PREDICTIVE ADMET

PREDICTIVE ADMET

Integrative Approaches in Drug Discovery and Development

Edited by

Jianling Wang

Metabolism and Pharmacokinetics
Novartis Institutes for Biomedical Research
Cambridge, MA, USA

Laszlo Urban

Preclinical Safety Profiling
Novartis Institutes for Biomedical Research
Cambridge, MA, USA

WILEY

Published by John Wiley & Sons, Inc., Hoboken, New Jersey.
Published simultaneously in Canada.

Library of Congress Cataloging-in-Publication Data:

Predictive ADMET : integrative approaches in drug discovery and development / edited by Jianling Wang, Laszlo Urban.
p. ; cm.
Includes bibliographical references and index.
ISBN 978-1-118-29992-0 (cloth)
I. Wang, Jianling, 1958- editor of compilation. II. Urban, Laszlo, 1951- editor of compilation.
[DNLM: 1. Drug Discovery–methods. 2. Pharmacokinetics. QV 745]
RM301.25
615.1′9–dc23

2013038054

Printed in the United States of America

10 9 8 7 6 5 4 3 2 1

CONTENTS

PREFACE

When we decided to put this book together, we asked ourselves whether it was possible to address absorption, distribution, metabolism, excretion, and toxicity (ADMET) from a different angle and provide novel information and guidance on the present pharmaceutical and academic practice. This was not an easy question to answer, considering numerous books and review articles on the topic. Although many ADMET aspects are addressed during the early phases of drug discovery, it is still common that projects treat data in isolation. Thus, we started with adding "integrated" that we believe is a most significant aspect of the drug discovery process. Although poor solubility and poor permeability are recognized as significant alerts for *in vitro* safety pharmacology assays, it is less obvious that already mitigated scaffold-based, off-target activity could resurface in metabolites. Therefore, testing both parent and metabolite in the *in vitro* safety pharmacology assays will provide a better chance to avoid unexpected adverse drug reactions (ADRs).

The other major element is the predictive, translational value of the ADMET efforts. We made all efforts to incorporate clinical cases into this book to demonstrate the relevance of reverse translation for the design of early safety assessment of molecules during the drug discovery and development process. It is invaluable to connect clinical observations with ADME properties, proteins, and gene targets using information obtained from marketed compounds. This is extremely valuable to correlate ADMET features of lead molecules in the absence of clinical proof of efficacy and safety with safety performance of marketed drugs. We are considering a simple equation that translates "hazard" into "risk."

The poor translational value of profiling efforts could come basically from two sources: (1) poor correlation of the target or measured chemical/biological characteristic of compounds with clinical features and (2) issues with assay quality. Many assays fall victim to either or both of the above and do not support optimization or mitigation. For example, hepatotoxicity assays performed on poor quality cells or within an inadequate physiological environment will have a poor predictive value and the data do not give any guidance for finding the right clinical candidate.

Another major mistake by drug discovery teams is that they look at the potency of compounds at off-targets without considering the expected therapeutic concentration. IC_{50} values on their own do not mean much; they have to be put in context with the available (free) concentration of the drug at the site of action. The best way to approach this is to consider the free effective therapeutic plasma concentration, which can be calculated in the preclinical setting from the determination of C_{max} or area under concentration (AUC) and plasma protein binding. Once all of these data are available, the project team can make an estimate on the safety margin. Caution has to be taken that this is not a therapeutic index that will be only available in the clinical setting; however, it can give a good estimate of the expected clinical safety of the compound. Therefore, we tried to provide a broad array of approaches from very early *in vitro* testing for bioavailability and safety in drug discovery to late-phase pharmacokinetic and *in vivo* safety pharmacology approaches in preclinical development phase continuing into the time of clinical trials. Finally several clinical examples will reinforce the message on the value of preclinical ADMET testing.

To meet the expectation, we recruited authors who are highly regarded specialists in various aspects of ADMET or clinical aspects of drug development; they understand and practice the integrative effort and are prepared to provide a critical view on their subject. We encouraged all contributors to think "translational" and highlight the most effective and novel methods and strategies for ADMET profiling of compounds. We hope that this book will be as enjoyable as useful for a broad range of scientists working in the pharmaceutical-biotechnology industries or in academia.

Last but not least we would thank the chapter authors for their high-quality contributions and the staff at Wiley for their patient support. We also would like to thank our loved ones for being patient with us during the time of editing this book.

Jianling Wang
Laszlo Urban

CONTRIBUTORS

Shirley A. Aguirre, Drug Safety Research and Development, Pfizer Global Research and Development, La Jolla Laboratories, San Diego, CA, USA

Praveen V. Balimane, Bristol–Meyers Squibb, Princeton, NJ, USA

Kevin Beaumont, Pharmacokinetics, Dynamics and Metabolism – New Chemical Entity, Pfizer Worldwide Research and Development, Cambridge, MA, USA

Jeremy Beck, Global Discovery Chemistry, Novartis Institutes for Biomedical Research, Cambridge, MA, USA

David Beckman, Preclinical Safety, Novartis Institutes for Biomedical Research, Basel, Switzerland

Dallas Bednarczyk, Metabolism and Pharmacokinetics, Novartis Institutes for Biomedical Research, Cambridge, MA, USA

Leslie Bell, Metabolism and Pharmacokinetics, Novartis Institutes for Biomedical Research, Cambridge, MA, USA

Eileen R. Blasi, Drug Safety Research and Development, Pfizer Global Research and Development, La Jolla Laboratories, San Diego, CA, USA

Christina Capacci-Daniel, Chemical Pharmaceutical Profiling, Novartis Institutes for Biomedical Research, Cambridge, MA, USA

Edmund J. Champness, Optibrium Ltd., Cambridge, UK

Donna M. Dambach, Genentech, South San Francisco, CA, USA

Andrew M. Davis, AstraZeneca R&D Charnwood, Loughborough, UK

Frank Dieterle, Molecular Diagnostics, Novartis Pharma AG, Basel, Switzerland

Stephanie Dodd, Chemical Pharmaceutical Profiling, Novartis Institutes for Biomedical Research, Cambridge, MA, USA

Jennifer B. Dressman, Institute of Pharmaceutical Technology, Goethe University Frankfurt/Main, Frankfurt/Main, Germany

Heidi J. Einolf, Department of Drug Metabolism and Pharmacokinetics, Novartis Institutes for Biomedical Research, East Hanover, NJ, USA

Gul Erdemli, Preclinical safety Profiling, Center for Proteomic Chemistry, Novartis Institutes for Biomedical Research, Cambridge, MA, USA

Ferry A. L. M. Eskens, Department of Medical Oncology, Erasmus University Medical Center, Rotterdam, The Netherlands

Urban Fagerholm, Lännavägen 7, Huddinge, Sweden

Bernard Faller, Metabolism and Pharmacokinetics, Novartis Institutes for Biomedical Research, Basel, Switzerland

Rakesh Gollen, Department of Pharmaceutical Sciences, Long Island University, Brooklyn, NY, USA

James R. Gosset, Pharmacokinetics, Dynamics and Metabolism – New Chemical Entity, Pfizer Worldwide Research and Development, Cambridge, MA, USA

F. Peter Guengerich, Department of Biochemistry and Center in Molecular Toxicology, Vanderbilt University School of Medicine, Nashville, TN, USA

Imad Hanna, Department of Drug Metabolism and Pharmacokinetics, Novartis Institutes for Biomedical Research, East Hanover, NJ, USA

Handan He, Department of Drug Metabolism and Pharmacokinetics, Novartis Institutes for Biomedical Research, East Hanover, NJ, USA

Tycho Heimbach, Department of Drug Metabolism and Pharmacokinetics, Novartis Institutes for Biomedical Research, East Hanover, NJ, USA

Keith Hoffmaster, Metabolism and Pharmacokinetics, Novartis Institutes for Biomedical Research, Cambridge, MA, USA

Hannah M. Jones, Department of Pharmacokinetics, Dynamics, and Metabolism, Pfizer Worldwide Research and Development, Cambridge, MA, USA

Kenneth I. Kaitin, Tufts Center for the Study of Drug Development, Tufts University School of Medicine, Boston, MA, USA

Teresa Kaserer, Computer-Aided Molecular Design Group, Institute of Pharmacy/Pharmaceutical Chemistry and Center for Molecular Biosciences University of Innsbruck, Innsbruck, Austria

Chris E. Keefer, Pharmacokinetics, Dynamics and Metabolism – New Chemical Entity, Pfizer Worldwide Research and Development, Groton, CT, USA

Estelle Marrer, Preclinical Safety, Novartis Institutes for Biomedical Research, Basel, Switzerland

Ruth L. Martin, GPRD, Abbott, Department R46R, Abbott Park, IL, USA

Virginie Y. Martiny, Molécules Thérapeutiques In Silico, Sorbonne Paris Cité, Université Paris Diderot, INSERM UMR-S 973, Paris, France; INSERM U973, Paris, France

Hans-Jörg Martus, Preclinical Safety, Novartis Institutes for Biomedical Research, Basel, Switzerland

Maria A. Miteva, Molécules Thérapeutiques In Silico, Sorbonne Paris Cité, Université Paris Diderot, INSERM UMR-S 973, Paris, France; INSERM U973, Paris, France

Lutz Mueller, Non-Clinical Safety, Pharma Research and Early Development, Hoffmann-La Roche Ltd., Basel, Switzerland

Patrick Y. Mueller, Preclinical Safety, Novartis Institutes for Biomedical Research, Cambridge, MA, USA

Ilza Pajeva, Institute of Biophysics and Biomedical Engineering, Bulgarian Academy of Sciences, Sofia, Bulgaria

Riccardo Panicucci, Chemical Pharmaceutical Profiling, Novartis Institutes for Biomedical Research, Cambridge, MA, USA

Neil Parrott, Pharmaceuticals Division, Non-Clinical Safety, Hoffmann-La Roche Ltd., Basel, Switzerland

Sebastien Ronseaux, Metabolism and Pharmacokinetics, Novartis Institutes for Biomedical Research, Cambridge, MA, USA

Patricia Schroeder, Modeling and Simulation – Drug Metabolism and Pharmacokinetics, EMD Serono Research and Development Institute, Billerica, MA, USA

Daniela Schuster, Computer-Aided Molecular Design Group, Institute of Pharmacy/ Pharmaceutical Chemistry and Center for Molecular Biosciences University of Innsbruck, Innrain, Innsbruck, Austria

Matthew D. Segall, Optibrium Ltd., Cambridge, UK

Hong Shen, Bristol–Meyers Squibb, Princeton, NJ, USA

M. P. S. Sie, Department of Medical Oncology, Erasmus University Medical Center, Rotterdam, The Netherlands

Nicole H. Siegel, Department of Ophthalmology, Boston University, Boston, MA, USA

Suzanne Skolnik, Metabolism and Pharmacokinetics, Novartis Institutes for Biomedical Research, Cambridge, MA, USA

Clayton Springer, Global Discovery Chemistry, Novartis Institutes for Biomedical Research, Cambridge, MA, USA

Manju L. Subramanian, Department of Ophthalmology, Boston University, Boston, MA, USA

Kiyohiko Sugano, Formulation Research, Asahi Kasei Pharma Corporation, Shizuoka, Japan

Veronika Temml, Computer-Aided Molecular Design Group, Institute of Pharmacy/ Pharmaceutical Chemistry and Center for Molecular Biosciences University of Innsbruck, Innrain, Innsbruck, Austria

Christopher Towler, Chemical Pharmaceutical Profiling, Novartis Institutes for Biomedical Research, Cambridge, MA, USA

Christian F. Trendelenburg, Preclinical Safety, Novartis Institutes for Biomedical Research, Basel, Switzerland

Laszlo Urban, Preclinical Safety profiling, Center for Proteomic Chemistry, Novartis Institutes for Biomedical Research, Cambridge, MA, USA

Jean-Pierre Valentin, AstraZeneca R&D, Innovative Medicines, Global Safety Assessment, Macclesfield, UK

Donna A. Volpe, Center for Drug Evaluation and Research, Food and Drug Administration, Silver Spring, MD, USA

Christian Wagner, Institute of Pharmaceutical Technology, Goethe University Frankfurt/Main, Frankfurt/Main, Germany

Jianling Wang, Metabolism and Pharmacokinetics, Novartis Institutes for Biomedical Research, Cambridge, MA, USA

Michael Wiese, Pharmaceutical Institute, Pharmaceutical Chemistry II, University of Bonn, Bonn, Germany

I

INTRODUCTION TO THE CURRENT SCIENTIFIC, CLINICAL, AND SOCIAL ENVIRONMENT OF DRUG DISCOVERY AND DEVELOPMENT

1

CURRENT SOCIAL, CLINICAL, AND SCIENTIFIC ENVIRONMENT OF PHARMACEUTICAL R&D

Laszlo Urban, Jean-Pierre Valentin, Kenneth I Kaitin, and Jianling Wang

1.1 THE CHANGING LANDSCAPE OF EPIDEMIOLOGY AND MEDICAL CARE

The rapidly changing landscape of epidemiology and associated medical requirements poses a significant challenge for the pharmaceutical industry, as well as for health care providers. Although the demand for new medicines is high in both developed and developing countries, the medical requirements differ considerably. In developed countries, the prevalence of diseases associated with lifestyle (e.g., obesity and type 2 diabetes) and aging (e.g., cancer and osteoporosis) is far greater than that in developing countries, whereas the latter suffer from a high frequency of infectious diseases, a large proportion of which is transmitted by vectors.

The demographic change in the United States and the developed Western world is biased toward older persons. The percentage of people 55 years and older in the United States has grown from 16.8% in 1950 to 23.6% in 2008. It is even more striking when considering real numbers, rising from 26 million to 73 million during the same time period. Moreover, as average life expectancy has increased, so have health-related issues. The major disease groups affecting this population include cardiovascular,

Predictive ADMET: Integrative Approaches in Drug Discovery and Development, First Edition.
Edited by Jianling Wang and Laszlo Urban.

respiratory and inflammatory diseases, cancer, neurodegenerative diseases, and psychiatric disorders. These diseases and disorders are associated with high overall drug costs. To support this, we looked at the statistics on arthritis and hypertension, which have high incidence in older persons. More specifically, according to government statistics, in 2010, more than 50 million adults of all ages in the United States suffered from doctor-diagnosed arthritis; that number is expected to exceed 60 million by 2015 [1]. However, whereas the overall prevalence of hypertension in people 18 years and older in the United States was 28.6%, it was 66.7% in people 60 years and older [2a].

Last, but not least, lifestyle, diet, and other behaviors in developed countries are associated with epidemics of diabetes, obesity, and stress-induced psychological/psychiatric disorders. There is emerging evidence that the combination of genetic and environmental factors could contribute to the development of particular mental disorders [2b]. In the United States, approximately 3.2% of boys between 6 and 17 years of age were found to have autism (National Health Statistics Report). Moreover, in the 15–17 year-old age group, the prevalence of parent-reported attentional deficit hyperactivity disorder (AD/HD) reached 13.6% in 2007, and these numbers are growing rapidly [3,4].

More and more diseases are controlled by chronic noncurative, symptomatic treatments, which contribute to persistently high medical expenses. Osteoporosis, high blood pressure, pain, atherosclerosis, diabetes, arthritis, and cancer, although not necessarily associated with an aging population, also require lifetime medications.

During the past 20–30 years, indications associated with large patient populations have led to increased consumption of particular drugs and have helped create a new class of "blockbuster" products. We have learned during the past decade that this strategy has not panned out well, as new blockbusters do not come easily, and the so-called patent cliff [5] could seriously affect the financial performance of companies.

Focus on particularly "profitable" areas in health care also meant that pharmaceutical companies neglected other segments of health care where return on investment was less assured. This was in part because little was known about the pathomechanisms of diseases, so efforts failed to produce viable drugs, for example, for Alzheimer's disease (AD), or because the disease was well managed by existing, relatively inexpensive medicines. This latter case is particularly relevant in the development of new agents to treat infectious diseases. The lack of investment into this area is resulting in an alarming dearth of therapies to treat multidrug-resistant bacterial strains.

Another contribution to the increase of expenditure on medicines is the parallel use of several medicines, either to affect various targets of the same disease, or to treat concomitant diseases. Approximately 37% of people 60 years and older in the United States takes five or more prescription medications at any time (Center for Disease Control and Prevention [CDC] statistics). Treatment regimens might also require further drugs to treat the side effects of the drugs used to treat the original disease. Cultural components are also pushing for more medications based on the commonly held belief that "pills" can provide easy and fast treatments in place of less expensive and sometimes more effective alternative methods, such as quitting bad habits, switching to healthy diet, physical exercise, and mental relaxation. An added benefit of these behavioral approaches would be the absence of adverse drug reactions (ADRs).

Use of antidepressants and antipsychotics has also skyrocketed during the past decades. For example, at the time of the publication of this book, one in ten U.S. adults reported having depression (CDC; [6]). The population 18 years of age and older reporting prescription antidepressant drug use in a single month has shown a significant increase between the periods 1988–1994 and 2005–2008. The increase is stunning from 5% to 22%. In the absence of appropriate medical care, the off-label use of psychotropic drugs has increased in the older population, particularly in senile dementia [7]. These aspects, however, did not facilitate more investment into drug development for psychiatric disorders. Instead, many companies discontinued investing in that disease area. A major reason for this trend is the fact that little is known about the pathomechanisms of mental disorders, which makes drug discovery highly speculative with a significant chance of failure.

We believe that the previous cases reflect a culture that nurtures a consumer-based approach to medicine in developed countries. Interestingly, the contribution of drug prices relative to total medical expenses remains relatively low, i.e., 10% in the United States, and similarly in the developed world.

Cost becomes even more of a pressing issue in the developing world, where there is likely to exist a large burden of epidemics of infectious, parasitic, and vector-carried diseases; lifestyle-associated diseases (e.g., smoking); or inappropriate nutrition and exposure to environmental hazards. Just to highlight the important contribution of infectious diseases to general health care issues, we looked at two diseases that overwhelmingly are associated with the developing world: dengue fever and drug-resistant tuberculosis. According to the World Health Organization (WHO), nine countries had reported severe dengue epidemics prior to 1970. Today the disease is endemic and affects more than 100 countries, with most cases reported in Southeast Asia and the western Pacific [8]. More recently, the number of reported cases has continued to rise. Because there is no treatment, and the most severe cases are life threatening or cause permanent disability, dengue fever remains a major neglected disease.

As another example, tuberculosis (TB) does not just remain a burden on its own, but more often it manifests in a multidrug-resistant format (MDR-TB). TB is the second highest cause of mortality worldwide as a result of a single infectious agent, resulting in an estimated 1.4 million death in 2010 alone (for further details and combination of HIV and TB, see Reference 8). A major problem is that MDR-TB does not respond to first-line standard anti-TB treatments. Second-line treatments are costly, significantly depend on compliance, and carry serious side effects. Alarmingly "extensively drug-resistant TB" (XDR-TB) is spreading and does not respond to any existing treatment. Of the greater than half million cases of MDR-TB in 2010, an estimated 9% progressed to XDR-TB, with an estimated annual death toll of 150,000. These numbers highlight the demand for action from health care professionals, governmental authorities, and the pharmaceutical industry.

1.2 COST OF DRUG DEVELOPMENT

The pharmaceutical industry faces challenges to develop affordable medications for diseases with high prevalence, regardless of whether they affect the population of a developed or developing country. Although hypertension, hyperlipidemia, and arthritis could be well controlled with relatively low-cost, generic products, new drugs for cancer with breakthrough results often demand high prices, which are likely to be unsustainable to health plans and other payers. Furthermore, diabetes, asthma, and mental disorders, to name a few conditions, are poorly managed even at higher cost.

During the last three decades, the cost of health care increased significantly worldwide, becoming a major strain on government resources [9]. National health expenditures in the United States skyrocketed during the past 10 years, from US$1,377.2B in 2000 to US$2,495.8B in 2009, according to CDC statistics [10]. The proportional rise in prescription drug spending during the same period (from US$120.9B to US$256.1B, respectively) contributed to increasing health care costs as well. The rise in prescription drug prices is driven by several factors, with a main contribution from three sources: the decline in overall industry productivity; the rise in safety issues during clinical testing and after approval; and the increase in regulatory requirements necessary to obtain market approval. We will look at each of these factors individually.

1.2.1 Decline in Industry Productivity

There has been considerable concern over the past two decades about the productivity of the pharmaceutical industry [10–13]. A recent analysis reveals that the average cost to bring a new molecular entity (NME) to the clinic today is US$1.3B [14]. Some estimates place the figure substantially higher, at an average of between US$5.5B and US$6B. Even the most productive companies, such as Amgen (Thousand Oaks, CA) and Novartis (Basel, Switzerland), spend more than US$3.5B to bring a novel drug to the clinic. In the worst case, this figure is greater than US$10B [15]. This latter calculation is based on simple mathematics: basically dividing research & development (R&D) expenditures by the number of new NMEs brought to the clinic by a company.

Indeed, the number of new medicines registered with the health authorities seriously declined during the first decade of the 21st century [12] followed by criticism of work practices, innovative spirit, and inadequate R&D strategies of the pharmaceutical companies [16–18]. The high expectations for the Human Genome Project did not seem to materialize and led to impatient cries of mismanagement. The slow utilization of new information was blamed for sluggish industry output. Now that the figures on new medicines have increased, it is obvious that there has been a certain time lag resulting from the adaptation and translation of new discoveries into drug discovery practices. In general, an increase in pharmaceutical productivity is expected to push down development costs, with the hope of producing less expensive medicines. Although the above is an important factor in drug prices, the development of novel cancer treatments and some special new treatments for rare diseases have been achieved by large investments. For example, the development of small-molecule kinase inhibitors took a long time because of the development of new tools for this class of targets and understanding the role and mode of action of kinases in disease pathomechanisms, particularly in cancer [19]. New technologies and work processes, such as high-throughput screening (HTS), helped to examine millions of compounds for interaction with protein targets and exert therapeutic effects. Although HTS made it possible to discover hits with considerable chemical diversity, it also flooded the assays with compounds with poor physicochemical properties. Molecules with low solubility, high lipophilicity, and poor stability cause major problems in absorption, distribution, metabolism, and excretion (ADME) assays, which result in inaccuracies, discrepancies between biochemical and cellular assays, etc. [20]. These properties lead to extended timelines and a higher cost during the discovery process and could be the reason for project termination or production of a poor quality clinical candidate [21]. To avoid these problems, profiling of biopharmaceutical properties has been included in early drug discovery road maps with effective results demonstrated by decreased attrition rates as a result of poor drug metabolism and pharmacokinetic (DMPK) properties [22]. Thus, ADME profiling proved to be highly effective particularly as it is done in parallel with testing for pharmacological potency and efficacy. Implementation of early profiling for ADME and toxicology (T) proved that simple assays could have great translational value at a low cost [23]. Another benefit of these efforts was the development and implementation of *in silico* tools, which greatly enhanced the predictive power of ADME profiling and gave direction for chemistry laboratories to generate compounds with better physicochemical properties.

1.2.2 Rise in Safety Issues

Another main factor contributing to increased drug-related expenses also has roots in early drug discovery; the cost, however, will appear in the clinic. This relates to safety issues, which are realized during clinical trials and in some cases postapproval only. Safety assessment of drug candidates is a late event in drug discovery, whereas early hazard identification was completely neglected in the past. The side effects of individual drugs or drug combinations generate a significant expense in health care either by rehospitalization; secondary treatments, such as management of concomitant hypertension; or number of lost working days. There are two categories of side effects: those that are associated with the primary target, and those associated with unintentionally encountered off-targets.

Many serious side effects are associated with primary targets and are difficult to avoid or manage. Recent withdrawals of anti-diabetic drugs (rosiglitazone), obesity treatment (rimonabant), and black box labels for many new drugs, in particular in cancer management, highlight this problem. For example, almost all kinase inhibitors carry box warnings for cardiovascular safety matters. In addition to acute effects, delayed post-treatment cardiological symptoms have been observed after treatment with tyrosine kinases inhibitors (TKIs; [24, 25]). We will address these new aspects and more issues associated with TKI treatment in the following chapters.

Cardiac and hepatic toxicities have remained the leading reasons for attrition and labeling for decades [26, 27]. These side effects largely develop because of the off-target activity of drugs and are a result of their metabolism by common enzymes. Idiosyncratic drug-induced liver injury (DILI) has been a dreaded, unpredictable threat until recently. Today newly developed *in vitro* and *in silico* technologies can assess the major metabolic route of compounds and predict possible drug–drug interactions in the liver, detect toxic effects of reactive metabolites, and link hepatobiliary hepatotoxicity to several bile transporters. Nevertheless other, currently not well-understood pathomechanisms or species differences in safety pharmacology keep hepatotoxicity prediction difficult. In the meantime, cardiotoxicity predictions improved significantly with the recognition of off-target effects at cardiac ion channels, particularly at the human ether a go-go related gene (hERG) potassium channel [28]. The hERG channel is a notoriously promiscuous protein with easy access for many compounds, and the block of its pore causes long QT (LQT) which can lead to life threatening ventricular arrhythmias, known as Torsades de Pointes. This off-target effect is well understood, and the regulatory assessment has high predictive value [29–31]. However, although rarely, we still encounter cases of hERG-related arrhythmias with compounds that are used for treatment of life-threatening diseases where the risk is accepted and managed by thorough examination of patients for prevalence of arrhythmia and monitoring during treatment.

In general, preclinical assessment of cardiac ion channel inhibition is well established and supports mitigation prior to clinical candidate selection and basically weeds out compounds with effects at these channels. Although mitigation of arrhythmias has become routine in preclinical development, the acute and delayed cardiotoxic effects of the TKIs has not been addressed satisfactorily [25]. This is in part a result of the poor preclinical assessment of cardiac inotropy, which could signal pathological changes in cardiomyocyte contraction and predict congestive heart failure [26, 32].

It is certainly more challenging to bring drug candidates to the clinic than a decade ago. The estimated average time of the preclinical phase is approximately 6 years, which includes preclinical safety studies. There is more expectation for both better efficacy than existing medications for a particular indication, and importantly, safety became a more sensitive issue. Although some drugs make it through the safety tests, labels can cripple their clinical applications, with disappointing return on investment. In general, there has been a notable move toward preclinical safety mitigation by the introduction of broad-scale safety profiling, including both target- and pathway-based safety assessments. There is good agreement within the pharmaceutical industry concerning target-based profiling practices [33] and associated predictive *in silico* tools [34]. Although this move could provide clinical candidates with higher safety margins, it costs more and prolongs timelines.

1.2.3 Increasing Regulatory Requirements

Regulatory requirements surrounding clinical trials have become significantly more stringent in recent years, demanding larger patient populations and including more diverse investigations [35, 36]. As an example, off-targets linked with suicidality can trigger demand for clinical evaluation of suicidal intent, which will add extra expense and prolong the trial. Large-scale trials to prove superior activity in comparison with existing treatments could demand thousands of patients in multiple centers, with the uncertainty that the result would be favorable for the drug (e.g., the case of Avandia; GlaxoSmithKline, London, U.K.).

1.3 THE NEW PARADIGM OF ADME/PK ASSESSMENT

1.3.1 Recent Advancement of ADME/PK Assessment of Drug Candidates in Discovery and Development

The introduction in the 1990s of the "rule of five," or drugability concept, by Lipinski et al. [37, 38] greatly facilitated the early assessment of the ADME properties of NMEs with a wealth of newly developed *in silico*, *in vitro*, and *in vivo* tools [39]. Not only does the availability of integrated application of *in vitro* and *in vivo* ADME/PK data enhance the optimization of efficacy for the designated therapeutic targets [40], but it also offers a wealth of knowledge for the potential development of tomorrows' more predictive *in silico* assay-free tools [41]. More predictive models require intelligent evaluation of dynamics, associations and disconnects within the historic data collection.

Today, tiered approaches are the standard, eliminating an earlier stepwise routine. First-line ADME assays provide "sentinel" for bad behavior and identify those factors that would prevent the advance of molecules toward clinical candidate status. These assays are relevant for structure activity relationship (SAR) support at a high-throughput rate. However, they mostly lack the sophistication for deeper analysis of individual components of ADME characteristics. High-throughput solubility, permeability, and microsomal stability tests would come under this category [42]. Mechanistic studies regularly are used as follow-up assays (e.g., tier 2,3…).

The application of state-of-the-art laboratory automation and bioanalytical technologies have resulted in high-quality ADME/PK data and have enhanced the predictability of downstream assays [41–43]. Centralization of assay facilities in pharmaceutical companies and contract research organizations (CROs) largely improved the overall efficiency and cost effectiveness of absorption, distribution, metabolism, excretion, and toxicity (ADMET) assessment. Multitiered PK tools range from fast snap-shot PK or cassette dosing all the way through to detailed mechanistic studies, such as sampling via the bile duct, sampling/dosing to the hepatic portal vein, or use of humanized or gene-knocked out animal models to address specific questions in early drug discovery.

Today, researchers in the pharmaceutical and biotechnology industry are trained for fundamental understanding of ADMET/PK principles that are typically incorporated into the drug discovery process for fast and efficient selection of the highest quality clinical candidate.

1.3.2 New Challenges and Emerging Fields of ADME/PK Development

From an overall ADMET perspective, early ADMET models offer a set of comprehensive "surrogate" filters to enable potentially problematic new chemical entities to be weeded out in the early discovery phase. Although effective in reducing ADME-related attrition in clinical phases, the current ADMET "justice" system is far from perfect and may transgress against "innocent" or "imperfect but manageable" candidates. As a matter of fact, a reasonable fraction of currently marketed drugs might have never qualified for clinical use if they were simply applied to rigid ADMET filters.

With the increasing costs and challenges in discovering new drug targets and the more crowded intellectual property space, the pharmaceutical industry can no longer afford to abandon a promising drug candidate simply as a result of a single unfavorable ADMET property. False positives derived from current ADMET models as a result of lack of robust *in silico–in vitro–in vivo* (ISIVIV) links have a further negative impact on the discovery of new drugs, although this imperfect approach may be valuable to reject the worst offenders. As an example, the assay applied to determine covalent binding, despite being a gold standard *in vitro* approach for identifying reactive metabolites, cannot reliably predict the occurrence of liver toxicity. Its correct role should be defined as flagging potential toxic structural alerts. Most of the existing *in silico* ADMET models seem to work well when used to predict simple physicochemical properties, such as ionization constant, LogP, and polar surface area, which are primarily governed by the intrinsic molecular descriptors of scaffolds and molecules. They, however, are less applicable to more complex interactions within the physiological environment (e.g., gastrointestinal solubility, clearance, bioavailability, etc.; [39]).

During optimization, overinterpreting ADMET data or overemphasizing the role of a single ADMET parameter is damaging. With tons of data generated in multiple ADMET models, the "more is better" approach is less of a dream and more of a nightmare for drug discovery teams. They may get buried in a flood of data and lost within seemingly "contradictory" messages. The strategy of "box-checking" or simple "all-around" profiling passed its time as it failed to deliver better candidates, faster to the clinic. For instance, one shortcoming of single-minded heat map tables is that a similar weight gets applied to each pixel of colors (e.g., red). A further limitation is that the possible interplay between remote pixels that impacts a pathway and may be potentially important in driving an undesired physiological effect is not reflected in the heat map. Therefore, integrating data within heat maps with consideration of biological networks, or pathways that interact, will make it easier and more reliable for the user to identify and mitigate risk. Clearly, success seems to rely not just on volume of the data but also on the meaningful integration of emerging patterns with translational relevance as a project matures (e.g., human exposure, PK/pharmacodynamics (PD)/efficacy in humans, therapeutic index, etc.). Data should be used to generate knowledge and drive testable hypotheses to enable right, reliable, and decisive conclusions. Thus, it is beneficial to connect all ISIVIV models together.

Gaps in the ISIVIV correlations (ISIVIVCs), within a single scaffold or across series, can be valuable to identify missing pieces of the puzzle and to generate a new and reiterated hypothesis.

An inferred algorithm based on experiences drawn from the chemotypes' history may offer greater advantage. Meanwhile, it should be kept in mind that the existing models might have been developed from historic or global chemical spaces and they might not necessarily apply to drug candidates from new series or local chemical space. Sometimes, global empirical rules might be misleading and their applications should be optimized in nontraditional ranges of physical properties. Although multiple statistics tools such as quantitative structure activity relationship (QSAR), principal component analysis (PCA), and partial least squares (PLS) are frequently used in the analysis of ADMET data, their limitation is that they need to be trained against a set of truly relevant data. In the early days of a project, there may be little alternative to reliance on more basic guidance such as the "rule of five," "absorption model," and some ADMET diagnostic models that are mostly applicable on a "global" basis. As measured ADMET data accumulate and get enriched, a better understanding of the chemotype emerges with the consequence of an increased confidence level for "local" models. This certainly extends to sophisticated modeling of exposure and effect relationships during preclinical and early clinical assessment with a better determination of safety margins.

In the current book, various data integration tools are introduced ranging from ADME diagnosis model (Chapter 3), PATH—a hypothesis generation tool (Chapter 4), PK-Matrix (Chapter 5), QSAR/PCA/chemoinformatics (Chapters 6 and 7); multiparameter optimization (Chapter 8), PBPK models (Chapters 9 and 10), and PK/PD models (Chapter 11). The comprehensive utility of the data integration models in real drug discovery and development projects to address specific ADMET issues are also presented (Chapters 12–21) along with successful stories and lessons learned (Chapters 22–26).

1.4 INCREASED SAFETY EXPECTATIONS

1.4.1 Early Awareness of Safety Hazards

Preclinical safety assessment of compounds has been significantly enriched during the past decades [44]. *In vivo* safety testing now routinely incorporates genetic toxicity, LQT-related (long QT interval) studies, and refined biomarkers for both DILI and nephrotoxicity are available (see Sections 2.7–2.10 of Chapter 2). Also, early assessment of hazards associated with well-defined molecular targets and pathways are now gaining ground and entering the domain of registration expectations [45,46]. Although *in vivo* safety pharmacology provides the backbone of preclinical safety evaluation of drug candidates, it has had considerable "misses" in the past. The best known cases are associated with 5-HT_{2B} receptor agonism and inhibition of 3',5'-cyclic-nucleotide phosphodiesterase (PDE3) inhibition (see Section 3.4 of Chapter 3), which are well publicized [47, 48]. The reasons for these undetected ADRs are various; some are the result of species differences, whereas others occurred because side effects associated with targets were not known and not detected by routine regulatory assays. Mechanisms that were not known prior to clinical observations include selective stimulation of the 5-HT_{2B} receptor that is associated with the gradual, time-dependent development of general fibrosis that manifests in cardiac valvular disease (CVD) and serious heart conditions [48]. Because of the slow development of this disease and the need for echocardiography for its early detection, CVD is difficult to diagnose *in vivo*. To the contrary, *in vitro* detection of 5-HT_{2B} agonism is easy and highly predictive for CVD. Therefore, compounds with agonism at 5-HT_{2B} can be de-prioritized before candidate selection or the activity engineered out of molecules [49]. Another example is inhibition of the biliary salt export protein (BSEP), which is associated with hepatobiliary DILI [50]. Although there is high homology between rat and human BSEP, the function of this transporter in the rodent differs from that in humans. BSEP in rats does not have a significant contribution to bile secretion, unlike in humans [51, 52]. These examples highlight the challenges surrounding the predictive value of the *in vitro* safety pharmacology assessment of drug candidates. Furthermore, the cost and speed of the *in vitro* assays provide a very powerful tool for hazard detection and mitigation and give guidance for the *in vivo* safety pharmacology studies in the regulatory panel [33, 49]. Certainly, once the cause is known and the link between the human molecular target and the adverse reaction has been established, simple *in vitro* assays can identify compounds acting on these targets. For these reasons, it was introduced into the drug discovery process during lead selection and lead optimization. As an added benefit, data accumulated by *in vitro* safety profiling gave way to the development of *in silico* tools with the predictive power of ADRs [34].

It is important to emphasize that *in vitro* safety pharmacology data provide alerts for safety hazards. The potency at the off-targets should be considered in the context of exposure, which is defined by the maximum free available drug concentration at the site of action in the organism. Most often the free C_{max} is used for this purpose, but area under concentration (AUC) can be considered if more relevant [53]. This simple calculation will define whether the hazard would represent a safety risk during clinical application. Once the risk was established, the *in vitro* profiling assays can lead mitigation by available SAR during lead optimization and, in case the hazard still persists to some extent at candidate selection, guide the preclinical safety assessment to explore no observed effect level (NOEL) for the associated ADR. Using the 5-HT_{2B} example again, if a clinical candidate with high potency at the therapeutic target shows weak agonist activity at this off-target, then a long-term rodent experiment for echocardiography should be considered within the preclinical assessment [54].

One particular area that is difficult to predict from nonclinical *in vivo* observations is suicidal ideation. It needs dedicated expert analysis and complex *in vivo* evaluation. However, once critical evidence emerges from clinical observations—in particular, from the U.S. Food and Drug Administration (FDA) adverse event recording system (AERS)—on high incidence of suicidal ideation of various drugs with a common central nervous system (CNS) target, one can establish links between the target and the observed ADR. Testing compounds at this target will give an opportunity for early warning for suicidal ideation and guide to surrogate *in vivo* testing and, if necessary, mitigation.

1.4.2 Logistics for *In Vitro* Safety Profiling

To set up the *in vitro* safety profiling panels is not trivial: Target selection should be guided by clinical information that is by observation of side effects during clinical trials and postapproval. Ranging from the discovery of the association of QT prolongation, to ventricular arrhythmias with hERG inhibition, to the suicidal ideation linked with vesicular monoamine transporter-2 (VMAT2) inhibition, a large volume of data has been derived from clinical observations and used in a reverse translational format to identify targets for *in vitro* safety assessment panels [33, 49]. Obviously, this is possible only if there are drugs that enter clinical development and/or proceed to full registration. When no drug-related data are available, one can use information obtained from human genetic diseases, pharmacological data in animal experiments, and knock-out/knock-in (KO/KI) animal models. Once targets associated with ADRs are identified, the next step is to establish panels for safety assessment.

The second most important criterion is to investigate the incidence and severity of the side effects associated with the targets and the hit rate of compounds at the selected targets, a phenomenon that is referred to as pharmacological promiscuity. By these standards a relatively slim panel can be identified and used for regular, iterative safety assessment [33] as an integral component of the design-make-test-analyze drug discovery cycle. Targets not included into such a panel still could be screened at the stage when clinical candidates are nominated. As this "secondary" or follow-up panel contains targets with links to less serious side effects and with a very low hit rate, it is unlikely that clinical candidates would be seriously affected.

Once the panels are assembled, the next question is concerned with the application of the panels to the different requirements during the drug discovery process and what is the most relevant assay format. These panels are used most extensively during lead selection and optimization when there is a chance for risk mitigation and produce molecules with the least off-target effects by clinical candidate stage. This requires rapid iterative cycles, which could be performed at the time of the primary target assay. *In vitro* safety profiling requires multiparallel assembly of assays, where a large number of compounds (from hundreds to several thousand/year) can be screened in an iterative manner. The assay cycle is set to the synthetic chemistry cycle, which allows testing for safety at the same frequency as for primary activity; thus, it does not introduce delays for the projects.

The cost should be minimal and the assays should not consume a large volume or quantity of material as it is rarely available at this stage. These readouts can indicate hazards, which should be considered in conjunction with other data, such as *in vitro* and *in vivo* data predicting human exposure at efficacious doses (see subsequent discussion).

Assay formats vary, but whenever possible robust biochemical assays are used for primary profiling, particularly if both receptor agonist and antagonist effects are associated with ADRs. As the panels are used for SAR, it is important to generate high-quality data that can be relied on for compound prioritization; therefore, only IC_{50} or *Ki* values are acceptable. Selection of the appropriate ligands and most full-length proteins will have an impact on the data and are crucial for both biochemical and functional assays. Functional assays are mostly considered for the primary panel when the modality is important: In some cases, only antagonist or agonist effects are associated with serious ADRs; in these instances, biochemical assays do not provide the crucial information.

In vitro safety profiling is primarily executed during lead selection and optimization when compounds have not been optimized for physicochemical properties and some synthetic impurities may exist. Therefore quality control (QC) and solubility testing is an essential component of the *in vitro* safety assessment.

1.4.3 Relevance and Confidence in Profiling Data

Relevance of preclinical safety profiling has two major aspects: (1) Predicted association with ADRs and (2) predicted manifestation. Early *in vitro* data are available at the time when no *in vivo* information exists and no clear safety window can be determined. However, iterative testing of compounds can determine pharmacological promiscuity and provide SAR for off-targets. These properties rarely warrant a Go/No-Go decision; however, they will initiate safety-related mitigation. This can be done at high confidence if an *in vivo* assay is available that links the *in vitro* data to clinical ADRs and provides a preclinical readout at a relatively low cost. Take hERG channel inhibition as an example: The hERG channel is a particularly promiscuous potassium channel, which attracts a large proportion of molecules and is associated with QT prolongation and ventricular arrhythmia. Potency at hERG only identifies the hazard; it does not predict the real risk in clinical settings. Although most disease models used rats or mice, rodents are not appropriate to evaluate the hERG-mediated pro-arrhythmic risk because the rodent hERG equivalent is not expressed in the adult cardiac tissue. Therefore, a nonrodent *in vivo* cardiotoxicity model (usually the dog or nonhuman primate) has been introduced [55] to translate the *in vitro* observation to clinical relevance [29]. The information from the nonrodent model also integrates any other off-target effects associated with cardiotoxicity, which can be the feature of the compound. Thus, this model provides integration of any cardiovascular effect and a predictive safety index. Still, clinical pathological conditions might modify the predicted index, which can be assessed, at least for the time being, only in the clinical setting.

Although nonrodent telemetry is incorporated into the regulatory safety assessment, many off-target effects could slip unnoticed into the clinical phase, such as PDE3 inhibition or 5-HT_{2B} receptor agonism. Positive finding at these targets should be followed up separately once the signal is identified. In case of positive findings in these assays, the predicted ADR will be high and could trigger a Go/No-Go decision.

1.5 TRANSLATIONAL VALUE OF *IN VITRO* PROFILING DATA

The true value of *in vitro* profiling manifests during integration. Target-based safety profiling, together with ADME, PK, and SAR or QSAR data, will provide the full picture and generate knowledge about clinical expectations. For example PK issues, such as blood-brain barrier (BBB) penetration, will determine which effect at a CNS site would manifest. This was clearly demonstrated during the development of histamine H1 receptor antagonists. Once these drugs enter the CNS, they produce somnolence and dizziness, in addition to their anti-allergic effects in the periphery [56]. Early ADME assays and *in silico* tools can predict BBB penetration; however, quantitative whole-body autoradiography (QWBA) will provide a much more refined picture.

Toxicology assessment is based on organ specific effects with few exceptions of genetic toxicity. In this respect, the relation of the off-target potency to the effective therapeutic free plasma C_{max} is determinant and used for finding the allowable therapeutic index (TI). Thus, simple IC_{50} values need to be translated into organ-specific read-outs in both animal experiments and clinical use. First, most of the modern preclinical *in vitro* assays use human targets and materials that have direct relevance for clinical applications. The importance of species specificity is highlighted when a predicted ADR does not manifest during the *in vivo* safety evaluation. In this case, the assay should be repeated with the target relevant for the *in vivo* species or the animal species should be changed to a species more relevant to human.

Second, the system organ class (SOC) nomenclature is used to determine correlation of targets with organ-specific toxicity. For example, inhibition of BSEP will translate into hepatobiliary DILI and will be recognized as cholestasis (preferred term). Although the pathomechanisms of many drug-related side effects are known, there remain a reasonable number of cases where ADRs appear unexpectedly.

It is important to note that the landscape of clinical ADRs can change within the same organ toxicity domain. Cardiotoxicity remained a leading reason for withdrawals and black box labels regardless of whether prediction and mitigation of arrhythmias improved significantly [25]. Arrhythmias are not the leading reason for cardiotoxicity anymore, although a more relaxed approach for toxic side effects in cancer indications led to the registration and marketing of several drugs that have hERG blockade. Instead, other cardiac ADRs, such as congestive heart failure, have appeared with new medications, such as tyrosine kinase inhibitors [57, 58].

A major aspect of clinical safety is concerned with co-administration of drugs. Medicines used in combination might converge on common metabolic pathways in the liver, or interference in absorption could contribute to serious alterations of PK. PD effects could be modified by drugs synergistically acting at the same target (off-target), within the same pathway or system (e.g., co-administration of monoamine oxidase (MAO) inhibitors with dopamine transporter (DAT) and norepinephrine transporter (NET) inhibitors) leading to hypertensive crisis [59]. Medicines with high pharmacological promiscuity might act at undetected off-targets, which could increase the incidence and/or severity of ADRs in combination therapy.

1.6 SUMMARY

The research-based pharmaceutical industry experienced a low output period between 2001 and 2010, with an average annual approval of 23 NMEs in the United States. There was a peak of 36 NME approvals in 2004. Although the figures were steady during these years, they alerted analysts to major productivity issues in the pharmaceutical industry, as this trend represented a decline in comparison with the

1990s when the annual output was greater than 40. Indeed, return on investment tipped the balance toward high expenses with low output, and led to questions about development strategy and practice in the pharmaceutical and biopharmaceutical industries. However, the output for the past two years improved considerably, with 30 registrations in 2011 (including 8 drugs for cancer; 10 targeting orphan diseases; 2 each for hepatitis C and chronic obstructive pulmonary disease (COPD)), and the upward trend continued in 2012, with 39 NME registrations; however, it is difficult to predict whether such a trend will continue. Drug development performance metrics for new product approvals indicate that, whereas total approvals were at a 25-year low between 2000 and 2009, the percentage of priority products was about 50% of the total—a 30-year high [11]. However, the duration of the clinical phases indeed increased during this period of time, with the longest in CNS (average 8.1 years) and oncology (average 6.9) [11]. Although this news reflects better on the state of the pharmaceutical and biotechnology sectors, the cost to bring a compound to registration remains high, the timelines do not support short-term gains on investment, and profitability, it is argued, is nonsustainable [12].

Available funds for health care in the face of the increasing 60-years-and-older population, with a high incidence of chronic diseases in developed countries, and a large demand for medications in the area of neglected diseases in developing countries, pose a big challenge for the industry, which is in need of radical changes in bio-innovation. Major innovations during the past two to three decades included the introduction of HTS, a large variety of ADME, efficacy, and safety profiling capabilities into the preclinical drug discovery process. However, the basic infrastructure and the drug discovery process have not changed significantly, and the timelines for target-to-clinic have remained about the same. It is encouraging that attrition rates resulting from ADME aspects diminished during this period of time, presumably because of the introduction of early ADME assessment [22]. Although it is too early to judge, there is a hope that safety-related attrition would follow in these steps as a result of prevention or mitigation of expected ADRs. However, this latter aspect might be compromised by the introduction of novel therapeutic targets with unknown side effect profiles, such as occurred with TK or PDE3 inhibitors.

Still on the bright side, better understanding of translational aspects of drug discovery seems to open opportunities for both industry and academia to work together in a more coherent manner to focus on integrated assessment of therapeutic targets and diminish attrition rates that are currently too high as a result of lack of efficacy. Better established target identification packages (target IDs) with translational focus prior to the initiation of full discovery projects are necessary to avoid disappointment at the stage of cellular and system biology evaluations or at later phases of drug development. Introduction of more rigorous target validation would certainly increase the costs of early drug discovery; however, this can be shared between industry (biotechnology, pharmaceutical companies, and CROs) and academia [60]. This is expected to provide a more solid basis for lead identification and optimization, which can be done in parallel without further expanding the discovery period.

Clinical development, however, is expected to remain a major challenge. First, there is increased awareness for both safety and efficacy, particularly when drugs compete for the same indication. This occurs already in oncology, where prices are high and new drugs have to show a better safety profile with improved survival rates in comparison with marketed drugs. Competition in therapeutic areas, where high-quality generic drugs are available, probably will hinder innovation for a period of time at least until a breakthrough novel drug candidate appears. Finally, in areas where the health care demand is high, such as in neuropsychiatric disorders, only novel, innovative approaches can generate new drugs [61]. This is more likely to happen during the next couple of years as various "omics" science matures and the move from academic laboratories into the development support by the biopharmaceutical sector accelerates. Early signs are already producing benefits, such as stratification of diseases and patients, particularly tumors, and providing better design for clinical trials.

Taken together, the pharmaceutical industry is facing many significant challenges, with slow improvement of productivity and increasing demand for safe and efficacious drugs in a rapidly changing medical and social landscape. This book provides a glimpse at those technologies and scientific advances that, at least to some extent, will help the industry reach its goals.

REFERENCES

1. http://www.cdc.gov/arthritis/data_statistics/national_nhis.htm.
2. (a) http://www.cdc.gov/nchs/data/databriefs/db107.htm. (b) Chaste, P., Leboyer, M. (2012). Autism risk factors: Genes, environment, and gene-environment interactions. *Dialogues in Clinical Neuroscience*, *14*, 281–292.
3. http://www.cdc.gov/Features/dsADHD.
4. Blumberg, S.J., Bramlett, M.D., Kogan, M.D., Schieve, L.A., Jones, J.R., Lu, M.C. (2013). Changes in prevalence of parent-reported autism spectrum disorder in school-aged US children: 2007 to 2011–2012. *National Health Statistics Report*, *65*.
5. Harrison, C. (2011). Patent watch: The patent cliff steepens. *Nature Reviews Drug Discovery*, *10*, 12–13.
6. http://www.cdc.gov/features/dsdepression.
7. Wetzels, R.B., Zuidema, S.U., de Jonghe, J.F., Verhey, F.R., Koopmans, R.T. (2011). Prescribing pattern of psychotropic drugs in nursing home residents with dementia. *International Psychogeriatrics*, *23*(8), 1249–1259.
8. http://www.who.int/mediacentre/factsheets/fs104/en/index.html.
9. Milne, C.P., Kaitin, K.I. (2010). Impact of the new US health-care-reform legislation on the pharmaceutical industry: Who are the real winners? *Clinical Pharmacology & Therapeutics*, *88*(5), 589–592.
10. http://www.cdc.gov/nchs/hus/contents2011.htm#128.
11. Kaitin, K.I. (2012). translational research and the evolving landscape for biomedical innovation. *Journal of Investigative Medicine*, *60*(7), 995–998.
12. Kaitin, K.I., DiMasi, J.A. (2011). Pharmaceutical innovation in the 21st century: New drug approvals in the first decade, 2000-2009. *Clinical Pharmacology & Therapeutics*, *89*(2), 183–188.

13. Milne, C.P., Kaitin, K.I. (2012). FDA review divisions: Performance levels and the impact on drug sponsors. *Clinical Pharmacology & Therapeutics*, *91*(3), 393–404.
14. DiMasi, J.A., Grabowski, H.G. (2007). The cost of biopharmaceutical R&D: Is biotech different? *Managerial and Decision Economics*, *28*, 469–479.
15. http://www.forbes.com/sites/matthewherper/2012/02/10/the-truly-staggering-cost-of-inventing-new-drugs.
16. Munos, B. (2009). Lessons from 60 years of pharmaceutical innovation. *Nature Reviews Drug Discovery*, *8*(12), 959–968.
17. Munos, B.H., Chin, W.W. (2011). How to revive breakthrough innovation in the pharmaceutical industry. *Science Translational Medicine*, *3*(89), 89.
18. Paul, S.M., Mytelka, D.S., Dunwiddie, C.T., Persinger, C.C., Munos, B.H., Lindborg, S.R., Schacht, A.L. (2010). How to improve R&D productivity: The pharmaceutical industry's grand challenge. *Nature Reviews Drug Discovery*, *9*(3), 203–214.
19. Zhang, J., Yan, P.L., Gray, N.S. (2009). Targeting cancer with small molecule kinase inhibitors. *Nature Reviews Cancer*, *9*, 28–39.
20. Di, L., Kerns, E.H. (2006). Biological assay challenges from compound solubility: Strategies for bioassay optimization. *Drug Discovery Today*, *11*, 446–451.
21. Chen, X.Q., Antman, M.D., Gesenberg, C., Gudmundsson, O.S. (2006). Discovery pharmaceutics—challenges and opportunities. *AAPS Journal*, *8*(2), E402–E408.
22. Kola, I., Landis, J. (2004). Can the pharmaceutical industry reduce attrition rates? *Nature Reviews Drug Discovery*, *3*(8), 711–715.
23. Faller, B., Urban, L. Integrative risk assessment, in B. Faller and L. Urban (eds.), *Hit and Lead Profiling*, Wiley-VCH Verlag, Hoboken, NJ, 2009, pp. 41–54.
24. Schmitz, K.H., Prosnitz, R.G., Schwartz, A.L., Carver, J.R. (2012). Prospective surveillance and management of cardiac toxicity and health in breast cancer survivors. *Cancer*, *118*, 2270–2276.
25. Mellor, H.R., Bell, A.R., Valentin, J.P., Robert, R.R. (2011). Cardiotoxicity associated with targeting kinase pathways in cancer. *Toxicological Sciences*, *120*(1), 14–32.
26. Laverty, H.G., Benson, C., Cartwright, E.J., Cross, M.J., Garland, C., Hammond, T.G., Holloway, C., et al. (2011). How can we improve our understanding of cardiovascular safety liabilities to develop safer medicines? *British Journal of Pharmacology*, *163*(4), 675–693.
27. Redfern, W.S., Valentin, J.P. (2011). Trends in safety pharmacology: Posters presented at the annual meetings of the Safety Pharmacology Society 2001–2010. *Journal of Pharmacological and Toxicological Methods*, *64*(1), 102–110.
28. Abbott, G.W., Sesti, F., Splawski, I., Buck, M.E., Lehmann, M.H., Timothy, K.W., Keating, M.T., Goldstein, S.A. (1999). MiRP1 forms IKr potassium channels with HERG and is associated with cardiac arrhythmia. *Cell*, *97*(2), 175–187.
29. Sarazan, R.D., Mittelstadt, S., Guth, B., Koerner, J., Zhang, J., Pettit, S. (2011). Cardiovascular function in nonclinical drug safety assessment: Current issues and opportunities. *International Journal of Toxicology*, *30*(3), 272–286.
30. Gintant, G. (2011). An evaluation of hERG current assay performance: Translating preclinical safety studies to clinical QT prolongation. *Pharmacology & Therapeutics*, *129*, 109–119.

31. Wallis, R.M. (2010). Integrated risk assessment and predictive value to humans of non-clinical repolarization assays. *British Journal of Pharmacology*, *159*, 115–121.
32. Harmer, A., Abi-Gerges, N., Morton, M., Pullen, G., Valentin, J.P., Pollard, C.E. (2012). Validation of an in vitro contractility assay using canine ventricular myocytes. *Toxicology and Applied Pharmacology*, *260*(2), 162–172.
33. Bowes, J., Brown, A.J., Hamon, J., Jarolimek, W., Sridhar, A., Waldron, G., Whitebread, S. (2012). Reducing safety-related drug attrition: The use of in vitro pharmacological profiling. *Nature Reviews Drug Discovery*, *11*(12), 909–922.
34. Lounkine, E., Keiser, M.J., Whitebread, S., Mikhailov, D., Hamon, J., Jenkins, J.L., Lavan, P., Weber, E., Doak, A.K., Côté, S., Shoichet, B.K., Urban, L. (2012). Large-scale prediction and testing of drug activity on side-effect targets. *Nature*, *486*(7403), 361–367.
35. Anon. (2008). *Guidance for Industry Diabetes Mellitus—Evaluating Cardiovascular Risk in New Antidiabetic Therapies to Treat Type 2 Diabetes*. U.S. Department of Health and Human Services. Food and Drug Administration Center for Drug Evaluation and Research (CDER). http://www.fda.gov/downloads/Drugs/GuidanceCompliance RegulatoryInformation/Guidances/ucm071627.pdf.
36. Anon. (2012). *Guidance for Industry Suicidal Ideation and Behavior: Prospective Assessment of Occurrence in Clinical Trials—Draft Guidance*. U.S. Department of Health and Human Services. Food and Drug Administration Center for Drug Evaluation and Research (CDER). http://www.fda.gov/downloads/Drugs/GuidanceCompliance RegulatoryInformation/Guidances/UCM225130.pdf.
37. Lipinski, C.A., Lombardo, F., Dominy, B.W., et al. (1997). Experimental and computational approaches to estimate solubility and permeability in drug discovery and development settings. *Advanced Drug Delivery Reviews*, *23*, 3–25.
38. Lipinski, C.A., Lombardo, F., Dominy, B.W., Feeney, P.J. (2001). Experimental and computational approaches to estimate solubility and permeability in drug discovery and development settings. *Advanced Drug Delivery Reviews*, *46*(1–3), 3–26.
39. Lesson, P.D., Springthorpe, B. (2007). The influence of drug-like concepts on decision making in medicinal chemistry. *Nature Reviews Drug Discovery*, *6*(11), 881–890.
40. Wang, J., Urban, L., Bojanic, D. (2007). Maximising use of in vitro ADMET tools to predict in vivo bioavailability and safety. *Expert Opinion on Drug Metabolism & Toxicology*, *3*(5), 641–665.
41. Gavaghan, C.L., Arnby, C.H., Blomberg, N., Strandlund, G., Boyer, S. (2007). Development, interpretation and temporal evaluation of a global QSAR of hERG electrophysiology screening data. *Journal of Computer-Aided Molecular Design*, *21*(4), 189–206.
42. Wang, J. (2009). Comprehensive assessment of ADMET risks in drug discovery. *Current Pharmaceutical Design*, *15*(19), 2195–2219.
43. Kamendi, H.W., Brott, D.A., Chen, Y., Litwin, D.C., Lengel, D.J., Fonck, C., Bui, K.H., Gorko, M.A., Bialecki, R.A. (2010). Combining radio telemetry and automated blood sampling: A novel approach for integrative pharmacology and toxicology studies. *Journal of Pharmacological and Toxicological Methods*, *62*, 30–39.
44. Ewart, L., Gallacher, D.J., Gintant, G., Guillon, J.M., Leishman, D., Levesque, P., McMahon, N., Mylecraine, L., Sanders, M., Suter, W., Wallis, R., Valentin, J.P. (2012). How do the top 12 pharmaceutical companies operate safety pharmacology? *Journal of Pharmacological and Toxicological Methods*, *66*(2), 66–70.

45. Anon. (2001). *ICH Topic S7A Safety Pharmacology Studies for Human Pharmaceuticals*. Note for Guidance on Safety Pharmacology Studies for Human Pharmaceuticals (CPMP/ICH/539/00) European Medicines Agency (EMEA). http://www.emea.europa.eu/docs/en_GB/document_library/Scientific_guideline/2009/09/WC500002831.pdf.
46. Anon. (2010). *Guidance for Industry Assessment of Abuse Potential of Drugs*. U.S. Department of Health and Human Services (DHHS), Food and Drug Administration (FDA), Center for Drug Evaluation and Research (CDER). http://www.fda.gov/downloads/Drugs/GuidanceComplianceRegulatoryInformation/Guidances/UCM198650.pdf.
47. Aguirre, S.A., Heyen, J.R., Collette, W. 3rd, Bobrowski, W., Blasi, E.R. (2010). Cardiovascular effects in rats following exposure to a receptor tyrosine kinase inhibitor. *Toxicologic Pathology*, *38*, 416–428.
48. Roth, B.L. (2007). Drugs and valvular heart disease. *New England Journal of Medicine*, *356*, 6–9.
49. Whitebread, S., Hamon, J., Bojanic, D., Urban, L. (2005). Keynote review: In vitro safety pharmacology profiling: An essential tool for successful drug development. *Drug Discovery Today*, *10*(21), 1421–1433.
50. Thompson, R.A., Isin, E.M., Li, Y., Weidolf, L., Page, K., Wilson, I., Swallow, S., Middleton, B., Stahl, S., Foster, A.J., Dolgos, H., Weaver, R., Kenna, J.G. (2012). *In vitro* approach to assess the potential for risk of idiosyncratic adverse reactions caused by candidate drugs. *Chemical Research in Toxicology*, *25*(8), 1616–1632.
51. Dawson, S., Stahl, S., Paul, N., Barber, J., Kenna, J.G. (2012). In vitro inhibition of the bile salt export pump correlates with risk of cholestatic drug-induced liver injury in humans. *Drug Metabolism and Disposition*, *40*(1), 130–138.
52. Jemnitz, K., Veres, Z., Vereczkey, L. (2010). Contribution of high basolateral bile salt efflux to the lack of hepatotoxicity in rat in response to drugs inducing cholestasis in human. *Toxicological Sciences*, *115*(1), 80–88.
53. Muller, P.Y., Milton, M.N. (2012). The determination and interpretation of the therapeutic index in drug development. *Nature Reviews Drug Discovery*, *11*(10), 751–761.
54. Droogmans, S., Franken, P.R., Garbar, C., Weytjens, C., Cosyns, B., Lahoutt, T., Caveliers, V. (2007). In vivo model of drug-induced valvular heart disease in rats: Pergolide-induced valvular heart disease demonstrated with echocardiography and correlation with pathology. *European Heart Journal*, *28*, 2156–2162.
55. Anon (2005). ICH harmonized tripartite guideline S7B.Safety pharmacology assessment of the potential for delayed ventricular repolarization (QT interval prolongation) by human pharmaceuticals. Available at: http://www.ich.org/cache/compo/276-254-1.html.
56. Obradovic, T., Dobson, G.G., Shingaki, T., Kungu, T., Hidalgo, I.J. (2007). Assessment of the first and second generation antihistamines brain penetration and role of P-glycoprotein. *Pharmaceutical Research*, *24*(2), 318–327.
57. Force, T., Kolaja, K.L. (2011). Cardiotoxicity of kinase inhibitors: The prediction and translation of preclinical models to clinical outcomes. *Nature Reviews Drug Discovery*, *10*(2), 111–126.
58. Mellor, H.R., Bell, A.R., Valentin, J.P., Robert, R.R. (2011). Cardiotoxicity associated with targeting kinase pathways in cancer. *Toxicological Sciences*, *120*(1), 14–32.
59. Beasley, C.M. Jr., Masica, D.N., Heiligenstein, J.H., Wheadon, D.E., Zerbe, R.L. (1993). Possible monoamine oxidase inhibitor-serotonin uptake inhibitor interaction: Fluoxetine

clinical data and preclinical findings. *Journal of Clinical Psychopharmacology*, *13*(5), 312–320.

60. Melese, T., Lin, S.M., Chang, J.L., Cohen, N.H. (2009). Open innovation networks between academia and industry: An imperative for breakthrough therapies. *Nature Medicine*, *15*, 502–507.
61. Akil, H., Brenner, S., Kandel, E., Kendler, K.S., King, M.C., Scolnick, E., Watson, J.D., Zoghbi, H.Y. (2010). Medicine. The future of psychiatric research: Genomes and neural circuits. *Science*, *327*, 1580–1581.

2

POLYPHARMACOLOGY AND ADVERSE BIOACTIVITY PROFILES PREDICT POTENTIAL TOXICITY AND DRUG-RELATED ADRs

Teresa Kaserer, Veronika Temml, and Daniela Schuster

2.1 INTRODUCTION

For the last several decades, the pharmaceutical industry has suffered from a permanent productivity problem: Although an immense amount of money has been spent on research and development (R&D) (an estimated US$50 billion per year across all pharmaceutical companies combined), the approval rates for new chemical entities (NCEs) have been declining. This makes drug development a costly business, with approximately US$1.8 billion in expenses per drug launch [1]. Especially high late-stage attrition rates are a persisting problem in drug development, despite the rising efforts made to minimize the risk of failure at late stages. In the period from 1991 to 2000, drug attrition rates in phase II and phase III clinical trials were reported 62% and 45% [2], respectively, and since then, the numbers have not considerably improved (66% and 30%, respectively) [1]. The two major reasons identified for drug withdrawal from the market during a similar period were missing pharmacological

Predictive ADMET: Integrative Approaches in Drug Discovery and Development, First Edition.
Edited by Jianling Wang and Laszlo Urban.

efficacy and clinical safety, two factors that traditionally are investigated very late in the drug development procedure [2, 3]. In phases I–III of clinical trials, insufficient efficacy and toxicity concerns contributed with 43% and 33% to drug development termination, respectively. In addition, toxic events are the major reason (>90%) held responsible for market withdrawal after approval [3]. Early assessment of pharmacokinetics, e.g., absorption, distribution, metabolization, excretion properties, and toxicity (together called ADMET parameters), is therefore considered a promising strategy to improve success rates [4, 5]. Parallel investigation of pharmacologic activity on the desired target and ADMET profiling already during early drug discovery and design can help to eliminate compounds with unfavorable properties and to concentrate on NCEs with a higher probability to succeed. Pharmacokinetic profiling therefore can help to guide through key decisions throughout the drug developing pipeline and prevent the termination of a drug development project at a stage, where already a lot of money has been invested.

In addition, the importance of another aspect is getting more and more in focus: Although designed to be highly selective, most xenobiotics seem to bind to multiple other targets in addition to the intended one. Recent analyses suggest that on average a single drug interacts with six different targets [6]. This has many implications for the biological effects of a drug as most of the adverse drug reactions are caused by off-target binding and drug–drug interactions.

However, although predominantly considered as a highly undesired property, target promiscuity also offers new perspectives for therapy optimization strategies. In various complex diseases like cancer, diabetes, or HIV, but also in microbial infections, polypharmacology seemed beneficial for the outcome of a disease. In cancer therapy, the development of mechanism-based therapies has been a breakthrough compared with conventional chemotherapy, which led to generation of toxic effects and damage of all tissues. But soon, drug resistance occurred via the activation of compensational pathways, which helped to circumvent the effects of the targeted therapy. So efforts are made toward modulating multiple targets of the involved pathways to prevent escape mechanisms [7, 8]. One prominent example is imatinib [8, 9], which also inhibits other kinases like platelet-derived growth factor receptor (PDGFR) and KIT besides the intended target kinase Bcr-Abl. Also, other chemotherapeutic agents like sorafenib and sunitinib have shown multiple modes of actions [8]. It is similar in antimicrobial therapy, where acquired drug resistance is an important topic. So recent trends move away from a single-target based strategy towards network pharmacology, which takes complex processes in biological systems better into account [10].

As mentioned, drug development is a very costly and risky issue, and polypharmacology investigations can help to decrease expenses for drug approval. Identifying novel targets for existing drugs is an emerging strategy to extend the application of a medication to additional clinical pictures, a procedure called drug-repurposing or drug-repositioning. For these compounds, pharmacological and safety-related data are already available, which helps to accelerate approval and lowers the financial burden [11]. Several examples have already been described where an existing drug successfully was applied for another disease: For instance, aprepitant, a neurokinin 1 (NK-1) receptor antagonist used for the treatment of nausea was found to be effective in drug-resistant HIV-1 [12, 13] and the antibiotic nitroxoline was proved to prevent human umbilical vein endothelial cell (HUVEC) proliferation and angiogenesis [14]. Even thalidomide, which was withdrawn from the market during the 1960s because it caused congenital abnormalities and birth defects in children whose mother had taken the drug during pregnancy, was found to be active in the treatment of erythema nodosum leprosum (ENL), a complication occurring in leprosy patients. Still, as a result of the reasons mentioned previously, the administration of thalidomide for the treatment of leprosy is not recommended by the World Health Organization (WHO) [15]. However, the U.S. Food and Drug Administration (FDA) and the European Medicines Agency (EMA) both recently approved thalidomide in combination with other drugs (melphalan and prednisone in Europe [16] and dexamethasone in the United States [17]) for the treatment of multiple myeloma.

Pharmacological profiling comprises the estimation and prediction of a compound's pharmacokinetic and toxicological properties. ADME screening deals with a broad range of unspecific and multifactorial processes. Absorption, for instance, is not only determined by one variable, but it also is a complex procedure influenced by compound solubility, lipophilicity, efflux and influx pumps, transporters, and much more [18]. Although unspecific toxic events like genotoxicity or the disruption of cell membranes also emerge as important issues, we will focus mainly on mechanism-based toxicity and investigation of adverse drug effects caused by off-target binding in this chapter.

Bioactivity profiling, no matter if in the sense of determining pharmacokinetic and toxicologic properties or exploring the target spectrum of a certain compound, is an important strategy that improves the drug discovery and development process in many ways. Safety and risk-assessment remains a highly important aspect, where there is still room for improvement. The more that knowledge about a drug's behavior in the human body can be acquired, the higher the chance to avoid unexpected and severe adverse drug reactions and the higher the chance to successfully design a highly effective and safe drug.

In general, profiling of the compound of interest can be conducted on several different ways. *In vivo* testing could theoretically cover most of the issues to be addressed, but profiling costs would definitely exceed the financial scope of every research institution and, in particular, ethical concerns remain. Instead, a plethora of different *in vitro* assays has been developed to address the different aspects of interest. They are much less costly and time consuming and allow for general investigations as well as for specific proof of principles on isolated targets. Therefore, they are widely applied in research and development. Understanding the molecular mechanisms underlying the biological effects enables a rational approach and the *in silico* prediction of a compound's behavior in a complex biological system. The latter two can be applied in a medium-to-high throughput scale and will be discussed in detail in this chapter.

2.2 *IN VITRO* ADMET PROFILING

In times of high-throughput screening (HTS) and combinatorial chemistry, there was rising need for reliable, cost-effective, and easy-to-handle assays that allowed medium-to-high throughput profiling of compounds. ADMET profiling methods should both address cost-related issues and be predictive biological models for *in vivo* systems [19]. The first challenge was to identify parameters and/or targets that are responsible for pharmacokinetic properties of compounds and have the best predictive value.

Physicochemical properties, e.g., molecular weight, logP, pKa, solubility, chemical stability etc., represent the simplest measures that can already help very early in the decision-making process. These parameters are relatively straightforward to investigate and provide early information about absorption rates and distribution, two important factors that largely contribute to the efficacy of a drug. The logP is determined via the distribution coefficient between octanol and water (or a suitable buffer). The classic determination of a compound's concentration in the aqueous layer after addition of the organic phase with the shake-flask [20] is still applied, but several other methods like titration or high-performance liquid chromatography (HPLC)-based investigations of lipophilicity can be conducted at a higher scale [21]. Similar to that, solubility assays are usually performed via addition of a stock solution of the sample diluted in dimethylsulfoxide (DMSO) to buffer. After the incubation time, the buffer is filtrated and the concentration of solved compound is again determined via HPLC [22].

Investigating absorption is already much more challenging because biomembranes are complex compositions of various different macromolecules. The permeability of a substance is therefore not only dependent on physicochemical properties like solubility and lipophilicity, but also the activities of several different transporters and efflux pumps can largely influence the absorption rate of a single compound [18]. Human epithelial colorectal adenocarcinoma cells (Caco-2 cells) show the same phenotype as enterocytes of the small intestine when they are cultured under the right conditions. A confluent monolayer of these cells grown on a transwell plate can be used for investigating compound permeability. A transwell plate consists of two chambers that are separated via a membrane. If cells are cultured on this membrane, all the compounds of chamber one have to pass the cells to reach chamber two. Concentration of the sample in chamber two is therefore directly correlated with permeability [23, 24]. Instead of cell cultures, artificial membranes are used in the parallel artificial membrane permeability assay (PAMPA), where the barrier between the two compartments is formed by a lipid-infused membrane [25]. The permeability in this assay is only dependent on passive transport, in contrast to the Caco-2 assay, where also the activity of transporters and pumps is taken into account. The major efflux pump responsible for unfavorable permeability properties and drug resistance is P-glycoprotein (P-gp, also called multidrug resistance protein 1, MDR1). It exports molecules from the cell and is often responsible for the lack of efficacy and missing biological activity of a potential drug [26]. In addition to P-gp, the EMA [27] and the FDA [28] also recommend investigating biological interactions with six other transporters, namely organic anion-transporting polypeptide B1 (OATP1B1) and 3 (OATP1B3), organic cation transporter 2 (OCT2), organic anion transporter 1 (OAT1) and 3 (OAT3), and breast cancer resistant protein (BCRP, adenosine triphosphate (ATP) binding cassette protein G2, ABCG2). The EMA also suggests the investigation of organic cation transporter 1 (OCT1), while this transporter is neglected in the FDA guideline so far. A lot of research is done at the moment on transporters and on their role in pharmacokinetics, so this list is expected to be updated in regular intervals. In principle, several different transporter assay systems are available to the scientist. ATPase assays are based on the assumption that export is an active transport that relies on ATP hydrolysis, and the extent of ATP consumption therefore correlates with the interaction of the transporter with substrates and inhibitors [26]. Another method is the membrane vesicle transport assay, where transport is inverted and the concentration of a drug is measured in the inside of the vesicle. Of course, cell-culture–based assays with transporter protein expressing cells can also be used in a similar way as described earlier [26].

Distribution is very much dependent on the parameters mentioned so far, but an additional aspect has to be taken into account. Only unbound drugs can exert biological effects, so the rate of plasma protein binding determines the concentration of available molecules. The most popular methods to investigate plasma protein binding *in vitro* are based on size exclusion of bound molecules. In the equilibrium dialysis method (EDM), two chambers are separated by a semipermeable membrane that restricts shuttling to a relatively small molecular weight. So only unbound molecules can diffuse into the second compartment and concentration can be determined then. Similar to that, the unbound drug is filtered through a size-restricted membrane via centrifugation in the ultrafiltration method or separated via centrifugation in the ultracentrifugation method [29]. More sophisticated approaches include capillary electrophoresis, liquid chromatography, spectroscopic and calorimetric methods, and PAMPA [30].

A factor that not only influences bioavailability and efficacy of a drug but also is often involved in toxic effects and drug–drug interactions is metabolism. At the beginning of a drug development program, metabolic stability should be determined. This can be done via experiments with tissue samples, primary hepatocytes, liver microsomes, or with an S9 mixture (which is the supernatant gained when liver tissue is centrifuged at 9.000 g [31]). The drug is incubated, and the loss of substance is measured at a certain time point. Next, the drug-metabolizing enzymes responsible for the degradation are determined with purified recombinant proteins. Important enzymes comprise members of the cytochrome P450 (CYP) family, monoamine oxidases (MAOs), xanthin oxidases (XOs), aldehyde oxidases (AOs), flavin monoxidases (FMOs), uridine diphosphate glucose (UDP)-glucuronosyl transferases (UGTs), sulfotransferases (STs), glutathione transferases (GSTs), and acetyl transferases. CYPs are the most important drug metabolizing enzymes that are responsible for the elimination of a large number of xenobiotics and endogenous ligands. The most important isoforms involved in metabolism of xenobiotics are 1A2, 2C9, 2C19, 2D6, 2E1, and 3A4. Together they metabolize approximately 90% of human drugs and are therefore important determents of systemic clearance [32]. Xenobiotics can largely influence CYP activity, thereby changing the pharmacokinetic properties of additional drugs. Some drugs, e.g., clotrimazole, rifampicin [33], or natural products from St. John's wort (*Hypericum perforatum*) [34], are known to bind to the pregnane X receptor (PXR) and other nuclear receptors responsible for CYP mRNA expression. Induction of CYP activity causes an enhanced metabolic rate of co-administered drugs and a decreased bioavailability or accumulation of toxic metabolites (reviewed in Reference 35). In contrast, direct inhibition of CYPs leads to reduced metabolization and higher plasma concentration levels, thereby causing toxic effects, especially in drugs with a narrow therapeutic range [36]. This effect is not only limited to medication, but also food intake can influence CYP activity. Grapefruit juice, for instance, was incidentally discovered to inhibit intestinal CYP 3A4 and to increase the bioavailability of several drugs including some calcium channel blockers, antimicrobials, and benzodiazepines (reviewed in Reference 37). For *in vitro* assessment of CYP activity modulation, the potential drug is incubated together with the specific enzyme isoform (inhibition), human liver microsomes (inhibition) or human primary hepatocytes (inhibition [36], and induction [35]), and a known CYP substrate, and then the concentration of the substrate metabolites is measured. Although the use of human primary hepatocytes is considered a "gold standard" for *in vitro* CYP induction investigation, gene reporter assays have been developed as a supplement to this method [33, 35].

Besides the metabolic turnover of a specific drug, its way of administration can have additional effects on its bioavailability.

General and unspecific toxicity assays focus on the mutational potential and genotoxicity of xenobiotics. Three well-established methods (Ames bacterial reverse mutation test, mouse lymphoma *tk* gene mutation assay, and the micronucleus clastogenicity assay) are available for this purpose [38]. In the Ames test, amino-acid–dependent bacteria are incubated with the sample. Genotoxic agents lead to a reverse mutation that allows the bacteria to reexpress the missing amino acid and to grow again under amino-acid depletion conditions [39]. Resistance against cytotoxic agents is explored in the mouse lymphoma mutation assay. The cytotoxic effects of pyrimidin analogous are mediated via thymidine kinase (TK), which phosphorylates and recycles thymidines. If toxic agents are metabolized via this enzyme, the cell will be killed. Mutations in the TK-locus caused by genotoxic drugs prevent the phosphorylation of the pyrimidine derivatives and the cells survive [40]. The micronucleus clastogenicity assay is a cellular imaging method that investigates morphological changes of chromosomes after exposure to a test substance [41]. Additional incubation together with S9 supernatant to the samples in all three assays provides information about potential genotoxic products generated during metabolization [38].

Mechanism-based toxicologic assays, beside others, address the role of ion channels in adverse drug reactions, with the human ether-a-go-go–related gene potassium channel (hERG channel) as its most prominent representative. The hERG channel is responsible for ventricle repolarization and thereby essential for normal heart activity. Inhibition can lead to severe cardiovascular arrhythmia, and even cases of sudden death have been reported. As it is a highly promiscuous protein, it is considered a classic antitarget for safety profiling [42]. A lot of methods like binding-, ion efflux-, and fluorescence-based assays as well as patch-clamping are employed for this purpose [38]. Several other toxicity-associated targets have been identified, which should be tested, including G-protein–coupled receptors like the serotonin 2b receptor [43], and targets associated with endocrine disruption like 17β-hydroxysteroid dehydrogenases type 3 (17β-HSD3 [44, 45], 11β-hydroxysteroid dehydrogenases type 2 (11β-HSD2

[46]), or the estrogen receptor (ER) [47, 48]. However, in principle, every drug–protein interaction in addition to the desired one could lead to adverse drug effects and toxic events, as it interferes with physiological processes not associated with the disease.

The least investigated aspect in pharmacokinetics so far is excretion, where again intrinsic clearance and organ uptake rates play an important role. Intrinsic clearance is of course determined by metabolization rate and reaction phenotype, whereas organ uptake depends on passive transport (lipophilicity), blood flow, plasma protein binding (availability of xenobiotics), and transporters [26, 49]. Watanabe et al. used isolated hepatocytes and kidney slices to determine organ uptake rates and successfully correlated it with biliary and renal clearance [50]. However, although they covered the effect of transporters, mainly organic anion transporters, the mechanisms of reabsorption and backflux from the kidney tubules to the organ has to be elucidated [50].

Establishing comprehensive ADMET testing probably exceeds the capabilities of most research institutions, so several contract research organizations offer ADMET profiling and/or ADMET profiling tools. A selection of them can be found in Table 2.1.

2.3 COMPUTATIONAL METHODS PREDICTING ADMET PROPERTIES

Obviously, investigating ADMET parameters and factors with *in vitro* experimental techniques, even if additional off-target effects are neglected, is a very cost- and time-consuming process. Additional approaches are therefore investigated to accelerate the risk profiling procedure and to reduce costs to a reasonable extent. The increase in computational power together with the amount of available data concerning ligands and protein structures make more rational approaches possible. Nowadays, a plethora of ADMET parameters can be predicted with *in silico* tools. Computational methods can aid in saving resources in biological testing in two ways. On the one hand, data mining and analysis tools can help to extract relevant information from the vast amount of bioactivity data that have already been created in research but is often spread among a myriad of publications. Thorough literature search and computational analysis allow for seeing patterns, thereby making the data interpretable and making links and correlations visible. The most prominent example is the deviation of Lipinski's rule of five from the World Drug Index, stating that bioavailable compounds usually do not violate more than one of the following criteria: They have no more than 5 hydrogen bond donors, 10 hydrogen bond acceptors, a molecular mass of less than 500 daltons, and a logP not greater than 5 [51].

On the other hand, it has been a long-standing goal of computational chemistry to predict bioactivity including drug metabolism, pharmacokinetics, and toxicity from the two-dimensional (2D) structure of a compound and thereby providing a filter to determine the most promising candidates for *in vitro* testing.

Table 2.1 Companies providing *in vitro* ADMET screening

Company	Assays	Homepage
admescope	gerneral ADME / toxicity screening	http://www.admescope.com/
APREDICA	gerneral ADME / toxicity screening	http://www.apredica.com/services.php
BD Biosciences	gerneral ADME / Toxicity screening	http://www.bdbiosciences.com/eu/research/admetox/
BioFocus	gerneral ADME / toxicity screening	http://biofocus.com/offerings/adme-pk-laboratory.htm
Caliper Life Sciences	gerneral ADME / toxicity screening	http://www.caliperls.com/products/contract-research/in-vitro/admetox/
Cerep	gerneral ADME / toxicity screening	http://www.cerep.fr/Cerep/Users/index.asp
COVANCE	gerneral ADME / toxicity screening	http://www.covance.com/products/nonclinical/drug-metabolism/index.php
Cyprotex	gerneral ADME / toxicity screening	http://www.cyprotex.com/home/
Enzo Lifesciences	gerneral ADME / toxicity screening	http://www.enzolifesciences.com/landing-page/fcode/90/
Invitrogen	gerneral ADME / toxicity screening	http://www.invitrogen.com/site/us/en/home/Products-and-Services/Applications/Drug-Discovery/adme-tox.html
MDS Pharma Services	gerneral ADME / toxicity screening	http://www.mdsps.com/
Millipore	gerneral ADME / toxicity screening	http://www.millipore.com/services/flx4/serviceshome
SolveBiotech	Transporter	http://www.solvobiotech.com
TCG Lifesciences/Chembiotek	gerneral ADME / toxicity screening	http://www.tcgls.com/dmpk.php

Over the last four decades, a variety of tools have become available to calculate molecular descriptors and correlate them with biological data to model quantitative structure activity/property relationships (QSARs, QSPRs). Physicochemical properties like lipophilicity, surface, number of rotable bonds, solubility, and number of hydrogen bond donors/acceptors already give important indications about the uptake and mobility of a drug within the body [52]. More complex problems like the prediction of toxicity or potential multiple targets for a drug are often approached with similarity searches, where functional groups and molecular fragments are compared with data from already examined compounds [53].

Many computational methods directly focus on the ligand/target interactions that underlie the activity of a compound. In structure-based modeling, information about the three-dimensional (3D) structure of the protein is gained from X-ray crystallography, nuclear magnetic resonance (NMR), or homology modeling. These data are then used to find the points of interaction between ligand and protein binding site following the lock-and-key concept. If not enough structural information is available, ligand-based methods searching for common chemical features in known active compounds can be employed.

In the docking approach, a small molecule is directly fitted to the binding pocket of the protein by calculating the energetically most favorable pose inside the pocket.

In structure-based pharmacophore modeling, the molecular interactions between a protein and its observed co-crystallized small-molecule ligand are mapped as a 3D pattern of abstracted physicochemical features (hydrogen bonds, ionic interactions, hydrophobic area, etc.) [54]. A virtual screening algorithm then fits query compounds to the resulting pharmacophore and estimates their fit, thus, yielding molecules with a high probability of binding to the protein [55].

A deeper mechanistic understanding can be gained with molecular dynamics (MD) simulations, where flexible ligand binding to a target is simulated over a period of time.

Shape and pharmacophore-based virtual screening can be performed very quickly once a model is established. In so-called parallel screening, a compound gets screened against a previously generated set of models for several targets to derive a bioactivity profile. The limitation of being only specific to one target can be overcome.

2.3.1 QSAR, QSPR, and Descriptor-based Methods

QSAR is the computational search for statistically significant correlations between the 2D or the 3D structure of a compound and its bioactivity. Thereby, the structural information is represented by different descriptors (simple examples: molecular weight, number of certain atom types, or number of rotatable bonds), which are then computationally correlated with observable properties of that structure [52]. If this approach is employed in drug design, such properties could be binding affinity, toxicity, percent enzyme inhibition, or cell permeability. If the function does not relate to a specific activity but to a general property (such as lipophilicity), the term “QSPR” applies. Once such a structure activity relation is observed in the data the model was created from, it can be used to make predictions of activities that are then experimentally validated. Once established, a QSAR model is a valuable tool both in drug design and in the evaluation of environmental and medicinally used chemicals, like, e.g., to predict the cytotoxicity of a series of metal oxide nanoparticles [56].

A wide variety of molecular descriptors have been published over the history of QSAR, and correlation methods have been redefined and refined to give now several underclasses of QSAR models, sorted by dimensionality: Although one-dimensional (1D)-QSAR focuses on global variables like logP or the pKa value, 2D- and 3D-models focus on the structural information contained in the respective depictions. In 2011, a study comparing competitive CYP2C9 inhibitors by 3D-QSAR was published [57]. Higher dimensional QSAR models add information from multiple conformations and external factors like solvation. The QSAR methods can also be classified by the correlation technique employed to create them into linear methods (e.g., linear regression) and nonlinear methods like artificial neural networks [58].

There are also fragment-based QSAR methods [59], which calculate descriptors and related properties for specific molecular fragments (fingerprints) that can be rearranged to new compounds.

Many computational ADMET prediction tools work with similar methods, calculating an array of molecular descriptors for a structure or structural element and using them to find correlating ADMET properties.

Similarity ensemble approach (SEA) [53] is an open-source Web tool, which is not specialized solely on pharmacokinetics but focuses on general bioactivity profiling. It uses molecular descriptors of more than a million already known ligands from the CHEMBL database (version 10) [60] to identify new targets for a query molecule. Very recently, it successfully elucidated the mode of action behind an adverse effect observed with the application of chlorotrianisene, a synthetic ER modulator. In the context of large-scale *in silico* profiling, chlorotrianisene was predicted to bind to cyclooxygenase-1. This hypothesis was confirmed with biological testing, thereby explaining the mechanism behind the occurrence of abdominal pain associated with this drug [61].

Prediction of Activity Spectra for Substances (PASS) [62] is another software program that predicts biological activities for an unknown query molecule on the basis of 2D structures of known ligands. It is a private computational tool, but a reduced version of the program is freely accessible online [63].

Other examples for free Web-based tools would be the OSIRIS property explorer that compares descriptors and fragments of query compounds with known drugs or toxic compounds [64], or the ADMET Predictor [65]. Another tool, Volsurf [66], calculates 2D molecular descriptors specifically designed for the optimization of pharmacokinetic properties from 3D molecular interaction energy grid maps.

Although these predictors are already optimized for general ADMET properties, it might be feasible to find a specific set of descriptors to analyze properties characteristic for a compound class in question. In this case, machine learning and neural networks can be employed to find the most suitable descriptors and correlations. For example, Molecular Networks (Erlangen, Germany) provides the program ADRIANA.Code [67] to calculate molecular descriptors that can in turn be analyzed with SONNIA, a self-organizing neural network tool [68].

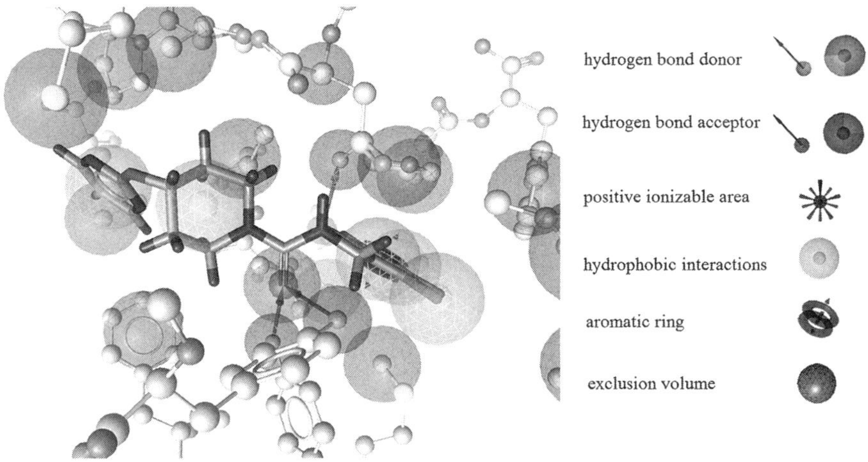

Figure 2.1 An exemplary pharmacophore model of an inhibitor (stick style) interacting with a target protein (ball-and-stick style). The example shows soluble epoxide hydrolase bound with an inhibitor and their protein–ligand interactions (Protein Data Bank accession number 3koo). (*See insert for color representation of the figure.*)

2.3.2 Molecular Interaction- and Shape-based Approaches

Molecular interaction fields encode the interaction energies between a target macromolecule, usually an enzyme or receptor, and a 3D conformation of a compound. This is the basis of pharmacophore modeling. According to the International Union of Pure and Applied Chemistry (IUPAC), a pharmacophore is defined as "an ensemble of steric and electronic features that is necessary to ensure the optimal supramolecular interactions with a specific biological target and to trigger (or block) its biological response" [69]. A pharmacophore feature can represent a hydrogen bond donor or acceptor, hydrophobic interactions, aromatic interactions, positively/negatively ionizable functional groups, and metal complexing groups (Figure 2.1). The features are mapped three-dimensionally, using experimental structural data (X-ray crystallography, NMR, and homology models) for a structure-based approach. In the ligand-based approach, the 3D pattern is derived from the 3D alignment of known active compounds.

Purely shape-based methods, where an active query molecule is used to provide a spatial template for searching similarly shaped molecules, have been shown to be most powerful in combination with pharmacophore interaction features. Pharmacophore modeling programs like LigandScout [70], Discovery Studio [71], and Phase [72] use exclusion volumes or shapes to mimic the binding pocket to prevent steric clashes, whereas the initially shape-based screening program ROCS [73] is now additionally using pharmacophore features to refine its screening algorithm.

A pharmacophore model describes the chemical, sterical, and electrostatic properties of a ligand that are required for drug–target interactions and can be used for virtual screening purposes. Of course, these feature patterns can also be used in ADMET prediction.

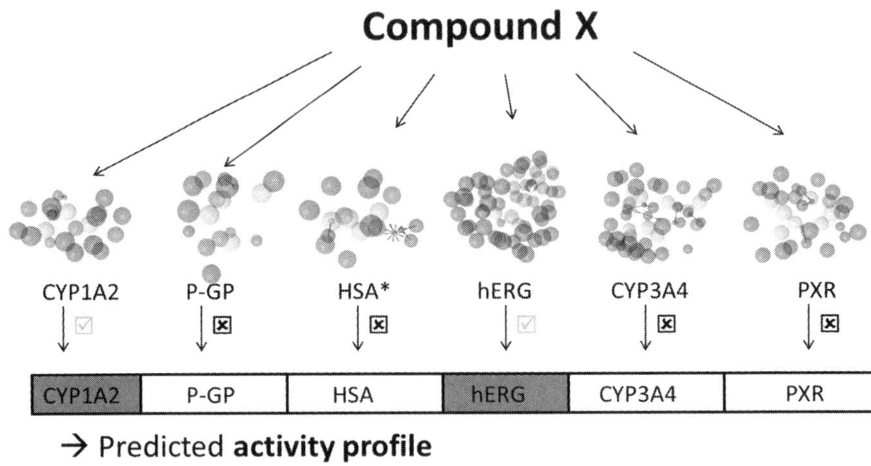

Figure 2.2 An array of pharmacophore models for different targets can be used to get a bioactivity profile, predicting the targets most interesting for biological testing. *Human serum albumin. (*See insert for color representation of the figure.*)

Pharmacophore-based virtual screening has been successfully used to identify active compounds for specific targets; for example, Waltenberger et al. used pharmacophore models for cyclooxygenase inhibitors to identify the components responsible for COX-inhibition in Prasaplai, a herbal remedy used in Thai medicine to treat inflammation [74].

Pharmacophore modeling software like Discovery Studio and LigandScout allows the users to develop their own pharmacophore collections and to use them for parallel screening (Figure 2.2). With this approach, targets for a specific compound can be identified. Such a database was used to screen leoligin, the major lignin from Edelweiß (*Leontopodium alpinum* C.). Leoligin was found to activate the cholesteryl ester transfer protein *in silico*, *in vitro*, and *in vivo* [75].

Virtual screening can also aid in finding unknown effects of xenobiotics; e.g., some chemical ultraviolet (UV) filters from the benzophenone class have been revealed by virtual screening to be active on 17β-HSD3, which was confirmed in *in vitro* testing. Thereby, UV filters could inhibit testosterone synthesis *in vivo* [45].

In the meantime, also several commercial pharmacophore databases have become available. Inte:Ligand (Maria Enzersdorf, Austria) provides the pharamacophore model database HypoDB [76] for use in Discovery Studio. Accelrys (San Diego, CA) recently implemented an automatically generated model database (PharmaDB) for its software package [71]. In addition, there is the freely accessible pharmacophore database PharmMapper [77] available, which can be used for pharmacophore-based activity profiling.

2.3.3 Docking

Molecular docking uses exclusively 3D structural data of a target for virtual screening. With the aim to predict binding modes and bioactive ligand conformations, molecules of a database are fitted into the binding site and the estimated free energy of binding is calculated. The molecules are then ranked according to their “fitness,” computed by a scoring function. Alternatively, also one compound can be docked into several different target binding pockets for the identification of putative off-target effects. Some popular docking programs are AutoDock [78], GOLD [79], FLExX [80], and GLIDE [81]. Docking is frequently used in combination with other methods; for example, a docking pose can be the basis for a pharmacophore model of a compound that has not been crystallized with its target so far.

Table 2.2 [46,82–96] provides a few examples of *in silico* modeling success stories.

2.4 OUTLOOK

In general, the limits of *in silico* models for a certain disease-related target are of course valid for ADMET-related targets as well. Pharmacophore-based and rapid overlay of chemical structures (ROCS)-based screening and QSAR models [97] will always be limited to the chemical space used for model generation. In contrast, docking may exert a greater variety in predicted scaffolds, but the problems of assessing the relevance of water molecules and the challenges of developing significant scoring functions remain [98]. In addition, the generation of *in silico* models is based on biological data, which thereby determine the quality and predictive value of a model. More high-quality and confidential experimental data are therefore needed to improve model quality [99]. An additional aspect has to be considered: Although the biological activity of disease-related targets is limited to specific functions integrated in a molecular pathway or network, ADMET-related targets are intended to exert effects on a much broader chemical variety of different endogenous and exogenous ligands. This is highlighted on the example of CYPs, where six isoforms, as mentioned earlier, are responsible for the metabolization of approximately 90% of all human drugs [32]. So they must be able to interact with a lot of chemically diverse entities and to cover a broad chemical space. As a consequence, they often are highly flexible and adopt multiple binding site conformations, which makes *in silico* predictions of this highly promiscuous targets even more challenging. This also has implications for the needs that predictive ADMET models have to fulfill: Because a lot of different chemical entities can interact with these proteins, more general models have to be created to cover all possible interactions [97]. In contrast, the focus on targeting disease-related proteins lies in the generation of rather strict models that predict a small number of ligands with a large probability to bind. So the strategy, how *in silico* models are generated, has to be adjusted to its intended application.

Although the modeler predominately adopted one method of choice on which he or she was an expert, the need to address the difficulties mentioned earlier move trends toward integrating different *in silico* approaches [97, 98]. Applying multiple techniques that are based on different principles helps to gain a better understanding of complex interaction patterns and addresses the problem in more than one way. This can not only be observed in the advancements made for established software programs, where producers try to integrate complementary methods into existing

Table 2.2 Selected examples of *in silico* ADMET profiling

	Example	Methods	
1.	Predicting new molecular targets for known drugs	2-D similarity search	[82]
2.	Computational models for drug inhibition of the human apical sodium-dependent bile acid transporter	QSAR, 2D-similarity search, pharmacophore	[83]
3.	Site of metabolism prediction on cytochrome P450 2C9: a knowledge-based docking approach	Docking	[84]
4.	Exploration of human serum albumin binding sites by docking and molecular dynamics flexible ligand–protein interactions	Docking, QSAR, MD-simulations	[85]
5.	Use of historic metabolic biotransformation data as a means of anticipating metabolic sites using MetaPrint2D and Bioclipse	Fingerprinting, datamining	[86]
6.	SMARTCyp: a 2D method for prediction of cytochrome P450-mediated drug metabolism	QSAR	[87]
7.	Pharmacophore modeling, 3D-QSAR studies, and in-silico ADME prediction of pyrrolidine derivatives as neuraminidase inhibitors	QSAR, Volsurf, Pharmacophore	[88]
8.	Parallel screening and activity profiling with HIV protease inhibitor pharmacophore models	Bioactivity Profiling Pharmacophore	[89]
9.	Pharmacophore modeling, molecular docking, QSAR, and in silico ADMET studies of gallic acid derivatives for immunomodulatory activity	Pharmacophor modeling, docking, QSAR	[90]
10.	Network pharmacology moves into the clinic	neural networks	[91]
11.	Development and validation of an in silico P450 profiler based on pharmacophore models	pharmacophore modeling	[92]
12.	A critical assessment of combined ligand- and structure-based approaches to hERG channel blocker modeling	QSAR, docking, pharmacophore modeling	[93]
13.	The identification of ligand features essential for PXR activation by pharmacophore modeling	pharmacophore modeling	[94]
14.	Data-driven homology modelling of P-glycoprotein in the ATP-bound state indicates flexibility of the transmembrane domains	homology modelling	[95]
15.	QSAR model for human pregnane X receptor (PXR) binding: Screening of environmental chemicals and correlations with genotoxicity, endocrine disruption and teratogenicity	QSAR, pharmacophore modeling	[96]
16.	Virtual screening as a strategy for the identification of xenobiotics disrupting corticosteroid action	Pharmacophore modeling	[46]

computational tools to improve the quality of the predictions, but also many recent publications [100, 101]. MD simulation represents a promising approach to investigate the dynamic interaction of a compound with its target, but as a result of the immense demands concerning time and computational power, the application is still very limited. However, with the increase in computational performance expected in the next time, MD simulation may be integrated in a standard procedure workflow, thereby dramatically increasing predictive accuracy [97]. Not only the parallel application of different techniques, but also the parallel screening of different targets for comprehensive bioactivity profiling is an advantageous strategy. Although already mentioned in the Methods section, it has to be admitted that this concept is still in its infancy and will be of interest in the future [102, 103].

Despite the ongoing refinements concerning software performance and model quality, *in silico* drug discovery will probably never be able to replace *in vitro* and *in vivo* experiments. It will provide a useful complement to existing and established methods to optimize the output and accelerate drug discovery [104, 105]. Combinatorial chemistry has enabled access to an immense chemical space [106], but the number of newly synthesized molecules exceeds the amount of compounds that can be experimentally tested by far. *In silico* screening can help here to prioritize available substances, minimize the size of the screening library to a reasonable extent, and increase the quality of the library at once [107]. Computational investigation of virtual libraries provides a selection of compounds even before they are synthesized, thereby saving resources during lead identification, but also throughout the hit optimization process. Concerning ADMET properties, computational tools can propose *in vitro* assays that should preferentially be accomplished very early because of the predicted properties of a certain compound. Different chemical properties are responsible for different problematic aspects that can be discovered during drug development (e.g., lipophilic substances have a larger probability to bind to the hERG channel and to cross the blood-brain barrier, while plasma protein binding might be increased and oral absorption decreased for hydrophilic molecules). Therefore, *in silico* methods can act as a filter that canalizes and prioritizes screening efforts to determine the benefits and risks of a potential new drug [108]. The ability to distinguish from active and inactive/binding and nonbinding molecules on the basis of their chemical structure can help to elucidate the mechanism underlying the biological function and the pharmacokinetic properties. Understanding the modes of action supports a rational drug design concept, which has proved to be much more effective than the application of the simple trial-and-error strategy of HTS alone [109, 110].

Extrapolating the actual *in vivo* effects of a drug from *in vitro* experiments is a difficult task that can result in promising drug candidates being excluded early on from further development, whereas ill-suited compounds are further pursued. Although *in vitro* models in general do yield valuable data for predicting *in vivo* activity, the results rarely translate quantitatively. Too many *in vivo* conditions cannot be mimicked in the *in vitro* experiment. For example, the occurrence of a nonspecific protein binding can result in a much smaller active site concentration of an inhibitor *in vivo* caused by the interaction with other proteins. Differences between *in vitro* and *in vivo* data also stem from the buffer/solvent conditions in the *in vitro* experiment, which often distort the activity of the tested enzyme. These effects are well researched for CYP metabolism, where *in vitro* models play an importing role in predicting *in vivo* drug–drug interaction [111].

To overcome these difficulties, *in silico* tools can be employed to take into account *in vivo* conditions (i.e., membrane permeability) and thus act as additional filters for *in vitro* screenings.

The perception that a single-targeting drug is more the exception than the rule has provided scientists with the possibility to gain mechanistic insight into adverse drug reactions and toxic side effects caused by additional drug–protein interactions. Understanding complex biological interactions and processes not only makes it possible to intentionally target multiple nodes of a pathway to prevent network robustness and design more effective drugs but also to estimate the risk of potential adverse drug effects in advance [112]. Further exploration of network pharmacology will identify additional protein targets that contribute to adverse drug effects or a certain disease pattern, thereby leading to an enlargement of the druggable space [113]. Having in mind that until now, approximately 300 different G-protein–coupled receptors are considered "targetable," but only 10% of them are known to be addressed with ligands on the market [114], it gets obvious that lots of target space remains to be explored. As data volumes are increasing exponentially and, in parallel, demands to the drug developing community are raising, a consolidation of *in silico*, *in vitro*, and *in vivo* efforts is probably the most effective strategy to address successfully future requirements and challenges [104, 105, 115].

ACKNOWLEDGMENTS

This work was supported by the Verein zur Förderung der wissenschaftlichen Ausbildung und Tätigkeit von Südtirolern an der Landesuniversität Innsbruck and the Erika Cremer Habilitation Program at the University of Innsbruck.

REFERENCES

1. Paul, S.M., Mytelka, D.S., Dunwiddie, C.T., Persinger, C.C., Munos, B.H., Lindborg, S.R., Schacht, A.L. (2010). How to improve R&D productivity: The pharmaceutical industry's grand challenge. *Nature Reviews Drug Discovery*, *9*, 203–214.
2. Kola, I., Landis, J. (2004). Can the pharmaceutical industry reduce attrition rates? *Nature Reviews Drug Discovery*, *3*, 711–715.
3. Schuster, D., Laggner, C., Langer, T. (2005). Why drugs fail–a study on side effects in new chemical entities. *Current Pharmaceutical Design*, *11*, 3545–3559.
4. Wang, J., Skolnik, S. (2009). Recent advances in physicochemical and ADMET profiling in drug discovery. *Chemistry & Biodiversity*, *6*, 1887–1899.
5. Wang, J., Urban, L., Bojanic, D. (2007). Maximising use of in vitro ADMET tools to predict in vivo bioavailability and safety. *Expert Opinion on Drug Metabolism & Toxicology*, *3*, 641–665.

6. Mestres, J., Gregori-Puigjané, E., Valverde, S., Solé, R.V. (2009). The topology of drug-target interaction networks: Implicit dependence on drug properties and target families. *Molecular Biosystems*, *5*, 1051–1057.
7. Petrelli, A., Giordano, S. (2008). From single- to multi-target drugs in cancer therapy: When aspecificity becomes an advantage. *Current Medicinal Chemistry*, *15*, 422–432.
8. Knight, Z.A., Lin, H., Shokat, K.M. (2010). Targeting the cancer kinome through polypharmacology. *Nature Reviews: Cancer*, *10*, 130–137.
9. Bantscheff, M., Eberhard, D., Abraham, Y., Bastuck, S., Boesche, M., Hobson, S., Mathieson, T., Perrin, J., Raida, M., Rau, C., Reader, V., Sweetman, G., Bauer, A., Bouwmeester, T., Hopf, C., Kruse, U., Neubauer, G., Ramsden, N., Rick, J., Kuster, B., Drewes, G. (2007). Quantitative chemical proteomics reveals mechanisms of action of clinical ABL kinase inhibitors. *Nature Biotechnology*, *25*, 1035–1044.
10. Hopkins, A.L. (2008). Network pharmacology: The next paradigm in drug discovery. *Nature Chemical Biology*, *4*, 682–690.
11. Wermuth, C.G. (2006). Selective optimization of side activities: The SOSA approach. *Drug Discovery Today*, *11*, 160–164.
12. Wang, X., Douglas, S.D., Lai, J.P., Tuluc, F., Tebas, P., Ho, W.Z. (2007). Neurokinin-1 receptor antagonist (aprepitant) inhibits drug-resistant HIV-1 infection of macrophages in vitro. *Journal of NeuroImmune Pharmacology*, *2*, 42–48.
13. Manak, M.M., Moshkoff, D.A., Nguyen, L.T., Meshki, J., Tebas, P., Tuluc, F., Douglas, S.D. (2010). Anti-HIV-1 activity of the neurokinin-1 receptor antagonist aprepitant and synergistic interactions with other antiretrovirals. *AIDS*, *24*, 2789–2796.
14. Shim, J.S., Matsui, Y., Bhat, S., Nacev, B.A., Xu, J., Bhang, H.E., Dhara, S., Han, K.C., Chong, C.R., Pomper, M.G., So, A., Liu, J.O. (2010). Effect of nitroxoline on angiogenesis and growth of human bladder cancer. *Journal of the National Cancer Institute*, *102*, 1855–1873.
15. World Health Organization (WHO), Use of thalidomide in leprosy. http://www.who.int/lep/research/thalidomide/en/index.html last accessed 10-02-2012.
16. European Medicines Agency (EMEA), http://www.ema.europa.eu/docs/en_GB/document_library/EPAR_-_Summary_for_the_public/human/000823/WC500037052.pdf last accessed 10-02-2012.
17. National Cancer Institute (NCI), http://www.cancer.gov/cancertopics/druginfo/fda-thalidomide last accessed 10-02-2012.
18. Pang, K.S. (2003). Modeling of intestinal drug absorption: roles of transporters and metabolic enzymes (for the Gillette Review Series). *Drug Metabolism and Disposition*, *31*, 1507–1519.
19. van de Waterbeemd, H., Gifford, E. (2003). ADMET in silico modelling: Towards prediction paradise? *Nature Reviews: Drug Discovery*, *2*, 192–204.
20. Leo, A., Hansch, C., Elkins, D. (1971). Partition coefficients and their uses. *Chemical Reviews*, *71*, 525–616.
21. Avdeef, A., Voloboy, D., Foreman, A. Dissolution and solubility, in B. Testa and H. Van de Waterbeemd (eds.), *Comprehensive Medicinal Chemistry II*, Elsevier, Oxford, UK, 2007, pp. 399–423.
22. Kerns, E.H., Di, L., Carter, G.T. (2008). In vitro solubility assays in drug discovery. *Current Drug Metabolism*, *9*, 879–885.

23. Hidalgo, I.J., Raub, T.J., Borchardt, R.T. (1989). Characterization of the human colon carcinoma cell line (Caco-2) as a model system for intestinal epithelial permeability. *Gastroenterology*, *96*, 736–749.
24. Shah, P., Jogani, V., Bagchi, T., Misra, A. (2006). Role of Caco-2 cell monolayers in prediction of intestinal drug absorption. *Biotechnology Progress*, *22*, 186–198.
25. Kansy, M., Senner, F., Gubernator, K. (1998). Physicochemical high throughput screening: Parallel artificial membrane permeation assay in the description of passive absorption processes. *Journal of Medicinal Chemistry*, *41*, 1007–1010.
26. Giacomini, K.M., Huang, S.M., Tweedie, D.J., Benet, L.Z., Brouwer, K.L., Chu, X., Dahlin, A., Evers, R., Fischer, V., Hillgren, K.M., Hoffmaster, K.A., Ishikawa, T., Keppler, D., Kim, R.B., Lee, C.A., Niemi, M., Polli, J.W., Sugiyama, Y., Swaan, P.W., Ware, J.A., Wright, S.H., Yee, S.W., Zamek-Gliszczynski, M.J., Zhang, L., Consortium, I.T. (2010). Membrane transporters in drug development. *Nature Reviews Drug Discovery*, *9*, 215–236.
27. European Medicines Agency (EMA), Guideline on the Investigation of Drug Interactions. http://www.ema.europa.eu/docs/en_GB/document_library/Scientific_guideline/2010/05/WC500090112.pdf last accessed 09-26-2012.
28. (FDA), F. a. D. A., Drug Interaction Studies—Study Design, Data Analysis, Implications for Dosing, and Labeling Recommendations. http://www.fda.gov/Drugs/GuidanceComplianceRegulatoryInformation/Guidances/default.htm last accessed 09-29-2012.
29. Barton, P., Austin, R.P., Fessey, R.E. In vitro models for plasma binding and tissue storage, in B. Testa, H. Van de Waterbeemd (eds.), *Comprehensive Medicinal Chemistry II*, Elsevier, Oxford, UK, 2007, pp. 21–340.
30. Vuignier, K., Schappler, J., Veuthey, J.L., Carrupt, P.A., Martel, S. (2010). Drug-protein binding: A critical review of analytical tools. *Analytical and Bioanalytical Chemistry*, *398*, 53–66.
31. Duffus, J.H., Nordberg, M., Templeton, D.M. (2007). Glossary of terms used in toxicology, 2nd ed. (IUPAC Recommendations 2007). *79*, 1153–1344.
32. Singh, D., Kashyap, A., Pandey, R.V., Saini, K.S. (2011). Novel advances in cytochrome P450 research. *Drug Discovery Today*, *16*, 793–799.
33. Lehmann, J.M., McKee, D.D., Watson, M.A., Willson, T.M., Moore, J.T., Kliewer, S.A. (1998). The human orphan nuclear receptor PXR is activated by compounds that regulate CYP3A4 gene expression and cause drug interactions. *Journal of Clinical Investigation*, *102*, 1016–1023.
34. Moore, L.B., Goodwin, B., Jones, S.A., Wisely, G.B., Serabjit-Singh, C.J., Willson, T.M., Collins, J.L., Kliewer, S.A. (2000). St. John's wort induces hepatic drug metabolism through activation of the pregnane X receptor. *Proceedings of the National Academy of Sciences U S A*, *97*, 7500–7502.
35. Lin, J.H. (2006). CYP induction-mediated drug interactions: In vitro assessment and clinical implications. *Pharmaceutical Research*, *23*, 1089–1116.
36. Fowler, S., Zhang, H. (2008). In vitro evaluation of reversible and irreversible cytochrome P450 inhibition: Current status on methodologies and their utility for predicting drug-drug interactions. *The AAPS Journal*, *10*, 410–424.
37. Kiani, J., Imam, S. Z. (2007). Medicinal importance of grapefruit juice and its interaction with various drugs. *Nutrition Journal*, *6*, 33.

38. Houck, K.A., Kavlock, R.J. (2008). Understanding mechanisms of toxicity: Insights from drug discovery research. *Toxicology and Applied Pharmacology*, *227*, 163–178.
39. Ames, B.N., Mccann, J., Yamasaki, E. (1975). Methods for detecting carcinogens and mutagens with the Salmonella/mammalian-microsome mutagenicity test. *Mutation Research*, *31*, 347–364.
40. Clive, D., Spector, J. F. (1975). Laboratory procedure for assessing specific locus mutations at the TK locus in cultured L5178Y mouse lymphoma cells. *Mutation Research*, *31*, 17–29.
41. Kirsch-Volders, M., Elhajouji, A., Cundari, E., Van Hummelen, P. (1997). The in vitro micronucleus test: A multi-endpoint assay to detect simultaneously mitotic delay, apoptosis, chromosome breakage, chromosome loss and non-disjunction. *Mutation Research*, *392*, 19–30.
42. Sanguinetti, M.C., Tristani-Firouzi, M. (2006). hERG potassium channels and cardiac arrhythmia. *Nature*, *440*, 463–469.
43. Roth, B.L. (2007). Drugs and valvular heart disease. *New England Journal of Medicine*, *356*, 6–9.
44. Schuster, D., Kowalik, D., Kirchmair, J., Laggner, C., Markt, P., Aebischer-Gumy, C., Ströhle, F., Möller, G., Wolber, G., Wilckens, T., Langer, T., Odermatt, A., Adamski, J. (2011). Identification of chemically diverse, novel inhibitors of 17β-hydroxysteroid dehydrogenase type 3 and 5 by pharmacophore-based virtual screening. *Journal of Steroid Biochemistry and Molecular Biology*, *125*, 148–161.
45. Nashev, L.G., Schuster, D., Laggner, C., Sodha, S., Langer, T., Wolber, G., Odermatt, A. (2010). The UV-filter benzophenone-1 inhibits 17beta-hydroxysteroid dehydrogenase type 3: Virtual screening as a strategy to identify potential endocrine disrupting chemicals. *Biochemical Pharmacology*, *79*, 1189–1199.
46. Nashev, L.G., Vuorinen, A., Praxmarer, L., Chantong, B., Cereghetti, D., Winiger, R., Schuster, D., Odermatt, A. (2012). Virtual screening as a strategy for the identification of xenobiotics disrupting corticosteroid action. *PLoS ONE*, *7*, e46958.
47. Clere, N., Lauret, E., Malthiery, Y., Andriantsitohaina, R., Faure, S. (2012). Estrogen receptor alpha as a key target of organochlorines to promote angiogenesis. *Angiogenesis*, *15*, 745–760.
48. Hao, C.J., Cheng, X.J., Xia, H.F., Ma, X. (2012). The endocrine disruptor diethylstilbestrol induces adipocyte differentiation and promotes obesity in mice. *Toxicology and Applied Pharmacology*, *263*, 102–110.
49. Wu, C.Y., Benet, L.Z. (2005). Predicting drug disposition via application of BCS: Transport/absorption/ elimination interplay and development of a biopharmaceutics drug disposition classification system. *Pharmaceutical Research*, *22*, 11–23.
50. Watanabe, T., Maeda, K., Kondo, T., Nakayama, H., Horita, S., Kusuhara, H., Sugiyama, Y. (2009). Prediction of the hepatic and renal clearance of transporter substrates in rats using in vitro uptake experiments. *Drug Metabolism and Disposition*, *37*, 1471-1479.
51. Lipinski, C.A., Lombardo, F., Dominy, B.W., Feeney, P.J. (1997). Experimental and computational approaches to estimate solubility and permeability in drug discovery and development settings. *Advanced Drug Delivery Reviews*, *23*, 3–25.
52. Gasteiger, J. (2006). Of molecules and humans. *Journal of Medicinal Chemistry*, *49*, 6429–6434.

53. Keiser, M.J., Roth, B.L., Armbruster, B.N., Ernsberger, P., Irwin, J.J., Shoichet, B.K. (2007). Relating protein pharmacology by ligand chemistry. *Nature Biotechnology*, *25*, 197–206.
54. Kier, L.B. *MO Theory in Drug Research*, Academic Press. New York, 1971, pp. 164–169.
55. Rester, U. (2008). From virtuality to reality - Virtual screening in lead discovery and lead optimization: A medicinal chemistry perspective. *Current Opinion on Drug Discovery and Development*, *11*, 559–568.
56. Puzyn, T., Rasulev, B., Gajewicz, A., Hu, X., Dasari, T.P., Michalkova, A., Hwang, H.M., Toropov, A., Leszczynska, D., Leszczynski, J. (2011). Using nano-QSAR to predict the cytotoxicity of metal oxide nanoparticles. *Nature Nanotechnology*, *6*, 175-178.
57. Lather, V., Fernandes, M.X. (2011). Comparative QSAR analyses of competitive CYP2C9 inhibitors using three-dimensional molecular descriptors. *Chemical Biology & Drug Design*, *78*, 112–123.
58. Verma, J., Khedkar, V.M., Coutinho, E.C. (2010). 3D-QSAR in drug design–a review. *Current Topics in Medicinal Chemistry*, *10*, 95–115.
59. Salum, L.B., Andricopulo, A.D. (2009). Fragment-based QSAR: Perspectives in drug design. *Molecular Diversity*, *13*, 277–285.
60. Rabal, O., Oyarzabal, J. (2012). Using novel descriptor accounting for ligand-receptor interactions to define and visually explore biologically relevant chemical space. *Journal of Chemical Information and Modeling*, *52*, 1086–1102.
61. Lounkine, E., Keiser, M.J., Whitebread, S., Mikhailov, D., Hamon, J., Jenkins, J. L., Lavan, P., Weber, E., Doak, A.K., Côté, S., Shoichet, B.K., Urban, L. (2012). Large-scale prediction and testing of drug activity on side-effect targets. *Nature*, *486*, 361–367.
62. Poroikov, V., Filimonov, D., Lagunin, A., Gloriozova, T., Zakharov, A. (2007). PASS: Identification of probable targets and mechanisms of toxicity. *SAR and QSAR in Environmental Research*, *18*, 101–110.
63. PASSonline. http://www.pharmaexpert.ru/passonline (accessed 09-20-2012).
64. OSIRIS property Explorer. http://www.organic-chemistry.org/prog/peo/ (accessed 10–04-2012).
65. *ADMET predictor*, http://www.simulationsplus.com/Products.aspx?grpID=1&cID=11&pID=13 09-25-2012.
66. Crivori, P., Cruciani, G., Carrupt, P.A., Testa, B. (2000). Predicting blood-brain barrier permeation from three-dimensional molecular structure. *Journal of Medicinal Chemistry*, *43*, 2204–2216.
67. *ADRIANA.Code*, Molecular Networks GmbH, Erlangen, Germany, http://www.molecular-networks.com/products/adrianacode 09-25-2012.
68. *SONNIA-Self-Organizing Neural Network Package*, Molecular Networks GmbH, Erlangen, Germany, http://www.molecular-networks.com/products/sonnia 09-25-2012.
69. Wermuth, C.G., Ganellin, C.R., Lindberg, P., Mitscher, L.A. (1998). Glossary of terms used in medicinal chemistry (IUPAC Recommendations 1998). *Pure and Applied Chemistry*, *70*, 1129–1143.
70. Wolber, G., Langer, T. (2005). LigandScout: 3-D pharmacophores derived from protein-bound ligands and their use as virtual screening filters. *Journal of Chemical Information and Modeling*, *45*, 160–169.

71. *Discovery Studio*, Accelrys Inc., San Diego, CA, USA http://accelrys.com/products/discovery-studio/ 10-02-2012.
72. Dixon, S.L., Smondyrev, A.M., Rao, S.N. (2006). PHASE: A novel approach to pharmacophore modeling and 3D database searching. *Chemical Biology & Drug Design*, *67*, 370–372.
73. Rush, T.S., Grant, J.A., Mosyak, L., Nicholls, A. (2005). A shape-based 3-D scaffold hopping method and its application to a bacterial protein-protein interaction. *Journal of Medicinal Chemistry*, *48*, 1489–1495.
74. Waltenberger, B., Schuster, D., Paramapojn, S., Gritsanapan, W., Wolber, G., Rollinger, J.M., Stuppner, H. (2011). Predicting cyclooxygenase inhibition by three-dimensional pharmacophoric profiling. Part II: Identification of enzyme inhibitors from Prasaplai, a Thai traditional medicine. *Phytomedicine*, *18*, 119–133.
75. Duwensee, K., Schwaiger, S., Tancevski, I., Eller, K., van Eck, M., Markt, P., Linder, T., Stanzl, U., Ritsch, A., Patsch, J.R., Schuster, D., Stuppner, H., Bernhard, D., Eller, P. (2011). Leoligin, the major lignan from Edelweiss, activates cholesteryl ester transfer protein. *Atherosclerosis*, *219*, 109–115.
76. HypoDB. Inte:Ligand GmbH, Maria Enzersdorf, Austria, 2006-2012 http://www.inteligand.com/pharmdb/, last accessed 10-05-2012.
77. Liu, X., Ouyang, S., Yu, B., Liu, Y., Huang, K., Gong, J., Zheng, S., Li, Z., Li, H., Jiang, H. (2010). PharmMapper server: A web server for potential drug target identification using pharmacophore mapping approach. *Nucleic Acids Research*, *38*, W609–614.
78. Morris, G.M., Goodsell, D.S., Halliday, R.S., Huey, R., Hart, W.E., Belew, R.K., Olson, A.J. (1998). Automated docking using a Lamarckian genetic algorithm and an empirical binding free energy function. *Journal of Computational Chemistry*, *19*, 1639–1662.
79. *Gold*, CCDC, Cambridge, UK, http://www.ccdc.cam.ac.uk/products/life_sciences/gold/ last accessed 10-02-2012.
80. Sousa, S.F., Fernandes, P.A., Ramos, M.J. (2006). Protein-ligand docking: current status and future challenges. *Proteins*, *65*, 15–26.
81. Friesner, R.A., Banks, J.L., Murphy, R.B., Halgren, T.A., Klicic, J.J., Mainz, D. T., Repasky, M.P., Knoll, E.H., Shelley, M., Perry, J.K., Shaw, D.E., Francis, P., Shenkin, P.S. (2004). Glide: A new approach for rapid, accurate docking and scoring. 1. Method and assessment of docking accuracy. *Journal of Medicinal Chemistry*, *47*, 1739–1749.
82. Keiser, M.J., Setola, V., Irwin, J.J., Laggner, C., Abbas, A.I., Hufeisen, S.J., Jensen, N.H., Kuijer, M.B., Matos, R.C., Tran, T.B., Whaley, R., Glennon, R.A., Hert, J., Thomas, K.L., Edwards, D.D., Shoichet, B.K., Roth, B.L. (2009). Predicting new molecular targets for known drugs. *Nature*, *462*, 175–181.
83. Zheng, X., Ekins, S., Raufman, J.P., Polli, J.E. (2009). Computational models for drug inhibition of the human apical sodium-dependent bile acid transporter. *Molecular Pharmaceutics*, *6*, 1591–1603.
84. Tarcsay, A., Kiss, R., Keseru, G.M. (2010). Site of metabolism prediction on cytochrome P450 2C9: A knowledge-based docking approach. *Journal of Computer-Aided Molecular Design*, *24*, 399–408.
85. Deeb, O., Rosales-Hernández, M.C., Gómez-Castro, C., Garduño-Juárez, R., Correa-Basurto, J. (2010). Exploration of human serum albumin binding sites by docking and molecular dynamics flexible ligand-protein interactions. *Biopolymers*, *93*, 161–170.

86. Carlsson, L., Spjuth, O., Adams, S., Glen, R.C., Boyer, S. (2010). Use of historic metabolic biotransformation data as a means of anticipating metabolic sites using MetaPrint2D and Bioclipse. *BMC Bioinformatics*, *11*, 362.
87. Rydberg, P., Gloriam, D.E., Zaretzki, J., Breneman, C., Olsen, L. (2010). SMARTCyp: A 2D method for prediction of Cytochrome P450-mediated drug metabolism. *ACS Medicinal Chemistry Letters*, *1*, 96–100.
88. Zhang, J., Pan, X., Wang, C., Wang, F., Li, P., Xu, W., He, L. (2012). Pharmacophore modeling, 3D-QSAR studies, and in-silico ADME prediction of pyrrolidine derivatives as neuraminidase inhibitors. *Chemical Biology & Drug Design*, *79*, 353–359.
89. Steindl, T.M., Schuster, D., Laggner, C., Chuang, K., Hoffmann, R.D., Langer, T. (2007). Parallel screening and activity profiling with HIV protease inhibitor pharmacophore models. *Journal of Chemical Information and Modeling*, *47*, 563–571.
90. Yadav, D.K., Khan, F., Negi, A.S. (2012). Pharmacophore modeling, molecular docking, QSAR, and in silico ADMET studies of gallic acid derivatives for immunomodulatory activity. *Journal of Molecular Modeling*, *18*, 2513–2525.
91. (2012) News in brief: Network pharmacology moves into the clinic. *Nature Reviews: Drug Discovery*, *11*, 589.
92. Schuster, D., Laggner, C., Steindl, T.M., Langer, T. (2006). Development and validation of an in silico P450 profiler based on pharmacophore models. *Current Drug Discovery Technologies*, *3*, 1–48.
93. Du-Cuny, L., Chen, L., Zhang, S. (2011). A critical assessment of combined ligand- and structure-based approaches to hERG channel blocker modeling. *Journal of Chemical Information and Modeling*, *51*, 2948–2960.
94. Schuster, D., Langer, T. (2005). The identification of ligand features essential for PXR activation by pharmacophore modeling. *Journal of Chemical Information and Modeling*, *45*, 431–439.
95. Stockner, T., de Vries, S.J., Bonvin, A.M., Ecker, G.F., Chiba, P. (2009). Data-driven homology modelling of P-glycoprotein in the ATP-bound state indicates flexibility of the transmembrane domains. *FEBS Journal*, *276*, 964–972.
96. Dybdahl, M., Nikolov, N.G., Wedebye, E.B., Jónsdóttir, S., Niemelä, J.R. (2012). QSAR model for human pregnane X receptor (PXR) binding: Screening of environmental chemicals and correlations with genotoxicity, endocrine disruption and teratogenicity. *Toxicology and Applied Pharmacology*, *262*, 301–309.
97. Moroy, G., Martiny, V.Y., Vayer, P., Villoutreix, B.O., Miteva, M.A. (2012). Toward in silico structure-based ADMET prediction in drug discovery. *Drug Discovery Today*, *17*, 44–55.
98. Kitchen, D.B., Decornez, H., Furr, J.R., Bajorath, J. (2004). Docking and scoring in virtual screening for drug discovery: methods and applications. *Nature Reviews: Drug Discovery*, *3*, 935–949.
99. Dearden, J.C. (2007). In silico prediction of ADMET properties: How far have we come? *Expert Opinion on Drug Metabolism & Toxicology*, *3*, 635–639.
100. von Grafenstein, S., Mihaly-Bison, J., Wolber, G., Bochkov, V.N., Liedl, K.R., Schuster, D. (2012). Identification of novel liver X receptor activators by structure-based modeling. *Journal of Chemical Information and Modeling*, *52*, 1391–1400.
101. Sato, T., Yuki, H., Takaya, D., Sasaki, S., Tanaka, A., Honma, T. (2012). Application of support vector machine to three-dimensional shape-based virtual screening using

comprehensive three-dimensional molecular shape overlay with known inhibitors. *Journal of Chemical Information and Modeling*, *52*, 1015–1026.

102. Schuster, D. (2010). 3D pharmacophores as tools for activity profiling. *Drug Discovery Today: Technologies*, *7*, 205–211.
103. Allen, J.A., Roth, B.L. (2011). Strategies to discover unexpected targets for drugs active at G protein-coupled receptors. *Annual Reviews of Pharmacology and Toxicology*, *51*, 117–144.
104. Bajorath, J. (2002). Integration of virtual and high-throughput screening. *Nature Reviews: Drug Discovery*, *1*, 882–894.
105. Guido, R.V., Oliva, G., Andricopulo, A.D. (2011). Modern drug discovery technologies: Opportunities and challenges in lead discovery. *Combinatorial Chemistry and High Throughput Screening*, *14*, 830–839.
106. López-Vallejo, F., Caulfield, T., Martínez-Mayorga, K., Giulianotti, M.A., Nefzi, A., Houghten, R.A., Medina-Franco, J.L. (2011). Integrating virtual screening and combinatorial chemistry for accelerated drug discovery. *Combinatorial Chemistry and High Throughput Screening*, *14*, 475–487.
107. Babaoglu, K., Simeonov, A., Irwin, J.J., Nelson, M.E., Feng, B., Thomas, C.J., Cancian, L., Costi, M.P., Maltby, D.A., Jadhav, A., Inglese, J., Austin, C.P., Shoichet, B. K. (2008). Comprehensive mechanistic analysis of hits from high-throughput and docking screens against beta-lactamase. *Journal of Medicinal Chemistry*, *51*, 2502–2511.
108. Gleeson, M.P., Hersey, A., Hannongbua, S. (2011). In-silico ADME models: A general assessment of their utility in drug discovery applications. *Current Topics in Medicinal Chemistry*, *11*, 358–381.
109. Schuster, D., Spetea, M., Music, M., Rief, S., Fink, M., Kirchmair, J., Schütz, J., Wolber, G., Langer, T., Stuppner, H., Schmidhammer, H., Rollinger, J.M. (2010). Morphinans and isoquinolines: Acetylcholinesterase inhibition, pharmacophore modeling, and interaction with opioid receptors. *Bioorganic & Medicinal Chemistry*, *18*, 5071–5080.
110. Doman, T.N., McGovern, S.L., Witherbee, B.J., Kasten, T.P., Kurumbail, R., Stallings, W.C., Connolly, D.T., Shoichet, B.K. (2002). Molecular docking and high-throughput screening for novel inhibitors of protein tyrosine phosphatase-1B. *Journal of Medicinal Chemistry*, *45*, 2213–2221.
111. Wienkers, L.C., Heath, T.G. (2005). Predicting in vivo drug interactions from in vitro drug discovery data. *Nature Reviews: Drug Discovery*, *4*, 825–833.
112. Xie, L., Bourne, P.E. (2011). Structure-based systems biology for analyzing off-target binding. *Current Opinion in Structural Biology*, *21*, 189–199.
113. Merino, A., Bronowska, A.K., Jackson, D.B., Cahill, D.J. (2010). Drug profiling: Knowing where it hits. *Drug Discovery Today*, *15*, 749–756.
114. Brown, J.B., Okuno, Y. (2012). Systems biology and systems chemistry: New directions for drug discovery. *Chemistry & Biology*, *19*, 23–28.
115. Wang, J., Collis, A. (2011). Maximizing the outcome of early ADMET models: Strategies to win the drug-hunting battles? *Expert Opinion on Drug Metabolism & Toxicology*, *7*, 381–386.

II

INTELLIGENT INTEGRATION AND EXTRAPOLATION OF ADMET DATA

3

ADMET DIAGNOSIS MODELS

Bernard Faller, Suzanne Skolnik, and Jianling Wang

3.1 INTRODUCTION

Preclinical *in vitro* absorption, distribution, metabolism, excretion, and toxicity (ADMET) assays have been useful to filter out molecules with potential liabilities and thus reduce attrition in later drug development. That said, when a flag appears in a given assay, additional information related to what structural element or physicochemical property is responsible for the liability would help design the next molecule and reduce the iterative optimization cycle. Incorporating diagnostic tools in the ADME optimization cycle helps move away from a pure trial-and-error approach toward a hypothesis-based optimization approach and can significantly shorten the compound optimization phase. Frequently, identifying the most influencing physicochemical properties of a specific parameter will also help selection of relevant assay(s) to monitor success.

3.2 SOLUBILITY DIAGNOSIS

Solubility and permeability largely determine the fraction absorbed; therefore, if one or both of these two parameters is unfavorable, then the fraction of the dose entering the systemic circulation might be insufficient to elicit the desired pharmacological effect and be prone to significant interindividual variability. Solubility limitations can be partially overcome via formulation, but this approach is not always successful

Predictive ADMET: Integrative Approaches in Drug Discovery and Development, First Edition.
Edited by Jianling Wang and Laszlo Urban.

and significantly increases the overall drug development time; therefore, the best approach is to address the molecule itself. Solubility is mainly controlled by three factors:

- Lipophilicity
- Ionization
- Solid state properties

Although these three parameters are not completely independent, one is usually dominant in controlling solubility. Therefore, if one could identify which of these three descriptors is dominant for a given chemotype, a suitable MedChem strategy could be put in place to improve solubility.

3.2.1 Lipophilicity and Maximum Solubility Concept

The importance of lipophilicity as a contributor to solubility has long been recognized [1], and many relationships linking solubility and octanol–water partition coefficients are described in the literature. Initial studies were restricted to liquids to avoid the complicating effect of crystal structure.

Hansch et al. reported on the following relationship for 156 liquid solutes using calculated logP values [1]:

$$\log S_0 = -1.33 \log P + 0.98 \tag{3.1}$$

Later, Valvani recalculated Equation 3.1 starting with experimental logP values [2] and found

$$\log S_0 = -1.01 \log P + 0.51 \tag{3.2}$$

Scientists have tried to extend the approach of Hansch by including crystalline solutes. For example, Isnard and Lambert obtained [3]

$$\log S_0 = -1.38 \log P + 1.17 \tag{3.3}$$

A minor improvement was obtained by the addition of a melting point term, perhaps because only a small proportion of the compounds in the training set had melting temperature (mp) > 125°C. In this respect, the molecules used to build the previous equations poorly reflect the current physicochemical property space [4], and therefore, it is unlikely that the correlations obtained in these studies can be reproduced with today's molecules.

As shown in earlier studies, lipophilicity largely defines the upper limit of intrinsic solubility [5]. Figure 3.1 shows the relationship between intrinsic solubility and

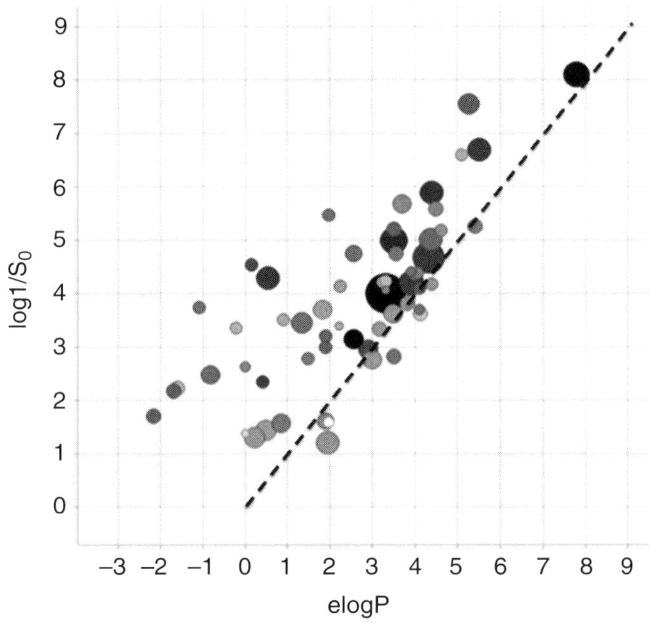

Figure 3.1 Intrinsic solubility–lipophilicity relationship. The dot size increases with increasing number of rotatable bonds. The dot color is by MW (the higher the MW, the darker the color). Few molecules fall below the dotted line indicating that logP controls the upper solubility limit. The solubility of compounds much above the dotted line is controlled by crystal lattice energy (ΔSL >2).

lipophilicity for 59 generic drugs. Data plotted in Figure 3.1 were essentially taken from the dataset compiled by Avdeef [6]. One can see that lipophilicity largely determines the maximal solubility with $S_0 \leq 10^{-\log P}$ as a generic rule as only very few molecules fall above the unity line. For example, the upper intrinsic solubility value of a compound of logP = 4 is 10^{-4} M. The fact that intrinsic solubility (S_0) can be much lower than anticipated by logP means that other factors than lipophilicity, such as crystal lattice energy, can further decrease intrinsic solubility as these interactions are additive. Simple descriptors such as molecular weight (MW) or the number of rotatable bonds are not sufficient to explain why the solubility of some molecules is substantially lower than anticipated from their logP value. Wassvik et al. [7] analyzed the respective contribution of lipophilicity and crystal packing on intrinsic solubility. As in Figure 3.1, they observed that the contribution of crystal packing was more important for molecules with a calculated octanol–water partition coefficient (clogP) < 2. In Wassvik et al.'s study, the fraction of compounds above the unity line ($S_0 > 10^{-\log P}$) was higher than in our Figure 3.1. We have observed the same trend if we substitute experimental octanol–water partition coefficient (elogP) for clogP and repeat the analysis with new molecules (outside the training set of prediction software). With currently available technology, it is not possible to anticipate the impact of solid state properties reliably from the molecular structure, and as a consequence, initial solubility assessment can only give an upper value.

3.2.2 Estimating the Impact of the Solid State in the Absence of Crystalline Material

Most high-throughput solubility assays start with dimethylsulfoxide (DMSO) stock solutions [8] because handling of liquids is a lot more efficient than powders, meaning that the property of the solid is lost during the dissolution phase. At the same time, early batches of new chemical entities are rarely crystalline, and therefore, a direct, experimental assessment of the properties of the solid is anyway rarely possible. That said, in some cases, it is possible to get an early estimate of the impact of the solid state on solubility. In earlier work [5], we defined the $\log 1/S_0$-logP (ΔSL) parameter: This is the difference in log units between the intrinsic solubility and lipophilicity. A ΔSL value greater than 2 usually correlates with a melting point above 200°C. ΔSL is, however, not always a good indicator of crystal packing, and in some cases, the solubility of molecules with a low ΔSL value might still be limited by the solid state properties, depending on whether the compound crystallizes during the incubation time of the solubility assay.

3.2.3 When Is the Maximal Effect of Ionization Reached?

Increasing ionization is a powerful way to improve solubility at physiological pH. Equations governing the impact of ionization on solubility are shown in Figure 3.2. For a monoprotic acid, solubility increases by 1 log unit / pH unit once the pH is above the pKa by at least 0.5 units. This is illustrated with a weak acid (Diclofenac) in Figure 3.3. Looking at the fraction ionized (FI) as such can be misleading. For this reason, we defined an ionization parameter (IP), which allows for quickly assessing the impact of ionization on solubility at physiological pH:

$$\mathrm{IP} = -\log(1 - \mathrm{FI}) \tag{3.4}$$

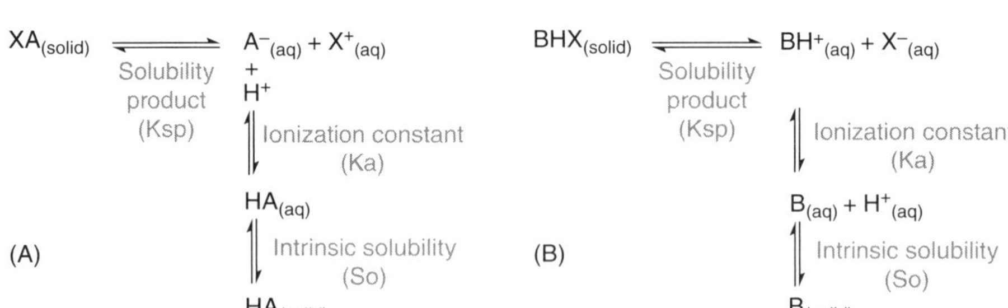

Figure 3.2 Equilibrium of a free acid (A) and base (B) and their corresponding salts; X^+ and X^- are the counterions.

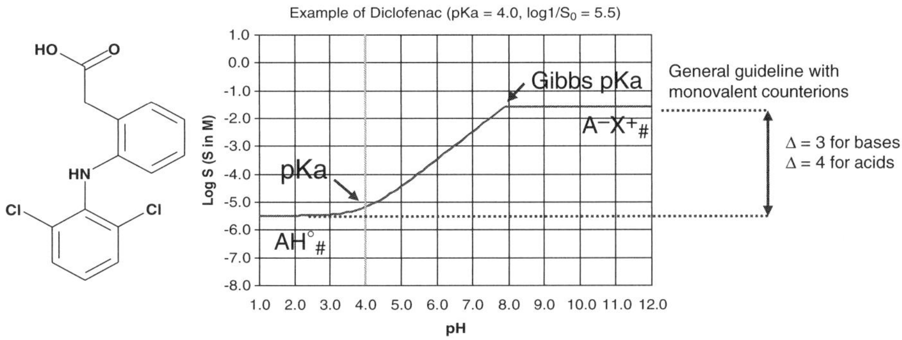

Figure 3.3 Solubility pH profile of Diclofenac. The lower plateau is defined by the intrinsic solubility and the upper plateau by the pKsp value (see equations in Figure 3.2). In the absence of other phenomena like aggregation, the slope is +1 for a monoprotic acid and –1 for a monoprotic base.

An $IP_{6.8}$ value of 2 means that the solubility at pH 6.8 is 2 log units above the intrinsic solubility (Table 3.1). There are, however, two points to consider carefully when using ionization to improve solubility:

- Increasing ionization can translate to a reduced passive permeability.
- There is a threshold IP value above which the solubility no longer increases because solubility becomes limited by the solubility product.

Table 3.1 Solubility and ionization[a]

pH	FI	IP	–logSo(M)
3.0	0.091	0.04	5.5
4.0	0.5	0.30	5.2
5.0	0.909	1.04	4.5
6.0	0.990	2.00	3.5
7.0	0.999	3.00	2.5

[a]Both FI and IP values refer to pH6.8 conditions.

3.2.4 The Solubility Diagnosis Matrix

The diagnosis matrix puts together parameters useful to dissect down solubility. An example is provided in Table 3.2.

Column A: This is the solubility number you want to optimize. Note that the solubility values are expressed in negative log units, and therefore, a lower number means a higher solubility value.

Columns F and G: These parameters report on the contribution of crystal lattice energy to intrinsic solubility (column C). Tm values above 200°C become prohibitive for a good solubility. ΔSL is obtained by subtracting the value in columns C and E.

Table 3.2 The solubility diagnosis matrix

ID	A $\text{Log1/S}_{6.8}$ (measured)	B $\text{Log1/S}_{6.8}$ (extrapolated)	C Log1/S_0	D $\text{IP}_{6.8}$	E logP	F ΔSL	G T_m
Compound 1	5.0	5.1	4.9	0.0	4.5	0.4	n/a
Compound 2	4.9	4.0	5.1	1.5	3.9	1.2	145
Compound 3	5.3	5.3	5.4	0	4.5	0.9	280
Compound 4	5.4	5.2	5.1	0.2	2.9	2.2	n/a

It can be used as a substitute for Tm in the absence of crystalline material. A ΔSL value greater than 2 is a sign of an unfavorable crystal lattice energy, but the reverse is not true.

Column C: This is the intrinsic solubility. This parameter dials out the ionization contribution to solubility.

Column D: This parameter describes the contribution of ionization to the solubility at pH 6.8 (see Table 3.1 for the definition of IP).

Column B: This is the solubility at pH 6.8 extrapolated from $\text{IP}_{6.8}$ and the intrinsic solubility (S_0). This value is equal to or higher than the number in column A. A higher number in column B versus column A indicates that the solubility product (Ksp) starts to become the limiting factor and that the maximum effect of ionization has been reached.

3.2.5 Diagnosis Examples (Molecules in Table 3.2)

Compound 1: Columns C and E are roughly equal. This means that intrinsic solubility is mainly controlled by lipophilicity. The compound is not ionized at pH 6.8 ($\text{IP}_{6.8} = 0$), and therefore, columns A, B, and C are equal. Solubility optimization strategy: Decrease lipophilicity and/or increase ionization.

Compound 2: Here the difference between columns C and E is a bit higher but still lower than 2 log units, suggesting that lipophilicity at least partially controls solubility. The melting point value at 145°C is not indicative of a particularly high crystal lattice energy. With this compound, $\text{IP}_{6.8}$ is 1.5, which means it is significantly ionized at pH 6.8. The value in column B is significantly lower than in column A, suggesting a solubility product limitation or aggregation.

Compound 3: Columns C and E are roughly identical, but the mp is high (column G). In this case, ΔSL is not a good surrogate for crystal lattice energy. This happens when (1) the compound does not crystalize in solution during the incubation of the solubility experiment and (2) the compound crystallizes in a different solid form.

Compound 4: ΔSL is high, suggesting solubility is limited by crystal lattice energy. Breaking the crystal lattice and/or introducing an ionizable group to the molecule can improve solubility.

3.3 DIAGNOSING PERMEABILITY

Permeation of a drug substance across the gastrointestinal (GI) membrane is a complicated process involving multiple mechanisms. Whereas passive diffusion may predominate for most drugs, active or facilitated transport may also occur. Therefore, identification of the most influencing physicochemical parameters affecting permeability will unsurprisingly be more complex than solubility. In this report, we have conducted a statistical analysis on Novartis Pharmaceuticals (Basel, Switzerland) drug discovery compounds from multiple projects and therapeutic indications to reveal which parameters mostly differentiate high- from low-permeable molecules. Since permeability is primarily driven by passive mechanism (that is, efflux is an additional burden for low permeable compounds) [9], the current permeability diagnosis model will focus on parameters affecting passive permeability. The analysis was conducted on both in-house Caco-2 and PAMPA datasets giving consistent results, while statistical details are reported for Caco-2 only.

The principal component analysis (PCA) loading plot in Figure 3.4 differentiates high- and low-passive permeability from an agreed binning dataset between in-house Caco-2 and PAMPA assays. Similar to those in solubility diagnosis, these properties may not be completely independent. Nonetheless, these properties identified by the trend analysis can be used for diagnosing key issues limiting permeability, proposing the most relevant experiments to test the hypothesis and prioritizing chemistry in drug discovery.

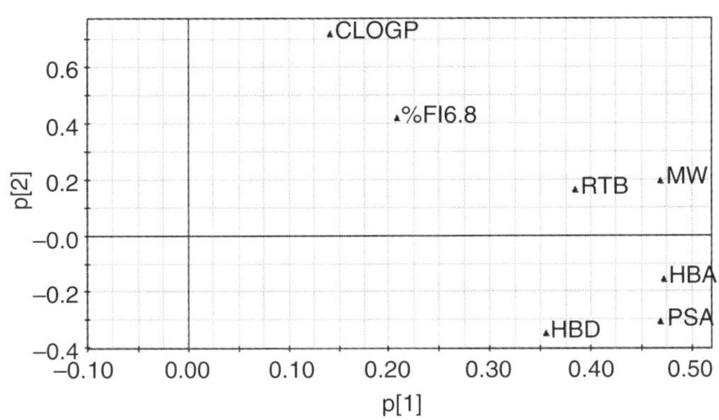

Figure 3.4 Key parameters related to permeability as illustrated via principal component analysis (PCA) of Novartis discovery NCEs that exhibited high- or low-permeability rank in both in-house PAMPA and Caco-2 experiments.

3.3.1 LogP, PSA, Absorption Model, and Polarity–Lipophilicity Line (PLL)

Despite the reported impact of cLogP and Polar Surface Area (PSA) on permeability and confirmed in our PCA analysis, we could not completely separate high- from

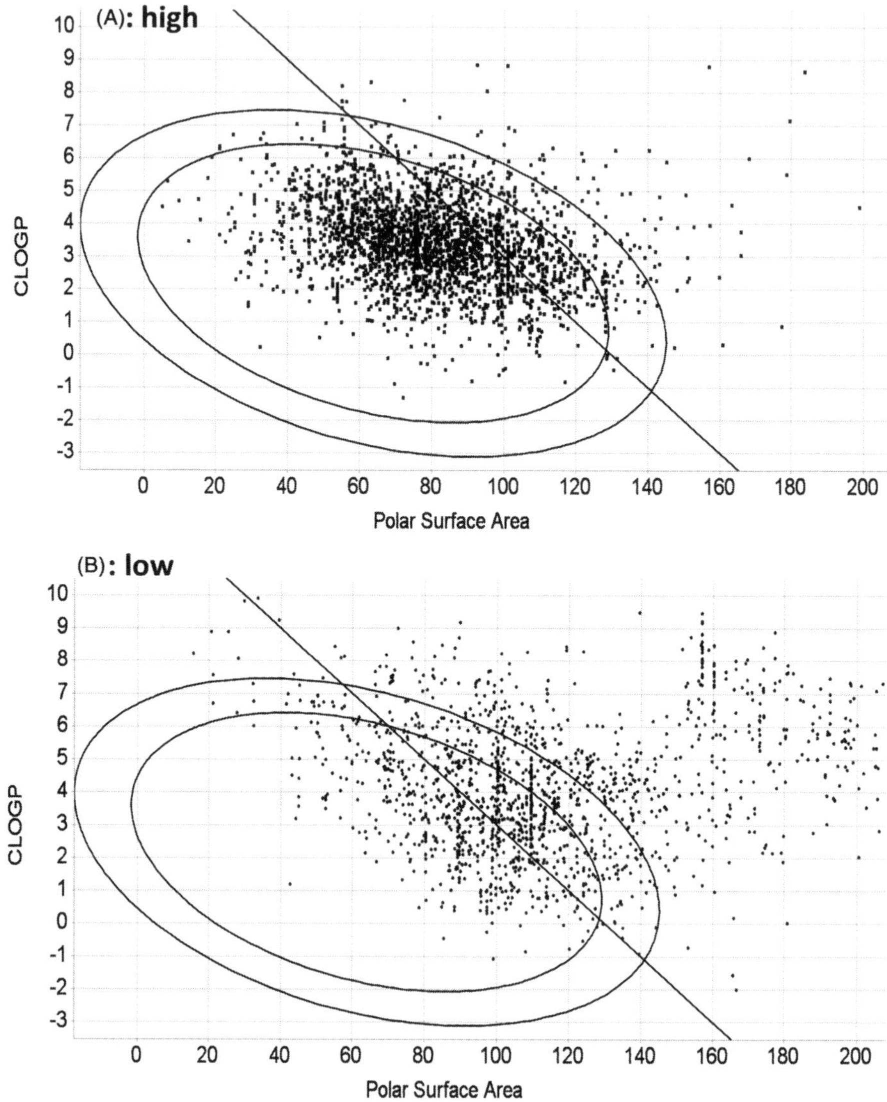

Figure 3.5 The projection of high- (A) and low-permeable NCEs (B) onto reported absorption model [10]. Data were collected from in-house Caco-2 assay (n = 7244). The newly proposed PLL can evenly differentiate NCEs with high and low permeability.

low-permeable compounds in the Caco-2 model from approximately 4400 in-house new chemical entities (NCEs) using the plot proposed by Egan et al. [10], which differentiated highly absorbed drugs from poorly absorbed ones (Figure 3.5). Here we populate the high- (A) and low-permeable NCEs (B) onto the absorption model (using cLogP, Biobyte) to evaluate its ability to differentiate high- and low-permeable NCEs. First, most of the high-permeable compounds cluster inside the absorption model (A), implying a high diagnostic power for high-permeable compounds. In contrast, most of the low-permeable NCEs also scatter INSIDE the absorption model, leading to a poor diagnostic power for these NCEs. In this case, it is difficult to attribute the low-permeable NCEs to either unfavorable PSA or cLogP values. Unfortunately, the diagnosis of low-permeable (rather than high-permeable) NCEs or series is far more critical for the chemistry optimization of ADME/pharmacokinetic (PK) properties of NCEs in drug discovery.

The physicochemical property space of NCEs shifts away from the original absorption model compounds that were mostly generic drugs [4], and therefore, semi-empirical rules that were derived from observations using generic drugs might become invalid with today's molecules. To strengthen the diagnostic power of PSA and clogP for this new chemical space, a semi-empirical boundary, the polarity–lipophilicity line (PLL), is thus proposed as shown:

$$\text{PLL boundary}: \ \text{cLogP} + 0.1 \times \text{PSA} - 13 \leq 0 \tag{3.5}$$

Using PLL as defined in Equation 3.5, 75% of high-permeable NCEs have PLL ≤ 0 in Caco-2, whereas 69% of NCEs with low Caco-2 A-B permeability show PLL > 0. Despite its ability to contribute to differentiating low- from high-permeable NCEs, the PLL boundary is not expected to explain in full all low- or high-permeable compounds. For the band of low-permeable compounds below PLL, additional physicochemical descriptors are necessary to diagnosis the poor permeability. Indeed, N_{RTB} and ionization flags were approximately twice as frequent for low-permeable NCEs compared with high-permeable NCEs below the PLL. Therefore, PLL may serve as the first filter when diagnosing permeability issues.

3.3.1.1 Number of Rotational Bonds and Molecular Weight

Smaller molecules tend to be leaky across biological membranes, whereas increased molecular size or flexibility may lead to retarded permeation. As PSA increases with the addition of nitrogen and oxygen atoms, the MW sometimes also increases. However, the correlation between PSA and MW is not strong for subsets of low-permeable compounds. The rise of N_{RTB} also coincides with the expansion of molecular weight, possibly implicating the impact of molecular size on the permeability [11]. We considered N_{RTB} and MW to be mostly redundant and found that Novartis NCEs tended to be more permeable when $N_{\mathrm{RTB}} \leq 6$ (~73% of high Caco-2 bin had 6 or fewer) and less permeable when $N_{\mathrm{RTB}} > 6$ (69% of low Caco-2 had > 6), confirming that a boundary of 6 gave an adequate separation between the two permeability bins.

3.3.1.2 Impact of Ionization on Permeability

It is generally known that unionized molecules tend to be favorable for permeation across phospholipid bilayers or GI membrane [12]. Our analysis on Novartis discovery NCEs supports this trend, despite being somewhat weak. First, the distribution of ionization for NCEs is quite binary, with most being either mostly charged or uncharged at neutral pH. Using 30% as a $\mathrm{FI\%}_{6.8}$ boundary, our analysis revealed that most of the highly permeable

Table 3.3 The permeability diagnosis matrix

Compound	A PLL	A-1 PSA	A-2 HBD	A-3 HBA	A-4 logP	B N_{RTB}	C $FI\%_{6.8}$	D-1 $FA\%_{Caco2}$	D-2 $FA\%_{PAMPA}$	E pKa[a]
Compound 5	–2.7	92.9	5	5	0.98	6	**100**	**28**	**54**	9.3, 8.2
Compound 6	–1.8	133.0	5	8	–1.8	**12**	**100**	**28**	n/a	6.7, 3.3
Compound 7	**7.5**	201.6	5	13	0.34	7	**100**	**24**	**18**	3.3, 2.9
Compound 8	**5.0**	141.3	3	9	3.9	6	**100**	**30**	n/a	8.7, 2.3
Compound 9	**8.4**	166.8	6	11	4.7	**13**	20	**16**	n/a	6.2, 3.5

[a]MoKa software (moldiscovery.com).

NCEs (61%) were not strongly ionized, suggesting a negative impact on the passive permeation process.

3.3.2 The Permeability Diagnosis Matrix (see Table 3.3)

Columns A, A-1, A-2, A-3, and A-4: Column A is the PLL as described by PSA (A-1) and LogP (A-4) in Equation 3.4. PSA is most affected by HBD (A-2) and HBA (A-3).

Column B: Number of rotational bonds or N_{RTB}.

Column C: Fraction of ionization at pH 6.8 ($FI\%_{6.8}$). All parameters in columns A–C contribute to passive permeability.

Columns D-1 and D-2: Measured membrane permeability as presented by the calculated fraction-absorbed (FA%) validated by marketed drugs with reported oral absorption in humans [13].

Column E: Calculated pKa values from commercial software.

3.3.3 Diagnosis Examples (Molecules in Table 3.3)

Compound 5 (Fenoterol) is a relatively small molecule (MW: 303.36) that permeates primarily via passive mechanisms, with low permeability in Caco-2 (Calc FA: 28%) [13]. From a diagnostic perspective, $FI\%_{6.8}$ is the only limiting factor as it is fully ionized at neutral pH. The chemistry optimization strategy, if any, may include the removal of the ionization center(s) and/or the shift of pKas, to reduce its ionized fraction.

Compound 6 (Lisinopril) is primarily featured by its high N_{RTB} (12) as a result of aliphatic side chain(s) in addition to its high $FI\%_{6.8}$. Although it has many HBAs (8) and therefore high PSA (133), overall it does not violate PLL guidance as a result of low lipophilicity (clogP: –1.8). It has a poor permeability in Caco-2 (Calc FA: 28%), predominantly via passive mechanisms [13].

Compound 7 (Aztreonam) is characteristic of the PLL issue, in contrast to compound 6. Even with low logP, its PSA is quite high (201.6) mainly because of high number of hydrogen bond acceptors (HBA) or hydrogen bond donors (HBD). The high PLL, along with high $FI\%_{6.8}$, is possibly responsible for its poor permeability in Caco-2 (Calc FA: 24%, [13]) and oral absorption in humans (FA: 1%, [14]). As its permeability is principally driven by the passive process, it is not surprising that similar permeability is reported in Caco-2 and PAMPA (Calc FA: 24% and 18%, respectively). The optimization strategy may start with reduction of total numbers for HBD and/or HBA and modification of ionization.

Compound 8 (Sulfasalazine) suffers from poor passive permeability and high $FI\%_{6.8}$. Apparently the low passive permeability may primarily be attributed to high PLL (5.0) along with high $FI\%_{6.8}$.

Compound 9 (Saquinavir) is an example where too many violations may contribute to the poor permeability. First, the molecular size (and volume) is quite high, with MW of 670.86 and N_{RTB} of 13, which greatly reduced mobility across the GI membrane. In addition, it has high PSA (166.8) and logP (4.7), thereby causing high PLL (8.4). The permeability issues for this type of molecule should be purposefully mitigated in early drug discovery.

As shown in the previous examples, it may be difficult to identify a single physicochemical parameter that is responsible for low permeability. In our study, 75% of low-permeable NCEs are associated with two to three violations from the permeability diagnosis matrix, whereas 75% of high-permeable NCEs feature only zero to one violation. Whereas the "number" of violations increases from high to low permeability, there is no clear overall trend in the "type" of violations for the change. Therefore, it is practical to mitigate low permeability for a series of compounds by first observing the trend with the total number of violations and then with each matrix parameter systematically.

3.4 GENERAL STRATEGY TO APPLY ADME DIAGNOSIS MODELS

The current permeability diagnosis model was derived after analyzing a large collection of discovery NCEs from multiple projects and chemical scaffolds. It is a rule-based semi-empirical model applicable to the chemical space covered by today's MedChem. Therefore, it may be valuable in early projects when limited experimental ADME (e.g., solubility and permeability) data are available, especially to direct chemistry optimization by prioritizing modifications in one to two areas/functional groups. For the cases where the general ADME diagnosis model does not lead to satisfactory improvement in the specific ADME property, a local model (e.g., solubility and/or permeability) may be established. For solubility optimization in early discovery phase where the final forms of NCEs are not yet available, solubility diagnosis is nevertheless valuable to monitor the interplay between the various factors governing solubility. At this stage, the measured solubility may be overestimated if crystal packing is a major contributor. Typically, the solubility of the final physical form is always lower than what is obtained with early batches, and therefore, the early measurements represent the upper solubility limit. For permeability, as active efflux only practically affects the effective permeability and consequently oral absorption of a small fraction of discovery NCEs, it is highly recommended to grant higher priority to diagnose and optimize key properties limiting passive permeability rather than efflux transport (e.g., via PLL, $FI\%_{6.8}$, and N_{RTB}). Most importantly, it should be kept in mind that some chemical modifications generally affect more than one ADME parameter and therefore one needs to consider the interplay between the various properties and find

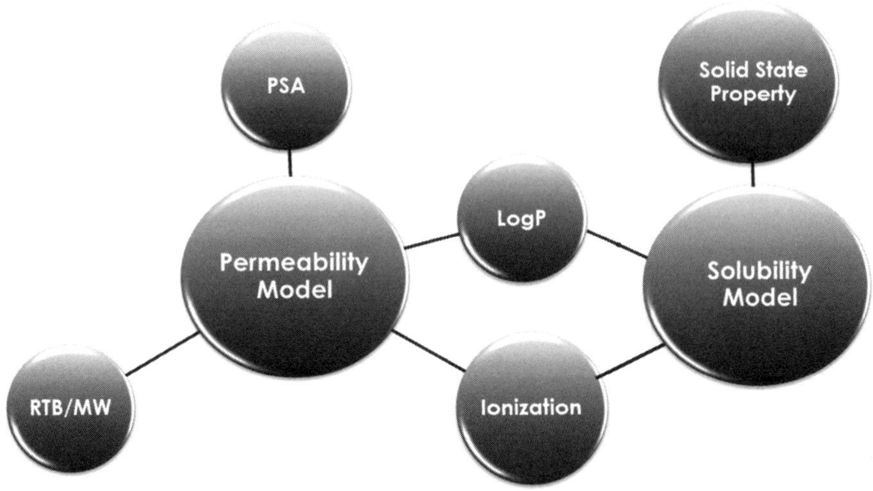

Figure 3.6 Demonstration of most critical properties affecting solubility and permeability.

an optimal balance. Figure 3.6 demonstrates the most influencing molecular properties affecting solubility and permeability together. First, improvement of solubility by reducing lipophilicity and enhancing ionization at neutral pH may concurrently jeopardize the permeability, where one needs to assess the overall impact of oral absorption.

3.5 CONCLUDING REMARKS

The introduction of diagnostic tools for ADME assays helps understand what structural elements are responsible for the unfavorable property and establishes a more adequate optimization strategy. One should, however, not overestimate the reliability of these tools, and it is advisable to analyze more than a single compound within a chemical series before drawing a solid conclusion.

REFERENCES

1. Hansch, C., Quinlan, J.E., Lawrence, G.L. (1968). The linear free energy relationships between partition coefficients and the aqueous solubility of organic liquids. *Journal of Organic Chemistry*, *33*, 347–350.
2. Valvani, S.C., Yalkowcky, S.H., Roseman, T.J. (1981). Solubility and partitioning IV: Aqueous solubility and octanol-water partition coefficients of liquid nonelectrolytes. *Journal of Pharmaceutical Sciences*, *70*, 502–507.
3. Isnard, P., Lambert, S. (1989). Aqueous solubility and n-octanol/water partition coefficients correlation. *Chemosphere*, *18*, 1837–1853.

4. Faller, B., Ottaviani, G., Berellini, G, Collis, A. (2011). Evolution of the physicochemical properties of marketed drugs: Can history foretell the future? *Drug Discovery Today*, *16*, 976–984.
5. Faller, B, Ertl, P. (2007). Computational approaches to determine drug solubility. *Advanced Drug Delivery Reviews*, *59*(7), 533–545.
6. Avdeef, A. *Absorption and Drug Development: Solubility, Permeability and Charge State*, Wiley, Hoboken, NJ, 2003.
7. Wassvik, C.M., Holmen, A.G., Draheim, R., Artursson, P., Bergstroem, C.A.S. (2008). Molecular characteristics for solid-state limited solubility. *Journal of Medicinal Chemistry*, *51*(10), 3035–3039.
8. Zhou, L., Yang, L., Tilton, S., Wang, J. (2007). Development of a high throughput equilibrium solubility assay using miniaturized shake-flask method in early discovery. *Journal of Pharmaceutical Sciences*, *96*(11), 3052–3071.
9. Lin, X., Skolnik, S., Chen, X., Wang, J. (2011). Attenuation of intestinal absorption by major efflux transporters: Quantitative tools and strategies using Caco-2 model. *Drug Metabolism and Disposition*, *39*(2), 265–274.
10. Egan, W., Merz, K. Jr., Baldwin, J. (2000). Prediction of drug absorption using multivariate statistics. *Journal of Medicinal Chemistry*, *43*, 3867–3877.
11. Veber, D.F., Johnson, S.R., Cheng, H.-Y., Smith, B.R., Ward, K.W., Kopple, K.D. (2002). Molecular properties that influence the oral bioavailability of drug candidates. *Journal of Medicinal Chemistry*, *45*, 2615–2623.
12. Schanker, L.S., Tocco, D.J., Brodie, B.B., Hogben, C.A.M. (1958). Absorption of drugs from the rat small intestine. *Journal of the American Chemical Society*, *123*, 81–88.
13. Skolnik, S., Lin, X., Wang, J., Chen, X., He, T., Zhang, B. (2010). Towards prediction of in vivo intestinal absorption using a 96-well Caco-2 assay. *Journal of Pharmaceutical Sciences*, *99*(7), 3246–3265.
14. Zhao, Y., Le, J., Abraham, M., Hersey, A., Eddershaw, P., Luscombe, C., Butina, D., Beck, G., Sherborne, B., Cooper, I., Platts, J. (2001). Evaluation of human intestinal absorption data and subsequent derivation of a quantitative structure–activity relationship (QSAR) with the Abraham descriptors. *Journal of Pharmaceutical Sciences*, *90*, 749–784.

4

PATH (PROBE ADME AND TEST HYPOTHESES): A USEFUL APPROACH ENABLING HYPOTHESIS-DRIVEN ADME OPTIMIZATION

Leslie Bell, Suzanne Skolnik and Dallas Bednarczyk

4.1 INTRODUCTION

Over the years, the cumulative industry experience with *in vitro* tools to predict drug absorption, distribution, metabolism, and elimination (ADME) has illuminated notable *in vitro–in vivo* correlation (IVIVC) gaps. The relatively high frequency with which primary *in vitro* models fail to predict *in vivo* outcome has tested our basic assumptions behind the most frequently employed *in vitro* models. This has challenged the industry's agility for interrogating the diverse mechanisms that impact clearance, absorption, and disposition in early drug discovery. A common route to identifying key drivers of emerging IV-IV disconnects is, frequently, process of elimination, which is a time-consuming and sometimes resource-intensive practice. It is risky, though, not to address disconnects when they arise. Poor IVIVC can diminish the apparent value of screening assays and lead project

Predictive ADMET: Integrative Approaches in Drug Discovery and Development, First Edition.
Edited by Jianling Wang and Laszlo Urban.

teams to abandon *in vitro* models as "failed" in favor of *in vivo* "screening." Without a viable plan to interrogate disconnects, some scaffolds may be abandoned, altogether, in favor of more highly correlated chemotypes. This unfortunate outcome can result in missed opportunities for exploiting precious novel chemical space. The ability to leverage profiling strategies that integrate physicochemical trends and diverse cues gathered from *in vitro* and *in vivo* ADME can enable more nimble and efficient diagnosis of drug pharmacokinetic (PK) behaviors. Within the pharmaceutical industry, the investment to anticipate and manage both typical and atypical PK events supports our essential endeavor to exploit new and patentable chemical matter.

Long-standing, industry-wide attention to physicochemical and ADME characterization has established a wealth of data from which to draw guidance around IVIVC. Historically, IVIVC has been based loosely on a matched-binning strategy. Relative "high," "moderate," or "low" metabolic stability *in vitro*, for example, may correlate with "high," "moderate," or "low" clearance *in vivo*. For projects benefitting from a fairly straightforward correlation, this approach can be effective in moving the project in the right direction. Frequently, however, simple and direct correlations are not the reality. Furthermore, definition and alignment of bins between *in vitro* and *in vivo* can be ambiguous. Without a clear foundation for making IV-IV data comparisons, IV-IV trends can be overstated, or even missed, and the role of important mechanisms can be overlooked.

Drawing on information gathered from the literature and data trend analysis of the Novartis Pharmaceutical (Basel, Switzerland) discovery library, this chapter presents a so-called "Probe ADME and Test Hypotheses" (PATH) approach to IVIVC interrogation. A rational, clearly defined expectation for IVIVC serves as a baseline for the PATH approach. From this foundation, the quality of a correlation may be described by IVIVC "zone." Hypothesis development and follow-up strategies from each zone are prescribed from the ADME and PK cues provided by the compounds/series of interest. The emphasis on the rat as an appropriate *in vivo* model to drive drug discovery chemistry is tenuous, despite acceptance as common practice. The goal of the PATH approach is not necessarily to focus on IV-IV correlation optimization. Rather, the emphasis is on interrogation of IV-IV disconnects to develop suitable progression strategies for otherwise attractive series. An objective review of early *in vitro* predictions against preclinical pharmacokinetic models can help identify trends for IVIVC gaps that may be useful considerations for scalability or projection of ADME in humans. Although most of the IVIVC analysis discussed in this article will focus on correlations with rodents, some appearance of similar behavioral trends of marketed drugs in humans will also be discussed.

The PATH models presented here focus on developing and interrogating IVIVC for drug clearance (CL, mL/min/kg) and oral bioavailability (%F) from the basic *in vitro* starting points of cytochrome P450 (CYP)-mediated metabolic stability in liver microsomes, parallel artificial membrane permeability assay (PAMPA) permeability, and aqueous equilibrium solubility. An examination of the ADME toolbox to assess CL and %F is offered. Some perspective on the impact of mechanistic ADME models will be shared in the following sections as we navigate the "PATH" forward from baseline IVIVC to elucidation of mechanisms as mapped by the available ADME profile.

4.2 ASSUMPTIONS AND LIMITATIONS

4.2.1 *In vitro*

The scaling approach described for hepatic CYP metabolism relies, first, on an assumption that hepatic CYPs limit *in vivo* clearance rates. Metabolism from other drug metabolizing enzymes (DMEs) (e.g., esterases, UDP-glucuronosyltransferase, etc.) commonly represented in microsomal preparations may not be adequately scaled by the previous approach [1, 2]. The utility of microsomal intrinsic clearance (CL_{int}) extrapolation for *in vivo* clearance estimations assumes, of course, that standard criteria for clearance measurements have been satisfied *in vitro* and *in vivo* (e.g., low microsomal protein concentration and nonsaturated enzyme kinetics) [3]. When reaction kinetics are very efficient, with $K_m \ll 1$ μM, an apparent *in vitro* CL_{int} obtained with even 1 μM of drug may underestimate true CL_{int}. When this is suspected, one may proceed by evaluating whether lowering the *in vitro* test article concentration produces a higher observed CL_{int}.

In reality, poor solubility of chemical matter can limit the extent and rate of absorption *in vitro* and *in vivo*. *In vitro* estimates of gut permeability should be derived from sufficiently low concentrations of test compound so as to minimize solubility concerns and resulting underestimation of f_a. Conversely, saturation of active transport processes *in vitro* with a very high concentration of drug can overestimate gut permeability. The *in vitro* estimates of f_a forming the basis for %F IVIVC largely assume nonsaturated transport and nonlimiting solubility. The role of solubility and transporters as critical determinants of bioavailability will be discussed in the development of hypotheses addressing IV-IV disconnects.

Another caveat of *in vitro* permeability assessment is that, although one of the primary *in vitro* assays is a nonbiological, species-independent model (PAMPA), the common biological, cell-based alternative (Caco-2) is human in origin [4, 5]. Caco-2, and its associated human drug transporters, has been widely accepted as predictive of drug absorption even in preclinical species, such as mouse, rat, and dog. Although it has been noted that there is a linear correlation of f_a between humans and rats, for example, this may not hold true for a diverse set of molecules [6]. Even relating the human Caco-2 model specifically to human drug absorption can pose challenges because the *in vitro* transporter activities and metabolic capacities cultivated in Caco-2 monolayers can be skewed relative to the intact gastrointestinal tract. Such fundamental differences between Caco-2 and *in vivo* models may be problematic for some projects.

In vitro systems inherently run the risk of poorly reflecting physiologically relevant drug distribution. The *in vivo* scenario for protein binding equilibria and physicochemical drug partitioning is complex, and there are many ways in which drug distribution within and between tissues can impact apparent clearance [7]. Studies have shown that *in vitro* protein binding can attenuate microsomal CL_{int} [8]. Although the equilibrium between microsomal and plasma protein binding is considered for hypothesis development in the following sections, the development of baseline IVIVC relies, largely, on an assumption that *in vitro* protein binding does not significantly limit the apparent rate of microsomal CYP metabolism or the permeability of the drug across Caco-2 monolayers.

4.2.2 *In vivo*

Although some chemical biotransformations are reproduced across species, few CYP isoforms are conserved between species [9]. Even when CYPs are conserved, abundances relative to other isoforms can vary dramatically between species. Similar observations have been made for other DMEs and drug transporters involved in clearance and distribution [10]. For these reasons, IVIVC should strive to compare only *in vitro* and *in vivo* clearance data from the same species. Less is understood about strain and gender differences. It is not uncommon for Sprague–Dawley and Wistar rats, for example, to be used indiscriminately for *in vivo* PK studies, and to be compared with *in vitro* studies derived from only one strain. When initiating IVIVC, it may be reasonable to group strains or genders within a species, especially if the combined data help to provide critical mass for beginning to identify data trends. However, it is worth considering that some disconnects may be strain- or gender-specific and it may become necessary to tease these dependencies apart. The use of diseased or genetically modified animal models could introduce additional physiological variables that impact IVIVC. In particular, models that potentially impact pathways of elimination (e.g., hepatic disease), blood flows to eliminating organs (e.g., drugs that affect heart rate), gastrointestinal function (e.g., gastric emptying and intestinal motility), or drug distribution to clearance tissues (e.g., obese animal models) can confound IVIVC.

Ideally, routine *in vivo* %F and CL estimates should come from fairly low doses ($\leq$ approximately 3 mg/kg), to minimize solubility limitations and avoid kinetic saturation of enzymatic and transporter pathways. Cassette dosing, often employed to conserve in-life resources in drug discovery, can introduce potential for metabolic drug interactions yielding *in vivo* clearance values that are lower, or intestinal absorption values that are higher, than might be observed with single drug administration. Lastly, to minimize underestimation of clearance and area under concentration (AUC) from drug concentration time profiles, arterial sampling intervals during PK studies must be sufficient to define clearance rates accurately and bioanalytical sensitivity must adequately estimate terminal phases of elimination [11, 12].

4.3 CLEARANCE IVIVC

Drug elimination or clearance generally refers to the irreversible removal of drugs or drug products from the body, primarily by two routes: excretion and metabolism (or biotransformation). The mechanisms driving these routes can be varied [13].

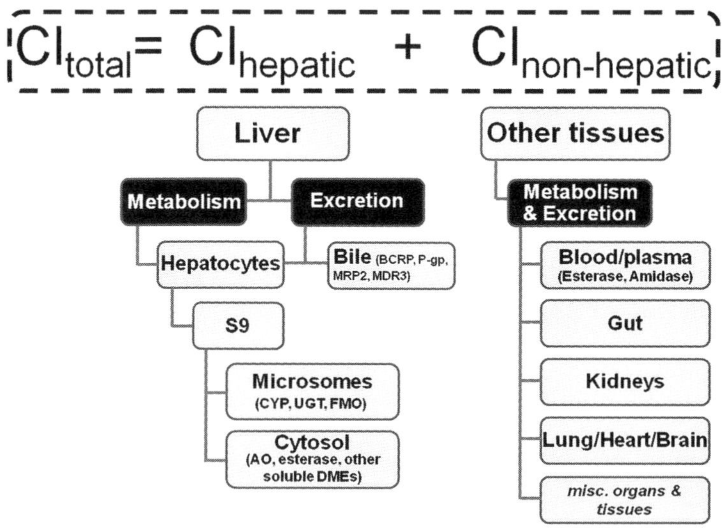

Figure 4.1 Overview of *in vivo* clearance mechanisms.

Total *in vivo* clearance will often reflect one or two prominent pathways, but it is, ultimately, determined by the sum of all hepatic (metabolism and biliary elimination) and extrahepatic clearance mechanisms (Figure 4.1). Today's *in vitro* capabilities, and most pharmaceutical drug discovery flowcharts, still heavily emphasize first-pass hepatic CYP metabolism for scaffold prioritization. This is not unreasonable because marketed drug analysis predicts hepatic CYP metabolism to dominate the rate of *in vivo* clearance for about half of all clinical drugs [13].

4.3.1 Establishing a Baseline for Clearance Correlation Analysis

Fractionation of liver yields a preparation, termed "microsomes," that is enriched for membrane-associated DMEs such as hepatic CYPs and UDP-glucuronosyltransferases (UGTs). The use of microsomes to predict hepatic metabolic stability for new chemical entities (NCEs) is widely accepted. This model is amenable to high-throughput investment and conveniently provides drug discovery with the earliest estimates of clearance for exposure optimization. The *in vitro* intrinsic clearance of drugs (CL_{int}, expressed as μL/min/mg microsomal protein) from an incubation of liver microsomes supplemented with CYP co-factor (β-nicotinamide adenine dinucleotide phosphate (NADPH)) provides a useful readout to triage against high clearance compounds and to identify the most promising candidates to take into *in vivo* studies. Historically, $CL_{int} < 50$ μL/min/mg has been a common threshold to reflect a low metabolism risk, whereas $CL_{int} > 150$ μL/min/mg flags high risk for metabolism. For rat, these cutoffs hold some merit. For instance, a CL_{int} of 50 μL/min/mg in rat liver microsomes, extrapolated to approximate *in vivo* hepatic clearance (CL_h) (IV-IV scaling discussed further in the next section), projects an *in vivo* clearance liability of <50% hepatic extraction of drug. That is, less than half of the drug dose delivered to the liver is expected to be irreversibly eliminated (in this case metabolized) upon one hepatic pass. In other species, however, $CL_{\text{int}} = 50$ μL/min/mg can translate to as much as 70% hepatic extraction. This misalignment of *in vitro* readouts across species further highlights the need for more robust guidelines in evaluating clearance.

Extrapolation of *in vitro* CL_{int} to *in vivo* hepatic clearance (CL_h, mL/min/kg) is routinely practiced, most often using a well-stirred model, as summarized in Equation 4.1 [3]:

$$CL_h = \frac{Q_h \bullet \dfrac{f_{u,b}}{f_{u,mic}} \bullet CL_{\text{int}} \bullet SF_1 \bullet SF_2}{Q_h + \dfrac{f_{u,b}}{f_{u,mic}} \bullet CL_{\text{int}} SF_1 \bullet SF_2} \tag{4.1}$$

The unbound fraction of the test compound in blood and microsomes is represented by $f_{u,b}$ and $f_{u,mic}$, respectively. SF_1 approximates mg microsomal protein per gram liver. Q_h and SF_2 represent species-specific portal blood flow (mL/min/kg) and liver size (g/kg), respectively. Recommended values for SF_1, SF_2, and Q_h are summarized in Table 4.1 [14].

The term "$f_{u,b}/f_{u,mic}$" in Equation 4.1 serves to capture the effect of protein binding differentials between blood and liver microsomes influencing drug distribution into metabolic tissue. This term is somewhat inconvenient for routine IVIVC. Protein binding measurements are sparsely acquired during early drug discovery. Additionally, pharmacokinetic screening often relies on plasma, rather than on whole blood, for drug concentration time profiles. As such, $f_{u,p}$ (fraction unbound in plasma) frequently substitutes for $f_{u,b}$ in Equation 4.1, with only occasional determination

Table 4.1 Recommended scaling factors for extrapolation of *in vitro*, hepatic microsomal CL_{int}

Species	SF_1 (microsomes) mg microsomal protein • g liver^{-1}	SF_1[a] (hepatocytes) 10^6 hepatocytes • g liver^{-1}	SF_2 g liver • kg body weight^{-1}	Q_h mL blood flow • min^{-1} • kg body weight^{-1}
Rat	45	125	40	55
Mouse	45	125	88	90
Dog	45	125	32	31
Monkey	45	125	30	44
Human	32	125	26	21

[a]Intrinsic clearance derived from hepatocyte suspensions is usually expressed as μL/mL/10^6 hepatocytes and can be scaled, similar to liver microsomes, with a simple substitution of hepatocellularity (10^6 hepatocytes • g liver^{-1}) in Equation 4.1 [15, 16].

of blood-to-plasma partitioning to investigate distribution concerns [17]. Initially assuming that $f_{u,p} = f_{u,mic}$ in Equation 4.1 yields a useful simplification that is more conducive to early drug discovery [3]:

$$CL_h = \frac{Q_h \bullet CL_{\text{int}} \bullet SF_1 \bullet SF_2}{Q_h + CL_{\text{int}} SF_1 \bullet SF_2} \tag{4.2}$$

The predicted hepatic extraction ratio (ER_h) considers CL_h in context with species-specific Q_h (Table 4.1), providing an intuitive reference for comparing metabolic clearance potential across species:

$$ER_h = \frac{CL_h}{Q_h} \tag{4.3}$$

Between the two extrapolated values (CL_h and ER_h), ER_h is less sensitive to the assumed Q_h value and is, thus, a more robust starting point for initiating IV-IV comparisons.

Hepatic CYP metabolism represents the rate-limiting metabolic step to *in vivo* clearance for most drug candidates. No other single *in vitro* clearance model can be expected to match the success of hepatic CYP metabolism for predicting *in vivo* CL across broad chemical space. Numerous modifications of *in vitro* metabolism models and scaling practices, including concomitant support of conjugative metabolism, inclusion of plasma protein, and correction for ionized fraction, have been investigated for improved clearance predictions among small subsets of compounds [18–20]. However, routine "correction" of microsomal clearance by one to two potentially relevant variables is unlikely to address IVIVC sufficiently across diverse chemical space and may, conversely, confound clearance hypotheses. Modified scaling practices are better suited to afford mechanistic investigation of disconnects rather than substitute as first-line models for clearance IVIVC. For "worst-case" estimates of metabolic liabilities, some groups prefer to screen metabolic stability in liver microsomes by dually supplementing the microsomal activities of hepatic CYPs and UGTs [21]. This option may impede rather than enhance optimization efficiency. The structure-property relationship (SPR) driving oxidative metabolism is distinct from the SPR for conjugative metabolism, and so the assay readout with this approach may not be instructive to medicinal chemists.

4.3.2 Clearance IVIVC Zones

4.3.2.1 Zones 1 and 2: Regions of Strong Clearance IVIVC

For the treatment of small molecules, it is reasonable to begin IVIVC with a direct comparison between the extent of hepatic extraction (ER_h) predicted from *in vitro* liver microsomes versus total *in vivo* clearance of drug from blood or plasma (CL_{total}, mL/min/kg). Typically, guidelines for evaluating the quality of an IVIVC are minimal. As a simple rule-of-thumb, *in vitro* clearance projections within twofold of *in vivo* clearance may be treated as "good" correlation. The danger here is that this model describes a correlation window with low tolerance for clearance differences in the lower range of clearance and relative insensitivity to clearance differences in the higher range. This can lead to missed opportunities for recognizing mechanistically relevant disconnects. By contrast, a line of unity, where complete CYP-based hepatic extraction (*in vitro* $ER_h = 1$) equals species-specific liver blood flow, provides a sound correlation expectation across the range of clearance values. Broadening the line of unity to account for ±30% variability defines a window that tolerates differences in animal physiology and experimental error (Figure 4.2A) [14, 22].

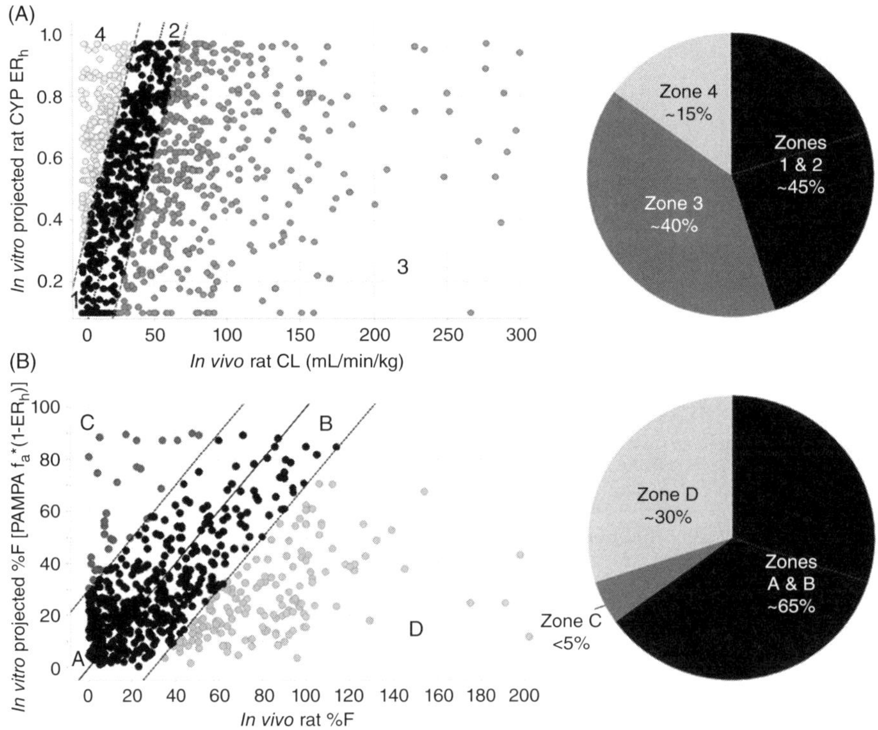

Figure 4.2 IVIVC for chemically diverse NCEs discretely dosed to male, Sprague–Dawley or Wistar rats. (A) Compares total clearance of intravenous dose from plasma versus *in vitro* microsomal projections of CYP-mediated hepatic extraction (ER_h). Solid and dashed lines represent unity (100% hepatic extraction equals clearance of 55 mL/min/kg) ± 30% deviation from unity, respectively. Clearance well predicted by liver microsomes (black circles) is divided into Zone 1 ($ER_h < 0.5$) and Zone 2 ($ER_h > 0.5$). Zone 3 is clearance underestimated by microsomal CYPs (dark gray circles), and Zone 4 is clearance overestimated by microsomal CYPs (light gray circles); (B) Compares oral %F versus *in vitro* projections of %F from PAMPA f_a and CYP ER_h estimates of f_h. Solid and dashed lines represent unity correlation ±30%, respectively. Well-predicted %F (black circles) is divided into Zone A (%F < 50) and Zone B (%F > 50). Zone C is %F overestimated by PAMPA and microsomal CYPs (dark gray circles), and Zone D is %F underestimated by PAMPA and microsomal CYPs (light gray circles); pie charts summarize % distribution of discovery NCEs across all four zones.

When scaled microsomal CYP ER_h correlates well with *in vivo* clearance, hepatic CYP metabolism is presumed to rate-limit clearance *in vivo* (PATH Clearance IVIVC "Zones 1 and 2"). In terms of follow-up strategies, it is useful to divide the window of strong IVIVC into two zones to distinguish the region for which CYP metabolic stability is more manageable ("Zone 1": $ER_h < 0.5$) from the space for which it is likely to compromise necessary exposures and where medicinal chemistry to improve metabolic stability may be warranted ("Zone 2": $ER_h > 0.5$).

4.3.2.2 *Zone 3: In Vivo Clearance Underestimation*

Often, *in vivo* clearance is much greater than anticipated from liver microsomes ("Zone 3"). In humans, nearly half of marketed drugs are cleared through pathways other than CYP metabolism [13]. Further contributing to the disconnect, CYPs and other DMEs are known to be present in extrahepatic tissues such as lung [23], heart [24], brain [25], intestine [26], and kidney [27] where they may play a significant role in drug clearance [28]. When *in vivo* clearance is dramatically underpredicted by hepatic CYPs (as is the case for, so-called, PATH "Zone 3" compounds), it is often challenging to deduce whether excretory mechanisms or alternative metabolic processes dominate *in vivo.* Activities triggered by clearance underprediction can range from extensive metabolite identification, to *in vivo* collection of bile or urine, to *in vitro* sampling of diverse metabolism models in an effort to uncover additional sources of biotransformation.

4.3.2.3 *Zone 4: In Vivo Clearance Overestimation*

There is a smaller, still relevant, fraction of compounds whose clearance predicted by *in vitro* CYPs is significantly greater than that measured *in vivo* ("Zone 4"). Although compounds prove to be very efficiently metabolized *in vitro* by hepatic CYPs, *in vivo* clearance can turn out to be surprisingly low, suggesting that the kinetic capacity for metabolism is not necessarily rate-limiting. Rather, the *in vivo* distribution of a drug to reach metabolic enzymes may represent the rate-limiting event for systemic clearance. The primary concern with Zone 4 disconnects is that microsomal CYPs can overstate the metabolic liability, leading project teams to invest too much in metabolic stabilization or discard highly promising structures without *in vivo* follow-up. Identifying root causes of distribution-limited clearance could help teams prioritize medicinal chemistry efforts and anticipate whether similar events are likely to influence IVIVC in higher species.

4.3.2.4 *Overview of Clearance IVIVC for Chemically Diverse NCEs*

The *in vitro* and *in vivo* clearance observations from 900 diverse Novartis compounds covering 10 disease areas and more than 50 different projects have been overlaid on the four-zone PATH clearance IVIVC map [14] (Figure 4.2A). At first assessment, hepatic CYP metabolism accurately projects *in vivo* clearance for ~45% of drug candidates (Zones 1 and 2 in Figure 4.2A). The remaining IVIVC lays, disproportionately, in the direction of *in vivo* clearance underestimation (Zone 3). Less frequently, clearance is significantly overpredicted by liver microsomes (Zone 4). As such, a hepatic CYP "snapshot" without *in vivo* confirmation can be grossly misleading for compound prioritization.

4.3.3 Trends, Hypotheses, and Strategies for Clearance Interrogation by Zone

4.3.3.1 Zones 1 and 2 Although CYP isoforms differ between species, some molecular features tend to be common. For instance, the collective CYPs represented in both human and rat livers exhibit a clear substrate lipophilicity optimum, with rapid CYP metabolism being highly associated with calculated (*in silico*) LogP (cLogP) in the range of ~3–7 (Figure 4.3A). Rat CYPs also display a striking preference for compounds having polar surface area (PSA) < 100 [14]. As a result of shared substrate features, frequently, although not always, modifications that stabilize compounds for human CYP metabolism also lead to stabilization in rat. It is important to monitor SPR between species, especially whenever significant medicinal chemistry efforts toward CYP stabilization will be invested.

Leveraging hydrophilicity to disfavor interaction with CYPs may reduce hepatic metabolism and yet concomitantly redirect clearance toward other pathways (hepatic and extrahepatic). Figure 4.3B illustrates that compounds with lower lipophilicity are more frequently associated with excretory clearance than metabolism. Conversely, marketed drugs characterized with significant renal elimination tend to be both stable against hepatic CYPs (*in vitro* ER_h < 0.5) and relatively hydrophilic (cLogP < 2.5) [29]. In many cases, a chemical strategy that succeeds in decreasing metabolism by increasing hydrophilicity can be a relatively productive outcome, especially if passive renal elimination is engaged as a dominant clearance mechanism. Passive renal elimination tends to remain largely associated with PATH CL module Zone 1

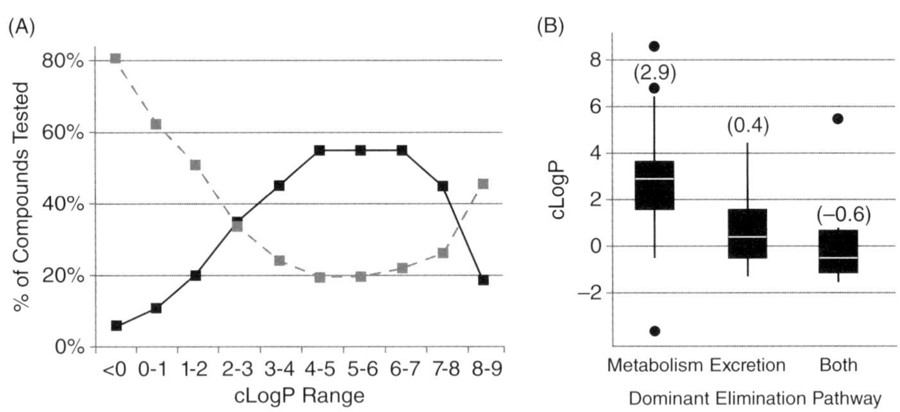

Figure 4.3 Clearance-lipophilicity trends. (A) Incidence of measured "high" (CL_{int} > 150 μL/min/mg; solid line) versus "low" (CL_{int} < 50 μL/min/mg; dashed line) rat CYP microsomal clearance *in vitro* across cLogP (*N* = 5420 discovery compounds). (B) Box plot illustrating the relationship between lipophilicity and *in vivo* human elimination pathways (*N* = 89 marketed compounds). Value in parentheses is median cLogP for subset.

as a result of both the CYP stability and the very low rate of glomerular filtration that limits passive renal elimination [14]. Other redirected modes of elimination, however, may be less benign.

4.3.3.2 Zone 3 Of greater concern is when lipophilic character evokes active elimination mechanisms, as shown in PATH clearance Zone 3. For both human (marketed) and rat (discovery) IVIVC, compounds with low cLogP tend to exhibit Zone 3 clearance disconnects [14]. When diagnosing Zone 3 disconnects, typically three major hypotheses drive follow-up activities: CYP-independent hepatic metabolism, biliary elimination, and extrahepatic metabolism.

4.3.3.2.1 CYP-independent Hepatic Metabolism Often, hydrophilicity or polar surface area are conferred through addition of groups such as hydroxyls, carboxyls, esters, ethers, and amines that serve as ready handles for metabolic conjugation [30]. The complementary array of metabolic enzymes, in addition to CYPs, presented in hepatocytes may help reconcile the IVIVC gaps, particularly when structures offer the types of conjugation handles just described. At Novartis, hepatocytes offer significantly improved clearance estimates over microsomal CYPs for ~10% of sampled Zone 3 clearance disconnects [14]. This rate is consistent with estimates of contribution from glucuronidation in humans [13]. Combined with CL_{int} estimates from hepatocytes, qualitative bioanalytical confirmation of common conjugation pathways, such as glucuronidation and sulfation, can be instructive for deducing relevant metabolic soft spots. When conjugation is coupled with significant *in vivo* enterohepatic recirculation, contribution from enzymatic conjugation can go undetected in the baseline IVIVC. Raloxifene (EVISTA; Eli Lilly, Indianapolis, IN) for example, is reported to undergo extensive first-pass glucuronidation with enterohepatic recycling in humans [31]. Rather than exhibiting a Zone 3 clearance disconnect, the total clearance that is observed is reasonably consistent with hepatic CYP projections (Zone 2). As the efficiency of enterohepatic recycling can differ across species, Raloxifene highlights the importance of *in vitro* and *in vivo* mechanistic assays for testing IVIVC-based clearance hypotheses. To evaluate enterohepatic recirculation *in vivo*, bile-duct cannulated rats can be useful to validate the mechanistic hypothesis.

4.3.3.2.2 Biliary Elimination Deducing whether one should invest in metabolic or excretory clearance characterization is not always easy. Exploration of the wrong hypothesis can cost a team weeks of delayed project progression. Clearance of parent drug through biliary excretion, without prior metabolism, is a major clearance pathway being identified more frequently with discovery compounds. Sandwich-cultured hepatocytes (SCHs) are a useful *in vitro* model for assessing biliary excretion [32,33]. Based on *in vivo* (bile-duct cannulated rats) and *in vitro* (SCH) models for biliary clearance, nearly half of sampled Zone 3 disconnects are found to be associated with measured (*in vivo*) or predicted (SCH) biliary excretion. Prime candidates for biliary excretion include compounds with poor passive permeability and demonstrated interaction with drug transporters [33, 34]. Considering the overlap of major drug transporters (e.g., P-glycoprotein (P-gp), multidrug resistant protein (MRP-2), and breast cancer resistant protein (BCRP)) in the intestine, liver, and kidney of both rats and humans, efflux in Caco-2 monolayers offers a potentially useful flag for broader transporter involvement *in vivo* [10, 35]. Furthermore, biliary elimination is most associated with compounds that are relatively CYP-stable (*in vitro* $ER_h < 0.5$). Too often, *in vivo* clearance studies using bile-duct cannulated rats are pursued without clear guidance regarding likelihood for involvement of biliary transporters. When (1) *in vivo* CL (within liver blood flow) for a compound is underpredicted by hepatic CYPs, (2) CYP metabolic liability is relatively low and (3) there is evidence for interaction with drug transporters; follow up with *in vivo* or *in vitro* biliary excretion studies is a particularly rational next step. Rarely, a Zone 3 IVIVC disconnect of very polar compounds has been attributed to transporter-mediated renal secretion, which can potentially contribute clearance rates approaching total kidney blood flow. There are currently no well-validated *in vitro* models to anticipate renal elimination, and this can only be confirmed through *in vivo* urine collection.

4.3.3.2.3 Extrahepatic Metabolism *In vivo* clearance in rats >100 mL/min/kg may be considered to be in excess of reasonable estimates of hepatic blood flow and warrants consideration of extrahepatic mechanisms. Rarely, total clearance exceeds cardiac output in rats (~300 mL/min/kg), attributable to chemistry occurring in the central compartment, for example, plasma-mediated hydrolysis, chemical instability, or both of the drug substance. When sampled across broad chemical space, plasma instability accounts for <10% of Zone 3 disconnects [14]. It is not uncommon for plasma instability to be associated with extreme microsomal instability, as well. This is, in part, a result of trace levels of hepatic esterase activity in microsomal preparations. A combination of plasma and hepatocytes may be needed to effectively interrogate the contribution of esterase activities to total clearance. When elimination is contributed from multiple organs, in particular, when processes are efficient and perfusion of these tissues is significant, clearance in excess of Q_h may be encountered. Together, gut, liver, and kidneys receive about half of the total cardiac output [22], so their collective contribution to clearance could be substantial. Several xenobiotics have been identified as substrates for renal glucuronidation, such as propofol, for which very high clearance in humans ($\sim 1.7 \bullet Q_{\mathrm{h}}$) reflects both high hepatic extraction and significant renal glucuronidation [27, 36, 37].

Interestingly, ~30% of sampled Novartis Zone 3 compounds exhibit NADPH-independent instability in hepatic S9, a liver fraction containing both microsomal and soluble DMEs, such as aldehyde oxidase that are also broadly expressed in other tissues [38, 39]. Despite the commercial availability of extrahepatic microsomes and S9, these systems are challenging to scale. Nevertheless, a few attempts have been described [40–42]. A "pan-CYP" inhibitor such as 1-aminobenzotriazole (ABT) can be an effective tool *in vitro* or *in vivo* to deduce non-CYP mechanisms of clearance. The caveat, however, is that some CYP isoforms are insensitive to ABT inactivation and it is difficult to validate whether requisite ABT exposure to extrahepatic tissues is achieved *in vivo* [43]. *Ex vivo* organ isolation, such as isolated perfused kidney and liver models, may offer alternative routes to dissect extrahepatic contribution to total clearance [44, 45].

4.3.3.3 Zone 4 A remarkable observation is that compounds characterized with a fairly low volume of distribution *in vivo* ($V_d \leq 1$) are highly associated with PATH Zone 4 for both humans and rats [14]. Although low V_d does not necessarily warrant exclusion of the drug from the liver, or any other specific tissue for that matter, it is known that hepatic exclusion of the drug can be associated with a dramatic impact on V_d [46]. In principle, NCEs with distribution limitations may be less "available" to catalytic machinery *in vivo* relative to the *in vitro* microsomal scenario, resulting in lower than expected *in vivo* clearance. When a Zone 4 disconnect remains undiagnosed, project teams can spin wheels working to correct a metabolic liability that, ultimately, does not critically limit *in vivo* exposure. When addressing Zone 4 IVIVC, typically two major hypotheses drive follow-up activities: restrictive protein binding and hepatocyte uptake.

4.3.3.3.1 Restrictive Protein Binding High-equilibrium plasma protein binding ($f_{u,p} \leq 0.02$) is encountered almost twice as often in Zone 4 (~60% of sampled compounds) relative to Zones 1–3 [14]. To evaluate the impact of protein binding on metabolic clearance rate projections, measured plasma protein binding and estimated microsomal protein binding can be, fairly easily, incorporated for *in vitro* into *in vivo* scaling, according to Equation 4.1. This practice tends to overwhelmingly attenuate clearance projections for most compounds, yielding mixed results with respect to IVIVC [3]. Although these "corrections" seem to work very nicely to improve clearance predictions within Zone 4, they generally deteriorate IVIVC across the other three zones.

4.3.3.3.2 Hepatocyte Uptake Although typically reserved for investigating Zone 3, *in vitro* hepatocytes can also provide an effective alternative tool for projects struggling with Zone 4 IVIVC disconnects. The potential for hepatocytes to discriminate effectively *in vivo* clearance for Zone 4 compounds may be partially attributed to cellular protein binding effects, but it may also reflect permeability barriers unique to intact hepatocytes. In a comparison of clearance projections obtained using hepatocytes versus microsomes, two major trends emerge. Lower ER_h projections obtained from hepatocytes are largely associated with Caco-2 efflux and poor passive permeability. This suggests that rate-limiting hepatic uptake could be a key physiological barrier for drug distribution. In some instances, though, hepatocytes project a lower clearance that actually worsens IVIVC relative to microsomal CYP. Here, *in vivo* rates of tissue perfusion, kinetics of distribution, saturation of transport processes, or all of the above may not be adequately represented in the *in vitro* scenario. Therefore, the incorporation of hepatocytes into a discovery ADME strategy should be monitored to ensure that hepatocytes appropriately reflect both high and low clearance for scaffold/series of interest.

4.4 ORAL BIOAVAILABILITY (%F) IVIVC

Ideally, every *in vitro* assay would have an output that could be correlated with a corresponding, readily measured *in vivo* parameter such as the practice for drug clearance. By contrast, the *in vivo* measure of oral bioavailability (%F) reflects the sum of several biological barriers that influence processes ranging from the fraction of dose absorbed in the gut (f_a) to the fraction of absorbed drug that is eliminated via first-pass, hepatic extraction (ER_h). Although, f_a can technically be measured *in vivo*, this parameter is not easily obtained and is seldom available in the discovery stage. As an alternative, we seek to relate *in vitro* measures of f_a and ER_h to generate our earliest estimates of *in vivo* %F. Additional factors, including gastrointestinal (GI) metabolism and drug solubility, provide additional considerations for interrogating the ability of orally administered compounds ultimately to reach systemic circulation.

4.4.1 Establishing a Baseline for %F Correlation Analysis

In vitro permeability assessments are frequently made across membranes on permeable supports. These membranes can be artificial, as is the case in PAMPA, or biological, as is the case in Caco-2. Both types of membrane have been used to provide an estimate of the fraction of drug absorbed from an oral dose [4, 5]. Both models are amenable to high-throughput investment and conveniently provide drug discovery projects with the early estimates of drug permeability (i.e., calculated fraction absorbed). Relative to PAMPA, however, Caco-2 is more resource intensive and requires greater expertise to generate consistent and sustainable monolayers. Furthermore, most drugs and NCEs permeate the gut primarily via passive mechanisms. Therefore, it is common for PAMPA to be surveyed for high-throughput estimates of f_a, with Caco-2 supplementing profiling portfolios for mechanistic follow-up.

Similar to the metabolic stability assay, the *in vitro* permeability value (log P_e, usually expressed as cm/s) from a PAMPA monolayer provides a useful means of triaging poorly permeable compounds and identifying the most promising candidates to move forward into *in vivo* studies. From this model, log $P_e < -5.3$ roughly correlates with gut $f_a < 0.35$ and reflects low permeability, whereas log $P_e > -4.8$ translates to $f_a > \sim 0.75$ and suggests high permeability. As described in the subsequent discussion, these data, when combined with *in vitro* clearance, can be used to estimate the oral bioavailability.

Oral bioavailability represents the fraction of orally dosed drug that is absorbed across the intestine (f_a); escapes intestinal, first-pass metabolism (f_g); and escapes hepatic first-pass elimination (f_h) [47]:

$$\text{In vivo } \%\text{F} = f_a \times f_g \times f_h \times 100\% \tag{4.4}$$

Each term in this equation can be estimated with *in vitro* methodologies. f_a is derived from the permeability measurements in PAMPA or Caco-2. f_g is generally assumed to be minimal and is ignored for the purpose of establishing baseline %F IVIVC. f_h is not measured directly but is inferred from *in vitro* CYP estimates of hepatic extraction ($f_h = 1 - ER_h$). The equation to estimate oral bioavailability can then be rearranged to:

$$\text{estimated } \%\text{F} = \text{PAMPA} f_a \times (1 - ER_h) \times 100\% \tag{4.5}$$

The hexadecane layer of the PAMPA model is well validated as a physicochemical mimetic of intestinal endothelium, and calibration of PAMPA P_e to *in vivo* f_a has previously been established [4]. Similar extrapolation has also been made from Caco-2 derived P_{app} values [5]. Most oral drug absorption occurs from the small intestine. However, theoretically, with permissive permeability, the drug could be absorbed at any point along gastrointestinal transit, from the very acidic conditions of the stomach to the relatively neutral pH of the jejunum. One advantage of the PAMPA model is that it allows for the effect of pH on permeability to be considered. Depending on pK_a and ionization of drug substances, the overall absorptive potential of NCEs can be misrepresented by permeability measured at only a single pH. Conventionally, the P_e readout for PAMPA reflects the maximum permeability observed across the pH range examined *in vitro* (typically from pH ~4–8). Biological permeability models, like Caco-2, on the other hand, are less tolerant to pH variation.

4.4.2 %F IVIVC Zones

4.4.2.1 Zones A and B: Regions of Strong %F IVIVC As both measured and predicted %F are reported in similar terms, a perfect, unity correlation will have a slope of 1. Allowing for ±30% deviation from unity defines a correlation window that simulates the common practice of ranking discovery drugs as "high," "moderate," or "low" %F using a "one-third" rule, where %F < 33% and %F > 67% approximate "low" and "high" oral bioavailability, respectively (Figure 4.2B). %F that is well predicted is separated into zones similar to the clearance module, where Zone A and Zone B represent the window of strong IVIVC associated with poor and good bioavailability, respectively.

4.4.2.2 Zone C: In vivo %F Overestimation Sometimes *in vitro* PAMPA and microsomal stability models fail to predict poor bioavailability. The particular challenge with this scenario is that, with predictive failure of the front-line *in vitro* models, a project team may resort to *in vivo* "screening" for optimization of oral exposure. Neither is this an efficient use of resources nor does it offer mechanistic insight or direction for chemistry optimization. An understanding of the interplay between determinants of oral bioavailability is required to outline a modified *in vitro* strategy to meet the needs of these projects.

4.4.2.3 Zone D: In vivo %F Underestimation Analogous to the outcome for clearance Zone 4 compounds, the molecules in %F Zone D are predicted *in vitro* to show poor bioavailability; yet *in vivo* bioavailability is medium to high. These compounds carry the same risk of misguided deprioritization for follow-up. Some discussion of features that might be associated with *in vitro* %F underestimation is shared to enable anticipation of this type of disconnect.

4.4.2.4 Overview of %F IVIVC for Chemically Diverse NCEs The overlay of >800 Novartis compounds on the four-zone PATH %F IVIVC map shows that the integration of PAMPA-predicted f_a and f_h derived from CYP ER_h as described by

Equation 4.5 effectively predicts *in vivo* %F for ~65% of discovery data (Zones A and B in Figure 4.2B). Infrequently, these *in vitro* data lead to overestimation of %F (<5%) (Zone C). When *in vivo* %F is poorly predicted, most often, these *in vitro* models underpredict oral bioavailability (~30%) (Zone D). The interplay between the major determinants of oral exposure must be considered to understand fully these IVIVC disconnects.

4.4.3 Trends, Hypotheses, and Strategies for %F Interrogation by Zone

4.4.3.1 Zones A and B Despite being well predicted by *in vitro* data, the compounds presented in Zone A demonstrate poor-to-moderate bioavailability. Low bioavailability in this zone may be a result of poor f_a, high ER_h, or a combination of both. When the estimated f_a from the *in vitro* assays is >50% (especially with clearance IVIVC Zone 2), clearance is likely the primary limiting factor and investment in optimized structural changes to mitigate the metabolic liability should be considered. Alternatively, when Zone A is associated with PAMPA $f_a < 50\%$, there could be significant payoffs with permeability optimization. The permeability diagnosis tool as described in Chapter 3, can instruct medicinal chemistry on modulation of physicochemical properties for improved compound absorption.

Zone B is an area that is well correlated with the *in vivo* data that predict medium-to-high bioavailability. The bioavailability in this zone generally reflects good permeability and favorable metabolic stability. There is typically limited opportunity to improve compounds in this zone, but, again, the permeability diagnosis tool (Chapter 3) may be of value in providing some incremental improvement in the permeability of compounds.

4.4.3.2 Zone C The overestimation of bioavailability from PAMPA f_a and CYP ER_h may arise from a few different sources. One caveat with estimating %F according to Equation 4.5 is that it assumes *in vitro* ER_h to predict accurately *in vivo* elimination. Many clearance IVIVC Zone 3 compounds, in particular, those that are associated with significant non-CYP, hepatic elimination mechanisms, may fall into bioavailability IVIVC Zone C simply because total hepatic contribution to elimination (f_h) has been underestimated. For these instances, it may be informative to substitute the apparent *in vivo* $ER_{h,\mathrm{app}}$ ($ER_{h,\mathrm{app}} = CL_{\mathrm{total}} / Q_h$) into Equation 4.5 and reevaluate bioavailability IVIVC. This may provide some insight as to the predictive value of PAMPA for permeability estimates. %F IVIVC disconnects attributable to gastrointestinal processes are generally investigated based on four primary hypotheses. These include overestimation of paracellular flux, involvement of transporters, gut metabolism, and drug solubility.

4.4.3.2.1 Paracellular Flux One source of overestimation from the PAMPA assay is the projection of absorption via paracellular flux. Paracellular flux is not truly mechanistically captured in the PAMPA assay and is instead calculated based on molecular properties [48, 49]. For some projects/scaffolds, this calculation may poorly estimate the paracellular flux relative to the intestine *in vivo*. The Caco-2 cell monolayer culture is considered a reasonably good model of intestinal permeability and does biologically mimic a paracellular component to absorption. However, even this model may not accurately mirror paracellular permeation through the small intestine for all projects.

4.4.3.2.2 Involvement of Transporters A key deficit of the PAMPA permeability model is the lack of drug transporters. Although PAMPA can be very effective for ranking and optimizing drug series for oral absorption governed by passive permeability, profiling must progress to biological models, such as Caco-2, for investigating active transport mechanisms and pathways of absorption. When physicochemical properties seem to be favorable for permeability, then it may be appropriate to investigate biological barriers to absorption. Mature Caco-2 monolayers, because of their expression of the major intestinal drug transporters, P-gp, BCRP, and MRP-2, offer utility as a secondary screen for investigating transporter impact on absorption.

It is important to note that the expression and relative proportion of various transporters may differ between *in vitro* cultures of Caco-2 cells and the small intestine [50]. Nevertheless, Caco-2 efflux provides good evidence for an involvement of drug transporters that might result in lower projection of f_a relative to PAMPA. In general, the *in vivo* impact of intestinal efflux is most profound for molecules having poor passive permeability (Figure 4.4). In particular, compounds with efflux ratios of >3 and low-to-moderate permeability in the Caco-2 assay are at risk for *in vivo* absorption issues. For such compounds, an *in vitro* kinetic study may merit consideration as a means to assess potential for transporter saturation and devise a dosing strategy to overcome restrictive efflux *in vivo* [51].

4.4.3.2.3 Gut Metabolism The amount of drug that is available for gut permeation *in vivo* may well be limited by gut metabolism, a process that is not at all addressed in the passive permeability estimate from PAMPA. Although Caco-2 cultures lack significant CYP 3A4 activity, one of the known DMEs with significant gut expression, some phase II (UGT and sulfotransferase) metabolic capacity seems to be retained by these cells. This may enable them, to some extent, to capture gut metabolism and, correspondingly, reduce the permeability value [52]. However, it is generally expected that metabolic capacity is underrepresented by this model, so it may be premature to assume that Caco-2 can significantly improve f_a projections on this basis. When restrictive gut metabolism is suspected, this hypothesis may be tested by *in vivo* portal vein sampling after oral administration of the drug to monitor the appearance of prehepatic metabolites.

4.4.3.2.4 Drug Solubility Good solubility is desirable for absorption, as compounds must be in solution to permeate the intestinal membrane. Early *in vivo* studies are often administered at low doses of compound, in a solution form, or both, presumably, to remove solubility limitations from the resulting drug-concentration time curves. However, even low doses can encounter solubility issues. Defining

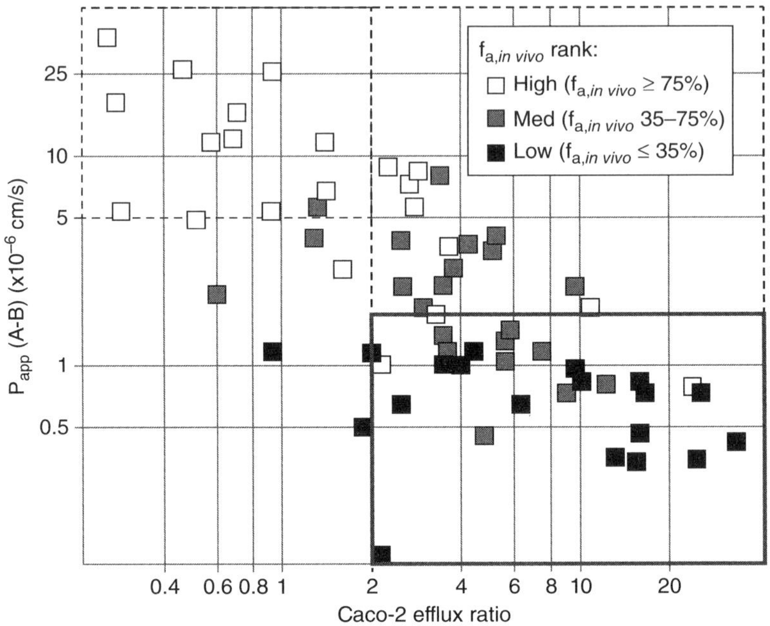

Figure 4.4 Impact of transporters on intestinal absorption. Compounds with measured human f_a were plotted using Caco-2 P_{app} (apical-to-basal transport) against Caco-2 efflux ratios in log-log scale. The solid box denotes the population of compounds where the balance between passive permeability and efflux most impacts f_a and where mechanistic studies to understand the net outcome of the two competing processes may be warranted. (Figure adopted from Reference 51 with permission.)

"sufficient" solubility for a given dose is best done in the context of permeability. The maximum absorbable dose (MAD) model integrates solubility with permeability and dose to anticipate the limits of oral drug exposure. This relatively simple concept introduced by Johnson and Swindell [53] and conceptualized further by Yu [54] is as follows:

$$\mathrm{MAD} = S \times P_{\mathrm{eff}} \times \mathrm{SIWV} \times \mathrm{SITT} \tag{4.6}$$

where S is the solubility (mg/mL) in the intestine, P_{eff} is the effective intestinal permeability coefficient (min^{-1}), SIWV is the small intestinal water volume (11 mL for rat), and SITT is the small intestinal transit time (90 min for rat) [22]. For purposes of discovery IVIVC, we generally consider the equilibrium (i.e., after 24 hours) solubility of the drug, measured in aqueous buffer (pH 6.8) or simulated intestinal fluid, as a measure of "S," whereas the P_{eff} is estimated from the PAMPA assay based on internal calibration between in-house PAMPA P_e and intestinal P_{eff} reported in the literature [55]:

$$P_{\mathrm{eff}} = 10^{(\log Pe+0.5)} \times 60\,\mathrm{s/min} \tag{4.7}$$

Reconsidering Equation 4.6, we can, conversely, assess the minimum acceptable solubility (MAS) required to achieve complete absorption of a predetermined dose. The MAS (g/L), then, is calculated according to Equation 4.8 and serves to answer the question "how much solubility does one need to accommodate the dose?":

$$\text{MAS} = \text{Dose}/(P_{\text{eff}} \times \text{SIWV} \times \text{SITT}) \tag{4.8}$$

From the MAS equation, it is observed that, as permeability decreases, the required solubility increases. When measured solubility is much less than the resultant MAS, solubility is likely to be limiting for gut absorption.

Figure 4.5 illustrates the MAS concept for Zones B and C. Compounds are plotted on a log-log plot of solubility in fasted-state simulated intestinal fluid (FaSSIF) and PAMPA permeability. The diagonal line is the MAS estimated for a MAD of 0.75 mg (~3 mg/kg) dosed orally in rats as permeability increases (left to right on x-axis). For any compound having measured solubility < approximately 0.004 g/L, even the highest theoretical permeability can no longer compensate for the low solubility and a 0.75-mg dose is unlikely to be completely absorbed. The horizontal dashed line in Figure 4.5 is this MAS at the highest achievable permeability. By comparing suspension-dosed compounds found in Zones B and C, where good %F is well predicted or falsely predicted, respectively, the impact of solubility can be observed. In Zone B, most compounds have solubilities above the MAS. Conversely, in Zone C, compounds are largely characterized with solubilities below the MAS, most likely having solubility-limited absorption. In these cases, dose-reduction studies to work within the theoretical MAD for a given compound may be informative toward understanding the role of solubility *in vivo*.

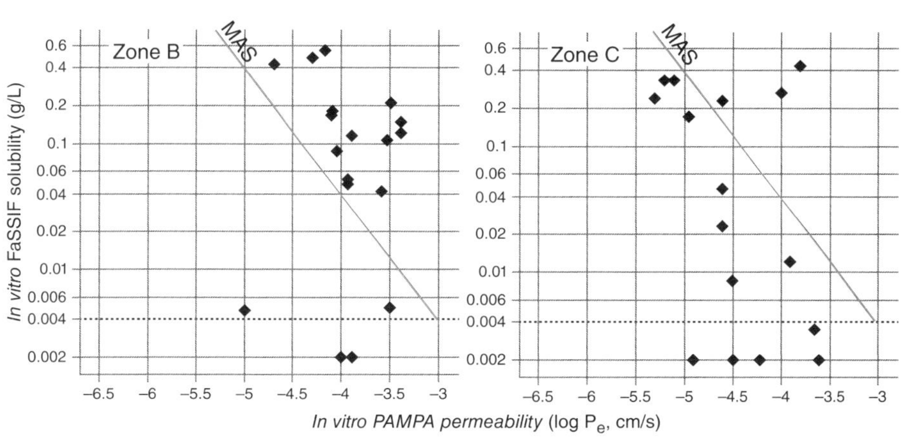

Figure 4.5 FaSSIF solubility of suspension-dosed NCEs relative to the theoretical MAS for a 3-mg/kg oral dose in rats. Graphs correspond to %F IVIVC Zones B (left) and C (right). Diagonal solid line represents the corresponding MAS across the range of permeability and denotes where solubility is likely to become restrictive for permeation. The dotted line represents the minimum solubility that must be satisfied at the highest achievable permeability.

Since the introduction of the MAD/MAS concept, solubility has increasingly been measured in FaSSIF over aqueous buffer, in an effort to improve IVIVC [56–58]. Most compounds have similar equilibrium solubility in pH 6.8 buffer and FaSSIF (70% within threefold), but overall, FaSSIF improves wetting and micellular solubilization of hydrophobic, lipophilic compounds [59]. Occasionally, drug dissolution in the gut is defined, not by the limit of equilibrium solubility, but by the initial solid-state character (e.g., crystallinity) of the chemical matter, a property that generally does not influence conventional high-throughput equilibrium solubility measurements. Particularly when MAD/MAS does not satisfactorily address suspicions of poor oral drug dissolution, the impact of solid-state character may be better understood through comparison of solution versus suspension formulations.

4.4.3.3 Zone D Not surprisingly, there is notable overlap between compounds with Zone D %F disconnects and Zone 4 clearance disconnects. A random subset of clearance Zone 4 compounds overlaid on the PATH %F IVIVC map shows that the majority lie toward the right of unity for measured %F, with >60% having significantly higher %F than would be expected from *in vitro* estimates (Figure 4.6). As discussed, the disconnect between *in vitro* and *in vivo* metabolic clearance may be adequately addressed in the developing %F IVIVC by substituting $ER_{h,\text{app}}$ into Equation 4.5. Where the disconnect is a result of absorptive processes, two primary hypotheses are typically considered, including absorptive transport (or active uptake) and enterohepatic recirculation.

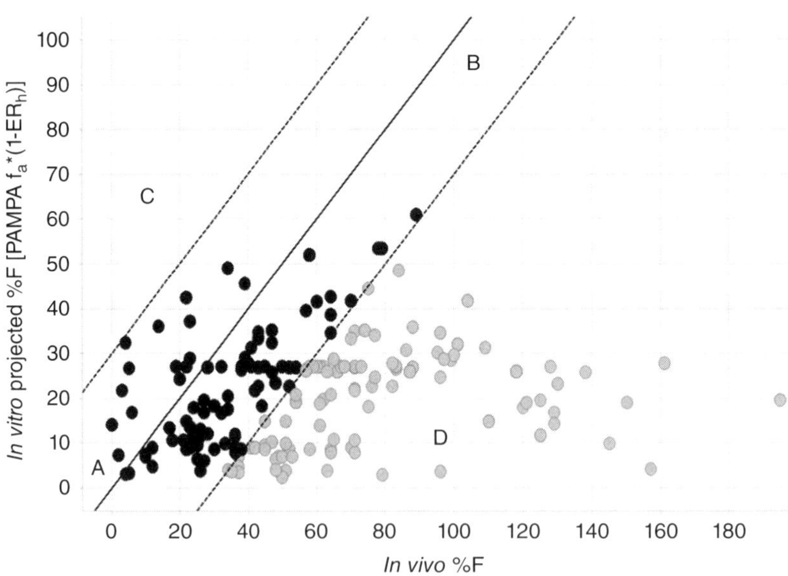

Figure 4.6 %F IVIVC for PATH Clearance Zone 4 compounds. A random subset of clearance Zone 4 compounds (n = 235) overlaid on the PATH %F IVIVC map.

4.4.3.3.1 Absorptive Transport In addition to efflux transporters, the gut expresses several transporters capable of active drug uptake from the gut [50]. Some of these transporters, such as PEPT1, are also expressed in Caco-2. *In vitro*, substances that undergo absorptive transport can be identified in instances where the efflux ratio is <0.33 as has been observed for the peptidomimetic, and PEPT1 substrate, bestatin [60]. *In vitro–in vivo* discontinuity reflected by an underestimation of the *in vivo* %F may arise when the transport capacity *in vivo* substantially exceeds that *in vitro*. Where both PAMPA and Caco-2 f_a estimates are available for Zone D compounds, Caco-2 typically predicts higher (if not comparable) f_a, which dramatically improves %F IVIVC where estimates between the two systems are most disparate.

4.4.3.3.2 Enterohepatic Recirculation When mechanisms of drug elimination are irreversible, the resulting AUC from oral drug absorption can be considered a quantitative reflection of the fraction of total drug dose that reached systemic circulation. Indeed, once metabolized, most drug products are excreted into urine or back into the gut and do not see systemic circulation again. Occasionally, the metabolic products of drug conjugation can undergo enterohepatic recirculation, whereby the metabolite (a drug-glucuronide, for example) is eliminated into the gut and cleaved to re-release the parent drug that, in turn, is reabsorbed back into circulation. This process is most substantial when permeability of the parent drug is moderate to high. Bile-duct cannulated rodents can be a useful *in vivo* model to investigate whether drug conjugates are significantly returned to the gut to potentiate recirculation.

4.5 PAYOFFS FOR INTELLIGENT DATA INTEGRATION IN EARLY DRUG DISCOVERY

The common strategy for optimizing *in vivo* PK in early drug discovery may be to rely, first, on the highest throughput *in vitro* models to gather estimates of clearance, absorption, and solubility. When baseline IVIVC is strong, it is not difficult for project teams to maintain a clear and constant strategy for monitoring progress and prioritizing compounds in series. However, when *in vivo* observations do not reconcile with *in vitro* expectations, the course of progression frequently becomes winding until a productive *in vitro–in vivo* strategy is defined. Engaging in productive next steps, when guided only by sparse compound information, can be challenging, and inefficient use of mechanistic models can diminish their apparent impact as studies fail time and again to deliver desired answers. In extreme cases, *in vitro* ADME models may eventually be abandoned altogether in favor of *in vivo* PK screening for promising NCEs. From our experience, this neither leads to accelerated project progression nor introduces efficient and cost-effective optimization.

In the "PATH" approach to IVIVC, the IVIVC gaps illuminate mechanistic insights crucial for the development of the scaffold SPR. Appreciation for the interplay between ADME properties is critical for identifying the most viable hypothesis describing IVIVC disconnects for the project at hand. A full ADME profile for each compound is not needed for this approach. Rather, the goal is, ultimately, to recognize general trends associated with particular IVIVC zones that may be applied to IVIVC described by subsets of carefully selected molecules of a series or scaffold to deduce relevant mechanisms accounting for disconnects and to engage mechanistic, follow-up studies (Figure 4.7).

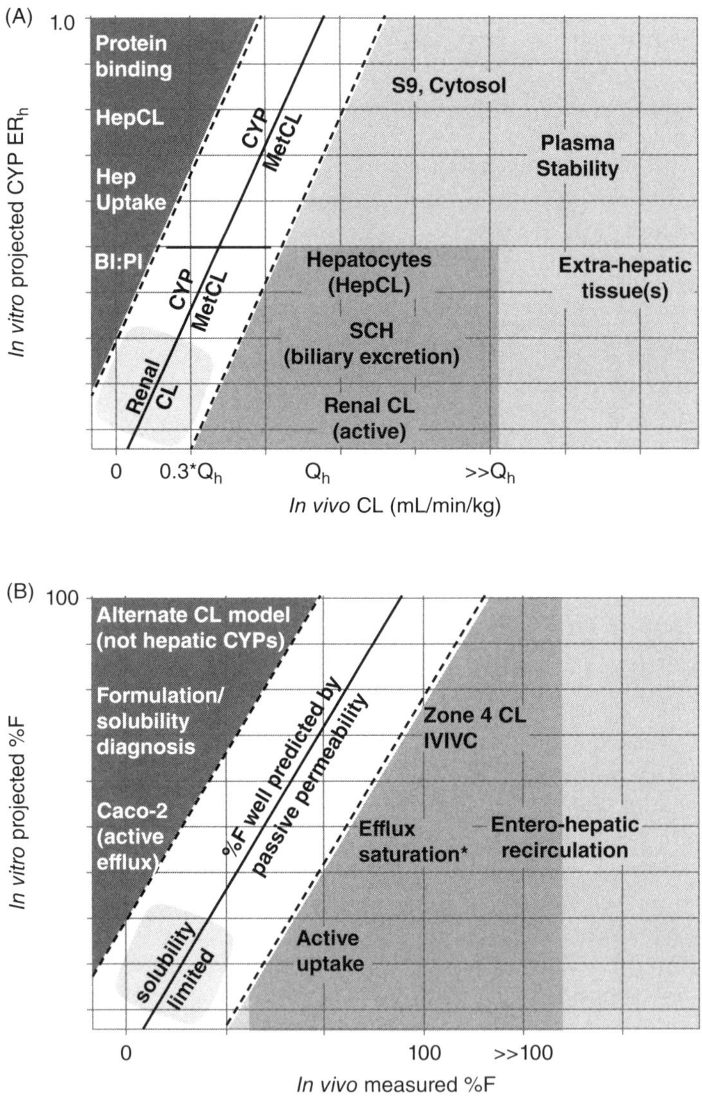

Figure 4.7 Summary of zonal hypotheses to guide follow-up activities relating to (A) clearance or (B) %F IVIVC disconnects. (*If %F is approximated from Caco-2 rather than from PAMPA.)

The general considerations presented in this chapter are derived from an ample database of rat PK. There is no guarantee that clearance or absorption mechanism(s) dominating in preclinical species may be transferrable to humans. For this reason, the approach described here does not prescribe optimization of preclinical IVIVC. Rather, PATH offers a method for interrogating emerging IVIVC and emphasizes diagnosis and testing of root causes for an IVIVC connection or gap. Once identified, the relevance of specific mechanisms to higher species can receive due consideration. This integrated approach to IVIVC development should ultimately elevate the overall return on investment for mechanistic *in vitro* and *in vivo* follow-up by refining rational criteria for engaging in these studies. In time, this approach is also expected to highlight critical gaps in the *in vitro–in vivo* mechanistic toolbox where additional investment can have the greatest impact.

REFERENCES

1. Lin, J.H., Wong, B.K. (2002). Complexities of glucuronidation affecting in vitro-in vivo extrapolation. *Current Drug Metabolism*, *3*, 623–646.
2. Crow, J.A., Borazjani, A., Potter, P.M., Ross, M.K. (2007). Hydrolysis of pyrethroids by human and rat tissues: Examination of intestinal, liver and serum carboxylesterases. *Toxicolology and Applied Pharmacology*, *221*, 1–12.
3. Obach, R.S. (1999). Prediction of human clearance of twenty-nine drugs from hepatic microsomal intrinsic clearance data: An examination of in vitro half-life approach and nonspecific binding to microsomes. *Drug Metabolism and Disposition*, *27*, 1350–1359.
4. Wohnsland, F., Faller, B. (2001), High-throughput permeability pH profile and high-throughput alkane/water log P with artificial membranes. *Journal of Medicinal Chemistry*, *44*, 923–930.
5. Skolnik, S., Lin, X., Wang, J., Chen, X.H., He, T., Zhang, B. (2010). Towards prediction of in vivo intestinal absorption using a 96-well Caco-2 assay. *Journal of Pharmaceutical Sciences*, *99*, 3246–3265.
6. Chiou, W.L., Barve, A. (1998). Linear correlation of the free fraction of oral dose absorbed of 64 drugs between humans and rats. *Pharmaceutical Research*, *15*, 1792–1795.
7. Smith, D.A., Di, L., Kerns, E.H. (2010). The effect of plasma protein binding on in vivo efficacy: Misconceptions in drug discovery. *Nature Reviews Drug Discovery*, *9*, 929–939.
8. Obach, R.S. (1997). Nonspecific binding to microsomes: Impact on scale-up of in vitro intrinsic clearance to hepatic clearance as assessed through examination of warfarin, imipramine, and propranolol. *Drug Metabolism and Disposition*, *25*, 1359–1369.
9. Lewis, D.F., Ioannides, C., Parke, D.V. (1998). Cytochromes P450 and species differences in xenobiotic metabolism and activation of carcinogen. *Environmental Health Perspectives*, *106*, 633–641.
10. Shitara, Y., Sato, H., Sugiyama, Y. (2005). Evaluation of drug-drug interaction in the hepatobiliary and renal transport of drugs. *Annual Review of Pharmacology and Toxicology*, *45*, 689–723.
11. Weiss, M. (1997). Errors in clearance estimation after bolus injection and arterial sampling: Nonexistence of a central compartment. *Journal of Pharmacokinetics and Biopharmaceutics*, *25*, 255–260.

12. Berezhkovskiy, L.M. (2011). On the accuracy of estimation of basic pharmaceutic parameters by the traditional noncompartmental equations and the prediction of the steady-state volume of distribution in obese patients based upon data derived from normal subjects. *Journal of Pharmaceutical Sciences*, *100*, 2482–2497.
13. Wienkers, L.C., Heath, T.G. (2005). Predicting in vivo drug interactions from in vitro drug discovery data. *Nature Reviews Drug Discovery*, *4*, 825–833.
14. Bell, L.C., Wang, J. (2012). Probe ADME and Test Hypotheses: A PATH beyond clearance in vitro-in vivo correlations in early drug discovery. *Expert Opinion on Drug Metabolism & Toxicology*, *8*, 1131–1155.
15. Sohlenius-Sternbeck, A.-K. (2006). Determination of the hepatocellularity number for human, dog, rabbit, rat and mouse livers from protein concentration measurements. *Toxicology in Vitro*, *20*, 1582–1586.
16. Marcos, R., Monteiro, R.A.F., Rocha, E. (2006). Design-based stereological estimation of hepatocyte number, by combining the smooth optical fractionator and immunocytochemistry with anti-carcinoembryonic antigen polyclonal antibodies. *Liver International*, *26*, 116–124.
17. Hinderling, P.H. (1997). Red blood cells: A neglected compartment in pharmacokinetics and pharmacodynamics. *Pharmacological Reviews*, *49*, 279–295.
18. Fisher, M.B., Campanale, K., Ackermann, B.L., VandenBranden, M., Wrighton, S.A. (2000). In vitro glucuronidation using human liver microsomes and the pore-forming peptide alamethicin. *Drug Metabolism and Disposition*, *28*, 560–566.
19. Berezhkovskiy, L.M., Khojasteh, S.C., Halladay, J.S., Hop, C.E. (2009). On the prediction of hepatic clearance using the diluted plasma in metabolic stability assay. *Journal of Pharmaceutical Sciences*, *98*, 1922–1927.
20. Berezhkovskiy, L.M. (2011). The corrected traditional equations for calculation of hepatic clearance that account for the difference in drug ionization in extracellular and intracellular tissue water and the corresponding corrected PBPK equation. *Journal of Pharmaceutical Sciences*, *100*, 1167–1183.
21. Yan, Z., Caldwell, G.W. (2003). Metabolic assessment in liver microsomes by co-activating cytochrome P450s and UDP-glycosyltransferases. *European Journal of Drug Metabolism and Pharmacokinetics*, *28*, 223–232.
22. Davies, B., Morris, T. (1993). Physiological parameters in laboratory animals and humans. *Pharmaceutical Research*, *10*, 1093–1095.
23. Aoki, M., Okudaira, K., Haga, M., Nishigaki, R., Hayashi, M. (2010). Contribution of rat pulmonary metabolism to the elimination of lidocaine, midazolam and nifedipine. *Drug Metabolism and Disposition*, *38*, 1183–1188.
24. Chaudhary, K.R., Batchu, S.N., Seubert, J.M. (2009). Cytochrome P450 enzymes and the heart. *IUBMB Life 61*, 954–960.
25. Ferguson, C.S., Tyndale, R.F. (2011). Cytochrome P450 enzymes in the brain: Emerging evidence of biological significance. *Trends in Pharmacological Sciences*, *32*, 708–714.
26. Cubitt, H.E., Houston, J.B., Galetin, A. (2009). Relative importance of intestinal and hepatic glucuronidation-impact on the prediction of drug clearance. *Pharmaceutical Research*, *26*, 1073–1083.
27. Lohr, J.W., Willsky, G.R., Acara, M.A. (1998). Renal drug metabolism. *Pharmacological Reviews*, *50*, 107–141.

28. Krishna, D.R., Klotz, U. (1994). Extrahepatic metabolism of drugs in humans. *Clinical Pharmacokinetics*, *26*, 144–160.
29. Tang, H., Mayersohn, M. (2006). A global examination of allometric scaling for predicting human drug clearance and the prediction of large vertical allometry. *Journal of Pharmaceutical Sciences*, *95*, 1783–1799.
30. Remmel, R., Nagar, S., Argikar, U. Conjugative metabolism of drugs, in D. Zhang, M. Zhu, and W.G. Humphreys (eds.), *Drug Metabolism in Drug Design and Development*, Wiley, Hoboken, NJ, 2008, pp. 37–88.
31. Morello, K.C., Wurz, G.T., DeGregorio, M.W. (2003). Pharmacokinetics of selective estrogen receptor modulators. *Clinical Pharmacokinetics*, *42*, 361–372.
32. Liu, X., LeCluyse, E.L., Brouwer, K.R., Lightfoot, R.M., Lee, J.I., Brouwer, K.L. (1999). Use of Ca2+ modulation to evaluate biliary excretion in sandwich-cultured rat hepatocytes. *Journal of Pharmacology and Experimental Therapeutics*, *289*, 1592–1599.
33. Ghibellini, G., Vasist, L.S., Leslie, E.M., Heizer, W.D., Kowalsky, R.J., Calvo, B.F., Brouwer, K.L. (2007). In vitro-in vivo correlation of hepatobiliary drug clearance in humans. *Clinical Pharmacology & Therapeutics*, *81*, 406–413.
34. Pan, G., Boiselle, C., Wang, J. (2012). Assessment of biliary clearance in early drug discovery using sandwich-cultured hepatocyte model. *Journal of Pharmaceutical Sciences*, *101*, 1898–1908.
35. Kusuhara, H., Sugiyama, Y. (2009). In vitro-in vivo extrapolation of transporter-mediated clearance in the liver and kidney. *Drug Metabolism and Pharmacokinetics*, *24*, 37–52.
36. Raoof, A.A., Van Obbergh, L.J., De Ville De, G.J., Verbeeck, R.K. (1996). Extrahepatic glucuronidation of propofol in man: Possible contribution of gut wall and kidney. *European Journal of Clinical Pharmacology*, *50*, 91–96.
37. Hiraoka, H., Yamamoto, K., Miyoshi, S., Morita, T., Nakamura, K., Kadoi, Y., Kunimoto, F., Horiuchi, R. (2005). Kidneys contribute to the extrahepatic clearance of propofol in humans, but not lungs and brain. *British Journal of Clinical Pharmacology*, *60*, 176–82.
38. Kitamura, S., Sugihara, K., Ohta, S. (2006). Drug-metabolizing ability of molybdenum hydroxylases. *Drug Metabolism and Pharmacokinetics*, *21*, 83–98.
39. Pryde, D.C., Dalvie, D., Hu, Q., Jones, P., Obach, R.S., Tran, T.D. (2010). Aldehyde oxidase: An enzyme of emerging importance in drug discovery. *Journal of Medicinal Chemistry*, *53*, 8441–8460.
40. Behera, D., Damre, A., Varghese, A., Addepalli, V. (2008). In vitro evaluation of hepatic and extra-hepatic metabolism of coumarins using rat subcellular fractions: Correlation of in vitro clearance with in vivo data. *Drug Metabolism and Drug Interactions*, *23*, 329–350.
41. Zientek, M., Jiang, Y., Youdim, K., Obach, R.S. (2010). In vitro-in vivo correlation for intrinsic clearance for drugs metabolized by human aldehyde oxidase. *Drug Metabolism and Disposition*, *38*, 1322–1327.
42. Cubitt, H.E., Houston, J.B., Galetin, A. (2011). Prediction of human drug clearance by multiple metabolic pathways: Integration of hepatic and intestinal microsomal and cytosolic data. *Drug Metabolism and Disposition*, *39*, 864–873.
43. Linder, C.D., Renaud, N.A., Hutzler, J.M. (2009). Is 1-Aminobenzotriazole an appropriate in vitro tool as a nonspecific cytochrome P450 inactivator? *Drug Metabolism and Disposition*, *37*, 10–13.
44. Taft, D.R. (2004). The isolated perfused rat kidney model: A useful tool for drug discovery and development. *Current Drug Discovery Technologies*, *1*, 97–111.

45. Liu, Y., Weber, S.J., Onua, E.T. (2004), Hepatic clearance and drug metabolism using isolated perfused rat liver. *Current Protocals in Pharmacololgy*, 7.9.1–7.9.10.

46. Grover, A., Benet, L.Z. (2009). Effects of drug transporters on volume of distribution. *AAPS J*, *11*, 250–261.

47. Kwan, K.C. (1997). Oral bioavailability and first-pass effects. *Drug Metabolism and Disposition*, *25*, 1329–1336.

48. Adson, A., Raub, T.J., Burton, P.S., Barsuhn, C.L., Hilgers, A.R., Audus, K.L., Ho, N.F. (1994). Quantitative approaches to delineate paracellular diffusion in cultured epithelial cell monolayers. *Journal of Pharmaceutical Sciences*, *83*, 1529–1536.

49. Adson, A., Burton, P.S., Raub, T.J., Barsuhn, C.L., Audus, K.L., Ho, N.F. (1995). Passive diffusion of weak organic electrolyte across caco-2 cell monolayers: Uncoupling the contributions of hydrodynamic, transcellular, and paracellular barriers. *Journal of Pharmaceutical Sciences*, *84*, 1197–1204.

50. Hilgendorf, C., Ahlin, G., Seithel, A., Artursson, P., Ungell, A.L., Karlsson, J. (2007). Expression of thirty-six drug transporter genes in human intestine, liver, kidney, and organotypic cell lines. *Drug Metabolism and Disposition*, *35*, 1333–1340.

51. Lin, X., Skolnik, S., Chen, X., Wang, J. (2011). Attenuation of intestinal absorption by major efflux transporters: Quantitative tools and strategies using a Caco-2 model. *Drug Metabolism and Disposition*, *39*, 265–274.

52. Jeong, E.J., Lin, H., Hu, M. (2004). Disposition mechanisms of raloxifene in the human intestinal Caco-2 model. *Journal of Pharmacology and Experimental Therapeutics*, *310*, 376–385.

53. Johnson, K.C., Swindell, A.C. (1996). Guidance in the setting of drug particle size specifications to minimize variability in absorption. *Pharmaceutical Research*, *13*, 1795–1798.

54. Yu, L.X. (1999). An integrated model for determining causes of poor oral drug absorption. *Pharmaceutical Research*, *16*, 1883–1887.

55. Winiwarter, S., Bonham, N.M., Ax, F., Karlen, A.A. (1998). Correlation of human jejunal permeability (*in vivo*) of drugs with experimentally and theoretically derived parameters. A multivariate data analysis approach. *Journal of Medicinal Chemistry*, *41*, 4939–4949.

56. Dressman, J., Reppas, C. (2000). In vitro-in vivo correlations for lipophilic, poorly water-soluble drugs. *European Journal of Pharmaceutical Sciences*, *11*, S73–S80.

57. Fagerberg, J.H., Tsinman, O., Sun, N., Tsinman, K., Avdeef, A., Bergström, C.A.S. (2010). Dissolution rate and apparent solubility of poorly soluble drugs in biorelevant dissolution media. *Molecular Pharmaceutics*, *7*, 1419–1430.

58. Zaki, N.M., Artursson, P., Bergström, C.A.S. (2010). A modified physiological BCS for prediction of intestinal absorption in drug discovery. *Molecular Pharmaceutics*, *7*, 1478–1487.

59. Ottaviani, G., Gosling, D., Patissier, C., Rodde, S., Zhou, L., Faller, B. (2010). What is modulating solubility in simulated intestinal fluids? *European Journal of Pharmaceutical Sciences*, *41*, 452–457.

60. Saito, H., Inui, K. (1993). Dipeptide transporters in apical and basolateral membranes of the human intestinal cell line Caco-2. *American Journal of Physiology*, *265*, G289–G294.

5

PK-MATRIX—A PERMEABILITY: INTRINSIC CLEARANCE SYSTEM FOR PREDICTION, CLASSIFICATION, AND PROFILING OF PHARMACOKINETICS AND DRUG–DRUG INTERACTIONS

Urban Fagerholm

5.1 INTRODUCTION

Established relationships between permeability (P_e) and uptake/reuptake in various important organs and understanding of the interplay between P_e and hepatic metabolic intrinsic clearance (CL_{int}) are desirable/valuable for predictions of pharmacokinetics (PK) and drug–drug interactions (DDIs).

Recently, Fagerholm [1] established relationships between passive *in vitro* P_e and *in vivo* fraction absorbed (f_a) (for the unbound fraction (f_u) during each passage

Predictive ADMET: Integrative Approaches in Drug Discovery and Development, First Edition.
Edited by Jianling Wang and Laszlo Urban.

through an organ) in the human intestines, liver, and brain, and between passive *in vitro* P_e and *in vivo* fraction reabsorbed in the human renal tubuli.

Based on these relationships, a P_e-Classification System (PCS) with four P_e-classes was developed—very high (I), high (II), intermediate (III), and low (IV) P_e [1]. According to defined limits, compounds of class I (without solubility/dissolution limitations) are or are predicted to be well absorbed and reabsorbed, and mainly eliminated by metabolism. Class III and IV compounds have limited passive uptake/reuptake in the intestines, liver, and renal tubuli, and their uptake and elimination are potentially most sensitive to involvement and changes of active transport.

Strengths with classification systems include improved overview and reduced systematic error potential when applied for *in vitro* to *in vivo* predictions. Such systems will automatically compensate for down(or up)regulation of enzymatic activities during *in vitro* conditions and for erroneous assumptions regarding, for example, organ weight, time, flow rates, dispersion, number of cells involved in the PK processes, and regional activity differences. The significant underpredictions of *in vivo* CL_{int}, hepatic CL (CL_{H}), and CL that is normally observed when using human microsome and hepatocyte CL_{int} data [2] could be avoided using such an approach.

Limitations include potential for and impact of incorrect classification. With an increasing number of classes, the former will increase, the latter will decrease, and more information is possible to obtain/show. Uncertain/erroneous *in vivo* data and poorly functioning prediction methods will also jeopardize their applicability. It is important that classification systems are well balanced and built based on available data. An extensive range of *in vivo* data enable more classes, a wider span (ratio between upper and lower limits) of classes, or both. The BCS (Biopharmaceutics Classification System) and BDDCS (Biopharmaceutics Drug Disposition Classification System) have two categories (high:low) for each parameter. The PCS has four.

The interplay between transport and metabolism is not fully understood. The development of PCS was an attempt to improve the understanding of it. By incorporating a CL_{int} dimension into the PCS (replacing f_a by CL_{int}), further improvements are achievable. Such a P_e- and CL_{int}-based matrix (called PK-MATRIX) enables PK classification, prediction, and profiling of drugs and drug candidates and is useful in risk assessment, lead optimization, and for optimization and rationalization of drug discovery and development programs. It is also suitable for test validation of various *in vitro* and *in silico* prediction tools (microsomes, hepatocytes, and permeability cell cultures inclusively).

The development and applicability of PK-MATRIX for PK and DDIs in humans (the approach is also useful for animal data) are demonstrated in this chapter. It includes (1) setting up of classes and limits, (2) *in vivo* classification of drugs (P_e vs. hepatic metabolic CL_{int}; including fraction excreted (f_e) for each class), (3) evaluation of the role and distribution of renal and biliary CL_{int}, (4) distribution of types and magnitudes of DDIs along P_e classes, and (5) *in vitro* to *in vivo* prediction and classification using available and suitable literature data of *in vitro* P_e and metabolic and excretion CL_{int}. The few available human *in vitro* renal and biliary CL_{int} data and studies limited the evaluation. However, as will be demonstrated, *in vitro* P_e

classification alone is useful for predicting excretion potential and routes and DDIs, and additional *in vitro* hepatic metabolic CL_{int} classification will increase the confidence and improve the predictability of drug excretion and corresponding DDIs.

5.2 SETTING UP THE PK-MATRIX

When setting up PK-MATRIX data ranges for *in vivo* P_e and hepatic metabolic CL_{int}, commonly used quality measures for prediction quality, and the pros and cons of having too few and too many classes were considered.

It seems that prediction errors of less than twofold are generally assumed to be reasonably good, and it was decided that upper and lower limits for P_e and CL_{int} classes should differ by a factor of at least 2 (except for class 8 where lower limits are zero).

An extensive set of plasma CL, CL_H (assumed to equal nonrenal CL), renal CL (CL_R), and f_u *in vivo* data in humans collected by Varma et al. [3] was used for estimation of hepatic metabolic CL_{int}. Blood hepatic metabolic CL_{int} was estimated using additional blood f_u and blood-to-plasma concentration data (collected from References 4–11). CL_{int} was calculated using the well-stirred model and a hepatic blood flow rate (Q_H) of 1500 mL/min. Many compounds in the dataset by Varma et al. are eliminated via bile. Corrections were made for these. Compounds having an apparent biliary CL (CL_{bile}; data taken from Reference 12) corresponding to more than 5% of total CL were excluded. The reason was the potential/unknown impact of gastrointestinal reabsorption on the true CL_{bile}. Calcium blockers, which have been demonstrated to increase the Q_H significantly (by 23% to 36% on average [13]), were also excluded.

Based on the range for *in vivo* hepatic metabolic CL_{int} (0 to 1,170,000 mL/min), it was decided that the PK-MATRIX should be built on eight CL_{int} classes with a threefold span (exceptions are class 8 with zero as the lowest limit and class 1 with infinity as the highest upper limit; see Table 5.1). The upper limit of class 8 (= lower limit of class 7) was set to 125 mL/min, which equals the glomerular filtration rate (GFR).

In vivo f_a after oral drug administration is used as a surrogate for *in vivo* P_e. When setting up the P_e part of the system, the (established) sigmoidal relationship between P_e and f_a [1] and data ranges for P_e and f_a were considered. F_a data were extracted from Reference 3. Compounds with apparent solubility/dissolution-limited uptake were excluded. The *in vitro* passive P_e dataset ($N = 124$) produced by Willmann et al. ([14]; P_e was estimated from membrane affinity at pH 7.4, defined as the equilibrium partition coefficient between water and immobilized lipid bilayers, and effective molecular weight) that was used for setting up the PCS was also used for PK-MATRIX. On the basis of these data, it was decided that the PK-MATRIX should be built on eight P_e classes (as for CL_{int}) with a twofold difference between lower and upper P_e limits (exceptions are class 8 with zero as the lowest limit and class 1 with infinite P_e as the highest upper limit; see Table 5.2). The lower limit of class 4 ($f_a = 0.9$) is similar to the high P_e limit of the BCS. High P_e classes (1–3) have narrow

Table 5.1 CL_{int} classes, CL_{int} limits, and suggested reference compounds

CL_{int} class	Lower *in vivo* CL_{int} limit (mL/min)	Reference compounds[a]	Alternatives
1	150,000	Propafenone	Diclofenac[b]
2	50,000	Midazolam	Omeprazole
3	15,000	Prazosin	Nortriptyline
4	5000	Triazolam	Methylprednisolone
5	1500	Dexamethasone	Acetaminophen
6	500	Tolbutamide	Warfarin
7	125	Amobarbital	Theophylline
8	0	—	—

[a]Represent blood hepatic metabolic CL_{int}.
[b]Does not fulfill f_u criteria for CL_{int} reference compounds (should be ≥ 0.01; see text).

f_a limits (a result of the nonlinear relationship between P_e and f_a), and therefore, *in vitro* (or *in silico*) P_e data are required for correct *in vivo* P_e-classification of such compounds. Compounds with $f_a \geq 0.98$ (P_e classes I–III) and without *in vitro* (or *in silico*) P_e were allocated to P_e class 3.

According to the PK-MATRIX, the most permeable high CL_{int} compounds belong to class PC11 and the least permeable low CL_{int} drugs belong to PC88.

Reference compounds for each class were selected (Tables 5.1 and 5.2). For P_e, the aim was to find and select substances without known significant active transport or solubility/dissolution limitations. For CL_{int}, the objective was to choose reference drugs with high passive P_e (low excretion potential), low f_e, $f_u \geq 0.01$, and without known influence on Q_H.

Table 5.2 P_e classes, P_e and f_a limits, and suggested reference compounds

P_e-class	Lower *in vitro* P_e limit[a] ($\bullet 10^{-6}$ cm/s)	Lower *in vivo* f_a limit[b]	Reference compounds
1	20	1.00	Carbamazepine[c]
2	12	0.99	Verapamil[c]
3	7.5	0.98	Metoprolol[c]
4	2.5	0.90	Scopolamine
5	0.66	0.50	Atenolol
6	0.33	0.25	Mannitol
7	0.15	0.10	Enalaprilat
8	0	0	—

[a]Based on *in vitro* passive P_e data from Reference 14. Other limits apply for other datasets.
[b]Calculated from established *in vitro* passive P_e versus *in vivo* f_a relationship [1].
[c]Known substrate for human transport protein(s) [24]. Ultimately reference compounds are not transported actively.
Note: Permeation-rate limited uptake is assumed.

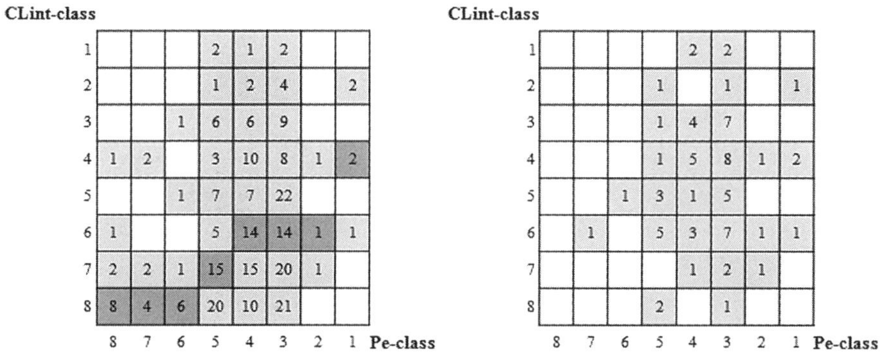

Figure 5.1 (A) *In vivo* P_e: plasma hepatic metabolic CL_{int} distribution of 261 drugs. (B) *In vivo* P_e: blood hepatic metabolic CL_{int} distribution of 71 drugs. The number of drugs for each category is presented. Median plasma hepatic metabolic CL_{int} for each P_e class is located in dark-shaded categories.

5.3 PK-MATRIX DISTRIBUTION/CLASSIFICATION OF DRUGS

5.3.1 P_e versus Hepatic Metabolic CL_{int}

Figure 5.1A shows the *in vivo* P_e: hepatic metabolic CL_{int} (plasma CL_{int}) distribution of 261 drugs. Corresponding results for blood hepatic metabolic CL_{int} for 71 compounds are demonstrated in Figure 5.1B.

The highest hepatic metabolic CL_{int} in this dataset is 1,170,000 mL/min. The absence of compounds in the upper left PK-MATRIX corner could probably be explained by the limited uptake capacity of the liver to absorb passively low P_e drugs [1, 15] and low capacity of liver enzymes to metabolize comparably hydrophilic drugs.

5.3.2 P_e versus Fraction Excreted

The median and maximum observed f_e ($f_e = CL_R/CL$; [3]) for each category are presented in Figure 5.2. Negligible f_e was found for P_e and CL_{int} classes 1 and 2. $F_e \geq 0.75$ was generally observed for compounds with low/intermediate P_e (P_e classes 4–8) and low hepatic metabolic CL_{int} (CL_{int} classes 7 and 8; $CL_{int} < 500$ mL/min). Apparently, intermediate passive P_e and absorption seems to be favorable for renal excretion potential. Many P_e class 5 drugs have comparably high f_e and renal CL_{int} (see below). Major renal elimination has been observed for a high hepatic metabolic CL_{int} drug (PC53).

Five compounds with elimination mainly via bile (estimated using CL and CL_{bile} data from References 3 and 12) were found. These all have a comparably high molecular weight (436 to 646 g/mole), which is in agreement with previous knowledge about properties important for bile excretion [12]. These belong to P_e classes 8 and 5 (P_e and f_a data were not available for all these compounds) and hepatic metabolic CL_{int} classes 7 or 8.

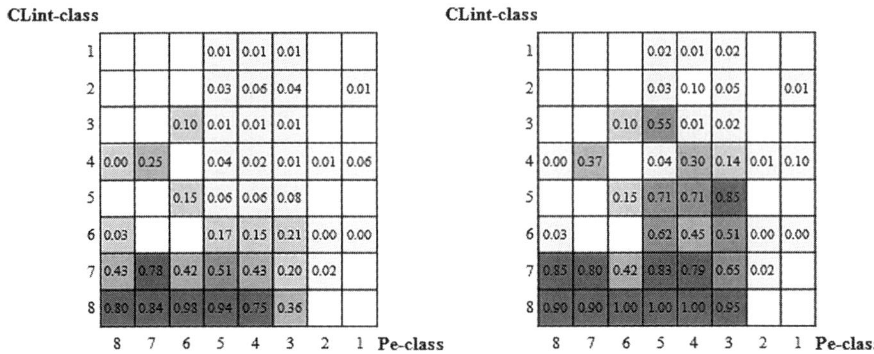

Figure 5.2 (A) Median and (B) maximum Fe for each category with *in vivo* P_e and plasma hepatic metabolic CL_{int} data. Classes with higher Fe are presented with a darker background.

5.3.3 P_e versus Renal and Biliary CL_{int}

CL_R, CL_{bile}, and plasma f_u data collected by References 3 and 12 enabled estimation of renal CL_{int} for 274 compounds and bile CL_{int} for 48 compounds (f_a is available for these).

Renal CL_{int} (includes the GFR; values > GFR for net secretion; values < GFR for net reabsorption) was estimated using the well-stirred model approach and by setting the GFR to 125 mL/min and renal Q to 1200 mL/min. Bile CL_{int} was calculated in analogy with hepatic metabolic CL_{int} and without considering intestinal reabsorption (which implies an overestimation potential). Both renal and bile CL_{int} were calculated using plasma f_u.

Figures 5.1, 5.2, and 5.3 clearly demonstrate the dominant role of hepatic metabolism, in particular at high P_e, f_a, and hepatic metabolic CL_{int}. They also show favorable windows (intermediate P_e and f_a) for excretion and that the renal route generally is dominant for drug excretion.

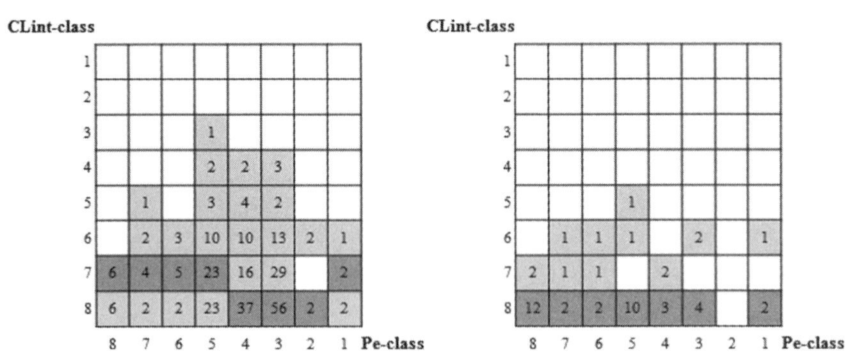

Figure 5.3 PK-MATRIX distribution patterns for (A) renal ($N = 274$) and (B) biliary plasma CL_{int} ($N = 48$). The number of drugs for each category is presented. Median CL_{int} for each P_e class is located in dark-shaded categories.

The PK-MATRIX distribution patterns for renal and biliary excretion are demonstrated in Figs. 5.3A and B, respectively. The highest renal and biliary CL_{int}-values, both observed for P_e class 5 compounds are 36,000 and 5600 mL/min, respectively. Lowest estimates were found for drugs with low and high P_e and, in particular, for the highly permeable compounds (which also possess high reabsorption potential). For P_e classes 5–8, most compounds show net renal secretion, whereas most compounds of a higher P_e class demonstrate net reabsorption. For high P_e compounds excreted via bile, a considerable portion of the extent excreted is likely to be reabsorbed from the intestines and undergo enterohepatic circulation. Therefore, their apparent biliary CL_{int} (generally estimated after collection of bile) is most probably higher than the true value.

5.3.4 Two Candidate Drug Examples

A compound predicted to belong to PC32 is anticipated to be more or less completely absorbed after oral administration (unless limited solubility/dissolution is expected) and to be eliminated mainly by metabolism. Preclinical and clinical investigations of excretion and related DDIs for such a compound are expected to be unnecessary (under the assumption that *in vitro* and/or *in silico* prediction tools are reliable). Such investigations are, however, required for a PC57 candidate drug, which is predicted to have intermediate f_a and anticipated to be eliminated partly renally (and potentially via bile if molecular weight is > approximately 400 g/mole).

5.4 DISTRIBUTION OF DDIS ACCORDING TO P_e CLASSES

Types and magnitudes (change in area under the time-plasma concentration curve for victim drugs) of DDIs were collected from References 16 and 17, and data were allocated to *in vivo* P_e classes. Only data observed with the strongest known inhibitors and inducers were used, including CYP3A4-inhibitor ketoconazole and CYP3A4-inducer rifampicin. CYP3A4 inhibition obtained with grapefruit juice was included as these results reflect the CYP3A4 inhibition specific for the gut-wall mucosa.

Figure 5.4 shows the maximum observed exposure change for different types of DDIs and grapefruit-juice–drug interactions for each P_e class. *Note that the distribution pattern is incomplete because data for some* $\mathrm{P_e}$ *classes are missing.* This figure clearly demonstrates that metabolic DDIs are of greatest general importance, especially for high P_e drugs and CYP3A4 substrates. Furthermore, it shows that inhibition of gut-wall CYP3A4 contributes significantly to the overall effect observed with CYP3A4-inhibition drugs and that this effect is P_e dependent. Inhibitors, inducers, and victim drugs with low P_e will be absorbed to a greater extent from the lower regions of the gastrointestinal tract where the levels and activity of CYP3A4 are lower. Thus, more pronounced mucosal CYP3A4-inhibition/induction effect is anticipated for high P_e compounds.

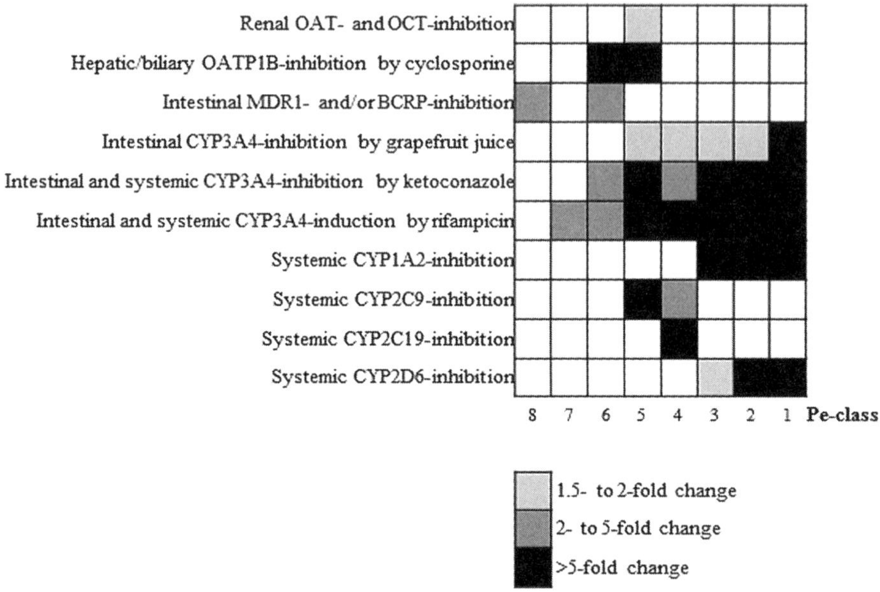

Figure 5.4 DDI types and magnitudes for victim drugs of different P_e classes. The maximum average x-fold change of CL or exposure with the strongest known inhibitors and inducers are presented. Blank categories lack data or represent compounds with <1.5-fold changes.

Figure 5.4 also shows that transporter DDIs occur for compounds with limited passive P_e and that these are most pronounced for the hepatobiliary route and of least importance for renal elimination. P_e classes for the most significant renal (inhibition by probenecid) and biliary (inhibition by cyclosporine and ritonavir) DDIs are consistent with those with highest excretion CL_{int}.

In contrast, clinical transporter DDIs for high P_e compounds (loperamide and verapamil) have been observed at the blood-brain barrier [1, 18, 19]. Explanations for this discrepancy include a comparably short residence time and right-shifted P_e versus f_a relationship at the blood-brain barrier [1, 20].

The PC32 compound used as an example earlier might therefore be sensitive to inhibition and induction of metabolism, in particular if it is a CYP3A4 substrate. Its brain uptake could potentially be sensitive to inhibition of efflux in case it is a substrate for important transporters at the human blood-brain barrier (in particular, MDR1 and BCRP). Transporter DDIs at the hepatobiliary, gastrointestinal, and renal levels are less likely. These are more likely to occur and be clinically important for the PC57 candidate drug.

5.5 *IN VITRO* VERSUS *IN VIVO* PREDICTION AND CLASSIFICATION

PK-MATRIX is thought to be primarily used for predictions of PK and DDIs (*in vivo* classification and profiling are secondary), and its applicability is highly dependent on the performance of *in vitro* (and *in silico*) prediction tools. Although poor prediction methods could limit the usefulness of PK-MATRIX, the system is useful for testing, validating, and comparing these.

When used for predictions, two different approaches are applicable—classification versus data for selected reference compounds (ranking) and classification based on established *in-vitro–in-vivo* relationships (functions/equations).

To demonstrate the applicability of PK-MATRIX and performance of *in vitro* prediction tools, the literature was searched for studies with P_e and CL_{int} datasets containing suitable reference compounds (predefined and shown in Tables 5.1 and 5.2).

5.5.1 *In Vitro* versus *In Vivo* P_e

For P_e predictions, passive *in vitro* P_e and *in vivo* f_a ($N = 124$) from Reference 14 were used. Correct *in-vitro–in-vivo* ranking of reference compounds was found, which is crucial for the ranking approach (but not when using a function that describes the *in-vitro–in-vivo* relationship). *In vivo* P_e (f_a) of 44% of compounds was correctly classified using *in vitro* passive P_e data. Sixty-nine percent had a classification error (difference between predicted and observed classes) of maximally 1, and 16% were misclassified by 3 to 5 classes. Compounds that were incorrectly classified by 3 to 5 classes were mainly those with complete or nearly complete absorption (classification most sensitive to small differences in f_a), and those with low/moderate passive P_e + active uptake (antibiotics and angiotensin-converting-enzyme (ACE) inhibitors). At least 13 (11%) of the selected test compounds are known to use active transport to some extent.

Thomas et al. [21] produced an extensive set with Caco-2 data ($N = 82$; obtained at pH 7.4). It lacks three of the preselected reference compounds. For this dataset, the P_e versus f_a relationship method was applied. Forty percent of compounds were correctly classified, and 82% had a classification error of maximally 1.

5.5.2 *In Vitro* versus *In Vivo* Hepatic Metabolic CL_{int}

In most of the explored reports with *in vitro* CL_{int}-based predictions, there were only a few of the preselected reference compounds, and some studies presented data obtained at different laboratories, with different methodologies, or both. A human liver microsome CL_{int}-based prediction study by Obach [7] was selected among studies with human liver microsomes and hepatocytes. This dataset contains 29 basic, neutral, and acidic drugs, including 6 of the preselected reference compounds (a seventh suitable was selected). CL_{int} (ranging by more than 300-fold) was estimated using the substrate depletion method and corrected for microsome binding. Correct *in-vitro–in-vivo* ranking of reference compounds was found. Fifty percent and 75% of classifications were correct or had an error of maximally 1, respectively. A twofold maximum CL_{int} classification error was found.

5.5.3 *In Vitro* versus *In Vivo* P_e and Hepatic Metabolic CL_{int}

Among the 20 test compounds with *in vivo* and *in vitro* CL_{int} data, 11 also had reported *in vivo* f_a and *in vitro* P_e data. For 7 of these 11 compounds (64%), *in vitro* and *in vivo* classes were similar or adjacent to each other. The *in vivo* class of two compounds (18%) was correctly predicted.

5.5.4 *In Vitro* versus *In Vivo* Excretion CL_{int}

Watanabe et al. [22] predicted the *in vivo* renal CL_{int} (maximum 1100 mL/min) from *in vitro* CL_{int} (using freshly prepared human kidney slices) for 10 anions. A correlation coefficient (R^2) of 0.8 was obtained between log *in vivo* CL_{int} and log *in vitro* CL_{int}, and the ranking was reasonably consistent. Eight of these compounds belong to renal *in vivo* CL_{int} class 7 and two belong to *in vivo* CL_{int} class 6, which limits the evaluation. According to *in-vitro*-to-*in-vivo* predictions, all compounds belong to CL_{int} class 7.

Attempts have also been made to predict *in vivo* biliary CL_{int} from *in vitro* data. Jones et al. [23] measured *in vitro* biliary CL_{int} for seven compounds using sandwich-cultured human hepatocytes for this purpose. *In vitro* and *in vivo* estimates correlated poorly. For example, rosuvastatin, which belongs to *in vivo* biliary CL_{int} class 5 (1600 mL/min; highest value of all compounds in this selection), had a comparably low *in vitro* biliary CL_{int} (class 8).

5.6 CASE STUDIES

So far, PK-MATRIX has been applied for selecting drug candidates, guiding lead optimization, and predicting PK, doses, exposures, and DDI potential in humans in two AstraZeneca (London, U.K.) projects. In both projects, lead compounds belong to P_e classes 1 or 2 and CL_{int} classes 3–6. Based on PK-MATRIX classification, none of these compound are anticipated to be significantly excreted in urine or bile (see Figures 5.2 and 5.3) or to demonstrate systemic PK sensitive to inhibition and induction of gastrointestinal, renal, or biliary transporters (see Figure 5.4). The compounds are substrates for CYP3A4, which makes their PK sensitive to inhibition and induction of this enzyme, not least in the small intestinal mucosa. Significant MDR1-efflux was observed for some substances belonging to CL_{int} class 3. Thus, it was preferred to select compounds of CL_{int} classes 4–6 (and P_e classes 1 or 2). These compounds have not been administered to and evaluated in healthy volunteers, which means that prospective PK-MATRIX–based predictions have not yet been evaluated for AstraZeneca compounds.

Retrospective analysis of *in vitro* P_e and CL_{int} data of an AstraZeneca compound given orally to humans correctly predicted significant renal excretion potential.

5.7 DISCUSSION

Every classification and prediction system has its limitations and strenghts.

Limitations/weaknesses with PK-MATRIX include the impact of data uncertainty and incorrect classification. A classification and prediction system is not stronger than the quality of the data. For example, there is a comparably high uncertainty of f_u (used for estimation of CL_{int}) for highly bound compounds, saturation might have influenced f_a-values, nonexcretion *in vivo* CL_{int} (used as a surrogate for hepatic metabolic *in vivo* CL_{int}) might not truly reflect *in vivo* hepatic metabolic CL_{int}, and the lower levels of quantification of commonly used *in vitro* P_e and CL_{int} methods could sometimes exceed the lower PK-MATRIX limits [6]. Another limitation is that intervenous and oral PK data and mass-balance results are required for *in vivo* classification of P_e and CL_{int}—and *in-vitro*-to-*in-vivo* comparison—of new compounds. There is also a requirement for additional *in vitro* or *in silico* data for *in vivo* P_e classing of high P_e compounds.

The strengths and advantages of PK-MATRIX include its integration of PK parameters and DDI risk potential, improved overview and understanding of PK and DDIs, automatic correction of systematic prediction errors and applicability for automatic validation, and comparison of *in vitro* and *in silico* prediction methods.

5.8 SUMMARY

It seems that reasonably good predictions can be obtained with commonly used *in vitro* hepatic metabolic CL_{int} and P_e methods and PK-MATRIX, and that a PK-MATRIX with 64 (8×8) classes is useful for its indended purposes (*in-vitro*-to-*in-vivo* extrapolation and prediction; test validation of *in vitro* and *in silico* prediction methods; PK and DDI classification and profiling; and improved overview and understanding of PK and DDIs). This together with user-friendliness and improved integration and overview make it an attractive tool for selection and profiling of candidate drugs, optimization of lead compounds, and rationalization of drug development programs.

REFERENCES

1. Fagerholm, U. (2008). The role of permeability in drug ADME/PK, interactions and toxicity – Presentation of a permeability-based classification system (PCS) for prediction of ADME/PK in humans. *Pharmaceutical Research, 25*, 625–638.
2. Fagerholm, U. (2007). Prediction of human pharmacokinetics – Evaluation of methods for prediction of hepatic metabolic clearance. *Journal of Pharmacy and Pharmacology*, *59*, 803–828.
3. Varma, M.V., Obach, R.S, Rotter, C., et al. (2010). Physicochemical space for optimum oral bioavailability: contribution of human intestinal absorption and first-pass elimination. *Journal of Medicinal Chemistry*, *53*, 1098–1108.
4. Shibata, Y., Takahashi, H., Chiba, M., Ishii, Y. (2002). Prediction of hepatic clearance and availability by cryopreserved human hepatocytes: an application of serum incubation method. *Drug Metabolism and Disposition*, *30*, 892–896.

5. Brown, H.S., Griffin, M., Houston, J.B. (2007). Evaluation of cryopreserved human hepatocytes as an alternative *in vitro* system to microsomes for the prediction of metabolic clearance. *Drug Metabolism and Disposition*, *35*, 293–301.
6. Stringer, R., Nicklin, P.L., Houston, J.B. (2008). Reliability of human cryopreserved hepatocytes and liver microsomes as *in vitro* systems to predict metabolic clearance. *Xenobiotica*, *38*, 1313–1329.
7. Obach, R.S. (1999). Prediction of human clearance of twenty-nine drugs from hepatic microsomal intrinsic clearance data: an examination of *in vitro* half-life approach and nonspecific binding to microsomes. *Drug Metabolism and Disposition*, *27*, 1350–1359.
8. Youdim, K.A., Zayed, A., Dickins, M., et al. (2008). Application of CYP3A4 *in vitro* data to predict clinical drug-drug interactions; predictions of compounds as objects of interaction. *British Journal of Clinical Pharmacology*, *65*, 680–692.
9. Kato, M., Shitara, Y., Sato, H., et al. (2008). The quantitative prediction of CYP-mediated drug interaction by physiologically based pharmacokinetic modeling. *Pharmaceutical Research*, *25*, 1891–1901.
10. De Buck, S.S., Sinha, V.K., Fenu, L.A., et al. (2007). Prediction of human pharmacokinetics using physiologically based modeling: A retrospective analysis of 26 clinically tested drugs. *Drug Metabolism and Disposition*, *35*, 1766–1780.
11. Paixão, P., Gouveia, L.F., Morais, J.A.G. (2009). Prediction of drug distribution within blood. *European Journal of Pharmaceutical Sciences*, *36*, 544–554.
12. Yang, X., Gandhi, Y.A., Duignan, D.B., Morris, M.E. (2009). Prediction of biliary excretion in rats and humans using molecular weight and quantitative structure-pharmacokinetic relationships. *AAPS Journal*, *11*, 511–525.
13. Bengtsson-Hasselgren, B., Ronn, O., Blychert, L.O. (1990). Acute effects of felodipine and nifedipine on hepatic and forearm blood flow in healthy men. *European Journal of Clinical Pharmacology*, *46*, 529–533.
14. Willmann, S., Schmitt, W., Keldenich, J., et al. (2004). A physiological model for the estimation of the fraction dose absorbed in humans. *Journal of Medicinal Chemistry*, *47*, 4022–4031.
15. Fagerholm, U. (2009). Presentation of a Modified Dispersion Model (MDM) for hepatic drug extraction and new methodology for prediction of the rate-limiting step in hepatic metabolic clearance. *Xenobiotica*, *39*, 57–71.
16. Giacomini, K.M. Huang, S.M., Tweedie, D.J. et al. (2010). Membrane transporters in drug development. *Nature Reviews*, *9*, 215–236.
17. Obach, R.S., Walsky, R.L., Venkatakrishnan, K., et al. (2006). The utility of *in vitro* cytochrome P450 inhibition data in the prediction of drug-drug interactions. *Journal of Pharmacology and Experimental Therapeutics 161*, 336–348.
18. Sadeque, A.J.M., Wandel, C., He, H., et al. (2000). Increased drug delivery to the brain by P-glycoprotein inhibition. *Clinical Pharmacology and Therapeutics*, *68*, 231–237.
19. Hsiao, P., Bui, T., Ho, R.J., Unadkat, J.D. (2008). *In vitro*-to-*in vivo* prediction of P-glycoprotein-based drug interactions at the human and rodent blood-brain barrier. *Drug Metabolism and Disposition*, *36*, 481–484.
20. Fagerholm, U. (2007). The highly permeable blood-brain barrier: An evaluation of current opinions about brain uptake capacity. *Drug Discovery Today*, *12*, 1076–1082.

21. Thomas, S., Brightman, F., Gill, H., et al. (2008). Simulation modelling of human intestinal absorption using Caco-2 permeability and kinetic solubility data for early drug discovery. *Journal of Pharmaceutical Sciences*, *10*, 4557–4574.
22. Watanabe, T., Kushuhara, H., Watanabe, T., et al. (2011). Prediction of the overall renal tubular secretion and hepatic clearance of anionic drugs and a renal drug-drug interaction involving organic anion transporter 3 in humans by *in vitro* uptake experiments. *Drug Metabolism and Disposition*, *39*, 1031–1038.
23. Jones, H.M., Barton, H.A., Lai, Y. et al. (2012). Mechanistic pharmacokinetic modeling for the prediction of transporter-mediated disposition in humans from sandwich culture human hepatocyte data. *Drug Metabolism and Disposition*, *40*, 1007–1017.
24. Pedersen, J.M., Matsson, P., Bergström, C.A., et al. (2008). Prediction and identification of drug interactions with the human ATP-binding cassette transporter multidrug-resistance associated protein 2 (MRP2; ABCC2). *Journal of Medicinal Chemistry*, *51*, 3275–3287.

6

MAXIMIZING THE POWER OF A LOCAL MODEL FOR ADMET-PROPERTY PREDICTION

Sebastien Ronseaux, Jeremy Beck, and Clayton Springer

6.1 INTRODUCTION

In silico models such as quantitative structure activity relationship modeling (QSAR models) have been used for more than 50 years and are a mature technology. The International Union of Pure and Applied Chemistry defines QSAR as "mathematical relationships linking chemical structure and pharmacological activity in a quantitative manner for a series of compounds. Methods which can be used in QSAR include various regression and pattern recognition techniques" [1]. In 1963, Hansch et al. published one of the first QSAR models to predict the effect of various chemical compounds on plant growth. The effect of the tested compound on the plant growth had been demonstrated to be dependent on the Hammett constant and partition coefficient into the cell, and soon after, relationships between lipophilicity and drug activity were formulated [2]. Subsequently, in 1972, Hansch et al. developed one of the first and most important pharmaceutical QSAR models: the clogP model. This early model was based on PLS model using the Hammett constant σ and the free energy-related constant π [3].

Semi-empirical rules such as the Lipinski "Rule-of-Five" were developed to identify compounds that are unlikely to progress in the clinic [4]. The "Rule-of-Five"

Predictive ADMET: Integrative Approaches in Drug Discovery and Development, First Edition.
Edited by Jianling Wang and Laszlo Urban.

states that if two or more of the following conditions were reached: MW greater than 500, more than 10 hydrogen bond acceptors (oxygens or nitrogens), more than 5 hydrogen bond donors (NH or OH), and logP greater than 5.0, then compounds are less likely to progress perhaps as a result of poor absorption or permeability. This rule became implemented in the Pfizer (New York, NY) registration system as an alert, and this particular model has been applied across the pharmaceutical industry [4, 5]. A different approach, based on two physical-chemical parameters (polar surface area (PSA) and AlogP98), and a statistical pattern recognition model, has been developed by Egan et al. to predict passive intestinal absorption. By using a Caco-2 cell permeability dataset from Pharmacopeia, Inc. (Cranbury, NJ), they developed a model that could perform between 74% and 92% accuracy for different subclass of compounds [6]. Other analyses have been carried out to study the differences among "lead-like," "drug-like," and "nondrug-like" compounds, leading to the development of global QSAR models. One of the earliest analyses was reported by Ajay et al. in 1998 that used the Comprehensive Medicinal Chemistry (CMC) database to define the drug chemical space and the Available Chemicals Database (ACD) to define the general chemical space [7]. The chemicals were described using nondimensional (0D) and one-dimensional (1D) parameters [8]. The analysis revealed that lead-like compounds fell in a different region of chemical space than "nondrug-like" compounds. Based on these descriptors, a Bayesian neural network was trained to recognize "drug-like" and "nondrug-like" compounds. Similar approaches were implemented by other pharmaceutical companies like BASF Corporation (Florham Park, NJ) and AstraZeneca (London, U.K.) [9, 10].

The application of *in silico* models is more extensive than predicting the likelihood of new molecular entities to be potent ligands. QSAR models are used on a regular basis by medicinal chemists. The most common parameters predicted by chemists are the logP, logD, logS, pKa, and pKb using partial least-squares (PLS) models [11]. Other *in silico* models more oriented for ADME have been developed, such as the prediction of metabolic liability using packages such as MetaSite (Molecular Discovery Ltd., UK) or StarDrop (Optibrium Ltd., UK) [12, 13]. Other significant software vendors in this arena include BioByte (BioByte Corp., USA), CompuDrug (CompuDrug International, Inc., USA), Simulations Plus (Simulations Plus, Inc., USA), Lhasa limited (Lhasa limited, UK), Accelrys Topkat (Accelyrs, Inc., USA), and Leadscope (Leadscope, Inc., USA) [14–19]. Most of the QSAR models developed by the vendors are global models with the exception of ACDLab's (Advanced Chemistry Development, Inc., Canada). This latest model provides pKa predictions by looking up templates and applying a local adjustment from a global database of experimental values [11]. These local models, which only apply in a limited chemical space, serve an important role in drug discovery. Many local models are developed to predict potency; however, in this review, we show local models for absorption, distribution, metabolism, and excretion (ADME) measurements. This review focuses on the use of *in silico* models for predicting absorption, distribution, metabolism, excretion, and toxicity (ADMET) properties and on highlighting the differences between global and local models.

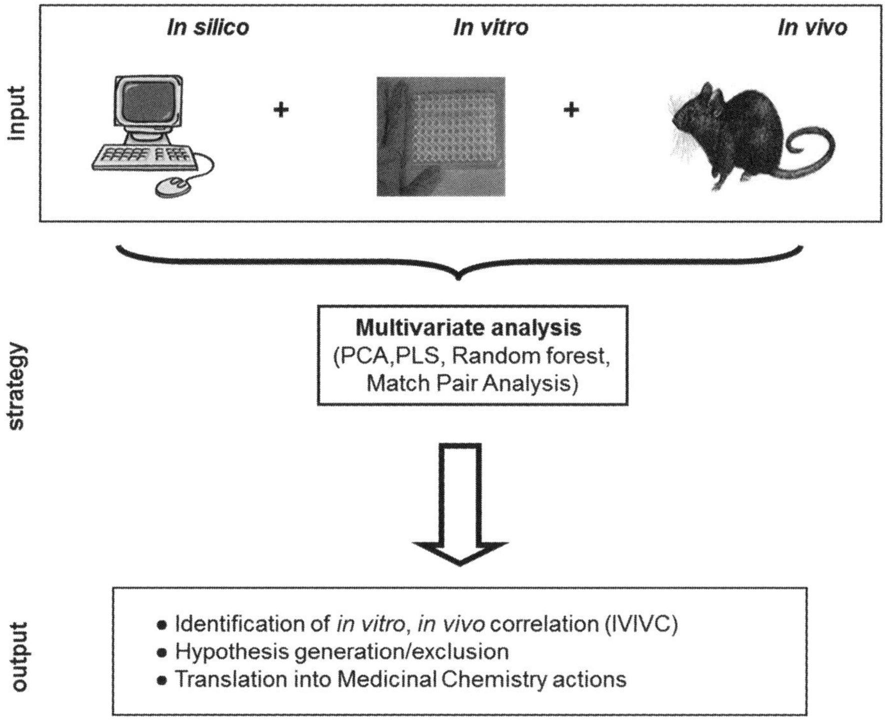

Figure 6.1 Flowchart illustrating the process used in ADME to build an *in silico* model.

6.2 SAR METHODOLOGY

We divide creating *in silico* models into three main components: an input containing experimental *in vivo* or *in vitro* data, *in silico* descriptors for each molecule, and then an appropriate statistical fitting algorithm to produce the model. These models quantify the formulation of a specific hypothesis, identify *in-vivo–in-vitro* correlation, and provide information to the medicinal chemist (Figure 6.1). Models will be able to provide an estimated value for each prospective compound. After construction, the model represents a hypothesis about how the descriptors are related to measured values and some models can be interpreted to provide chemical insight. Any model is dependent on the quality of the experimental data on which it is built, and extensive effort should be invested to obtain sufficient information and to ensure the quality of the dataset.

6.2.1 Data and Descriptors Selections

Numerous free data repositories exist, such as the National Institute of Standards and Technology (NIST) [20], the National Cancer Institute (NCI) [21], the National

Library of Medicine (NLM) [22], the U.S. Environmental Protection Agency (EPA) [23], the British National Formulation [24], PubChem [25], ChemID-Plus [26], and ChemSpider [27]. All of these databases are maintained by public funding, whereas other databases such as WOMBAT [28], GVK [29], and Drug Bank [30] are maintained by private funding. These databases and the primary literature contain a large amount of data. However, extensive care needs to be used, as differences in protocols between laboratories and even units have been found to be inconsistent in some datasets. To add to the complexity of the process, some studies revealed issues with the WOMBAT database, showing data and structure error rates as high as 8% [31, 32]. To improve the accuracy of WOMBAT, the database was curated to form the WOMBAT-PK database. Another study from Young et al. investigated databases from several public and commercial sources and showed that both public and private databases contained structural information with significant errors related to target engagement. Young et al. also reported that small errors in the structural representations could lead to significant errors in predictions. On average, they found that error rates ranged from 0.1% to 3.4% depending on the database [33].

In contrast to the publically available database, large pharmaceutical companies maintain their own extensive confidential databases containing data from well-defined high-throughput methods such as assays measuring solubility, logP, or permeability parameters using the parallel artificial membrane permeability assay (PAMPA) and Caco-2. Even in these situations, establishing a reliable global model can be difficult. In particular, the data for a specific high-throughput assay could relate to several closely defined series of compounds, rather than encompassing all regions of chemical space. As a result of competitiveness and the need to preserve intellectual property, the disclosure of information by pharmaceutical companies into the public literature has been limited. In an attempt to circumvent this problem, several initiatives have formed, such as Vitic, eTOX, and the Cytochrome P450 Consortium. In each case, data from a consortium of pharmaceutical companies and public organizations are compiled by an independent institution that has unrestricted access to the entire dataset. In return, the consortium can provide more global QSAR models such as the one provided by the Cytochrome P450 consortium or chemical database that can recognize and search for similarities in new chemical structures.

In conjunction with data curation and new algorithms, the number of descriptors available to represent compounds is large; in 2000, Todeschini and Consonni described ~1800 descriptors [8]. This increase in the number of descriptors was mostly driven by the assumption that more descriptors to characterize molecules would improve the understanding and relationship between compounds and the assayed value. Despite the large number of available descriptors, some have called for the development of new descriptors [34], whereas others believe that there are already more than enough descriptors [35]. Programs like Dragon (TALETE srl., Italy) [36] alone can generate up to 4885 molecular descriptors, but other software like MOE (Chemical Computing Group, Canada) [37], ADAPT (Jurs research group, USA) [38], VOLSURF (Molecular Discovery, Italy) [39], and RDKit [40] can calculate hundreds of descriptors. Descriptors fall into many classes: physicochemical (measured or calculated), geometrical, constitutional (including group contributions), topological, electro-topological, quantum chemical, and molecular fingerprints. To organize these myriad descriptors, we classify them into four different groups: 0D-descriptors provide general information on the molecule such as the number and type of atoms and molecular weight; 1D-descriptors are used to describe structural fragments; and 2D-descriptors explain how the atoms are interconnected in terms of the presence and nature of chemical bonds. In contrast to the 2D-descriptors, the 3D-descriptors describe not only the atom's connectivity but also the overall spatial configuration of the molecule and include 3D-MoRSE descriptors; WHIM descriptors; GETAWAY descriptors; quantum-chemical descriptors; and size, steric, surface, and volume descriptors. Finally, the 4D-descriptors or grid-based descriptors, such as those obtained with VolSurf, map a molecular interaction with a probe such as steric, electrostatic, and hydrophilic to the molecular surface [41].

With increasing computational power, larger numbers of descriptors can be calculated for increasingly large datasets. The challenge with obtaining a multitude of descriptors is selecting the subset of descriptors with the highest information content for separating compounds according to the desired ADMET value. A recent paper by Kjeldahl and Bro discusses several common misunderstandings and pitfalls in practical multivariate data analysis that one should be aware of. Among these are the selection of relevant samples and variables as well as diagnosis and interpretation of the models [42]. Sprous et al. emphasizes the importance of eliminating redundancy in the selection process to define the training set and in the selection of descriptors used to produce the model [43]. They assert that descriptors should be uncorrelated, have a reasonable variance, and possess physical relevance. Various solutions to selecting descriptors have been employed, such as the transformation of the descriptors into an uncorrelated set obtained via a principal component analysis where linear combinations of the original descriptors set are used to describe the system.

6.2.2 Methods for Multivariate Analyses

Forward and backward selections of the relevant descriptors have also been used by various authors [44–46]. These two techniques either add or eliminate a subset of descriptors to find those that obtain the best QSAR model performance. To achieve a high-performance linear model, the desired performance statistics or prespecified desired numbers of descriptors need to be defined. Then an iteration to add or remove descriptors is carried out until the optimal conditions are achieved. These methods were evaluated by Xu and Zhang using the toxicity of 35 nitrobenzenes, and they showed that the backward selection exhibited better performance than the forward selection [47]. However, other studies have found the opposite conclusion: Forward selection resulted in a more robust QSAR model [48]. Neither approach has been clearly demonstrated as superior to the other. A multitude of other methods to reduce the number of descriptors is available such as the one described by Kompany-Zareh and Omidikia where they used a filter approach [49]. In this particular study, the Gram–Schmidt algorithm was used as a descriptor thinning method to remove the collinear and redundant descriptors, which was followed by a variable selection using the jackknifing approach [50]. Only a few methods are available to remove nonrelevant descriptors in QSAR modeling; more information related to this topic can be found in various papers and reviews such as Stanton, Goodarzi et al., and Sprous et al. [43, 51, 52]. Another method for feature selection is the lasso (and the related technique of elastic net). This is a linear method where the fitting function is linear (L1) with the residual error rather than the usual quadratic (L2). Because of the linear loss function, the final optimized model can have some of its coefficients set to zero, which eliminates them from the model [53].

6.2.3 Strategy for *In Silico* Model

6.2.3.1 *Principal Component Analysis (PCA) and Partial Least-Squares (PLS) Methods* To process the large number of descriptors used to describe individual compounds, various approaches have been developed to build QSAR models that relate the quantitative chemical descriptors to biological activities. As described previously, the refinement of the descriptor set used to construct the model is important and will facilitate the interpretation of the QSAR model. However, the selection of the algorithm to build the QSAR model is another important component. Different modeling approaches can be applied, which will result in different types of QSAR models. Linear models, such as the PLS model, were originally developed for modern econometrics [54]. It consists of finding a fundamental relationship between two matrices X and Y. The linear regression model is obtained by projecting the observable variables (X) and predicted variables (Y) to a new space. The projection on a new space is required to eliminate concern of colinearity between the physicochemical parameters. In contrast, multiple linear regression requires a dependent variable. Another widely used model in the social and physical sciences is PCA, which uses a linear technique. It is one of the oldest and most common latent variable projection methods [55, 56]. PCA was originally developed by Pearson, and the details of extracting components from a data matrix were presented in an extensive two-part paper by Hotelling [57, 58]. Subsequently, the interpretation of PCA has been greatly improved through the work of Jeffers, who calculated, plotted, examined, and interpreted the role of eigenvalues and eigenvectors [59]. It was further emphasized by Baxter, who mapped principal component scores as an illustration of the possibilities offered by computer graphics [60]. This transformation is defined in such a way that the first principal component has the largest possible variance (that is, it accounts for as much of the variability in the descriptors as possible), and each succeeding component in turn has the highest variance possible under the constraint that it be orthogonal to (i.e., uncorrelated with) the preceding components. Principal components are guaranteed to be independent only if the dataset is jointly normally distributed. PCA is sensitive to the relative scaling of the original variables. It can be used to examine the correlation between variables of a selected set, to reduce the variable dimensionality in the measured set, to eliminate variables that have low relevance, and to examine the grouping of new chemical entities in n-dimensional space. Additionally, an interpretation from the PCA output can be achieved by comparing the score and loading plots. This visual interpretation will allow for defining the weight of each multivariate for a specific chemical space defined by the dataset.

6.2.3.2 Decision Tree or Random Forest Methodology A completely distinct method for QSAR modeling is the decision tree. A decision tree starts with a set of data points with labels (for example, active or inactive) and a pool of descriptors. The algorithm then identifies a descriptor and criteria that best separates the data labels (i.e., for drug-like molecules, MW $\leq$ 500), the data are split into two separate sets, and the process is repeated for the two resultant sets. In this way, the data are partitioned into smaller subsets until they can either no longer be separated, or a stopping criterion is met. The decision tree method will therefore identify a set of criteria (decisions) for separating molecules, resulting in a deterministic model that may be applied to new datasets. The user must exercise caution when constructing decision tree models as they are prone to overfitting on the training set if too many criteria are used to split the data. A common process to limit overfitting is pruning of the tree using either model complexity parameters or cross-validation. Generally, a single decision tree does not provide a high-performance model. Ho proposed the use of an ensemble of decision trees each created using a subset of the total descriptor set to increase the variance of the predictions, which he called the random decision forest [61]. Ensemble learners are better than an individual constituent learner and benefit from variability in the ensemble members, and so they take advantage of the variance of decision trees. The modern adaptation of the random decision forest, the random forest algorithm developed by Breiman, introduced bagging and subset selection at each node of the decision tree to increase further prediction variance [62]. We briefly describe the technique here. To obtain a diverse set of decision tree committee members, the random forest technique only uses a fraction (by default two thirds) of the data in each tree. To increase further the variability between trees, only a small fraction ($n^{0.5}$ for classification) of the features is considered at each split. Each member of the committee that is overfit has less prospective predictive power than a regular decision tree. However, committees of these trees have been shown to have excellent machine learning performance. The random forest generally results in a robust and accurate model. In the examples presented, we used the implementation found in the random Forest package of Liaw and Wiener [63].

6.2.3.3 Match Pair Analysis Method Matched molecular pairs analysis (MMPA) is another recently proposed method for analyzing SAR trends in a dataset [64, 65]. MMPA provides a framework for analyzing SAR within a dataset by identifying structurally similar molecules that differ at a single point (a chemical transformation), and then calculating a property/activity change associated with that transformation [66]. Thus, the analysis is based on comparing the changes between pairs of molecules—the assumption being that a similar change between two pairs of molecules should result in a similar change in an activity/property. Previous work has used a fragment-based algorithm by Hussain and Rea for identification of matched pairs [67]. An alternative to standard QSAR modeling was recently described, which uses the concept of matched molecular pairs to make prospective predictions [66]. This protocol involves finding matched pairs between prospective compounds and assayed compounds and simply applying activity changes from known chemical transformations to those pairs. By averaging predicted activities across all prospective matched pairs, the matched pairs QSAR applied to a potassium ion channel assay exhibited increased performance over a nearest-neighbors algorithm and random forest algorithm [66]. This approach will be applied to our Caco-2 dataset. The matched molecular pairs for the training set compounds were determined using the "matched pairs detectors" node in KNIME (KNIME.com AG, Switzerland) [68]. The pairs were aggregated by their associated chemical transformation, and the mean activity change for each transformation was stored. For each test compound, matched pairs with training set molecules were determined, and the associated chemical transformation was looked up in the repository of chemical transformations from the training set. This approach calculated the permeability for each test set compound from the combination of the known permeability of the training set compound and the predicted property change for the known chemical transformation.

Calculation of matched pairs using molecular fingerprint distance (also called fingerprint pairs herein) is a less stringent definition of molecular pairs that allows for not only a single R-group chemical transformation but also highly similar structures (that is the transformation that may not be a clean matched pair transformation) and cases where two small groups are altered, to be classified as pairs. Loosening this constraint increases the population of molecular pairs within a dataset, but it may complicate the application of chemical transformations to prospective compounds, as multiple changes may contribute to a fingerprint pair. In the example presented, fingerprint pairs were identified by calculating the Manhattan distance between the 1024-bit Morgan fingerprints and defining a fingerprint pair as any two compounds having a distance of less than 50. To perform QSAR modeling on the matched pairs, pairwise descriptors were constructed by taking the difference between descriptors for the initial and final molecules. Additionally, descriptors for the initial "known" molecule of a pair were included in the model, as well as the permeability of the initial molecule.

6.3 CASES IN DRUG DISCOVERY

To illustrate the use of *in silico* models to predict the ADMET property, two proprietary datasets were used. The first one was an in-house project where the large volume of distribution at steady state (Vss) needed to be addressed. Because measuring Vss is resource-intensive, *in silico* approaches to predict Vss guide the project. The second example, predicting Caco-2 permeability, is based on the proprietary Novartis (Basel, Switzerland) dataset and illustrates how random forest and match pair analyses have been used to build a QSAR model predicting the apical to basolateral permeability in a Caco-2 cell line.

6.3.1 Use of PCA and PLS to Guide Permeability and Vss

To explore the chemical space of this in-house project, 73 compounds selected for this project were projected in the ChemMAP (Figure 6.2A), a PCA-based tool that has been developed specifically to navigate the physicochemical space of bioactive molecules [69]. Additionally, based on the loading plot from ChemMAP's PCA, all compounds possessed an overall high hydrophobicity and a low number of H-bonding

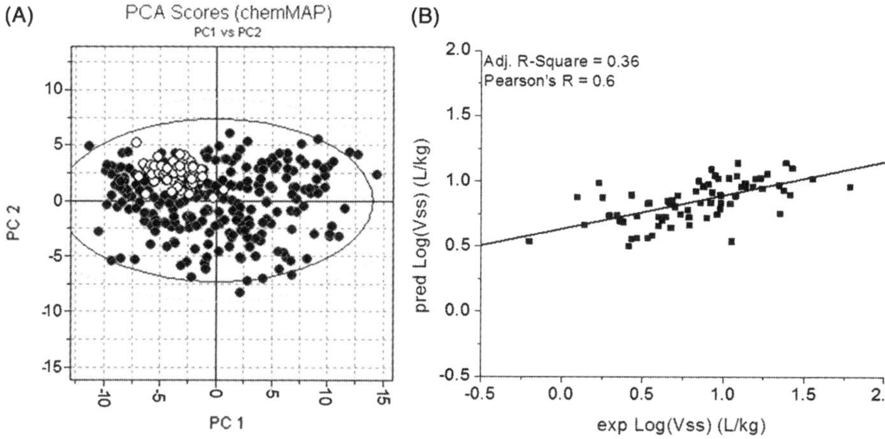

Figure 6.2 Projection of the data from an in-house project in the ChemMAP chemical space [74] (white circles = in-house project, black circle = commercial compounds used to define the ChemMAP chemical space) (A) and volume of distribution at the steady state in rat predicted using an in-house global model in comparison with the experimental data, where the linear regression shows the efficiency of the model (B).

acceptors and donors and poor hydrophilicity. After this chemical space analysis, the volume of distribution of these 73 compounds was predicted using an in-house global Vss model based on VolSurf parameters and described by Lombardo et al. (Figure 6.2B) [70]. The correlation between the measured and predicted values shows a relatively low R^2 and Pearson R value indicating a larger chance to have a poor ranking. For this specific class of compounds, the global model seemed to underperform. To respond to the needs of the project, a local model was developed using the following *in vitro* data: high-throughput equilibration solubility assay, PAMPA, and fraction unbound at pH 7.4, as well as 14 *in silico* descriptors generated in VolSurf. After compiling the data, a PLS model was constructed using all 16 descriptors (Figure 6.3A). The results for the local model indicate a significantly improved prediction compared with the global model. Analysis of the coefficient plot from the PLS regression show that 4 descriptors (exp FU7.4 = experimental fraction of neutral compound at pH 7.4, clogP = calculated logP, Pi = polarizability, and alpha = capacity to donate H-bonds) contributed to a much greater extent than the other 12 descriptors (Figure 6.3B). In an attempt to simplify the local model and to improve understanding regarding the role of the descriptors on the Vss, a PLS model with only the four most important descriptors defined in the previous local model was evaluated (Figure 6.3C).

The results of this analysis show that this *in silico* model with fewer parameters performs as well as the previous model, with an R^2 of 0.5 and a Pearson R of 0.71 (Figure 6.3D). A follow-up analysis revealed that the percentage of compounds correctly predicted to within a twofolds margin of error was equal to 85% (Figure 6.4). The remaining poorly predicted compounds correspond to high plasma protein binding > 99.5%. Despite having included the descriptors fraction of the neutral compound at pH 7.4 and clogP, the current local model is still underperforming and shows over-predicted values for low Vss compounds (Figure 6.4). Nonetheless, one advantage of

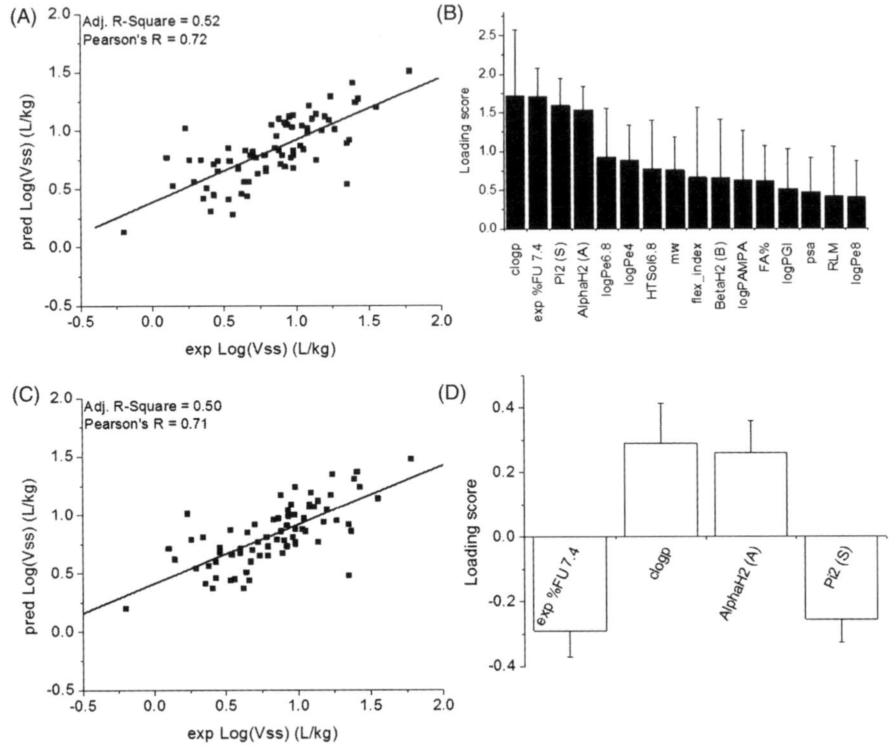

Figure 6.3 (A) The predicted Vss values in rat using the local PLS model based on 16 descriptors were compared with the experimental values, and a linear regression was used to evaluate the accuracy of the model. (B) The respective loading score for the 16 descriptors used in the PLS model. (C) The predicted Vss values in rat using the local PLS model based on 4 descriptors were compared with the experimental values, and a linear regression was used to evaluate the accuracy of the model. (D) The respective loading score for the 4 descriptors used in the PLS model.

this local model compared with the global model is the improved interpretability. For example, an increase of the clogP or H-bond donor capacity increases the volume of distribution at steady state, whereas a decrease of the Vss should be observed when the fraction of neutral compound at pH = 7.4 or polarizability is lowered (Figure 6.3D).

6.3.2 Caco-2 Permeability via Random Forest and Match Pairs Analysis

6.3.2.1 Random Forest The Novartis Caco-2 dataset contains 23,427 compounds. The dataset was chronologically split into a training set of 17,500 compounds measured prior to March 2010 and into 5927 test set compounds measured from March 2010 to late 2012. A random forest was constructed to predict the apical to

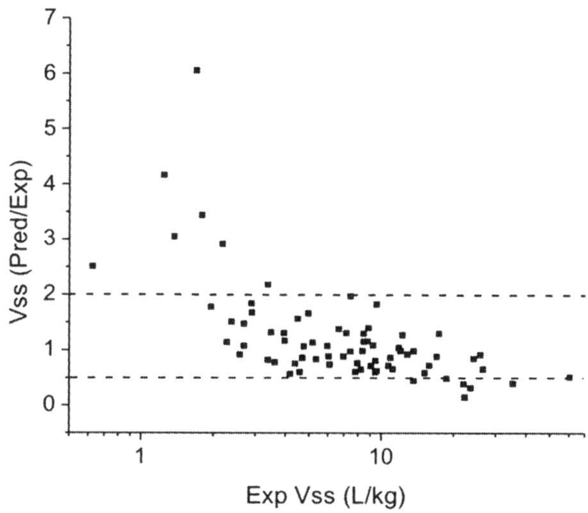

Figure 6.4 The ratio of the volume of distribution at steady-state predicted data versus experimentally measured values plotted against experimentally measured, showing the predictability of the model within a twofold interval of confidence for most compounds except those with low Vss (defined by the dotted black lines).

basolateral permeability in log scale, using the chiral Morgan fingerprint calculated in RDKit (calculated with the "rdMolDescriptors.GetMorganFingerprint()" command with the radius of 3 and then hashed to 1024 bits), a total of 51 2D descriptors from MOE (version 2011.10), and descriptors for the ionization states of acidic and basic centers determined using pKas calculated in MoKa (version 1.1.0) at pH 7.4 and the Henderson–Hasselbalch equation.

The random forest ensemble was constructed from 400 decision trees. The resulting QSAR model accounted for 61% of the variance in the experimental permeability for the training set compounds (root-mean-square error (RMSE) = 0.40, $R^2 = 0.61$) (Figure 6.5A). When applied to the test set compounds, the performance of the model dropped significantly, with only 39% of the variance in the permeability explained (Figure 6.5C: RMSE = 0.47, $R^2 = 0.39$). The reason for the performance decrease becomes apparent when the Tanimoto fingerprint similarity between the training and the test set compounds is calculated. Most test set compounds have an extremely low Tanimoto similarity to their nearest neighbor in the training set, suggesting that they are either outside the domain of the model or in sparsely populated regions of the training set chemical space. The model RMSE plotted versus time was observed to increase concomitantly with the decrease in average chemical similarity. By selecting the training set using a random sampling of the data rather than a chronological split, we obtained a model with comparable performance to the chronological model, and equal performance between the training ($R^2 = 0.59$) and the test ($R^2 = 0.61$) sets (Figure 6.5B and 6.5D). The modeler and subsequent users should exercise caution when using QSAR models to ensure that predictions fall within the domain of the

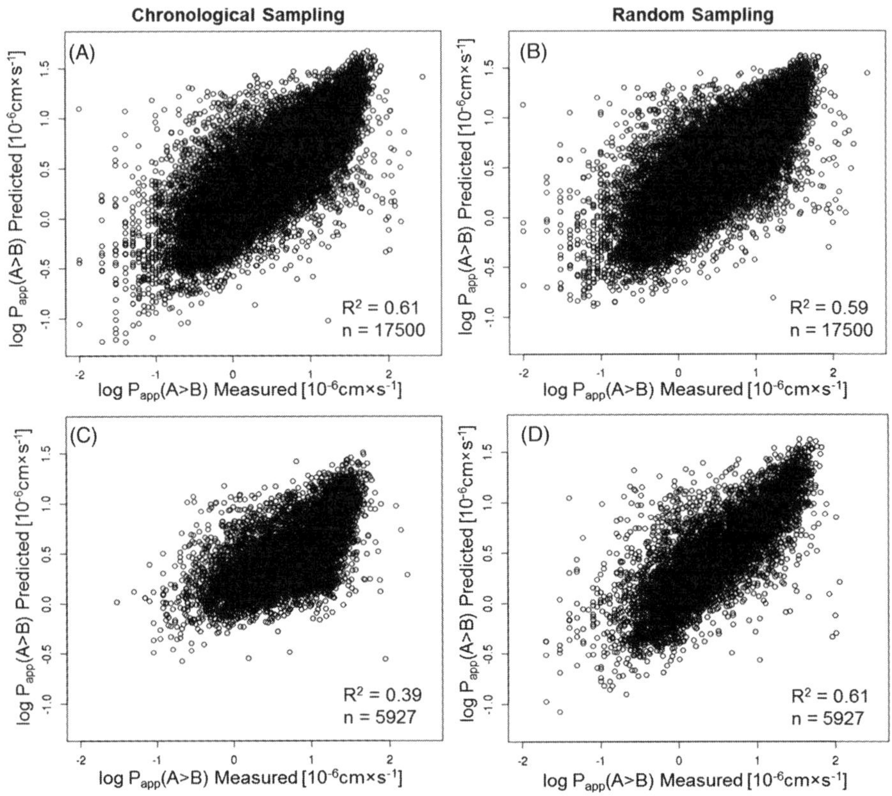

Figure 6.5 Predicted versus measured Caco-2 P_{app}(A > B) for the training set (A,B) and test set (C,D) using chronological sampling (A,C) and random sampling methods (B,D).

model, and periodically, they should update models to counteract inaccuracies that appear as a result of the shifting chemical space.

The random forest algorithm in *R* provides an internal estimate for the relative importance of each descriptor to the final model, which is reported in what is called a variable importance plot, although the variable importance plot does not provide any indication of how those descriptors contribute to the model. A second technique, the partial dependence plot, estimates the effect of each descriptor on the model predictions. The partial dependence plot represents the range of the descriptor on the *x*-axis and the relative model response on the *y*-axis (Figure 6.6). The highest five descriptor contributions to the Caco-2 permeability model may be grouped into three criteria, corresponding to the size (weight, van der Waals (VdW) area, and VdW volume), the fraction ionization, and the number of hydrogen bond donors. The partial dependence plot for molecular weight indicates that as weight increases, the permeability decreases rapidly. Very few compounds have a molecular weight higher than 600, and so the model response above MW = 600 is relatively flat. The partial

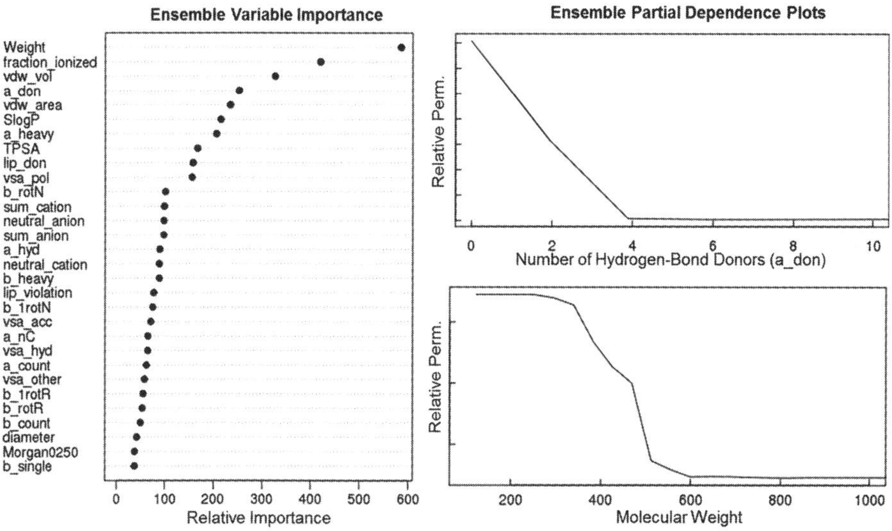

Figure 6.6 Caco-2 variable importance measure (left) depicting the algorithm's measure of how significantly each descriptor contributes to the model predictions and variable partial dependence plots (right) for hydrogen bond donors (a_don) and molecular weight, depicting the model response across the variable's range.

dependence plot for the number of hydrogen bond donors presents a similar case—the permeability is predicted to decrease linearly with the number of hydrogen-bond donors up to a total of 4, at which point increasing the number of hydrogen bonds is no longer predicted to have a significant effect on permeability.

6.3.2.2 Matched Pairs Analysis As a result of the low prospective performance of the chronological random forest model stemming from applicability domain issues, the concept of matched molecular pairs was applied to predicting Caco-2 permeability. The rationale for this approach is that chemical transformations should have similar changes in activity regardless of the exact molecule on which they occur, and therefore, transformations observed in the training set should be able to predict the corresponding activity changes between training and prospective compounds, and ideally between prospective compounds. The Caco-2 permeability rates calculated using direct application of known chemical transformations to prospective pairs underperforms standard QSAR by a significant measure. The lack of similarity between the training ($N = 17{,}500$) and test sets ($N = 5927$) results in only 662 test set compounds forming pairs to training compounds with known chemical transformations, resulting in no predictions for 5265 compounds. For those 662 prospective compounds, the molecular matched pair (MMP) model only accounts for 23% of the variance ($R^2 = 0.41$, RMSE $= 0.52$) in observed permeability (Figure 6.7A) versus 56% of the variance ($R^2 = 0.52$, RMSE $= 0.40$) for the subset of compounds in the standard QSAR model (Figure 6.7B). Two issues are demonstrated by applying the matched pairs methodology to Caco-2: First, the low similarity between the training

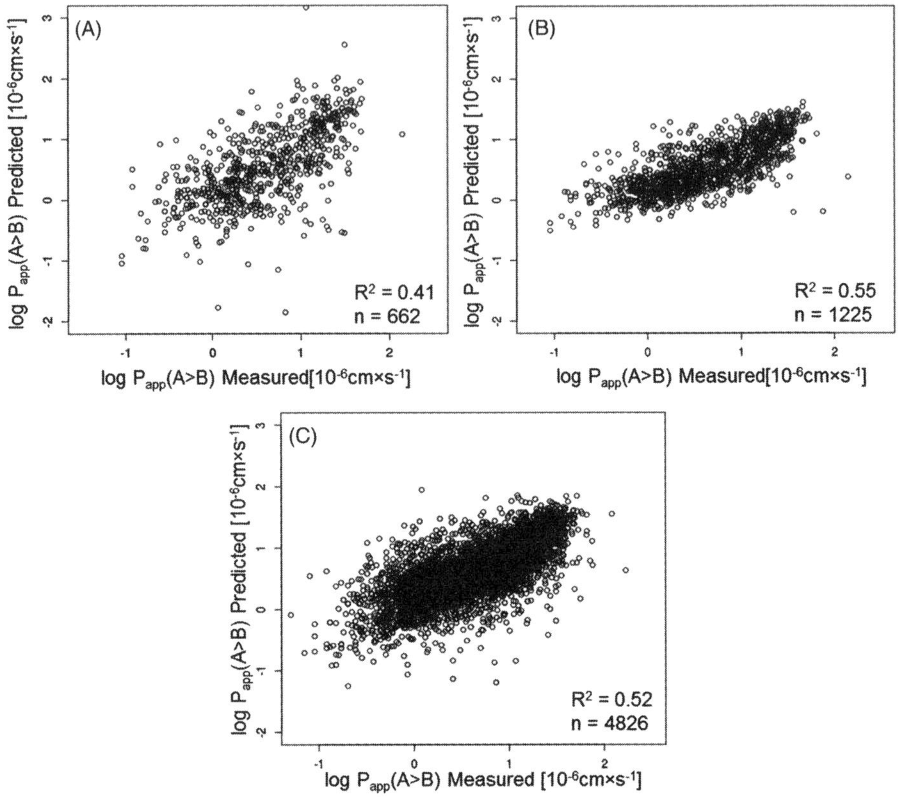

Figure 6.7 Average prediction for P_{app}(A > B) from QSAR by matched pairs (A) and from transformation QSAR models using pairs with the training set (B) or purely prospective pairs (C).

and the test sets results in a low number of pairs, and second, the observed chemical transformations are not strictly transferrable.

As a result of the limited number of test set compounds involved in pairs with known chemical transformations and low model performance, an alternative method of modeling pairs in datasets was employed as described in the methodology. The fingerprint difference between the two compounds will also incorporate the local environment of each transformation, allowing one to consider the local chemical environment of the transformation. QSAR modeling on the chemical transformation of each fingerprint pair can then be carried out to predict the activity change associated with each pair, a method that was successfully demonstrated for modeling the structure activity landscape index (SALI) of several datasets [71]. This methodology is denoted as transformation QSAR here.

Although transformation QSAR is based on the same principles as QSAR-by-MMPA, there are several notable exceptions. First, pairwise descriptors are calculated by taking the difference in molecular descriptors for each training pair, and using those descriptors, a QSAR model is constructed to predict the activity change.

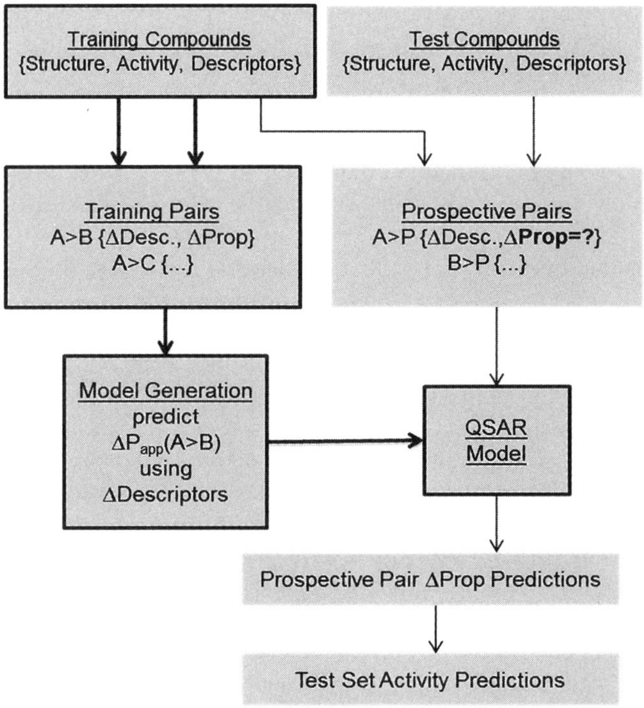

Figure 6.8 The schematic for transformation QSAR. The bolded steps comprise the transformation QSAR model generation and occur first, followed by evaluation of prospective pairs.

Prospective pairs are formed by pairing the test set compounds with the training set molecules. The activity change for each prospective pair can then be calculated using the QSAR model. By adding the predicted activity change to the known activity for the training compound, the activity of the test set compound can be estimated. Multiple predictions for a single test set compound will be aggregated, and the mean activity will be reported. The basic process for this procedure is similar to the QSAR-by-MMPA methodology, although activity changes are determined using a QSAR model as in prospective SALI predictions by Guha [66,71]. The process is depicted in Figure 6.8.

At the Manhattan distance cutoff of 50 for fingerprint pairs, predictions were obtained for 1225 fingerprint pairs as the test set. The model predictions accounted for 55% of the variance in permeability for this set of compounds ($R^2 = 0.55$, RMSE = 0.39) (Figure 6.7B). For this subset of 1225 compounds, the standard QSAR model is more accurate (53% of the variance explained) than for the entire test set (39% of the variance explained). In the Caco-2 dataset, transformation QSAR performs comparably with the standard random forest algorithm, but predictions are obtained for test set compounds that have overall lower errors in the standard model.

The success of transformation QSAR at predicting the activity of test set compounds using known transformations from the training set suggests that the methodology is sound, although once again performance is hampered as a result of the low similarity between training and test set molecules. However, the test set molecules share higher structural similarity to one another than to the training set, and if the chemical transformations are transferable, then one should be able to predict the activity change between two unknown compounds. For the sake of demonstration, we have repeated the transformation QSAR using prospective pairs involving two test set molecules, rather than one test set and one training set compound. Using this methodology, predictions are obtained for 4826 test set compounds. The transformation QSAR model predictions account for 50% of the variance in the permeability (RMSE = 0.38, $R^2 = 0.52$). The same subset of compounds in the standard QSAR model account for 38% of the variance in the permeability (RMSE = 0.46, $R^2 = 0.38$) (Figure 6.7C). Here, even when the compounds themselves are outside the applicability domain of the standard QSAR model, information on the known chemical transformations and how they affect Caco-2 permeability can be applied to the test set compounds to obtain a significant improvement to model performance.

The low performance observed using QSAR-by-MMPA, and the minimal performance change between transformation QSAR and a standard random forest QSAR model, is a side effect of the low structural similarity between the training set and the test set chemical space of the Novartis Caco-2 dataset. We have demonstrated that when test set compounds may be outside the domain of a standard QSAR model to predict its activity, a transformation QSAR model may not suffer the same limitations. The matched pairs methodology and transformation QSAR apply previously observed chemical transformations and their associated activity changes to unknown compounds, resulting in a model that occurs in a chemical space associated with structural changes rather than with the chemical entity itself. In the case of our Caco-2 model, the transformation QSAR model returned predictions for a substantially larger subset of compounds than the matched pairs implementations, and it did not suffer from the same applicability domain-related performance decrease from which the standard random forest QSAR model suffered.

6.4 POTENTIAL CAVEATS FOR QSAR ANALYSES

The *in silico* methodology provides scientific tools that have been used and developed for several years to explore the ADME chemical space and predict the activity of new chemical species. However, this approach presents some caveats like the one mentioned previously regarding the gathering of information to build robust QSAR models. The process of data curation is challenging and could be a limiting factor especially when data were compiled from multiple sources with different experimental protocols, methods of calculation, or simply scaling factors. Obviously these problems are minimized when the data originate from a single source, but even in this case, variation can occur and data curation issues cannot be disregarded. In ADME, one of the common problems encountered with high-throughput assays is the solubility of the new chemical entities. Multiple studies have demonstrated that compounds in early phases of drug discovery tend to have a higher logP and lower solubility, which tends to improve as the compounds progress to clinical trials.

Additionally, concern around the limitation of prediction has to be considered and it could have some implication on the overall SAR approach. This concern was voiced very early in the development of SAR (1977) by Corwin Hansch; "Turning now to the problem of direct predicting of the biological activity of untested molecules from known QSAR, it is instructive to divide the problem into two parts: 1) predictions within the [chemical space]; 2) prediction outside of the [chemical space]" [72]. In response to these concerns, much research has been carried out to develop better physicochemical descriptors, more complex algorithms to improve SAR implementation, and access to larger datasets to span more chemical space. From these initiatives, numerous global models were published, such as the partition coefficient, aqueous solubility, blood-brain barrier (BBB) penetration, plasma protein binding, and volume of distribution [73–76]. These models have been applied successfully to a certain degree but have also shown their limitations. One such example was the global volume of distribution model described by Berellini et al. [77]. This model seemed to underperform with the specific set of compounds selected from the in-house project described here, triggering the need to develop a local model to address its shortcomings. This local model should be expected to perform better on the in-house project set, as it was tailored to a smaller chemical space. However, this local model still failed in providing accurate predictions for compounds with a high plasma protein binding. This latest observation seemed less marked in the global model. The simplicity of the local model allowed user interpretability, which is not always possible for more complex global models, especially when embedded in software packages. For this specific series of compounds, the Vss prediction was affected by four properties (see earlier), which provided valuable information for the development of the next series of compounds.

The assumption that global models were too limited and unable to provide useful information for ADME properties would be inaccurate, but care must be taken to ensure the applicability to novel chemical entities. Multiple packages employ global SAR models such as BBB penetration, P-glycoprotein (P-gp) efflux, metabolism, PBPK (Physiologically Based PharmacoKinetics), and more complex models such as the advance compartmental absorption transit (ACAT) model or the advance dissolution, absorption, and metabolism (ADAM) model developed by GastroPlus (Simulations Plus, Inc., USA) and Simcyp (Simcyp Limited, UK), respectively [15–18,78,79]. In contrast, the local model could have a small advantage regarding the interpretability of the data and the predictive accuracy but will lack the ability to explore new chemical space. Despite the imperfections of the global and local SAR models, computational methods have enormous potential in narrowing the range of compounds to consider by eliminating the ones predicted to have poor ADME properties [80]. It is clear that a good understanding of confidence of a predictive model is critical and that to translate the computational results into interpretable information will be important for the development of this field. Further development in the understanding of the confidence of predictions for global and local models, and the translation of models into tangible knowledge, such as that provided by matched pairs analyses, will be necessary to show their applicability to a larger community.

REFERENCES

1. McNaught, A.D., Wilkinson, A. *IUPAC. Compendium of Chemical Terminology, 2nd ed. (the "Gold Book")*, Wiley, New York, 1997.
2. Hansch, C., Muir, R.M., Fujita, T., Maloney, P.P., Geiger, F., Streich, M. (1963). The correlation of biological activity of plant growth regulators and chloromycetin derivatives with hammett constants and partition coefficients. *Biological Activity of Plant Growth Regulators*, *85*, 2817–2824.
3. Hansch, C. (1972). Structure-activity relationship in the chymotrypsin hydrolysis of p-nitrophenyl esters. *Journal of Organic Chemistry*, *37*, 92–95.
4. Lipinski, C.A., Lombardo, F., Dominy, B.W., Feeney, P.J. (1997). Experimental and computational approaches to estimate solubility and permeability in drug discovery and development settings. *Advenced Drug Delivery Reviews*, *23*, 3–25.
5. Lipinski, C.A. (2003). Chris Lipinski discusses life and chemistry after the Rule of Five. *Drug Discovovery Today*, *8*, 12–16.
6. Egan, W.J., Merz, K.M., Jr., Baldwin, J.J. (2000). Prediction of drug absorption using multivariate statistics. *Journal of Medicinal Chemistry*, *43*, 3867–3877.
7. Ajay, A., Walters, W.P., Murcko, M.A. (1998). Can we learn to distinguish between "drug-like" and "nondrug-like" molecules? *Journal of Medicinal Chemistry*, *41*, 3314–3324.
8. Todeschini, R., Consonni, V. *Handbook of Molecular Descriptors*, Wiley-VCH, New York, 2000.
9. Sadowski, J., Kubinyi, H. (1998). A scoring scheme for discriminating between drugs and nondrugs. *Journal of Medicinal Chemistry*, *41*, 3325–3329.
10. Oprea, T.I., Davis, A.M., Teague, S.J., Leeson, P.D. (2001). Is there a difference between leads and drugs? A historical perspective. *Journal of Chemical Information and Computer Sciences*, *41*, 1308–1315.
11. http://www.acdlabs.com.
12. http://www.optibrium.com/stardrop/.
13. http://www.moldiscovery.com/soft_metasite.php.
14. http://biobyte.com.
15. http://www.compudrug.com.
16. http://www.simulations-plus.com/.
17. https://www.lhasalimited.org/.
18. http://accelrys.com/products/discovery-studio/admet.html.
19. http://www.leadscope.com.
20. http://webbook.nist.gov/chemistry/.
21. http://www.cancer.gov/drugdictionary.
22. http://www.epa.gov/.
23. http://www.epa.gov/.
24. http://www.bnf.org/.
25. http://pubchem.ncbi.nlm.nih.gov/.
26. http://chem.sis.nlm.nih.gov/chemidplus/.
27. http://www.chemspider.com.
28. http://www.sunsetmolecular.com/.

29. https://gostardb.com/gostar/.
30. http://www.drugbank.ca/.
31. Olah, M., Bologa, C., Oprea, T.I. (2004). An automated PLS search for biologically relevant QSAR descriptors. *Journal of Computer-Aided Molecular Design*, *18*, 437–449.
32. Olah, M., Rad, R., Ostopovici, L., Bora, A., Hadaruga, N., Hadaruga, D., Moldovan, A., Fulias, M., Mracec, M., Oprea, T.I. WOMBAT and WOMBAT-PK: Bioactivity Database for Lead and Drig Discovery. In *Chemical Biology: From Small Molecules to Systems Biology and Drug Design*, Wiley-VCH, New York, 2007, pp. 760–786.
33. Young, D., Martin, D., Venkatapathy, R., Harten, P. (2008). Are the chemical structures in your QSAR correct? *QSAR Combinatorial Science*, *27*, 1337–1345.
34. Gola, J., Obrezanova, O., Segall, M. (2006). ADMET property prediction: The state of the art and current challenges. *QSAR Combinatorial Science*, 1172–1180.
35. Lombardo, F., Gifford, E., Shalaeva, M.Y. (2003). In silico ADME prediction: Data, models, facts and myths. *Mini Reviews in Medicinal Chemistry*, *3*, 861–875.
36. http://www.talete.mi.it/products/dragon_description.htm.
37. https://www.chemcomp.com/.
38. http://www.netsci.org/Resources/Software/Modeling/CADD/adapt.html.
39. http://www.moldiscovery.com/soft_volsurf.php.
40. http://www.rdkit.org/.
41. Todeschini, R., Ballabio, D., Consonni, V., Manganaro, A., Mauri, A. (2009). Canonical measure of correlation (CMC) and canonical measure of distance (CMD) between sets of data. Part 1. Theory and simple chemometric applications. *Analytica Chimica Acta*, *648*, 45–51.
42. Kjeldahl, K., Bro, R. (2010). Some common misunderstanding in chemometrics. *Journal of Chemometrics*, *24*, 558–564.
43. Sprous, D.G., Palmer, R.K., Swanson, J.T., Lawless, M. (2010). QSAR in the pharmaceutical research setting: QSAR models for broad, large problems. *Current Topics in Medicinal Chemistry*, *10*, 619–637.
44. Li, H., Ung, C.Y., Yap, C.W., Xue, Y., Li, Z.R., Cao, Z.W., Chen, Y.Z. (2005). Prediction of genotoxicity of chemical compounds by statistical learning methods. *Chemical Research in Toxicology*, *18*, 1071–1080.
45. Horvath, D., Bonachera, F., Solov'ev, V., Gaudin, C., Varnek, A. (2007). Stochastic versus stepwise strategies for quantitative structure-activity relationship generation–how much effort may the mining for successful QSAR models take? *Journal of Chemical Information and Modeling*, *47*, 927–939.
46. Dutta, D., Guha, R., Wild, D., Chen, T. (2007). Ensemble feature selection: consistent descriptor subsets for multiple QSAR models. *Journal of Chemical Information and Modeling*, *47*, 989–997.
47. Xu, L., Zhang, W.-J. (2001). Comparison of different methods for variable selection. *Analytica Chimica Acta*, *446*, 477–483.
48. Acevedo-Martinez, J., Escalona-Arranz, J.C., Villar-Rojas, A., Tellez-Palmero, F., Perez-Roses, R., Gonzalez, L., Carrasco-Velar, R. (2006). Quantitative study of the structure-retention index relationship in the imine family. *Journal of Chromatography A 1102*, 238–244.

49. Kompany-Zareh, M., Omidikia, N. (2010). Jackknife-based selection of Gram-Schmidt orthogonalized descriptors in QSAR. *Journal of Chemical Information and Modeling*, *50*, 2055–2066.
50. Bjornstad, A., Westad, F., Martens, H. (2004). Analysis of genetic marker-phenotype relationships by jack-knifed partial least squares regression (PLSR). *Hereditas*, *141*, 149–165.
51. Stanton, D.T. (2012). QSAR and QSPR model interpretation using partial least squares (PLS) analysis. *Current Computer Aided Drug Design*, *8*, 107–127.
52. Goodarzi, M., Dejaegher, B., Vander, H.Y. (2012). Feature selection methods in QSAR studies. *Journal of AOAC (Association of Official Analytical Chemists) International*, *95*, 636–651.
53. Tibshirani, R. (1996). Regression shrinkage and selection via the lasso. *Journal of the Royal Statistical Society, B 58*, 267–288.
54. Wold, S. PLS for multivariate linear modeling. *QSAR:Chemometric Methods in molecular Design. Methods and Principles in Medicinal Chemistry*, Weinheim,Verlag-Chemie, 1994.
55. Jackson, J.E. *User's Guide to Principal Components*, 1991.
56. Wold, S., Esbensen, K., Geladi, P. (1987). Principal component analysis. *Chemometrics and Intelligent Laboratory Systems*. *2*, 37–52.
57. Pearson, K. (1901). On lines and planes of closest fit to systems of points in space. *Philosophical Magasine 6*, 559–572.
58. Hotelling, H. (1933). Analysis of a complex of statistical variables into principal components. *Journal of Education Psychology*, *24*, 417–441.
59. Jeffers, J. (1967). Two case studies in the application of principal component. *Applied Statistics*, *16*, 225–236.
60. Baxter, M.J. (1995). Standardization and transformation in principal component analysis, with applicatiosn to archaeometry. *Applied Statistics*, *44*, 513–527.
61. Tin, Kam Ho. (1995). Document analysis and recognition. *Proceedings of the Third International Conference on Document Analysis and Recognition*, *1995*, 278-282.
62. Breiman, L. (2001). Random forests. *Machine Learning*, *45*, 5–32.
63. Liaw, A., Wiener, M. (2002). Classification and regression by random forest. *R News*, *2*, 18–22.
64. Sheridan, R., Hunt, P., Culberson, J. (2006). Molecular transformations as a way of finding and exploiting consistent local QSAR. *Journal of Chemical Information and Modeling*, *46*, 180–192.
65. Wassermann, A.M., Dimova, D., Iyer, P., Bajorath, J. (2012). Advances in computational medicinal chemistry: Matched molecular pair analysis. *Drug Development Research*, *73*, 518–527.
66. Warner, D.J., Bridgland-Taylor, M.H., Sefton, C.E., Wood, D.J. (2012). Prospective prediction of antitarget activity by matched molecular pairs analysis. *Molecular Informatics*, *31*, 365–368.
67. Hussain, J., Rea, C. (2010). Computationally efficient algorithm to identify matched molecular pairs (MMPs) in large data sets. *Journal of Chemical Information and Modeling*, *50*, 339–348.
68. http://www.knime.org/.

69. Faller, B., Ottaviani, G., Ertl, P., Berellini, G., Collis, A. (2011). Evolution of the physicochemical properties of marketed drugs: can history foretell the future? *Drug Discovery Today*, *16*, 976–984.

70. Lombardo, F., Obach, R.S., Shalaeva, M.Y., Gao, F. (2002). Prediction of volume of distribution values in humans for neutral and basic drugs using physicochemical measurements and plasma protein binding data. *Journal of Medicinal Chemistry*, *45*, 2867–2876.

71. Guha, R. (2012). Exploring uncharted territories: predicting activity cliffs in structure-activity landscapes. *Journal of Chemical Information and Modeling*, *52*, 2181–2191.

72. Hansch, C. On the predictive value of QSAR. *Biologial Activity and Chemical Structure*, 1977, pp. 47–61.

73. Shi, J.Q., Cheng, J., Wang, F.Y., Flamm, A., Wang, Z.Y., Yang, X., Gao, S.X. (2012). Acute toxicity and n-octanol/water partition coefficients of substituted thiophenols: determination and QSAR analysis. *Ecotoxicology and Environmental Safety*, *78*, 134–141.

74. Ali, J., Camilleri, P., Brown, M.B., Hutt, A.J., Kirton, S.B. (2012). Revisiting the general solubility equation: in silico prediction of aqueous solubility incorporating the effect of topographical polar surface area. *Journal of Chemical Information and Modeling*, *52*, 420–428.

75. Obrezanova, O., Gola, J.M., Champness, E.J., Segall, M.D. (2008). Automatic QSAR modeling of ADME properties: blood-brain barrier penetration and aqueous solubility. *Journal of Computer-Aided Molecular Design*, *22*, 431–440.

76. Zhivkova, Z., Doytchinova, I. (2012). Prediction of steady-state volume of distribution of acidic drugs by quantitative structure-pharmacokinetics relationships. *Journal of Pharmaceutical Sciences*, *101*, 1253–1266.

77. Berellini, G., Springer, C., Waters, N.J., Lombardo, F. (2009). In silico prediction of volume of distribution in human using linear and nonlinear models on a 669 compound data set. *Journal of Medicinal Chemistry*, *52*, 4488–4495.

78. http://chemsilico.com/.

79. http://www.simcyp.com/.

80. Selick, H.E., Beresford, A.P., Tarbit, M.H. (2002). The emerging importance of predictive ADME simulation in drug discovery. *Drug Discovery Today*, *7*, 109–116.

7

CHEMOINFORMATIC AND CHEMOGENOMIC APPROACH TO ADMET

Virginie Y. Martiny, Ilza Pajeva, Michael Wiese, Andrew M. Davis, and Maria A. Miteva

7.1 INTRODUCTION

Evaluation of the absorption, distribution, metabolism, excretion, and toxicity (ADMET) profile of lead compounds has been used for a long time at the later stages of drug discovery campaigns. Today it is recognized that predicting ADMET properties as early as possible helps to reduce poor bioavailability and toxicity problems [1]. Although a large variety of experimental *in vitro*, *in vivo*, and high-throughput ADMET assays are available nowadays [2], computational approaches are very valuable for ADMET prediction [3,4].

Two large groups of chemoinformatic approaches, molecular modeling and data-derived modeling, have been successfully applied to ADMET prediction. Among the data-derived methods, knowledge-based modeling and quantitative structure activity/property relationships (QSAR/QSPR) have been widely used to predict various ADMET properties by employing statistical and learning techniques [5, 6] on available experimental data and molecular descriptors. More than 6000 molecular descriptors can represent molecular structures (e.g., physicochemical, topological, quantum chemical, etc.) [7]. Overall drug-likeness and oral bioavailability of drug candidates can be evaluated based on statistically derived rules (e.g., "Rule of five") for physicochemical properties (polar surface area, rotatable bonds, logP, pKa, etc.) [8,9]. Several

Predictive ADMET: Integrative Approaches in Drug Discovery and Development, First Edition.
Edited by Jianling Wang and Laszlo Urban.

Table 7.1 ***In silico*** **ADMET properties and techniques prediction**

ADMET properties	*in silico* techniques
solubility, pKa	**data-derived based**
oral absorption	QSAR/QSPR
intestinal absorption	machine learning
blood-brain-barrier permeability	ligand-based pharmacophore
plasma protein binding	
renal clearance	**molecular modeling**
enzyme metabolism (CYP, transferases)	ligand-protein docking
inhibition of drug metabolizing enzymes	quantum mechanics
inhibition of hERG	protein structure-based pharmacophore
affinity to transporters	
affinity to nuclear receptors	**proteochemometrics**
mutagenicity, carcinogenicity	matching ligand (2D, 3D based) space and protein space (sequences, structures based)
acute toxicity, hepatotoxicity	

computer packages can perform such calculations, such as FAF-Drugs2 [10] (MTi, Inserm U973 – University Paris Diderot, France), Screening-Assistant [11] (UMR CNRS 6005, University Orleans, France), or Filter (OpenEye Scientific Software, USA). Models representing "safe" chemical space [12], or predicting more complex chemical biology processes (e.g., permeability, blood-brain barrier (BBB), intestinal absorption [5], and toxicity of molecules [13, 14]), have been developed during the last few decades (Table 7.1).

More recently, the early homology models of human cytochrome P450 (CYPs) [15, 16] have enabled the first employment of chemoinformatic approaches for ADMET prediction based on three-dimensional (3D) structures of proteins that can complement the data-derived models to move toward a deeper mechanistic prediction [17, 18]. Yet, the structure-based approach is limited to the experimentally known 3D structures of ADMET proteins.

In this chapter, we introduce current chemoinformatic and chemogenomic methods used for *in silico* prediction of ADMET properties of small molecules and discuss their advantages and disadvantages. Furthermore, we demonstrate the utility of such an approach by presenting different *in silico* techniques for the prediction of small-molecule interactions with the adenosine triphosphate (ATP)-binding cassette transporter (ABC-transporter) P-glycoprotein (P-gp).

7.2 QSAR TO ADMET

7.2.1 QSAR Modeling

QSAR is an old science, with its origins stretching back to the 19th century, but popularized in medicinal chemistry by the work of Hansch and Fujita in the 1960s

[19]. It is an empirical modeling method, where an observed relationship between descriptors of structural changes across a series of compounds and the response they show in an experiment is used to infer an underlying physical model, from which one can predict the response for new compounds. The "relationship" between descriptors and response is often identified by use of a machine learning algorithm or statistical regression approach. Many such mathematical methods have been investigated. Each has strengths and weaknesses, as well as supporters and critics, although an objective and widespread evaluation of commonly used methods across many datasets has not been reported. Published validations often rely on performance comparisons on single or small numbers of datasets. The earliest used methods, linear regression and multiple linear regression, are still commonly used, as a result of the widespread availability of the mathematical algorithms, and ease of interpretation. But tools such as partial least-squares regression, neural network methods, random forests, Gaussian processes, and support vector machines, genetic algorithms (see details in References 4–6), to name but a few, are often used in QSAR studies. A detailed understanding of the underlying mathematics of the varying approaches is usually little help in choosing what method to apply to a particular dataset. Experience teaches us that often if one method can find a model, other methods can also find similar quality models. The improvement of one method over another assessed on a chosen test set may be marginal. The major weakness in many QSAR approaches more often seems to be the choice of descriptors rather than the choice of machine learning method. To some extent, though, some datasets are more suitable to one machine learning method rather than another. For example, if the relationship between response and descriptors is inherently nonlinear, methods that apply a linear model, such as multiple linear regression, partial least-squares regression, and simple support vector machines, may be at a disadvantage. Modern automation may provide an objective approach to choosing the best machine learning method for a particular dataset.

QSAR is now firmly established as one tool in the computational chemistry toolbox. The field of computational chemistry has evolved to embody more physics-based approaches attempting to describe the molecular structure and its functional relationship to activity, from the underlying mathematics of the system. QSAR approaches may seem so passé in light of such deep science. But QSAR, carefully executed and applied, can often exceed the accuracy and precision of physics-based approaches. The "Achilles heel" of physics-based methods is often the degree of understanding of the physics of the system and our ability to encode this mathematically, which limits the accuracy and precision to fit and predict. But the strength of the physics-based approach is that information, which is encoded within the model, can probably be applied broadly across the chemical space. The "Achilles heel" of the QSAR approaches is the degree of confidence we have that the built model indeed encodes the important underlying controlling properties, and even if it does, that the model can extrapolate away from the set of compounds used to define those controlling properties, to new compounds. At their worst, empirical QSAR models may only identify chance relationships in the training set, and at their best, they can capture the underlying physics of the system being modeled, which may be extrapolated broadly, but this is a rare event, and most QSAR models fall somewhere between those extremes.

QSAR models have had a particularly important impact in the past 10–15 years. With many of the important mechanisms of ADMET being relatively more bulk property and less structure dependent, even when transporters and metabolizing enzymes are involved, empirical QSAR models have proved particularly powerful in guiding compound design. The screens for ADMET properties, such as solubility, logD, Caco-2 permeability, microsomal and hepatocyte determination, clearance, cytochrome P450 (CYP) inhibition, phospholipidosis, and human ether a go-go related gene (hERG) interactions, and so on, are generic to all medicinal chemistry projects, so many companies have built large datasets of diverse compounds profiled through these consistent assays. For example, in a recent publication, GlaxoSmithKline (London, U.K.) has described simple interpretable QSAR models containing between 10,000 and 50,000 data points across many assays [20].

7.2.2 Derivation of Robust QSAR Models

The literature is full of QSAR models, but not all QSAR models are born equal. Models with apparently appealing statistics can in reality encode little more than chance relationships (as a result of the power of chance, and the speed of machine learning algorithms to find random correlations in large data matrices). Models that describe the compounds used to define the model well (training set) may suffer when trying to predict compounds unlike the training set [21]. Superficially appealing statistics can easily mislead the unwary, and when used for prediction, they can waste time and money on the synthesis of compounds subsequently found not to have improved properties. On the contrary, a good QSAR model can very quickly guide a chemist toward optimal compounds. For example, the hERG models [22] used by AstraZeneca (London, U.K.) chemists have contributed to the reduction in compounds showing hERG potency $< 1\,\mu M$, from 25.8% of all compounds tested in 2003 to only 6% in 2010. But the problem of identifying a robust QSAR model is so great, and the need for predictive models so high, that the Organization for Economic and Commercial Development (OECD) has defined good practice in the validation QSAR models, with the aim of increasing regulatory body acceptance of QSAR models for toxicity and environmental impact (http://www.oecd.org/env/hazard/qsar). The OECD has also developed a "QSAR toolbox" for use by regulatory authorities and companies to allow look-up, near-neighbor analysis, and gap-filling (robust prediction) of toxicity and environmental impact of new compounds based on its database of structures and responses. Indeed the European Community in its Registration, Evaluation, Authorisation, and Restriction of Chemicals (REACH) now accepts validated QSARs as part of the complete dossier on the physicochemical biological and technological properties, which is required for every chemical circulating in Europe (Anon, "Regulatory Use of (Q)SARs under REACH, 2009," URL: http://echa.europa.eu/doc/press/webinars/regulatory_use_of_qsars_under_reach_doris_hirmann_echa.pdf). The OECD guidelines are recommended reading for any scientist involved in the derivation of QSAR models.

The publication policies of journals have also changed, and many now explicitly require that all data and molecular structures used to carry out a QSAR/QSPR study should be reported in the paper in its supporting information or should be readily available without infringements or restrictions. This allows the reader to reproduce and validate the analysis in the paper, and it further adds to improving the quality of published work (see, for instance, http://pubs.acs.org/paragonplus/submission/jmcmar/jmcmar_authguide.pdf).

Modern computing power allows many repetitive tasks to be automated with the tasks being performed in consistent and objective fashion. Recently, we have seen automation being applied to QSAR model development. This offers several advantages. As stated previously, many companies have large datasets of compounds measured in consistent, well-defined assays. Using these large datasets, so-called "global" QSAR models are built, in the hope that the breadth of chemical space contained within the model will allow it to be applied to a wide range of future chemistries. Many commercially available QSAR models can be described as global models. But as drug discovery projects continue, the chemistries evolve and move away from the domain of applicability of the model, meaning that the accuracy and precision of predictions is liable to degrade. Models can be manually kept up to date, but this is a laborious task. Automation can be used to keep these models up-to-date.

It is often found that "local" QSAR models, focused only on the chemistry of a particular subseries, can perform better than predictions from a global model. But where multiple projects are being prosecuted in parallel, the task of building many local models, together with the global models, becomes so onerous that models are not kept up-to-date. Automation can build these models and keep them up-to-date. The choice of which model to use ("old" vs. "new," "global" vs. "project," method A vs. method B) requires many decisions and comparisons, and automation can apply decision rules consistently and objectively.

To manage the automation of building and validation of hundreds or thousands of QSAR models, Leahy and colleagues at Cyprotex (U.K.) built a system called "Discovery Bus" [23]. Discovery Bus allows new machine learning agents, datasets, or descriptors to be plugged in and be available for model building and model comparison within "a competitive workflow." In this manner, good models can be found and kept up-to-date.

The comparisons across models does increase the risk of chance correlations, but the continuous feedback of measurements from predicted compounds builds confidence in the success of these approaches. AstraZeneca (U.K.) has also embraced the concept of automation and in the "AutoQSAR" system [24]. AutoQSAR also includes comparison of chemical series versus project series versus global modeling to find the best model for the next new compound. Discovery Bus and AutoQSAR apply good modeling and validation practice to maximize the robustness of models built and selected for use. In a two-year temporal validation across nine active projects, AstraZeneca demonstrated that the hierarchical and updating modeling method of AutoQSAR outperforms updating global models, which outperform static global models [24]. AutoQSAR is now being used routinely in project teams across AstraZeneca. Many projects experience remarkable improvements in the predictive

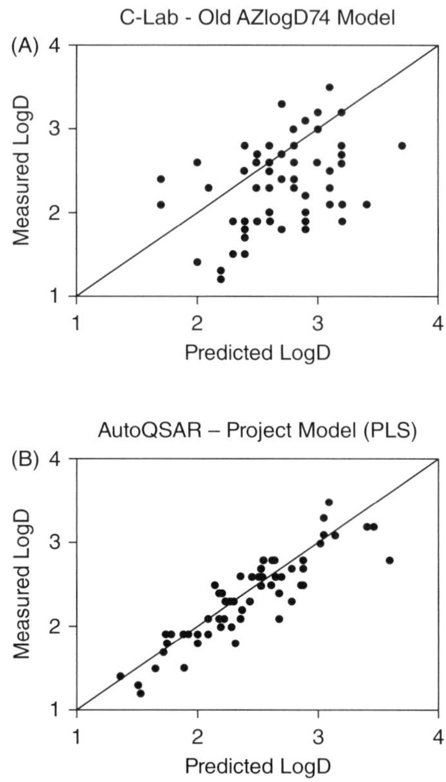

Figure 7.1 A comparison of predictive ability of the global AstraZeneca logD7.4 model (A) with the AutoQSAR logD7.4 model (B) for a current oncology project.

power of their local project QSAR models compared with the global models. Figure 7.1 shows the prediction of n-octanol-water logD7.4 for a current oncology project. The global AstraZeneca logD7.4 model is itself a QSAR model using the complete AstraZeneca logD7.4 database with a combination of physical properties and commercial logP and logD7.4 predictions as input. But the AZlogD7.4 model does a very poor job for this project. Their own AutoQSAR model is good enough to guide chemistry.

Developments such as Discovery Bus and AutoQSAR help to minimize many major problems associated with QSAR modeling. They apply QSAR modeling in a consistent fashion using objective decision-making criteria; they allow for the updating of QSAR models with the most recent data, which in active drug discovery programs ensures the distance between the compound predicted and the domain of applicability of the QSAR model is minimized; and finally they apply the best model, whether it is a local chemical series model or the global x-project model, based on using temporal predictions to make the decision. A weakness in the competitive workflow approach is that as multiple comparisons are being made, this increases the chance that a spurious model could be selected. But AstraZeneca's experience shows that in practice, more accurate and precise predictions are made routinely across projects, so the benefit clearly outweighs the risk, and improved models are available to project teams. AutoQSAR and Discovery Bus also offer the hope that, the value of the plethora of machine learning methods and descriptors previously mentioned can be objectively evaluated.

7.3 STRUCTURE-BASED APPROACHES TO ADMET

7.3.1 Modeling Methods

Several recent reviews [17, 18, 25] reported different strategies exploiting 3D structures of proteins to predict ADMET properties of compounds to move toward understanding of the involved molecular mechanisms. Structure-based (SB) molecular modeling approaches are being employed for such goals. Molecular docking computations, widely used for ADMET proteins, aim at finding low-energy conformations of a ligand bound in a protein binding site. Various docking search algorithms, as stochastic, systematic, and deterministic, are implemented in free or commercial software [26, 27]. Stochastic algorithms (e.g., Monte Carlo, genetic algorithms, etc.) involve random translation, rotation, and modification of torsion angles of the ligand to generate different conformations (e.g., software AutoDOCK [28], etc.). Systematic (or pseudo-systematic) algorithms attempt to explore all degrees of freedom of a ligand and often employ incremental construction of the ligand in the binding site (e.g., software DOCK [29], etc.). Deterministic methods generate a conformation that is dependent on the previous simulation step and has to be energetically more preferable than the previous one, such as the molecular dynamics approach. In general, different generated conformations are evaluated by scoring functions to identify the most energetically favorable pose of the ligand through an estimation of the binding energy by taking into account several energetic terms as intermolecular hydrophobic and polar interactions, intramolecular energy, solvation, and entropy contributions to binding. Three main types of scoring functions, force field-based, knowledge-based, and empirical, have been developed [30]. The first ones estimate the binding energy as a sum of the contribution of several molecular mechanics force field potentials, the second ones are derived from statistical analyses of atom-pairs frequencies in protein–ligand complexes, and the last ones represent the binding energy as a weighted sum of interaction terms based on experimental binding energy data. When approximating the binding energy, scoring methods suffer from insufficient precision. For example, explicit water molecules are often neglected; only configurational entropy contribution is taken into consideration in most cases [30]. For a short list of compounds, more rigorous thermodynamics approaches can be used to calculate binding free energies of structurally diverse or structurally similar ligands [31] for ADMET prediction purposes. Furthermore, quantum chemistry-based calculations offer the opportunity to explore in detail the nature of interactions between molecules, and mechanical and hybrid quantum mechanical/molecular mechanical (QM/MM) methods have been used to calculate the reactivity of compounds for drug metabolism prediction [25]. For example, CypScore is a tool that predicts the exact atomic positions of CYP oxidative attack on small drug molecules; it is now a routine tool in drug discovery at Bayer Schering Pharma (Berlin, Germany) [32]. Its application to diazepam demonstrated its utility for the regioselectivity prediction for lead optimization goals. A structure-based approach and QSAR analysis have been used to guide the metabolism-directed optimization of several thrombin inhibitors recently developed by Merck (Whitehouse Station, NJ) [33].

Applying the structure-based molecular modeling approach for ADMET proteins can be considered more challenging than for other proteins because such ADMET proteins interact with a large range of diverse small molecules; i.e., they are promiscuous. Most ADMET proteins have flexible and large binding sites naturally ensuring its promiscuity. To investigate such problems, molecular dynamics or other simulation methods can be used to generate different conformations of the protein prior docking [34] or to explore the conformational landscape of protein–ligand complexes [35].

7.3.2 Pharmacophore Modeling

Key properties of a small molecule required for a pharmacological activity are called pharmacophores. Pharmacophore models based on a 3D protein structure of the binding site can be very useful to determine the potency of a ligand to interact with a protein receptor by identifying specific interactions. Pharmacophore hypotheses can be explored by identification of similar atoms or groups in terms of chemical and physical properties and approximation of their 3D positions. Many pharmacophore models have been developed to predict the effects of small molecules on transporters (see the review in Reference 36), metabolizing enzymes [37], or off-targets [38].

7.3.3 Structure-based Approach Applied to Metabolizing Enzymes

Historically, structure-based methods to ADMET started with exploring CYPs [15, 16]. CYPs are major phase I drug metabolizing enzymes. More than 50% of marketed drugs are metabolized by CYPs. Their natural role is to detoxify the organism from xenobiotics by oxidation. However, the oxidation sometimes leads to activating procarcinogens or to creating highly reactive metabolites [39]. Furthermore, CYPs' inhibition is often a reason for drug–drug interaction with possible toxicity. Numerous X-ray structures of human CYP proteins are now available from the Protein Data Bank (PDB) that enable prediction of ligand binding to CYP for metabolism or inhibition mechanism prediction. Vedani and Smiesko combined flexible docking and multidimensional QSAR to investigate the inhibitory potential of several ligands of CYP3A4 [18]. Recently, studies combining docking calculations with molecular dynamics (MD) simulations and quantum chemistry to predict regioselectivity of some ligands have been reported [25]. MD simulations and simulated annealing have been performed to generate different conformations of CYP2D6 (CYP2D6 is the second-most studied drug-metabolizing enzyme), and a neural network model was used to identify relevant conformations for binding affinity prediction based on docking scores [40]. MD simulations combined with docking calculations and binary decision trees have also been performed on the crystal structure of CYP2D6 apo-form, to find the three most essential structures enabling the accurate prediction of the metabolism site of known substrates [41]. It was previously demonstrated that better prediction with docking-scoring on CYP2D6 could be obtained by adding water molecules in the binding site [42]. Indeed, 12 hydration sites in the active site of CYP2D6 (PDB ID: 2F9Q) have been found by using MD simulations [43]. Other studies on CYP2C9 (PDB ID: 1OG2) showed the important role of water molecules in the substrate–enzyme reaction [44]. Recently, Stjernschantz and Oostenbrink combined docking, MD simulations, and free energy calculations with the linear interaction energy (LIE) approach to improve the prediction of ligand affinity. Finally, a computational procedure predicting ligand binding based on the similarity between molecular interaction fields of the active site and substrates combined with substrate reactivity enabled MetaSite to predict successfully CYP-specific metabolites [45].

Although phase II metabolizing enzymes (UDP-glucuronosyltransferases (UGTs), sulfotransferases (SULTs), etc.) also play an important role in drug metabolism, atomic-level predictions of drug candidates have been rare up to now. SULTs are a supergene family catalyzing sulfoconjugation of numerous compounds including drugs via the co-factor 3′–phosphoadenosine 5′-phosphosulfate (PAPS). In some cases, SULTs create highly reactive metabolites that are mutagenic and carcinogenic [46]. The analysis of its active site and diverse substrates [47] suggests an important plasticity of the binding site, and a single experimental 3D structure of a SULT shows multiple co-crystallized ligands [48]. Based on docking computations and MD simulations, Stjernschantz et al. explored SULT1E1 to find selective inhibitors and to explain their specificity [49]. Substrate selectivity profiles of SULT1A3 and SULT1E1 were investigated by virtual screening experiments, and it was shown that the structure-based approach is useful for prediction of preferential substrate classes [50]. Recently, we developed a new protocol to distinguish binders from nonbinders for the SULT1A1 (PDB ID: 1LS6) family. Based on MD simulations and hierarchical protein conformation clustering, we selected different representative conformations of the binding site. The performance of the employed structure-based approach was validated by assessing the discrimination between binders and nonbinders for a large number of diverse SULT1A1 ligands (substrates and inhibitors) and the obtained results were very promising [17].

7.3.4 Structure-based Approach Applied to Toxicity Prediction

The hERG potassium ion channel is a key player in the cardiac rhythm regulation. Drug-induced blocking of the hERG channel can cause the prolongation of the Q-T interval and ultimately potential life-threatening ventricular tachyarrhythmia. A wide variety of compounds prolong the Q-T interval through action on the hERG channel. hERG is a tetramer, and each subunit consists of six transmembrane helices. The drug-binding site is within the cavity formed by four S6 helices. The entire structure has not been solved, and several homology models have been built based on other channels [51]. Several studies based on docking methods have identified key residues implicated in the hERG binding mechanism and drug interactions [52].

Prediction of 20 hERG blockers by docking has been found to be consistent with experimental data. Thus, a model predicting drug blockage of hERG was obtained using docking computations combined with 3D-QSAR [53]. MD simulations have also been undertaken to optimize docking solutions and to evaluate relative binding affinities by the LIE approach [51].

Other types of proteins can interfere with drug toxicity. Nuclear receptors (NRs), which are ligand-regulated transcription factors, can induce drug–drug interactions by upregulating transcription of metabolizing enzymes. They are also capable of binding endocrine disruptors leading to abnormal functions (e.g., reproductive, developmental, etc.). They are composed of a ligand-binding domain and a DNA-binding domain controlling gene expression. Pregnane X receptor (PXR) is implicated in the regulation of the CYP3A, UGT, and ABC transporter genes through activation by a wide range of ligands. Two structures of the human apo PXR and seven structures of the human ligand-bound domain of PXR are known. Ekins et al. combined docking, hybrid scoring strategies, and 3D-QSAR modeling to improve the prediction of PXR agonists among the ToxCast database [54]. The constitutive androstane receptor (CAR) activates the transcription of genes including some CYP members, UGT, SULTs, and ABC transporters, thus, influencing drug response. Two structures of the human CAR ligand-binding domain have been solved with a general fold similar to the PXR one. The flexibility of the CAR binding site has been explored by MD simulations combined with molecular docking based on homology models in the complexed and uncomplexed states. Küblbeck et al. identified 19 compounds as CAR agonists by virtual screening performed on a CAR structure [55]. A study exploring the flexibility of 13 NRs by collecting multiple NR conformations from crystallographic structures and by docking of 157 diverse NR ligands into the multiple ensemble showed the gain in accuracy when taking into account the plasticity of the receptor [56].

7.4 CHEMOGENOMICS TO ADMET

Among various concepts and algorithms used for *in silico* prediction to ADMET, chemogenomic approaches attempt to match fully ADMET-related targets and ligands spaces. In such an approach, the targets are analyzed not according to sequence or fold but according to the similarity of ligands, e.g. via reverse docking [57] or by structure-based pharmacophores [38]. Chemogenomics can help to prioritize chemical classes by mapping out ADMET-relevant areas of chemical space for a particular target. Along this line, so-called proteochemometric modeling has been developed to extrapolate the activity of known ligands for known protein targets to novel targets that could cause toxicity problems [58]. Recently, a novel model was developed to navigate drug–target interactions on a genome-wide scale [59]. This method allows for predicting molecules with too many potential off-target interactions or specific off-targets that can lead to severe side effects. In the same direction, we can notice the *in silico* network methods mapping adverse drug reactions on the chemical space [60]. Yet, several difficulties are present with such modeling, as the involvement of a large number of parameters to model both small molecules and binding proteins spaces, which can reduce the prediction performance, or also the interpretability and compatibility of ligand and protein descriptors.

7.5 COMBINATION OF VARIOUS APPROACHES EXEMPLIFIED FOR P-GP

P-gp is one of the most important antitargets in drug discovery projects. This fact is related to the ability of the protein to recognize and actively export out of the cells numerous compounds with a totally different structure [61]. Among them are many drugs, whose ADMET properties and intracellular bioavailability can be strongly influenced by the P-gp transport [62]. P-gp is one of the main transport proteins involved in multidrug resistance (MDR) in cancer cells; its substrates are many cytotoxic anticancer drugs (anthracylines, Vinca alkaloids, taxanes, and epipodophyllotoxins) and various MDR modulators (Figure 7.2) that can inhibit the transport function of the protein [63].

The structure of P-gp is composed of the functional units common to all ABC transporters acting as exporters: two transmembrane domains (TMDs), each containing six transmembrane helices, and two cytoplasmatic nucleotide binding domains (NBDs). During the transport cycle, the protein undergoes several functional states that, generally, vary from the ATP-free one, corresponding to the open-to-inside or inward-facing conformation, to the nucleotide-bound that corresponds to the open-to-outside or outward-facing conformation. The closed one is considered competent for drug binding [64]; in contrast, the open conformation is associated with the drug release from the protein [65]. The transition of the protein from the nucleotide-free to the bound functional state is accompanied by large conformational changes that simultaneously affect the TMDs and NBDs.

Aiming at better understanding the structure-function relationships of the protein and the ways to overcome MDR by developing selective and potent P-gp inhibitors, numerous studies have been performed applying both experimental and *in silico* approaches. Being more time and cost effective, the *in silico* methods have increasingly been used for P-gp in recent years. A variety of approaches, including ligand-based, structure-based, and a combination of both, have been applied.

7.5.1 Ligand-based Methods

The numerous data accumulated from rational synthesis and pharmacological testing of compounds of different structural scaffolds provide excellent opportunities for using ligand-based methods. In the past several years, these studies have shifted from classic QSAR and 3D-QSAR methods (reviewed in References 66 and 67) to more sophisticated techniques like pharmacophore identification [68–70] and machine learning methods for compounds' recognition as P-gp substrates and non-substrates/inhibitors [20, 71, 72] with ongoing studies in these directions [73, 74].

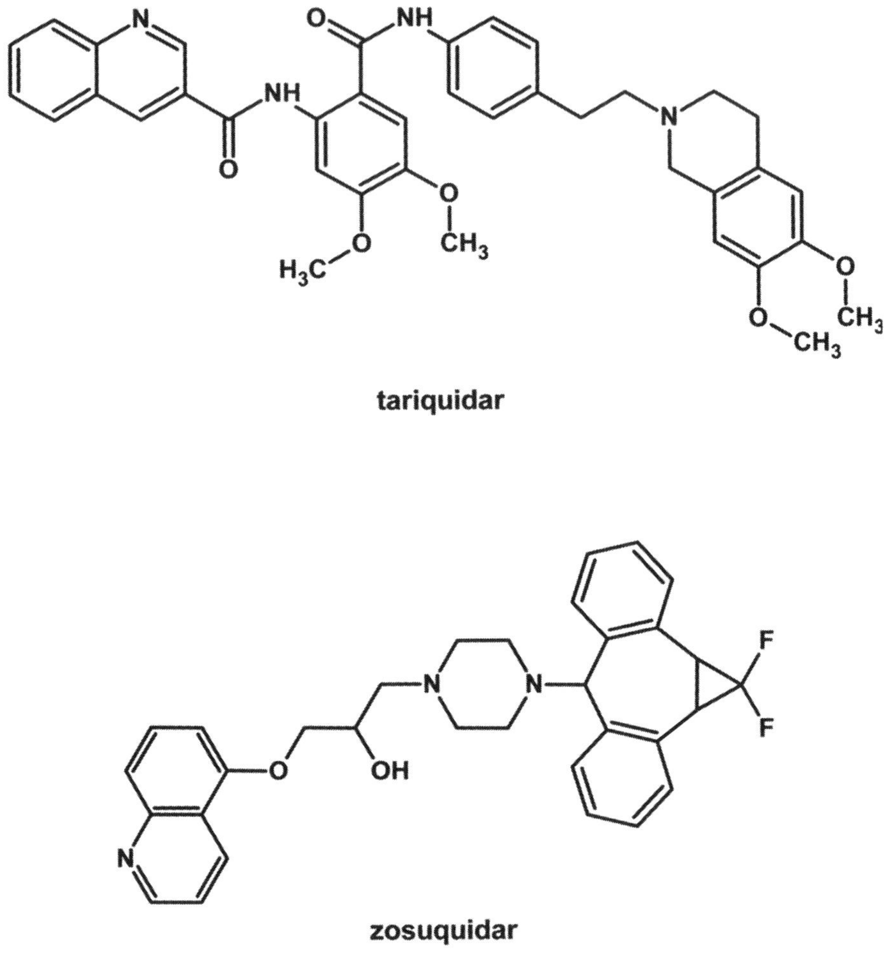

Figure 7.2 Structures of the third-generation MDR modulators tariquidar and zosuquidar composed of structural features preselected on rational QSAR analysis.

7.5.2 Structure-based Methods

The recently solved physiologically relevant crystallographic structures of MDR transporters [75], including mouse P-gp [64](PDB ID: 3G60, 3G61), gave a strong impetus to the application of structure-based methods in the P-gp studies. Homology models of human P-gp have been developed corresponding to the open and closed conformations of the protein [76–80]. Two main directions are currently explored using these models: binding site identification and characterization and docking of various ligands [77, 81–84]. A combination of both ligand- and structure-based methods seems like a promising approach in studying P-gp [81, 85]. Recently, using

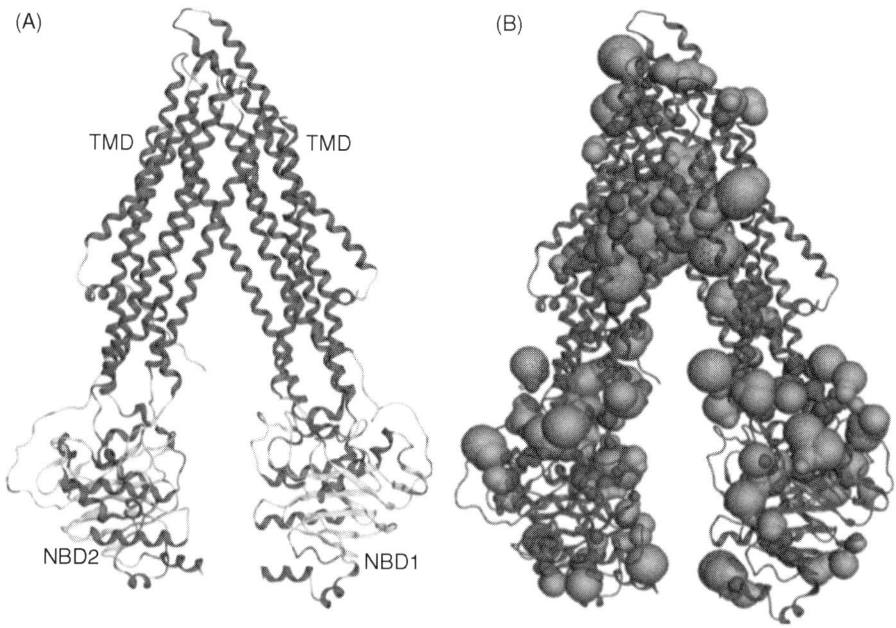

Figure 7.3 Homology model of the human P-glycoprotein corresponding to the inward-open conformation (ATP-free functional state) [80]. (A) A 3D view of the protein structure: The protein backbone is rendered as a line ribbon and colored according to the secondary structure: loop – gray; helix – red; strand – yellow; turn – blue; (B) Potential binding sites of the protein filled with alpha spheres (each sphere is classified as either "hydrophobic" (gray) or "hydrophilic" (red) depending on whether the sphere is a good hydrogen bonding point in the receptor. The radius of the sphere is proportional to the convex hull of the point set and is invariant to rotation of the atomic coordinates, generated with the "Site Finder" module in MOE (http://www.chemcomp.com). (*See insert for color representation of the figure.*)

targeted MD Wise generated several possible conformations of the protein that can be used for docking and virtual ligand screening purposes [86].

The results on P-gp modeling accumulated so far contribute undoubtedly to the better understanding of the protein behavior and to the interactions with its substrates and inhibitors. The protein promiscuity as well as the conformational flexibility of its ligands and of the protein itself make the modeling tasks rather complex. For example, when deciding on possible binding sites of the P-gp interacting drugs, one encounters a vast amount of possibilities. This is illustrated in the example of the sites allocated in the closed (inward-facing) homology model of the human P-gp (Figure 7.3) that can potentially be used for docking and virtual ligand screening tasks. Despite its complexity, the *in silico* modeling of P-gp has achieved significant progress in recent years, and its impact on application-oriented studies is permanently improving. The future progress of its use for the purposes of *in silico* ADMET screening will depend on the most optimal combination of the results of both theoretical and experimental investigations.

ACKNOWLEDGMENTS

Financial support from the Inserm Institute is greatly appreciated. V.Y.M. thanks the doctoral school "MTCE" of the Universities Paris Descartes and Paris Diderot. I.P. thanks the National Science Fund of Bulgaria (Grant DTK 02/58).

REFERENCES

1. Giacomini, K.M., Krauss, R.M., Roden, D.M., Eichelbaum, M., Hayden, M.R., Nakamura, Y. (2007). When good drugs go bad. *Nature*, *446*, 975–977.
2. Hamon, J., Whitebread, S., Techer-Etienne, V., Le Coq, H., Azzaoui, K., Urban, L. (2009). In vitro safety pharmacology profiling: What else beyond hERG? *Future Medicinal Chemistry*, *1*, 645–665.
3. Nigsch, F., Lounkine, E., McCarren, P., Cornett, B., Glick, M., Azzaoui, K., Urban, L., Marc, P., Muller, A., Hahne, F., Heard, D.J., Jenkins, J.L. (2011). Computational methods for early predictive safety assessment from biological and chemical data. *Expert Opinion on Drug Metabolism & Toxicology*, *7*, 1497–1511.
4. Lagorce, D., Reynes, C., Camproux, A.-C., Miteva, M.A., Sperandio, O., Villoutreix, B.O. In Silico ADME/Tox Predictions, in K. Tsaioun and S.A. Kates (eds.), *ADMET for Medicinal Chemists*, Wiley, Hoboken, NJ, 2011, pp. 29–124.
5. Michielan, L., Moro, S. (2010). Pharmaceutical perspectives of nonlinear QSAR strategies. *Journal of Chemical Information and Modeling*, *50*, 961–978.
6. Varnek, A., Baskin, I. (2012). Machine learning methods for property prediction in chemoinformatics: Quo vadis? *Journal of Chemical Information and Modeling*, *52*, 1413–1437.
7. Helguera, A.M., Combes, R.D., Gonzalez, M.P., Cordeiro, M.N. (2008). Applications of 2D descriptors in drug design: A DRAGON tale. *Current Topics in Medicinal Chemistry*, *8*, 1628–1655.
8. Lipinski, C.A., Lombardo, F., Dominy, B.W., Feeney, P.J. (2001). Experimental and computational approaches to estimate solubility and permeability in drug discovery and development settings. *Advanced Drug Delivery Reviews*, *46*, 3–26.
9. Mitra, R., Shyam, R., Mitra, I., Miteva, M.A., Alexov, E. (2008). Calculating the protonation states of proteins and small molecules: Implications to ligand-receptor interactions. *Current Computer-Aided Drug Design*, *4*, 169–179.
10. Lagorce, D., Maupetit, J., Baell, J., Sperandio, O., Tuffery, P., Miteva, M.A., Galons, H., Villoutreix, B.O. (2011). The FAF-Drugs2 server: A multistep engine to prepare electronic chemical compound collections. *Bioinformatics*, *27*, 2018–2020.
11. Monge, A., Arrault, A., Marot, C., Morin-Allory, L. (2006). Managing, profiling and analyzing a library of 2.6 million compounds gathered from 32 chemical providers. *Molecular Diversity*, *10*, 389–403.
12. Peironcely, J.E., Reijmers, T., Coulier, L., Bender, A., Hankemeier, T. (2011). Understanding and classifying metabolite space and metabolite-likeness. *PLoS One*, *6*, e28966.
13. Low, Y., Uehara, T., Minowa, Y., Yamada, H., Ohno, Y., Urushidani, T., Sedykh, A., Muratov, E., Kuz'min, V., Fourches, D., Zhu, H., Rusyn, I., Tropsha, A. (2011). Predicting

drug-induced hepatotoxicity using QSAR and toxicogenomics approaches. *Chemical Research in Toxicology*, *24*, 1251–1262.

14. Benigni, R., Bossa, C., Alivernini, S., Colafranceschi, M. (2012). Assessment and validation of US EPA's OncoLogic(R) expert system and analysis of its modulating factors for structural alerts. *Journal of Environmental Science and Health - Part C: Environmental Carcinogenesis & Ecotoxicology Reviews*, *30*, 152–173.
15. de Groot, M.J., Vermeulen, N.P., Kramer, J.D., van Acker, F.A., Donne-Op den Kelder, G.M. (1996). A three-dimensional protein model for human cytochrome P450 2D6 based on the crystal structures of P450 101, P450 102, and P450 108. *Chemical Research in Toxicology*, *9*, 1079–1091.
16. Ekins, S., de Groot, M.J., Jones, J.P. (2001). Pharmacophore and three-dimensional quantitative structure activity relationship methods for modeling cytochrome p450 active sites. *Drug Metabolism and Disposition*, *29*, 936–944.
17. Moroy, G., Martiny, V.Y., Vayer, P., Villoutreix, B.O., Miteva, M.A. (2012). Toward in silico structure-based ADMET prediction in drug discovery. *Drug Discovery Today*, *17*, 44–55.
18. Vedani, A., Smiesko, M. (2009). In silico toxicology in drug discovery - concepts based on three-dimensional models. *Alternatives to Lab Animals*, *37*, 477–496.
19. Hansch, C.M., Fujita, T., Muir, R.M. (1962). Correlation of biological activity of phenoxyacetic acids with hammett substituent constants and partition coefficients. *Nature*, *194*, 178–180.
20. Gleeson, M.P. (2008). Generation of a set of simple, interpretable ADMET rules of thumb. *Journal of Medicinal Chemistry*, *51*, 817–834.
21. Stouch, T.R., Kenyon, J.R., Johnson, S.R., Chen, X.Q., Doweyko, A., Li, Y. (2003). In silico ADME/Tox: Why models fail. *Journal of Computer-Aided Molecular Design*, *17*, 83–92.
22. Gavaghan, C.L., Arnby, C.H., Blomberg, N., Strandlund, G., Boyer, S. (2007). Development, interpretation and temporal evaluation of a global QSAR of hERG electrophysiology screening data. *Journal of Computer-Aided Molecular Design*, *21*, 189–206.
23. Cartmell, J., Enoch, S., Krstajic, D., Leahy, D.E. (2005). Automated QSPR through competitive workflow. *Journal of Computer-Aided Molecular Design*, *19*, 821–833.
24. Wood, D.J., Buttar, D., Cumming, J.G., Davis, A.M., Norinder, U., Rodgers, S.L. (2011). Automated QSAR with a hierarchy of global and local models. *Molecular Informatics*, *30*, 960–972.
25. Sun, H., Scott, D.O. (2010). Structure-based drug metabolism predictions for drug design. *Chemical Biology & Drug Design*, *75*, 3–17.
26. Leach, A.R., Shoichet, B.K., Peishoff, C.E. (2006). Prediction of protein-ligand interactions. Docking and scoring: Successes and gaps. *Journal of Medicinal Chemistry*, *49*, 5851–5855.
27. Sperandio, O., Villoutreix, B.O., Miteva, M.A. Structure-based virtual screening, in M.A. Miteva ed.), *In Silico Lead Discovery*, 2011, pp. 20–46.
28. Morris, G.M., Huey, R., Lindstrom, W., Sanner, M.F., Belew, R.K., Goodsell, D.S., Olson, A.J. (2009). AutoDock4 and AutoDockTools4: Automated docking with selective receptor flexibility. *Journal of Computational Chemistry*, *30*, 2785–2791.
29. Moustakas, D.T., Lang, P.T., Pegg, S., Pettersen, E., Kuntz, I.D., Brooijmans, N., Rizzo, R.C. (2006). Development and validation of a modular, extensible docking program: DOCK 5. *Journal of Computer-Aided Molecular Design*, *20*, 601–619.

30. Moitessier, N., Englebienne, P., Lee, D., Lawandi, J., Corbeil, C.R. (2008). Towards the development of universal, fast and highly accurate docking/scoring methods: A long way to go. *British Journal of Pharmacology*, *153*(1), S7–S26.
31. Stjernschantz, E., Oostenbrink, C. (2010). Improved ligand-protein binding affinity predictions using multiple binding modes. *Biophysical Journal*, *98*, 2682–2691.
32. Hennemann, M., Friedl, A., Lobell, M., Keldenich, J., Hillisch, A., Clark, T., Goller, A.H. (2009). CypScore: Quantitative prediction of reactivity toward cytochromes P450 based on semiempirical molecular orbital theory. *ChemMedChem*, *4*, 657–669.
33. Burgey, C.S., Robinson, K.A., Lyle, T.A., Sanderson, P.E., Lewis, S.D., Lucas, B.J., Krueger, J.A., Singh, R., Miller-Stein, C., White, R.B., Wong, B., Lyle, E.A., Williams, P.D., Coburn, C.A., Dorsey, B.D., Barrow, J.C., Stranieri, M.T., Holahan, M.A., Sitko, G.R., Cook, J.J., McMasters, D.R., McDonough, C.M., Sanders, W.M., Wallace, A.A., Clayton, F.C., Bohn, D., Leonard, Y.M., Detwiler, T.J., Jr., Lynch, J.J., Jr., Yan, Y., Chen, Z., Kuo, L., Gardell, S.J., Shafer, J.A., Vacca, J.P. (2003). Metabolism-directed optimization of 3-aminopyrazinone acetamide thrombin inhibitors. Development of an orally bioavailable series containing P1 and P3 pyridines. *Journal of Medicinal Chemistry*, *46*, 461–473.
34. Sperandio, O., Mouawad, L., Pinto, E., Villoutreix, B.O., Perahia, D., Miteva, M.A. (2010). How to choose relevant multiple receptor conformations for virtual screening: A test case of Cdk2 and normal mode analysis. *European Biophysics Journal*, *39*, 1365–72.
35. Rossato, G., Ernst, B., Smiesko, M., Spreafico, M., Vedani, A. (2010). Probing small-molecule binding to cytochrome P450 2D6 and 2C9: An in silico protocol for generating toxicity alerts. *ChemMedChem*, *5*, 2088–2101.
36. Yamashita, F., Hashida, M. (2004). In silico approaches for predicting ADME properties of drugs. *Drug Metabolism and Pharmacokinetics*, *19*, 327–338.
37. Smith, P.A., Sorich, M.J., Low, L.S.C., McKinnon, R.A., Miners, J.O. (2004). Towards integrated ADME prediction: Past, present and future directions for modelling metabolism by UDP-glucuronosyltransferases. *Journal of Molecular Graphics and Modelling*, *22*, 507–517.
38. Meslamani, J., Li, J., Sutter, J., Stevens, A., Bertrand, H.O., Rognan, D. (2012). Protein-ligand-based pharmacophores: Generation and utility assessment in computational ligand profiling. *Journal of Chemical Information and Modeling*, *52*, 943–955.
39. Shimada, T. (2006). Xenobiotic-metabolizing enzymes involved in activation and detoxification of carcinogenic polycyclic aromatic hydrocarbons. *Drug Metabolism and Pharmacokinetics*, *21*, 257–276.
40. Bazeley, P.S., Prithivi, S., Struble, C.A., Povinelli, R.J., Sem, D.S. (2006). Synergistic use of compound properties and docking scores in neural network modeling of CYP2D6 binding: Predicting affinity and conformational sampling. *Journal of Chemical Information and Modeling*, *46*, 2698–2708.
41. Hritz, J., de Ruiter, A., Oostenbrink, C. (2008). Impact of plasticity and flexibility on docking results for cytochrome P450 2D6: A combined approach of molecular dynamics and ligand docking. *Journal of Medicinal Chemistry*, *51*, 7469–7477.
42. de Graaf, C., Oostenbrink, C., Keizers, P.H., van der Wijst, T., Jongejan, A., Vermeulen, N.P. (2006). Catalytic site prediction and virtual screening of cytochrome P450 2D6 substrates by consideration of water and rescoring in automated docking. *Journal of Medicinal Chemistry*, *49*, 2417–2430.

43. Santos, R., Hritz, J., Oostenbrink, C. (2010). Role of water in molecular docking simulations of cytochrome P450 2D6. *Journal of Chemical Information and Modeling*, *50*, 146–154.
44. Shi, R., Li, J., Cao, X., Zhu, X., Lu, X. (2011). Exploration of the binding of proton pump inhibitors to human P450 2C9 based on docking and molecular dynamics simulation. *Journal of Molecular Modeling*, *17*, 1941–1951.
45. Cruciani, G., Carosati, E., De Boeck, B., Ethirajulu, K., Mackie, C., Howe, T., Vianello, R. (2005). MetaSite: Understanding metabolism in human cytochromes from the perspective of the chemist. *Journal of Medicinal Chemistry*, *48*, 6970–6979.
46. Bojarova, P., Williams, S.J. (2008). Sulfotransferases, sulfatases and formylglycine-generating enzymes: A sulfation fascination. *Current Opinion in Chemical Biology*, *12*, 573–581.
47. Gamage, N., Barnett, A., Hempel, N., Duggleby, R.G., Windmill, K.F., Martin, J. L., McManus, M.E. (2006). Human sulfotransferases and their role in chemical metabolism. *Toxicological Sciences*, *90*, 5–22.
48. Gamage, N.U., Duggleby, R.G., Barnett, A.C., Tresillian, M., Latham, C.F., Liyou, N.E., McManus, M.E., Martin, J.L. (2003). Structure of a human carcinogen-converting enzyme, SULT1A1. Structural and kinetic implications of substrate inhibition. *The Journal of Biological Chemistry*, *278*, 7655–7662.
49. Stjernschantz, E., Reinen, J., Meinl, W., George, B.J., Glatt, H., Vermeulen, N.P., Oostenbrink, C. (2010). Comparison of murine and human estrogen sulfotransferase inhibition in vitro and in silico—implications for differences in activity, subunit dimerization and substrate inhibition. *Molecular and Cellular Endocrinology*, *317*, 127–140.
50. Campagna-Slater, V., Schapira, M. (2009). Evaluation of virtual screening as a tool for chemical genetic applications. *Journal of Chemical Information and Modeling*, *49*, 2082–2091.
51. Osterberg, F., Aqvist, J. (2005). Exploring blocker binding to a homology model of the open hERG K +channel using docking and molecular dynamics methods. *FEBS Letters*, *579*, 2939–2944.
52. Fernandez, D., Ghanta, A., Kauffman, G.W., Sanguinetti, M.C. (2004). Physicochemical features of the HERG channel drug binding site. *The Journal of Biological Chemistry*, *279*, 10120–10127.
53. Durdagi, S., Duff, H.J., Noskov, S.Y. (2011). Combined receptor and ligand-based approach to the universal pharmacophore model development for studies of drug blockade to the hERG1 pore domain. *Journal of Chemical Information and Modeling*, *51*, 463–474.
54. Ekins, S., Kortagere, S., Iyer, M., Reschly, E.J., Lill, M.A., Redinbo, M.R., Krasowski, M.D. (2009). Challenges predicting ligand-receptor interactions of promiscuous proteins: The nuclear receptor PXR. *PLoS Computational Biology*, *5*, e1000594.
55. Kublbeck, J., Jyrkkarinne, J., Poso, A., Turpeinen, M., Sippl, W., Honkakoski, P., Windshugel, B. (2008). Discovery of substituted sulfonamides and thiazolidin-4-one derivatives as agonists of human constitutive androstane receptor. *Biochemical Pharmacology*, *76*, 1288–1297.
56. Park, S.J., Kufareva, I., Abagyan, R. (2010). Improved docking, screening and selectivity prediction for small molecule nuclear receptor modulators using conformational ensembles. *Journal of Computer-Aided Molecular Design*, *24*, 459–471.

57. Kinnings, S.L., Jackson, R.M. (2011). ReverseScreen3D: A structure-based ligand matching method to identify protein targets. *Journal of Chemical Information and Modeling*, *51*, 624–634.
58. van Westen, G.J., Wegner, J.K., Geluykens, P., Kwanten, L., Vereycken, I., Peeters, A., Ijzerman, A.P., van Vlijmen, H.W., Bender, A. (2011). Which compound to select in lead optimization? Prospectively validated proteochemometric models guide preclinical development. *PLoS One*, *6*, e27518.
59. Yamanishi, Y., Pauwels, E., Saigo, H., Stoven, V. (2011). Extracting sets of chemical substructures and protein domains governing drug-target interactions. *Journal of Chemical Information and Modeling*, *51*, 1183–1194.
60. Panagiotou, G., Taboureau, O. (2012). The impact of network biology in pharmacology and toxicology. *SAR QSAR Environmental Research*, *23*, 221–235.
61. Glavinas, H., Krajcsi, P., Cserepes, J., Sarkadi, B. (2004). The role of ABC transporters in drug resistance, metabolism and toxicity. *Current Drug Delivery*, *1*, 27–42.
62. Marquez, B., Van Bambeke, F. (2011). ABC multidrug transporters: Target for modulation of drug pharmacokinetics and drug-drug interactions. *Current Drug Targets*, *12*, 600–620.
63. Chen, K.G., Sikic, B.I. (2012). Molecular pathways: Regulation and therapeutic implications of multidrug resistance. *Clinical Cancer Research*, *18*, 1863–1869.
64. Aller, S.G., Yu, J., Ward, A., Weng, Y., Chittaboina, S., Zhuo, R., Harrell, P.M., Trinh, Y.T., Zhang, Q., Urbatsch, I.L., Chang, G. (2009). Structure of P-glycoprotein reveals a molecular basis for poly-specific drug binding. *Science*, *323*, 1718–1722.
65. Gutmann, D.A., Ward, A., Urbatsch, I.L., Chang, G., van Veen, H.W. (2010). Understanding polyspecificity of multidrug ABC transporters: Closing in on the gaps in ABCB1. *Trends in Biochemical Sciences*, *35*, 36–42.
66. Wiese, M., Pajeva, I.K. (2001). Structure-activity relationships of multidrug resistance reversers. *Current Medicinal Chemistry*, *8*, 685–713.
67. Wiese, M., Pajeva, I.K. Algorithms to predict affinity for transporters, in B. Testa, L. Turski, eds., *Virtual ADMET Assessment in Target Selection and Maturation*, Solvay Pharmaceuticals Conferences Series, Amsterdam, 2006, pp. 187-208.
68. Ekins, S., Kim, R.B., Leake, B.F., Dantzig, A.H., Schuetz, E.G., Lan, L.B., Yasuda, K., Shepard, R.L., Winter, M.A., Schuetz, J.D., Wikel, J.H., Wrighton, S.A. (2002). Application of three-dimensional quantitative structure-activity relationships of P-glycoprotein inhibitors and substrates. *Molecular Pharmacology*, *61*, 974–981.
69. Pajeva, I.K., Wiese, M. (2002). Pharmacophore model of drugs involved in P-glycoprotein multidrug resistance: Explanation of structural variety (hypothesis). *Journal of Medicinal Chemistry*, *45*, 5671–5686.
70. Penzotti, J.E., Lamb, M.L., Evensen, E., Grootenhuis, P.D. (2002). A computational ensemble pharmacophore model for identifying substrates of P-glycoprotein. *Journal of Medicinal Chemistry*, *45*, 1737–1740.
71. Bakken, G.A., Jurs, P.C. (2000). Classification of multidrug-resistance reversal agents using structure-based descriptors and linear discriminant analysis. *Journal of Medicinal Chemistry*, *43*, 4534–4541.
72. Gombar, V.K., Polli, J.W., Humphreys, J.E., Wring, S.A., Serabjit-Singh, C.S. (2004). Predicting P-glycoprotein substrates by a quantitative structure-activity relationship model. *Journal of Pharmaceutical Sciences*, *93*, 957–968.

73. Ferreira, R.J., dos Santos, D.J., Ferreira, M.J., Guedes, R.C. (2011). Toward a better pharmacophore description of P-glycoprotein modulators, based on macrocyclic diterpenes from Euphorbia species. *Journal of Chemical Information and Modeling*, *51*, 1315–1324.
74. Poongavanam, V., Haider, N., Ecker, G.F. (2012). Fingerprint-based in silico models for the prediction of P-glycoprotein substrates and inhibitors. *Bioorganic & Medicinal Chemistry*, *20*, 5388–5395.
75. Dawson, R.J., Locher, K.P. (2006). Structure of a bacterial multidrug ABC transporter. *Nature*, *443*, 180–185.
76. Becker, J.P., Depret, G., Van Bambeke, F., Tulkens, P.M., Prevost, M. (2009). Molecular models of human P-glycoprotein in two different catalytic states. *BMC Structure Biology*, *9*, 3.
77. Globisch, C., Pajeva, I.K., Wiese, M. (2008). Identification of putative binding sites of P-glycoprotein based on its homology model. *ChemMedChem*, *3*, 280–295.
78. O'Mara, M.L., Tieleman, D.P. (2007). P-glycoprotein models of the apo and ATP-bound states based on homology with Sav1866 and MalK. *FEBS Letters*, *581*, 4217–4222.
79. Stockner, T., de Vries, S.J., Bonvin, A.M., Ecker, G.F., Chiba, P. (2009). Data-driven homology modelling of P-glycoprotein in the ATP-bound state indicates flexibility of the transmembrane domains. *FEBS Journal*, *276*, 964–972.
80. Pajeva, I.K., Globisch, C., Wiese, M. (2009). Comparison of the inward- and outward-open homology models and ligand binding of human P-glycoprotein. *FEBS Journal*, *276*, 7016–7026.
81. Bikadi, Z., Hazai, I., Malik, D., Jemnitz, K., Veres, Z., Hari, P., Ni, Z., Loo, T.W., Clarke, D.M., Hazai, E., Mao, Q. (2011). Predicting P-glycoprotein-mediated drug transport based on support vector machine and three-dimensional crystal structure of P-glycoprotein. *PLoS One*, *6*, e25815.
82. Pajeva, I.K., Globisch, C., Wiese, M. (2009). Combined pharmacophore modeling, docking, and 3D QSAR studies of ABCB1 and ABCC1 transporter inhibitors. *ChemMedChem*, *4*, 1883–1896.
83. Klepsch, F., Chiba, P., Ecker, G.F. (2011). Exhaustive sampling of docking poses reveals binding hypotheses for propafenone type inhibitors of P-glycoprotein. *PLoS Computational Biology*, *7*, e1002036.
84. Tarcsay, A., Keseru, G.M. (2011). Homology modeling and binding site assessment of the human P-glycoprotein. *Future Medicinal Chemistry*, *3*, 297–307.
85. Kroemer, R.T. (2007). Structure-based drug design: Docking and scoring. *Current Protein & Peptide Science*, *8*, 312–328.
86. Wise, J.G. (2012). Catalytic transitions in the human MDR1 P-glycoprotein drug binding sites. *Biochemistry*, *51*, 5125–5141.

8

MULTIPARAMETER OPTIMIZATION OF ADMET FOR DRUG DESIGN

Matthew D. Segall and Edmund J. Champness

8.1 INTRODUCTION

Experience has shown that finding a new drug is a difficult, time consuming, and expensive process. The estimated research and development cost per marketed drug continues to escalate, and the attrition rate of compounds, even in clinical development, remains stubbornly high; recent estimates suggest that the average cost per new marketed drug exceeds US$1.8B and that the success rate of compounds that enter development is approximately 12% [1]. One major cause of attrition is the validity of the biological mechanism targeted by the candidate drug, but even when the biological target is well validated, the challenge of finding a compound with appropriate potency, pharmacokinetics (PK), and safety is enormous.

The impact of poor PK properties on clinical success rates was noted in the 1990s [2], and this led to an increased interest in characterization of the absorption, distribution, metabolism, and elimination (ADME) properties of compounds early in the drug discovery process, to eliminate compounds with poor PK as early as possible. High-throughput *in vitro* and computational (*in silico*) methods were developed to measure or calculate ADME properties that can be used to identify potential PK issues [3–6]. The impact of this can be observed in a reduction in the proportion of clinical failures as a result of poor PK, from an estimated 39% in 1991 [2] to an approximate 10% in 2000 [7]. However, the introduction of additional testing and

Predictive ADMET: Integrative Approaches in Drug Discovery and Development, First Edition.
Edited by Jianling Wang and Laszlo Urban.

more "hurdles" that a compound must overcome has increased the cost and time taken in drug discovery and reduced productivity [8]. Meanwhile, over the same period, the attrition resulting from toxicity issues has increased and the overall success rate in development has not changed significantly—from 11% in 2000 [7] to 12% in 2010 [1]. The response to this has been, in turn, to push assessments of toxicity earlier in the drug discovery process in an attempt to eliminate further risky compounds.

The result of these trends is that that the quantity and complexity of the data generated on compounds in early drug discovery has increased dramatically. This has introduced a new challenge: how to make effective use of these data to guide the design and selection of compounds with a better balance of properties and, hence, a better chance of downstream success. This goal is made even more challenging by the fact that the data that are routinely generated in early discovery have significant uncertainty, both regarding their relevance to the ultimate success of a compound and the property values themselves. The former source of uncertainty comes from the fact that all discovery data are generated using models of the ultimate human patient, whether *in silico*, *in vitro*, or *in vivo*. The uncertainty in the property values is the result of experimental variability and statistical errors in predictions. Therefore, appropriate weight must be given to these data when making decisions about the progression or rejection of a compound. The goal is to save time and effort that is wasted by pursuing compounds that are unlikely to succeed; however, where opportunities to find a good drug molecule are limited, the cost of opportunities missed as a result of incorrectly discarding compounds, while harder to quantify, is likely to be similar in magnitude to late-stage failures.

Even experienced scientists find it difficult to make good decisions based on complex data, particularly when there is significant uncertainty and there is a lot at stake, which is a good description of the situation in drug discovery. The impact of this can be observed in several tendencies, such as oversensitivity to early signs of problems, leading to inappropriate rejection of compounds and missed opportunities or, conversely, wasted effort caused by the desire to be "certain" before moving on to a new idea, chemistry, or project. For a detailed discussion of these so-called "cognitive biases" in drug discovery decision making, see Reference 9.

One commonly observed tendency is to focus first on identifying chemistry with a high degree of potency against the intended therapeutic target before giving significant consideration to the physicochemical, ADME, and safety properties required in the ultimate drug. This may come from a natural inclination to simplify the problem by focusing on the most critical requirement of a drug in the expectation (or hope?) that any problems that may be encountered downstream can be overcome, for example, during lead optimization or with an appropriate formulation. The risk with this unidimensional approach is that it can lead to a premature narrowing of the chemistry explored because typically only small changes to a highly potent compound can be tolerated without an "unacceptable" loss of potency. Furthermore, the compound characteristics required to achieve a high potency can often conflict directly with those for appropriate ADME and safety. For example, for many targets, increasing the lipophilicity of a compound can increase the potency, but a high lipophilicity is also associated with poor solubility and absorption [10] and a greater tendency for toxicity [11]. The result is often many, slow design–test–redesign iterations in lead optimization or the need to "hop" to a new lead series in the search for appropriate properties, adding further time and cost. One common approach to mitigate this risk is to progress two or more chemical series in parallel. However, even with this strategy, it would be ideal to consider a broad range of required properties as early as possible in the drug discovery process, to ensure that effort is placed on chemistries with an appropriate balance of potency with the other factors required in a high-quality lead or candidate drug.

The opposite extreme from a uni-dimensional approach is to insist on identifying the "perfect" compound in all respects, setting hard criteria for all compound properties in an effort to "fail fast, fail cheap." However, as discussed, the ideal properties are often conflicting and the impact of this may be too rapid attrition of compounds leading to missed opportunities. Although difficult to measure, it is likely that the opportunity cost of inappropriately failed projects is comparable with that of late-stage failures. It is important to define property requirements that are consistent with each project's therapeutic goals, rather than simply to "tick the boxes," and consider acceptable compromises if an ideal property profile cannot be achieved.

The process of simultaneously optimizing multiple factors is commonly described as multiparameter optimization (MPO) or equivalently multidimensional optimization (MDO) or multiobjective optimization (MOOP). For simplicity we will use the term "MPO" to describe all of the approaches discussed in this chapter. Fortunately, this problem has previously been studied in many fields, including engineering, economics, and quality management; in fact, some approaches date back as far as the early 1900s. Many of the solutions developed in these domains may be adapted for the specific environment of drug discovery, which differs from domains such as engineering most notably in the level of uncertainty in the data, as discussed earlier.

In addition to the need to take account of the uncertainty in the data, other requirements should ideally be satisfied by a MPO method for drug discovery. It is important that the method is flexible, so that the property criteria can be adjusted to reflect the specific objectives of each project; for example, the properties of a good orally available drug for a chronic central nervous system (CNS) indication will be very different from those of an inhaled drug for an acute respiratory indication. Furthermore, it should be possible to weight the individual property criteria because, as discussed, the property requirements for a good drug often conflict and it is necessary to define acceptable trade-offs in the search for the best possible balance. Finally, it is essential that the output from an MPO method is easily interpretable; "black boxes" that produce an output with no explanation are likely to be disregarded, and a good understanding of the implications of the results will help a scientist to apply his or her expertise to overcoming potential issues and designing improved compounds.

In this chapter, we will explore a number of approaches to MPO and their application to drug discovery. In the next section we will discuss the starting point for any MPO process, namely how we can establish the property criteria we wish to optimize for a given therapeutic objective. The following section will describe the most common MPO methods used in drug discovery. Finally, we will present a brief case study before drawing some conclusions.

8.2 ESTABLISHING THE RULES

Before any MPO method may be applied, the property criteria that define the optimization goal must first be defined. This will be based on the target product profile for the project, for example, the intended route of administration, dosing regimen, biology of the targeted disease process, and potentially other factors such as the limitations of existing therapies that could be improved. The relationship between these ultimate therapeutic goals and the compound properties that may be measured or predicted in early discovery will inform the definition of success criteria for these properties. In the following sections, we will discuss several approaches to establishing these criteria.

8.2.1 Expert Opinion

The most common basis for determining the property criteria for a high-quality lead or candidate is the expertise of the scientists on the drug discovery project team or their peers in their organization. Years of personal or institutional experience can build up an understanding of the properties of successful compounds, and it is now common to define up front the criteria for an acceptable lead or candidate drug. An example of a typical "checklist" for a lead compound in a project with the goal of identifying an orally dosed drug is shown in Table 8.1.

Table 8.1 Example criteria for a lead compound intended for oral dosing

Property	Criterion
Pharmacology	
Potency against target (K_i)	<100 nM
Selectivity against related off-targets	>100 ×
Physicochemical	
LogP	<4
Solubility	>100 μM
MW	<450 Da
ADME	
Caco-2* permeability (P_{app})	$>10 \times 10^{-6}$ cm/s
Intrinsic Clearance in Human Liver Microsomes (Cl_{int})	<25 μL/min/mg protein
Absence of P-glycoprotein transport (Caco2 BA:AB)	<3
Safety	
Avoid Cytochrome P450-mediated drug-drug interactions (K_i for CYP3A4, CYP2C9, CYP2D6, CYP1A2, CYP2B6, CYP2C19, CYP2C8)	>1 μM
Avoid interaction with hERG potassium ion channel (IC_{50})	>10 μM
Cytotoxicity in HepG2† cells (LD_{50})	>1 mM

*Human epithelial colorectal adenocarcinoma cell line [12].
†Hepatocellular carcinoma cell line [13].

The input of experts who understand the biology and chemistry of a project is very important to defining appropriate criteria with which to identify high-quality compounds; however, some caveats should be taken into account. Personal opinion brings a high degree of subjectivity, colored by individual experience. Therefore, as a project progresses, it is important periodically to check that the chosen criteria for early data correlate with outcomes in downstream studies, such as *in vivo* PK, toxicity, or efficacy studies. This should include testing "negative" examples that do not meet all criteria, to ensure that potentially good compounds are not being inappropriately rejected.

Furthermore, given the long time scales of pharmaceutical R&D, it is difficult for any individual or even many institutions to accumulate a statistically significant sample of successful and unsuccessful projects to identify the criteria that best discriminate these outcomes. Shared knowledge from a statistical analysis of successful and unsuccessful projects, combined with an expert's understanding of the specific requirements of an individual project, is essential to achieving best practice.

8.2.2 "Drug-like" Properties

One source of commonly applied criteria is the analysis of successful drugs to identify simple compound characteristics that they have in common. The most well-known set of criteria of this type are Lipinski's Rule of Five (RoF) [10], which define limits on four easily calculated compound properties that are obeyed by most orally absorbed compounds:

- Molecular weight (MW) < 500
- Logarithm of the octanol:water partition coefficient (logP) < 5
- Number of hydrogen bond donors (HBDs) < 5
- Number of hydrogen bond acceptors (HBAs) < 10

Many other rules have also been defined for other compound properties. For example, publications by Veber et al. [14] and Lu et al. [15] derived criteria for oral bioavailability based on the number of rotatable bonds (ROTBs) and polar surface area (PSA). Johnson et al. identified rules linking the logarithm of the octanol:water partition coefficient at pH 7.4 (logD) and MW with membrane permeability and metabolic stability. In this case, instead of expressing these as simple criteria, the authors found it more effective to define a "golden triangle" in (MW,logD) space in which most permeable and stable compounds were found [16].

Other criteria have been defined for objectives such as "developability" or clinical success, based on the "flatness" of compounds, defined in terms of either the number of aromatic rings (AROMs) or the fraction of sp^3 carbons (Fsp^3) [17,18]. Furthermore, Hughes et al. [11] noted an increased likelihood of *in vivo* toxicity for compounds with logP > 3 and PSA < 75 Å.

These simple criteria have the advantage that they are easy to understand and apply. Compounds with "nondrug-like" properties lie in regions with poor precedence of finding successful drugs, which may indicate an additional risk, and therefore, these criteria help to avoid potential pitfalls. However, some important caveats should be noted: The simple characteristics on which these criteria are based are only weakly predictive of the biological properties of compounds; hence, even "drug-like" compounds have only a very small chance of ultimate success, and over-rigorous application of these rules may miss valuable compounds. This risk is discussed in more detail in the later section on filtering, and some approaches to address this are described in the desirability functions section.

It is also important to recognize that many of the sets of criteria are derived for one specific objective, most commonly oral availability, and should only be applied in this context. Compounds intended for other routes of administration may obey very different rules [19]. Unfortunately, criteria such as the RoF are sometimes applied as general definitions of "drug-likeness," which may lead to incorrect rejection of compounds intended for nonoral applications. It is important to choose appropriate criteria for the objective of the project, and where sufficient historical data are available on compounds intended for a similar objective, a statistical analysis could yield tailored criteria.

Furthermore, rules for "drug-like" properties are, by definition, derived from the analysis of successful drugs for precedented therapeutic targets. As new target classes are explored, such as protein–protein interactions or epigenetic targets, it may be necessary to find compounds that break the existing rules, accepting that this will increase the risk of finding poor compound properties in addition to the risk associated with an unprecedented biological target.

8.2.3 Bump Hunting

Manual analysis of high-dimensional, historical data to find multiparameter criteria with which to identify successful compounds is very difficult. One computational approach to automate this process, described as "bump hunting," has been developed by scientists in the field of machine learning. Bump hunting methods search for regions of a multidimensional parameter space with a greater probability of success [20]. Many machine learning methods result in "black box" models (a black box model will predict an outcome with some degree of accuracy, but it does not provide any insight into the underlying criteria for this prediction), but one advantage of bump hunting is that the output is a set of interpretable criteria, expressed as desirable property ranges. This allows an expert to understand the output and, if necessary, adjust the criteria based on his or her detailed knowledge. This achieves a good balance between the power of a computer to analyze complex data efficiently and the input of an expert scientist.

Bump hunting is not limited to the analysis of simple compound characteristics such as those described earlier in the "Drug-like Properties" section. It can be applied to any predicted or experimental data that are available on compounds for which the outcome is known. This highlights another advantage of bump hunting, in that the importance of each criterion can also be calculated, which indicates the properties that provide the greatest value in distinguishing successful and unsuccessful outcomes.

Therefore, experimental efforts can be focused on generating the most important data and saves effort wasted on measuring data that add little value to the decision-making process. Furthermore, the relative importance of the criteria informs the acceptable trade-offs when an ideal compound cannot be found.

The application of bump hunting to drug discovery data is still in its early stages [21], and there are particular challenges resulting from the sparseness and uncertainty in the available data compared with other fields in which it has previously been applied.

8.2.4 Pareto Optimization

When the appropriate selection criteria are not known *a priori*, either based on experience or on statistical analysis of historical data, Pareto optimization provides an approach to explore different "optimal" combinations of properties. Originally published in 1906 by Vilfredo Pareto [22], an Italian economist, this approach defines an optimal solution (in our case, a compound) as one for which no other solution is better in all properties. Therefore, Pareto optimization does not provide a single optimal solution, but a family of solutions that each represent different, optimal balances of properties, described as "nondominated" solutions.

Figure 8.1 illustrates a simple two-dimensional case, exploring the balance between potency and metabolic stability, with the ideal compound exhibiting both high potency and good metabolic stability. However, this is not always possible and it may not be clear whether it is more important to optimize potency or metabolic stability to achieve sufficient efficacy. Exploring the Pareto optimal compounds and generating downstream data, e.g., *in vivo* efficacy, will identify the most appropriate balance of these two factors and help to guide the optimization of further compounds.

The concept of Pareto optimality may be generalized to rank solutions based on the number of compounds by which they are dominated (i.e., that are better in all properties). Nondominated solutions are given a rank of 0, compounds dominated by a single compound a rank of 1, etc.

A limitation of Pareto optimization is that the method does not distinguish between compounds with the same rank and the number of Pareto optimal solutions increases exponentially with the number of properties considered simultaneously. This means that for a larger number of properties, e.g., above approximately 4, the number of rank 0 compounds often becomes intractable. One approach to this is to combine Pareto optimization with another MPO method to reduce the dimensionality of the search space. For example, several related properties may be combined in a single score reflecting a single factor, e.g., appropriate ADME properties, and this score is balanced with another factor such as potency.

A further caveat to the use of Pareto optimization is that it does not explicitly consider the uncertainty in the property values. As a result of this uncertainty, the number of compounds with a significant probability of being nondominated is often very high and apparently optimal compounds have a significant chance of being dominated. Therefore, given these limitations, Pareto optimization is probably most

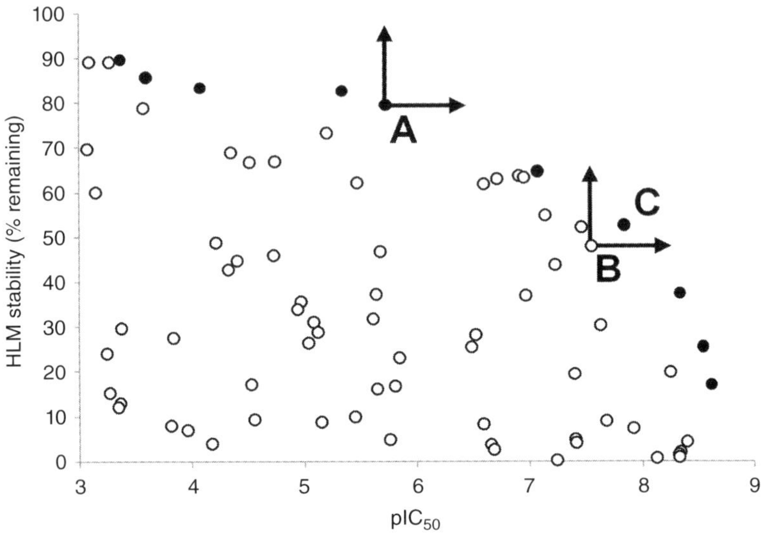

Figure 8.1 An illustrative example of Pareto optimization, exploring the trade-off between potency (pIC_{50}) and metabolic stability measured as percent of parent compound remaining after a 40-minute incubation with human liver microsomes (HLMs % remaining). An ideal goal, indicated by the star, is a compound with both high potency and metabolic stability. The compounds tested so far are shown by the points on the graph of pIC_{50} against HLM % remaining. The solid circles, such as the point labeled A, represent Pareto optimal compounds; i.e. there are no compounds that are better in both properties. The open circles represent nonoptimal compounds. For example, the compound represented by the point labeled B is "dominated" by the compound labeled C.

appropriate as a method for exploration of the optimization space rather than for rigorous prioritization of compounds.

8.3 APPROACHES TO MPO

Having determined the target criteria for optimization, the next step is to apply an appropriate methodology to assess new compounds for their potential to be successful against this profile. In the following sections, we describe some methods that are regularly used to apply MPO within drug discovery.

8.3.1 Filtering

The process of filtering compounds is the simplest way to apply MPO to optimize a set of compounds against multiple requirements. Often this approach is used when trying to reduce the number of compounds below a predetermined limit, for example, to meet a practical limit on the available capacity for screening or synthesis. However, filtering can also be used independent of this goal, with the aim of finding compounds that meet all the specified criteria. A typical approach would be to apply the chosen criterion for each property in sequence as a means of discarding compounds that fail to meet the criterion. Often this will be carried out by starting with the most important parameter, with the following filters for each parameter applied in order of the parameters' relative importance, until all the filters have been applied or an appropriate size of compound set has been achieved.

Although this process is simple, it is important to give a word of caution about some of the risks associated with filtering compounds in this way. First, applying criteria in a "black-and-white" fashion could easily result in compounds being discarded that are negligibly different than those considered acceptable. As an example, consider two compounds that have very similar property profiles, one of which has a molecular weight of 499 and the other 501. Using a filter that only selects compounds with a molecular weight of less than 500, only the lower molecular weight compound would be retained. However, it is unrealistic to believe that there is any significant difference between these compounds in terms of their ultimate potential for success.

Perhaps more importantly, it is worth also recalling that the data used within drug discovery are typically associated with significant statistical or experimental error. Often these errors are not taken into consideration when filtering compounds, and as a result, the filtering system can produce poor results. Applying filters sequentially will accumulate the errors in the individual filters, resulting in a filtering system that, as a whole, may be more likely to discard "good" compounds or, worse, select "poor" ones. For example, consider a process that applies filters sequentially using criteria for 10 parameters. If each filter is 90% accurate (which may be optimistic for many predictive models or data generated in high-throughput assays), by the time all 10 filters have been applied, the system as a whole has an accuracy of 35%. This means that the filtering process is more likely to discard "good" compounds than to retain them.

One common variation on filtering is a "traffic light" or "heat map" visualization, in which a spreadsheet of data is displayed on which the data values are colored green for property values that pass the corresponding criterion, red for those that fail, and yellow for those that are "close." These views can be very helpful in getting an overview of a dataset and in spotting the rare occasions when a compound's properties are all "green." However, in the more common case, when a perfect compound cannot be identified, it is difficult to use these views to prioritize compounds confidently. For example, is a compound with two "red" properties better or worse than a compound with one "red" and three "yellow" properties? The answer will depend on the relative importance of each property criterion to the project's objective and the confidence with which we can assign the property values to red, yellow, and green.

Of course, the advantages of filtering are that it is very easy to apply and to interpret the results. However, this may be more than offset by chance of missing valuable opportunities or incorrectly progressing poor compounds leading to wasted effort and late-stage failures.

8.3.2 Ligand Efficiency Metrics

Another simple approach to applying MPO within drug discovery is through the use of calculated metrics that combine multiple parameters into a single parameter for optimization. Well-known examples of this are ligand efficiency (LE) [23] and ligand lipophilicity efficiency (LLE) [24]:

$$\mathrm{LE} = \frac{RT \times p\mathrm{IC}_{50}}{N_H} = \frac{1.4 \times p\mathrm{IC}_{50}}{N_H} \tag{8.1}$$

$$\mathrm{LLE} = p\mathrm{IC}_{50} - \log P \tag{8.2}$$

where R is the gas constant, T is the absolute temperature, $p\mathrm{IC}_{50}$ is the negative log of the molar concentration at half of the maximum inhibition, and N_H is the number of nonhydrogen atoms in the molecule.

Low LE is associated with an increased risk of physicochemical and ADME properties (for equipotent compounds), whereas low LLE has also been shown to increase the risk of safety issues [25]. Motivated by these two metrics, additional, more complicated metrics have also been proposed; for a review of these, see Reference 26.

The key advantage to working with ligand efficiency metrics is that their results are easy to interpret because there is only one value to monitor to guide optimization; they help to optimize potency while maintaining attention on well-known risk factors such as large compound size or high lipophilicity. However, as discussed in relation to drug-like properties, the simple characteristics used to calculate these efficiency metrics, such as number of heavy atoms or logP, have only a weak correlation with the biological properties of a compound, and therefore, the results should not be overinterpreted. In addition, the uncertainty in the calculated efficiency metric will be a combination of the uncertainties in the properties used in the calculation, e.g., potency or logP, and hence, a small change in value of a calculated metric is unlikely to be statistically significant. Therefore, these models give an indication toward optimization of potency without sacrificing good "drug-like" properties, but hard cut-offs based on efficiency metrics should not be applied too rigidly. Finally, as with any methods based on achieving good drug-like properties, the objective depends on the ultimate therapeutic goal, and a metric derived for one goal, e.g., an orally absorbed compound, may not be applicable in other scenarios.

8.3.3 Desirability Indices

Desirability indices are a more general approach for applying MPO that can be used to give an indication as to the overall "quality" of a compound that is tailored to a specific objective [27]. A desirability index is made up of a combination of individual "desirability functions" for each property that link the value of a compound property with the desirability of that outcome. A desirability function may be as simple as a threshold or filter, whereby a compound would score 1 (the ideal value) if its property meets the criterion or 0 (the worst possible outcome) if the property fails to meet the criterion. However, a more realistic approach is to tailor each desirability function to reflect the relative importance of the parameter. Used in this way, an ideal property value would still score 1, but property values that do not meet the criterion would score a value between 0 and 1, depending how critical the property

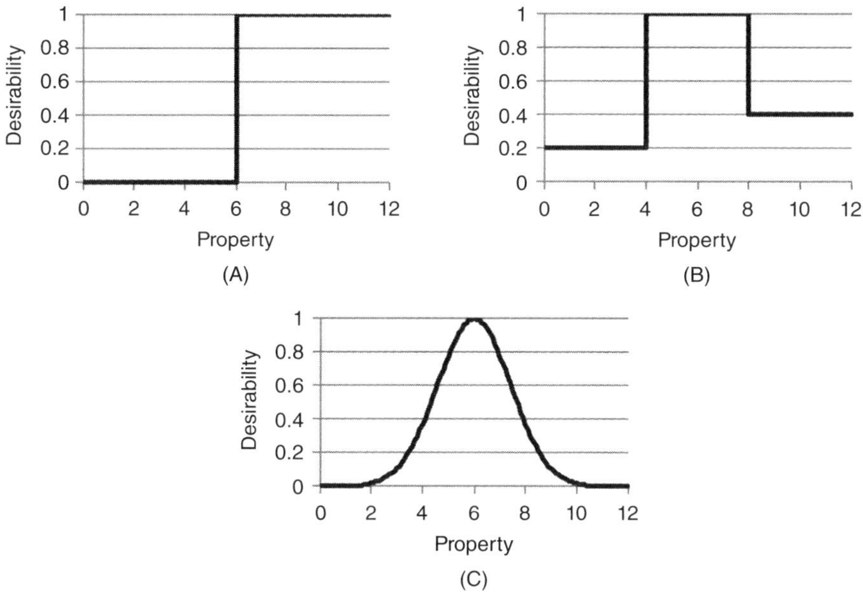

Figure 8.2 Three examples of desirability functions: (A) represents a simple filter; compounds with a property value greater than 6 will be accepted, and those with a value less than 6 will be rejected. (B) represents an ideal property range of 4 to 8; compounds with a value below this range are less desirable than those with a property value above this range. (C) shows an example of a nonlinear desirability function, in this case, a Gaussian distribution with an ideal property value of 6 and a standard deviation of 2.

is to the overall multiparameter objective. The more critical a property, the lower the score for failing to meet the criterion. In this way, compounds will be penalized if they have suboptimal parameter values, but the penalties will reflect the effect of this "failure" on the overall project goals. Furthermore, desirability functions are not limited to simple thresholds and may encode more subtle trade-offs, including desirable ranges, trends in desirability, or even nonlinear relationships. Figure 8.2 shows some examples of desirability functions.

A desirability index, reflecting the overall quality of a compound, may then be calculated by combining the values of the desirability functions of the individual properties. The most common approaches for calculation of a desirability index are additive (Equations 8.3 and 8.4):

$$D = \frac{d_1(Y_1) + d_2(Y_2) + \cdots + d_n(Y_n)}{n} \tag{8.3}$$

and multiplicative:

$$D = \sqrt[n]{d_1(Y_1) \times d_2(Y_2) \times \ldots d_n(Y_n)} \tag{8.4}$$

where d_i is the desirability function for property i, Y_i is the value of property i, and n is the total number of properties.

An interesting example of the use of a desirability index to give an indication of the suitability of a compound as an oral drug for a target in the CNS was published by Wager et al. [28]. In this paper, the authors described a function that combines the desirabilities of six compound properties: calculated logP (clogP), topological PSA (TPSA), logD, HBD, MW, and the acid dissociation constant of the most basic site (pK_a). For each property, a function was constructed that indicated the desired range of that property. Figure 8.3 shows the desirability functions that were used for each property. The overall desirability of a compound, referred to as the CNS MPO score, was then calculated as the sum of the desirabilities for each of that compound's

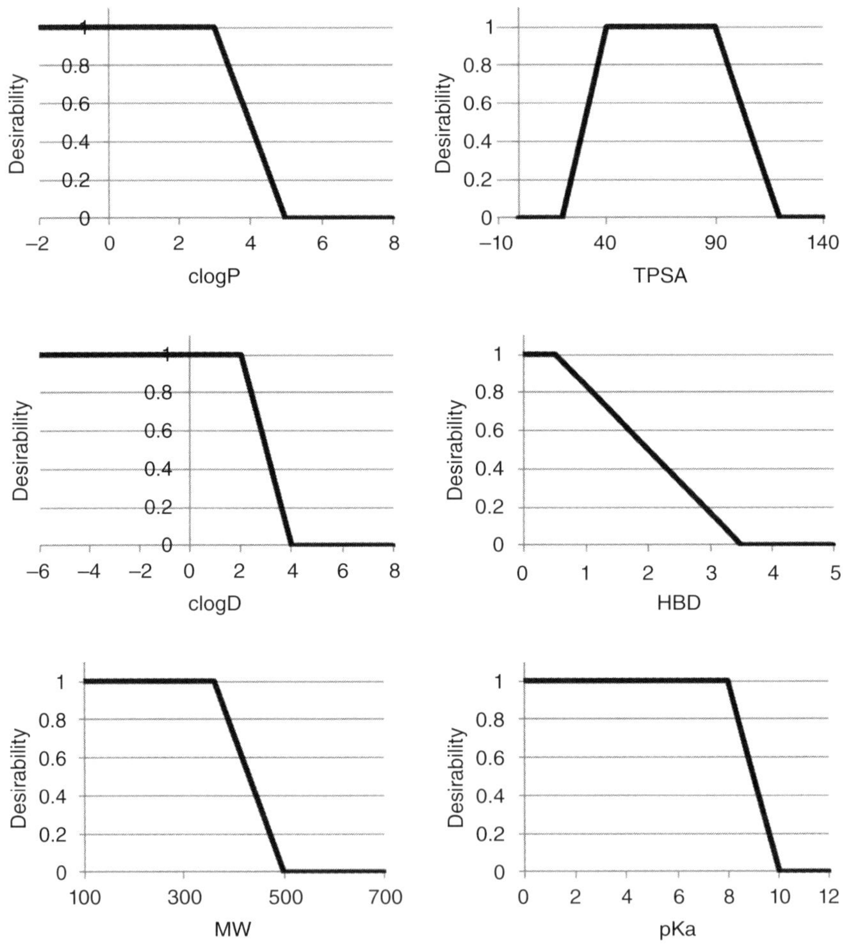

Figure 8.3 The desirability functions defined in Wager et al. [28] for selection of compounds intended for a CNS target on the basis of six properties, as described in the text.

properties. As each property contributed a maximum value of 1 to the total, the maximum possible value for the CNS MPO score was 6. Wager et al. showed that 74% of marketed CNS drugs achieved a CNS MPO index > 4 versus 60% of Pfizer (New York, NY) candidate compounds. They also observed correlations between a high CNS MPO index and good *in vitro* ADME properties such as MDCK P_{app}, human liver microsome stability and P-glycoprotein transport. At the same time, it was noted that in general, as compounds' CNS MPO scores increased, incidences of cytotoxicity and human ether a go-go related gene (hERG) inhibition decreased, enabling the index to be used for assisting in the reduction of safety risks. Importantly, although several properties have been combined to produce this index, it is still easy to interpret because of the simplicity of the functions and the parameters involved.

Another example use of a desirability index, to give an assessment of a compound's similarity to marketed, oral drugs, is the quantitative estimate of drug-likeness (QED) published by Bickerton et al. [29]. In this method, eight compound characteristics—MW, logP, HBD, HBA, PSA, ROTB, AROM, and counts of alerts for undesirable substructures (ALERT)—are combined. For each property, a function was fitted to the distribution of the values of the property for marketed oral drugs, working on the premise that a compound with similar properties to most of these drugs would have a greater probability of success (see Figure 8.4 for an example).

By combining the desirabilities using a multiplicative approach, with weights given to each property, the QED can then be calculated as follows (Equation 8.5):

$$\mathrm{QED}_w = \exp\left(\frac{\sum_{i=\mathrm{MW,LOGP,HBA,HBD,PSA,ROTB,AROM,ALERT}} w_i \ln d_i}{\sum_{i=\mathrm{MW,LOGP,HBA,HBD,PSA,ROTB,AROM,ALERT}} w_i}\right) \quad (8.5)$$

where d_i is the desirability function and w_i is the weighting applied to that property.

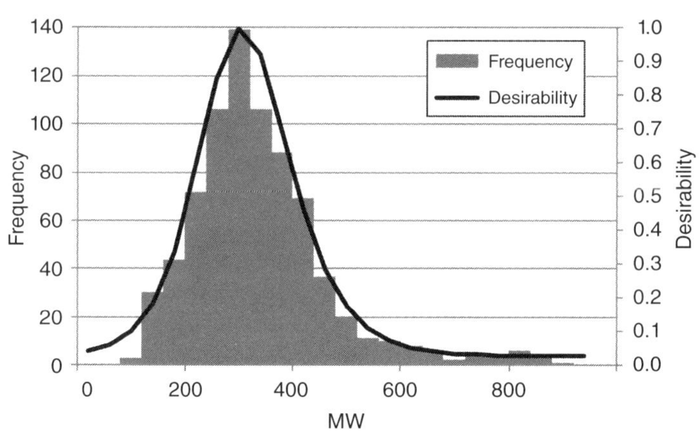

Figure 8.4 Frequency plot of molecular weight (MW) values for a set of 771 oral drugs. This was used by Bickerton et al. [29] to fit a desirability function, shown as a solid line, for MW values as a contribution to the QED desirability index.

One limitation of this approach is that a compound with a value of a property that is similar to known drugs does not necessarily have a better chance of being a drug. This is only the case if the probability distribution of that property for drugs is significantly different from the distribution of all compounds that might be considered. If the distribution of a property is the same for drugs as nondrugs, the property provides no information about the difference between drugs and nondrugs. With this in mind, an extension to this approach was proposed by Yusof et al. [30] in which a metric called the relative drug-likelihood (RDL) is calculated by applying the Bayes' theorem to estimate the relative probability of a compound being a drug, given the value of a property (Equation 8.6):

$$\frac{P(\text{Drug}|X)}{P(\text{not Drug}|X)} = \frac{P(X|\text{Drug})}{P(X|\text{not Drug})} \times \frac{P(\text{Drug})}{P(\text{not Drug})} \tag{8.6}$$

where X is the property of interest, $P(\text{Drug}|X)$ is the probability of a compound being a drug given the value of a property X, $P(X|\text{Drug})$ is the probability of a compound having a value of a property X given that it is a drug, and $P(\text{Drug})$ is the probability of a compound being a drug given no information about its properties (known as the prior probability). The terms related to "not drugs" are probabilities similarly defined for nondrug compounds.

Within this equation, $P(\text{Drug})$ and $P(\text{not Drug})$ are unknown constants. However, although the ratio of these values is likely to be very small, it is also constant, and therefore, the value does not need to be known to calculate a value that is proportional to the relative likelihood of a compound becoming a drug. Therefore, the desirability of a value x of property X can be defined as:

$$d(x) = \frac{P(X = x|\text{Drug})}{P(X = x|\text{not Drug})} \tag{8.7}$$

where $P(X = x|\text{Drug})$ is the probability of a drug having value x of property X.

The desirabilities of the individual parameters can then be combined to give the overall relative drug likelihood (see Equation 8.8):

$$\text{RDL} = \exp\left(\frac{1}{n}\sum_{i=1}^{n} \ln(d_i(x_i))\right) \tag{8.8}$$

As an example, Yusof et al. constructed an RDL metric using a set of known oral drugs and a sample from the ChEMBL database (https://www.ebi.ac.uk/chembl/) as a representative set of typical "nondrug" medicinal chemistry compounds. Figure 8.5 shows an example desirability function for MW, relative to the distributions of the known drugs and the representative set. The resulting RDL was shown to enrich significantly the selection of known drugs from an independent set of "med chem" compounds in ChEMBL.

The three examples of desirability indices described earlier demonstrate how this approach can be used to enrich a selection of compounds. However, it is also important to remember that their utility is dependent on the question being asked. If the objective

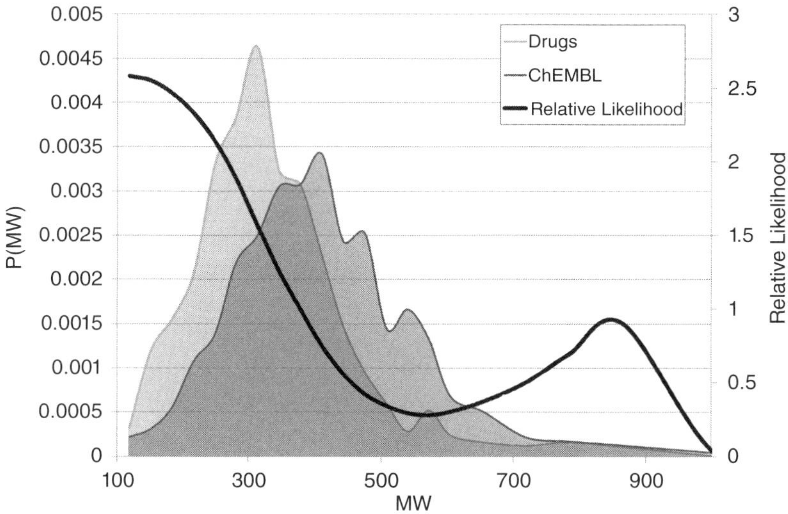

Figure 8.5 Graph illustrating the basis for calculation of the relative drug likelihood for MW, shown as a solid black line, derived by comparing the distribution of property values of a set of 771 oral drugs (blue) with the distribution for >650k compounds from the ChEMBL database (red).

is not consistent with that for which the methods were conditioned—in these cases, the selection of small oral or CNS drugs from other compounds explored in drug discovery—-then they are less likely to be able to distinguish "good" from "bad" compounds in that context. However, the flexibility of the desirability index approach means that it can be easily adapted to create similar indices for different objectives or therapeutic classes (e.g., antibiotics, nonoral drugs, etc.). The straightforward way in which the overall score for a compound is built up from contributions resulting from individual properties means that it is easy for a scientist or project team to construct manually a desirability metric. Furthermore, the results are easily interpretable to understand the critical issues that impact the quality of a compound.

8.3.4 Probabilistic Scoring

One key consideration that has not been addressed by the methods described earlier is the uncertainty in the data that are typically available in drug discovery. Nearly all data generated and used as the basis for decision making during drug discovery will have associated uncertainties; there are statistical errors in all *in silico* predictions, and *in vitro* and *in vivo* experiments are typically subject to experimental variability that will vary depending on the experimental protocol. The probabilistic scoring approach [31] builds on desirability indices to include explicitly these sources of uncertainty and allow the confidence with which compounds may be distinguished to be considered.

Probabilistic scoring combines individual desirability functions, as in a desirability index approach, to define the properties of an "ideal" compound and the acceptable trade-offs, sometimes referred to as a scoring profile. The method then calculates the likelihood of a compound achieving the required profile (a number between 0 and 1) given the uncertainty in the data and the relative importance of the individual properties. The uncertainty in the overall score resulting from the uncertainty in the underlying data is also calculated (Figure 8.6 shows some example scores and associated error bars for a compound series scored against the profile shown). By using this approach, data with varying levels of uncertainty can be combined without the risk of a highly uncertain value inappropriately biasing the result. The explicit handling of uncertainty enables this approach to highlight when scores for individual compounds are significantly different and when they are not, which is important for making decisions about which compounds may be progressed or discarded with confidence and for avoiding missed opportunities.

The probabilistic scoring approach also benefits from the ease of interpretation of desirability indices because the contribution of each property to the overall score can be identified, as illustrated by the histogram inset in Figure 8.6. This shows the properties with the greatest impact on the likelihood of success of a compound and guides optimization to address the most critical issues first. This analysis takes into account not only the property values but also the relative importance of the criteria and the confidence in the data; the most critical issue is one where a property value fails to meet an important criterion with high confidence.

8.4 CASE STUDY: COMPOUND PRIORITIZATION FOR A BALANCE OF PROPERTIES

The following case study highlights how simultaneous consideration of multiple properties and their uncertainties can result in better outcomes than rigid filtering based on individual properties.

The case study relates to an early-stage lead optimization project in which the project team was pursuing an orally dosed therapy. The project team had generated *in vitro* data for a set of 75 compounds, including potency, selectivity, solubility, and microsomal stability in human and rat liver microsomes. Using a traditional filtering approach, the project team had filtered compounds on the basis of selectivity (>eightfold) and potency (<1 μm), resulting in only 17 compounds that met both criteria. Unfortunately, when tested *in vivo,* these compounds had no measurable oral activity and poor efficacy in animal tests after intraperitoneal dosing. These results suggested issues caused by poor solubility and metabolic instability, leading to poor bioavailability.

To compare the outcome of filtering on potency and selectivity with a true MPO approach, Optibrium's StarDrop software (www.optibrium.com/stardrop) was first used to generate probabilistic scores for the compounds on the basis of their selectivity and potency, taking into account the experimental error. The scoring profile used reflected the thresholds employed within the filtering approach and assigned a higher importance to selectivity, reflecting the fact that the data were first filtered on selectivity. This scoring profile is shown in Figure 8.7A along with the results

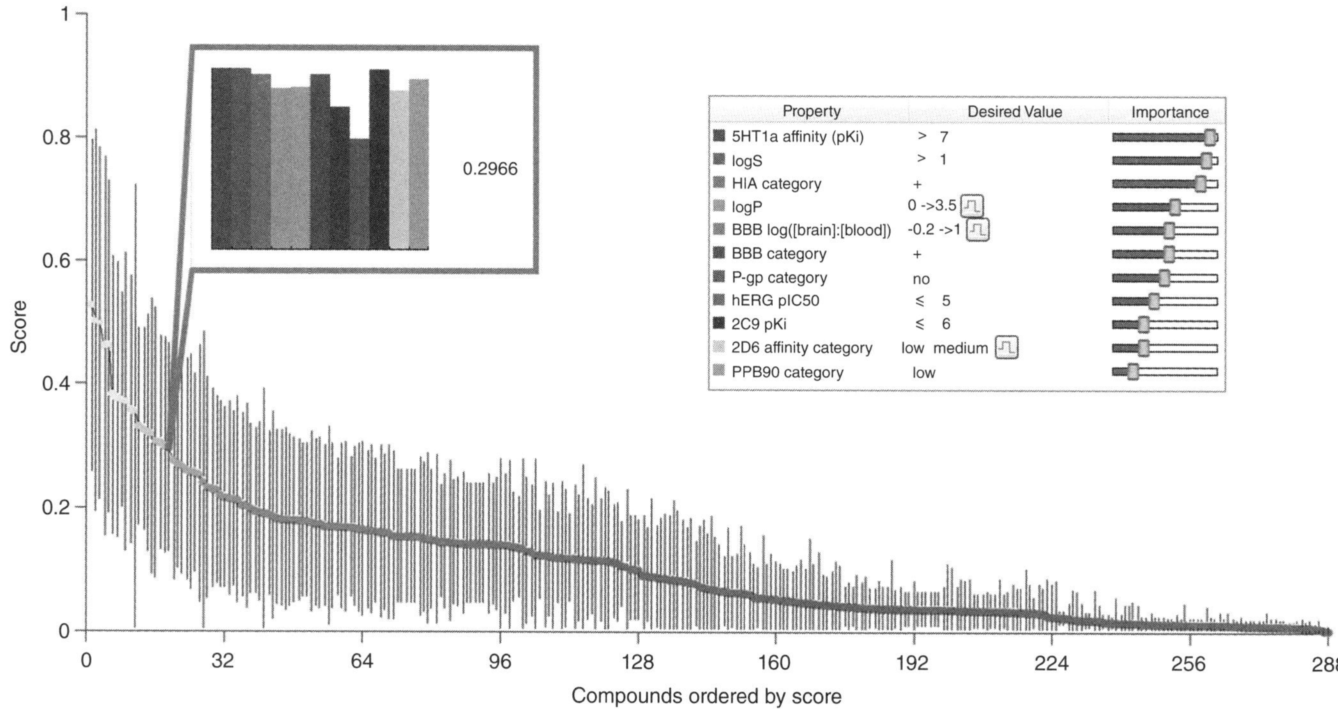

Figure 8.6 Example results from scoring 288 compounds using the probabilistic scoring method and a scoring profile designed to prioritize compounds with good potency against the $5HT_{1a}$ target and appropriate properties for an orally dosed compound with good brain penetration. The compounds are ordered by score along the *x*-axis, and the score is plotted on the *y*-axis. The error bars indicate the confidence in the score as a result of the uncertainty in the underlying data. Note that, given the uncertainty in the scores, there are a number of compounds that cannot be confidently distinguished from the "best" compound. The inset scoring profile shows the property criteria that were specified, along with their importance of each criterion to the overall success of the project. A histogram for an example compound shows the contribution of each property to the overall score for that compound, clearly identifying the most important issues that should be addressed to improve significantly the compound's chance of success. (*See insert for color representation of the figure.*)

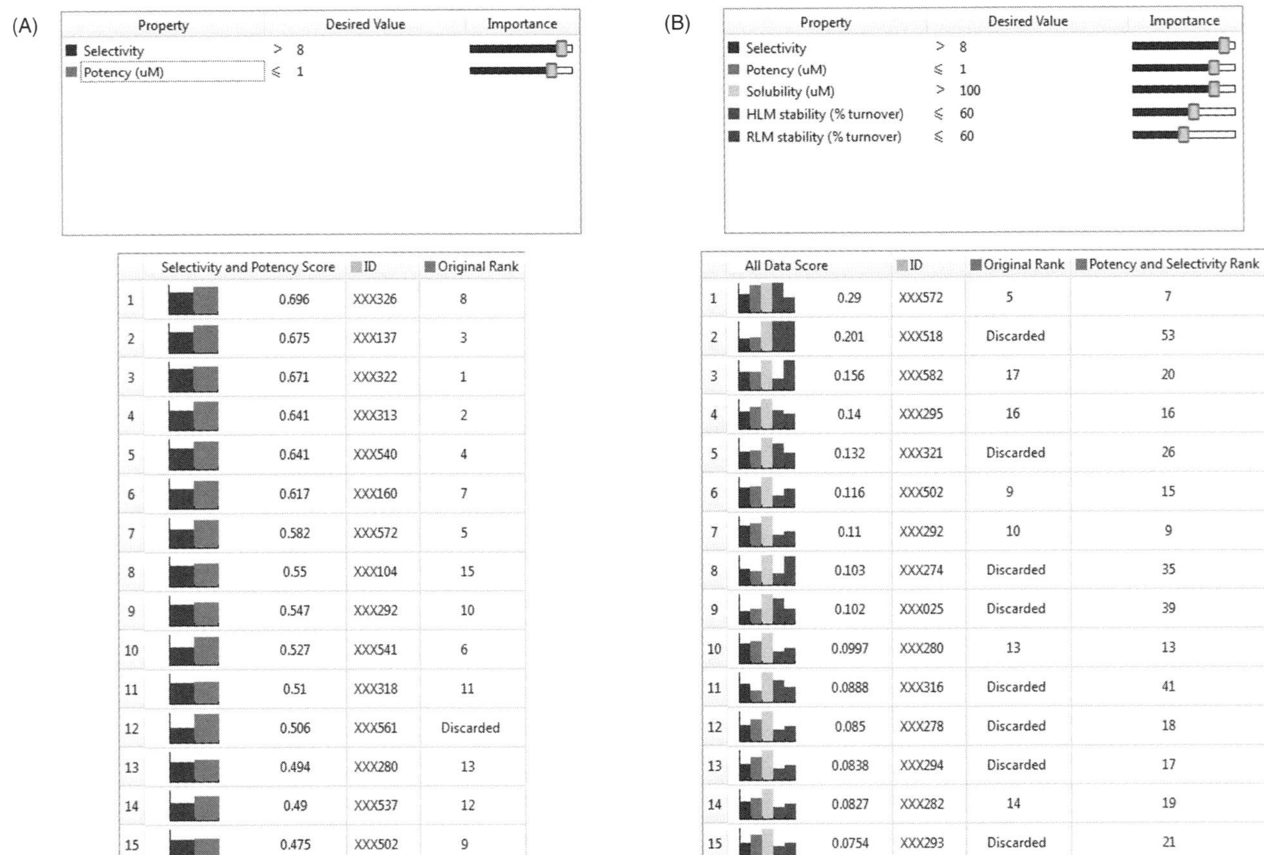

(A)

Property	Desired Value	Importance
Selectivity	> 8	
Potency (uM)	⩽ 1	

	Selectivity and Potency Score	ID	Original Rank
1	0.696	XXX326	8
2	0.675	XXX137	3
3	0.671	XXX322	1
4	0.641	XXX313	2
5	0.641	XXX540	4
6	0.617	XXX160	7
7	0.582	XXX572	5
8	0.55	XXX104	15
9	0.547	XXX292	10
10	0.527	XXX541	6
11	0.51	XXX318	11
12	0.506	XXX561	Discarded
13	0.494	XXX280	13
14	0.49	XXX537	12
15	0.475	XXX502	9

(B)

Property	Desired Value	Importance
Selectivity	> 8	
Potency (uM)	⩽ 1	
Solubility (uM)	> 100	
HLM stability (% turnover)	⩽ 60	
RLM stability (% turnover)	⩽ 60	

	All Data Score	ID	Original Rank	Potency and Selectivity Rank
1	0.29	XXX572	5	7
2	0.201	XXX518	Discarded	53
3	0.156	XXX582	17	20
4	0.14	XXX295	16	16
5	0.132	XXX321	Discarded	26
6	0.116	XXX502	9	15
7	0.11	XXX292	10	9
8	0.103	XXX274	Discarded	35
9	0.102	XXX025	Discarded	39
10	0.0997	XXX280	13	13
11	0.0888	XXX316	Discarded	41
12	0.085	XXX278	Discarded	18
13	0.0838	XXX294	Discarded	17
14	0.0827	XXX282	14	19
15	0.0754	XXX293	Discarded	21

Figure 8.7 Comparison of the results of probabilistic scoring applied to a set of 75 compounds. (A) shows the profile used to score the compounds based only on selectivity and potency, taking into account the uncertainty in the experimental data, and the resulting scores for the top 15 compounds. The original rank shows the result of filtering based on potency and selectivity using the same cut-offs. Note that some compounds that had previously been discarded by filtering have a greater chance of success than some of the originally selected compounds. (B) shows a scoring profile that also takes into account the available *in vitro* ADME data in addition to the selectivity and potency and the results of applying this profile to the same 75 compounds. The original ranking and those achieved when scoring based only on potency and selectivity alone are shown for comparison. Note that most of the highest

of this process compared with that of filtering. From this, it can be observed that, although the top 10 compounds remain largely unchanged, there are some notable differences. For example, compound XXX561, which had previously been discarded because of its sevenfold potency, appeared as the 12th ranked compound. This is because it was the most potent compound measured to date, and the high uncertainty in its selectivity measurement means that it cannot be confidently rejected based on a cut-off of eightfold. Therefore, in terms of the balance between these properties, this compound is worthy of consideration.

However, the *in vivo* issues that were encountered by the project were thought to be a result of poor solubility and stability. Therefore, an additional scoring profile was created that included these parameters in addition to potency and selectivity (see Figure 8.7B). The solubility and stability criteria were assigned lower importance than selectivity and potency, but they were still significant. The results of scoring the data using this new scoring profile can also be observed in Figure 8.7B. From this, it is notable that most of the top scoring compounds had previously been discarded by the original filtering process. For example, compound XXX518 had originally been discarded and, when considering only potency and selectivity, was ranked outside the top 50 as a result of selectivity (fivefold) and potency (1.67 μm) values that did not meet the specified criteria. However, given the uncertainty in these values, the very good solubility and stability of this compound meant that it was one of the more likely compounds to have an overall balance of the required properties. *In vivo* testing showed that XXX518 had a superior pharmacokinetic profile and oral activity. Furthermore, it was the only representative of a chemical series that had previously be discarded by the filtering process. Therefore, the MPO approach opened a new route forward for the project team, which could further optimize the potency and selectivity of this chemotype with the knowledge that acceptable ADME/PK properties were accessible.

The results from this case study highlight the importance of taking uncertainty into consideration to avoid missed opportunities and the importance of simultaneously considering all relevant parameters. By initially focusing only on a subset of the properties, the project was left trying to develop compounds that did not have the balance of properties necessary to succeed. With the data that were available, this would have been avoided if all the parameters had been used within the decision-making process from the beginning.

8.5 CONCLUSION

It is widely recognized that a successful, safe, and efficacious drug must bring together many biological and physicochemical properties. Achieving this delicate balance is a major challenge, but if a project can focus on appropriate chemistry early in the process, downstream risks can be minimized and a high-quality lead or candidate compound can be identified more quickly. In this chapter, we have compared and contrasted several approaches for determining the most appropriate balance of properties and then for applying this proactively to the design and selection of new compounds with an improved chance of success. These MPO approaches are being increasingly adopted in drug discovery to integrate the complex *in silico* and *in vitro* data that are available throughout the drug discovery process.

Drug discovery is necessarily a collaborative endeavor, bringing together scientists from many disciplines such as chemistry, biology, ADME/PK, and toxicology. These backgrounds will bring different perspectives on the available data, colored by the individuals' experiences, and it is tempting for an individual to focus on the data with which he or she has the greatest familiarity, perhaps giving insufficient weight to other factors. MPO methods have the added benefit of providing a framework for collaborative discussions between different disciplines and then an objective way to apply the consensus opinion regarding the profile of properties required to identify the most appropriate compounds.

However, to achieve the maximum impact, it is important that these methods are easily accessible and interpretable to scientists from all backgrounds. This facilitates interactive exploration, allowing new ideas to be proposed with rapid feedback to stimulate the search for optimization strategies. If new ideas must be subjected to analysis by computational specialists and the results must be interpreted and reported back, the resulting delays mean that the results are less likely to be taken into account and reduce the chance to learn from earlier ideas to develop better strategies. Therefore, software to support MPO should be user friendly and intuitive to users from many disciplines.

The ultimate goal of MPO is to maximize the impact of *in silico* and *in vitro* technologies on the decision-making process in drug discovery to improve timelines, reduce wasted effort and missed opportunities, and thereby contribute to improvements in productivity.

REFERENCES

1. Paul, S.M., Mytelka, D.S., Dunwiddie, D.T., Persinger, C.C., Munos, B.H., Lindborg, S.R., Schacht, A.L. (2010). How to improve R&D productivity: The pharmaceutical industry's grand challenge. *Nature Reviews Drug Discovery*, *9*, 203–214.
2. Kennedy, T. (1997). Managing the drug discovery/development interface. *Drug Discovery Today*, *2*, 436–444.
3. Van de Waterbeemd, H., Gifford, E. (2003). ADMET in silico modelling: Towards prediction paradise? *Nature Reviews Drug Discovery*, *2*, 192–204.
4. Ekins, S., Boulanger, B., Swaan, P.W., Hupcey, M.A. (2001). Towards a new age of virtual ADME/TOX and multidimensional drug discovery. *Journal of Computer-Aided Molecular Design*, *16*, 381–401.
5. White, R.E. (2000). High-throughput screening in drug metabolism and pharmacokinetic support of drug discovery. *Annual Review of Pharmacology and Toxicology*, *40*, 133–157.
6. Roberts, S.A. (2001). High-throughput screening approaches for investigating drug metabolism and pharmacokinetics. *Xenobiotica*, *31*, 557–589.

7. Kola, I., Landis, J. (2004). Can the pharmaceutical industry reduce attrition rates. *Nature Reviews Drug Discovery*, *3*, 711–716.
8. Morgan, S., Grootendorst, P., Lexchin, J., Cunningham, C., Greyson, D. (2011). The cost of drug development: A systematic review. *Health Policy*, *100*, 4–17.
9. Chadwick, A.T., Segall, M.D. (2010). Overcoming psychological barriers to good discovery decisions. *Drug Discovery Today*, *15*, 561–569.
10. Lipinski, C.A., Lombardo, F., Dominy, B.W., Feeney, P.J. (1997). Experimental and computational approaches to estimate solubility and permeability in drug discovery and development settings. *Advanced Drug Delivery Review*, *23*, 3–25.
11. Hughes, J.D., Blagg, J., Price, D.A., et al. (2008). Physicochemical drug properties associated with in vivo toxicological outcomes. *Bioorganic & Medicinal Chemistry Letters*, *18*, 4872–4875.
12. Hidalgo, I.J., Raub, T.J., Borchardt, R.T. (1989). Characterization of the human colon carcinoma cell line (Caco-2) as a model system for intestinal epithelial permeability. *Gastroenterology*, *96*, 736–749.
13. Schoonen, W.G., de Roos, J.A., Westerink, W.M., Débiton, E. (2005). Cytotoxic effects of 110 reference compounds on HepG2 cells and for 60 compounds on HeLa, ECC-1 and CHO cells. II mechanistic assays on NAD(P)H, ATP and DNA contents. *Toxicology In Vitro*, *19*, 491–503.
14. Veber, D.F., Johnson, S.R., Cheng, H.Y., Smith, B.R., Ward, K.W., Kopple, K.D. (2002). Molecular properties that influence the oral bioavailability of drug candidates. *Journal of Medicinal Chemistry*, *45*, 2615–2623.
15. Lu, J.J., Crimin, K., Goodwin, J.T., et al. (2004). Influence of molecular flexibility and polar surface area metrics on oral bioavailability in the rat. *Journal of Medicinal Chemistry*, *47*, 6104–6107.
16. Johnson, T.W., Dress, K.R., and Edwards, M. (2009). Using the Golden Triangle to optimize clearance and oral absorption. *Bioorganic & Medicinal Chemistry Letters*, *19*, 5560–5564.
17. Lovering, F, Bikker, J., Humblet, C. (2009). Escape from flatland: Increasing saturation as an approach to improving clinical success. *Journal of Medicinal Chemistry*, *52*, 6752–6756.
18. Ritchie, T.J., Macdonald, S.J.F. (2009). The impact of aromatic ring count on compound developability—are too many aromatic rings a liability in drug design? *Drug Discovery Today*, *14*, 1011–1020.
19. Choy, Y.B., Prausnitz, M.R. (2011). The rule of five for non-oral routes of drug delivery: Ophthalmic, inhalation and transdermal. *Pharmaceutical Research*, *28*, 943–948.
20. Friedman, J.H., Fisher, N.I. (1999). Bump hunting in high-dimensional data. *Statistics and Computing*, *9*, 123–143.
21. Hashimoto, T., Segall, M.D. Finding drug discovery "rules of thumb" with bump hunting, in *American Chemical Society Fall National Meeting*, Boston, MA, 2010.
22. Jaffe, W. (1972). Pareto translated: A review article. *Journal of Economic Literature*, *10*, 1190–1201.
23. Hopkins, A.L., Groom, C.R., Alexander, A. (2004). Ligand efficiency: A useful metric for lead selection. *Drug Discovery Today*, *9*, 430–431.
24. Leeson, P.D., Springthorpe, B. (2007). The influence of drug-like concepts on decision-making in medicinal chemistry. *Nature Reviews Drug Discovery*, *6*, 881–890.

25. Edwards, M.P., Price, D. (2010). Role of physicochemical properties and ligand lipophilicity efficiency in addressing drug safety risks. *Annual Reports in Medicinal Chemistry*, *45*, 381–391.
26. Abad-Zapatero, C. (2007). Ligand efficiency indices for effective drug discovery. *Expert Opinion on Drug Discovery*, *2*, 469–488.
27. Harrington, E.C. (1965). The desirability function. *Industrial Quality Control*, *21*, 494–498.
28. Wager, T.T., Hour, X., Verhoest, P.R., Villalobos, A. (2010). Moving beyond rules: The development of a central nervous system multiparameter optimization (CNS MPO) approach to enable alignment of druglike properties. *ACS Chemical Neuroscience*, *1*, 435–449.
29. Bickerton, G.R., Paolini, G.V., Besnard, J., Muresan, S., Hopkins, A.L. (2012). Quantifying the chemical beauty of drugs. *Nature Chemistry*, *4*, 90–98.
30. Yusof, I., Segall, M.D. (2012). Beauty is relative: Going beyond "drug-like" properties to identify compounds with an improved chance of success. *Drug Discovery Today*, in press.
31. Segall, M.D., Champness, E., Obrezanova, O., Leeding, C. (2009). Beyond profiling: Using ADMET models to guide decisions. *Chemistry and Biodiversity*, *6*, 2144–2151.

9

PBPK: INTEGRATING *IN VITRO* AND *IN SILICO* DATA IN PHYSIOLOGICALLY BASED MODELS

Hannah M. Jones and Neil Parrott

9.1 WHAT IS PHYSIOLOGICALLY BASED PHARMACOKINETIC MODELING?

9.1.1 Modeling Drug Pharmacokinetics

To assist with interpretation of drug pharmacokinetics (PK) in animals and humans, modeling is often applied to drug concentration data obtained after intravenous or oral dosing. Usually the measurements consist of concentrations determined in plasma samples taken at different time points and a commonly used model takes the form of a sum of exponential functions where parameters are adjusted to fit the observed concentration versus time profiles. Such a model can be represented as Eq. 9.1:

$$C(t) = \sum_{i} C_i e^{-k_i t} \tag{9.1}$$

where $C(t)$ represents the concentrations at different times and C_i and k_i are model parameters. Interpretable and useful pharmacokinetic descriptors such as clearance or volume of distribution can then be obtained by transformation of the fitted model parameters [1]. Such models are useful because they allow a concise and standard

Predictive ADMET: Integrative Approaches in Drug Discovery and Development, First Edition.
Edited by Jianling Wang and Laszlo Urban.

representation of experimental results. For example, preclinical determinations of the clearance, volume, and half-life for different molecules after dosing in rats can be linked to physicochemical or structural properties of the molecule by medicinal chemists and can be used to guide optimization of pharmacokinetics for new compounds. Alternatively, in the clinical setting, the pharmacokinetic parameters determined for different individuals may be related to the subject characteristics or the models fitted to one dataset may be used to simulate the plasma concentrations expected under different doses or dosing regimens.

However, when using this type of empirical modeling, the ability to link pharmacokinetics to drug or physiological properties is limited because the fitted model is purely descriptive of the measured data and bears no relationship to the physiological system being studied. Furthermore, such a modeling approach ignores prior information available on both the drug and the physiology under study, restricting the possibility to forecast pharmacokinetics for a similar drug or to extrapolate the pharmacokinetics to different physiological conditions. For example, there would be no way with such modeling to predict pharmacokinetics in obese subjects given the pharmacokinetics in healthy volunteers.

9.1.2 PHYSIOLOGICALLY BASED MODELS

A distinguishing feature of physiologically based pharmacokinetic (PBPK) models is that modeling starts from a mathematical structure representing the physiology of interest. Compartments in the model correspond to separate organs or tissues and are connected by links that parallel the flow of body fluids (Figure 9.1). The

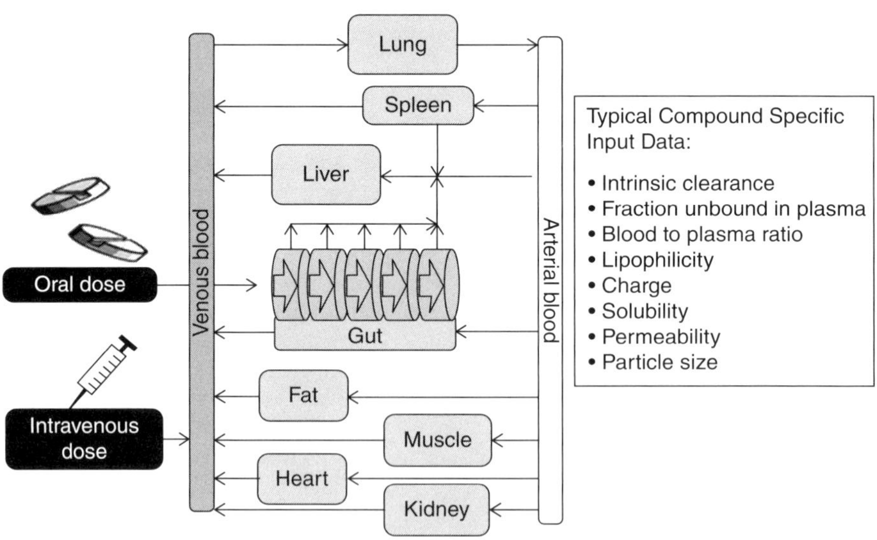

Figure 9.1 Schematic of a typical PBPK model and relevant input data.

model is highly parameterized with parameter values based on measured anatomical/physiological data. A whole-body PBPK model integrates different submodels describing the simultaneously occurring processes of absorption, distribution, metabolism, and excretion (ADME). In addition to the physiological data relevant for each process, each submodel requires numerous compound specific data that adapt the model to the drug being studied. Such a model allows for prediction of plasma and tissue concentration time profiles after intravenous or oral administration.

9.1.2.1 Absorption Submodel Oral absorption can be modeled by a series of linked compartments corresponding to the different segments of the gastrointestinal tract. The transit of the drug from the stomach through the small intestines to the large intestine is modeled by matching the rate of transfer between compartments to the actual transit times for different regions of the gut. Such models rely on a variety of *in vitro* or *in silico* input data, or both, such as solubility, permeability, particle size, lipophiliity, and charge, to model the kinetics associated with dissolution, precipitation, uptake, and absorption of a compound as it transits through the different segments of the digestive tract.

9.1.2.2 Distribution Submodel Drug distribution throughout the body is modeled by tracking the drug concentrations in the tissue compartments and the transfer between these compartments via the blood. Estimation of the extent of drug accumulation within individual tissues relies on knowledge of the specific makeup of each tissue in terms of lipids, proteins, and water and on drug-specific data on lipophilicity, charge, and protein binding. Methods for prediction of tissue partitioning of drugs based on their physicochemical properties has developed considerably in recent years, and this has been a driving force in the growth of PBPK in the pharmaceutical industry as it has replaced the need for laborious *in vivo* experiments that would otherwise be needed to measure distribution into tissues. A detailed description of the mechanistic derivation of these equations can be found in the literature [2, 3].

9.1.2.3 Metabolism and Excretion Submodels Metabolism predominantly occurs in the liver and can be scaled up from measurements made *in vitro* using liver tissue extracts (e.g., microsomes or hepatocytes) or recombinant enzymes. This scaling process accounts for the relative concentrations of each enzyme *in vitro* and in the whole liver and employs a submodel for the liver that accounts for physiological factors such as blood flow and binding to proteins in blood. Although metabolism may occur in tissues other than the liver, this is usually a minor contribution for most drugs and can be added on an ad hoc basis applying similar principles to those used for the liver.

Excretion of drug into urine occurs in the kidney via a combination of filtration and active secretion and into feces via active elimination into bile in the liver or via secretion into the gut from the blood. These processes can be added to the model although the scaling from *in vitro* data for active transport is not currently as well established as metabolism scaling.

9.2 PBPK HISTORY AND CURRENT STATUS

9.2.1 Industrial Adoption of PBPK Models and Commercial Software Tools

PBPK models are implemented as a series of linked differential equations that are solved numerically on the computer to track the time course of drug concentrations in blood and tissues. This requires substantial computing power, so although many of the basic concepts of PBPK were outlined much earlier, their widespread application has been dependent on the availability of powerful personal computers that occurred in the early 1980s. Most early publications on PBPK described its application for risk assessment of environmental chemicals [4]. In this area, controlled trials in humans are not ethical and PBPK offers the only scientifically sound method of extrapolating animal-to-human toxicology. Despite the numerous advantages of PBPK, application for therapeutic drug development, both within the pharmaceutical industry and by regulatory agencies, was initially limited [5]. Adoption by the industry has been comparatively slow, and empirical methods, such as allometry, are still widely employed. Explanations for this include the level of mathematical and computer expertise needed to develop such models as well as the requirement for *in vitro* inputs to feed the models. Recently, however, an increase in the number of PBPK publications coming from the pharmaceutical industry has been noted [6]. This acceleration has been aided by the rapid advances in *in silico* prediction that can often replace *in vivo* measurements [3, 7, 8] and by the availability of powerful and user-friendly PBPK software such as Cloe PK (Cyprotex, Macclesfield, U.K.; http://www.cyprotex.com/cloepredict/physiological_modelling/cloe-pk/), PKSim http://www.systems-biology.com/products/pk-sim.html), SimCYP Population-Based ADME Simulator (Simcyp Ltd, Blades Enterprise Centre, John Street, Sheffield S2 4SU UK, http://www.simcyp.com), and the advanced absorption, pharmacokinetic, and pharmacodynamic simulation software GastroPlus (Simulations Plus, Inc., Lancaster, USA, http://www.simulations-plus.com/).

9.3 PBPK APPLICATION AT DIFFERENT STAGES OF PHARMACEUTICAL RESEARCH AND DEVELOPMENT

PBPK models can be useful guides in the research and development of new drugs from the early discovery phase through to clinical development. Early in discovery, uncertainties are large and the models are used more to forecast the impact of changes in key properties than to make precise quantitative prediction. As molecules advance through preclinical development, the models can integrate numerous *in vitro* and *in silico* data and add value by transforming these properties into possible *in vivo* outcomes. Once compound-specific PBPK models have been verified with *in vivo* animal data, they provide reliable prediction of human pharmacokinetics [9, 10] and by integrating relevant *in vitro* data are also useful for simulations of drug–drug interaction (DDI) potential [11]. After further refinement based on phase 1 clinical data, the models can be applied to simulate age dependency [12, 13] and disease dependency [14,15] and to explore the variability expected in different patient populations [16,17]. The models can also provide valuable guidance in the design of clinical drug–drug interaction studies [18].

Drug discovery is increasingly data rich with high-throughput chemistry generating numerous compounds that are rapidly screened for pharmacological and pharmacokinetic properties. These data drive the decision-making process leading to the selection of the best molecules for development. The job of a drug metabolism and pharmacokinetic (DMPK) scientist within drug discovery is to select the compounds with human pharmacokinetics appropriate to achieve the desired efficacy/safety profile in humans. Considering the multifactorial nature of this decision, tools that assist by integrating the data and weighing their combined relevance are essential. PBPK models provide such capability by incorporating preclinical ADME data into a mechanistic framework. Furthermore, by linking to pharmacodynamic models incorporating pharmacology data, PK/PD relationships can be modeled facilitating a quantitative, science-driven decision.

9.3.1 Application at the Optimization Stage

One application of PBPK modeling is to ensure the best use of *in vivo* studies in animals. During drug discovery, considerable resources are required to assess the pharmacokinetic properties of potential drug candidates *in vivo* in rodents. By feeding PBPK models with *in vitro* and *in silico* data inputs, concentration versus time profiles can be predicted before any *in vivo* experiment is performed and if sufficiently reliable can be used to prioritize compounds for the more costly *in vivo* testing. Equally importantly, for those molecules tested *in vivo*, when comparison of the predictive simulations shows a large discrepancy with measured data, this indicates knowledge gaps and the need for further experiments to quantify additional processes. This use of a rat PBPK model for early drug discovery was evaluated at F. Hoffmann LaRoche, Basel, CH [19]. Plasma profiles after intravenous and oral dosing were simulated in rats for 68 compounds from six chemical classes. Input data consisted of structure-based predictions of lipophilicity, ionization, and protein binding plus measured intrinsic clearance in rat hepatocytes, aqueous solubility, and artificial membrane permeability. Overall, the simulations were able to predict differences in pharmacokinetic parameters of more than twofold although verification of the simulations with *in vivo* data for a few compounds of each compound class was recommended because the accuracy of simulations was class dependent. This step could thus serve to verify the reliability of the model simulations and identify knowledge gaps.

At this early stage, it is not necessarily expected that an accurate predictive model can be obtained. However, even when a perfect simulation is not possible, a model sensitivity analysis can be useful to explore the factors that might be causes for gaps between simulation and observed data. This exploration may lead to insights that trigger future experiments, eventually bringing improved mechanistic understanding. This is illustrated in Figure 9.2, which shows a sensitivity analysis plot of the variation in fraction of dose absorbed for changes in drug solubility, drug permeability, and drug

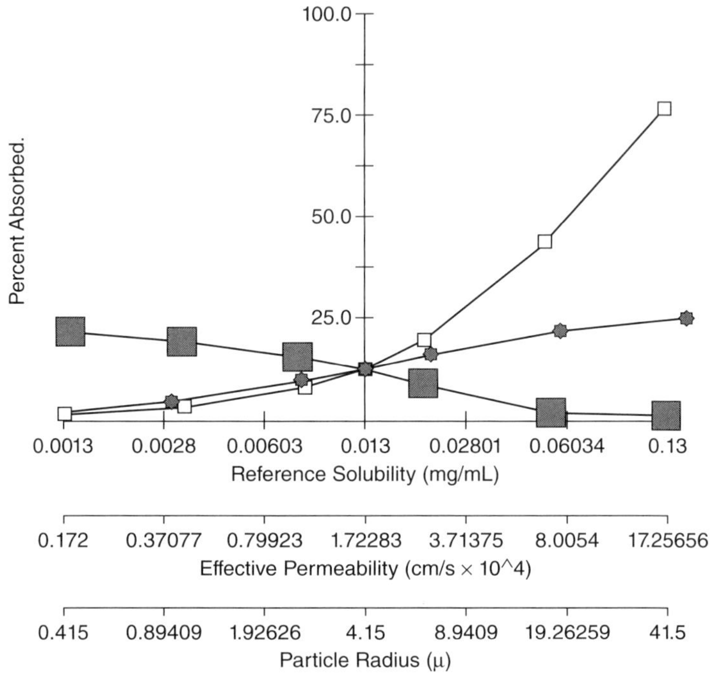

Figure 9.2 Plot of a parameter sensitivity analysis showing how simulated fraction of dose absorbed varies for changes in solubility (small open squares), permeability (closed circles), and particle radius (large filled squares).

particle size for a molecule where simulation underpredicted the oral bioavailability by threefold. A poor estimation of any, or a combination, of these factors could lead to the simulation mismatch, but the plot indicates that solubility is most likely to be the cause because simulated exposures are more sensitive to small changes in this factor. Subsequent measurements supported this prediction when solubility in a medium simulating the intestinal fluids showed a value of 47 ug/mL versus the original aqueous buffer solubility of 13 ug/mL.

An analysis at Pfizer (New York, NY) has shown that PBPK methods can be effectively used within the discovery environment to predict absorption in preclinical species. Typically, available input data for several diverse soluble analogs were used to predict accurately *in vivo* oral PK in rats. Confidence gained by accurate prediction of oral PK for the test compounds allowed the teams to rely totally on simulation for further analogs, thus, helping to reduce the number of animals used [20].

9.3.2 Application at the Clinical Candidate Selection Stage

The phase when a compound is prepared for the first administration to humans is relatively time and resource demanding because it involves several *in vivo* animal experiments to assess safety and much of the work is subject to regulatory guidelines. Such characterization is only possible for a limited number of compounds per project so the selection of optimal compounds is a critical step. The decision on the clinical candidate must consider numerous factors, and the data related to the pharmacokinetics and pharmacodynamics (PD) of the compounds need to be combined and compared in a rational way. The benefits of the use of PBPK at this stage were illustrated at Roche [21]. PK/PD models were built using *in vitro* and *in vivo* data for five potential clinical candidates and were applied to make estimations of the effective human doses and associated exposures, thus, aiding in the selection of the most promising compound. To ensure that the decision was based on significant differences between the compounds, estimates of variability and uncertainty in both the pharmacokinetics and the pharmacodynamics were carried through in the modeling. A similar approach has also been illustrated at Pfizer [22], where PBPK modeling was used to profile a competitor compound and to predict exposure and dose for potential internal clinical candidates without the need for preclinical *in vivo* data.

9.3.3 PBPK for Human PK Prediction

The application of PBPK modeling for human PK prediction has been more prevalent in recent years with several examples being published from the pharmaceutical industry [9, 10, 20, 23–30].

Workers at Roche [9] proposed and validated a strategy for the use of PBPK for human PK simulations using a set of 19 compounds. Initially, the PBPK simulation is performed in animals using animal PBPK models, animal *in vitro* data, and compound-specific physicochemical data. The animal simulation is compared with the *in vivo* data; if simulation in animals is reasonable, then the human simulation is performed using a human PBPK model and human *in vitro* data. If the simulation in animals is inaccurate, this would indicate a violation of one or more of the model assumptions. In this case, further experiments may be performed to understand the mismatch. In their evaluation, Jones et al. showed that for those compounds where good predictions of animal PK were achieved, 92%, 67%, and 100% of compounds had a prediction accuracy within twofold error in humans for area under concentration (AUC), Cmax, and Tmax, respectively. The prediction accuracy was reduced [76%, 47%, and 94% for AUC, Cmax, and Tmax respectively) when the compounds that were judged as having poor predictions in animals were included in the human analysis emphasizing the need for good understanding in animals prior to the simulation in humans. When compared against traditional allometric scaling approaches, the novel PBPK techniques were found to be superior (Figure 9.3). This iterative PBPK prediction approach has subsequently been further validated by several other authors using diverse drug datasets [10, 26, 31].

Figure 9.4 illustrates a real example where PBPK was compared with allometric scaling for human pharmacokinetic prediction. The molecule has a molecular weight of 579 and is weakly basic and highly lipophilic (LogP 6.5). A high volume of distribution (>12 L/Kg) and a moderate clearance (10–50% of liver blood flow) were observed in rats, dogs, and monkeys. Physiological clearance scaling was verified using *in vitro* data generated in hepatocytes, and as this gave reasonable prediction (within twofold) of total clearance in animals, human hepatocytes were used to predict human clearance. Use of allometry for clearance prediction was less successful

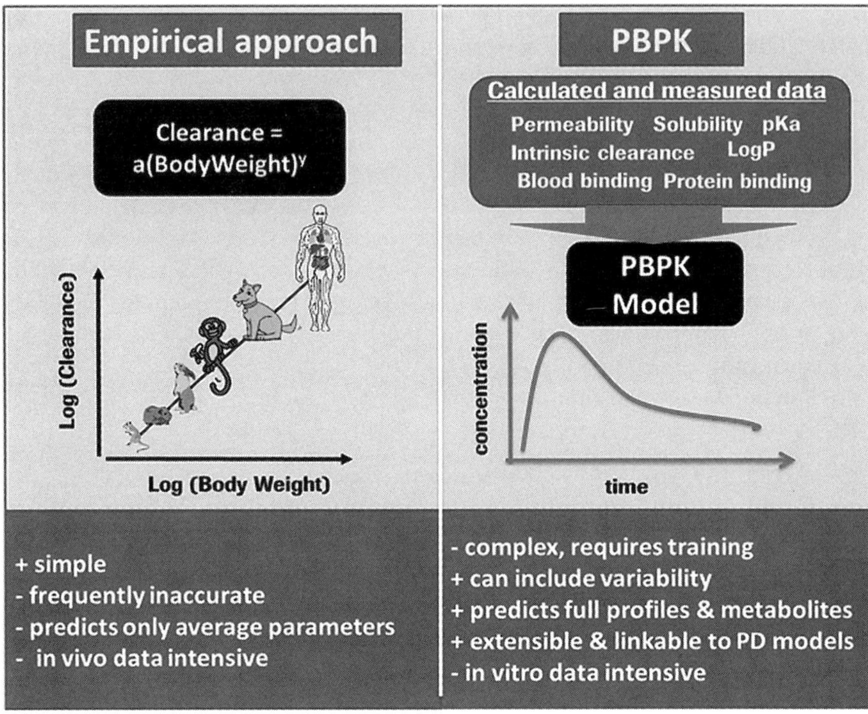

Figure 9.3 Comparison of prediction of human pharmacokinetics using empirical methods such as allometry (left) and PBPK modeling (right). (*See insert for color representation of the figure.*)

because a poor correlation was observed between animal species. On the other hand, allometric scaling for volume of distribution showed a good correlation for the preclinical species but led to a fourfold overprediction for humans. Having verified the physiological oral absorption simulations in rats and dogs based on measured permeability in the parallel artificial membrane permeability assay (PAMPA) and solubility in a biorelevant medium, the human PBPK model was applied to predict oral concentration versus time profiles in humans. Good agreement with observed data was observed for doses ranging from 10 mg to 450 mg.

Other examples of the prospective use of PBPK modeling for human PK prediction can be found in the literature [32–34]. These publications provide "real-life" examples of where PBPK modeling has successfully been used to predict exposure and design phase 1 clinical trials.

9.3.4 Application for Translational PBPK/PD

The combination of PBPK with PK/PD models provides the basis for quantitative comparisons of potency and safety of drug candidates. In particular, bridging of PK/PD relationships established preclinically to clinical studies results in improved

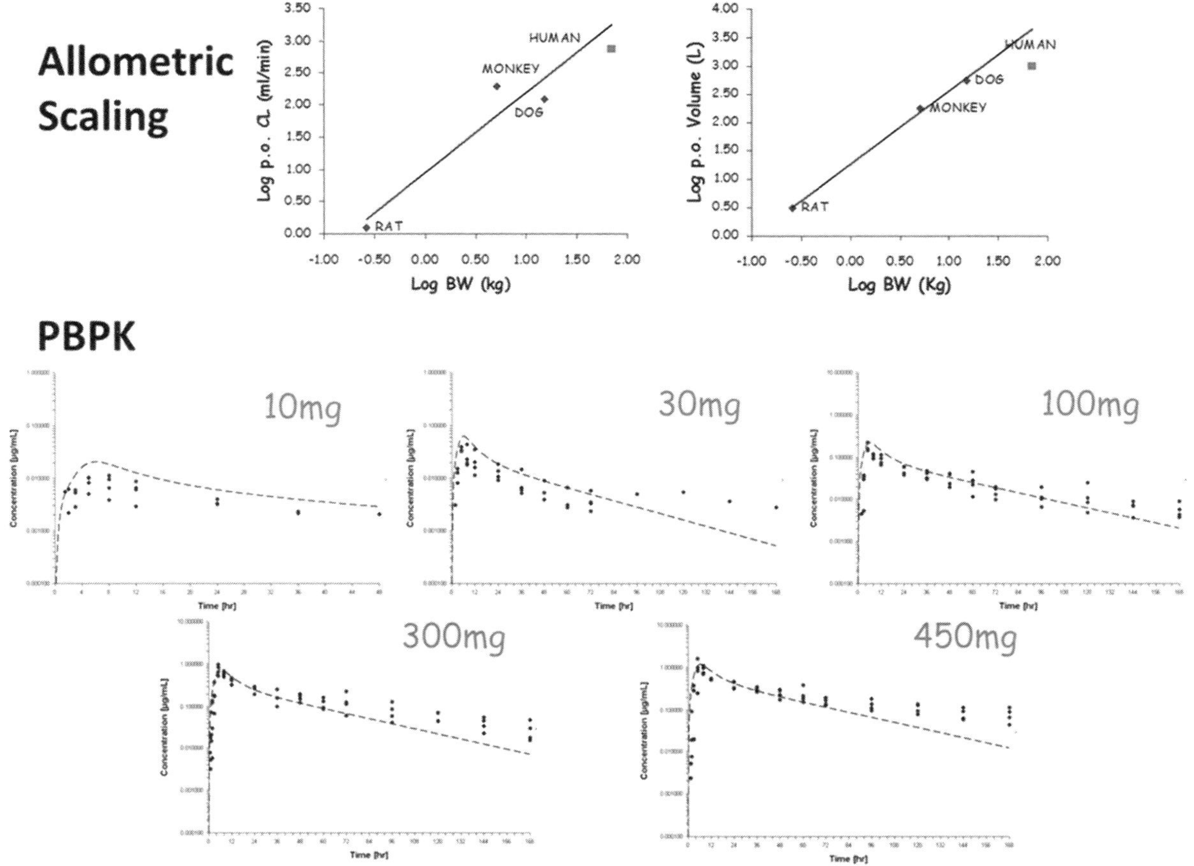

Figure 9.4 Comparison of allometry and PBPK methods for prediction of human pharmacokinetics. Allometric scaling predicts only PK parameters and cannot be used to assess the confidence in predictions as a good correlation in animals does not imply a reliable human prediction. PBPK methods may be verified in animals prior to prediction in humans and may predict full concentration versus time profiles including dose dependency.

decision making during the selection process [35]. A typical example of such an approach was provided for a Roche anti-obesity project in [36]. To get a better understanding of the determinants of efficacy in rats for compounds tested *in vivo* in a subchronic food intake model, PBPK simulations were extended to predict brain receptor occupancy for the target of interest. Receptor occupancy predictions used the *in vitro* Ki for binding to the rat receptor and unbound fraction in plasma assuming that unbound concentration in plasma is the same as unbound concentration in the brain. Interestingly, these PK/PD simulations indicated that *in vivo* active compounds exhibited high (>60%) and sustained (>8 h) occupancies while inactive compounds showed lower or less sustained occupancies. This work supported that a meaningful way to select compounds for the clinic was to employ a PBPK/PD human model using the same target for receptor occupancy predicted using human-relevant *in vitro* data.

9.3.5 Application for Planning of Toxicological Studies

Typically, early toxicology studies are performed over 7–14 days and a single dose toxicology study would be performed at three dose levels to help design the 7–14-day study. PBPK modeling can be used to reduce the need for the preliminary dose setting study, thus, improving efficiency and reducing compound requirement and animal usage. As described in Reference 22 at Pfizer, a retrospective analysis has been conducted using a cohort of 39 compounds for which toxicokinetic data from the 7–14-day toxicology study were available. In this analysis, PBPK methods were shown to predict more accurately AUC and Cmax than an approach assuming linearity from low-dose studies. Jones et al. described the validation of such techniques in an early discovery project, building confidence in the predictability of PBPK techniques for a particular chemical series. Subsequently, prospective application of PBPK techniques was made for follow-up compounds in the project, without the need for the dose setting study.

PBPK absorption modeling is also used at Roche to predict plateauing in exposures in toxicology studies as a result of absorption limitations. An example illustrating the advantages of PBPK over more simplistic approaches such as the maximum absorbable dose calculation [37, 38] is given in Reference 39. Advantages include that the simulated plot of predicted exposures versus dose conveys more information than a single number and helps to focus preclinical discussions and decisions around safety margins based on exposures rather than on doses.

9.3.6 PBPK for Drug–Drug Interaction Modeling

In clinical practice it is common for multiple drugs to be administered concomitantly either for more effective disease treatment or for treatment of multiple co-morbidities. A drug–drug interaction occurs when co-administration of drugs results in a pharmacological or clinical response different than the responses with the same drugs given alone. Many drug interactions are a result of changed pharmacokinetics when either increased or decreased exposures result and can, in the worst cases, lead to severe adverse reactions or loss of efficacy. Very often, drug interactions are related to inhibition or induction of metabolism, or both, but other mechanisms, such as changes in absorption, distribution, and excretion (e.g., via active transporters) are possible.

PBPK simulation provides several advantages over more simple approaches for assessment of drug–drug interactions. Advantages include the capability to consider the time-varying concentrations of the interacting drugs as opposed to the use of average concentrations and the ability to explore different dosing regimens and times of administration. These factors can have a significant effect on the magnitude of interaction predicted [18]. Other advantages are related to the capacity to account for the pharmacokinetic properties of interacting drugs in a more comprehensive and mechanistic way. For instance, active hepatic uptake of a metabolic inhibitor (or a perpetrator drug) increases its concentration and results in greater effect on the concentration of co-administered (victim) drugs metabolized by the inhibited enzyme. Such uptake is a result of transporter proteins expressed in the hepatocytes and can be included as a simple scaling factor based on *in vitro* data [40] or can be modeled more mechanistically [41]. As another example, drugs that are substrates of CYP3A4 may be subject to significant metabolism in the gut wall, which can be best captured in a PBPK model [42,43]. Metabolic drug interactions are often most severe when the inhibition is irreversible and recovery requires synthesis of new enzyme. This irreversible inhibition can be included in PBPK models [44]. Drug metabolites can also be included in a PBPK model, allowing for consideration of cases where multiple inhibitors all contribute to an interaction [45,46].

Several evaluations of PBPK techniques for DDI prediction have been reported [27,47–50]. Specific examples of the application of PBPK modeling for drug–drug pairs from Roche and Pfizer have been published [51,52]. SimCYP was used by Rakhit et al. to predict the pharmacokinetics of erlotinib in the presence and absence of ketoconazole. *In vitro* data indicate that erlotinib is metabolized via CYP3A4 (~70%) and CYP1A2 (~30%). Consistent with the clinical data, SimCYP predicted that ketoconazole caused a twofold increase in erlotinib plasma AUC curve and in maximum plasma concentration. Hyland et al. characterized the metabolism of maraviroc *in vitro* and used this information to develop a SimCYP model to forecast DDIs with CYP inhibitors. CYP3A4 was identified as the major CYP responsible for maraviroc metabolism. Simulations were performed with ketoconazole, ritonavir, saquinavir, and atazanvir. The simulated median AUC ratios were in good agreement with observed clinical changes (within twofold) in all cases. These examples show the retrospective use of PBPK modeling techniques for the simulation of DDIs. However, the real value of such techniques is to replace or optimize clinical DDI studies, and several examples of such application by the pharmaceutical industry have been recently published [53].

9.3.7 PBPK for Absorption and Formulation Development

Drug delivery via the oral route is usually preferable because of safety, convenience, and cost. Thus, small-molecule drugs are often considered viable only if they can be given orally. To aid the development of oral drugs, there have been many attempts to model drug absorption. The biopharmaceutics classification system (BCS) [54] is widely used as an aid in decision making in pharmaceutical formulation development. It recognizes that the key determinants of oral absorption are the drug properties of permeability and solubility. Physiologically based absorption models also integrate these key drug properties but go further by accounting for physiological factors such as drug transit, fluid volumes, and intestinal pH. This approach offers several advantages and allows for a more refined consideration of the possibilities of drug formulation to improve oral absorption [55, 56].

An estimate of intestinal permeability is a key input for absorption models. At the lead optimization stage, compound permeability is often predicted with *in silico* models or may be measured in a higher throughput assay such as PAMPA [57, 58]. For more advanced compounds, a measurement of cellular permeability such as that obtained with Caco-2 cells [59] is often preferred. For use of these *in vitro* data in PBPK simulations, it is necessary to scale to the *in vivo* situation. The reference data for this approach are usually the human jejunal permeability, which has been measured for many reference drugs [60].

Simulation of absorption for the poorly soluble lipophilic compounds common during drug discovery are often limited by the poor relevance of aqueous solubility for the situation *in vivo* where bile salts and lipids can enhance solubilization. It is possible to generate more relevant solubility data in media such as fasted state simulating intestinal fluid (FaSSIF) and fed state simulating intestinal fluid (FeSSIF) [61]. Such biorelevant solubility measurements have been shown to be essential for reliable PBPK simulation in both animals and humans [39, 62, 63].

Simulations in preclinical species offer the first opportunity to assess the relevance of *in vitro* data and PBPK modeling to predict oral absorption. This begins with the rat, typically dosed per gavage with a solution or suspension, and then, when solid formulations are developed, the dog becomes the model of choice. The validation of PBPK simulations in these preclinical species is believed to increase confidence in human simulations [9, 10, 63, 64], although several gaps remain to be filled such as species differences in intestinal fluid composition [64] and intestinal metabolism [65].

Many molecules selected for clinical development have less than ideal properties for oral absorption, and drug formulation plays an important role in their chances of success. The choice of an adequate clinical formulation is difficult, as it has to meet the general project timelines and must take the involved risks into consideration. Those risks can be considerable with biopharmaceutically challenging drugs like those in the BCS class II, III, or IV. A straightforward technical development could lead to a formulation that does not show adequate exposure in humans, whereas de-risking, in the form of parallel development of sophisticated formulations, is cost intensive. This is especially the case for bioavailability testing in animals. An extensive formulation screening would not only bind substantial resources but also is uncertain in terms of its relevance for the human situation. The use of PBPK modeling as an additional biopharmaceutical assessment program in the preclinical development phase at Roche was described by Kuentz et al. [55].

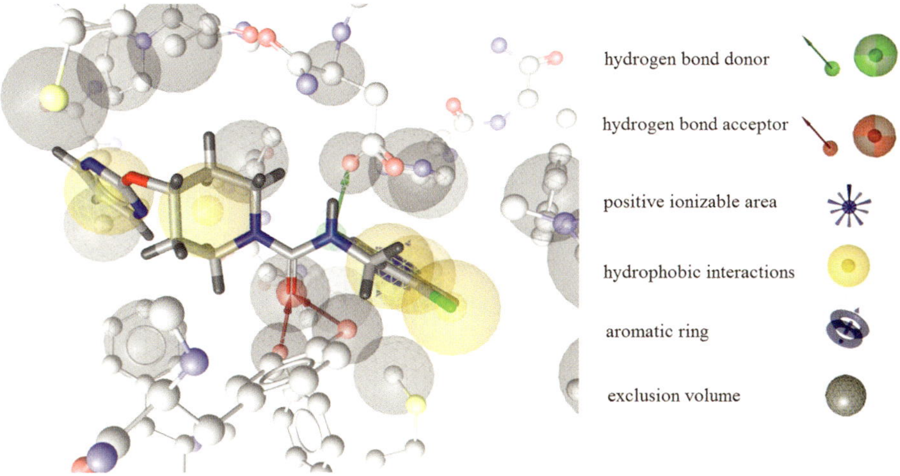

Plate 2.1 An exemplary pharmacophore model of an inhibitor (stick style) interacting with a target protein (ball-and-stick style). The example shows soluble epoxide hydrolase bound with an inhibitor and their protein–ligand interactions (Protein Data Bank accession number 3koo).

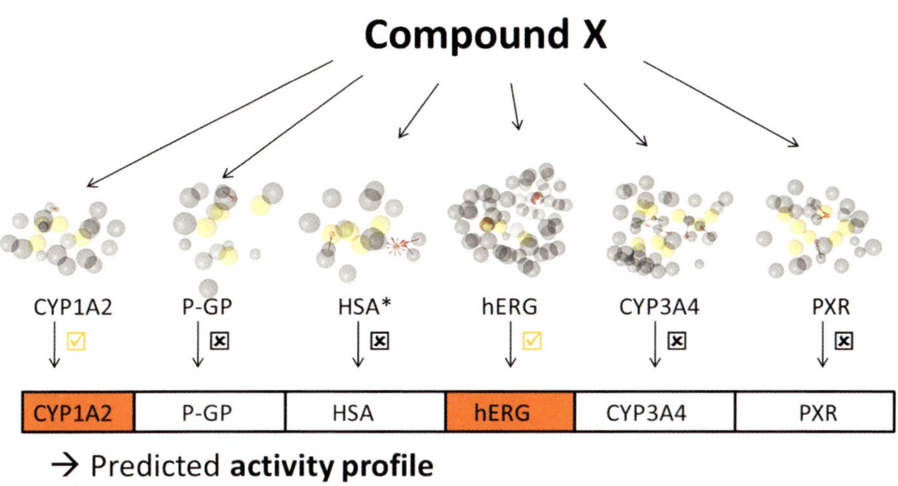

Plate 2.2 An array of pharmacophore models for different targets can be used to get a bioactivity profile, predicting the targets most interesting for biological testing. *Human serum albumin.

Predictive ADMET: Integrative Approaches in Drug Discovery and Development, First Edition.
Edited by Jianling Wang and Laszlo Urban.

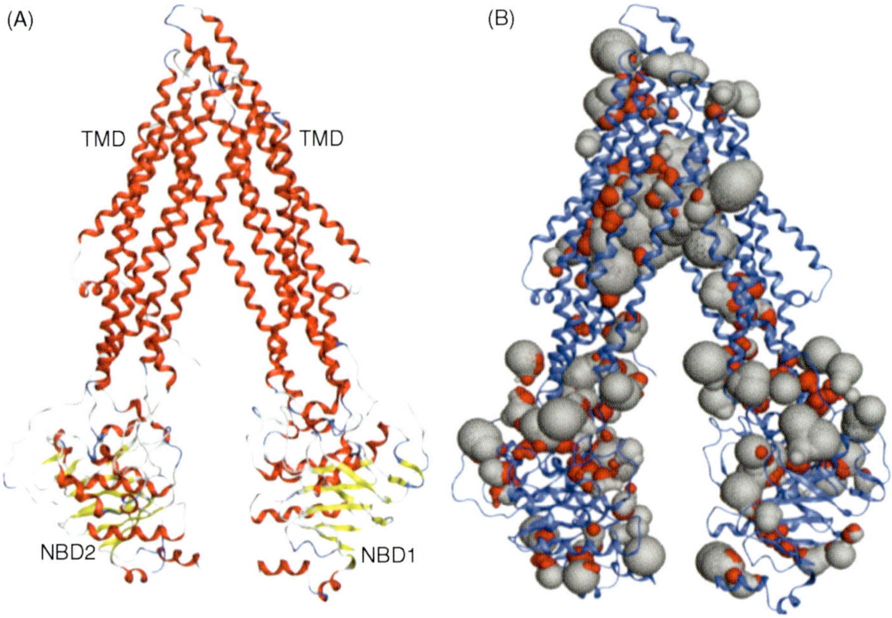

Plate 7.3 Homology model of the human P-glycoprotein corresponding to the inward-open conformation (ATP-free functional state) [80]. (A) A 3D view of the protein structure: The protein backbone is rendered as a line ribbon and colored according to the secondary structure: loop – gray; helix – red; strand – yellow; turn – blue; (B) Potential binding sites of the protein filled with alpha spheres (each sphere is classified as either "hydrophobic" (gray) or "hydrophilic" (red) depending on whether the sphere is a good hydrogen bonding point in the receptor. The radius of the sphere is proportional to the convex hull of the point set and is invariant to rotation of the atomic coordinates, generated with the "Site Finder" module in MOE (http://www.chemcomp.com).

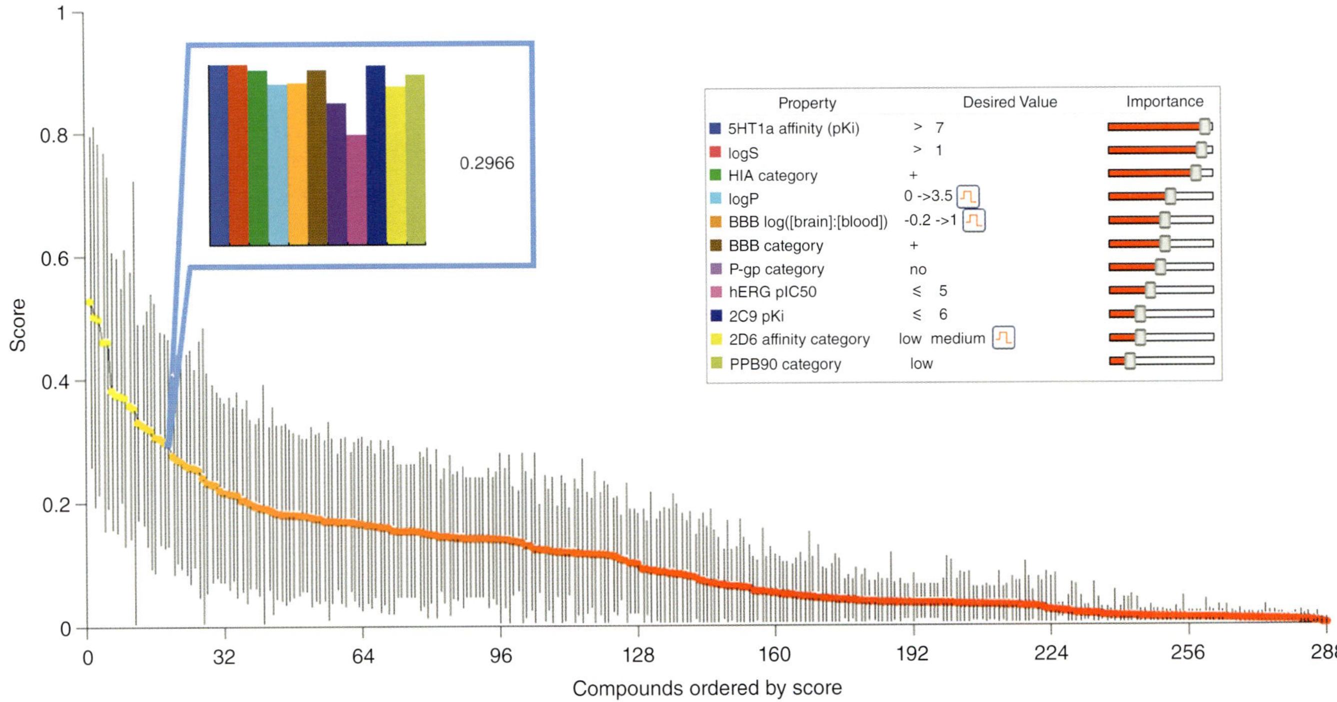

Plate 8.6 Example results from scoring 288 compounds using the probabilistic scoring method and a scoring profile designed to prioritize compounds with good potency against the $5HT_{1a}$ target and appropriate properties for an orally dosed compound with good brain penetration. The compounds are ordered by score along the *x*-axis, and the score is plotted on the *y*-axis. The error bars indicate the confidence in the score as a result of the uncertainty in the underlying data. Note that, given the uncertainty in the scores, there are a number of other compounds that cannot be confidently distinguished from the "best" compound. The inset scoring profile shows the property criteria that were specified, along with their importance of each criterion to the overall success of the project. A histogram for an example compound shows the contribution of each property to the overall score for that compound, clearly identifying the most important issues that should be addressed to improve significantly the compound's chance of success.

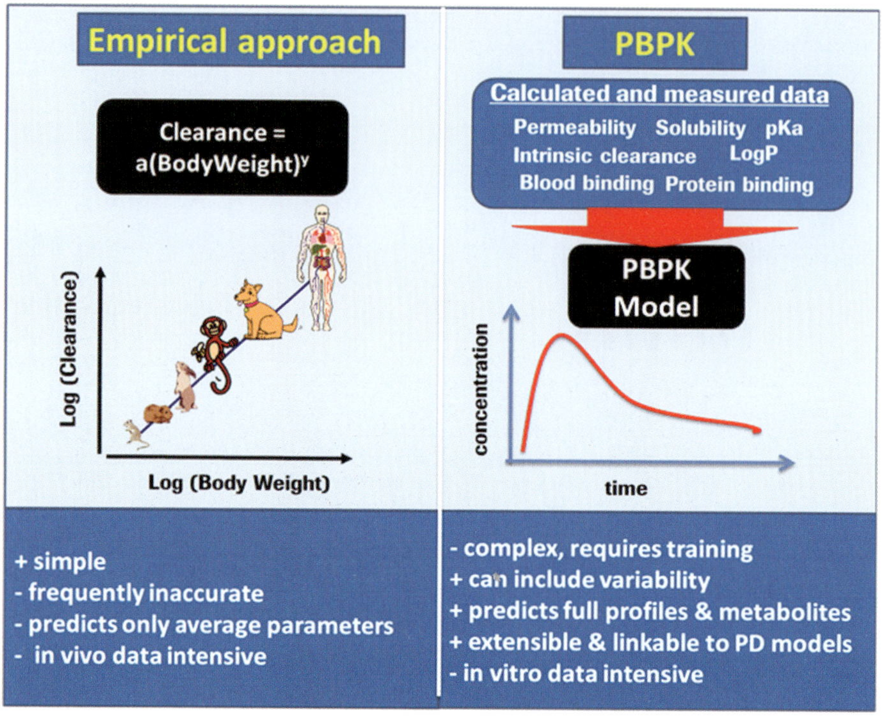

Plate 9.3 Comparison of prediction of human pharmacokinetics using empirical methods such as allometry (left) and PBPK modeling (right).

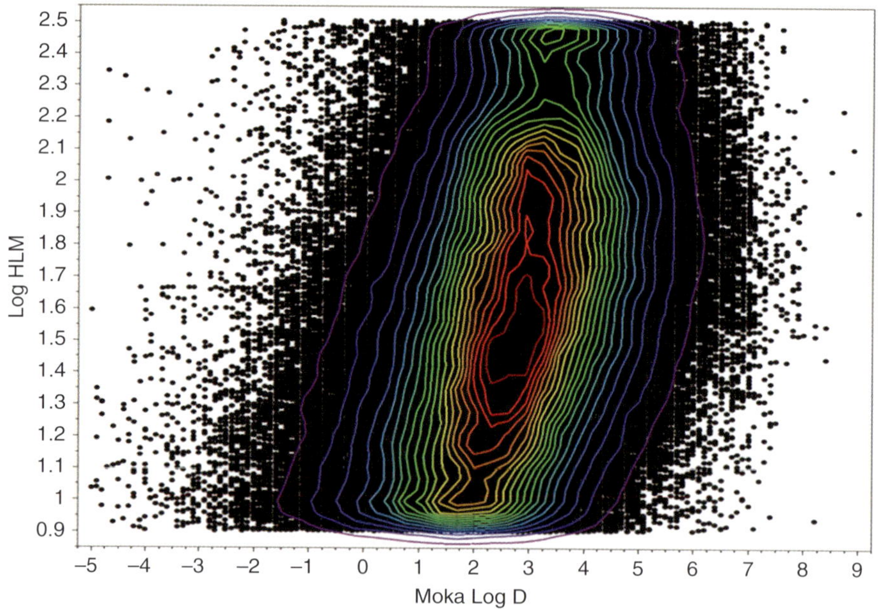

Plate 14.4 Log (HLM $Clint_{app}$) versus calculated Log $D_{(7.4)}$ values for a set of discovery compounds showing weak correlation and high variability (for any given value of HLM, the LogD spans > ±2 units). Colored lines are 5% nonparametric density lines.

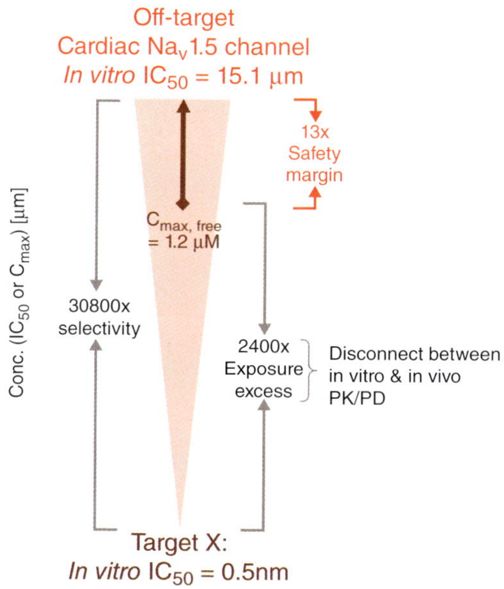

Plate 17.1 Drug candidate X failed as a result of off-target toxicity in humans. On- vs. off-target selectivity *in vitro* (calculated as Off-target IC_{50}/Target IC_{50}, outlined in Table 17.1) is reflected by the length of the vertical triangle with the upper end of the triangle at the IC_{50} value of the off-target and the lower end at the IC_{50} value of the target. The starting point of the vertical arrow corresponds to $C_{max,free}$ at efficacious dose. The length of the vertical arrow reflects the off-target safety margin (calculated as Off-target IC50/$C_{max,free}$ at efficacious dose, outlined in Table 17.1). For further details, refer to text.

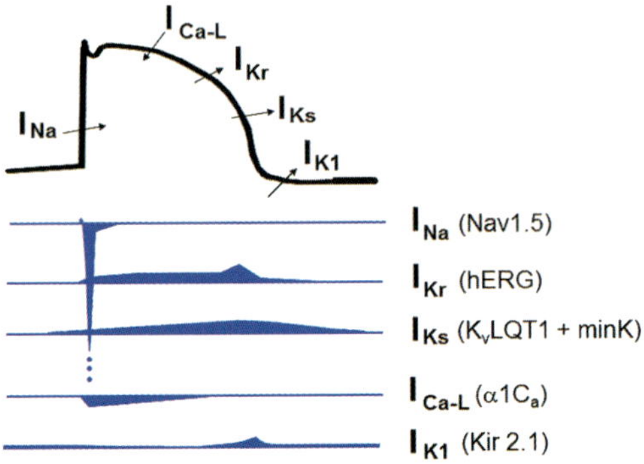

Plate 18.1 Cardiac ventricular action potential primary currents. Arrows indicate inward or outward currents. Not pictured are $I_{Ca,T}$, I_{to}, I_{Kur}, I_{KATP}, I_{KAch}, I_F, or I_{Cl}.

Table 18.1 Evidence, prevalence, and occurrence of safety liabilities relating to the cardiovascular and hepatic systems

Phase	Nonclinical	Phase I	Phase I-III	Phase III/ postapproval	Postapproval	Postapproval	Post approval
Information	Causes of attrition	Serious ADRs	Causes of attrition	ADRs on label	Serious ADRs	Withdrawal from sale	Withdrawa from sale
Source	Car 2006 [111]	Sibille et al. 1998 [112]	Olson et al 2000 [113]	BioPrint® 2006	Budnitz et al. 2006 [114]	Fung et al. 2001 [115]	Stevens & Baker 2009 [116]
Sample size	88 CDs stopped	1,015 subjects	82 CDs stopped	1,138 drugs	21,298 patients	121 drugs	47 drugs
Cardiovascular	27%	9%	21%	36%	15%	9%	45%
Hepatotoxicity	8%	7%	21%	13%	0%	26%	32%

0% (white) 1-10% (green) 1-20% (yellow) >20% (red)

Note: Information was collated from the commercially available database, BioPrint®, and published articles [111–116]. ADRS, Adverse Drug Reactions. CD, Candidate drug. (Reprinted from Laverty et al. [2] with permission.)

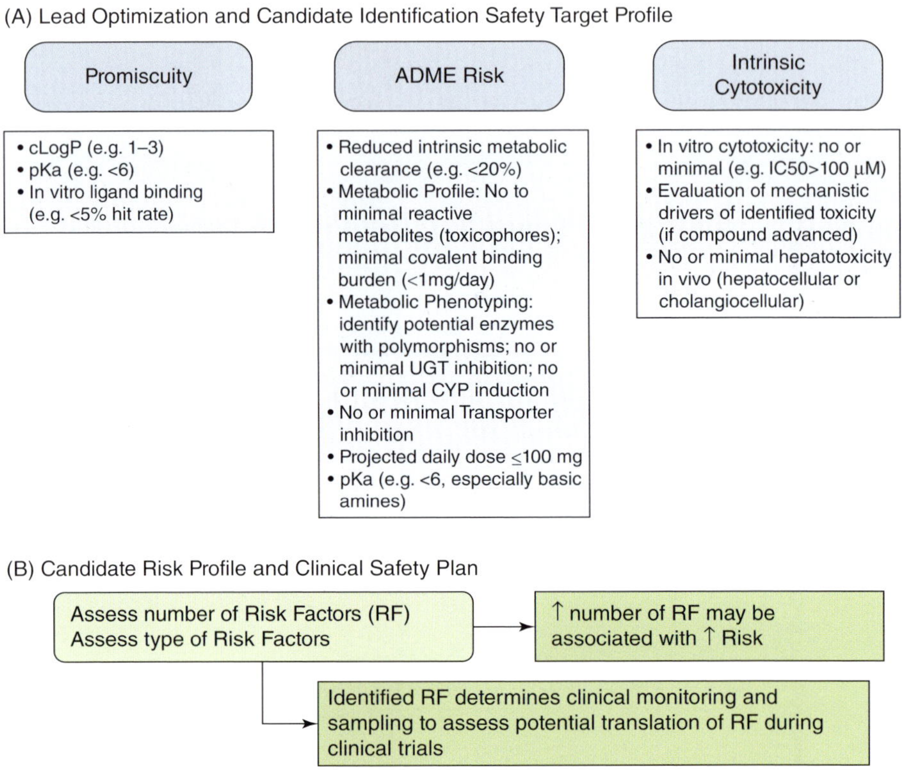

Plate 19.1 iDILI risk assessment and mitigation strategy components.

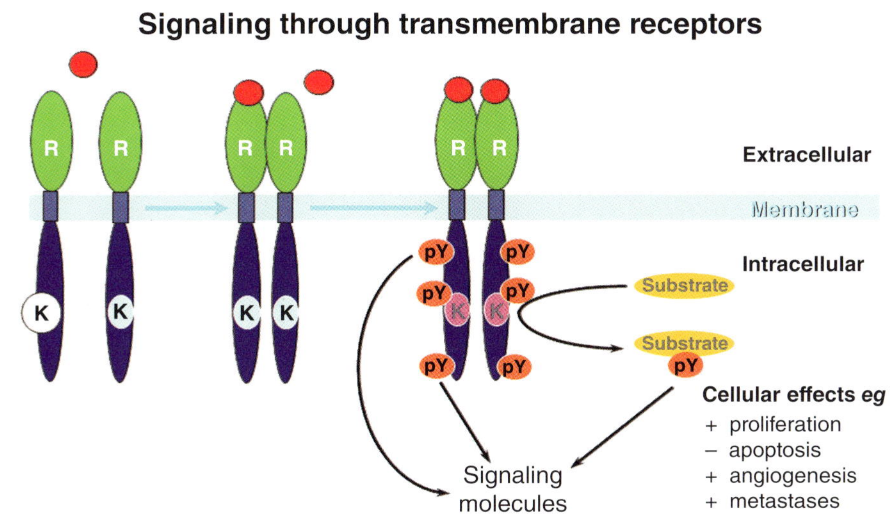

Plate 24.1 Signaling through transmembrane receptors. o = ligand; R = extracellular domain of the receptor; K = intracellular domain of the receptor; pY = phosphorylation domains of the receptor.

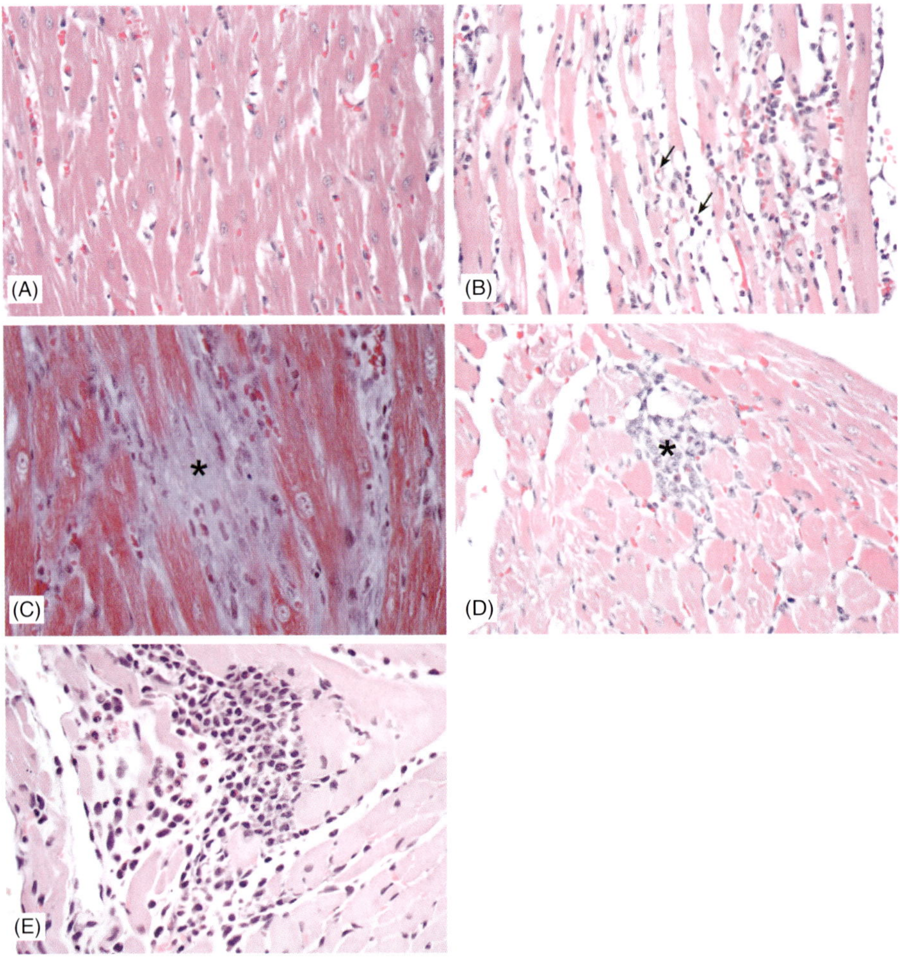

Plate 25.1 Microscopic findings of hematoxylin and eosin stained or Masson's Trichrome stained sections of rat myocardial interventricular tissue. (A) Hematoxylin and eosin staining of myocardial sections from a control rat in the TK study (400X). (B) Hematoxylin and eosin staining of myocardium sections from a rat in the TK study with 500-mg/kg PF-04254644 24 h after dosing, 400X. Moderate myocardial degeneration in the interventricular septum (arrows). (C) Masson's Trichrome staining of myocardial sections of a rat with 320/160-mg/kg PF-04254644 in the 7-day repeat dose toleration study after 7 days of dosing (400X). Fibrosis of the myocardial interventricular septum (asterisk). (D) Hematoxylin and eosin staining of myocardial sections of a rat given 80-mg/kg PF-04254644 in the 7-day repeat dose toleration study after 7 days of dosing (400X). Minimal myocardial degeneration in the interventricular septum (asterisk). (E) Hematoxylin and eosin staining of myocardial tissue taken from a rat 24 h after dosing with 10.5-mg/kg milrinone (400X). Note the minimal myocardial degeneration with macrophage and few neutrophil infiltrates.

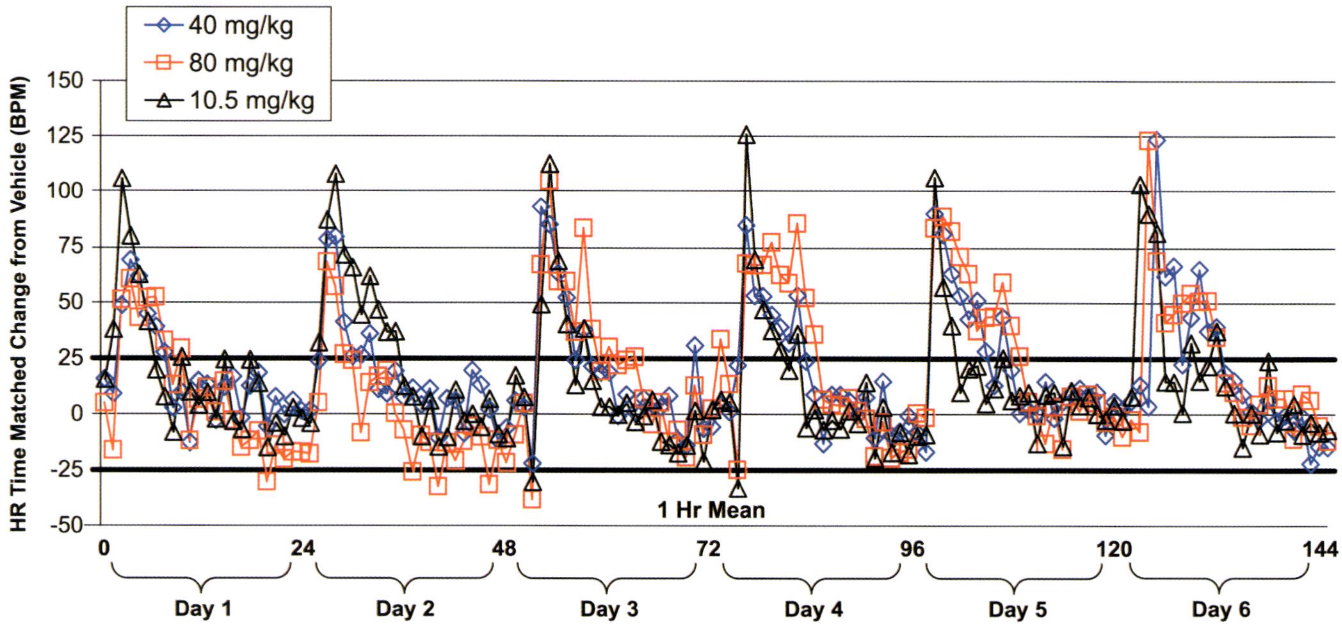

Plate 25.3 Heart rate profile in telemetry instrumented rats from the 6-day repeat-dose investigative study. Rats were given 40- or 80-mg/kg PF-04254644 or 10.5-mg/kg milrinone. Heart rate data represented as 1 h means of time-matched change from rats given vehicle. Dosing occurred at or during hours 2, 25, 49, 73, 96, and 121. Heavy black bars at 25 and -25 depict baseline HR range prior to dosing for male Sprague–Dawley rats ages 7 to 9 weeks.

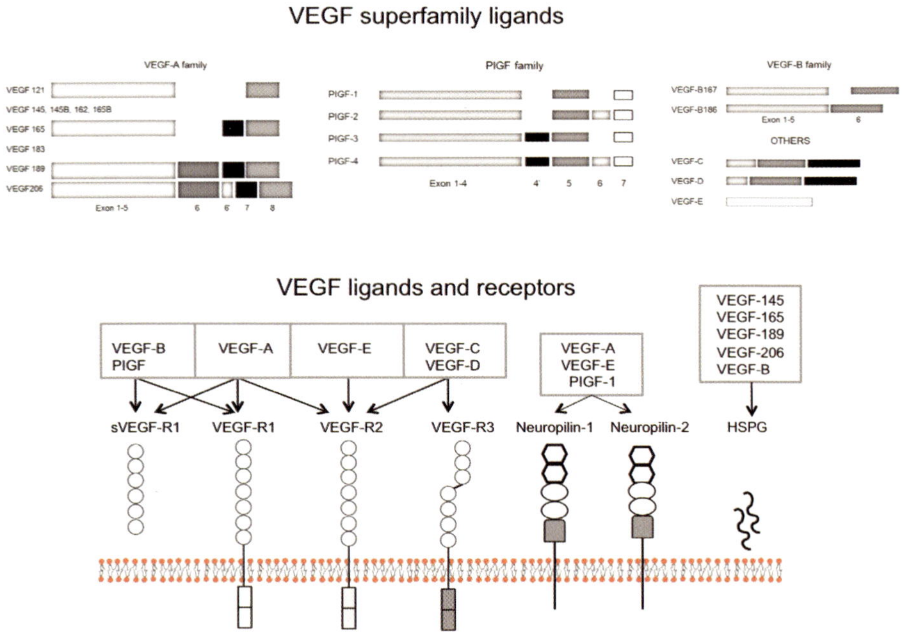

Plate 26.1 VEGF superfamily ligands and receptors. A schematic of VEGF ligands and their interactions with VEGF receptors is shown.

9.3.8 PBPK for Prediction of Variability in Different Populations (Pediatrics, Disease, and Ethnicity)

It has been shown that PBPK models verified with *in vivo* animal data provide reliable prediction of human pharmacokinetics [9, 10]. However, such predictions are always based on several assumptions, and scaling from *in vitro* data always entails uncertainty. Therefore, when the first clinical pharmacokinetic data are obtained in phase 1 studies, they should be compared with the simulations and the model assumptions and inputs reassessed, which is a process that may lead to model refinement. The refined model may then be usefully applied for further simulations as the compound advances through early clinical development. In this regard, the strategy outlined by Jones et al. [9] can be extended to include these human populations as suggested by Rowland [6]. Such applications would include forecasting of age dependency [12, 13] and disease dependency of pharmacokinetics [14, 15] and exploration of the variability expected in different patient populations [16, 17].

Application of clinical PBPK models to predict variability is an area of great potential. This so-called "bottom-up" modeling [66] can enable predictions to be made not just for an "average" person but also in different human populations of interest. This may be extended to diseased populations [67,68] and special populations such as pregnant women [69]. In addition, the requirement of companies to submit a pediatric investigation plan before completion of the phase I trial has led many companies to use PBPK techniques to predict PK and set doses in children [70].

9.4 ADVANTAGES, LIMITATIONS, AND FUTURE PERSPECTIVES OF PBPK MODELING

Physiologically based pharmacokinetic models are increasingly replacing more empirical approaches in pharmaceutical research. Their strength is the ability to incorporate, in a mechanistic framework, the various data generated during drug discovery and development. As more is learned via experiment, the models are refined and act as a knowledge repository. Importantly, the mechanistic basis of the models allows extrapolation to predict outcomes in untested situations. During the early stages, only limited measured data are available to feed the models, and although *in silico* tools can be used to obtain first estimates for inputs, there are limitations to the accuracy that can be expected. At this stage, parameter sensitivity analysis is a useful tool to guide additional experiments and explore scenarios. However, drug discovery is increasingly data rich and the portfolio of *in vitro* assays applied to generate predictive data prior to clinical development is expanding. Therefore, it is recommended to begin with PBPK modeling early. Prior to the first clinical studies, PBPK is the best approach to predict human pharmacokinetics and may be used to explore likely scenarios and guide early clinical studies. During clinical development, definitive data on the clinical pharmacokinetics and drug interactions will become available and the PBPK model should then be refined and used to provide more reliable extrapolation and guidance.

One barrier to more widespread application of PBPK is that scientists with the knowledge and experience to perform such modeling are currently in short supply. However, the increased availability of user-friendly PBPK software packages is helping to alleviate this. More fundamental limitations of PBPK arise as a result of a lack of detailed understanding of the physiological and biochemical basis of certain pharmacokinetic processes (e.g., active transport) or when *in vitro* systems are unable to capture the complexity of the *in vivo* situation. However, developments are ongoing in these areas, which should help to reduce some of these limitations in the near future.

Extrapolation of measurements obtained *in vitro* to the *in vivo* situation is a cornerstone of PBPK modeling [71] and is now routine for metabolism by the cytochrome P450 enzymes (CYPs) where knowledge of the expression levels in tissues and *in vitro* systems are established [7]. Until recently, the best technique for measurement of expression levels of proteins in biological samples was immunoquantification. This involves time-consuming preparation of specific antibodies is often challenging because of the high sequence similarity among protein isoforms. However, advanced proteomics techniques for the identification of proteins in biological samples have recently been developed and promise to accelerate growth in our knowledge of expression levels of many ADME-relevant proteins. For example, simultaneous quantification of 11 CYP isoforms in human liver tissue by multiplexed selected reaction monitoring (SRM) analysis with liquid chromatography tandem mass spectrometry has been demonstrated [72], and a similar technique was also successful in quantifying CYP enzyme expression in dog gut [65]. Furthermore, the SRM technique has been applied to quantifying membrane transporter proteins in the blood-brain barrier, liver, and kidney of a mouse [73].

Often, PBPK models assume that the dominant process determining the movement of a drug across cell membranes is passive diffusion and active transport is neglected. This assumption seems to be valid in many cases, but it is increasingly recognized, also by health authorities [74], that active transport can be important and should be included. With efforts to reduce clearance in *in vitro* systems during drug discovery, transporter-mediated PK is becoming more prevalent and the inclusion of such considerations into PBPK models is becoming a requirement. Transporters are known to be present at the plasma membrane domains of intestinal epithelia, hepatocytes, kidney proximal tubules, and brain capillary endothelial cells. These transporters can affect the absorption, distribution, metabolism, and excretions of drugs. Active transport in the hepatocytes has been included in PBPK models as a simple scaling factor based on *in vitro* data [40], and it has been modeled more mechanistically using a combination of scaling from *in vitro* and of fitting of *in vivo* data [41]. Most recently, scaling purely from *in vitro* data has been demonstrated for a range of compounds [75]. In the case of the liver, information is available on organic anion-transporting polypeptide (OATP) expression in human hepatocytes [76] and Jones et al [75] demonstrated that PBPK approaches were already more successful than predicting clearance from either *in vitro* metabolic data alone or from preclinical PK data. However, the approach still requires large relative activity factors to account for the differences *in vitro* and *in vivo,* and the basis for these scaling factors is unclear. A greater understanding of this is required before prospective application.

The difference between OATP expression in culture and *in vivo* is currently unknown and is likely to contribute to the scaling factors.

As in the liver, PBPK models can be used to simulate transporter processes in other tissues, such as the kidney, brain, and intestine. The major limitation to the inclusion of such processes in PBPK models currently is the availability of expression/abundance data *in vitro* and *in vivo*. As the proteomics techniques mentioned earlier advance, the availability of such data will rapidly increase as will our understanding of transporter mechanisms. Thus, the application of PBPK modeling in this area is certain to intensify.

First-pass metabolism in the intestinal mucosa can be an important determinant of oral bioavailability for many CYP3A substrates, and so the ability to predict the fraction of absorbed drug that escapes metabolism in the gut (Fg) is important for PBPK modeling. Several recent papers have described PBPK methods to model this [42,43,77], and a recent evaluation at Roche showed that the commercial PBPK tool, GastroPlus, may also be applied for this purpose [78]. GastroPlus Fg predictions for 20 CYP3A substrates were assessed using *in vitro* data for metabolic clearance and membrane permeation and resulted in good Fg prediction accuracy with no significant bias and 95% of predictions within twofold of the *in vivo* estimated Fg.

Development of absorption PBPK models has been rapid within the past 10 years, and this trend continues to be driven by workers in the industry [79,80], in academia [81–83], and in commercial software suppliers [84–86]. Relevant areas of recent development include the refinement of biorelevant methods for solubility and dissolution determination for forecasting of food effects in humans [82,87] as well as for application to advanced formulations such as lipid suspensions [88]. In addition, efforts are being devoted to characterize better the dog models used for preclinical formulation testing [89–91]. The motivation of this work is to use the preclinical species to verify *in vitro*-to-*in vivo* translation before making human predictions [64].

It is now clear that PBPK is well established in the pharmaceutical industry [92], and a recent analysis of new drug application (NDA) and investigational new drug (IND) application submissions to the U.S. Food and Drug Administration (FDA) [93] showed a pronounced increase in the number of applications where PBPK modeling is included. The FDA is also using commercial PBPK tools to check submitted work or to perform PBPK simulations in cases where this has not yet been included in the submission [93]. Both the FDA [74] and the European Medicines Agency (EMA) [94] make prominent mention of PBPK methods in recent guidance documents. Best practice guidance for documentation of PBPK models submitted to health authorities has also recently been published [53]. Given this acceptance of PBPK modeling by the pharmaceutical industry and the regulatory authorities, and the mechanistic nature of these models, the future for PBPK modeling is promising.

REFERENCES

1. Gabrielsson, J., Weiner, D. *Pharmacokinetic and Pharmacodynamic Data Analysis: Concepts and Applications*, 4th ed. Swedish Pharmaceutical Press, United States, 2007.

2. Poulin, P., Theil, F. (2002). Prediction of pharmacokinetics prior to in vivo studies I. Mechanism-based prediction of volume of distribution. *Journal of Pharmaceutical Sciences, 91*, 129–156.
3. Rodgers, T., Rowland, M. (2007). Mechanistic approaches to volume of distribution predictions: Understanding the processes. *Pharmaceutical Research, 24*(5), 918–933.
4. Reddy, M., Yang, R.S., Andersen, M.E., Clewell, H.J.I. *Physiologically Based Pharmacokinetic Modeling: Science and Applications*. New York, Wiley, 2005.
5. Rowland, M., Balant, L., Peck, C. (2004). Physiologically based pharmacokinetics in drug development and regulatory science: A workshop report. *AAPS PharmSci, 6*.
6. Rowland, M., Peck, C., Tucker, G. (2011). Physiologically-based pharmacokinetics in drug development and regulatory science. *Annual Review of Pharmacology and Toxicology, 51*(1), 45–73.
7. Rostami-Hodjegan, A., Tucker, G.T. (2007). Simulation and prediction of in vivo drug metabolism in human populations from in vitro data. *Nature Reviews Drug Discovery, 6*(2), 140–148.
8. Sugano, K. (2009). Introduction to computational oral absorption simulation. *Expert Opinion on Drug Metabolism & Toxicology, 5*(3), 259–293.
9. Jones, H., Parrott, N. Jorga, K., Lavé, T. (2006). A novel strategy for physiologically based predictions of human pharmacokinetics. *Clinical Pharmacokinetics, 45*(5), 511–542.
10. Jones, H.M., Gardner, I.B., Collard, W.T., Stanley, P.J., Oxley, P., Hosea, N.A., Plowchalk, D., Gernhardt, S., Lin, J., Dickins, M., Rahavendran, S.R., Jones, B.C., Watson, K.J., Pertinez, H., Kumar, V., Cole, S. (2011). Simulation of human intravenous and oral pharmacokinetics of 21 diverse compounds using physiologically based pharmacokinetic modelling. *Clinical Pharmacokinetics, 50*(5), 331–347.
11. Baneyx, G., Fukushima, Y., Parrott, N. (2012). Use of physiologically based pharmacokinetic modeling for assessment of drug–drug interactions. *Future Medicinal Chemistry, 4*(5), 681–693.
12. Bjorkman, S. (2005). Prediction of drug disposition in infants and children by means of physiologically based pharmacokinetic (PBPK) modelling: Theophylline and midazolam as model drugs. *British Journal of Clinical Pharmacology, 59*(6), 691–704.
13. Parrott, N., Davies, B., Hoffmann, G., Koerner, A., Lave, T., Prinssen, E., Theogaraj, E., Singer, T. (2011). Development of a physiologically based model for oseltamivir and simulation of pharmacokinetics in neonates and infants. *Clinical Pharmacokinetics, 50*(9), 613–623.
14. Yeo, R.K., Aarabi, M., Jamei, M., Rostami-Hodjegan, A. (2011). Modeling and predicting drug pharmacokinetics in patients with renal impairment. *Expert Review of Clinical Pharmacology, 4*(2), 261–274.
15. Edginton, A.N., Willmann, S. (2008). Physiology-based simulations of a pathological condition: Prediction of pharmacokinetics in patients with liver cirrhosis. *Clinical Pharmacokinetics, 47*(11), 743–752.
16. Cubitt, H.E., Yeo, K.R., Howgate, E.M., Rostami-Hodjegan, A., Barter, Z.E. (2011). Sources of interindividual variability in IVIVE of clearance: An investigation into the prediction of benzodiazepine clearance using a mechanistic population-based pharmacokinetic model. *Xenobiotica, 0*(0), 1–16.
17. Inoue, S., Howgate, E.M., Yeo, R.K., Shimada, T., Yamazaki, H., Tucker, G.T., Rostami-Hodjegan, A. (2006). Prediction of in vivo drug clearance from in vitro data. II: Potential inter-ethnic differences. *Xenobiotica, 36*(6), 499–513.

18. Zhao, P., Ragueneau-Majlessi, I., Zhang, L., Strong, J.M., Reynolds, K.S., Levy, R.H., Thummel, K.E., Huang, S.-M. (2009). Quantitative evaluation of pharmacokinetic inhibition of CYP3A substrates by ketoconazole: A simulation study. *The Journal of Clinical Pharmacology*, *49*(3), 351–359.
19. Parrott, N., Paquereau, N., Coassolo, P., Lave, T. (2005). An evaluation of the utility of physiologically based models of pharmacokinetics in early drug discovery. *Journal of Pharmaceutical Sciences*, *94*(10), 2327–2343.
20. Jones, H.M., Gardner, I.B., Watson, K.J. (2009). Modelling and PBPK simulation in drug discovery. *The AAPS Journal*, *11*(1).
21. Parrott, N., Jones, H., Paquereau, N., Lave, T. (2005). Application of full physiological models for pharmaceutical drug candidate selection and extrapolation of pharmacokinetics to man. *Basic & Clinical Pharmacology & Toxicology*, *96*, 193–199.
22. Jones, H.M., Dickins, M., Youdim, K., Gosset, J.R., Attkins, N.J., Hay, T.L., Gurrell, I.K., Logan, Y.R., Bungay, P.J., Jones, B.C., Gardner, I.B. (2012). Application of PBPK modelling in drug discovery and development at Pfizer. *Xenobiotica*, *42*(1), 94–106.
23. Poulin, P., Theil, F. (2002). Prediction of pharmacokinetics prior to in vivo studies II. Generic physiologically based pharmacokinetic models of drug disposition. *Journal of Pharmaceutical Sciences*, *91*, 1358–1370.
24. Theil, F., Haddad, S., Guentert, T., Poulin, P. (2003). Utility of physiologically based pharmacokinetic models to drug development and rational drug discovery candidate selection. *Toxicology Letters*, (138), 29–49.
25. Brightman, F.A., Leahy, D.E., Searle, G.E., Thomas, S. (2006). Application of a generic physiologically based pharmacokinetic model to the estimation of xenobiotic levels in human plasma. *Drug Metabolism and Disposition*, *34*(1), 94–101.
26. De Buck, S.S., Sinha, V.K., Fenu, L.A., Nijsen, M.J., Mackie, C.E., Gilissen, R.A.H.J. (2007). Prediction of human pharmacokinetics using physiologically based modeling: A retrospective analysis of 26 clinically tested drugs. *Drug Metabolism and Disposition*, *35*(10).
27. Sinha, V.K., Snoeys, J., Osselaer, N.V., Peer, A.V., Mackie, C., Heald, D. (2012). From preclinical to human—prediction of oral absorption and drug–drug interaction potential using physiologically based pharmacokinetic (PBPK) modeling approach in an industrial setting: A workflow by using case example. *Biopharmaceutics & Drug Disposition*, *33*(2), 111–121.
28. Chen, Y., Jin, J.Y., Mukadam, S., Malhi, V., Kenny, J.R. (2012). Application of IVIVE and PBPK modeling in prospective prediction of clinical pharmacokinetics: Strategy and approach during the drug discovery phase with four case studies. *Biopharmaceutics & Drug Disposition*, *33*(2), 85–98.
29. Peters, S.A., (2008). Evaluation of a generic physiologically based pharmacokinetic model for lineshape analysis. *Clinical Pharmacokinetics*, *47*(4), 261–275.
30. Edginton, A.N., Theil, F.-P., Schmitt, W., Willmann, S. (2008). Whole body physiologically-based pharmacokinetic models: Their use in clinical drug development. *Expert Opinion on Drug Metabolism & Toxicology*, *4*(9), 1143–1152.
31. Poulin, P., Jones, R.D.O., Jones, H.M., Gibson, C.R., Rowland, M., Chien, J.Y., Ring, B.J., Adkison, K.K., Ku, M.S., He, H., Vuppugalla, R., Marathe, P., Fischer, V., Dutta, S., Sinha, V.K., Björnsson, T., Lavé, T., Yates, J.W.T. (2011). PHRMA CPCDC initiative on predictive models of human pharmacokinetics, part 5: Prediction of plasma

concentration–time profiles in human by using the physiologically-based pharmacokinetic modeling approach. *Journal of Pharmaceutical Sciences*, *100*(10), 4127–4157.

32. Bungay, P., Tweedy, S., Howe, D., Gibson, K., Jones, H., Mount, N. (2011). Pre-clinical and clinical pharmacokinetics of PF-02413873, a non-steroidal progesterone receptor antagonist. *Drug Metabolism and Disposition*, *39*(8),1396–1405.
33. Yamazaki, S., Skaptason, J., Romero, D., Vekich, S., Jones, H.M., Tan, W., Wilner, K.D., Koudriakova, T. (2011). Prediction of oral pharmacokinetics of cmet kinase inhibitors in humans: Physiologically based pharmacokinetic model versus traditional one-compartment model. *Drug Metabolism and Disposition*, 383–393.
34. Allan, G., Davis, J., Dickins, M., Gardner, I., Jenkins, T., Jones, H., Webster, R., and Westgate, H. (2008). Pre-clinical pharmacokinetics of UK-453,061, a novel non-nucleoside reverse transcriptase inhibitor (NNRTI), and use of in silico physiologically based prediction tools to predict the oral pharmacokinetics of UK-453,061 in man. *Xenobiotica*, *38*(6), 620–640.
35. Lave, T., Parrott, N., Grimm, H.P., Fleury, A., Reddy, M. (2007). Challenges and opportunities with modelling and simulation in drug discovery and drug development. *Xenobiotica*, *37*(10), 1295–1310.
36. Berry, P., Parrott, N., Reddy, M., David-Pierson, P., Lave, T. putting it all together, in hit and lead profiling, in B. Faller and L. Urban, (eds.), *Identification and Optimization of Drug-like Molecules*, Wiley, Hoboken, NJ, 2009.
37. Johnson, K.C., Swindell, A.C. (1996). Guidance in the setting of drug particle size specifications to minimize variability in absorption. *Pharmaceutical Research*, *13*(12), 1795–1798.
38. Sun, D., Yu, L.X., Hussain, M.A., Wall, D.A., Smith, R.L., Amidon, G.L. (2004). In vitro testing of drug absorption for drug "developability" assessment: Forming an interface between in vitro preclinical data and clinical outcome. *Current Opinion in Drug Discovery & Development*, *7*(1), 75–85.
39. Parrott, N., Lave, T. (2008). Applications of physiologically based absorption models in drug discovery and development. *Molecular Pharmaceutics*, *5*(5), 760–775.
40. Einolf, H.J. (2007). Comparison of different approaches to predict metabolic drug–drug interactions. *Xenobiotica*, *37*(10), 1257–1294.
41. Poirier, A., Cascais, A.C., Funk, C., Lave, T. (2009). Prediction of pharmacokinetic profile of valsartan in human based on in vitro uptake transport data. *Journal of Pharmacokinetics and Pharmacodynamics*, *36*(6), 585–611.
42. Yang, J., Jamei, M., Yeo, K.R., Tucker, G.T., Rostami-Hodjegan, A. (2007). prediction of intestinal first-pass drug metabolism. *Current Drug Metabolism*, *8*, 676–684.
43. Bruyere, A., Decleves, X., Bouzom, F., Ball, K., Catie Marques, M.P., Treton, X., Valleur, P., Yoram Bouhnik, Y.P., Scherrmann, J.-M., Mouly, S. (2010). Effect of variations in the amounts of P-Glycoprotein (ABCB1), BCRP (ABCG2) and CYP3A4 along the Human Small Intestine on PBPK models for predicting intestinal first pass. *Molecular Pharmaceutics*, *7*(5), 1596–1607.
44. Wang, Y.-H. (2010). Confidence assessment of the simcyp time-based approach and a static mathematical model in predicting clinical drug-drug Interactions for mechanism-based CYP3A inhibitors. *Drug Metabolism and Disposition*, *38*(7), 1094–1104.
45. Yeo, R.K., Jamei, M., Yang, J., Tucker, G.T., Rostami-Hodjegan, A. (2010). Physiologically based mechanistic modelling to predict complex drug-drug interactions involving

simultaneous competitive and time-dependent enzyme inhibition by parent compound and its metabolite in both liver and gut–The effect of diltiazem on the time-course of exposure to triazolam. *European Journal of Pharmaceutical Sciences*, *39*(5), 298–309.

46. Lukacova, V., Parrott, N., Howard, M., Woltosz, W., Bolger, M. (2010). Prediction of omeprazole's disposition and drug-drug interactions using a physiologically-based pharmacokinetic model. Poster at ADMET Europe, 2010.
47. Peters, S.A., Schroeder, P.E., Giri, N., Dolgos, H. (2012). Evaluation of the use of static and dynamic models to predict drug-drug interaction and its associated variability: Impact on drug discovery and early development. *Drug Metabolism and Disposition*, *40*, 1495–1507.
48. Guest, E.J., Yeo, R.K., Rostami-Hodjegan, A., Tucker, G.T., Houston, J.B., Galetin, A. (2011). Assessment of algorithms for predicting drug–drug interactions via inhibition mechanisms: Comparison of dynamic and static models. *British Journal of Clinical Pharmacology*, *71*(1), 72–87.
49. Fahmi, O.A., Hurst, S., Plowchalk, D., Cook, J., Guo, F., Youdim, K., Dickins, M., Phipps, A., Darekar, A., Hyland, R., Obach, R.S. (2009). Comparison of different algorithms for predicting clinical drug-drug interactions, based on the use of CYP3A4 in vitro data: Predictions of compounds as precipitants of interaction. *Drug Metabolism And Disposition*, *37*(8), 1658–1666.
50. Youdim, K.A., Zayed, A., Dickins, M., Phipps, A., Griffiths, M., Darekar, A., Hyland, R., Fahmi, O., Hurst, S., Plowchalk, D.R., Cook, J., Guo, F., Obach R.S. (2008.) Application of CYP3A4 in vitro data to predict clinical drug–drug interactions; Predictions of compounds as objects of interaction. *British Journal of Clinical Pharmacology*, *65*(5), 680–692.
51. Hyland, R., Collins, M.D.C., Jones, H., Jones, B. (2008). Maraviroc: In vitro assessment of drug–drug interaction potential. *British Journal of Clinical Pharmacology*, *66* (4), 498–507.
52. Rakhit, A., Pantze, M., Fettner, S., Jones, H., Charoin, J.-E., Riek, M., Lum, B., Hamilton, M. (2008). The effects of CYP3A4 inhibition on erlotinib pharmacokinetics: Computer-based simulation (SimCYP) predicts in vivo metabolic inhibition. *European Journal of Clinical Pharmacology*, *64*(1), 31–41.
53. Zhao, P., Rowland, M., Huang, S.M. (2012). Best practice in the use of physiologically based pharmacokinetic modeling and simulation to address clinical pharmacology regulatory questions. *Clinical Pharmacology & Therapeutics*, *92*(1), 17–20.
54. Amidon, G.L., Lennernas, H., Shah, V.P., Crison, J.R. (1995). A theoretical basis for a biopharmaceutic drug classification: The correlation of in vitro drug product dissolution and in vivo bioavailability. *Pharmaceutical Research*, *12*(3), 413–420.
55. Kuentz, M., Nick, S., Parrott, N., Rothlisberger, D. (2006). A strategy for preclinical formulation development using GastroPlusTM as pharmacokinetic simulation tool and a statistical screening design applied to a dog study. *European journal of pharmaceutical sciences*, *27*, 91–99.
56. Kuentz, M. (2008). Drug absorption modeling as a tool to define the strategy in clinical formulation development. *The AAPS Journal*, *10*(3).
57. Kansy, M., Avdeef, A., Fischer, H. (2004). Advances in screening for membrane permeability: High-resolution PAMPA for medicinal chemists. *Drug Discovery Today: Technologies*, *1*(4), 349–355.
58. Avdeef, A. *Absorption and Drug Development—Solubility Permeability and Charge State*, Wiley-IEEE, Hoboken, NJ, 2003.

59. Alsenz, J., Haenel, E. (2003). Development of a 7-day, 96-well Caco-2 permeability assay with high-throughput direct uv compound analysis. *Pharmaceutical Research*, *20*(12), 1961–1969.
60. Lennernas, H. (1997). Human intestinal permeability. *Journal of Pharmaceutical Sciences*, *87*(4), 403–410.
61. Jantratid, E., Janssen, N., Reppas, C., Dressman, J.B. (2008). Dissolution media simulating conditions in the proximal human gastrointestinal tract: An update. *Pharmaceutical Research*, *25*(7), 1663.
62. Jones, H., Parrott, N., Ohlenbusch, G., Lavé, T. (2006). Predicting pharmacokinetic food effects using biorelevant solubility media and physiologically based modelling. *Clinical Pharmacokinetics*, *45*(12), 1213–1226.
63. Parrott, N., Lukacova, V., Fraczkiewicz G., Bolger, M. (2009). Predicting pharmacokinetics of drugs using physiologically based modeling—application to food effects. *The AAPS Journal*, *11*(1), 45.
64. Parrott, N., Lave, T. Computer models for predicting drug absorption, in J. Dressman and C. Reppas, (eds.), *Oral Drug Absorption*, Springer, New York, 2010.
65. Heikkinen, A.T., Friedlein, A., Lamerz, J., Jakob, P., Cutler, P., Fowler, S., Williamson, T., Tolando, R., Lave, T., Parrott, N. (2012). Mass spectrometry based quantification of cyp enzymes to establish in vitro-in vivo scaling factors for intestinal and hepatic metabolism in beagle dog. *Pharmaceutical Research*, *29*(7), 1832–1842.
66. Jamei, M., Dickinson, G.L., Rostami-Hodjegan, A. (2009). A framework for assessing inter-individual variability in pharmacokinetics using virtual human populations and integrating general knowledge of physical chemistry, biology, anatomy, physiology and genetics: A tale of "bottom-up" vs "top-down" recognition of covariates. *Drug Metabolism and Pharmacokinetics*, *24*(1), 53–75.
67. Johnson, T.N., Boussery, K., Yeo, R.K., Tucker, G.T., Rostami-Hodjegan, A. (2010). A semi-mechanistic model to predict the effects of liver cirrhosis on drug clearance. *Clinical Pharmacokinetics*, *49*(3), 189–206.
68. Yeo, R.K., Aarabi, M., Jamei, M., Rostami-Hodjegan, A. (2011). Modeling and predicting drug pharmacokinetics in patients with renal impairment. *Expert Review of Clinical Pharmacology*, *4*(2), 261–274.
69. Gaohua, L., Abduljalil, K., Jamei, M., Johnson, T.N., Rostami-Hodjegan, A. (2012). A pregnancy physiologically-based pharmacokinetic (p-pbpk) model for disposition of drugs metabolized by CYP1A2, CYP3A4 and CYP2D6. *British Journal of Clinical Pharmacology*, *74*(5), 873–885.
70. Johnson, T.N., Rostami-Hodjegan, A. (2010.) Resurgence in the use of physiologically based pharmacokinetic models in pediatric clinical pharmacology: Parallel shift in incorporating the knowledge of biological elements and increased applicability to drug development and clinical practice. *Pediatric Anesthesia*, *21*(3), 291–301.
71. Rostami-Hodjegan, A. (2012). Physiologically based pharmacokinetics joined with in vitro–in vivo extrapolation of ADME: A marriage under the arch of systems pharmacology. *CPET*, *92*(1), 50–61.
72. Kawakami, H., Ohtsuki, S., Kamiie, J., Suzuki, T., Abe, T., Terasaki, T. (2011). Simultaneous absolute quantification of 11 cytochrome P450 isoforms in human liver microsomes by liquid chromatography tandem mass spectrometry with in silico target peptide selection. *Journal of Pharmaceutical Sciences*, *100*(1), 341–352.

73. Kamiie, J., Ohtsuki, S., Iwase, R., Ohmine, K., Katsukura, Y., Yanai, K., Sekine, Y.U.Y., Ito, S., Terasaki, T. (2008). Quantitative atlas of membrane transporter proteins: Development and application of a highly sensitive simultaneous LC/MS/MS method combined with novel in-silico peptide selection criteria. *Pharmaceutical Research*, *25*(6), 1469–1483.
74. Administration, F.A.D. (2012). Guidance for industry drug interaction studies—Study design, data analysis, implications for dosing, and labeling recommendations, http://www.fda.gov/downloads/Drugs/GuidanceComplianceRegulatoryInformation/Guidances/ucm072101.pdf.
75. Jones, H.M., Barton, H.A., Lai, Y., Bi, Y., Kimoto, E., Kempshall, S., Tate, S.C., El-Kattan, A., Houston, J.B., Galetin, A., Fenner, K.S. (2012). Mechanistic pharmacokinetic modeling for the prediction of transporter-mediated disposition in human from sandwich culture human hepatocyte data. *Drug Metabolism and Disposition*, *40*(5), 1007–1017.
76. Bi, Y.-A., Kimoto, E., Sevidal, S., Jones, H.M., Barton, H.A., Kempshall, S., Whalen, K.M., Zhang, H., Ji, C., Fenner, K.S., El-Kattan, A.F., Lai, Y. (2012). In vitro evaluation of hepatic transporter-mediated clinical drug-drug interactions: Hepatocyte model optimization and retrospective investigation. *Drug Metabolism and Disposition*, *40*(6), 1085–1092.
77. Gertz, M., Houston, J.B., Galetin, A. (2011). Physiologically-based pharmacokinetic modeling of intestinal first-pass metabolism of CYP3A substrates with high intestinal extraction. *Drug Metabolism and Disposition*, *39*(9), 1633–1642.
78. Heikkinen, A.T., Baneyx, G., Caruso, A., Parrott, N. (2012). Application of PBPK modeling to predict human intestinal metabolism of CYP3A substrates—an evaluation and case study using GastroPlus™. *European Journal of Pharmaceutical Sciences*, *47*, 375–386.
79. Sugano, K., Kataoka, M., da Costa Mathews, C., Yamashita, S. (2010). Prediction of food effect by bile micelles on oral drug absorption considering free fraction in intestinal fluid. *European Journal of Pharmaceutical Sciences*, *40*(2), 118–124.
80. Takano, R. (2010). Quantitative analysis of the effect of supersaturation on in vivo drug absorption. *Molecular Pharmaceutics*, *7*(5), 1431–1440.
81. Lundahl, A., Hedeland, M., Bondesson, U., Lennernäs, H. (2011). In vivo investigation in pigs of intestinal absorption, hepatobiliary disposition and metabolism of the 5-alpha reductase inhibitor finasteride and the effects of co-administered ketoconazole. *Drug Metabolism and Disposition*, *39*(5), 847–857.
82. Shono, Y., Jantratid, E., Janssen, N., Kesisoglou, F., Mao, Y., Vertzoni, M., Reppas, C., Dressman, J.B. (2009). Prediction of food effects on the absorption of celecoxib based on biorelevant dissolution testing coupled with physiologically based pharmacokinetic modeling. *European Journal of Pharmaceutics and Biopharmaceutics*, *73*(1), 107–114.
83. Pang, K.S., Chow, E.C.Y. (2012). Prediction of flow effects on intestinal (FI) and systemic availability (Fsys) in PBPK intestine models: The traditional model (TM), segregated flow model (SFM) and QGut model. *Drug Metabolism and Disposition*, *40*(10), 1869–1877.
84. Thelen, K., Coboeken, K., Willmann, S., Burghaus, R., Dressman, J.B., Lippert, J. (2011). Evolution of a detailed physiological model to simulate the gastrointestinal transit and absorption process in humans, part 1: Oral solutions. *Journal of Pharmaceutical Sciences*, *100*(12), 5324–5345.
85. Jamei, M., Turner, D., Yang, J., Neuhoff, S., Polak, S., Rostami-Hodjegan, A., Tucker, G. (2009). Population-based mechanistic prediction of oral drug absorption. *The AAPS Journal*, *11*(2), 225–237.

86. Abuasal, B.S., Bolger, M.B., Walker, D.K., Kaddoumi, A. (2012). In silico modeling for the nonlinear absorption kinetics of UK-343,664: A P-gp and CYP3A4 substrate. *Molecular Pharmaceutics*, *9*(3), 492–504.
87. Shono, Y., Jantratid, E., Kesisoglou, F., Reppas, C., Dressman, J.B. (2010). Forecasting in vivo oral absorption and food effect of micronized and nanosized aprepitant formulations in humans. *European Journal of Pharmaceutics and Biopharmaceutics*, *76*(1), 95–104.
88. Jantratid, E., Janssen, N., Chokshi, H., Tang, K., Dressman, J.B. (2008). Designing biorelevant dissolution tests for lipid formulations: Case example—Lipid suspension of RZ-50. *European Journal of Pharmaceutics and Biopharmaceutics*, *69*(2), 776–785.
89. Arndt, M., Jantratid, E., Chokshi, H., Tang, K., Parrott, N.J., Reppas, C., Dressman, J.B. (2010). Characterization of homogenized dog meals used for food effect study in pre-clinical stage. *7th World Meeting on Pharmaceutics, Biopharmaceutics and Pharmaceutical Technology*, APV Malta.
90. Arndt, M., Wagner, C., Kambayashi, A., Otsuka, K., Chokshi, H., Tang, K., Parrott, N.J., Reppas, C., Dressman, J.B. (2012). Use of canine biorelevant media in prediction of dog pharmacokinetics for an oral celecoxib formulation. APV Istanbul 2012 Poster.
91. Haller, S., Schuler, F., Lazic, S.E., Dalila, B.-C., Stefanie, D.K., Neil, J.P., Guido, S., Sara, B. (2012). Expression profiles of metabolic enzymes and drug transporters in the liver and along the intestine of beagle dogs. *Drug Metabolism and Disposition*, *40*, 1603–1611.
92. Rostami-Hodjegan, A., Tamai, I., Pang, K.S. (2012). Physiologically based pharmacokinetic (PBPK) modeling: It is here to stay! *Biopharmaceutics & Drug Disposition*, *33*(2), 47–50.
93. Zhao, P., Zhang, L., Grillo, J.A., Liu, Q., Bullock, J.M., Moon, Y.J., Song, P., Brar, S.S., Madabushi, R., Wu, T.C., Booth, B.P., Rahman, N.A., Reynolds, K.S., Gil Berglund, E., Lesko, L.J., Huang, S.M. (2011). Applications of physiologically based pharmacokinetic (PBPK) modeling and simulation during regulatory review. *Clinical Pharmacology & Therapeutics*, *89*(2), 259–267.
94. EMA, *Guideline on the Investigation of Drug Interactions*, CHMP, Editor, 2012.

10

EMERGING FULL MECHANISTIC PHYSIOLOGICALLY BASED MODELING

Kiyohiko Sugano

10.1 INTRODUCTION

The prediction of clinical pharmacokinetics (PK) from preclinical data is still challenging. Recently, physiologically based pharmacokinetics (PBPK) modeling became available for drug discovery and development. Currently, the paradigm of PBPK modeling is moving from semi-empirical to full-mechanistic ones.

10.2 PARADIGM SHIFT

In the semi-empirical PBPK modeling paradigm, *apparent in vitro* data are directly used for modeling (Figure 10.1A). For example, *apparent* Caco-2 permeability (P_{app}) was directly used to predict *in vivo* P_{eff}. However, apparent *in vitro* data are not always appropriate for the full-mechanistic PBPK modeling, as they often consist of multiple processes and do not represent an intrinsic (primary) process. *To move forward to a full-mechanistic PBPK modeling paradigm, apparent* in vitro *data should be first decomposed to intrinsic (primary) PK parameters by applying a mechanistic mathematical model(s) to the* in vitro *data* (Figure 10.1B). The primary PK parameters

Predictive ADMET: Integrative Approaches in Drug Discovery and Development, First Edition.
Edited by Jianling Wang and Laszlo Urban.

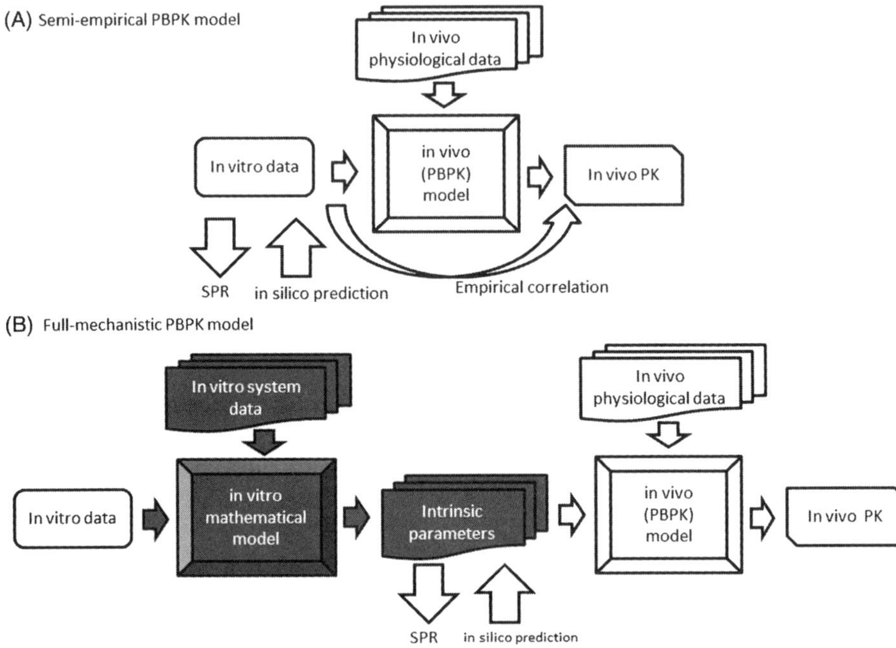

Figure 10.1 Current and future paradigms of PK prediction.

should be then used for the full-mechanistic PBPK models. This paradigm shift also enables the in-combo approach. *In silico* (prediction from chemical structural information), *in vitro*, and *in vivo* data are used in a combined manner to fill the gap of each method. The paradigm shift also changes the target of structure PK relationship (SPR) studies. The use of a primary PK parameter for SPR would result in clearer and simpler correlation with the chemical structure (Section 10.5.6).

10.3 PAST PARADIGM OF BIOPHARMACEUTICAL MODELING

Biopharmaceutical modeling is one area where human prediction has been extensively investigated [1]. The oral absorption of a drug is determined by the solubility, dissolution rate, precipitation rate and permeability of a drug in the *in vivo* gastrointestinal (GI) tract (Figure 10.2).

In a semi-empirical modeling paradigm, the apparent solubility of a drug in an artificial biorelevant media, such as the fasted state simulated intestinal fluid (FaSSIF [2]), was directly used to calculate the pH solubility profile and the dissolution rate of the drug by the Henderson–Hasselbalch (HH) and Noyes Whitney–Nernst Brunner (NWNB) equations, respectively. The *in vivo* effective intestinal membrane permeability (P_{eff}) of a drug was predicted from P_{app} using an empirical linear

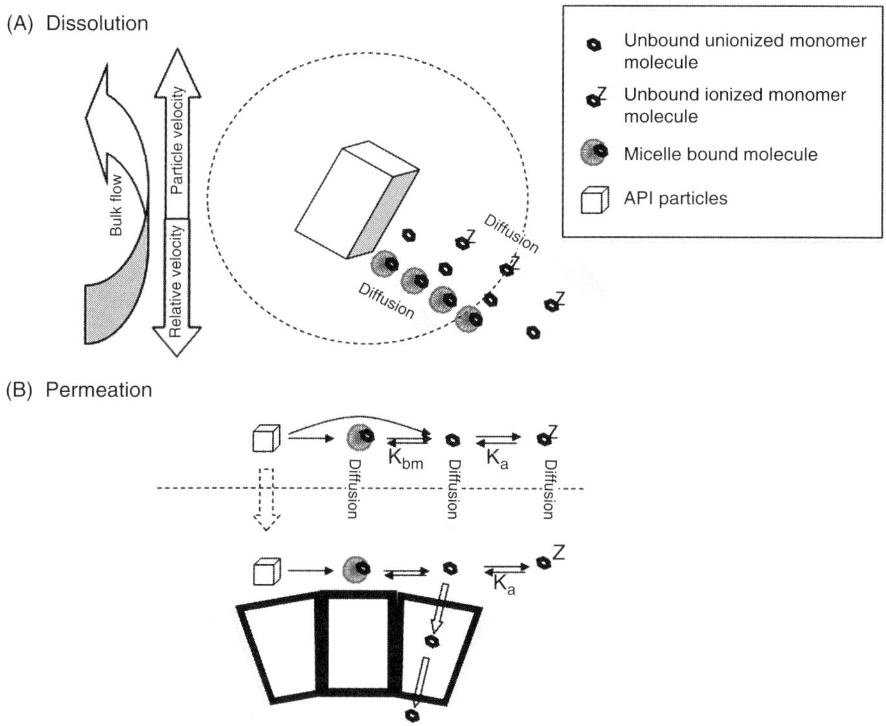

Figure 10.2 Schemes of oral absorption of a drug.

equation. Together with apparent solubility, this P_{eff} value was used to calculate the absorption rate of a drug.

10.4 FUTURE PARADIGM OF BIOPHARMACEUTICAL MODELING

Full-mechanistic biopharmaceutical modeling has been extensively investigated during the last several years [1, 3–22]. The framework of theoretical equations is called the gastrointestinal unified theoretical (GUT) framework in this chapter [1, 3]. The unified concept and description of drug concentration is one key feature of the GUT framework. Each molecular species of a drug in a media, i.e., unbound undissociated molecules, unbound dissociated molecules, bile micelle bound molecules, etc., is explicitly taken into account. In addition, the anatomical and physiological features of the GI tract are explicitly captured by the model (the plica and villi structures, the asymmetric cellular membrane surface area, diffusion through the unstirred water layer (UWL) and the subepithelial space, the villi blood flow, transporters, metabolisms, etc.). In addition to the GI compartment transit model [23], the physiological GI tube model and the simple approximate model are provided [3].

10.4.1 Solubility and Dissolution Rate

The solubility of a drug is a classic example of where a mechanistic model has been applied for *in vitro* data. The HH equation can be modified to incorporate the bile micelle partitioning [1, 24, 25]. For a base:

$$S_{dissolv} = S_0 \left(1 + \frac{[H^+]}{K_a} + \frac{C_{bile}}{C_{water}} \cdot K_{bm,0} + \frac{[H^+]}{K_a} \frac{C_{bile}}{C_{water}} \cdot K_{bm,+} \right) \quad (10.1)$$

For an acid:

$$S_{dissolv} = S_0 \left(1 + \frac{K_a}{[H^+]} + \frac{C_{bile}}{C_{water}} \cdot K_{bm,0} + \frac{K_a}{[H^+]} \cdot \frac{C_{bile}}{C_{water}} \cdot K_{bm,-} \right) \quad (10.2)$$

$S_{dissolv}$: solubility in bile micelle media
S_0: intrinsic solubility (of undissociated drug form)
K_a: dissociation constant
K_{bm}: bile micelle partition coefficient (subscripts indicate charge of molecule)
C_{bile}: bile concentration
C_{water}: water concentration (55.5 M)

K_{bm} can be calculated from the solubility data in FaSSIF and blank FaSSIF. The S_0, K_{bm}, and pK_a values are used to predict the solubility of a drug *in vivo*. The regional and species differences of pH and bile micelle concentration can be captured by this approach. In addition, these data are used to calculate the unbound fractions (f_u) and the undissociated fractions (f_n).

$$f_u = \frac{S_{blank}}{S_{dissolv}} \quad (10.3)$$

$$f_n = \frac{S_0}{S_{blank}} \quad (10.4)$$

S_{blank}: solubility of a drug in a buffer without bile micelle (same pH)

The dissolution rate coefficient of a drug is not proportional to the apparent equilibrium solubility of the drug as the diffusion coefficient of bile micelle bound molecules is ca. 1/10~1/80 of monomer drug molecules [16, 26]. In addition, for a dissociable drug, the solubility at the solid surface is significantly different from that in a bulk buffer as a result of the self-buffering effect of the dissolving drug [27, 28]. The NWNB equation can be modified to incorporate the effects of bile micelles and solid surface pH [1]. For monodispersed spherical particles with the particle diameter of <60 μm, the dissolution rate ($dX_{dissolve} / dt$) can be approximated as:

$$\frac{dX_{disssolv}}{dt} = k_{diss}X_{undissolv} = \frac{3D_{eff}}{r_p^2 \rho_p} \cdot S_{surface} \cdot \left(1 - \frac{X_{dissolv}}{V_{GI} \cdot S_{dissolv}}\right) X_{undissolv} \quad (10.5)$$

$$D_{eff} = D_{mono} \cdot f_u + D_{bm}(1 - f_u) \quad (10.6)$$

$X_{dissolv}$: dissolved drug amount
$X_{undissolv}$: undissolved drug amount
k_{diss}: dissolution rate coefficient
$S_{surface}$: solid surface solubility
D_{eff}: effective diffusion coefficient
D_{mono}: diffusion coefficient of monomer drug molecule
D_{bm}: diffusion coefficient of bile micelle bound drug molecule
ρ_p: true density
r_p: particle radius
V_{GI}: fluid volume

The solid surface solubility of a drug depends on S_0, pK_a, D_{mono}, bulk pH, buffer composition, and buffer concentration [27,28]. It should be noted that the dissolution test fluid usually has a higher buffer concentration than that *in vivo*, leading to overestimation of the dissolution rate for dissociable drugs [29].

Nucleation of the new solid form in the GI tract is another area where a mechanistic approach should be applied (Figure 10.3). Currently, supersaturation in the GI tract is under extensive investigation [30]. The nucleation theories would be the basis for mechanistic modeling [11].

10.4.2 Intestinal Membrane Permeability

10.4.2.1 Passive Permeation Recently, a full-mechanistic model for passive membrane permeation was proposed (Figure 10.4). The effect of microclimate pH, bile micelles (unbound fraction), UWL, the villi and fold (plica) structures, the intestinal tube shape, and the paracellular pathway are explicitly taken into account.

$$k_{perm} = \frac{2DF}{R_{GI}} P_{eff} \quad (10.7)$$

$$P_{eff}(or\ P_{app}) = \frac{PE}{\frac{1}{P_{UWL}} + \frac{1}{P'_{ep}}} = \frac{PE}{\frac{1}{P_{UWL}} + \frac{1}{VE \cdot Acc \cdot f_u P_{ep}}} \quad (10.8)$$

$$P_{ep} = f_n \cdot P_{trans,n} + P_{para} \quad (10.9)$$

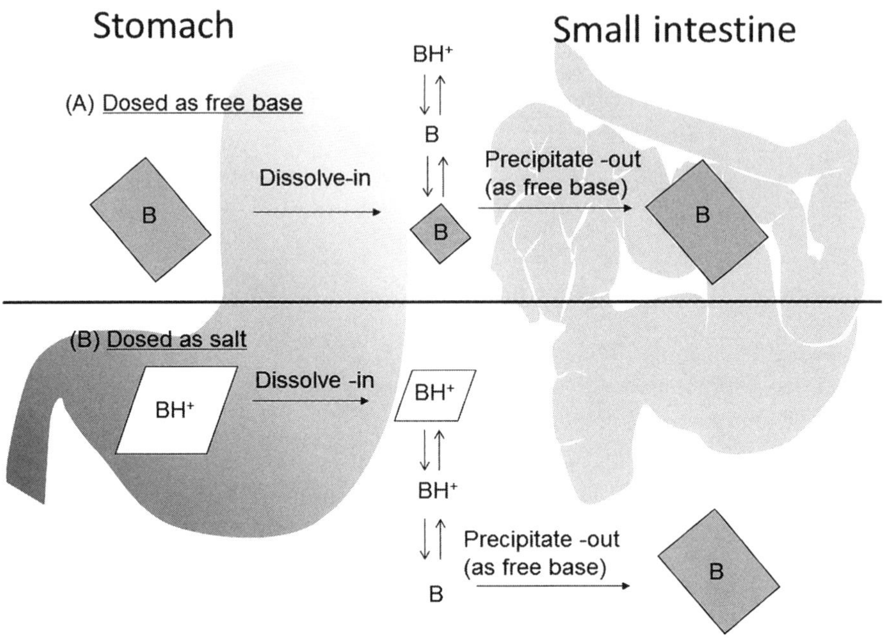

Figure 10.3 Dissolution and precipitation of free base and salt. (A) After dosing a free base, a portion of a drug dissolves in the stomach. However, some portion reaches the small intestine before completely dissolved and regrows in the small intestine. (B) After dosing a salt, the drug molecules precipitate out as a free base in the small intestine.

$$P_{UWL} = \frac{D_{eff}}{h_{UWL}} + P_{WC} = \frac{f_u \cdot D_{mono} + (1 - f_u)D_{bm}}{h_{UWL}} + P_{WC} \tag{10.10}$$

$$P_{para} = A_{para} \cdot \frac{1}{MW^{1/3}} RK \left(\frac{MW^{1/3}}{R_{MW}} \right) \sum^{z} f_z E(z) \tag{10.11}$$

$$RK(x) = (1 - x)^2 (1 - 2.104 \cdot x + 2.09(x)^3 - 0.95(x)^5) \tag{10.12}$$

$$E(z) = \frac{Z_{para} \cdot z}{1 - \exp(-Z_{para} \cdot z)} \qquad (for\ z = n, E(z) = 1) \tag{10.13}$$

k_{perm}: permeation rate coefficient
P_{ep}, P_{ep}': epithelial membrane permeability
$P_{trans,n}$: intrinsic passive transcellular permeability of undissociated species

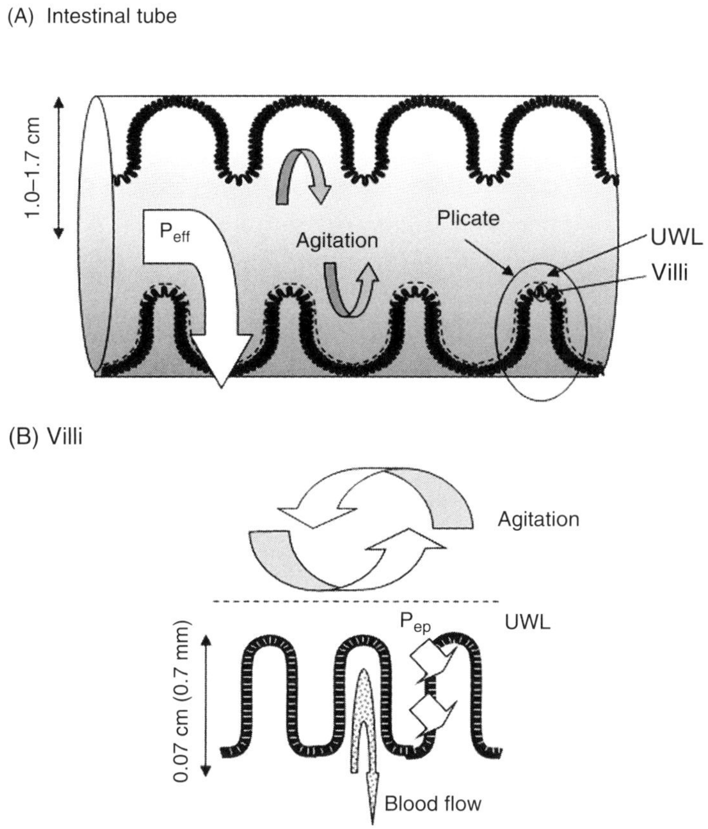

Figure 10.4 Intestinal membrane permeability.

P_{UWL}: UWL permeability
P_{para}: paracellular pathway permeability
h_{UWL}: effective thickness of UWL
R_{MW}: paracellular pore size based on MW
Z_{para}: paracellular electric potential factor
z: molecular charge (n (neutral: 0), + (mono cation: +1), ++ (di cation: +2), -, –, etc.)
A_{para}: paracellular area coefficient
DF: degree of flatness
PE: plica expansion coefficient
VE: villi expansion coefficient
Acc: villi surface accessibility (can be assumed ≈ 1 for many cases) [31]
P_{WC}: water conveyance

As a rule of thumb:

$$\log D_{oct,pH6.5} < 0 : P_{eff} \approx PE \cdot VE \cdot f_u \cdot P_{ep} \tag{10.14}$$

$$\log D_{oct,pH6.5} < 2 : P_{eff} \approx \frac{D_{eff}}{h_{UWL}} \cdot PE \tag{10.15}$$

The apparent permeability in Caco-2 (or another *in vitro* assay) can be converted to $P_{trans,n}$ (Figure 10.5). For drugs with $\log D_{oct} > 1.5$, membrane binding can significantly hinder the permeability of a drug, leading to misclassification to P_{ep} limited

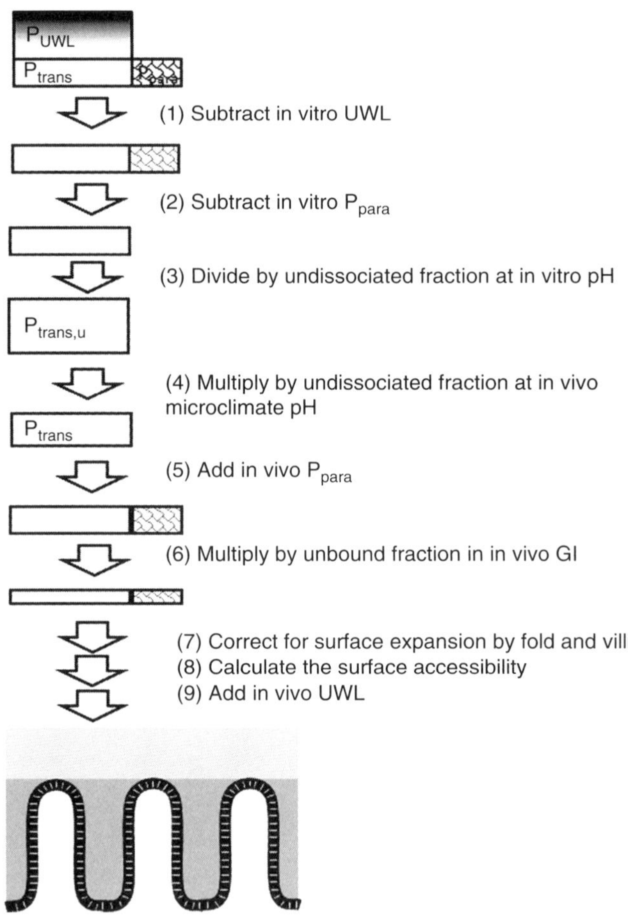

Figure 10.5 Procedure of P_{eff} prediction. *In vitro* hUWL can be obtained from the pH–P_{app} relationship of a high-permeability drug (e.g., ketoprofen) or the log$D$–P_{app} relationship. A_{para}, R_{MW}, and Z_{para} can be obtained by fitting Equations 7–13 to the *in vitro* permeability of several paracellular substrates.

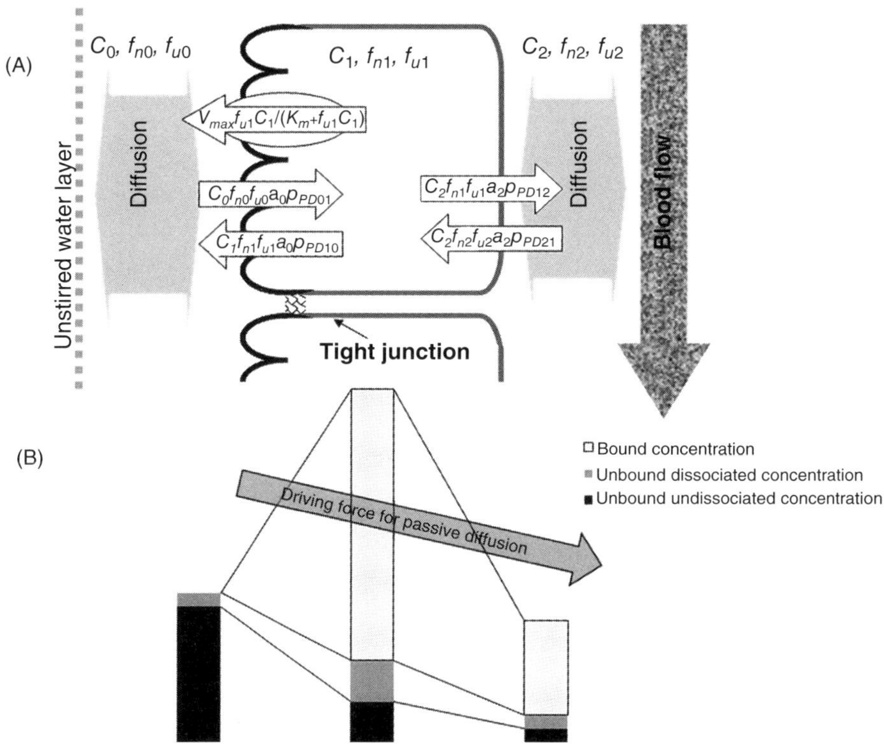

Figure 10.6 Schematic presentation of intestinal epithelial cells (A) and concentration gradient across the epithelial cell (B).

permeation [32]. Therefore, for this case, an estimation of $P_{trans,n}$ from $\log P_{oct}$ can be more appropriate (Equation 10.27). Usually, *in vivo* permeation becomes UWL-limited for this case and the estimation error of $P_{trans,n}$ does not become an issue.

10.4.2.2 Carrier-Mediated Permeation In the full-mechanistic enteric cell model (Figure 10.6) [4], it is critically important to consider explicitly the difference of surface areas of apical and basolateral membranes and the difference of pH in the apical, basolateral, and cytosol fluids. A mathematical model can be written as [4]:

$$P_{app,ep,A-B} = \frac{1}{C_0 A_{app}} f_{n1} a_2 p_{12}(C_1 f_{u1}) + P_{para} \tag{10.16}$$

$$P_{app,ep,B-A} = \frac{1}{C_2 A_{app}} (C_2 f_{n2} f_{u2} a_2 p_{21} - (C_1 f_{u1}) f_{n1} a_2 p_{12}) + P_{para} \tag{10.17}$$

$$C_1 f_{u1} = \frac{-b' + \sqrt{b'^2 - 4a'c'}}{2a'} \tag{10.18}$$

$$a' = f_{n1}a_0p_{10} + f_{n1}a_2p_{12} \tag{10.19}$$

$$b' = K_{m10}a' + V_{\max 10} + \frac{c'}{K_{m10}} \tag{10.20}$$

$$c' = -K_{m10}(C_0f_{n0}f_{u0}a_0p_{01} + C_2f_{n2}f_{u2}a_2p_{21}) \tag{10.21}$$

f_n: the fraction of undissociated (uncharged) species
f_u: the fraction of unbound species
p: ideal passive permeability
C: total dissolved drug concentration in each compartment
a: absolute surface area
X: compound amount in each compartment
PD: passive diffusion
$0,1,2$: compartment in Figure 10.6
A_{app}: apparent surface area

Even though it is counterintuitive, f_{u1} is not required to estimate the unbound drug concentration ($C_1 \cdot f_{u1}$) at the steady state because the concentration gradient of the unbound undissociated drug determines the permeation flux (Figure 10.6B).

The mechanistic enteric cell model can capture the asymmetric patterns of apical to basal (A-to-B) and B-to-A permeability (Figure 10.7) caused by the difference of the $p \times a$ values being threefold higher for the basolateral membrane [4, 33]. In addition, it can also capture the expression-level dependency of apparent K_m values [34]. The intrinsic K_m values can be obtained from the concentration dependency of P_{app} [4]. The model can be further improved to incorporate the effect of UWL [3].

The mechanistic enteric cell model can be also used to estimate the extent of intestinal first-pass metabolism (Fg). The anatomical Fg model was recently proposed [3], which considers the diffusion resistance of the subepithelial space in Fg calculation.

10.5 IMPACT OF FULL-MECHANISTIC MODELING ON DRUG DISCOVERY AND DEVELOPMENT

10.5.1 Prediction of Fraction of a Dose Absorbed (Fa%)

The fraction of a dose absorbed (Fa%) is the main target for biopharmaceutical modeling. Even though it is counterintuitive, once we obtain k_{perm}, k_{diss}, and $S_{dissolv}$, Fa% can be calculated as accurately by using a simple equation [1, 10] as by using a multiple compartment GI model [35] for the cases with no gastric pH effect and

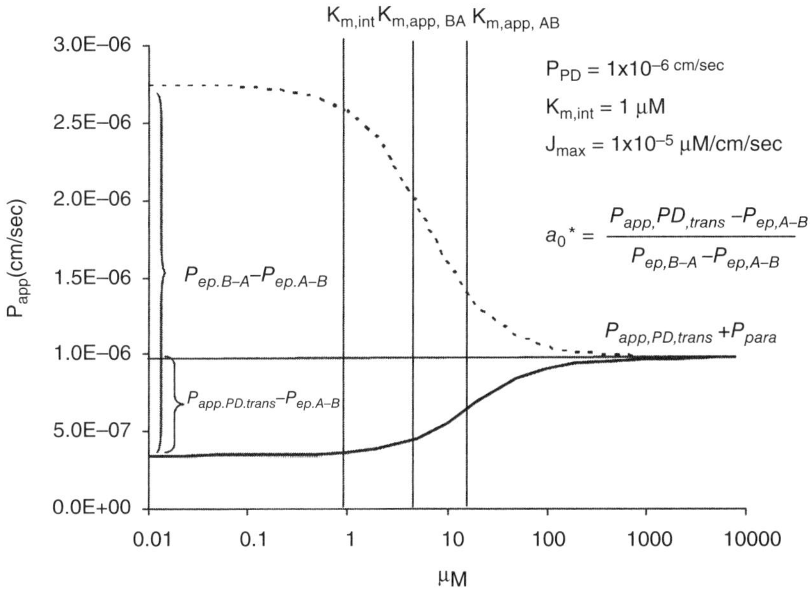

Figure 10.7 Asymmetry of A to B and B to A permeability with an efflux transporter in the apical membrane. (Adapted from Reference 4 with permission.)

supersaturation. By using the dimensionless parameters introduced by Amidon and co-workers [36], Fa% can be calculated as [1, 10]:

$$Fa = 1 - \exp\left(\frac{1}{\frac{1}{Dn} + \frac{Do}{Pn}}\right) \text{ ifDo} < 1, \text{ setDo} = 1. \tag{10.22}$$

$$Do = \frac{Dose}{S_{dissolv} \cdot V_{GI}}, \; Dn = k_{diss}T_{si}, \; Pn = k_{perm}T_{si} \tag{10.23}$$

T_{si}: small intestine transit time
Dn: dissolution number
Do: dose number
Pn: permeation number

Based on Equation 10.22, oral absorption of a drug can be categorized as [6, 17, 37]:

- Epithelial membrane permeability limited (PL-E): $Do < 1$, $Pn < Dn$, $P_{ep}{}' < P_{UWL}$
- UWL permeability limited (PL-U): $Do < 1$, $Pn < Dn$, $P_{ep}{}' > P_{UWL}$
- Dissolution rate limited (DRL): $Do/Pn < 1/Dn$

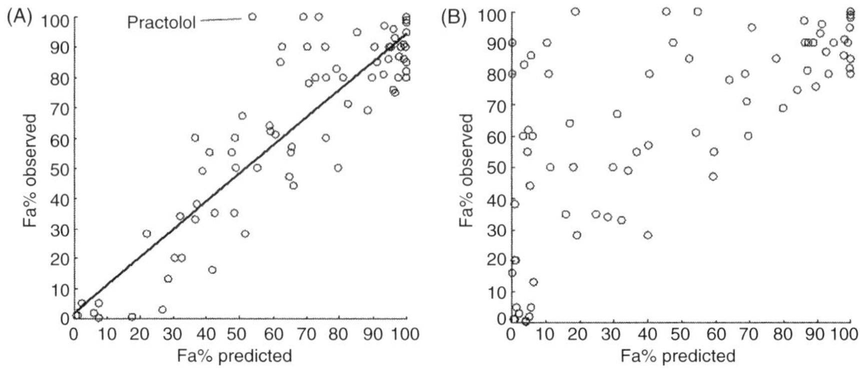

Figure 10.8 Fa% prediction from PAMPA data with (A) and without (B) P_{para} correction. Low-solubility drugs (*Do* > 1) and transporter substrates were excluded from the data set. (Adapted from Reference 22 with permission.)

- Solubility–epithelial membrane permeability limited (SL-E): $Do > 1$, $Do/Pn > 1/Dn$, $P_{ep}' < P_{UWL}$
- Solubility–UWL permeability limited (SL-U): $Do > 1$, $Do/Pn > 1/Dn$, $P_{UWL} < P_{ep}'$

Figure 10.8 shows the results of Fa% prediction for low permeability-high solubility drugs by parallel artificial membrane permeation assay (PAMPA) [38] with and without paracellular correction [21, 22]. The paracellular model can be also used in combination with Caco-2, MDCK (Madin-Darby canine kidney cells), and other *in vitro* data [20, 39].

Figure 10.9 shows the results of Fa% prediction for low-solubility drugs (undissociable, free acid and free base drugs without gastric pH effect). It is critically important to consider the effect of UWL on P_{eff} even though. The importance of UWL has been often overlooked [7, 40].

10.5.2 Understanding Rate-Limiting Steps

10.5.2.1 Solubility and Dissolution Rate The main reason for low Fa% of low-solubility drugs had often been speculated to be slow dissolution. However, recent analysis revealed that for most low-solubility drugs (ca. 70%), oral absorption was categorized as solubility-permeability limited (SL) but not dissolution rate limited (DRL) [5]. This finding explains the reason that level A IVIVC (*in vitro in vivo* correlation) has been hardly achieved for many low-solubility drugs by a conventional dissolution test and urges us to reconsider the design of dissolution tests as well as the biowaiver strategy [3]. The critical particle size ($r_{p,crit}$) discriminating DRL and SL can be calculated by rearranging $1/Dn = Do/Pn$ (for $Do > 1$) as [5]:

$$r_{p,crit} = \sqrt{\frac{3D_{eff} \cdot Dose \cdot R_{GI}}{2 \cdot V_{GI} \cdot DF \cdot P_{eff} \cdot \rho_p}} \tag{10.24}$$

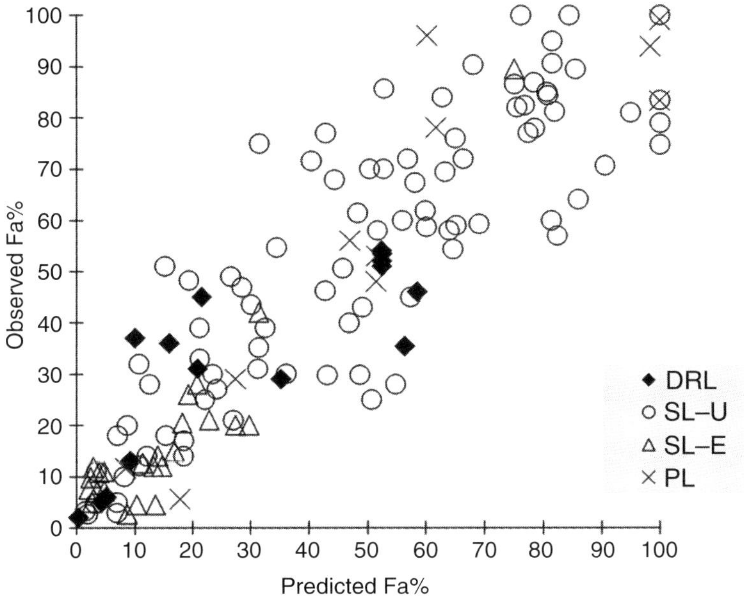

Figure 10.9 Fa% prediction using S_0, K_{bm} (FaSSIF), $P_{trans,n}$ (from Caco-2 permeability (for $\log D_{pH6.5} < 1.5$) or experimental $\log P_{oct}$ (for $\log D_{pH6.5} > 1.5$)), pK_a, and MW for low-solubility drugs. Undissociable and acid drugs are used as model drugs. Free bases were also used in the case with the high pH stomach. Salts were excluded from the dataset. Physiological parameters: $DF = 1.7$, $VE = 10$, $R_{mucus} = 2.9$ μm, $C_{pd} = 2.2$, and $h_{pd} = 285$ μm, and $h_{fam} = 15$ μm for both humans and dogs; $V_{GI} = 130$ and 19 mL, $P_{WC} = 0.23$ and 0.29×10^{-4} cm/s, $PE = 3$ and 1, $R_{GI} = 1.5$ and 0.5 cm, and $T_{si} = 3.5$ and 2 hours, for humans and dogs. V_{GI} in the fed state was set to 1.2-fold larger than that in the fasted state. See Reference 5 for details. (Adapted from Reference 5 with permission.)

$S_{dissolv}$ is canceled out from the equation, suggesting that $r_{p,crit}$ does not depend on $S_{dissolv}$ for $Do > 1$ cases. During drug development, particle size reduction is usually employed to remove dissolution rate limitation (can be reduced to <10 μm). The upper limit of P_{eff} would be ca. 5×10^{-4} cm/s (UWL limited). Therefore, in most cases, the oral absorption becomes SL. However, it should not be interpreted that particle size reduction would have no effect on Fa% when the particle radius is smaller than $r_{p,crit}$. It was experimentally observed that particle size reduction improved Fa% for the SL-U cases, e.g., danazol [41]. This point is later discussed.

10.5.2.2 *Permeability*

The transcellular pathway had been considered to be the main determinant of the intestinal membrane permeability for most drugs. However, recent analysis revealed that the paracellular pathway and UWL play more significant roles than originally thought (ca. 30% and 40% of drugs with reported P_{eff} values, respectively) [22, 39].

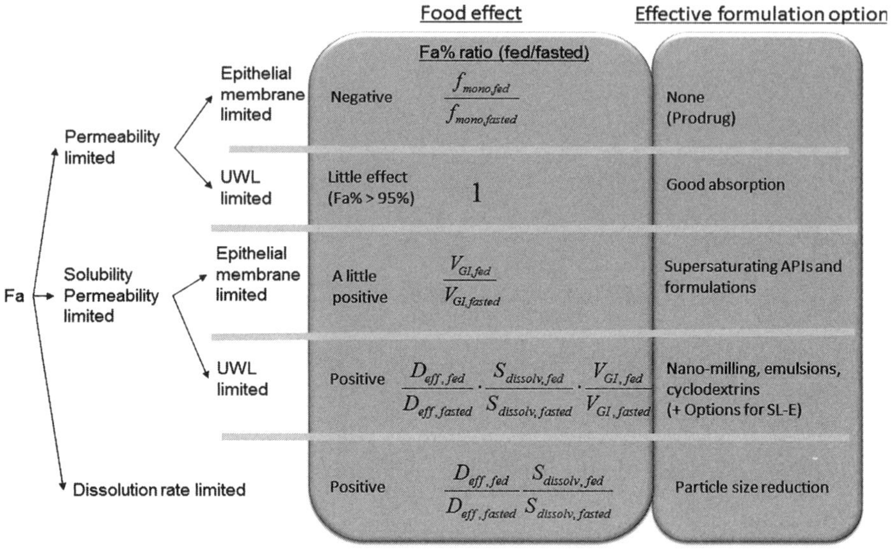

Figure 10.10 Simple decision trees to predict food effect and to select a formulation option to improve Fa% based on the GUT framework.

10.5.3 Effect of Food, Formulation, API Form, and Gastric pH

The food effect can be appropriately captured by the GUT framework [6, 42]. A simple decision tree to estimate the food and formulation effects was derived from the GUT framework (Figure 10.10) [6, 43].

Food intake induces the bile secretion and increases the bile concentration ca. fivefold compared with the fasted state [44]. To simulate the food effects, f_u and P_{UWL} have to be taken into account. Bile micelles increase $S_{dissolv}$ and k_{diss}; however, they reduce both P_{UWL} and P_{ep}' (hence, reduce P_{eff}).

The GUT framework suggests that bile micelle solubilization (binding) will increase Fa% for the DRL and SL-U cases, have little effect for SL-E cases, and reduce Fa% for PL-E cases [6] (Figure 10.10). This is in good agreement with experimental observations and empirical rules [45]. Reduction of P_{eff} by bile micelle binding has been suggested to be one of the main reasons for the negative food effect [6, 46]. For the SL-E cases such as pranlukast, the increase in solubility would be canceled out by the decrease in permeability, resulting in little food effect [47]. For the SL-U cases, as the bile micelle bound drug molecules can pass through the UWL, the net flux increases (cf. flux = concentration × permeability).

The effect of solubilizing formulations can be similarly explained [43]. Micelle solubilization will be effective for SL-U, whereas a supersaturating formulation will be required for SL-E as it can increase unbound drug concentration.

The drug particles drifting into the UWL would reduce the effective thickness of UWL and increase P_{eff}. This particle drifting effect can explain the reason why particle size reduction is effective for SL-U (including nano-milling) [9].

Prediction of the gastric pH effect has been investigated for free base drugs [8]. To simulate the pH shift from the stomach to the small intestine, a compartment transit model was used [23]. It is critically important to consider the solid surface solubility in the calculation of the dissolution rate in the stomach.

It is well known that the solid form of an active pharmaceutical ingredient (API) affects Fa%. When a drug is administered as a salt, before it precipitates out as a free form, a supersaturated concentration is induced, resulting in increased Fa% compared with after dosing the free form (Figure 10.3). Biopharmaceutical modeling for supersaturating APIs (salt, co-crystal, anhydrate, amorphous) and formulations needs further investigation.

10.5.4 Species Differences and Lab-to-lab Variation of Membrane Permeation

Species differences of passive permeability can be appropriately captured by the GUT framework [48]. As a result of the differences of the plica and villi structures, the P_{eff} values of passive transcellular permeability in rats and dogs would be 1/6 and 1/3 of that in humans (but the smaller R_{GI} cancels out the reduction of P_{eff}, resulting in similar Fa%. cf. $k_{perm} = 2 \cdot DF \cdot P_{eff} / R_{GI}$). For paracellular permeation, permeability in dogs becomes much higher than that in rats and humans [49].

Another interesting application is correction of lab-to-lab variation of *in vitro* P_{app} [50]. P_{para}, assay pH, and UWL in each laboratory can be virtually standardized by applying the model equations.

10.5.5 Effect of Transporters and Metabolism in the Intestinal Epithelial Cells

Simulation of the interplay of passive diffusion, carrier-mediated transport, and metabolism in the intestinal epithelial cells would be of great importance, especially for drug–drug interaction (DDI) prediction [51]. The mechanistic enteric cell model suggests that $C_l f_{ul}$ becomes 1/2, 1/3, and 1/10 of that in the apical side for acid, neutral, and basic drugs, respectively, in the *in vivo* situation (apical pH = 6.5, cytosol pH = 7, for passive PL-E cases)[4]. Interestingly, it has been empirically suggested that the 1/10 of apical concentration should be used to predict DDI in the enteric cells [52]. The permeation resistance of UWL would further reduce the ratio of cytosol/apical concentration [3].

Dose-supralinearity of *in vivo* Fa% can be also estimated by the epithelial cell model. Maraviroc is a P-gp substrate ($K_m = 37$ μM) [53]. The area under concentration (AUC) after oral administration was found to be dose-supra proportional at 3~100 mg, whereas it is dose-linear after intravenous administration [54, 55]. Maraviroc is mainly metabolized by CYP3A4 ($K_m = 21$ μM) [32]. The unbound drug concentration in the portal vein is estimated to less than 1 μM (= Dose × k_{abs} × Fg × $f_{u,p}$ / Qh). The intrinsic hepatic clearance is 44 mL/min/kg, suggesting that Fg is 0.91 [56]. Therefore, it was suggested that the dose supralinearity was caused by saturation of P-glycoprotein (P-gp) efflux. The estimated dose strength that saturates P-gp efflux was 25 mg (from the *in vitro* K_m value and V_{GI} of 130 mL), which is in good agreement with the clinical observation.

Linagliptin is another example (P-gp K_m = 187 μM (inversed vesicle assay))[57]. The estimated dose strength that saturates P-gp efflux was 115 mg. Linagliptin actually shows dose-supralinearity at 50~400 mg.

10.5.6 Structure–Primary PK Parameter Relationship

When a primary PK parameter is used as the dependent variable for structure-PK parameter relationship, a clear correlation with a simple physicochemical parameter can be obtained. The following equations are useful for rough estimation of S_0 [58], K_{bm} [25, 59, 60], $P_{trans,n}$ [3, 14], and D_{mono} [50]:

$$\log S_0(M) = -\log P_{oct} - 0.01(m.p. - 25) + 0.50 \quad (10.25)$$

$$\log K_{bm,0} = 0.74 \log P_{oct} + 2.29, \log K_{bm,+} = \log K_{bm,0} - 1,$$

$$\log K_{bm,-} = \log K_{bm,0} - 2 \quad (10.26)$$

$$\log \mathrm{P}_{trans,n}(\mathrm{cm/s}) = 1.1 \log P_{oct} - 5.6 \quad (\text{for MW} < 450) \quad (10.27)$$

$$D_{mono}(\mathrm{cm^2/s}) = 9.9 \times 10^{-5}\ \mathrm{MW}^{-0.453} \quad (10.28)$$

It is highly recommended to use experimental logP_{oct} values [61].

10.6 EPISTEMOLOGY AND GOOD SIMULATION PRACTICE

The greatest obstacle in improving the prediction sciences is not ignorance but the illusion of good prediction. Even for the simplest cases (i.e., passive permeation, no gastric pH effect, and no supersaturation), as shown in Figure10.8 and 10.9, a modest predictability would be the reality of current achievement. This reality should be the starting point for future improvements.

10.6.1 Transparency

To improve transparency, it is critically important to disclose all the equations, physiological parameters, and drug parameters. Unfortunately, publications using commercial software tend not to disclose sufficient information for peer review. There are excellent review articles summarizing the *in vivo* physiologies [62–65].

10.6.2 Comprehensiveness

As the mathematical models become more complicated, it becomes less comprehensive. A complicated GI transit model is usually employed in commercial software. To improve transparency and comprehensiveness, a simple model should be used when a complex model is not necessary. Categorization by the limiting step, biopharmaceutical classification system, or both [36, 66] would be of great help to grasp the whole picture of oral absorption and to check the sanity of computational calculation.

10.6.3 Validation Dataset

Sufficient number of drugs with suitable characteristics should be used for model validation. For mechanistic modeling, a model equation should be validated for both *in vitro* and *in vivo* levels.

10.6.4 Benefits and Limitations of GUT Framework

The GUT framework is transparent and comprehensive. All the equations and physiological data used in the GUT framework have been published, and the framework was validated using more than 100 Fa% data [3]. The Fa equation is simple but sufficient for undissociable and free acid cases. Dose and particle size dependency of Fa% [12], food effect [6], species differences [48], etc., can be predicted by the GUT framework.

Mechanistic prediction for supersaturable APIs and formulations is very challenging as nucleation and particle growth in the GI tract are complicated phenomena. Even though classic nucleation theory (CNT) has been incorporated in the GUT framework [1, 3, 11], the nucleation and particle growth in the real GI tract is complicated and CNT may not explain the real situation.

10.7 OUTLOOK

The full-mechanistic models have been applied for other *in vitro* and *in situ* models such as the suspended hepatocyte assay [67] and the brain perfusion assay [68]. Population-based prediction is another area where the full-mechanistic approach has been anticipated to be successful [69].

REFERENCES

1. Sugano, K. (2009). Introduction to computational oral absorption simulation. *Expert Opinion on Drug Metabolism & Toxicology*, *5*, 259–293.
2. Galia, E., Nicolaides, E., Horter, D., Lobenberg, R., Reppas, C., Dressman, J.B. (1998). Evaluation of various dissolution media for predicting in vivo performance of class I and II drugs. *Pharmaceutical Research*, *15*, 698–705.
3. Sugano, K. *Biopharmaceutics modeling and simulations: Theory, practice, methods and applications*, Wiley, Hoboken, NJ, 2012.

4. Sugano, K., Shirasaka, Y., Yamashita, S. (2011). Estimation of Michaelis-Menten constant of efflux transporter considering asymmetric permeability. *International Journal of Pharmaceutics*, *8*, 8.
5. Sugano, K. (2011). Fraction of a dose absorbed estimation for structurally diverse low solubility compounds. *International Journal of Pharmaceutics*, *405*, 79–89.
6. Sugano, K., Kataoka, M., Mathews, C.D.C., Yamashita, S. (2010). Prediction of food effect by bile micelles on oral drug absorption considering free fraction in intestinal fluid. *European Journal of Pharmaceutical Sciences*, *40*, 118–124.
7. Sugano, K. (2010). Aqueous boundary layers related to oral absorption of a drug: From dissolution of a drug to carrier mediated transport and intestinal wall metabolism. *Molecular Pharmaceutics*, *7*, 1362–1373.
8. Sugano, K. (2010). Computational oral absorption simulation of free base drugs. *International Journal of Pharmaceutics*, in press.
9. Sugano, K. (2010). Possible reduction of effective thickness of intestinal unstirred water layer by particle drifting effect. *International Journal of Pharmaceutics*, *387*, 103–109.
10. Sugano, K. (2009). Fraction of dose absorbed calculation: Comparison between analytical solution based on one compartment steady state approximation and dynamic seven compartment model. *CBI Journal*, *9*, 75–93.
11. Sugano, K. (2009). A simulation of oral absorption using classical nucleation theory. *International Journal of Pharmaceutics*, *378*, 142–145.
12. Sugano, K. (2009). Oral absorption simulation for low solubility compounds *Chemistry and Biodiversity*, *6*, 2014–2029.
13. Sugano, K. (2009). Theoretical investigation of passive intestinal membrane permeability using Monte Carlo method to generate drug like molecule population. *International Journal of Pharmaceutics*, *373*, 55–61.
14. Sugano, K. (2009). Estimation of effective intestinal membrane permeability considering bile micelle solubilisation. *International Journal of Pharmaceutics*, *368*, 116–122.
15. Sugano, K. (2008). Theoretical comparison of hydrodynamic diffusion layer models used for dissolution simulation in drug discovery and development. *International Journal of Pharmaceutics*, *363*, 73–77.
16. Okazaki, A., Mano, T., Sugano, K. (2008). Theoretical dissolution model of polydisperse drug particles in biorelevant media. *Journal of Pharmaceutical Sciences*, *97*, 1843–1852.
17. Sugano, K., Okazaki, A., Sugimoto, S., Tavornvipas, S., Omura, A., Mano, T. (2007). Solubility and dissolution profile assessment in drug discovery. *Drug Metabolism and Pharmacokinetics*, *22*, 225–254.
18. Sugano, K., Obata, K., Saitoh, R., Higashida, A., Hamada, H. (2006). Processing of biopharmaceutical profiling data in drug discovery. *Pharmacokinetic Profiling in Drug Research: Biological, Physicochemical, and Computational Strategies, [LogP2004, Lipophilicity Symposium], 3rd, Zurich, Switzerland, Feb. 29-Mar. 4, 2004*441-458.
19. Obata, K., Sugano, K., Saitoh, R., Higashida, A., Nabuchi, Y., Machida, M., Aso, Y. (2005). Prediction of oral drug absorption in humans by theoretical passive absorption model. *International Journal of Pharmaceutics*, *293*, 183–192.
20. Saitoh, R., Sugano, K., Takata, N., Tachibana, T., Higashida, A., Nabuchi, Y., Aso, Y. (2004). Correction of permeability with pore radius of tight junctions in Caco-2 monolayers improves the prediction of the dose fraction of hydrophilic drugs absorbed by humans. *Pharmaceutical Research*, *21*, 749.

21. Sugano, K., Nabuchi, Y., Machida, M., Aso, Y. (2003). Prediction of human intestinal permeability using artificial membrane permeability. *International Journal of Pharmaceutics*, *257*, 245–251.
22. Sugano, K., Takata, N., Machida, M., Saitoh, K., Terada, K. (2002). Prediction of passive intestinal absorption using bio-mimetic artificial membrane permeation assay and the paracellular pathway model. *International Journal of Pharmaceutics*, *241*, 241–251.
23. Yu, L.X., Lipka, E., Crison, J.R., Amidon, G.L. (1996). Transport approaches to the biopharmaceutical design of oral drug delivery systems: Prediction of intestinal absorption. *Advanced Drug Delivery Reviews*, *19*, 359–376.
24. Rippie, E.G., Lamb, D.J., Romig, P.W. (1964). Solubilization of weakly acidic and basic drugs by aqueous solutions of polysorbate 80. *Journal of Pharmaceutical Sciences*, *53*, 1346–1348.
25. Glomme, A., März, J., Dressman, J. Predicting the intestinal solubility of poorly soluble drugs, in B. Testa, S. Krämer, H. Wunderli-Allenspach, and G. Folkers (eds.), *Pharmacokinetic Profiling in Drug Research*, Wiley-VCH, Zurich, Switzerland, 2006, pp. 259–280.
26. Bakatselou, V., Oppenheim, R.C., Dressman, J.B. (1991). Solubilization and wetting effects of bile salts on the dissolution of steroids. *Pharmaceutical Research*, *8*, 1461–1469.
27. Mooney, K.G., Mintun, M.A., Himmelstein, K.J., Stella, V.J. (1981). Dissolution kinetics of carboxylic acids. II: Effects of buffers. *Journal of Pharmaceutical Sciences*, *70*, 22–32.
28. Mooney, K.G., Mintun, M.A., Himmelstein, K.J., Stella, V.J. (1981). Dissolution kinetics of carboxylic acids. I: Effect of pH under unbuffered conditions. *Journal of Pharmaceutical Sciences*, *70*, 13–22.
29. Sheng, J.J., McNamara, D.P., Amidon, G.L. (2009). Toward an in vivo dissolution methodology: A comparison of phosphate and bicarbonate buffers. *Molecular Pharmaceutics*, *6*, 29–39.
30. Carlert, S., Palsson, A., Hanisch, G., von Corswant, C., Nilsson, C., Lindfors, L., Lennernas, H., Abrahamsson, B. (2010). Predicting intestinal precipitation–a case example for a basic BCS class II drug. *Pharmaceutical Research*, *27*, 2119–2130.
31. Oliver, R.E., Jones, A.F., Rowland, M. (1998). What surface of the intestinal epithelium is effectively available to permeating drugs? *Journal of Pharmaceutical Sciences*, *87*, 634–639.
32. Wils, P., Warnery, A., Phung-Ba, V., Legrain, S., Scherman, D. (1994). High lipophilicity decreases drug transport across intestinal epithelial cells. *The Journal of Pharmacology and Experimental Therapeutics*, *269*, 654–658.
33. Trotter, P.J., Storch, J. (1991). Fatty acid uptake and metabolism in a human intestinal cell line (Caco-2): Comparison of apical and basolateral incubation. *The Journal of Lipid Research*, *32*, 293–304.
34. Tachibana, T., Kitamura, S., Kato, M., Mitsui, T., Shirasaka, Y., Yamashita, S., Sugiyama, Y. (2010). Model analysis of the concentration-dependent permeability of P-gp substrates. *Pharmaceutical Research*, *27*, 442–446.
35. Yu, L.X., Amidon, G.L. (1999). A compartmental absorption and transit model for estimating oral drug absorption. *International Journal of Pharmaceutics*, *186*, 119–125.
36. Oh, D.M., Curl, R.L., Amidon, G.L. (1993). Estimating the fraction dose absorbed from suspensions of poorly soluble compounds in humans: A mathematical model. *Pharmaceutical Research*, *10*, 264–270.

37. Yu, L.X. (1999). An integrated model for determining causes of poor oral drug absorption. *Pharmaceutical Research*, *16*, 1883–1887.
38. Kansy, M., Senner, F., Gubernator, K. (1998). Physicochemical high throughput screening: Parallel artificial membrane permeation assay in the description of passive absorption processes. *Journal of Medicinal Chemistry*, *41*, 1007–1010.
39. Avdeef, A., Tam, K.Y. (2010). How well can the Caco-2/Madin-Darby canine kidney models predict effective human jejunal permeability? *Journal of Medicinal Chemistry*, *53*, 3566–3584.
40. Youdim, K.A., Avdeef, A., Abbott, N.J. (2003). In vitro trans-monolayer permeability calculations: Often forgotten assumptions. *Drug Discovery Today*, *8*, 997–1003.
41. Liversidge, G.G., Cundy, K.C. (1995). Particle size reduction for improvement of oral bioavailability of hydrophobic drugs: I. Absolute oral bioavailability of nanocrystalline danazol in beagle dogs. *International Journal of Pharmaceutics*, *125*, 91–97.
42. Sugano, K. (2010). Computational oral absorption simulation of free base drugs. *International Journal of Pharmaceutics*, *398*, 73–82.
43. Kataoka, M., Sugano, K., da Costa Mathews, C., Wong, J.W., Jones, K.L., Masaoka, Y., Sakuma, S., Yamashita, S. (2011). Application of dissolution/permeation system for evaluation of formulation effect on oral absorption of poorly water-soluble drugs in drug development. *Pharmaceutical Research*, *29*(6), 1485–1494.
44. Jantratid, E., Janssen, N., Reppas, C., Dressman, J.B. (2008). Dissolution media simulating conditions in the proximal human gastrointestinal tract: An update. *Pharmaceutical Research*, *25*, 1663–1676.
45. Wu, C.-Y., Benet, L.Z. (2005). Predicting drug disposition via application of BCS: Transport/absorption/ elimination interplay and development of a biopharmaceutics drug disposition classification system. *Pharmaceutical Research*, *22*, 11–23.
46. Lennernaes, H., Regaardh, C.G. (1993). Evidence for an interaction between the b-blocker pafenolol and bile salts in the intestinal lumen of the rat leading to dose-dependent oral absorption and double peaks in the plasma concentration-time profile. *Pharmaceutical Research*, *10*, 879–883.
47. Pranlukast. Interview form. http://www.info.pmda.go.jp/go/pack/4490017M1036_1_11/
48. Sugano, K. (2009). Theoretical investigation of passive intestinal membrane permeability using Monte Carlo method to generate drug-like molecule population. *International Journal of Pharmaceutics*, *373*, 55–61.
49. Chiou, W.L., Jeong, H.Y., Chung, S.M., Wu, T.C. (2000). Evaluation of using dog as an animal model to study the fraction of oral dose absorbed of 43 drugs in humans. *Pharmaceutical Research*, *17*, 135–140.
50. Avdeef, A. (2010). Leakiness and size exclusion of paracellular channels in cultured epithelial cell monolayers interlaboratory comparison. *Pharmaceutical Research*, *27*, 480–489.
51. Cummins, C.L., Jacobsen, W., Benet, L.Z. (2002). Unmasking the dynamic interplay between intestinal P-glycoprotein and CYP3A4. *Journal of Pharmacology and Experimental Therapeutics*, *300*, 1036–1045.
52. Giacomini, K.M., Huang, S.M., Tweedie, D.J., Benet, L.Z., Brouwer, K.L., Chu, X., Dahlin, A., Evers, R., Fischer, V., Hillgren, K.M., Hoffmaster, K.A., Ishikawa, T., Keppler, D., Kim, R.B., Lee, C.A., Niemi, M., Polli, J.W., Sugiyama, Y., Swaan, P.W., Ware, J.A., Wright, S.H., Yee, S.W., Zamek-Gliszczynski, M.J., Zhang, L. (2010). Membrane transporters in drug development. *Nature Reviews Drug Discovery*, *9*, 215–236.

53. Hyland, R., Dickins, M., Collins, C., Jones, H., Jones, B. (2008). Maraviroc: In vitro assessment of drug-drug interaction potential. *British Journal of Clinical Pharmacology*, *66*, 498–507.

54. Abel, S., Russell, D., Whitlock, L.A., Ridgway, C.E., Nedderman, A.N., Walker, D.K. (2008). Assessment of the absorption, metabolism and absolute bioavailability of maraviroc in healthy male subjects. *British Journal of Clinical Pharmacology*, *65*(1), 60–67.

55. Abel, S., van der Ryst, E., Rosario, M.C., Ridgway, C.E., Medhurst, C.G., Taylor-Worth, R.J., Muirhead, G.J. (2008). Assessment of the pharmacokinetics, safety and tolerability of maraviroc, a novel CCR5 antagonist, in healthy volunteers. *British Journal of Clinical Pharmacology*, *65*(1), 5–18.

56. Kato, M. (2008). Intestinal first-pass metabolism of CYP3A4 substrates. *Drug Metabolism and Pharmacokinetics*, *23*, 87–94.

57. Linagliptin, Interview Form, http://www.info.pmda.go.jp/go/pack/3969014F1024_1_06/

58. Yalkowsky, S.H., Valvani, S.C. (1980). Solubility and partitioning. I: Solubility of nonelectrolytes in water. *Journal of Pharmaceutical Sciences*, *69*, 912–922.

59. Miyazaki, J., Hideg, K., Marsh, D. (1992). Interfacial ionization and partitioning of membrane-bound local anesthetics. *Biochimica et Biophysica Acta, Biomembranes*, *1103*, 62–68.

60. Avdeef, A., Box, K.J., Comer, J.E., Hibbert, C., Tam, K.Y. (1998). pH-metric logP 10. Determination of liposomal membrane-water partition coefficients of ionizable drugs. *Pharmaceutical Research*, *15*, 209–215.

61. Mannhold, R., Poda, G.I., Ostermann, C., Tetko, I.V. (2009). Calculation of molecular lipophilicity: State-of-the-art and comparison of log P methods on more than 96,000 compounds. *Journal of Pharmaceutical Sciences*, *98*, 861–893.

62. Mudie, D.M., Amidon, G.L., Amidon, G.E. (2010). Physiological parameters for oral delivery and in vitro testing. *Molecular Pharmaceutics*, *7*(5), 1388–1405.

63. Kararli, T.T. (1995). Comparison of the gastrointestinal anatomy, physiology, and biochemistry of humans and commonly used laboratory animals. *Biopharmaceutics & Drug Disposition*, *16*, 351–380.

64. DeSesso, J.M., Williams, A.L. (2008). Contrasting the gastrointestinal tracts of mammals: Factors that influence absorption. *Annual Reports in Medicinal Chemistry*, *43*, 353–371.

65. McConnell, E.L., Fadda, H.M., Basit, A.W. (2008). Gut instincts: Explorations in intestinal physiology and drug delivery. *International Journal of Pharmaceutics*, *364*, 213–226.

66. Amidon, G.L., Lennernas, H., Shah, V.P., Crison, J.R. (1995). A theoretical basis for a biopharmaceutic drug classification: The correlation of in vitro drug product dissolution and in vivo bioavailability. *Pharmaceutical Research*, *12*, 413–420.

67. Jigorel, E., Houston, J.B. (2012). Utility of drug depletion-time profiles in isolated hepatocytes for accessing hepatic uptake clearance: Identifying rate-limiting steps and role of passive processes. *Drug Metabolism and Disposition: The Biological Fate of Chemicals*, *40*, 1596–1602.

68. Avdeef, A., Sun, N. (2011). A new in situ brain perfusion flow correction method for lipophilic drugs based on the pH-dependent Crone-Renkin equation. *Pharmaceutical Research*, *28*, 517–530.

69. Rostami-Hodjegan, A., Tucker, G.T. (2007). Simulation and prediction of in vivo drug metabolism in human populations from in vitro data. *Nature Reviews Drug Discovery*, *6*, 140–148.

11

PHARMACOKINETIC/ PHARMACODYNAMIC MODELING IN DRUG DISCOVERY: A TRANSLATIONAL TOOL TO OPTIMIZE DISCOVERY COMPOUNDS TOWARD THE IDEAL TARGET-SPECIFIC PROFILE

Patricia Schroeder

11.1 INTRODUCTION

Drug discovery and development has been plagued by clinical attrition rates exceeding 90%, where the primary cause was identified by Kola and Landis as lack of efficacy and unexpected safety findings [1]. This finding demonstrates the urgency to improve the prediction of drug efficacy and safety in the clinic. In response, pharmacokinetic/pharmacodynamic (PK/PD) modeling is emerging as a key area of

Predictive ADMET: Integrative Approaches in Drug Discovery and Development, First Edition.
Edited by Jianling Wang and Laszlo Urban.

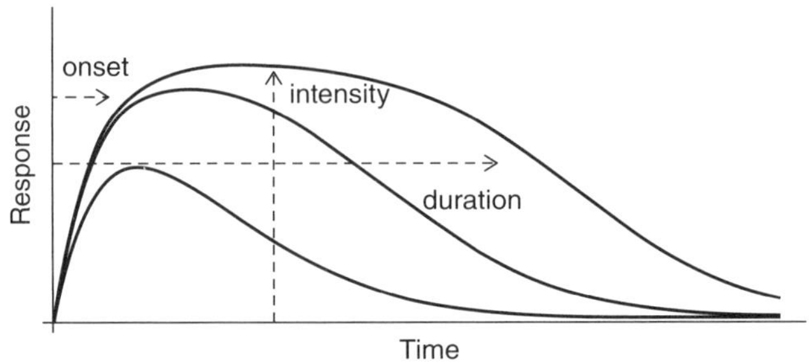

Figure 11.1 Response simulations describing what is meant by the onset, intensity, and duration of pharmacodyńamic effect.

investment in pharmaceutical research that plays a critical role in the drug discovery process [2, 3]. PK/PD modeling uses mathematics to describe the impact of drug concentration (PK) on the onset, intensity, and duration of the observed response (PD). Historically, PK/PD modeling has been used for the optimization of phase III clinical trials by using existing clinical proof of mechanism (POM) and PK data. However, PK/PD modeling also has great utility in early drug discovery where it should be considered a critical component in early chemistry asset optimization and lead prioritization. Following the target validation and hit identification stages of drug discovery, the temptation is to optimize chemistry leads toward a generic set of *in vitro* absorption, distribution, metabolism, and excretion (ADME) and safety endpoints. Although this is common practice, it is also a missed opportunity toward understanding the intensity, extent, and duration of target engagement required for meaningful efficacy for novel targets and emerging chemistry assets. Thus, within the drug discovery value chain, PK/PD modeling can play an important role in lead optimization, drug candidate selection, and the optimization of early proof of mechanism, principal, and concept clinical trials [4].

Translational drug PK/PD modeling often relies on the quantitative description of PK/PD relationships demonstrated in preclinical *in vivo* studies or *in vitro* bioassays. It has evolved from a descriptive approach to a mechanistic science incorporating physiological mechanisms and biology describing the PK/PD relationship, and these mechanism-based PK/PD models have a greater capability of translational extrapolations compared with empirical models [4]. Mechanism-based PK/PD modeling incorporates not only the exposure-response relationship, but also it can include the intra- and interindividual pharmacokinetic and pharmacodynamic variability, where exposure, as expressed by the concentration-time profile in plasma, is linked to the time course of the pharmacodynamic effect and includes the onset, intensity, and duration demonstrated in Figure 11.1. Here we will describe utility PK/PD modeling and simulation in the early phases of drug discovery, where an emerging understanding of the concentration-effect relationship can help guide chemistry lead optimization toward the ADME/PK that will generate the extent of target engagement required for clinical hypothesis testing.

11.2 THE DISTINCTION BETWEEN DRUG AND SYSTEM-SPECIFIC PARAMETERS

As shown in Figure 11.2, there is a distinction between drug- and biological-system-specific parameters. Drug-specific parameters describe the interplay between the compound and the biological system in terms of target affinity or activation, whereas biological-system-specific parameters describe the function of the biological system [5]. The distinction between drug and system parameters is a key component for the best estimate of response. A drug-specific parameter such as target affinity is often well predicted from *in vitro* binding bioassays and comparable between species; thus, they typically do not require scaling to clinical outcomes [6]. In contrast, the biological-system-specific parameters, including the expression of target protein or the transduction rate constants to effect, should be measured *in vivo*, and these values often vary among species, individuals within a species, the state of disease, and experimental conditions [5]. Thus, interspecies scaling of biological system parameters are typically required for simulating clinical response. The resulting drug and system parameters not only provide a description of the PD at the doses tested, but also they allow for simulation of other doses and dosing paradigms [7] providing an anchor for optimization of the clinical dosing regimen and informing on the extent of pharmacokinetic coverage, thereby highlighting the need for further chemistry asset optimization.

11.3 EXTRAPOLATION OF *IN VITRO* BIOASSAYS

In vitro bioassays are tools to prioritize discovery leads, aid in *in vivo* study design, and can help to identify the first time in human dose and schedule for clinical hypothesis testing of experimental drugs and drug combinations [8, 9]. *In vitro* bioassays often describe the potency of discovery compounds against the target and, in the absence of cross-species differences in target structure, have demonstrated to translate across species where the unbound plasma drug concentration required to generate an effect (often EC_{50}) is often comparable in experimental animals and humans [10]. Target binding kinetics can be described as a function of time and concentration as described in Equation 11.1:

$$\frac{dE}{dt} = k_{on} \cdot (E_{max} - E) \cdot C - k_{off} \cdot E \qquad (11.1)$$

Where k_{on} and k_{off} is the on and off rate, respectively, E_{max} is the maximal response, and C is the unbound plasma concentration. Although it has been documented that interspecies differences in receptor affinity exist [11], several examples show reasonable concordance of such properties between preclinical test species and humans. For instance, a linear relationship between the equilibrium dissociation constants between rats and humans of benzodiazepines was demonstrated by Ito and colleagues that spanned over four orders of magnitude [12]. A similar relationship between rats

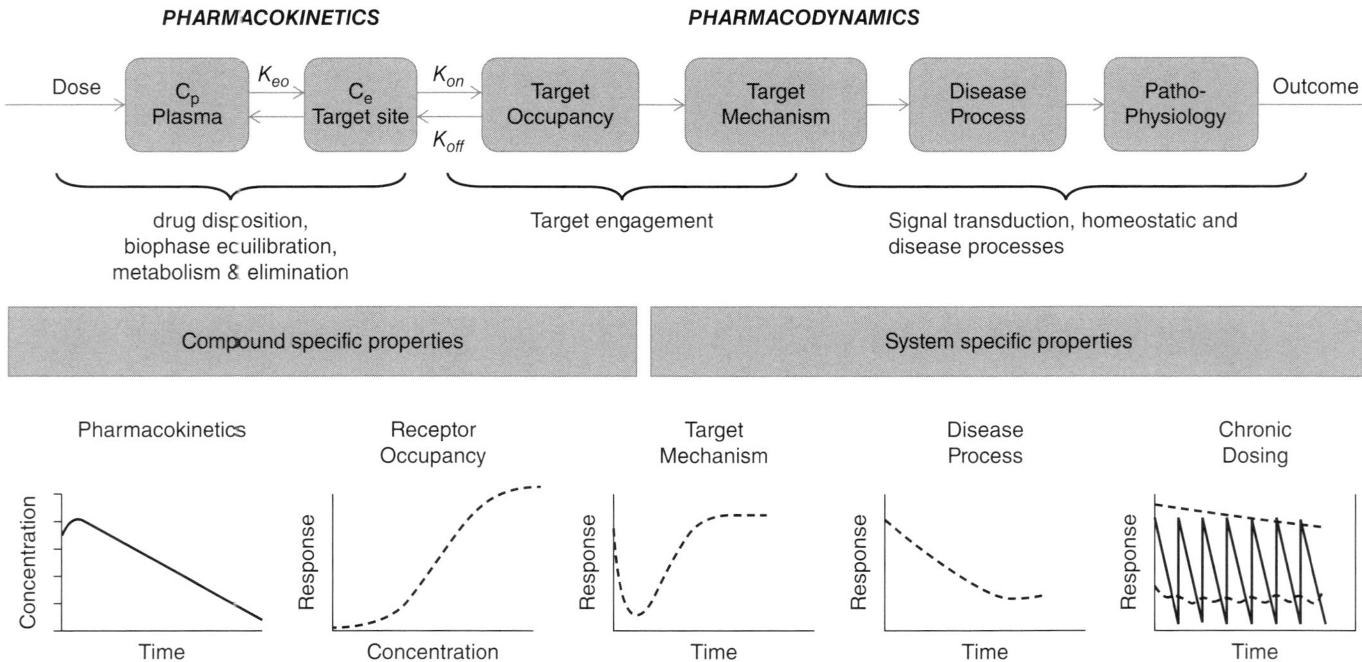

Figure 11.2 Distinction between compound and system-specific PK/PD properties of a compound, where k_{eo} is the biophase distribution rate constant; k_{on} and k_{off} are the receptor on and off rate, respectively.

and humans for the EC_{50} values of synthetic opioids was demonstrated by Cox and colleagues [13]. Kalvass et al. [14] showed that the unbound pharmacokinetic distribution of seven opioids into the brain allowed for a direct-effect correlation between unbound mouse brain EC_{50} and *in vitro* opiate receptor binding K_i. Mechanism-based PK/PD models that used *in vitro* estimates including cortisol suppression, neutrophil and lymphocyte trafficking, and *ex vivo* inhibition of lymphocyte proliferation all demonstrated a high correlation with their relative receptor affinity in a placebo-controlled crossover study, whose aim was to capture the dosing equivalency of four corticosteroids [15, 16].

11.4 CONCENTRATION-RESPONSE RELATIONSHIPS AND THE IMPACT OF TIME

Different target classes have different temporal relationships between the pharmacokinetic time course and the pharmacodynamic time course. For example, the antithrombotic effect demonstrates a rapid equilibria between concentration and effect [17], whereas significant clinical response of antipsychotics typically takes weeks to separate from placebo [18, 19]. Therefore, the pharmacokinetic profile required for optimal pharmacodynamic effect will differ substantially depending on the target and the proximity of the target to the biochemical pathway responsible for the effect. Here we observe that there is no such thing as a generic optimal human PK profile, and therefore, an integrated PK/PD approach should be considered for novel targets.

11.4.1 Direct Response

The impact of time is one of the most important considerations of PK/PD modeling, and it is often the most neglected one in the high-throughput environment of early drug discovery. When screening new chemical entities (NCEs), it is common practice to compare compounds via a dose response where pharmacodynamic endpoints and concentration are compared at a single time point (often t_{max}). The dataset is analyzed as a response to a function of concentration with the assumption that PD response tracks directly with pharmacokinetics, as depicted graphically in Figure 11.3. Here the assumption is that there is no duration of pharmacodynamic response that extends beyond the pharmacokinetic profile and that biophase and plasma concentrations are at instantaneous equilibrium.

Although this is often not the case, there are documented instances of direct PK/PD relationships. For example, Mitenko and Ogilvie [20] described the impact of theophylline on airway obstruction using a direct-effect E_{max} model and Singh et al. [21] used a direct-effect inhibition E_{max} model to build a relationship between timolol plasma concentrations and heart rate after exercise and at rest. In most instances, however, there is a measureable difference between the plasma and the pharmacodynamic maximum (C_{max} and PD_{max}, respectively) with the duration of the pharmacodynamic effect extending past the plasma pharmacokinetics. This may be a function of pharmacokinetic distribution into the biophase, an indirect relationship

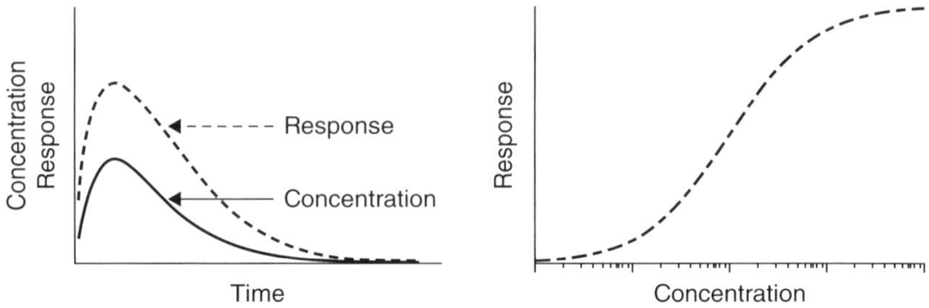

Figure 11.3 The direct-effect PK/PD relationship. Only in the instance where the onset and duration of pharmacodynamic response tracks with concentration (panel 1) can the PKPD relationship be called direct (panel 2).

between pharmacokinetics and pharmacodynamics, or both. Thus, mechanism-based PK/PD models quantitatively capture the path between plasma concentration and effect including (1) drug distribution to the target site, (2) drug target binding, and (3) transduction mechanisms [4] with the capability of distinguishing between drug effect and disease process [22].

11.4.2 Biophase Distribution

A common approach to capturing and modeling the impact of pharmacokinetic distribution to the biophase is the use of the effect compartment model [23]. The model assumptions are that the transport of drug to the target site occurs by passive diffusion and allows for the construction of a hypothetical effect compartment where the kinetics of diffusion is described by the following differential equation:

$$\frac{dC_e}{dt} = k_{1e} \cdot C_p - k_{e0} \cdot C_e \tag{11.2}$$

Here, k_{1e} and k_{e0} represent the first-order rate constants for distribution into and out of the hypothetical effect compartment, and C_p and C_e are the unbound drug concentrations in plasma and the hypothetical effect compartment, respectively.

As the amount of drug distributing into the effect compartment is negligible and does not significantly influence the overall systemic pharmacokinetic profile, the distribution rate parameters k_{1e} and k_{e0} are assumed to be the same (Equation 11.3). At a pharmacokinetic steady state, the drug concentration in the biophase is the same as unbound plasma, simplifying the equation describing the distribution into the biophase as demonstrated in Figure 11.4:

$$\frac{dC_e}{dt} = k_{e0} \cdot C_p - k_{e0} \cdot C_e \tag{11.3}$$

Sheiner et al. [23] showed this as a reasonable assumption for drugs that distribute to the site of action by passive diffusion. An early report of the biophase distribution

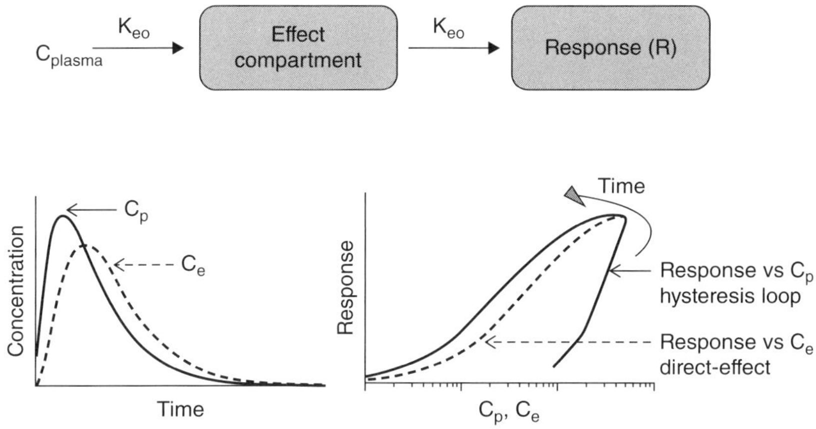

Figure 11.4 Schematic and graphic of a biophase model. Pharmacokinetic distribution into the hypothetical effect compartment C_e (biophase) fully describes the source of hysteresis, and PK/PD relationships in the biophase are direct at pharmacokinetic steady state.

was demonstrated by Galeazzi et al. showing that the change in Q-T interval (QT) effect after procainamide dose had a hysteresis loop when plotted as a function of plasma concentrations, but this hysteresis loop was not measureable when QT effect was plotted as a function of saliva procainamide concentration, leading the authors to recommend saliva as an appropriate biophase [24]. Several approaches of modeling distribution into the effect compartment have been proposed [6, 25–28] and have demonstrated success in deriving *in vivo* steady-state concentration-response relationships under acute dosing conditions.

Here we observe that effect compartment models have utility when the response delays relative to plasma concentration are clearly a function of distribution from the central blood compartment to the biophase. It is important to note that biophase effect modeling is not mechanistic and still presumes that the biophase concentration-effect relationship is direct. When using this sort of model, it is important to appreciate the proximity of the biomarker to the target and binding/functional kinetics to justify the use of this model for advanced PK/PD modeling in drug discovery and development.

11.4.3 Basic Indirect Models: More Mechanistic

Hysteresis in concentration-effect relationships is now widely regarded as the norm in *in vivo* pharmacology. However, conclusions, which have been recently editorialized by Van Der Graaf and Gabrielsson [3], include assuming the observed hysteresis is simply a function of biophase distribution and that a direct relationship between biophase PK and PD needs to be drawn with caution. A case study was highlighted by Gabrielsson and colleagues [29] demonstrating the likely inappropriate use of the effect compartment model. Here, the authors describe an analysis where the effects of 5-hydroxytryptamine reuptake inhibitors in mice could be explained by an effect compartment model [30]. However, as highlighted by Gabrielsson et al., the fact that

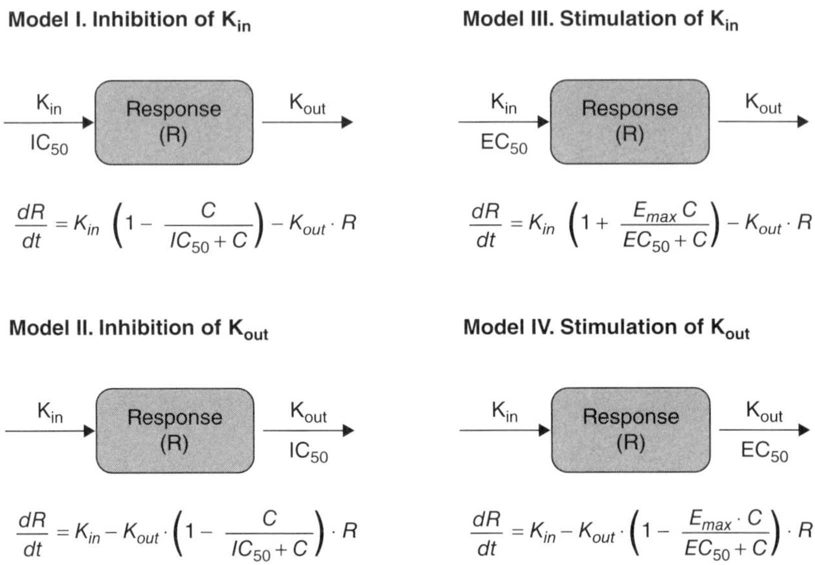

Figure 11.5 Four basic indirect response models characterized by either inhibition or stimulation of the response variable as first described by Dayneka and Jusko [31] (adopted with permission).

Kreilgaard and colleagues found dose-dependent values for the equilibration rate constant to the effect compartment (k_{e0}) calls into question the validity of the choice of the effect compartment to describe their data and suggests that target binding rather than drug distribution was the reason for the observed hysteresis.

Unlike the biophase distribution model that identifies the hysteresis between unbound plasma levels and pharmacodynamic effect as an exclusive function of pharmacokinetic distribution into the biophase, indirect pharmacodynamic models capture both reversible and irreversible effects. Indirect PK/PD models mathematically capture the onset, intensity, and duration of drug effect after biophase distribution. Four basic models that were initially introduced by Dayneka and colleagues [31] describe the indirect response, and they have been elegantly reviewed and expanded on by Jusko and colleagues [32, 33]. They are described as four distinct types as shown in Figure 11.5. Indirect response models describe the PK/PD of most targets and downstream biomarkers that can either inhibit or stimulate the baseline response.

A good example of an indirect PK/PD relationship is that demonstrated by warfarin. Warfarin is the most commonly used anticoagulant that acts via inhibition of VKORC1, exerting its activity by interrupting the regeneration of the active form of vitamin K, thereby decreasing carboxylation of the vitamin-K-dependent clotting factors [34, 35]. As carboxylation is necessary for the coagulation process, the blocking of the reductase activity by warfarin leads to the inhibition of coagulation [36]. Using the prothrombin time course as a measure of the coagulation process, which was first published by Nagashima et al. in 1969 [37], Jusko and Ko [32] captured the pharmacodynamic hysteresis by using indirect model I (Figure 11.6).

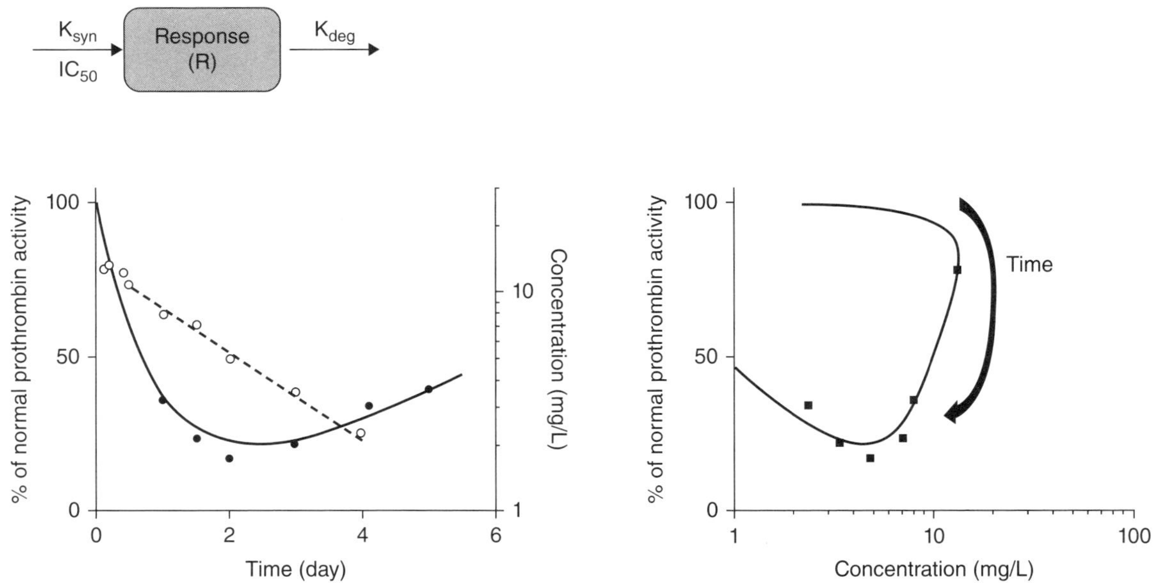

Figure 11.6 The composite graphic describes the pharmacokinetics of warfarin (open circles) and the pharmacodynamic time course of prothombin complex activity (solid circles) after an oral dose of 1.5 mg/kg of sodium warfarin. PK/PD composite graphic and analysis were obtained from Jusko [32] (adopted with permission) with dose day 1 PK/PD parameter estimates of 121%/day for k_{deg} and an IC_{50} of 1.26 mg/L.

Here, Jusko and Ko described k_{in} as k_{syn} as the zero-order rate constant for the synthesis of clotting factors. The first-order degradation process, generically described as k_{out}, is labeled here as (k_{deg}). The warfarin plasma pharmacokinetics was obtained after a 1.5-mg/kg oral dose of sodium warfarin, and the pharmacodynamic time course of prothombin complex activity was obtained by Jusko and Ko from the Nagashima et al. publication [37].

11.4.4 Irreversible Effects

Many anticancer and anti-infective agents are biologically active through irreversible interactions leading to cell/bacteria death. The classic model to describe cell kill was first described by Jusko [38] and is described in Equation 11.4:

$$\frac{dR}{dt} = k_g \cdot R - C_p \cdot R \tag{11.4}$$

In this instance, R is the cell number, k_g is the first-order growth rate constant observed in the system, and C_p is the concentration in plasma/biophase.

Modification of this model provides an elegant retrospective analysis of mouse tumor xenograft predicting clinical outcomes using an irreversible effect model by Wong and colleagues [39]. With the development of mouse tumor xenograft models, demonstrating an NCE antitumor effect *in vivo* has become a fundamental step in the preclinical development of oncology drug candidates. Despite being the workhorse efficacy model for most drug discovery programs, the capability of xenograft efficacy models to translate into meaningful clinical outcomes has been the subject of much debate [40–44]. Here, the investigators interrogated the predictive nature of eight molecularly targeted and cytotoxic chemotherapeutics in disease-relevant xenograft/allograft efficacy models with published clinical outcome data.

As shown in Figure 11.7, Wong et al. [39] modeled tumor growth inhibition using either an indirect response model (A) or a tumor transit (B) PK/PD model. The preclinical anticancer agent mediated tumor growth inhibition was mathematically linked to the pharmacokinetics of the anticancer agent providing the drug and system parameters allowing for clinical simulations. As demonstrated for Erlotinib in Figure 11.8, once the preclinical drug and system parameters for the anticancer agents were described from the preclinical model (Figure 11.8A), the authors incorporated the human pharmacokinetics into the model to simulate clinical outcomes (Figure 11.8B) and found an excellent correlation between simulated and observed clinical response [39]. Although xenograft models cannot fully capture patient populations, this study provides a framework for integrating preclinical data and predicted clinical outcomes in meaningful disease models to form a basis for a clinical testable hypothesis.

11.5 THE UTILITY OF TRANSLATION BIOMARKERS

Integration of translational biomarkers can aid in the description of the onset, extent, and duration of target engagement in preclinical PK/PD modeling at the

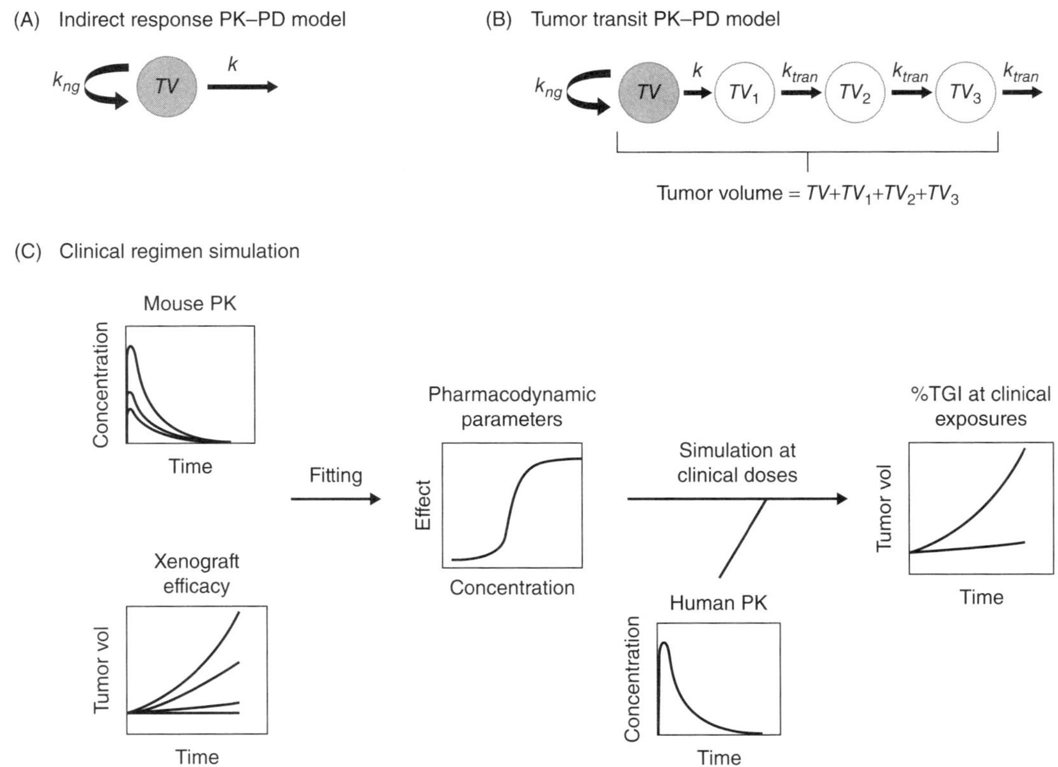

Figure 11.7 The PK/PD modeling and simulation process used to translate preclinical data is depicted. Tumor growth kinetics in xenograft/allograft studies were described using either a standard indirect response (panel A) or a tumor transit (panel B) PK–PD model. Where TV is the tumor volume, TV1, TV2, and TV3 are transit compartments representing tumor cells irreversibly damaged by treatment with an anticancer agent, k_{ng} (day^{-1}) is the net growth rate constant, K (day^{-1}) is the rate constant describing the antitumor effect of the anticancer agent, and k_{tran} (day^{-1}) is the rate constant describing movement of cells through the transit compartments. The simulation process integrating human pharmacokinetics is described in panel C. (Reprinted with permission from the American Association for Cancer Research [39].)

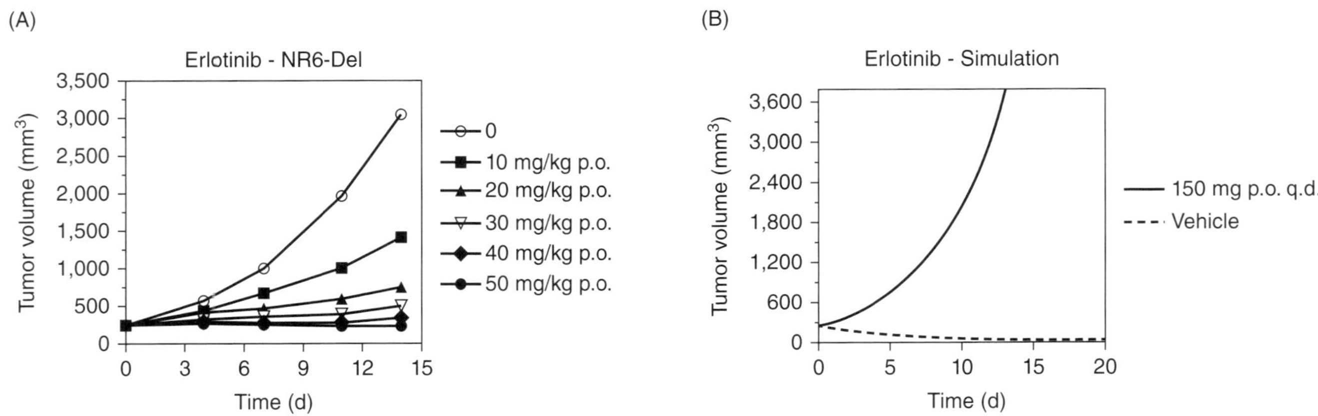

Figure 11.8 The PK/PD translational simulation for Erlotinib. The preclinical dose-response in the mouse xenograft (panel A) provided the drug and system parameters that allowed for clinical outcome response simulations (panel B).

earliest stages of a project's life. Identification of a mechanism-based biomarker whose modulation can be correlated with efficacy endpoints is of great utility for translational PK/PD/efficacy modeling. The emerging interest in the use of biomarkers in clinical drug development has been reviewed [45–49], where most of the discussion has focused on the utility of biomarkers in forming regulatory decision making [50–52]. For drug discovery projects with a novel biological mechanism of action, the identification and assessment of biomarkers is often examined in parallel with chemistry asset optimization. The successful integration of PK/PD modeling with biomarker data is anchored on four key aspects [46]: (1) The selection of a mechanism-based biomarker that can be linked with a clinically measureable endpoint. (2) Accurate quantification of drug, active metabolites, or both in biological fluids under good laboratory practice (GLP) conditions. (3) GLP-like assay methods for biomarkers. (4) Mechanism-based PKPD modeling and validation. Biomarker response to drugs should be reproducible, objective, and sensitive, with the anticipated efficacious response within its dynamic range, allowing for proof-of-concept (POC) studies to be dose anchored on biomarker response. In 2005, Danhof and colleagues proposed a new mechanistic classification system that identified seven types of biomarkers ranging from a patient genotype/phenotype that impacts patient response to treatment (type 0) to a measurable clinical response (type 6) [53].

The early stages of target identification and validation should include plans to develop a translational biomarker for linking PD/efficacy relationships identified preclinically with clinical POM/proof-of-principle(POP)/POC studies. For meaningful cross-species translation, quantitative analysis of functional biomarkers should be standardized from preclinical to clinical, and most importantly, study design should incorporate pharmacodynamic variability such that it is adequately powered to distinguish between treated and untreated study groups [45, 54, 55]. Ideally, a translational biomarker that is linked to efficacy greatly increases confidence in clinical hypothesis testing and the establishment of POM. In early drug discovery, and in some instances clinical development, these translational tools are either unavailable or do not have a sufficient dynamic range versus intrinsic animal-to-animal or patient-to-patient variability. In these instances and prior to establishing an appropriate biomarker tool, it is recommended to design the right sort of PK/efficacy study to guide chemistry asset optimization with the tools available for building PK/PD or PK/efficacy relationships.

11.6 PK/PD STUDY DESIGN: UNDERSTANDING PK DRIVERS FOR PD/EFFICACY RESPONSE

In early drug discovery, *in vivo* PD screening models are typically used to distinguish and prioritize compounds in a high-throughput manner for beneficial drug effect, as opposed to describing the drug effect on pathology of the human disease. Once a positive result is achieved, lead compounds are progressed for further testing, either through a more elegant study design or to a more stringent disease-like model when one exists [56]. The design of a PK/PD/efficacy study should complement the question it proposes to answer. Gabrielsson and colleagues have recommended a

PK/PD design/test cycle that captures the process of fully understanding the PK/PD relationship of chemistry leads [29] for future asset *in vitro* potency and ADME chemistry optimization.

11.6.1 Optimizing the Dose Route and Characterizing the Concentration Time Curve

When assessing the PK/PD relationship of different NCEs, stratification by ED_{50} (dose responsible for 50% of the maximal PD effect) is unlikely to be meaningful as it does not account for dose route, clearance, tissue distribution, and plasma protein binding. When performing preclinical PK/PD/Efficacy studies, it is common practice to aim to use the same dose route as that proposed for clinical development. This may not be the wisest choice in many instances and needs appropriate vetting to defend. For example, for many central nervous system (CNS) active compounds, there is a great body of evidence to support the use of rodents as the preclinical PD/efficacy screening species for behavioral or receptor occupancy studies. It is common for the disposition profile in rodents to be dramatically different from that observed in humans; primarily, a result of cross-species difference is hepatic clearance. The rapid clearance observed in preclinical test species results in poor and uncertain bioavailability and in shorter therapeutic half-lives compared with that observed in humans. Thus, oral dosing of an antidepressant preclinically, as it is done in the clinic, would require multiple doses per day that could ultimately impact PD or efficacy readouts (especially behavioral based) [57, 58]. In these instances, and when extended constant exposure to a test compound is required, an osmotic minipump can be a viable option [59] assuming the test compound is stable for the duration of the experiment.

Prior to an elaborate PK/PD study, it is useful to obtain bridging PK data after oral, subcutaneous (SC), or intraperitoneal (IP) dose route administration in the PD test species. Here the scientist can identify the dose route that will provide the PK target coverage that will allow for adequate and meaningful collection of PD endpoints for hypothesis testing. Depending on the pharmacokinetic distribution profile, an intravenous (IV) bolus preclinically can lead to C_{max} ranges in great excess to those likely to be clinically observed. Thus, any additional pharmacology can either add to or antagonize the primary pharmacology of interest, thereby potentially confounding attempts of preclinical PK/PD translation.

11.6.2 Interrogate the PK/Response Time Profile

Single time-point dose response data are often used in drug discovery screening and are likely one of the greatest missed opportunities of PK/PD study design. Single time-point dose response studies can only inform the response at the dose and time measured and do not inform the PD-time course over the dosing interval, which is often the goal of the experiment. The PD time course experiment allows for a better understanding of the onset, intensity, and duration of pharmacodynamic response and is superior to a single-time point dose response, which simply provides the capability to describe the PD response at a certain dose in that moment in time. Time course

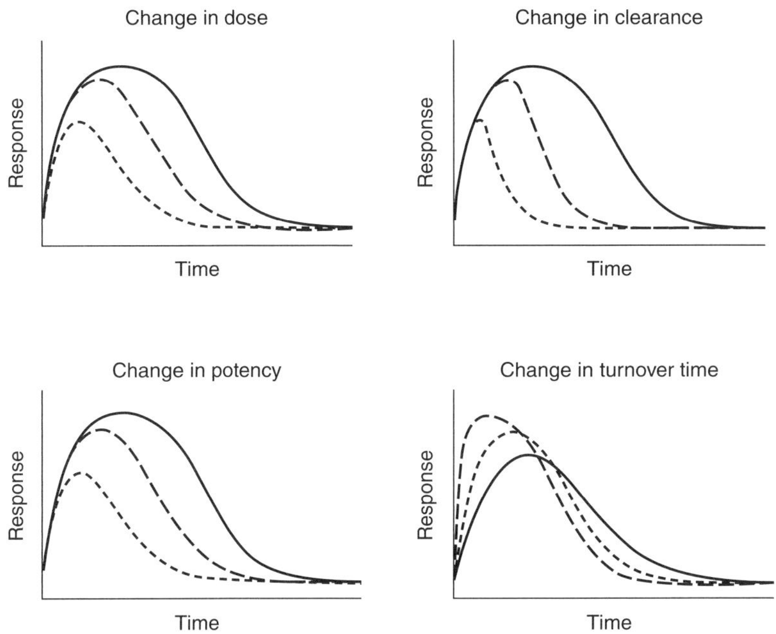

Figure 11.9 Simulations of the impact of changing dose, clearance, potency (EC_{50}), and turnover time using an indirect response time curve (indirect model III, stimulation of production). In each instance, only the stated parameter is changed to demonstrate the impact of each parameter on the PD time course. The top panels demonstrate the impact of changing plasma concentrations on the PD time course, where the top left includes three simulations with escalating dose and the top right demonstrates three simulations with different clearance. The bottom panels demonstrate the impact of changing drug potency (drug parameter) and the impact of changing the turnover time (system-specific parameter).

PK/PD relationships provide the mathematical relationship to describe the effect at a certain dose and concentration over time and may be used to simulate other dosing scenarios for further preclinical hypothesis testing.

Ideally, initial preclinical PK/PD experiments should be run with probe compounds selective against the pharmacology that is proposed to be the principal driver of pharmacodynamic/efficacy outcome. Using this approach, investigators can interrogate the patterns of PK and PD; the PD effects after a single dose or during steady state; the onset, intensity, and duration of pharmacodynamic effect; and the consequences of varying dosing and exposure-time patterns on pharmacodynamic readouts, as shown in Figure 11.9.

11.6.3 Examine the Impact of Controlled Absorption on Response Profile

Gabrielsson and colleagues recently published an elegant review on optimizing PK/PD study design and outlining factors impacting PK/PD relationships [29]. A

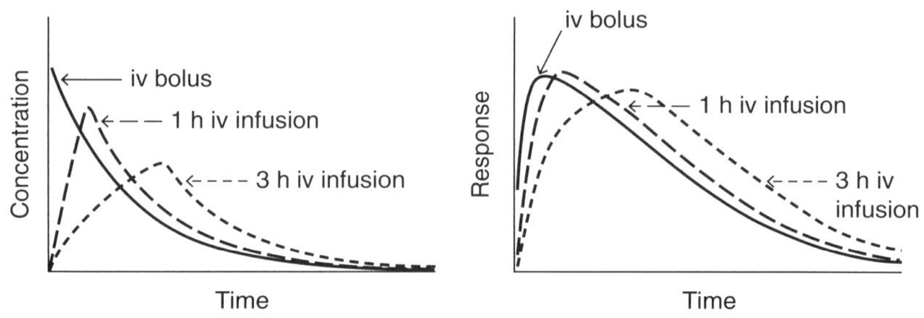

Figure 11.10 Impact of pharmacokinetic input on the pharmacodynamic time course. In all three instances, the dose and pharmacokinetic parameters were identical where only the input rate (infusion time) was changed.

key principal considered by the group was the importance of the impact of pharmacokinetic absorption, recognizing that cross-species differences can be profound and potentially confound PK/PD translation. To assess the impact of rapid or slow oral absorption in preclincial species, a constant rate infusion is proposed. This allows for a controlled means to describe both the impact of rate (t_{max}) and extent (C_{max}) with good accuracy and reproducibility. Testing different infusion rates with the same total dose would conserve the total exposure but significantly alter the maximum exposure, allowing a pharmacologist to begin to identify the pharmacodynamic contribution or safety liability of C_{max} with identical area under concentration $(AUC)_{0\text{-}t}$ between dose groups, as demonstrated in Figure 11.10. In most instances, the pharmacodynamic response is influenced by the initial rate of drug concentration (whether this is from absorption or IV infusion). It is also likely that should a compound be given orally, there is intra-animal variability in rate and extent of absorption, and this should be considered when examining the pharmacodynamic variability, especially if pharmacokinetic samples are not taken from the pharmacodynamic test group.

11.6.4 PK/PD Time Points: Frequency and Duration

As shown in the warfarin case study (Figure 11.6), it is advisable to collect PD readouts prior to C_{max} (after oral, IP, SC, or IV infusion studies) and during the elimination phase of the PK curve. This allows for the ability to identify hysteresis (whether this be target-based or a result of pharmacokinetic distribution into the biophase) and describe the onset, intensity, and duration of PD effect. It is also recommended to test more than one dose (ideally two to four) as doing so will not only increase the likelihood of capturing the maximum response but also identify onset of response patterns that can further describe the response-time curve for simulation. To that end, and in most instances, it is highly recommended to aim for a dose-response time-course of the test compound in the study design. Continuous PD/efficacy readouts as a function of dose and time in the instance where multiple pharmacodynamic measurements may be taken from the same animal are ideal. When a complete or partial PD time course may be taken from the same animal, the number of animals required is greatly diminished, where the number of animals per dose group should take into account the dynamic range of the response and accuracy of the PD readouts. In the common instance where PK is not sampled during PD measurements, pharmacokinetic variability should also be considered for appropriate powering of the study. There is an emergence of automation technologies to lessen the stress of PK sampling during readouts. In the absence of this possibility, using same-strain satellite animals (ideally on the same dose day and dose solution) to capture the PK profile as necessary is an acceptable alternative.

11.6.5 Translational Consideration: Cross-species Differences in Plasma Protein Binding

For preclinical PK/PD studies with the aim of estimating human response, a key aspect of characterizing the translational concentration-effect relationship of a specific compound can be the cross-species differences in plasma protein binding [2, 60]. Many NCEs demonstrate some extent of plasma protein binding where, in most instances, albumin and α1-acid glycoprotein are the two major drug-binding plasma proteins [61]. Despite that in many instances the pharmacodynamic biophase is not in blood or plasma, the translational pharmacodynamic effect is typically anchored on unbound plasma levels. The assumption here is that there is no cross-species difference in the extent of unbound distribution into the biophase. The total target-tissue levels of the compound describe both the unbound fraction and the compound that is nonspecifically bound to tissue, and thus, there are more caveats than with unbound plasma levels. Aside from the cross-species translational ease of anchoring PK/PD relationships to unbound plasma levels, in most instances, clinical sampling for quantification of the compound will be taken from the blood compartment and not from the biophase target tissue.

11.6.6 Study Design Summary

Despite the fact that PD/efficacy responses are rarely driven exclusively through plasma C_{max} or plasma $AUC_{0\text{-dosing interval}}$, it is common practice for discovery teams to describe an effect as either a C_{max} or AUC-driven. In most instances, is it more likely that the PD/efficacy response is a composite of both C_{max} and AUC; therefore, a superior analysis is to generate the best estimate of the PK/PD-time profile. Care must be taken not to extend conclusions past what the study design allows. Remember that a dose-response concentration-effect relationship gives information regarding the response observed at the dose/concentration measured at one time point. Unless there is precedence that the concentration/response relationship is direct, simulations of response as a function of time from these data are entirely inappropriate.

11.7 CAPTURING PHARMACODYNAMIC UNCERTAINTY

Describing preclinical pharmacodynamic uncertainty can be challenging in drug discovery. It is not uncommon for pharmacodynamic endpoints to be evaluated without complimentary pharmacokinetic data. Population PK/PD, first introduced by Sheiner [62], describes the concentration-effect relationship in populations, rather than the mean data or individual data, and allows for quantitating the sources of interindividual pharmacokinetic and pharmacodynamic variability. The emergence of user-friendly software to describe variability can allow for the characterization of preclinical PK/PD variability and parameter uncertainty. Describing pharmacodynamic variability, even qualitatively, is useful when describing an NCE PK/PD relationship.

11.8 PERSPECTIVES

PK/PD models are mathematical tools to describe the underlying biological and pharmacological mechanisms of actions that can be used to predict outcomes in alternate dosing paradigms and ultimately scale to the clinic [10, 46]. It is now recognized that a better understanding of the concentration-effect-time relationship in early drug discovery can help set the chemistry lead optimization criteria toward the ideal human pharmacokinetic profile. Failure to integrate PK/PD in drug discovery can lead to poor or incomplete study design that fails to demonstrate the concentration required to generate a pharmacodynamic response. Ultimately, should these flawed extrapolations extend into clinical candidate nomination, there can be a substantial over- or underprediction of the concentration-time profile required for clinical hypothesis testing resulting in the advancement of the wrong candidate(s), termination of projects, or the costly redesign of clinical trials. Integrative PK/PD modeling should be viewed as a core activity for translational methods for integrating several sources of preclinical and *in vitro* data on the predictions of dose and exposure requirements to generate the pharmacological response in humans [63].

REFERENCES

1. Kola, I., Landis, J. (2004). Can the pharmaceutical industry reduce attrition rates? *Nature Reviews Drug Discovery*, *3*, 711–715.
2. Gabrielsson, J., Green, A.R. (2009). Quantitative pharmacology or pharmacokinetic pharmacodynamic integration should be a vital component in integrative pharmacology. *Journal of Pharmacology and Experimental Therapeutics*, *331*, 767–774.
3. Van Der Graaf, P.H., Gabrielsson, J. (2009). Pharmacokinetic-pharmacodynamic reasoning in drug discovery and early development. *Future Medicinal Chemistry*, *1*, 1371–1374.
4. Danhof, M., de Jongh, J., De Lange, E.C., Della Pasqua, O., Ploeger, B.A., Voskuyl, R.A. (2007). Mechanism-based pharmacokinetic-pharmacodynamic modeling: Biophase distribution, receptor theory, and dynamical systems analysis. *Annual Review of Pharmacology and Toxicology*, *47*, 357–400.
5. Danhof, M., de Lange, E.C., Della Pasqua, O.E., Ploeger, B.A., Voskuyl, R.A. (2008). Mechanism-based pharmacokinetic-pharmacodynamic (PK-PD) modeling in translational drug research. *Trends in Pharmacological Sciences*, *29*, 186–191.

6. Van der Graaf, P.H., Danhof, M. (1997). Analysis of drug-receptor interactions in vivo: A new approach in pharmacokinetic-pharmacodynamic modelling. *International Journal of Clinical Pharmacology, Therapy*, *35*, 442–446.
7. Derendorf, H., Meibohm, B. (1999). Modeling of pharmacokinetic/pharmacodynamic (PK/PD) relationships: Concepts and perspectives. *Pharmaceutical Research*, *16*, 176–185.
8. Meno-Tetang, G.M., Lowe, P.J. (2005). On the prediction of the human response: A recycled mechanistic pharmacokinetic/pharmacodynamic approach. *Basic and Clinical Pharmacology and Toxicology*, *96*, 182–192.
9. Aarons, L., Karlsson, M.O., Mentre, F., Rombout, F., Steimer, J.L., van Peer, A., COST B15 Experts. (2001). Role of modelling and simulation in Phase I drug development. *European Journal of Pharmaceutical Sciences*, *13*, 115–122.
10. Levy, G. The case for preclinical pharmacodynamics, in: A. Yacobi, J. Skelly, V. Shah, and L.Z. Benet (eds.), *Integration of Pharmacokinetics, Pharmacodynamics, and Toxicokinetics in RationalDrug Development*, Plenum, New York, 1993.
11. Chien, J.Y., Friedrich, S., Heathman, M.A., de Alwis, D.P., Sinha, V. (2005). Pharmacokinetics/Pharmacodynamics and the stages of drug development: Role of modeling and simulation. *AAPS Journal*, *7*, E544–E559.
12. Ito, K., Asakura, A., Yamada, Y., Nakamura, K., Sawada, Y., Iga, T. (1997). Prediction of the therapeutic dose for benzodiazepine anxiolytics based on receptor occupancy theory. *Biopharmaceutics & Drug Disposition*, *18*, 293–303.
13. Cox, E.H., Langemeijer, M.W., Gubbens-Stibbe, J.M., Muir, K.T., Danhof, M. (1999). The comparative pharmacodynamics of remifentanil and its metabolite, GR90291, in a rat electroencephalographic model. *Anesthesiology*, *90*, 535–544.
14. Kalvass, J.C., Olson, E.R., Cassidy, M.P., Selley, D.E., Pollack, G.M. (2007). Pharmacokinetics and pharmacodynamics of seven opioids in P-glycoprotein-competent mice: Assessment of unbound brain EC50,u and correlation of in vitro, preclinical, and clinical data. *Journal of Pharmacology and Experimental Therapeutics*, *323*, 346–355.
15. Hong, Y., Mager, D.E., Blum, R.A., Jusko, W.J. (2007). Population pharmacokinetic/pharmacodynamic modeling of systemic corticosteroid inhibition of whole blood lymphocytes: Modeling interoccasion pharmacodynamic variability. *Pharmaceutical Research*, *24*, 1088–1097.
16. Mager, D.E., Lin, S.X., Blum, R.A., Lates, C.D., Jusko, W.J. (2003). Dose equivalency evaluation of major corticosteroids: Pharmacokinetics and cell trafficking and cortisol dynamics. *Journal of Clinical Pharmacology*, *43*, 1216–1227.
17. Kalantzi, K.I., Tsoumani, M.E., Goudevenos, I.A., Tselepis, A.D. (2012). Pharmacodynamic properties of antiplatelet agents: Current knowledge and future perspectives. *Expert Review of Clinical Pharmacology*, *5*, 319–336.
18. Khin, N.A., Chen, Y.F., Yang, Y., Yang, P., Laughren, T.P. (2012). Exploratory analyses of efficacy data from schizophrenia trials in support of new drug applications submitted to the US Food and Drug Administration. *Journal of Clinical Psychiatry*, *73*, 856–864.
19. Maher, A.R., Maglione, M., Bagley, S., Suttorp, M., Hu, J.H., Ewing, B., Wang, Z., Timmer, M., Sultzer, D., Shekelle, P.G. (2011). Efficacy and comparative effectiveness of atypical antipsychotic medications for off-label uses in adults: A systematic review and meta-analysis. *Journal of the American Medical Association*, *306*, 1359–1369.

20. Mitenko, P.A., Ogilvie, R.I. (1973). Rational intravenous doses of theophylline. *New England Journal of Medicine*, *289*, 600–603.
21. Singh, B.N., Williams, F.M., Whitlock, R.M., Collett, J., Chew, C. (1980). Plasma timolol levels and systolic time intervals. *Clinical Pharmacology & Therapeutics*, *28*, 159–166.
22. Post, T.M., Freijer, J.I., DeJongh, J., Danhof, M. (2005). Disease system analysis: Basic disease progression models in degenerative disease. *Pharmaceutical Research*, *22*, 1038–1049.
23. Sheiner, L.B., Stanski, D.R., Vozeh, S., Miller, R.D., Ham, J. (1979). Simultaneous modeling of pharmacokinetics and pharmacodynamics: Application to d-tubocurarine. *Clinical Pharmacology & Therapeutics*, *25*, 358–371.
24. Galeazzi, R.L., Benet, L.Z., Sheiner, L.B. (1976). Relationship between the pharmacokinetics and pharmacodynamics of procainamide. *Clinical Pharmacology & Therapeutics*, *20*, 278–289.
25. Fuseau, E., Sheiner, L.B. (1984). Simultaneous modeling of pharmacokinetics and pharmacodynamics with a nonparametric pharmacodynamic model. *Clinical Pharmacology & Therapeutics*, *35*, 733–741.
26. Unadkat, J.D., Bartha, F., Sheiner, L.B. (1986). Simultaneous modeling of pharmacokinetics and pharmacodynamics with nonparametric kinetic and dynamic models. *Clinical Pharmacology & Therapeutics*, *40*, 86–93.
27. Veng-Pedersen, P., Mandema, J.W., Danhof, M. (1991). A system approach to pharmacodynamics. III: An algorithm and computer program, COLAPS, for pharmacodynamic modeling. *Journal of Pharmaceutical Sciences*, *80*, 488–495.
28. Minematsu, T., Ohtani, H., Yamada, Y., Sawada, Y., Sato, H., Iga, T. (2001). Quantitative relationship between myocardial concentration of tacrolimus and QT prolongation in guinea pigs: Pharmacokinetic/pharmacodynamic model incorporating a site of adverse effect. *Journal of Pharmacokinetics and Pharmacodynamics*, *28*, 533–554.
29. Gabrielsson, J., Green, A.R., Van der Graaf, P.H. (2010). Optimising in vivo pharmacology studies–Practical PKPD considerations. *Journal of Pharmacological and Toxicological Methods*, *61*, 146–156.
30. Kreilgaard, M., Smith, D.G., Brennum, L.T., Sanchez, C. (2008). Prediction of clinical response based on pharmacokinetic/pharmacodynamic models of 5-hydroxytryptamine reuptake inhibitors in mice. *British Journal of Pharmacology*, *155*, 276–284.
31. Dayneka, N.L., Garg, V., Jusko, W.J. (1993). Comparison of four basic models of indirect pharmacodynamic responses. *Journal of Pharmacokinetics and Biopharmaceutics*, *21*, 457–478.
32. Jusko, W.J., Ko, H.C. (1994). Physiologic indirect response models characterize diverse types of pharmacodynamic effects. *Clinical Pharmacology & Therapeutics*, *56*, 406–419.
33. Sharma, A., Jusko, W.J. (1998). Characteristics of indirect pharmacodynamic models and applications to clinical drug responses. *British Journal of Clinical Pharmacology*, *45*, 229–239.
34. Garcia, A.A., Reitsma, P.H. (2008). VKORC1 and the vitamin K cycle. *Vitamins & Hormones*, *78*, 23–33.
35. Wallin, R., Sane, D.C., Hutson, S.M. (2002). Vitamin K 2,3-epoxide reductase and the vitamin K-dependent gamma-carboxylation system. *Thrombosis Research*, *108*, 221–226.

36. Whitlon, D.S., Sadowski, J.A., Suttie, J.W. (1978). Mechanism of coumarin action: Significance of vitamin K epoxide reductase inhibition. *Biochemistry*, *17*, 1371–1377.
37. Nagashima, R., O'Reilly, R.A., Levy, G. (1969). Kinetics of pharmacologic effects in man: The anticoagulant action of warfarin. *Clinical Pharmacology & Therapeutics*, *10*, 22–35.
38. Jusko, W.J. (1971). Pharmacodynamics of chemotherapeutic effects: Dose-time-response relationships for phase-nonspecific agents. *Journal of Pharmaceutical Sciences*, *60*, 892–895.
39. Wong, H., Choo, E.F., Alicke, B., Ding, X., La, H., McNamara, E., Theil, F.P., Tibbitts, J., Friedman, L.S., Hop, C.E., Gould, S.E. (2012). Antitumor activity of targeted and cytotoxic agents in murine subcutaneous tumor models correlates with clinical response. *Clinical Cancer Research*, *18*, 3846–3855.
40. Johnson, J.I., Decker, S., Zaharevitz, D., Rubinstein, L.V., Venditti, J.M., Schepartz, S., Kalyandrug, S., Christian, M., Arbuck, S., Hollingshead, M., Sausville, E.A. (2001). Relationships between drug activity in NCI preclinical in vitro and in vivo models and early clinical trials. *British Journal of Cancer*, *84*, 1424–1431.
41. Kelland, L.R. (2004). Of mice and men: Values and liabilities of the athymic nude mouse model in anticancer drug development. *European Journal of Cancer*, *40*, 827–836.
42. Kerbel, R.S. (2003). Human tumor xenografts as predictive preclinical models for anticancer drug activity in humans: Better than commonly perceived-but they can be improved. *Cancer Biology & Therapy*, *2*, S134–S139.
43. Peterson, J.K., Houghton, P.J. (2004). Integrating pharmacology and in vivo cancer models in preclinical and clinical drug development. *European Journal of Cancer*, *40*, 837–844.
44. Troiani, T., Schettino, C., Martinelli, E., Morgillo, F., Tortora, G., Ciardiello, F. (2008). The use of xenograft models for the selection of cancer treatments with the EGFR as an example. *Critical Reviews in Oncology /Hematology*, *65*, 200–211.
45. Colburn, W.A. (1997). Selecting and validating biologic markers for drug development. *Journal of Clinical Pharmacology*, *37*, 355–362.
46. Colburn, W.A., Lee, J.W. (2003). Biomarkers, validation and pharmacokinetic-pharmacodynamic modelling. *Clinical Pharmacokinetics*, *42*, 997–1022.
47. Jackson, D.H., Banks, R.E. (2010). Banking of clinical samples for proteomic biomarker studies: A consideration of logistical issues with a focus on pre-analytical variation. *Proteomics—Clinical Applications*, *4*, 250–270.
48. Lee, J.W., Hulse, J.D., Colburn, W.A. (1995). Surrogate biochemical markers: Precise measurement for strategic drug and biologics development. *Journal of Clinical Pharmacology*, *35*, 464–470.
49. Wagner, J.A., Williams, S.A., Webster, C.J. (2007). Biomarkers and surrogate end points for fit-for-purpose development and regulatory evaluation of new drugs. *Clinical Pharmacology & Therapeutics*, *81*, 104–107.
50. Kuhlmann, J., Wensing, G. (2006). The applications of biomarkers in early clinical drug development to improve decision-making processes. *Current Clinical Pharmacology*, *1*, 185–191.
51. Lesko, L.J., Atkinson, A.J., Jr. (2001). Use of biomarkers and surrogate endpoints in drug development and regulatory decision making: Criteria, validation, strategies. *Annual Review of Pharmacology and Toxicology*, *41*, 347–366.
52. Rolan, P., Atkinson, A.J., Jr., Lesko, L.J., Scientific Organizing Committee, Conference Report Committee. (2003). Use of biomarkers from drug discovery through clinical

practice: Report of the Ninth European Federation of Pharmaceutical Sciences Conference on Optimizing Drug Development. *Clinical Pharmacology & Therapeutics*, *73*, 284–291.

53. Danhof, M., Alvan, G., Dahl, S.G., Kuhlmann, J., Paintaud, G. (2005). Mechanism-based pharmacokinetic-pharmacodynamic modeling—a new classification of biomarkers. *Pharmaceutical Research*, *22*, 1432–1437.
54. Gao, J., Garulacan, L.A., Storm, S.M., Opiteck, G.J., Dubaquie, Y., Hefta, S.A., Dambach, D.M., Dongre, A.R. (2005). Biomarker discovery in biological fluids. *Methods*, *35*, 291–302.
55. Goodsaid, F., Frueh, F. (2007). Biomarker qualification pilot process at the US Food and Drug Administration. *AAPS Journal*, *9*, E105–E108.
56. Bueters, T.J., Hoogstraate, J., Visser, S.A. (2009). Correct assessment of new compounds using in vivo screening models can reduce false positives. *Drug Discovery Today*, *14*, 89–94.
57. Stevens, J., Suidgeest, E., van der Graaf, P.H., Danhof, M., de Lange, E.C. (2009). A new minimal-stress freely-moving rat model for preclinical studies on intranasal administration of CNS drugs. *Pharmaceutical Research*, *26*, 1911–1917.
58. Zuideveld, K.P., van Gestel, A., Peletier, L.A., Van der Graaf, P.H., Danhof, M. (2002). Pharmacokinetic-pharmacodynamic modelling of the hypothermic and corticosterone effects of the 5-HT1A receptor agonist flesinoxan. *European Journal of Pharmacology*, *445*, 43–54.
59. Abe, C., Tashiro, T., Tanaka, K., Ogihara, R., Morita, H. (2009). A novel type of implantable and programmable infusion pump for small laboratory animals. *Journal of Pharmacological and Toxicological Methods*, *59*, 7–12.
60. Gabrielsson, J., Dolgos, H., Gillberg, P.G., Bredberg, U., Benthem, B., Duker, G. (2009). Early integration of pharmacokinetic and dynamic reasoning is essential for optimal development of lead compounds: Strategic considerations. *Drug Discovery Today*, *14*, 358–372.
61. Howard, M.L., Hill, J.J., Galluppi, G.R., McLean, M.A. (2010). Plasma protein binding in drug discovery and development. *Combinatorial Chemistry and High Throughput Screening*, *13*, 170–187.
62. Sheiner, L.B. (1984). The population approach to pharmacokinetic data analysis: Rationale and standard data analysis methods. *Drug Metabolism Reviews*, *15*, 153–171.
63. Agoram, B.M., Martin, S.W., van der Graaf, P.H. (2007). The role of mechanism-based pharmacokinetic-pharmacodynamic (PK-PD) modelling in translational research of biologics. *Drug Discovery Today*, *12*, 1018–1024.

III

ASSESSMENT AND MITIGATION OF CRITICAL CLINICALLY RELEVANT ADMET RISKS IN DRUG DISCOVERY AND DEVELOPMENT

12

IN VITRO–IN SILICO TOOLS TO PREDICT PHARMACOKINETICS OF POORLY SOLUBLE DRUG COMPOUNDS

Christian Wagner and Jennifer B. Dressman

12.1 INTRODUCTION AND GENERAL CONSIDERATIONS

The main factors that influence the *in vivo* performance and exposure of an orally administered drug are (1) the amount of drug in solution at the site of absorption (which, of course, strongly depends on the solubility of the drug, on the one hand, and on the rate and extent of drug dissolution in the gastrointestinal tract, on the other hand); (2) inactivation by, e.g., complexation or acid- or enzyme-induced decomposition; (3) the ability of the drug to permeate the gut wall (also taking into consideration the possibility that the drug is a substrate for transporters, e.g., for P-glycoprotein (P-gp) or organic anion-transporting polypeptide (OATP)); and 4) enzymatic decomposition in the gut wall, liver, or both by, e.g., cytochrome P 450 (CYP) enzymes (the so-called presystemic or first-pass metabolism) [1]. For drugs that are not released instantaneously from the dosage form, e.g., for extended release dosage forms, the rate, duration, and mechanism of release are of additional importance for the *in vivo* behavior of the drug.

Predictive ADMET: Integrative Approaches in Drug Discovery and Development, First Edition.
Edited by Jianling Wang and Laszlo Urban.

Table 12.1 Overview of *in vitro* surrogate techniques to estimate absorption characteristics of drug compounds

In vivo process	*In vitro* surrogate technique[a]
Drug solubility/dissolution	Biorelevant dissolution/solubility testing [1, 17, 18]
Intestinal decomposition, including first-pass metabolism	Enterocytes; hepatocytes; liver microsomes; S9 [11–13, 16, 19]
Drug permeability, absorption	Caco 2 cell studies [6–8, 15]; PAMPA studies [4, 9]; PBPK modeling; (Q)SAR, physicochemical drug properties [2, 3, 10, 14]

[a]Caco-2 (human epithelial colorectal adenocarcinoma cells); PAMPA (Parallel Artificial Membrane Permeability Assay); PBPK (Physiologically Based Pharmacokinetic); (Q)SAR ((Quantitative) structure activity relationship).

In vivo studies, conducted in humans or animals, provide the most accurate information about the rate and extent of drug absorption. However, such studies are time- and cost-intensive, and so *in vitro* surrogate techniques are sought for this purpose [2–16].

Table 12.1 gives a brief overview of the factors that impact the *in vivo* absorption of a drug and possible surrogate techniques to estimate those factors *in vitro*.

In this chapter, we focus on the application of biorelevant dissolution conditions to estimate the *in vivo* solubility and dissolution behavior of poorly soluble drug compounds. The solubility of a drug in the gastrointestinal environment is one key factor that impacts the rate and extent of drug absorption and, thus, its bioavailability because only drug which is molecularly dissolved can cross the gut wall. However, biorelevant solubility and dissolution results do not always give a quantitative prediction of the drug's absorption. We thus additionally focus on application of physiologically based pharmacokinetic (PBPK) modeling; an *in silico* approach to predict the absorption, distribution, metabolism, and excretion (ADME) properties of drug compounds and the possible advantages of linking biorelevant dissolution test results with PBPK models. Other techniques for linking *in vitro* with *in vivo* data, e.g., classic *in vitro–in vivo* correlations (IVIVCs), the biopharmaceutics classification system (BCS), or the "biowaiver" approach, are not the subjects of this book chapter.

12.1.1 The Concept of Biorelevant Dissolution Conditions for Forecasting the *In Vivo* Dissolution of Poorly Soluble Drugs

As only a dissolved drug can cross the gut wall and enter into the systemic circulation, it is of utmost importance in preclinical and clinical development to get an estimate of the drug's dissolution and solubility behavior in the gastrointestinal tract, both in preclinical species like dogs and in humans. Biorelevant dissolution methods have been developed specifically for this purpose.

12.1.1.1 Media Selection

Plain buffer solutions or buffers containing artificial surfactants like sodium dodecyl sulfate (SDS) are described in the pharmacopeias, but they do not fully reflect the composition of natural gastric and intestinal juices and thus fall short of being "biorelevant." In particular, for poorly soluble drugs, these simple dissolution media are not an appropriate tool for forecasting the drug's behavior in the gastrointestinal tract [17, 18–23].

Briefly, the concept behind biorelevant dissolution media is to adjust the media composition as close as possible to the *in vivo* composition of pre- and postprandial gastric and intestinal juices. Of course, one should bear in mind that biorelevant media represent a compromise between reflecting the *complete* composition of gastrointestinal fluids on the one hand and the need for reproducibility in the laboratory setting on the other hand.

In general, there are three sets of biorelevant media, i.e., those media that simulate (1) gastric, (2) small intestinal, and (3) colonic fluids in both the fasted and the fed state, and these media are fasted and fed state simulated gastric fluids (FaSSGF, FeSSGF), fasted and fed state simulated intestinal fluids (FaSSIF, FeSSIF), and fasted and fed state simulated colonic fluids (FaSSCoF, FeSSCoF). Meanwhile, the composition of the small intestinal media has been updated to FaSSIF v2 and FeSSIF v2 (version 2), respectively [5, 24–26]. The media differ in their pH, osmolality, buffer capacity, and amounts of bile salts, lecithin, and lipolysis products. Thus, the media show a different solubilization capacity, e.g., for neutral, basic, and acidic drug compounds. For composition and preparation of these media, we refer to the primary literature [5, 24–26].

12.1.1.2 Media Volumes

In vivo, the amount of fluid available for drug dissolution in the gastrointestinal tract depends on the liquid intake, the secretion of digestion juices, and water absorption [1].

Basal preprandial gastric fluid volumes are often in the range between 20 and 30 mL [1, 27], and administration of a formulation together with a glass of water thus leads to a gastric fluid volume of 200 to 300 mL. As the stomach is not a typical organ for drug absorption, gastric absorption of most drugs can be considered negligible, and gastric dissolution thus can be simulated well *in vitro* using the Mini paddle setup. This setup is a scaled-down version of the United States Pharmacopeia (USP) paddle setup, whereby the hydrodynamics remain essentially similar [28].

Depending on the meal size, a medium volume up to 1 L might be reasonable for simulation of postprandial gastric conditions [24].

In 2005, Schiller et al. [27] reported on small intestinal liquid volumes of 105 mL in the fasted state and 54 mL in the fed state. This liquid was observed to be not distributed equally along the small intestine but to occur as liquid-filled pockets [27]. However, earlier researchers had observed markedly higher fluid volumes in the range of 120 mL up to 350 mL [29]. At that time it was also observed that the amount of fluid, especially in the small intestine, depends on the tonicity of the meal given (hypertonic/hypotonic) [30]. The small intestinal liquid volumes observed by Schiller et al. [27] are quite a bit lower than the values observed by, e.g., Dillard et al. [29]. One reason for this apparent discrepancy might be that in the study conducted by Schiller et al. [27], samples were taken 1 h after administration of the liquid, and it is likely that during this time, water absorption had already occurred and thus led to lower observed liquid volumes in the small intestine. It is reasonable that small intestinal liquid volumes are highly variable and depend on the measuring technique, the time of measurement, the volume of fluid co-administered, and for postprandial measurements, the size composition of the meal.

In contrast to small intestinal fluid volumes, colonic fluid volumes seemed to be unaffected by the prandial state and were reported as approximately 12 mL in the study conducted by Schiller et al. [27]. Other researchers report values between 22 and 30 mL [31, 32].

In most cases, at least for immediate release formulations, the paddle apparatus with a medium volume between 250 and 500 mL provides sufficient accuracy to give a good qualitative estimate about the *in vivo* dissolution behavior of poorly soluble drug compounds in the upper gastrointestinal tract.

One limitation of the paddle setup is the relatively high medium volume (250 mL for the Mini Paddle setup and up to 900 mL for the standard paddle setup [28]), which does not always reflect the human gastrointestinal volumes [27, 29, 30]. Additionally, there is no standard technique for simultaneously measuring drug dissolution and absorption in the same vessel, as drug dissolution is influenced by the absorption process *in vivo*.

The high liquid volumes typically used for simulation of small intestinal dissolution (typically 500 mL) can lead to a pronounced overestimation of the extent of dissolution of poorly soluble drugs in the small intestine. On the other hand, the dissolution rate in the *in vitro* setup decreases rapidly as soon as nonsink conditions are reached, which can lead to an underestimation of *in vivo* dissolution.

In vivo, the small intestinal dissolution is driven by the absorption rate of the drug, and at least for highly permeable drug compounds, it is quite feasible that dissolution of the drug takes place under sink conditions. By contrast, for poorly permeable compounds, the standard paddle setup can be a useful approach to simulate the *in vivo* dissolution of the drug because it is likely that *in vivo* dissolution will occur under nonsink conditions.

In conclusion, even though the paddle setup shows some limitations in predicting the intralumenal performance of immediate-release solid dosage forms [20, 33, 34], it is an appropriate tool for estimating the initial dissolution rates of the drug compound, and these dissolution rates can subsequently be coupled with PBPK software [18, 21, 23, 35–37].

12.1.1.3 Hydrodynamics Even though considerable progress has been made over the last several years in adjusting the composition of dissolution media to more "biorelevant" compositions, the progress in updating the dissolution setup to a more "biorelevant" method in terms of gastrointestinal hydrodynamics and sheer forces has been slower, and there are only a few apparatus that attempt to mimic the gastric or small intestinal hydrodynamics, e.g., TNO (Toegepast Natuurwetenschappelijk Onderzoek) testing [38], biorelevant stress testing [39], or the application of the dynamic gastric model (DGM) [40, 41]. Although these approaches seem promising as ways of significantly improving prediction of a formulation's behavior in the gastrointestinal tract, the data are still too limited to draw conclusions about the usefulness of these approaches. In the meantime, compendial apparatus, such as the paddle setup (USP apparatus 2), the BioDis (USP apparatus 3), and the flow-through tester (USP apparatus 4) have often proven to be sufficient for predicting a drug's *in vivo* dissolution behavior [42–46].

Usually, the standard paddle revolution speed is either 50 or 75 rpm. Depending on the physiological conditions to be simulated (e.g., simulating different phases of the preprandial gastric motility), one might decelerate the paddle revolution speed, as an example. However, coning of the drug should be avoided as this is an artifact of the setup and it is unlikely that coning occurs in the human stomach.

For the USP apparatus 3, dip rates can be adjusted individually in each row of vessels, and dip rates between 5 and 20 dpm have proven suitable in establishing *in vitro–in vivo* correlations for several poorly soluble dugs (unpublished data).

With the USP apparatus 4, flow rates between 4 and 16 mL/min are often used. As small intestinal flow rates *in vivo* are often in the range between 0 and 7 mL/min, it has been suggested that *in vitro* flow rates used for the USP apparatus 4 should also be selected in this range. However, better *in vivo* correlations might also be obtained with faster flow rates because the flow in the flow-through cell is laminar, whereas in the gut, it is probably turbulent [43–46].

To design a dissolution apparatus with optimal hydrodynamics in relation to gastrointestinal physiology, it would be necessary to conduct more experiments in volunteers to obtain more precise information about *in vivo* hydrodynamics.

12.1.1.4 Special Considerations for Modified Release Formulations For nondisintegrating modified release formulations, e.g., coated tablets, accurate simulation of gastrointestinal pH, hydrodynamics, and media composition is of great importance for simulating the *in vivo* performance of the formulation, and the BioDis setup provides great advantage for that purpose. For formulations with a pH-dependent release mechanism as well as for drugs with pH-dependent solubility, it is of utmost importance to simulate the passage of the formulation through the gastrointestinal tract as closely as possible, and gastric and intestinal residence times as well as the composition and pH of the gastrointestinal juices have to be reflected in the setup. A reasonable preprandial gastric residence time for monolithic formulations is 1 to 2 h, depending on the time-point when phase 3 of the myoelectric motor complex (MMC) starts. For simulating the fed state stomach, the residence time depends on the amount of calories administered with the meal (please note that even though a monolithic dosage form might be administered in the fed state, it leaves the stomach after the meal has been emptied and the gastric motility reverts to the *fasted* state, i.e., in phase 3 of the MMC). Therefore, the monolithic dosage form encounters the small intestine in the fasted state, independent of the timing of application (fasted or fed). A mean small intestinal resident time of 4 h is reasonable for both prandial states [47]. To simulate the small intestinal passage as close as possible, FaSSIF v2, with a pH-gradient, beginning with pH 6.2 (duodenum) and ending with pH 7.4 (distal ileum), is reasonable. Additionally, the concentrations of bile salts and lecithin should be decreased carefully, as these components are reabsorbed during their passage through the small intestine [1].

The residence time of formulations in the colon is highly variable. However, as a starting point, it is reasonable to use FaSSCoF or FeSSCoF as dissolution medium, with a test time of up to 12 h [47] to reflect time spent in the proximal colon.

The flow-through cell and the BioDis setup seem to be appropriate alternatives compared with the paddle setup to manage the changes in composition and hydrodynamics encountered by dosage forms with extended release. Several studies have reported that the BioDis and the flow-through tester seem to provide better predictions of the gastrointestinal behavior of extended release dosage forms [17, 20, 33, 48–50].

12.1.2 PBPK Modeling as a Tool for Predicting the Absorption, Distribution, Metabolism, and Excretion Properties of Poorly Soluble Drugs

In 1937, Torsten Teorell provided the first, physiologically based approach of the main determinants influencing the distribution kinetics of drug compounds [51]. However, at that time, it was not possible to solve the series of differential equations that were evolved because appropriate computational aids were not available. In the late 1960s and early 1970s, as a result of advances in computer technology, interest in the field of physiology-based prediction of a drug's pharmacokinetics was renewed [52,53], and these days, Pub Med lists hundreds of publications dealing with that topic.

PBPK modeling now constitutes a promising, computer-assisted, mathematical approach to predict the ADME properties of drug compounds. Of great interest is the prediction of the pharmacokinetics of poorly soluble drugs, as many of the drugs currently under development have this characteristic [54].

The most widespread and sophisticated PBPK models are GastroPlus™ (Simulations Plus, Lancaster, CA), PK Sim® (Bayer Technology Services, Leverkusen, Germany), and Simcyp® (Simcyp Ltd, Sheffield, U.K.). These software tools enable the prediction of the drug's permeability, based on the physicochemical properties of the drug, as well as drug distribution into the different tissues. Besides absorption, distribution, and excretion processes (both active and passive), enzymatic degradation of the drug, e.g., by CYP enzymes, can also be simulated.

The PBPK models currently available enable not only pharmacokinetic predictions in well-defined, human standard populations (e.g., the typical First-in-Man population) but also in special population groups (e.g., in obese patients, older people, and children) and nonhuman species, e.g., in rats, pigs, and dogs. For more detailed information, we refer to further literature about this topic [3, 22, 55–69].

12.1.3 Summary

Most of the drug compounds currently in pharmaceutical development possess suboptimal biopharmaceutical properties according to the BCS [70], e.g., poor drug solubility, permeability, or both [54].

For most poorly soluble drugs, biorelevant dissolution testing provides a prognostic tool for qualitatively predicting the *in vivo* performance of that drug or oral dosage form. In some cases, however, the biorelevant dissolution results have to be coupled with PBPK software to improve the predictability of the ADME properties of the drug [17, 23, 46, 62, 71, 72].

In some cases, such as osmotic pump formulations of metoprolol, simple dissolution tests can be correlated with *in vivo* data. These formulations are extended release dosage forms of a highly soluble drug and are robust to physiological variations in the gastrointestinal tract. Simple dissolution tests also suffice for defining "safe zones" for the formulation of highly soluble drugs in immediate release dosage forms (via biowaiver techniques). For poorly soluble but highly permeable drugs, dissolution in biorelevant test conditions may provide adequate qualitative discrimination for identifying promising formulations and for identifying food effects, but often a quantitative *in vitro–in vivo* correlation is not possible. On the other hand, using *in silico* techniques to predict *in vivo* behavior of drugs relies purely on drug characteristics, and although this may work quite well for the drug itself, this is not particularly useful for predicting formulation effects. To improve forecasting of formulation performance *in vivo*, it thus seems reasonable to combine the advantages of biorelevant dissolution tests and *in silico* models that feature the gastrointestinal tract but also address distribution, metabolism, and elimination.

The remainder of this chapter is devoted to case examples illustrating the benefits and limitations of combining *in vitro* and *in silico* techniques, and from these examples, a systematic approach to predicting the *in vivo* performance of poorly soluble drugs is proposed.

12.2 *IN VITRO–IN SILICO–IN VIVO* APPROACH

The aim of this section is to present examples of how to combine dissolution test results with PBPK modeling to establish *in vitro–in silico–in vivo* correlations (IVISIVCs) and thus predict better the *in vivo* performance of poorly soluble drug compounds. Additionally, the limitations of this approach are discussed.

This approach, i.e., calculating a drug's pharmacokinetic profile from dissolution data, can be considered to be a type of convolution. The opposite, deconvolution, is used for the classic IVIVC approach in which the fraction of drug absorbed from a drug's pharmacokinetic profile is calculated and compared with the dissolution profile.

12.2.1 Case Examples

12.2.1.1 Neutral Compounds In contrast to weakly acidic or basic drugs, neutral compounds do not show a pH-dependent solubility behavior. Rather, the solubility of the drug depends mainly on the concentration of solubility-enhancing agents, i.e., surfactants, fat, or proteins. In the fasted state, the concentration of surfactants like bile salts and lecithin is very low in comparison with the fed state, and there are no significant amounts of lipids in the gastrointestinal juices. As these components are crucial for the solubility of neutral drugs, the bioavailability of these drugs is often lower in the fasted state, compared with the fed state [73–78].

As the drug has to be dissolved in the small intestine to be available for absorption, it is of utmost importance to obtain a good estimate of drug solubility and dissolution in the upper gastrointestinal tract. Even though drug absorption is negligible in the stomach, simulation of gastric conditions can provide an advantage for predicting the absorption of those drugs for which both gastric and small intestinal dissolution contribute to the overall dissolution. It is thus recommended to use both sets of biorelevant media, i.e., those that simulate the stomach and the small intestine in both the fasted and fed states to obtain a complete picture of the drug's ability to dissolve and become available for absorption. In such cases, it is likely that the high permeability will compensate for low solubility to a large degree. However, the *in vivo* performance of the formulation will be strongly dependent on the rate of release of the drug, and so for drugs with these characteristics, it is particularly important to characterize the dissolution *rate* of the drug.

A classic example for a highly lipophilic, poorly water-soluble, neutral drug compound is fenofibrate. The compound shows a significantly higher exposure when the drug is administered in the fed state (please note that this is valid for the micronized drug; the nanosized drug, however, does not show an appreciable food effect) [75,76]. In a recently published study, the plasma profiles of both micronized and nanosized fenofibrate were predicted, and the authors coupled the results of biorelevant solubility and dissolution experiments with a PBPK model. The *in vitro* solubility results correctly predicted the food effect of fenofibrate qualitatively. However, the magnitude of the food effect was overestimated, which is an observation that has been frequently made in literature [23, 35–37]. To quantify the *in vivo* pharmacokinetics of fenofibrate, the authors coupled the *in vitro* results with a PBPK model (STELLA® software). The "complete" *in vitro* dissolution profile did not reflect the *in vivo* dissolution of fenofibrate accurately, and the authors thus used the *initial dissolution rates*, based on the Noyes–Whitney theory, together with the drug's solubility in biorelevant media, to describe the *in vivo* dissolution of fenofibrate [21]. In the case of fenofibrate, the standard paddle setup provided sufficient accuracy for estimating the *in vivo* dissolution rates of the drug on a quantitative basis.

In some cases, e.g., nifedipine, deconvolution of the *in vivo* plasma profiles reveals that the standard paddle setup overpredicts the *in vivo* dissolution rate (unpublished

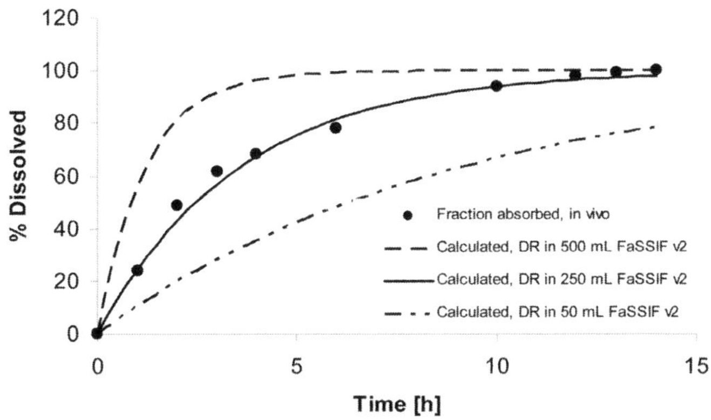

Figure 12.1 Comparison between the *in vivo* dissolution of a nifedipine formulation, obtained after deconvolution of the *in vivo* plasma profile [79], and the calculated dissolution of the same formulation, using dissolution rates (DRs) obtained in FaSSIF v2. To determine the dissolution rate, several media volumes were used.

data). Dissolution of a fast disintegrating nifedipine formulation (Adalat®, Bayer Schering Pharma, Leverkusen, Germany) was tested in FaSSIF v2 using the standard paddle setup. As changing the paddle speed between 25 and 100 rpm proved to have no substantial influence on the dissolution rate, the volume of medium was reduced from 500 to 250 mL and again to 50 mL [27, 29, 30]. The initial dissolution rates obtained from these *in vitro* experiments were then used to calculate the intralumenal dissolution of nifedipine from the formulation, and the results were compared with the observed *in vivo* data, obtained by deconvolution of the *in vivo* plasma profile [79]. Initial dissolution rates obtained in 250 mL media volume resulted in the best correlation between the dissolution rate of the drug *in vivo* [79] and *in vitro* (Figure 12.1). By contrast, media volumes of 500 and 50 mL over- and underestimated the *in vivo* dissolution rate of nifedipine, respectively. A volume of 250 mL lies within the physiological range for the small intestine [27, 29, 30]. These results suggest that reducing the medium volume from 900 or 500 mL to more physiological values might lead to a closer prediction of the intestinal dissolution behavior of a poorly soluble drug.

However, biorelevant dissolution testing is not always needed to predict the intralumenal dissolution behavior of a drug. In the case of cilostazol, the intralumenal dissolution of the drug could be predicted successfully solely on an *in silico* basis based on the particle size of the drug [34].

12.2.1.2 Weak Acids Weakly acidic drug compounds show pH-dependent solubility, with higher solubility at high pH values (e.g., in the small intestine) and lower solubility at low pH values (e.g., in the preprandial stomach). Additionally, one should bear in mind that ionization not only affects the solubility but also the permeability. Whereas the ionized form will be the more soluble, the nonionized form will be the more permeable.

Glibenclamide (glyburide), a weakly acidic sulfonylurea, shows very high interindividual variability *in vivo*. Sources for the high variability might theoretically be, on the one hand, variability in the dissolution behavior of the drug from the formulation and, on the other hand, variations in the patient physiology, e.g., different gastric and small intestinal fluid volumes, surfactant concentrations and pH, differences in the small intestinal surface area, variations in gastric emptying rates (GER), and variations in the metabolism (e.g., different enzyme concentrations and genetic polymorphisms).

For weakly acidic drug compounds, such as glibenclamide, biorelevant dissolution testing and PBPK modeling can be combined to examine to what extent solubility and dissolution issues might be a source of the high variability in the drug's pharmacokinetics.

To obtain a good estimate about the *in vivo* dissolution behavior, dissolution tests with a marketed glibenclamide formulation, containing 1.75 mg of the micronized drug (Semi Euglucon® N, Sanofi Aventis, Frankfurt, Germany), were performed using biorelevant media to simulate both the fasted and the fed state in the stomach and small intestine, i.e., FaSSGF pH 1.6, FeSSGF pH 5.0, FaSSIF v2 pH 6.5, and FeSSIF v2 pH 5.0. Additional dissolution tests in simple buffer media, i.e., simulated gastric fluid (SGF) pH 1.2 and simulated intestinal fluid (SIF) pH 6.8, were also performed.

To estimate the influence of variations of natural surfactants, i.e., bile salts and lecithin, on the dissolution behavior of the drug, dissolution tests in biorelevant media that contained varying amounts of sodium taurocholate and lecithin were also performed. To support these studies, the drug's solubility was also determined in the media used for the dissolution study.

The dissolution and solubility results revealed that glibenclamide dissolution does not depend significantly on the concentration of bile salts and lecithin, and that the pH of the medium has a much greater influence (Figure 12.2). Thus, variations in the concentrations of natural surfactants, i.e., bile salts and lecithin, can be excluded as sources of the high interindividual variability.

The solubility and dissolution studies showed that dissolution of glibenclamide in the fasted state stomach is negligible, and that dissolution of the drug mainly takes place in the small intestine. However, in the fed state, both gastric and small intestinal dissolution seem to play a role.

As a result of the high pH of FaSSIF v2 and SIF, the solubility and dissolution rate of glibenclamide in these media do not differ substantially (data not shown). For the media representing the fasted state stomach, i.e., FaSSGF and SGF, the solubility and dissolution rate were also similar (data not shown). In the case of glibenclamide, biorelevant media did not provide any significant advantage compared with simple buffers.

For explaining the high interindividual variability in glibenclamide pharmacokinetics, it is important to mention that, as gastric dissolution is negligible, the drug enters the fasted state small intestine as a suspension. As a result of the high solubility and fast dissolution rate of glibenclamide at small intestinal pH values (the dissolution rate in FaSSIF v2 is much faster than typical *in vivo* gastric emptying rates [80–82]), the gastric emptying rate becomes the rate-limiting step to the *in vivo* dissolution of

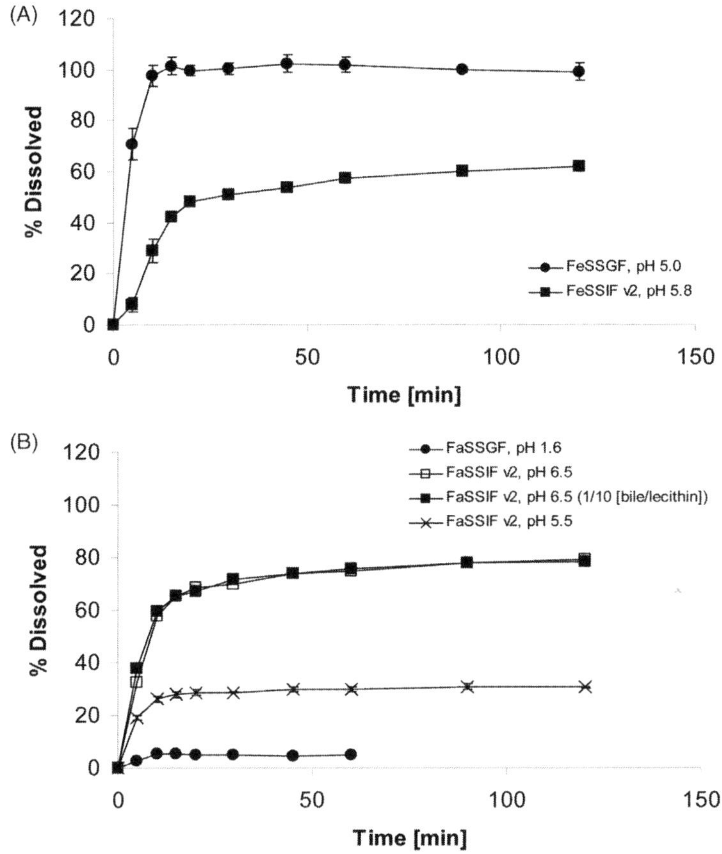

Figure 12.2 Dissolution of a marketed glibenclamide formulation, containing 1.75 mg micronized glibenclamide, in biorelevant media simulating fasted (panel A) or fed (panel B) gastrointestinal conditions. Please note that the dissolution curves for FaSSIF v2 and FaSSIF v2 (with only 1/10 bile and lecithin) in panel B superimpose. The curves are the mean of three experiments ($n = 3$; $\pm$ SD).

the drug, and *in vitro* dissolution tests thus *overestimate* the rate of drug absorption *in vivo*.

Thus, for weakly acidic drugs (in addition to glibenclamide, glimepiride, atorvastatin, and furosemide have been studied), biorelevant dissolution testing seems to provide no particular advantage compared with simple buffer solutions for predicting fasted state behavior. However, it is highly recommended that it should be a case-to-case decision whether biorelevant or compendial media are used to predict the *in vivo* performance of a drug compound.

To quantify the influence of variations in gastric emptying rates on glibenclamide's pharmacokinetics, a PBPK model was developed that describes oral glibenclamide pharmacokinetics, with special emphasis on the role of gastric emptying. In the literature, three distinct gastric emptying patterns are reported, and these are slow, fast, and biphasic. As a result of the complexity and extremely high variability in gastric emptying in the fasted state, we refer to adequate literature dealing especially with that topic [80–82].

An additional issue for the *in vivo* performance of glibenclamide is the drug's metabolism: For many drugs, genetic polymorphisms in the metabolizing enzyme system contribute to the high variability of the drug's pharmacokinetics. The most prominent enzyme systems that are subject to genetic polymorphisms are CYP 2C9, 2C19, and 2D6. Genetic polymorphisms in metabolizing enzyme systems result in poor (PM), intermediate (IM), and extensive (EM) metabolizing patterns [83–87]. *In vivo*, glibenclamide is metabolized via CYP 2C9 and is thus subject to genetic polymorphism.

Therefore, not only variations in gastric emptying but also genetic polymorphisms had to be taken into account for the PBPK simulations of glibenclamide. The computer simulations were performed using PK Sim®. To simulate the *in vivo* pharmacokinetics of glibenclamide accurately, subgroups with the three distinct gastric emptying patterns (slow, fast, biphasic) and the three distinct metabolism patterns (PM, IM, EM) were combined in a 3 × 3 matrix so that, in total, nine different subgroups were taken into consideration. The prevalence of each gastric emptying and metabolism pattern was taken from the literature [83–89]. By combining different gastric emptying patterns with different metabolism patterns, glibenclamide pharmacokinetics could be simulated accurately on a retrospective basis, as shown in Figure 12.3 [90].

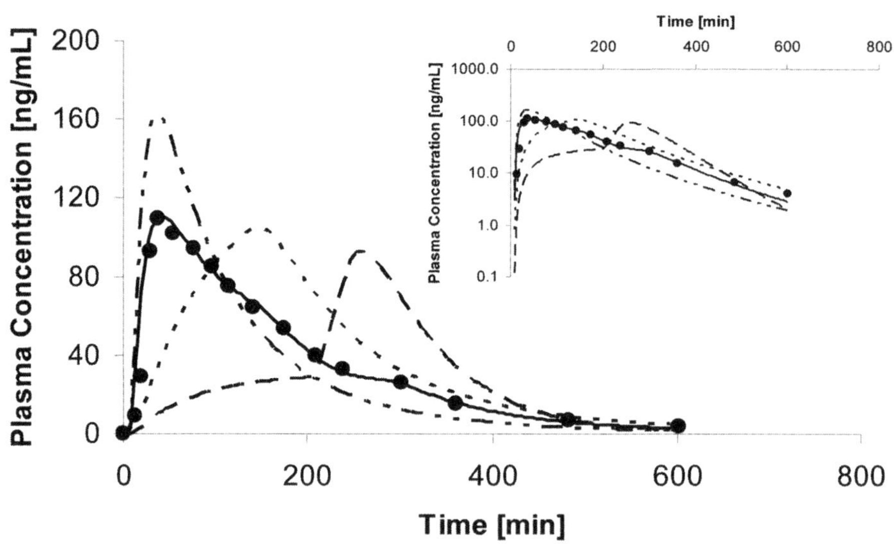

Figure 12.3 Observed *in vivo* [90] (•, mean data) and simulated (—) mean glibenclamide plasma profile after oral application of one tablet of the marketed formulation, containing 1.75 mg micronized glibenclamide, to postprandial healthy subjects ($n = 10$). The prevalence for each of the nine metabolism and gastric emptying pattern combinations used to construct the simulation was calculated from literature data. Hatched lines indicate fast (-—), delayed (- - -), and biphasic (——) gastric emptying patterns. Insert display the same plot on a semi-logarithmic scale. For further explanation, please see text.

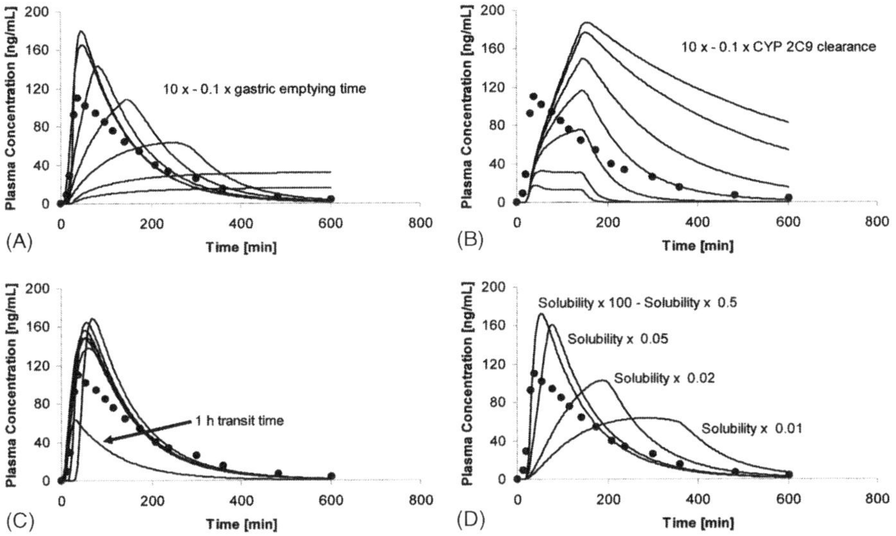

Figure 12.4 Sensitivity analysis to identify the crucial parameters for glibenclamide pharmacokinetics. The observed *in vivo* data are plotted for comparison (•) [90]. For the sensitivity analysis, the gastric emptying rate (panel A), the clearance (panel B), and the intestinal transit time (panel C) were varied in the range 0.1–10 times the corresponding mean value. Additionally, the drug's solubility was varied in the range between 0.01 and 10 times the reference solubility (panel D). The sensitivity analysis presented here was performed for the subgroup "slow gastric emptying/intermediate metabolizer" as a representative example.

To test the assumption that variations in gastric emptying rates and genetic polymorphism account for the high variability of glibenclamide pharmacokinetics, a sensitivity analysis was performed. For this purpose, parameters crucial for the *in vivo* performance of the drug were varied (Figure 12.4).

By combining both *in vitro* and *in silico* techniques, the results show that the large interindividual variations in pharmacokinetics of micronized glibenclamide derive from (1) the genetic polymorphisms in the CYP 2C9 enzyme system and (2) large variations in gastric emptying rates within the population, and that dissolution and solubility of micronized glibenclamide in the small intestine is not a major issue for the *in vivo* performance of the drug. By contrast, variations in the intestinal transit time of the drug do not influence the drugs pharmacokinetics markedly as long as it is longer than approximately 1 h. If the transit time is shorter than approximately 1 h, the time available to dissolve the complete dose is too short, and the extent of absorption decreases. Longer transit times do not influence the drug's pharmacokinetics markedly, as the drug is also absorbed in the colon.

Additionally, glibenclamide's pharmacokinetics are influenced by the drug's small intestinal solubility only when this parameter is decreased by a factor of approximately 50. This results from the drug's low dose combined with its relatively high solubility at small intestinal pH.

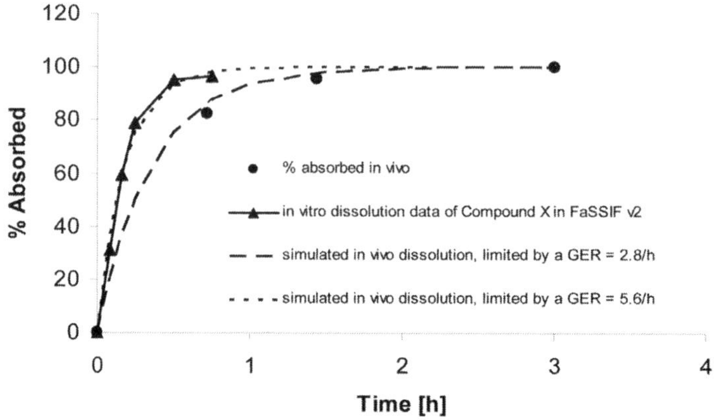

Figure 12.5 Comparison between the fraction absorbed of the weakly acidic Compound X *in vivo* and actual dissolution data of the compound in FaSSIF v2.

As glibenclamide pharmacokinetics strongly depend on both the gastric emptying time and the metabolic phenotype of each subject, it is obvious that dissolution profiles obtained from *in vitro* methods cannot be correlated on a stand-alone basis with the *in vivo* observed plasma profiles *prospectively*, and thus, an *in vitro–in vivo* correlation with an individual *a priori* prediction of pharmacokinetics seems not to be feasible for glibenclamide preparations. It is anticipated that these findings are not only valid for glibenclamide but also for a variety of other drug weakly acidic substances.

Figure 12.5 compares the fraction absorbed profile of a further highly permeable, weakly acidic drug (Compound X) with its *in vitro* dissolution data from the same formulation in FaSSIF v2. It is clear that the *in vitro* dissolution is much faster than the *in vivo* absorption. However, under the assumption that (1) gastric dissolution of Compound X is negligible (which was confirmed with dissolution tests in FaSSGF); (2) Compound X dissolves more or less instantaneously upon entry into the small intestine (which is reasonable as a result of the *pKa* of the drug as well as the dissolution test results in FaSSIF); and (3) Compound X is highly permeable (which was confirmed separately by *in vivo* data), it is proposed that the dissolution of the drug *in vivo* is driven primarily by the gastric emptying rate. By applying a typical mean population gastric emptying rate (and thus dissolution rate) of 2.8 h^{-1} [18] in *in silico* simulations, one obtains an absorption profile that matches well with the observed *in vivo* results. These results suggest that the *in vivo* performance of Compound X is determined by gastric emptying rather than by dissolution issues.

For weakly acidic, poorly soluble drugs like Compound X, a "safe zone" consists of dissolution performance under intestinal conditions that is fast or faster than the gastric emptying rate. If this can be established, physiological variables (e.g., gastric emptying or first-pass metabolism) will dictate the *in vivo* performance and formulation issues will be less important. Only when the formulation is changed so that the dissolution performance is no longer in the safe zone, for example, through an increase in dose or particle size or a change in the release mechanism of the drug product, dissolution issues would come into play.

12.2.1.3 Weak Bases As a result of their physicochemical characteristics, basic drug compounds show pH-dependent solubility, with high solubility at a low pH and low solubility at a high pH.

When administered in the fasted state, the large "pH gap" between the gastric and the small intestinal pH leads to precipitation of some weak bases when they enter the small intestine, leading to a decrease in drug bioavailability. In other cases, weak bases can show higher *in vivo* exposure when administered in the fasted state compared with the fed state. For these drugs, precipitation upon entry into the small intestine seems unlikely [23, 35, 91–93].

At the pH of the postprandial stomach, which is, depending on the meal administered, approximately pH 5 [1], the solubility of a weak base is lower than at preprandial gastric pH values. So based on the pH difference, one might expect lower *in vivo* dissolution and thus absorption of the drug in the fed state. However, the pH effect is often compensated by solubilization of the weak base, e.g., in fat contained in the meal co-administered with the drug, secretion of natural surfactants like bile salts and phospholipids in response to a meal intake, and lipolysis products.

The opposing effects of pH and solubilization between the fasted and the fed state make it difficult to predict food effects for poorly soluble, weakly basic drugs. In contrast to weakly acidic drug compounds, solubility and dissolution of weakly basic drugs in both the stomach *and* the small intestine have to be considered for predicting the overall *in vivo* performance of that drug. Additionally, the potential for drug precipitation in the fasted small intestine has to be taken into account.

One *in vitro* setup to test the precipitation tendency of weakly basic drug compounds is the transfer model described by Kostewicz and colleagues [94].

The driving force for drug precipitation is known to be the degree of supersaturation. This implies that, when the concentration of drug is kept low in the small intestine by continuing drug absorption, the extent of supersaturation will be kept lower. Highly permeable weak bases thus tend to show a lower *in vivo* precipitation tendency than poorly permeable weak bases [91, 93]. For highly permeable weak bases, it was shown that the precipitation tendency of the drug is often overestimated using *in vitro* tools like the transfer model, and highly permeable drugs that tend to precipitate *in vitro* do not always show a high precipitation tendency *in vivo* [23, 91, 93]. For those drugs, additional transfer model experiments seem to bring no significant advantage to prediction of *in vivo* performance, and the standard dissolution tests seem sufficient to predict the gastrointestinal performance of the drug.

For poorly permeable weak bases, precipitation of the drug upon entry into the fasted small intestine is an important factor that can limit both the rate and the extent of gastrointestinal drug absorption. For poorly permeable weak bases, a higher degree of supersaturation in the small intestine can be generated. As a result, poorly permeable weak bases show a higher tendency to precipitate when they are emptied into the fasted small intestine than highly permeable weak bases. The effect of precipitation on absorption may depend on whether the drug precipitates in its crystalline or

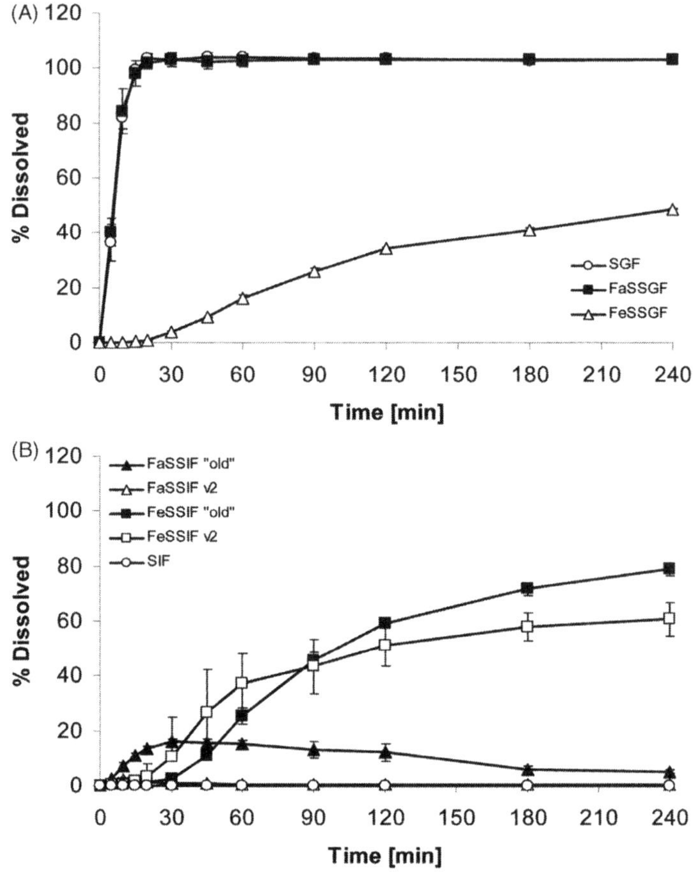

Figure 12.6 Dissolution profiles of Compound A, a weakly basic drug compound, in dissolution media simulating both the pre- and postprandial stomach (panel A) and the small intestine (panel B), respectively. (Reprinted from "Predicting the oral absorption of a poorly soluble, poorly permeable weak base using biorelevant dissolution and transfer model tests coupled with a physiologically based pharmacokinetic model, *Eur J Pharm Biopharm* 82 (1), Wagner C et al., 127–138 (2012)", with permission from Elsevier [23].)

amorphous form (with a higher solubility of the amorphous form compared with the crystalline drug), and whether precipitation is complete [23, 91, 92, 94].

For such drugs, it should be kept in mind that inhibition of first-pass metabolism as a result of meal components, increase in splanchnic blood flow, or lymphatic uptake in response to a meal might also contribute to the food effect [95, 96].

As a result of the larger pH gap between the preprandial stomach and the small intestine, precipitation of poorly permeable weak bases is usually more of a problem when the drug is administered in the fasted state. A good estimate of whether precipitation might also occur after fed state administration can be made by comparing the solubilities of the drug in the medium simulating the fed stomach, FeSSGF, with

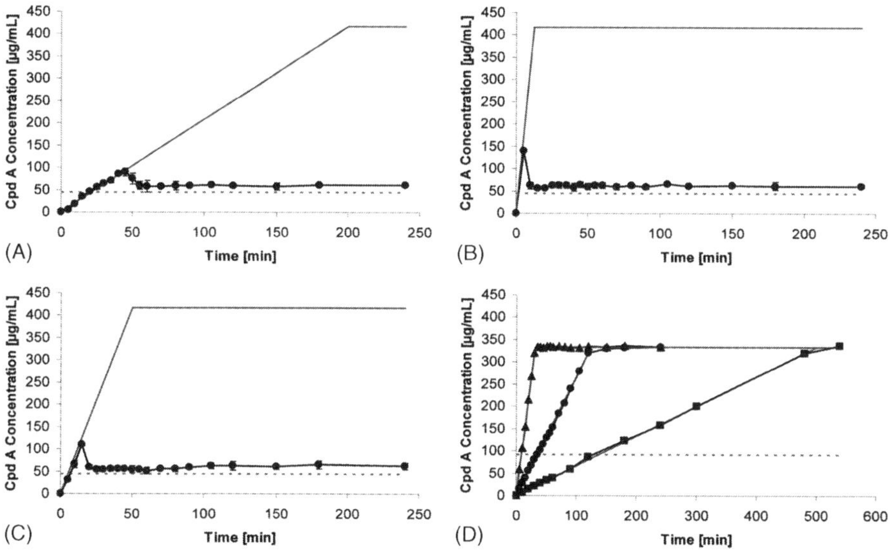

Figure 12.7 Results from the transfer model experiments, conducted with the weakly basic drug Compound A. The figure shows the results in the fasted state with three different transfer rates (A: 0.5 mL/min; B: 4.0 mL/min; C: 8.0 mL/min). Panel D shows the results for the fed state transfer model experiments (▲ = 8.0 mL/min; • = 4.0 mL/min; ■ = 0.5 mL/min). The light gray lines show the theoretical compound concentrations, and the dashed lines show the equilibrium solubility of the compound in FaSSIF v2 (A, B, C) and FeSSIF v2, respectively (D). (Reprinted from "Predicting the oral absorption of a poorly soluble, poorly permeable weak base using biorelevant dissolution and transfer model tests coupled with a physiologically based pharmacokinetic model, *Eur J Pharm Biopharm* 82 (1), Wagner C et al., 127–138 (2012)", with permission from Elsevier [23].)

the drug's solubility in FeSSIF v2. As the FeSSGF pH is quite similar to the pH in FeSSIF v2 (5.0 vs. 5.8) and the solubilities of many poorly soluble weak bases are similar in these two media, precipitation in the fed state is not anticipated in most cases.

Recently, the absorption of a poorly soluble, poorly permeable weak basic research compound (Compound A) was predicted by coupling the results of biorelevant dissolution tests with a PBPK model [23]. The *in vitro* dissolution test results revealed that the complete dose of the drug, i.e., 250 mg, can dissolve in both prandial states: in the fasted state, because the drug is highly soluble in the acidic gastric environment (even though solubility in the simulated intestinal fluid was very low), and in the fed state, because the solubility of the drug in the media simulating both the fed state stomach and the small intestine was relatively high (Figure 12.6).

However, the drug showed a 2.5-fold higher exposure in a clinical study when administered in the fed state. From the solubility and dissolution study alone, one could not explain the higher exposure of the drug in the fed state, and the absorption of the drug could not be predicted quantitatively. For this reason, transfer model experiments, as described by Kostewicz and colleagues [94], were conducted.

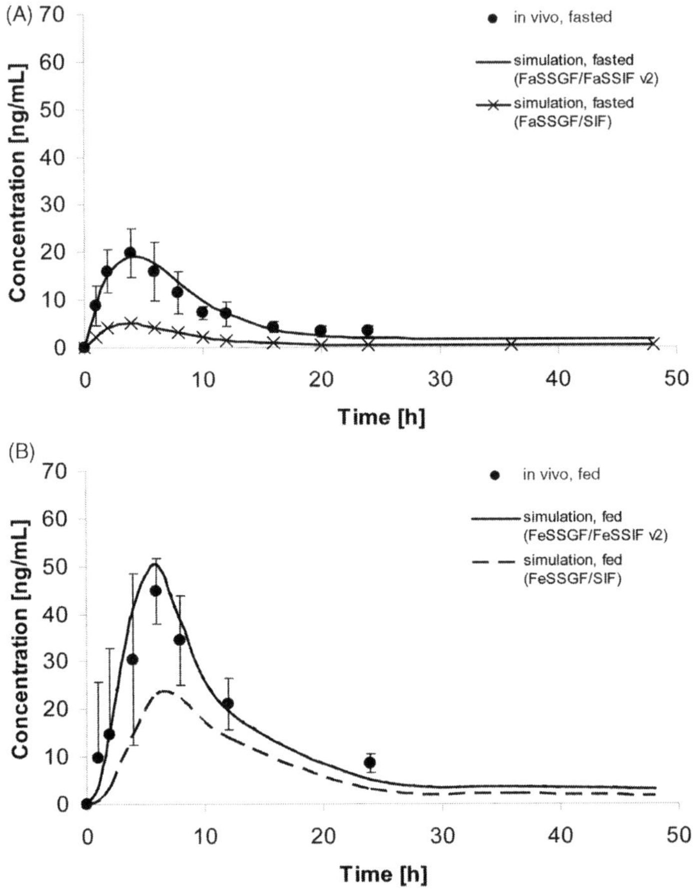

Figure 12.8 Comparison between observed *in vivo* and simulated plasma profiles for the poorly soluble, poorly permeable weak base Compound A. Panel A: fasted state; panel B: fed state. (Reprinted from "Predicting the oral absorption of a poorly soluble, poorly permeable weak base using biorelevant dissolution and transfer model tests coupled with a physiologically based pharmacokinetic model, *Eur J Pharm Biopharm* 82 (1), Wagner C et al., 127–138 (2012)", with permission from Elsevier [23].)

The pale gray, continuous lines in Figure 12.7 show the concentration in the acceptor phase with time if no precipitation occurs during or after transfer. In the fasted state experiments (panels A, B, and C), Compound A showed a very strong tendency to precipitate, with the final concentration close to the solubility of the drug in FaSSIF v2 as indicated by the light dashed lines. Thus, supersaturation is very short-lived for Compound A under fasted state conditions. In fact, the precipitation rates observed *in vitro* were much faster than the average gastric emptying rate of 2.8 h^{-1} *in vivo*, suggesting that the drug would precipitate more or less instantaneously upon entering the small intestine in the fasted state. By contrast, no precipitation was observed in the fed state (Figure 12.7D).

To predict *quantitatively* the absorption of that drug, the *in vitro* dissolution and transfer model results were coupled with a PBPK model. To describe the *in vivo* dissolution of the drug, the results from the *in vitro* experiments, i.e., (1) initial dissolution rates in the media simulating the fasted and the fed state stomach and small intestine, and (2) precipitation rates from the transfer model experiments, were used.

Besides drug precipitation in the fasted small intestine, the poor permeability of the drug, which resulted in pronounced supersaturation under intestinal conditions, became an issue for the *in vivo* performance of the drug [23]. To calculate the drug flux of the poorly permeable weak base, validated Caco-2 data were used in the PBPK model as an *in vitro* surrogate parameter for the permeability coefficient, with physiological values for the small intestinal surface area and small intestinal liquid volumes to convert the data into an estimate of *in vivo* flux. Using this approach, the rate and extent of absorption could be predicted accurately on an *a priori* basis, and it was shown that drug precipitation is the limiting factor in the *in vivo* performance of Compound A in the fasted state (Figure 12.8).

The simulations for Compound A pharmacokinetics illustrate that, for prediction of the ADME properties of weakly basic drugs, dissolution in both the stomach and the small intestine plays a role in determining the absorption of the drug. Simulation of only gastric or only small intestinal dissolution conditions will likely lead to misleading results. For weak bases, it is also highly recommended to use biorelevant media rather than simple buffers and to consider application of the transfer model to detect precipitation problems under fasted state conditions. For weak bases, prediction of the *in vivo* exposure of the drug (and thus its ADME properties) is best tackled by a combination of *in vitro* and *in silico* tools.

12.2.2 Summary

As shown in the various case examples, it was the aim of the authors to present examples of how to use biorelevant dissolution techniques for forecasting the *in vivo* dissolution behavior of poorly soluble drugs. To date, there is no “master setup” to predict the gastrointestinal dissolution of a drug compound adequately, and the selection of the appropriate setup, i.e., apparatus, media, and hydrodynamics, is still a case-to-case decision.

Human physiology is too complex and variable to reflect every single aspect with either *in vitro* or *in silico* approaches, and there is no ultimate method to predict the *in vivo* absorption of every drug compound. The methods presented herein can thus only give an estimate about how the drug or formulation will behave *in vivo*.

To improve the predictability of biorelevant dissolution tests, the *in vitro* results can be coupled with PBPK software to predict the absorption of poorly soluble drugs, and for neutral and weakly basic drug compounds, a prediction of the pharmacokinetic profile of the drug seems feasible. For poorly soluble weak acids, a “safe zone” for formulation can be identified.

Figure 12.9 summarizes the outcome of this chapter and proposes a systematic approach that can be used in preclinical and clinical development for predicting the absorption of poorly soluble drugs.

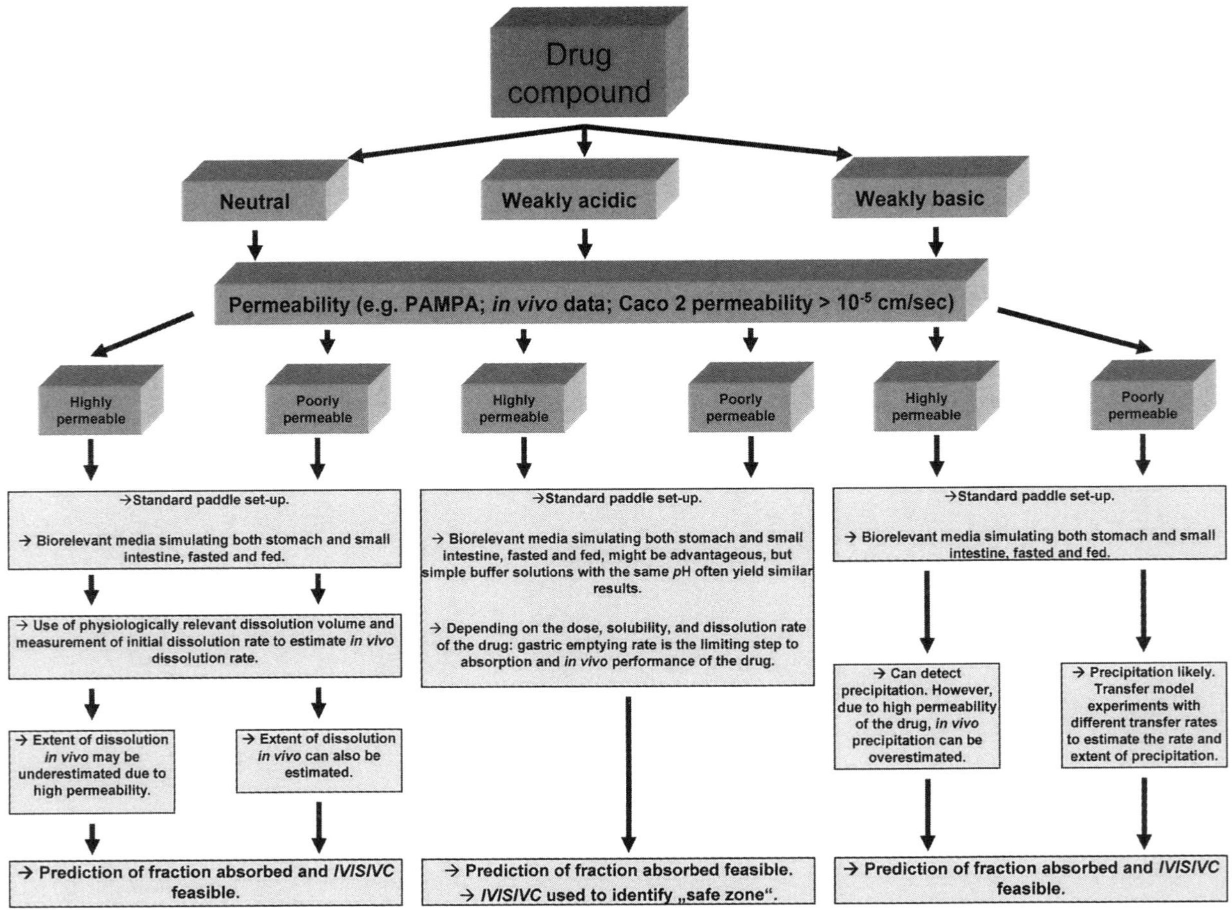

Figure 12.9 Proposed working scheme for estimating the *in vivo* solubility and dissolution of poorly soluble drug compounds from immediate release formulations.

REFERENCES

1. Dressman, J.B., Amidon, G.L., Reppas, C., Shah, V.P. (1998). Dissolution testing as a prognostic tool for oral drug absorption: Immediate release dosage forms. *Pharmaceutical Research, 15*, 11–22.
2. Akamatsu, M., Fujikawa, M., Nakao, K., Shimizu, R. (2009). In silico prediction of human oral absorption based on QSAR analyses of PAMPA permeability. *Chemistry and Biodiversity, 6*, 1845–1866.
3. Dearden, J.C. (2007). In silico prediction of ADMET properties: How far have we come? *Expert Opinion on Drug Metabolism & Toxicology, 3*, 635–639.
4. Faller, B. (2008). Artificial membrane assays to assess permeability. *Current Drug Metabolism, 9*, 886–892.
5. Galia, E., Nicolaides, E., Horter, D., Lobenberg, R., Reppas, C., Dressman, J.B. (1998). Evaluation of various dissolution media for predicting in vivo performance of class I and II drugs. *Pharmaceutical Research, 15*, 698–705.
6. Hidalgo, I.J., Raub, T.J., Borchardt, R.T. (1989). Characterization of the human colon carcinoma cell line (Caco-2) as a model system for intestinal epithelial permeability. *Gastroenterology, 96*, 736–749.
7. Hu, M., Li, Y., Davitt, C.M., Huang, S.M., Thummel, K., Penman, B.W., Crespi, C.L. (1999). Transport and metabolic characterization of Caco-2 cells expressing CYP3A4 and CYP3A4 plus oxidoreductase. *Pharmaceutical Research, 16*, 1352–1359.
8. Ingels, F., Beck, B., Oth, M., Augustijns, P. (2004). Effect of simulated intestinal fluid on drug permeability estimation across Caco-2 monolayers. *International Journal of Pharmaceutics, 274*, 221–232.
9. Kansy, M., Senner, F., Gubernator, K. (1998). Physicochemical high throughput screening: Parallel artificial membrane permeation assay in the description of passive absorption processes. *Journal of Medicinal Chemistry, 41*, 1007–1010.
10. Lipinski, C.A., Lombardo, F., Dominy, B.W., Feeney, P.J. (2001). Experimental and computational approaches to estimate solubility and permeability in drug discovery and development settings. *Advanced Drug Delivery Reviews, 46*, 3–26.
11. Paine, M.F., Khalighi, M., Fisher, J.M., Shen, D.D., Kunze, K.L., Marsh, C.L., Perkins, J.D., Thummel, K.E. (1997). Characterization of interintestinal and intraintestinal variations in human CYP3A-dependent metabolism. *Journal of Pharmacology and Experimental Therapeutics, 283*, 1552–1562.
12. Schmiedlin-Ren, P., Edwards, D.J., Fitzsimmons, M.E., He, K., Lown, K.S., Woster, P.M., Rahman, A., Thummel, K.E., Fisher, J.M., Hollenberg, P.F., Watkins, P.B. (1997). Mechanisms of enhanced oral availability of CYP3A4 substrates by grapefruit constituents. Decreased enterocyte CYP3A4 concentration and mechanism-based inactivation by furanocoumarins. *Drug Metabolism and Disposition, 25*, 1228–1233.
13. Williams, J.A., Ring, B.J., Cantrell, V.E., Jones, D.R., Eckstein, J., Ruterbories, K., Hamman, M.A., Hall, S.D., Wrighton, S.A. (2002). Comparative metabolic capabilities of CYP3A4, CYP3A5, and CYP3A7. *Drug Metabolism and Disposition, 30*, 883–891.
14. Winiwarter, S., Bonham, N.M., Ax, F., Hallberg, A., Lennernas, H., Karlen, A. (1998). Correlation of human jejunal permeability (in vivo) of drugs with experimentally and theoretically derived parameters. A multivariate data analysis approach. *Journal of Medicinal Chemistry, 41*, 4939–4949.

15. Yee, S. (1997). In vitro permeability across Caco-2 cells (colonic) can predict in vivo (small intestinal) absorption in man–fact or myth, *Pharmaceutical Research*, *14*, 763–766.
16. Zhang, Q.Y., Dunbar, D., Ostrowska, A., Zeisloft, S., Yang, J., Kaminsky, L.S. (1999). Characterization of human small intestinal cytochromes P-450. *Drug Metabolism and Disposition*, *27*, 804–809.
17. Jantratid, E., De Maio, V., Ronda, E., Mattavelli, V., Vertzoni, M., Dressman, J.B. (2009). Application of biorelevant dissolution tests to the prediction of in vivo performance of diclofenac sodium from an oral modified-release pellet dosage form. *European Journal of Pharmaceutical Sciences*, *37*, 434–441.
18. Nicolaides, E., Symillides, M., Dressman, J.B., Reppas, C. (2001). Biorelevant dissolution testing to predict the plasma profile of lipophilic drugs after oral administration. *Pharmaceutical Research*, *18*, 380–388.
19. Galetin, A., Houston, J.B. (2006). Intestinal and hepatic metabolic activity of five cytochrome P450 enzymes: Impact on prediction of first-pass metabolism. *Journal of Pharmacology and Experimental Therapeutics*, *318*, 1220–1229.
20. Jantratid, E., Janssen, N., Chokshi, H., Tang, K., Dressman, J.B. (2008). Designing biorelevant dissolution tests for lipid formulations: Case example–lipid suspension of RZ-50. *European Journal of Pharmaceutics and Biopharmaceutics*, *69*, 776–785.
21. Juenemann, D., Jantratid, E., Wagner, C., Reppas, C., Vertzoni, M., Dressman, J.B. (2010). Biorelevant in vitro dissolution testing of products containing micronized or nanosized fenofibrate with a view to predicting plasma profiles. *European Journal of Pharmaceutics and Biopharmaceutics*, *77*, 257–264.
22. Okumu, A., DiMaso, M., Lobenberg, R. (2009). Computer simulations using GastroPlus to justify a biowaiver for etoricoxib solid oral drug products. *European Journal of Pharmaceutics and Biopharmaceutics*, *72*, 91–98.
23. Wagner, C., Jantratid, E., Kesisoglou, F., Vertzoni, M., Reppas, C., Dressman, J.B. (2012). Predicting the oral absorption of a poorly soluble, poorly permeable weak base using biorelevant dissolution and transfer model tests coupled with a physiologically based pharmacokinetic model. *European Journal of Pharmaceutics and Biopharmaceutics*, *82*, 127–138.
24. Jantratid, E., Janssen, N., Reppas, C., Dressman, J.B. (2008). Dissolution media simulating conditions in the proximal human gastrointestinal tract: An update. *Pharmaceutical Research*, *25*, 1663–1676.
25. Vertzoni, M., Diakidou, A., Chatzilias, M., Soderlind, E., Abrahamsson, B., Dressman, J.B., Reppas, C. (2010). Biorelevant media to simulate fluids in the ascending colon of humans and their usefulness in predicting intracolonic drug solubility. *Pharmaceutical Research*, *27*, 2187–2196.
26. Vertzoni, M., Dressman, J., Butler, J., Hempenstall, J., Reppas, C. (2005). Simulation of fasting gastric conditions and its importance for the in vivo dissolution of lipophilic compounds. *European Journal of Pharmaceutics and Biopharmaceutics*, *60*, 413-417.
27. Schiller, C., Frohlich, C.P., Giessmann, T., Siegmund, W., Monnikes, H., Hosten, N., Weitschies, W. (2005). Intestinal fluid volumes and transit of dosage forms as assessed by magnetic resonance imaging. *Alimentary Pharmacology and Therapeutics*, *22*, 971–979.
28. Klein, S., Shah, V.P. (2008). A standardized mini paddle apparatus as an alternative to the standard paddle. *AAPS PharmSciTech*, *9*, 1179–1184.

29. Dillard, D., Eastman, H., Fordtan, J.S. (1965). Volume-flow relationship during the transport of fluid through the human small intestine. *Gastroenterology*, *49*, 58–66.

30. Fordtran, J.S., Locklear, T. (1966). Ionic constituents and osmolarity of gastric and small intestinal fluids after eating. *American Journal of Digestive Diseases*, *11*, 503–521.

31. Diakidou, A., Vertzoni, M., Goumas, K., Soderlind, E., Abrahamsson, B., Dressman, J., Reppas, C. (2009). Characterization of the contents of ascending colon to which drugs are exposed after oral administration to healthy adults. *Pharmaceutical Research*, *26*, 2141–2151.

32. Vertzoni, M., Goumas, K., Soderlind, E., Abrahamsson, B., Dressman, J.B., Poulou, A., Reppas, C. (2010). Characterization of the ascending colon fluids in ulcerative colitis. *Pharmaceutical Research*, *27*, 1620–1626.

33. Reppas, C., Vertzoni, M. (2012). Biorelevant in-vitro performance testing of orally administered dosage forms. *Journal of Pharmacy and Pharmacology*, *64*, 919–930.

34. Willmann, S., Thelen, K., Becker, C., Dressman, J.B., Lippert, J. (2010). Mechanism-based prediction of particle size-dependent dissolution and absorption: Cilostazol pharmacokinetics in dogs. *European Journal of Pharmaceutics and Biopharmaceutics*, *76*, 83–94.

35. Shono, Y., Jantratid, E., Dressman, J.B. (2011). Precipitation in the small intestine may play a more important role in the in vivo performance of poorly soluble weak bases in the fasted state: Case example nelfinavir. *European Journal of Pharmaceutics and Biopharmaceutics*, *79*(2), 349–356.

36. Shono, Y., Jantratid, E., Janssen, N., Kesisoglou, F., Mao, Y., Vertzoni, M., Reppas, C., Dressman, J.B. (2009). Prediction of food effects on the absorption of celecoxib based on biorelevant dissolution testing coupled with physiologically based pharmacokinetic modeling. *European Journal of Pharmaceutics and Biopharmaceutics*, *73*, 107–114.

37. Shono, Y., Jantratid, E., Kesisoglou, F., Reppas, C., Dressman, J.B. (2010). Forecasting in vivo oral absorption and food effect of micronized and nanosized aprepitant formulations in humans. *European Journal of Pharmaceutics and Biopharmaceutics*, *76*(1), 95–104.

38. Brouwers, J., Anneveld, B., Goudappel, G.J., Duchateau, G., Annaert, P., Augustijns, P., Zeijdner, E. (2011). Food-dependent disintegration of immediate release fosamprenavir tablets: In vitro evaluation using magnetic resonance imaging and a dynamic gastrointestinal system. *European Journal of Pharmaceutics and Biopharmaceutics*, *77*, 313–319.

39. Garbacz, G., Golke, B., Wedemeyer, R.S., Axell, M., Soderlind, E., Abrahamsson, B., Weitschies, W. (2009). Comparison of dissolution profiles obtained from nifedipine extended release once a day products using different dissolution test apparatuses. *European Journal of Pharmaceutical Sciences*, *38*, 147–155.

40. Vardakou, M., Mercuri, A., Barker, S.A., Craig, D.Q., Faulks, R.M., Wickham, M.S. (2011). Achieving antral grinding forces in biorelevant in vitro models: Comparing the USP Dissolution Apparatus II and the Dynamic Gastric Model with human in vivo data. *AAPS PharmSciTech*, *12*, 620–626.

41. Vardakou, M., Mercuri, A., Naylor, T.A., Rizzo, D., Butler, J.M., Connolly, P.C., Wickham, M.S., Faulks, R.M. (2011). Predicting the human in vivo performance of different oral capsule shell types using a novel in vitro dynamic gastric model. *International Journal of Pharmaceutics*, *419*, 192–199.

42. Jinno, J., Kamada, N., Miyake, M., Yamada, K., Mukai, T., Odomi, M., Toguchi, H., Liversidge, G.G., Higaki, K., Kimura, T. (2006). Effect of particle size reduction on

dissolution and oral absorption of a poorly water-soluble drug, cilostazol, in beagle dogs. *Journal of Controlled Release*, *111*, 56–64.

43. Jinno, J., Kamada, N., Miyake, M., Yamada, K., Mukai, T., Odomi, M., Toguchi, H., Liversidge, G.G., Higaki, K., Kimura, T. (2008). In vitro-in vivo correlation for wet-milled tablet of poorly water-soluble cilostazol. *Journal of Controlled Release*, *130*, 29–37.
44. Katori, N., Aoyagi, N., Terao, T. (1995). Estimation of agitation intensity in the GI tract in humans and dogs based on in vitro/in vivo correlation. *Pharmaceutical Research*, *12*, 237–243.
45. Scholz, A., Kostewicz, E., Abrahamsson, B., Dressman, J.B. (2003). Can the USP paddle method be used to represent in-vivo hydrodynamics? *Journal of Pharmacy and Pharmacology*, *55*, 443–451.
46. Sunesen, V.H., Pedersen, B.L., Kristensen, H.G., Mullertz A. (2005). In vivo in vitro correlations for a poorly soluble drug, danazol, using the flow-through dissolution method with biorelevant dissolution media. *European Journal of Pharmaceutical Sciences*, *24*, 305–313.
47. Davis, S.S., Hardy, J.G., Fara, J.W. (1986). Transit of pharmaceutical dosage forms through the small intestine. *Gut*, *27*, 886–892.
48. Fotaki, N., Aivaliotis, A., Butler, J., Dressman, J., Fischbach, M., Hempenstall, J., Klein, S., Reppas, C. (2009). A comparative study of different release apparatus in generating in vitro-in vivo correlations for extended release formulations. *European Journal of Pharmaceutics and Biopharmaceutics*, *73*, 115–120.
49. Fotaki, N., Symillides, M., Reppas, C. (2005). Canine versus in vitro data for predicting input profiles of L-sulpiride after oral administration. *European Journal of Pharmaceutical Sciences*, *26*, 324–333.
50. Fotaki, N., Symillides, M., Reppas, C. (2005). In vitro versus canine data for predicting input profiles of isosorbide-5-mononitrate from oral extended release products on a confidence interval basis. *European Journal of Pharmaceutical Sciences*, *24*, 115–122.
51. Teorell, T. (1937). Kinetics of distribution of substances administered to the body: I. The extravascular modes of administration. *Archives Internationales de Pharmacodynamie et de Therapie*, *57*, 205–225.
52. Bischoff, K.B., Dedrick, R.L. (1968). Thiopental pharmacokinetics. *Journal of Pharmaceutical Sciences*, *57*, 1346–1351.
53. Bischoff, K.B., Dedrick, R.L. (1970). Generalized solution to linear, to-compartment, open model for drug distribution. *Journal of Theoretical Biology*, *29*, 63–68.
54. Wu, C.Y., Benet, L.Z. (2005). Predicting drug disposition via application of BCS: Transport/absorption/ elimination interplay and development of a biopharmaceutics drug disposition classification system. *Pharmaceutical Research*, *22*, 11–23.
55. Ball, K., Bouzom, F., Scherrmann, J.M., Walther, B., Decleves, X. (2012). Development of a physiologically based pharmacokinetic model for the rat central nervous system and determination of an in vitro-in vivo scaling methodology for the blood-brain barrier permeability of two transporter substrates, morphine and oxycodone. *Journal of Pharmaceutical Sciences*, *101*(11), 4277–4292.
56. Corley, R.A., Gordon, S.M., Wallace, L.A. (2000). Physiologically based pharmacokinetic modeling of the temperature-dependent dermal absorption of chloroform by humans following bath water exposures. *Toxicological Sciences*, *53*, 13–23.

57. Fisher, J., Lumpkin, M., Boyd, J., Mahle, D., Bruckner, J.V., El-Masri, H.A. (2004). PBPK modeling of the metabolic interactions of carbon tetrachloride and tetrachloroethylene in B6C3F1 mice. *Environmental Toxicology and Pharmacology*, *16*, 93–105.

58. Grass, G.M. (1997). Simulation models to predict oral drug absorption from in vitro data. *Advanced Drug Delivery Reviews*, *23*, 199–219.

59. Grass, G.M., Sinko, P.J. (2002). Physiologically-based pharmacokinetic simulation modeling. *Advanced Drug Delivery Reviews*, *54*, 433–451.

60. Heikkinen, A.T., Baneyx, G., Caruso, A., Parrott, N. (2012). Application of PBPK modeling to predict human intestinal metabolism of CYP3A substrates—An evaluation and case study using GastroPlus. *European Journal of Pharmaceutical Sciences*, *47*, 375–386.

61. Jones, H.M., Gardner, I.B., Watson, K.J. (2009). Modelling and PBPK simulation in drug discovery. *AAPS Journal*, *11*, 155–166.

62. Jones, H.M., Parrott, N., Ohlenbusch, G., Lave, T. (2006) Predicting pharmacokinetic food effects using biorelevant solubility media and physiologically based modeling *Clinical Pharmacokinetics*, *45*, 1213–1226.

63. Kesisoglou, F., Wu, Y. (2008). Understanding the effect of API properties on bioavailability through absorption modeling. *AAPS Journal*, *10*, 516–525.

64. Pang, K.S., Chow, E.C. (2012). Prediction of flow effects on intestinal (FI) and systemic availability (Fsys) in PBPK intestine models: The traditional model (TM), segregated flow model (SFM) and QGut model. *Drug Metabolism and Disposition*, *40*(10), 1869–1877.

65. Parrott, N., Lave, T. (2002). Prediction of intestinal absorption: Comparative assessment of GASTROPLUS and IDEA. *European Journal of Pharmaceutical Sciences*, *17*, 51–61.

66. Parrott, N., Lukacova, V., Fraczkiewicz, G., Bolger, M.B. (2009). Predicting pharmacokinetics of drugs using physiologically based modeling–application to food effects. *AAPS Journal*, *11*, 45–53.

67. Thelen, K., Coboeken, K., Willmann, S., Burghaus, R., Dressman, J.B., Lippert, J. (2011). Evolution of a detailed physiological model to simulate the gastrointestinal transit and absorption process in humans, part 1: Oral solutions. *Journal of Pharmaceutical Sciences*, *100*, 5324–5345.

68. Thelen, K., Coboeken, K., Willmann, S., Dressman, J.B., Lippert, J. (2012). Evolution of a detailed physiological model to simulate the gastrointestinal transit and absorption process in humans, part II: Extension to describe performance of solid dosage forms. *Journal of Pharmaceutical Sciences*, *101*, 1267–1280.

69. Willmann, S., Edginton, A.N., Kleine-Besten, M., Jantratid, E., Thelen, K., Dressman, J.B. (2009). Whole-body physiologically based pharmacokinetic population modelling of oral drug administration: Inter-individual variability of cimetidine absorption. *Journal of Pharmacy and Pharmacology*, *61*, 891–899.

70. Amidon, G.L., Lennernas, H., Shah, V.P., Crison, J.R. (1995). A theoretical basis for a biopharmaceutic drug classification: The correlation of in vitro drug product dissolution and in vivo bioavailability. *Pharmaceutical Research*, *12*, 413–420.

71. de Campos, D.R., Klein, S., Zoller, T., Vieria, N.R., Barros, F.A., Meurer, E.C., Coelho, E.C., Marchioretto, M.A., Pedrazzoli, J. (2010). Evaluation of pantoprazole formulations in different dissolution apparatus using biorelevant medium. *Arzneimittelforschung*, *60*, 42–47.

72. Shah, V.P., Konecny, J.J., Everett, R.L., McCullough, B., Noorizadeh, A.C., Skelly, J.P. (1989). In vitro dissolution profile of water-insoluble drug dosage forms in the presence of surfactants. *Pharmaceutical Research*, *6*, 612–618.
73. Bailey, D.G., Spence, J.D., Munoz, C., Arnold, J.M. (1991). Interaction of citrus juices with felodipine and nifedipine. *Lancet*, *337*, 268–269.
74. Charman, W.N., Rogge, M.C., Boddy, A.W., Berger, B.M. (1993). Effect of food and a monoglyceride emulsion formulation on danazol bioavailability. *Journal of Clinical Pharmacology*, *33*, 381–386.
75. Guichard, J.P., Blouquin, P., Qing, Y. (2000). A new formulation of fenofibrate: Suprabioavailable tablets. *Current Medical Research and Opinion*, *16*, 134–138.
76. Sauron, R., Wilkins, M., Jessent, V., Dubois, A., Maillot, C., Weil, A. (2006). Absence of a food effect with a 145 mg nanoparticle fenofibrate tablet formulation. *International Journal of Clinical Pharmacology, Therapy*, *44*, 64–70.
77. Schug, B.S., Brendel, E., Chantraine, E., Wolf, D., Martin, W., Schall, R., Blume, H.H. (2002). The effect of food on the pharmacokinetics of nifedipine in two slow release formulations: Pronounced lag-time after a high fat breakfast. *British Journal of Clinical Pharmacology*, *53*, 582–588.
78. Welling, P.G. (1996). Effects of food on drug absorption. *Annual Review of Nutrition*, *16*, 383–415.
79. Pasanisi, F., Meredith, P.A., Reid, J.L. (1985). Pharmacokinetics of nifedipine. *International Journal of Clinical Pharmacology, Therapy*, *5*, 63–66.
80. Berger, T. (1969). Studies on the gastric emptying mechanism in healthy persons and patients after partial gastrectomy. *Acta Chirurgica Scandinavica—Supplementum*, *404*, 1–51.
81. Heading, R.C., Nimmo, J., Prescott, L.F., Tothill, P. (1973). The dependence of paracetamol absorption on the rate of gastric emptying. *British Journal of Pharmacology*, *47*, 415–421.
82. Petring, O.U., Flachs, H. (1990). Inter- and intrasubject variability of gastric emptying in healthy volunteers measured by scintigraphy and paracetamol absorption. *British Journal of Clinical Pharmacology*, *29*, 703–708.
83. Kirchheiner, J., Brockmoller, J., Meineke, I., Bauer, S., Rohde, W., Meisel, C., Roots, I. (2002). Impact of CYP2C9 amino acid polymorphisms on glyburide kinetics and on the insulin and glucose response in healthy volunteers. *International Journal of Clinical Pharmacology, Therapy*, *71*, 286–296.
84. Kirchheiner, J., Roots, I., Goldammer, M., Rosenkranz, B., Brockmoller, J. (2005). Effect of genetic polymorphisms in cytochrome p450 (CYP) 2C9 and CYP2C8 on the pharmacokinetics of oral antidiabetic drugs: Clinical relevance. *Clinical Pharmacokinetics*, *44*, 1209–1225.
85. Niemi, M., Cascorbi, I., Timm, R., Kroemer, H.K., Neuvonen, P.J., Kivisto, K.T. (2002). Glyburide and glimepiride pharmacokinetics in subjects with different CYP2C9 genotypes. *International Journal of Clinical Pharmacology, Therapy*, *72*, 326–332.
86. Yin, O.Q., Tomlinson, B., Chow, M.S. (2005). CYP2C9, but not CYP2C19, polymorphisms affect the pharmacokinetics and pharmacodynamics of glyburide in Chinese subjects. *International Journal of Clinical Pharmacology, Therapy*, *78*, 370–377.
87. Zhou, L., Naraharisetti, S.B., Liu, L., Wang, H., Lin, Y.S., Isoherranen, N., Unadkat, J.D., Hebert, M.F., Mao, Q. (2010). Contributions of human cytochrome P450 enzymes to glyburide metabolism. *Biopharmaceutics & Drug Disposition*, *31*, 228–242.

88. Coppack, S.W., Lant, A.F., McIntosh, C.S., Rodgers, A.V. (1990). Pharmacokinetic and pharmacodynamic studies of glibenclamide in non-insulin dependent diabetes mellitus. *British Journal of Clinical Pharmacology*, *29*, 673–684.

89. Ikegami, H., Shima, K., Tanaka, A., Tahara, Y., Hirota, M., Kumahara, Y. (1985). Pharmacokinetics of glibenclamide: Heterogeneity in its absorption. *Medical Journal of Osaka University*, *35*, 55–61.

90. Neugebauer, G., Betzien, G., Hrstka, V., Kaufmann, B., von Mollendorff, E., Abshagen, U. (1985). Absolute bioavailability and bioequivalence of glibenclamide (Semi-Euglucon N). *International Journal of Clinical Pharmacology, Therapy and Toxicology*, *23*, 453–460.

91. Carlert, S., Palsson, A., Hanisch, G., von Corswant, C., Nilsson, C., Lindfors, L., Lennernas, H., Abrahamsson, B. (2010). Predicting intestinal precipitation–a case example for a basic BCS class II drug. *Pharmaceutical Research*, *27*, 2119–2130.

92. Hsieh, Y.L., Ilevbare, G.A., Van Eerdenbrugh, B., Box, K.J., Sanchez-Felix, M.V., Taylor, L.S. (2012). pH-induced precipitation behavior of weakly basic compounds: Determination of extent and duration of supersaturation using potentiometric titration and correlation to solid state properties. *Pharmaceutical Research*, *29*(10), 2738–2753.

93. Psachoulias, D., Vertzoni, M., Goumas, K., Kalioras, V., Beato, S., Butler, J., Reppas, C. (2011). Precipitation in and supersaturation of contents of the upper small intestine after administration of two weak bases to fasted adults. *Pharmaceutical Research*, *28*, 3145–3158.

94. Kostewicz, E.S., Wunderlich, M., Brauns, U., Becker, R., Bock, T., Dressman, J.B. (2004). Predicting the precipitation of poorly soluble weak bases upon entry in the small intestine. *Journal of Pharmacy and Pharmacology*, *56*, 43–51.

95. Charman, W.N., Stella, V.J. (1991). Transport of lipophilic molecules by the intestinal lymphatic system. *Advanced Drug Delivery Reviews*, *7*, 1–14.

96. Svensson, C.K., Edwards, D.J., Mauriello, P.M., Barde, S.H., Foster, A.C., Lanc, R.A., Middleton, E. Jr., Lalka, D. (1983). Effect of food on hepatic blood flow: Implications in the "food effect" phenomenon. *International Journal of Clinical Pharmacology, Therapy*, *34*, 316–323.

13

EVALUATION OF THE COLLECTIVE IMPACT OF PASSIVE PERMEABILITY AND ACTIVE TRANSPORT ON *IN VIVO* BLOOD-BRAIN BARRIER AND GASTROINTESTINAL DRUG ABSORPTION

Donna A. Volpe*, Hong Shen, and Praveen V. Balimane

13.1 INTRODUCTION TO DRUG ABSORPTION

A successful drug possesses acceptable pharmacological properties to reach its intended target site of action regardless of its route of administration. Therefore, permeability and transport characteristics play an important role in the therapeutic efficacy of the drug product. As a result, there is a need for reliable and accurate data on a drug's passive and active absorptive properties during drug discovery and development phases for the selection of compounds most likely to be useful as therapeutic

**Disclaimer*: This chapter reflects the views of the author and should not be construed to represent views or policies of the U.S. Food and Drug Administration (FDA).

Predictive ADMET: Integrative Approaches in Drug Discovery and Development, First Edition.
Edited by Jianling Wang and Laszlo Urban.

drugs [1–3]. Early introduction of screening models enhances the probability of success in clinical trials with drugs having favorable absorptive properties [4].

Oral administration is the most common, preferential, and convenient dosing route for drug therapy. This means that a drug must have adequate solubility and permeability in the gastrointestinal tract to reach therapeutic levels as the intestinal epithelium provides a physiological and biochemical barrier from the lumen to the systemic blood supply. Therefore, lead optimization to identify new chemical entities with sufficient predicted oral absorption is of great importance.

The development of drugs targeting the central nervous system (CNS) is the fastest growing franchise in the pharmaceutical industry, although this growth has been tempered by the relatively poor success of novel candidates. One of the significant challenges in the treatment of CNS disorders is drug passage across the blood-brain barrier (BBB), a continuous layer of cerebrovascular endothelial cells connected with intercellular tight junctions that express numerous efflux transporters and drug-metabolizing enzymes. Thus, investigation of drug properties that are favorable for CNS and intestinal absorption can greatly improve efforts in drug discovery.

As drug absorption is an important selection criterion in drug discovery and development, there is a need for models to assess passive and active intestinal and brain permeability [3–6]. The purpose of this chapter is to (1) evaluate the collective impact of passive permeability and active transport on *in vivo* BBB and gastrointestinal drug absorption, and (2) provide an overview of existing tools as guidance or examples to predict the gastrointestinal or brain absorption of drugs. Models ranging from computer simulations to whole animals have importance in drug discovery and development to make reasonable estimates of a drug's absorption potential without human studies. Furthermore, the models allow a better understanding of the rate-limiting processes of drug intestinal and BBB absorption to improve delivery strategies.

13.2 GASTROINTESTINAL DRUG ABSORPTION

The gastrointestinal tract has a very large surface area that is a result of its visible circular folds that are covered by microscopic villi and epithelial microvilli. The role of the small intestine is to absorb nutrients into the bloodstream and to provide a barrier to digestive enzymes and ingested foreign materials. Approximately 90% of oral drug absorption occurs in the small intestine.

The intestinal mucosa consists of the epithelial layer, the lamina propria, and the muscularis mucosa. A single layer of cells covers the inner intestinal wall providing a physiological and biochemical barrier to drug absorption that is composed of absorptive enterocytes, mucus-secreting goblet cells, microfold (M) cells, pareth cells, and endocrine cells. The enterocytes are polarized columnar epithelial cells that have a brush border formed by microvilli. Tight junctions form a barrier in the epithelium that regulates the transport of ions and molecules through the paracellular pathway.

Permeability together with solubility, dissolution, and gastrointestinal transit are the key variables that determine the *in vivo* rate and extent of intestinal drug absorption. The transport of drugs via the intestinal membrane is a complex process that

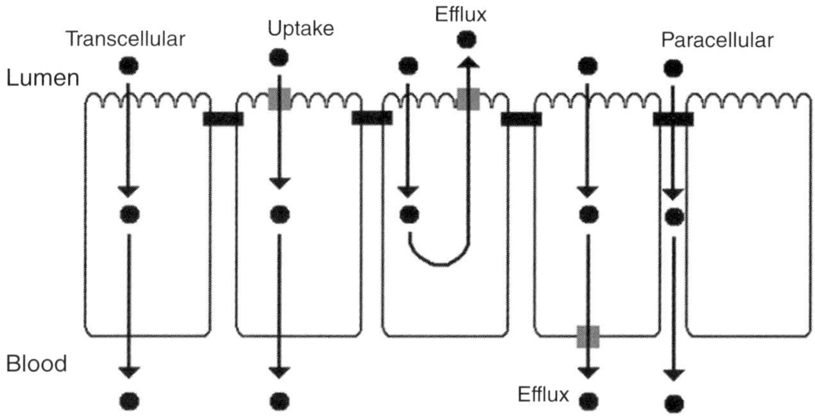

Figure 13.1 A schematic representation of the drug (•) transport systems formed by enterocytes at the intestinal epithelium. Gray bars represent tight junctions. Passive diffusion is represented by the paracellular and transcellular routes. Active transport is represented by carrier-mediated apical (e.g., P-gp, BCRP, MRP2) and basolateral (e.g., MRP3) efflux and uptake (e.g., OATP, OCTN, PepT1, MCT1) transporters.

includes the passage of compounds via several functional pathways in parallel. The intestinal barrier is composed of multiple passive and active transport processes (Figure 13.1) whose total functional activity affects net absorption [1]. Passive diffusion is gradient driven and composed of paracellular and transcellular components. Paracellular diffusion between cells is restricted by the tight junctions and is generally used by low-molecular-weight and hydrophilic compounds. More lipophilic compounds partition from the lumen into the cell membrane by the transcellular pathway through the enterocyte. Carrier-mediated uptake and efflux transport is also driven by membrane-bound transporters against a concentration gradient.

Transporter proteins play a relevant role in clinical drug absorption, distribution, and elimination of a variety of drugs. Drug transporters are present on the apical and basolateral surfaces of enterocytes and affect the uptake or efflux of compounds [7,8]. The ATP-binding cassette (ABC) transporters efflux substrate from cells using energy released from adenosine triphosphate (ATP) hydrolysis. These transporters pump substrates out of cells against a concentration gradient. Intestinal efflux transporters include P-glycoprotein (P-gp), breast cancer-resistant protein (BCRP), and multidrug resistance proteins (MRP) [7, 8]. The solute carrier (SLC) transporters enable the uptake of substrates into cells using ion gradients or co-transporters. Intestinal uptake transporters include monocarboxylic acid transporter (MCT1), oligopeptide transporters (PepT), organic anion transporting proteins (OATP) 1A2, OATP2B1, organic cation transporters (OCT), apical sodium-dependent bile acid transporter (ASBT), concentrative nucleoside transporter (CNT) 1, organic cation/zwitterion transporter (OCTN) 1, OCTN2s, equilibrative nucleoside transporter (ENT)1, and ENT2 [7, 8]. These efflux and uptake transporters regulate oral drug disposition by limiting absorption and bioavailability.

13.2.1 *In Silico* Models

The goal of *in silico* modeling is to predict *in vivo* oral drug absorption accurately using statistical tools to search for correlations between a given property (e.g., permeability and extent of absorption) and a set of molecular or structural descriptors of a drug candidate [9]. *In silico* models have been developed for predicting intestinal absorption based on molecular descriptors [9, 10], *in vitro* cell permeability [11, 12], *in vitro* artificial membrane permeability [13], *in vivo* animal absorption studies [14], and *in vivo* human absorption studies [15]. Physiological- and nonphysiological-based commercial software programs are also available to simulate human intestinal absorption from experimental data and molecular descriptors including GastroPlus™ (Simulations Plus, Lancaster CA), Intellipharm® PKCR (Niantic, CT), Cloe® HIA (Cyprotex, Cheshire, UK), SimCYP (Sheffield, UK), Accelrys® (San Diego, CA), and VolSurf (Tripos, St. Louis, MO).

In silico models of intestinal permeability and absorption have the advantages of relative ease in a high-throughput mode to provide predictions early in drug discovery. However, there are limitations to these models in that it is difficult to predict active transport processes and they are dependent on the quality and quantity of input data. The considerable inter- and intralaboratory variability in cell-assay–derived permeability values affects model development [12].

13.2.2 *In Vitro* Models

13.2.2.1 Artificial Membrane Studies Immobilized artificial membranes (IAMs) are phospholipids bound to a solid support in a chromatography column with polar head groups and lipid side chains. The measure of a drug's affinity, or capacity factor (k_{IAM}), in the columns allows ranking based on lipophilicity or affinity, which correlates with human intestinal absorption [16, 17].

Kotecha and colleagues optimized pH conditions and validated an IAM assay for routine analysis and were able to demonstrate a good correlation between k_{IAM} and human oral absorption [16]. Yen et al. built a stepwise regression model to predict human intestinal drug absorption based on retentions of drugs in an IAM column along with a drug's molecular descriptors and solubility [17]. Absorption was reciprocally correlated to the negative value of the capacity factor, and predictions improved with the addition of descriptors representing molecular size and shape, solubility, and polarity [17].

The parallel artificial membrane assay (PAMPA) is composed of two chambers separated by a porous filter that contains a lipid layer in an organic solvent. Drug transport is assessed by adding a drug to the donor chamber and measuring its appearance in the receiver chamber. By performing the assay in a 96-well microtiter format, it allows for high-throughput screening of passive permeability. Different types of PAMPA (e.g., biomimetic [BAMPA], hexadecane, and double-sink) have evolved with variable lipid compositions on the filter, and solvents have all been shown to influence resultant apparent permeability (P_{app}) values for drugs [18].

A validated 96-well PAMPA model constructed of a lipid trilayer in hexane was able to classify accurately drugs as having high or low permeability [19]. A microplate

BAMPA assay was combined with a prediction model of the paracellular pathway to predict better the total passive permeability of compounds with estimates of paracellular and transcellular permeability separately [20].

The artificial membrane models have the advantage of being rapid, simple, and automatable for high throughput. The assays can tolerates wider pH ranges and higher solubilizer concentrations than tissues or cells. The different artificial membrane assays are, however, limited in that they only model transcellular permeability and lack active processes.

13.2.2.2 Cell Assays

The reconstruction of the human differentiated epithelial cell monolayer by cell lines *in vitro* allows for the prediction of oral drug absorption in humans. Cells are grown on filters in an apparatus with apical and basolateral chambers to mimic the passage of drugs from the intestinal lumen to the blood supply. A drug is added to the apical or basolateral chamber, and its appearance is measured over time in the basolateral (absorption) or apical (secretion) chamber, respectively. Applications of cell-based *in vitro* models include permeability classification, mechanisms of absorption, luminal pH effects, and prediction of transporter-related drug interactions [21–26]. Advanced cell assays have been developed to increase the throughput of permeability studies including accelerated models to reduce monolayer growth time, microtiter assays to increase throughput, and automation for higher throughput [27–29].

Caco-2 cells cultured on permeable supports were introduced as an experimental tool for studies of drug transport and have become the most established cell model for the mechanistic study of intestinal drug absorption [3, 30]. The properties of Caco-2 cells that make them a useful model of drug absorption include their human origin, morphological and functional similarities to enterocytes, and formation of tight junctions and brush borders. Caco-2 cells model transcellular and paracellular permeability, active transport, carrier-mediated transport, and efflux systems [2, 23, 30].

Several investigators had found a qualitative relationship between intestinal absorption in humans and the P_{app} values for a drug in the Caco-2 cell assay [22, 28, 30]. Caco-2 cell's tighter paracellular route and smaller absorptive area than the intestine prevents a quantitative relationship between experimental permeability and *in vivo* absorption [31, 32]. Furthermore, there can be a lack of correlation for carrier-mediated absorption as the expression of important intestinal transporters in the Caco-2 cells is often lower than that observed *in vivo* in the intestine [31].

The Madin–Darby canine kidney (MDCK) epithelial cell line has been used as an alternative to Caco-2 cells in permeability studies because of its faster growing time, lower transepithelial electrical resistance (TEER) values, and low endogenous transporter expression [33]. P_{app} values from MDCK cell monolayers correlated with those from Caco-2 cells and ranked well to human intestinal absorption [21, 34].

The conditionally immortalized 2/4/A1 cell line from fetal rat intestine forms differentiated monolayers and has lower TEER values than Caco-2 [32]. In testing several model drugs, P_{app} values with the 2/4/A1 cells were lower than those for Caco-2 cells but correlated with human fraction absorbed [32]. TC-7 cells are clones of Caco-2 cells that have a greater homogeneity and higher enzyme content in their brush border, which have the potential for rank ordering compounds and predicting of human intestinal absorption [35]. The rat small intestinal crypt cell line, IEC-18, which originated from the ileum, more closely resembles the small intestine than Caco-2 cells. IEC-18 cells form a size-selective barrier for paracellular transport to screen for passive transcellular and paracellular transport of hydrophilic macromolecules [36].

The cell models can characterize paracellular and transcellular passive diffusion, active uptake, and active efflux of drug substances. The models are easy to use with both animal and human cell lines, and they can be automated to increase throughput. A significant disadvantage of the cell models is that the transport properties of the cells vary as the result of culture and transport conditions that require assay standardization [33]. The cell models are also limited by their lower expression of intestinal transporters, lack of mucus layer, genetic drift of cells, and concerns with the unstirred water layer [33].

13.2.2.3 Transporter Assays If a drug is subject to considerable efflux or uptake transport, then its absorption, distribution, elimination, or all of the above, may be altered by strong inhibitors of that transporter in drug–drug interactions (DDIs) [37]. Inhibition of an uptake transporter in the intestine by one drug can decrease the absorption of a second drug that is a substrate of that transporter. Alternatively, inhibition of an efflux transporter by one drug can increase the absorption of a second drug that is a substrate of that transporter.

To evaluate whether drugs are substrates of inhibitors of intestinal transporter proteins, cell-based assays are used. Accumulation assays are conducted in cell suspensions or monolayers to assess the uptake or efflux of a potential substrate or effects of an inhibitor. To study further the influence of efflux transporters on a drug's absorption, bidirectional assays with polarized cells cultured on a filter are considered to be the definitive method [37]. The transporter cell assays need to be characterized and validated with appropriate negative and positive controls in routine use along with setting acceptance criteria for transporter function [26, 37].

The presence of a large complement of influx and efflux transporters in Caco-2 cells makes it a relevant cell model to study the interaction between drugs and intestinal transporters [8]. Caco-2 cell assays can readily identify substrate specificity or potential drug interaction risk by comparing absorptive with secretive P_{app} (efflux ratios) for P-gp, BCRP, and MRP2 [23, 24, 29].

Transfected cell lines can target one or two transporters, thus, removing ambiguity in identifying whether a drug is a specific substrate or an inhibitor [38, 39]. Wild-type or parental cells are tested in parallel to control for the background effects of endogenous transporters. MDCK and porcine kidney LLC-PK1 cells are amenable to transfections as a result of their low background expression of endogenous transporters. The use of knockdown cells provides another means to study drug transporter interactions that eliminates the uncertainty associated with using inhibitors alone [40].

13.2.3 Animal Models

The prediction of human pharmacokinetic behavior of a compound is often based on data obtained from animals. Animal studies are used to extrapolate the extent of intestinal absorption in humans, although there are differences resulting from species variations in physiological or biochemical conditions including pH and the expression and localization of active drug transporters [41, 42]. Whole animal *in vivo*, *in situ* perfusion, and *ex vivo* excised tissue studies are used to determine a drug's extent of absorption and systemic bioavailability.

13.2.3.1 Whole Animal Studies The extent of absorption in humans and rats for more than 60 compounds has been found to be significantly correlated between the species [42]. The lower bioavailability of drugs in monkeys compared with humans makes them a poor predictor of human pharmacokinetics [43]. Absorption data for humans and dogs under fasted state had a relatively poor overall correlation [44]. Even though more than half of the drugs were essentially completely absorbed in both dogs and humans, several drugs showed much better absorption in dogs than in humans [44].

Orally administered drugs may have good absorption in the intestinal tract but poor systemic exposure or bioavailability. The rate and extent of absorption can be measured in animals with hepatic portal vein cannulas to determine intestinal first-pass effects. This model provides a reliable method to estimate intestinal drug absorption through the measurement of the fraction of dose entering the portal blood [45].

Knockout animals lacking genes encoding specific transporters provide a model to establish their role in intestinal drug absorption and tissue distribution of a drug candidate [46]. The model also provides a method to examine the effects of specific transporters on the pharmacokinetics (e.g., clearance, systemic exposure, plasma levels, and bioavailability) of orally administered drugs [47].

13.2.3.2 Perfusion Assays The *in situ* perfusion model measures drug permeability in isolated segments of the intact intestine in anesthetized animals. In the model, a section of intestine is cannulated proximally and distally, and it is perfused with a drug solution with serial measurements over time. Effective permeability (P_{eff}) is calculated from the difference between solute concentration entering and leaving the cannulated region.

A single-pass *in situ* perfusion study in rats with comparison to *in vivo* perfusion values in humans found that jejunal P_{eff} estimates for passively absorbed compounds correlated well and could be used to predict human *in vivo* absorption [48]. However, carrier-mediated transport required scaling between models as there were differences in transport maximums, substrate specificity, or both [48]. Kim et al. demonstrated the suitability of a rat single-pass perfusion method to classify twenty drugs as high or low permeability [49].

The advantage of the perfusion model is that it is closest to *in vivo* intestinal absorption containing intact blood flow and innervation with minimal interference on intestinal function and architecture [3]. The model complexity and animal usage requiring anesthesia and surgery does limit its use in the discovery setting.

13.2.3.3 Gut Sac Assays In the gut sac assay, a section of the intestine is prepared to measure drug transport from the lumen to the serosal compartment. Everted sacs are filled with oxygenated buffer and placed in a vessel containing the test drug. After a specified time period, the amount of drug in the sac is measured. With noneverted sacs, the drug solution is placed inside the sac, which is placed in a container with oxygenated buffer, and the appearance of the drug is measured over time in the container.

An everted gut sac assay compared the permeability of passively (paracellular, transcellular) and actively (efflux, uptake) transported molecules in different segments in the rat intestine [50]. The feasibility of a noneverted intestinal sacs method was validated and demonstrated for over 10 marketed compounds [51]. The study showed a good relationship between the permeability of the model drugs and their corresponding human fraction-absorbed data [51].

The advantages of the gut sac model are that it measures absorption at different sites along the intestinal tract, includes all intestinal cell types and the mucus layer, is easy and inexpensive to perform, and allows for mechanistic or directional studies. The gut sacs have limited tissue viability, and stirring conditions are often suboptimal during the experiment.

13.2.3.4 Diffusion Chamber Assays For diffusion chamber studies, a small section of the intestine is removed from an animal and mounted in a dual-chamber apparatus, with oxygenated buffer, to represent the serosal and mucosal compartments of drug transport. Drug solution is added to the buffer in the donor chamber, and its appearance is measured over time in the receiver chamber. The advantages of this model are that it retains gut architecture, provides a means to compare permeability in different intestinal regions, and represents all the absorption processes. It also allows for directional transport and can use human or animal tissue. However, the model has the drawback of limited tissue viability and low throughput.

Rat permeability (P_{app}) in jejunal, ileum, and colon segments in a diffusion chamber correlated well with human permeability and was able to classify compounds as high or low permeability [52]. In another diffusion chamber study with isolated rat jejunum, the resultant permeability values had similar correlations with human fraction-absorbed and P_{eff} values [53].

13.2.4 Integrated Approaches

As a result of the multiple processes involved in drug absorption in the intestine, the use of a single model makes it difficult to predict accurately a drug's *in vivo* permeability characteristics. Therefore, a tiered approached allows for estimation of the multiple processes involved in a drug's intestinal absorption [3]. A tier-based strategy features *in silico* screening, high-throughput *in vitro* permeability screening for lead selection and optimization, followed by *in vitro* and *in vivo* mechanistic studies.

First tier systems during the discovery phase involve *in silico* models to determine whether a drug may be permeable through the intestinal barrier based on its molecular properties. Next, by combining artificial membranes (e.g., IAM, PAMPA, and BAMPA) with *in vitro* cellular assays (e.g., Caco-2 and MDCK), the contribution of passive and active mechanisms to the drug's absorption potential is determined [54]. A second tier involves further studies in Caco-2, or other transporter-expressing cell lines, of selected candidates in bidirectional assays with inhibitor compounds for complete characterization of active transport mechanisms [54]. These studies may be complemented by mechanistic *ex vivo* studies with intestinal tissue. Third tier studies in various animal and cell models further investigate the potential for intestinal transporter and metabolism interplay along with transporter-mediated drug–drug interactions.

A tiered approach can also be applied for transporter specificity and the potential for transporter-based drug interactions. The primary tier would be a robust high-throughput model to identify substrates and inhibitors of relevant efflux and uptake intestinal transporters [23, 24, 26]. The secondary tier would use more refined cell models (e.g., transfected or knockdown) to look at substrate and inhibitor specificity and their impact on drug permeability [38–40]. The third tier may involve animal models including knock-out animals lacking specific transporters [46, 55].

The use of a combination of approaches to understand the absorption potential ensures recognition of the permeability mechanisms for decisions on the continued development of drug candidates. Several investigators have suggested strategies that combine Caco-2 cell and PAMPA assays to assess the permeability potential of new drugs in the discovery phase [4, 54, 56]. PAMPA serves as a fast and high-throughput tool for ranking transcellular passive permeability, while the Caco-2 model is used to delineate active (efflux and uptake) transport mechanisms. The differences between Caco-2 cell and PAMPA permeability coefficients ranking serve as an indicator for passive diffusion (PAMPA $\approx$ Caco-2), drug efflux (PAMPA > Caco-2), or absorptive transport (Caco-2 > PAMPA) [54, 56]. Balimane et al. combined a Caco-2 assay for transcellular, passive, and carrier-mediated transport with a PAMPA assay for transcellular permeability to bin drugs into broad categories of high, medium, and low permeability based on the P_{app} values [4].

Matsson et al. developed *in silico* models based on molecular descriptors and found that models based on *in vitro* permeability from 2/4/A1 cells (paracellular) performed better than those with hexadecane membranes (transcellular) [5]. A similar model by Fliszar et al. also compared permeability data from Caco-2 and 2/4/A1 cells along with hexadecane membranes [13]. Again 2/4/A1 yielded the highest degree of accuracy in predicting drug exposure and maximal plasma concentration [13].

13.3 BLOOD-BRAIN BARRIER DRUG ABSORPTION

Developing drugs for brain disorders, compared with non-CNS drugs, takes longer to get to market and has a greater attrition rate [6]. Despite successful examples of drug delivery to the CNS, more than 98% of all new drugs discovered for therapeutic

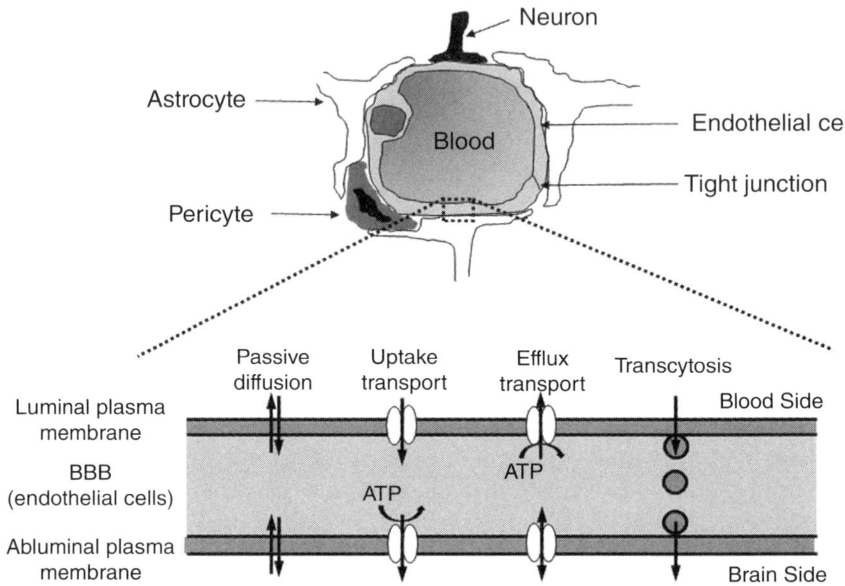

Figure 13.2 A schematic representation of the cell types and transporter systems at the blood-brain barrier (BBB). The endothelial layer coupled by tight junctions forms the barrier, surrounded by a basement membrane. Perivascular astrocytes and pericytes have associated roles including BBB phenotype in the endothelium. Other associated cells are neuron and microglia. The transport rate of a drug across BBB is a sum of active and passive processes in and out of the brain.

use in the CNS never reach the market because of their inherent inability to cross the BBB [57]. Most affected patients are not being adequately treated today. Brain disorders impose the largest disease burdens worldwide, and they cost more than cancer, cardiovascular disease, and diabetes to treat [58].

The BBB is a polarized layer of endothelial cells that physically separate circulating blood from the brain tissue (Figure 13.2). These endothelial cells are stitched together by tight junctions. Astrocyte cell projections called astrocytic feet surround the endothelial cells of the BBB, providing nutrient support to these cells. Drug delivery, distribution, and clearance within the brain involves the low-resistance twisting pathways associated with diffusion through the interstitial fluid (ISF) and cerebrospinal-fluid (CSF), which surrounds the neuron and other brain tissues [59]. Bulk flow of the ISF occurs, especially along perivascular spaces and axon tracts, draining through brain parenchyma and into the CSF. CSF is produced at a low rate from the choroid plexus, circulated, and then drained out through arachnoid villi into the lymphatics. Thus, ISF bulk flow likely contributes to the passive elimination of drugs from the brain parenchyma.

Because of restriction in the paracellular pathway through the BBB by tight junctions, exchange between the blood and the brain is dominated by the transcellular routes, making the endothelial cells the "gatekeepers" of the brain (Figure 13.2). Thus, passive diffusion through the barrier is the primary process of translocation from the bloodstream to the brain for most therapeutic compounds. The physicochemical properties of a drug including lipid solubility, polar surface area (PSA), ionization, molecular size, flexibility, and hydrogen-bonding potential govern its passive permeation through the BBB and entry into the brain [60,61]. Carrier-mediated uptake via SLC transporters expressed on the apical and basolateral membrane of endothelial cells transport many essential polar molecules such as glucose, amino acids, and nucleosides into the CNS. This is not surprising because there is a huge appetite that the CNS has for energy, and the brain constitutes just 2% of the body weight and consumes 25% of total body glucose utilization. SLC transporters suggested as relevant for drug transport at the BBB include glucose transporter 1 (GLUT1), L-type amino acid transporter, ENT1, organic anion transporter (OAT3), OATP1A2, and OATP2B1 [62–65]. Various efflux transporter proteins are polarized in their expression at apical or basolateral membrane of endothelial cells (Figure 13.2). By the mechanism of efflux, drugs are extruded from the brain, principally with P-gp [66,67]. In addition to P-gp, other ABC efflux transporters such as BCRP [68] and members of the MRP family [69] contribute to the restrictive function of BBB.

Once it has moved from the bloodstream to CNS interstitial fluid, the unbound drug is available to interact with its target, mostly neurotransmitter receptor, transporter, or enzyme. With sufficient exposure above appropriate concentrations, the drug exerts a physiological effect via receptor antagonism or enzyme inhibition, leading to a therapeutic effect. Therefore, it is important to determine the free drug concentration in the brain to predict *in vivo* efficacy based on *in vitro* activity data [70]. The free drug concentration in the brain can be affected by a variety of modulating factors: plasma or blood concentration (controlled by drug absorption, distribution, metabolism, and excretion (ADME)), plasma protein binding, permeability across the BBB (via passive diffusion, active transport, or both), brain metabolism, binding to lipids and proteins in the brain, and the “sink effect” of continual secretion and drainage of CSF and ISF, giving turnover of the fluids. The current approach to understanding and quantifying BBB absorption has moved on from expressing a single value representing permeability and partition extent to now focus on a greater understanding of the three major factors that control drug disposition in the brain: (1) passive membrane permeability, (2) active transport processes at the BBB (primarily focused on the efflux transporter P-gp), and (3) relative tissue binding between the plasma or blood and brain compartments.

13.3.1 *In Silico* Models

Several types of *in silico* models have been developed as tools for predicting drug brain penetration and distribution as with intestinal absorption. *In silico* prediction methods are gaining popularity in CNS drug discovery as they allow rapid and inexpensive screening. The key determinants of passive entry of a drug into the CNS are physicochemical properties such as hydrogen bonding, molecular weight, lipid solubility, and plasma-protein binding. Consequently, by quantifying such molecular features with *in silico* methods, it is possible to predict the *in vivo* BBB permeation, allowing conducting as much compound screening as possible without need to use *in vitro* and *in vivo* biology preparations. Several such *in silico* models are available, and they have evolved from simple regression models based on calculation of lipophilicity and polar surface area to complicated grid-based classification approaches [60, 71, 72]. However, modeling the BBB permeability of compounds is still challenging because of the quality and quantity of experimental data available and the difficult task of establishing a useful relationship between the molecular structure and the measured blood-brain partitioning. The absorption of compounds into the brain is heavily regulated by active transporter mechanisms, regulating both influx and efflux of drugs and toxins, so proper prediction models should also take into account the active transport phenomena.

Physiologically based pharmacokinetic (PBPK) CNS models include a multicompartmental representation of the blood-CNS barriers, allowing prediction of drug concentrations in the brain and CSF over time, that is, the rate and extent of drug absorption. PBPK models reduce biological and physiological features to a simplified framework that can then be described mathematically. Such models incorporate both compound- and species-specific inputs. The output of PBPK models enables investigators to predict prospectively the compounds that will be excluded from the brain and elucidate the salient factor responsible for the restriction. Furthermore, this feature can provide insights to the efficacy and safety of CNS drugs. Estimated concentrations of a drug in each brain segment can also be used as a driving force to assess any wanted or unwanted pharmacodynamic effects. Despite these benefits, few PBPK models have been developed for CNS penetration over the past two decades. Oyler et al. constructed a BBB model to simulate brain-uptakes for the compounds undergoing active transport [73]. Later, Kim and his colleagues described the distribution of the neurotoxin domoic acid into various regions of the brain as a function of time using a PBPK approach [74]. Subsequently, researchers implemented various model systems to assess the time required to achieve equilibrium, the extent of brain penetration [75, 76], and the impact of efflux transporters [77]. Recently, simCYP introduced a user-friendly PBPK brain model to provide various possibilities for simulating drug absorption and disposition in the central nervous system particularly by taking account of the effect of active transporters existing at the BBB and blood-cerebrospinal fluid barrier. The paucity of PBPK CNS absorption models can be attributive in part to the difficulty in obtaining key physiological and compound-specific input parameters. Because of the inherent complexity of the living system and the imperfect predictive power of existing PBPK model, a general, fully parameterized PBPK model for CNS penetration has yet to be developed, validated, and implemented.

13.3.2 *In Vitro* Methods

13.3.2.1 Methods for BBB Permeability Permeability is a well-established concept that represents the rate of permeation in physical chemistry. Several *in vitro* model systems are used to assess BBB permeability, including PAMPA, bovine, and human brain endothelial cells co-cultured with astrocytes, immortalized brain endothelial cell lines, along with models using cells not derived from endothelial cells, such as Caco-2 and MDCK cell lines. The key aspect of such assays is that they effectively and reliably predict the CNS penetration of drug candidates *in vivo*, so that they usefully inform structure–activity relationships.

The PAMPA system uses a specially designed artificial lipid-impregnated membrane. A high degree of automation compatibility, assay reproducibility, rapid evaluation, and substantially low cost makes it an attractive first-pass screening method for BBB passive permeability. Recently, PAMPA models based on porcine brain lipid extract was introduced. Mensch and his colleagues have compared the permeability of 19 structurally diverse, commercially available drugs and 14 internal drug candidates assessed in four different PAMPA models [78]. Both the PAMPA-black lipid membrane (BLM) model, a traditional paradigm for the prediction of gastrointestinal (GI) absorption, and the PAMPA-BBB model predicted the passive diffusion of the compounds reasonably well. The differences between these two models (BLM versus BBB) on the BBB classification of the tested compounds seem to be minimal although the porcine brain lipid was more discriminating [78]. Moreover, a major limitation of PAMPA systems versus cell-based and *in vivo* studies is its inability to produce active transport mechanisms and metabolic transformations incurred at the BBB endothelium. This may significantly affect the predictability of the bioavailability of a drug candidate on the targeted site, and thus, use of PAMPA alone would be insufficient.

Several *in vitro* cell-based BBB models have been developed with various types of cell source (human and animal, primary and immortalized, brain- and non-brain-derived, single and co-cultures), each having advantages and disadvantages, to improve permeability assessment. These models allow to some degree the ability to reproduce physiological expression of drug efflux systems and metabolizing enzymes that may facilitate the study of BBB absorption. However, although primary brain endothelial cell cultures are too cumbersome to be applied routinely, the immortalized brain endothelial cell lines usually lack the tightness required for permeability assessment, although some are promising [79]. The Caco-2 cell line has been used as a surrogate model in brain penetration research although this system was first highlighted for the study of intestinal absorption and was derived from human colon epithelial cancer cells [2]. Some CNS discovery groups in industry use MDCK cells to screen compounds for BBB permeability, primarily as a result of their short culture time and the tightness of the monolayer, resulting in permeability values that correlate well with *in vivo* brain permeation [80]. However, those surrogate epithelial cells lack representation of the whole spectrum of BBB transporters and enzymes. Variability in transporter expression has been observed between Caco-2 and MDCK cell subclones that were used in different laboratories [33]. Therefore, extrapolation of permeability data from these *in vitro* cell-based BBB models to *in vivo* conditions needs to be done cautiously.

13.3.2.2 Methods for Efflux Transporter Study Along with *in vitro* membrane permeability models, there has been great interest in using cell lines to express transporters of interest to evaluate the effect of an efflux transporter on BBB absorption. The cell lines expressing transporters of interest can be used to measure bidirectional transport across confluent cell monolayer grown on a semipermeable membrane in the absence or presence of an inhibitor to measure passive and active components. Transporter affinity is usually determined as an efflux ratio (basolateral-apical/apical-basolateral). The simplest and easiest systems use transfected immortalized adherent cell lines such as MDCK or LLC-PK1 to overexpress the major drug efflux transporters present at BBB including P-gp and BCRP. Using the MDCK-MDR1 cells expressing human P-gp, Mahar Doan and colleagues have compared both the passive permeability and the P-gp-mediated efflux of 48 CNS and 45 non-CNS drugs. The passive permeability was 43% greater in the CNS compounds, whereas P-gp efflux was 65% lower, indicating that the CNS compounds had a better brain absorption profile [80]. The use of MDCK and LLC-PK1 cell systems can be of increasing complexity as a result of their endogenous expression of efflux transporters. In addition, it is difficult to extrapolate from these assays to the *in vivo* BBB because allocation of "weighting factors" reflecting the relative importance of particular transporters is problematic, and overexpression of specific transporters in cell lines may distort transport kinetics.

13.3.2.3 Methods for Brain Tissue Binding Study The measurement of total drug levels in the brain (i.e., amount of drug per gram of brain) has historically been the most common method of brain tissue binding because concentrations of unbound drug in the interstitial fluid of the brain cannot be rapidly measured. But the total concentrations are not helpful in drug discovery programs relating to central drug effects. The ratio of the free brain concentration to the free plasma concentration is considered the "intrinsic" partition coefficient between brain and plasma and should represent a better parameter to assess the penetration. Two simple *in vitro* techniques using isolated brain tissues have been used to estimate quickly free brain drug concentrations. One binding method uses brain homogenate dialyzed against a drug-containing buffer solution mimicking ISF-CSF to estimate the fraction of unbound drug in the tissues ($f_{\text{u,brain}}$) [81, 82]. The fraction unbound in brain, like the fraction unbound in plasma, is an easily understood concept. However, it shares the limitation of a homogenate method from which it originates; there is no distinction made between intra- and extracellular distribution. The homogenization may change brain binding properties by unmasking binding sites that are not normally accessible to a drug *in vivo*. The other method uses brain slices incubated in drug-containing buffer to measure the volume of distribution of unbound drug in the slices ($V_{\text{u,brain}}$) [83, 84]. $V_{\text{u,brain}}$ describes the distribution of drug inside the brain regardless of brain-to-plasma distribution, and it is determined according to the following equation:

$$V_{\text{u,brain}} = \frac{A_{\text{brain}} - V_{\text{blood}} \times C_{\text{blood}}}{C_{\text{u,brain}}} \qquad (13.1)$$

A low value for $V_{\text{u,brain}}$, close to the volume of the interstitial space, would thus describe predominantly extracellular distribution, whereas a high value would indicate that the drug enters brain cells and binds to tissue components [84]. The free drug concentration in brain, the pharmacologically relevant concentration, would be calculated if total brain (i.e., A_{brain}) and blood (i.e., $V_{\mathrm{blood}} \times C_{\mathrm{blood}}$) drug levels are available. Estimations of unbound drug concentration in the brain using both methods agreed well with *in vivo* microdialysis measurement results, and the approaches are likely to be more successful with lipophilic drugs [85].

13.3.3 *Ex Vivo* and *In Vivo* Methods

13.3.3.1 Methods for BBB Permeability The permeability or rate of transport of a drug across the BBB can be estimated by several *in situ* and *in vivo* methods including brain-to-plasma concentration ratio and *in situ* brain perfusion. Historically, an *in vivo* determination of drug brain permeability in preclinical species especially rodents based on the measurement of a ratio of brain concentration over plasma concentration is the most commonly used parameter for measuring brain permeability in addition to brain tissue binding, which is referred as K_{p}. A large K_{p} is considered a favorable property of a good CNS compound. The ratio can then be used to calculate a P_{app} value, as described by Ohno and colleagues [86]. It provides a simple measure of permeability and partitioning but does not take into account the presence of drug in the brain vasculature.

The brain perfusion method represents an *in vivo* highly sensitive means of studying BBB permeability without the confounding factor of systemic disposition in whole animals [87]. This method uses catheterization of the common carotid artery in the anesthetized rat and the ligation of the external carotid artery. The brain is then perfused via the internal carotid using an oxygenated physiological saline buffer containing the test substance. Once perfusion is complete, the brain is removed for analysis and uptake is determined. *In situ* brain perfusion is a rapid technique that measures the total concentration of drugs in the brain at short time points after drug administration. It is assumed that BBB permeability is a rate-limiting step in determining the value of the *in situ* permeability constant and that any subsequent distribution processes will be faster. But with increasing lipophilicity, brain tissue binding and distribution seemed to influence *in situ* BBB permeability despite short brain perfusion time. The *in situ* brain infusion remains difficult to deconvolute one process from another.

13.3.3.2 Methods for Efflux Transporter Studies *In vivo* studies of efflux transporters are now a well-established part of modern CNS drug discovery. Knock-out mouse and rat studies demonstrate that these animal models are valuable tools to determine a role of P-gp in drug brain absorption. For instance, the absence of the mdr1a gene in *mdr1a* (–/–) mice, which encodes the predominant P-gp isoform expressed at BBB, leads to a significant increase in the brain penetration of ivermectin, vinblastine, digoxin, cyclosporin A, ondansetron, and lopermaide compared with wild-type animals [46, 88]. Recently, *mdr1a* (–/–) rat model using ZFN-mediated gene targeting was developed. $K_{\mathrm{p,brain}}$ in knock-out versus wild-type rats was 17- to 63-fold for loperamide and 9- to 30-fold for indinavir, respectively [55]. The increased brain levels of loperamide in P-gp knockout rat results in hallmark opioid central effects [47]. Similar findings were observed in mice and rats where P-gp has been chemically ablated with selective inhibitors of P-gp such as PSC833, GF120918, and LY335979 [89, 90]. In each case, the most striking effects were observed in the brain even though systemic exposures were not altered. A concern of use of the knock-out animals is species-dependent P-gp activity. The species differences in substrate and inhibition of P-gp-mediated transport could be observed for a particular structural class of compounds [91–93]. The differences in expression levels of BBB transporters between species exit. It was reported that BCRP showed the most abundant protein expression among drug transporters in human brain microvessels by means of a liquid chromatography-tandem mass spectrometric quantification method, and its expression level was 1.85-fold greater in humans than in mice. In contrast, the expression level of P-gp in humans was 2.3-fold lower than that of mdr1a in mice [94]. Hsiao et al. demonstrated that the brain-to-blood ratio of [^{11}C]verapamil was increased by only 79% by chemical inhibition on P-gp in humans, while that of [^{3}H]verapamil was increased by 1,460% in mice by gene deletion [95, 96]. Finally, the absence of P-gp alone does not enhance brain distribution of a drug that is a substrate for both P-gp and BCRP. But genetic or chemical knock out of both transporter significantly increases the brain uptake of dual P-gp/BCRP substrate drug [97]. Taken together, efflux transporter genetic or chemical knock-out animal studies cannot readily be extrapolated to humans.

13.3.3.3 Methods for Tissue Binding Studies The approach of CSF sampling was first introduced in 1992 [98] and has been applied extensively in CNS drug research since then because CSF concentrations are a reasonably good measure of free drug concentration in the brain [99]. CSF sampling can be performed postmortem or by repeated sampling of CSF from the cisterna *in vivo*. CSF concentrations follow the same profile as that of interstitial concentrations [100], but this is not always the case. It has been reported that in most cases, the ISF concentration is lower than CSF concentration; thus, CSF values will usually overpredict brain exposure [101]. Consequently, sampling time point (single versus multiple), sampling site (cisterna magna versus ventricles or lumbar thecal), and dosage (single versus multiple dosing) should be considered when performing CSF sampling studies.

Cerebral microdialysis offers a direct measure of free drug concentration in brain interstitial fluid over time. The continuous quantitative measure provides a solid basis to select drug candidates with the best brain exposure profile. Microdialysis also allows a level of anatomical precision that is not available with any other method. This is important because systemically administered compounds do not distribute evenly across the brain [102]. Another advantage of microdialysis is that it can be used to assess pharmacodynamic measures, which thus permits the neuropharmacokinetic profile of a compound to be related to its neuropharmacodynamic profile. For example, a microdialysis study using BMS-505130, a potent and selective serotonin transport inhibitor, showed that its interstitial concentrations mirrored the changes in the interstitial concentrations of serotonin [103]. This approach is, however, expensive, time consuming, and can be technically very challenging. In particular, the experimental conditions are critical to the integrity and reproducibility of microdialysate data. This method is of limited value unless the exact site of pharmacological action is known.

As direct sampling of human brain tissue is generally limited to postmortem subjects, noninvasive techniques have emerged to image drug concentrations within the human brain. Of these techniques, positron emission tomography (PET) has emerged as the standard platform used to assess the biodistribution and receptor occupancy of compounds in the human and animal CNS [95,96]. After administration of a radiolabeled compound into the body, the subject is placed in a counter that detects the emission of gamma photons emitted by the compound in the brain as a result of destruction of the positrons. With the use of computerized imaging techniques, two-dimensional images of the brain can be determined over time, allowing for a quantitative kinetic evaluation of brain uptake [104]. However, it is expensive and the preparation and stability of the tracers are matters of concern. In addition, no distinction can be made among parent compound, metabolites, and protein bound or unbound tracer in PET study.

13.3.4 Integrated Approach to Study Drug Absorption across BBB

CNS drug discovery and development is a time-consuming, resource-demanding, and costly endeavor. The selection of right compounds using the right models at the right time is critical for progressing through different phases and gain approval, and it is eventually financially rewarding for the companies. The utilization of right models ensures understanding drug brain absorption mechanisms and benefit/risk ratios, and making a decision for next step. The ideal strategy for CNS compound selection involves the use of a combination of *in silico* and *in vitro* models that are high throughput but less predictive with *in vivo* models that are low throughput but more predictive to evaluate effectively brain absorption for thousands of compounds during the lead selection and only a few leads during lead optimization process (Figure 13.3). The three distinct but interdependent aspects of brain absorption including passive permeability, active transport, and brain tissue binding must be examined and integrated into one coherent concept (Figure 13.3), and the integrated approach to brain absorption is briefly outlined in the following.

In line with routine high-throughput screening procedures at the early stage of drug discovery, a panel of *in vitro* ADME assays is run, e.g., physicochemical profiling, liver microsomes metabolic stability, PAMPA and Caco-2 permeability, CYP inhibition and induction, and plasma binding. In relation to brain absorption, this *in vitro* panel is often supplemented by the assays of bidirectional permeability in MDCK-MDR1 cells and equilibrium dialysis of brain homogenate from the pharmacological species. This early prospective assessment in conjunction with computational models allows quick identification of compounds with good CNS absorption properties in efficient and cost-effective manner. Although polar surface area is the most commonly cited physiochemical approaches with respect to the traditional brain-to-plasma concentration ratio measurements [105], recent studies suggested that efflux-mediated reduction in CNS exposure may be a more important factor. The MDR1-MDCK cell

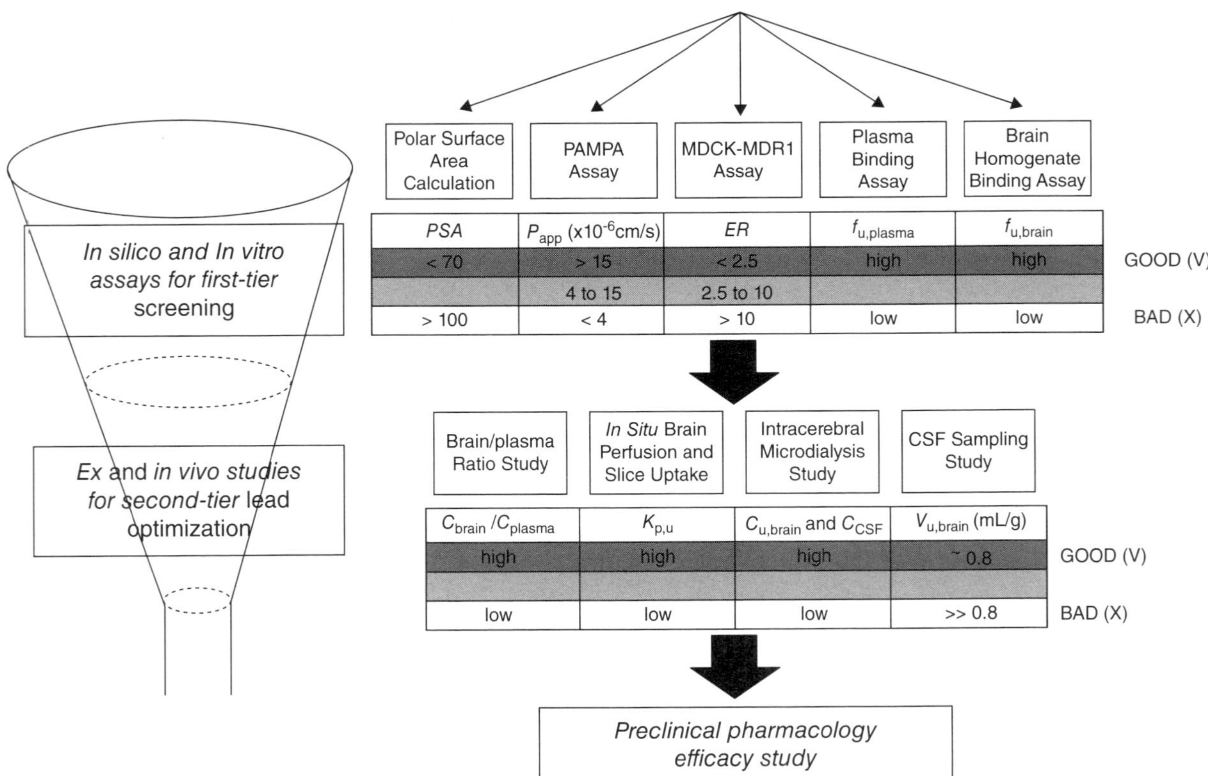

Figure 13.3 Schematic illustration of the integrated approach to study drug brain absorption using both the first and the second tier of *in vitro* assays and *in vivo* studies.

data are often used to assess the general permeation behavior of a compound in a cellular setup (P_{app} value), and its susceptibility to drug efflux by P-gp (ER). P_{app} value can be taken as surrogate for the rate of permeation across the BBB and should ideally have a P_{app} value of $>15 \times 10^{-6}$ cm/s [80]. The value of ER is indicative of potential limitations in the brain penetration by drug efflux at the BBB and should ideally have an ER < 2.5 [80]. However, it needs to be kept in mind that these thresholds are not rigid and may well be modified in supporting a CNS drug discovery program, as a poor P_{app} or a high ER value may both be compensated for by other properties of the compound. For practical reasons, information on ER may be taken alternatively from the bidirectional Caco-2 permeation assay, as in many instances, these data are being determined on a routine basis in all drug discovery projects. The values of unbound fraction (i.e., $f_{u,plasma}$ and $f_{u,brain}$) will be used to estimate the brain-to-plasma unbound concentration ratio ($K_{p,u}$) described in the next section. Kalvass et al. suggested that the ratio of steady-state unbound plasma over unbound brain concentration ratio ($f_{u,plasma}/f_{u,brain}$) could be used as a simple mechanism-independent alternative method for assessing the CNS distribution of compounds [106]. If the ratio is close to unity, passive diffusion across the BBB or any influx being offset by efflux can be assumed. In contrast, a ratio that is significantly greater or less than one implies for active efflux or uptake at the BBB.

Once potential drug candidates with favorable profiles have been identified through *in vitro* screening assays, *in vivo* studies will follow to examine the brain penetration of selected compounds in the whole animal before *in vivo* pharmacology efficacy studies (Figure 13.3).

Traditionally the total brain-to-plasma concentration ratio (K_p) is determined either at several time points after IV bolus and oral administration, or at a single time point where near steady state is assumed after IV constant infusion in animals. Because it is considered that only the unbound brain drug concentration ($C_{u,brain}$) is available for interaction with most CNS targets, it is essential to determine this parameter for all CNS drug candidates. $C_{u,brain}$ can be obtained in several ways. First, a brain microdialysis technique can directly determine the concentration of unbound drug in the brain interstitial fluid at the site of the probe placement. However, this technique is resource-intensive and not broadly applicable, as a result of probe recovery problems with lipophilic compounds. Consequently, brain microdialysis is usually reserved for drugs in the development stage. Second, the classic K_p is calculated using $f_{u,brain}$ and $f_{u,plasma}$ data from the first tier; K_p is then transformed into the brain-to-plasma unbound concentration ratio ($K_{p,u}$). Furthermore, the $C_{u,brain}$ as well as the unbound volume of distribution in brain ($V_{u,brain}$) can be estimated. The parameters of $K_{p,u}$ and $V_{u,brain}$ provide surrogate measures of free drug concentration within the parenchyma; this approach is more practical than brain microdialysis for routine screening. As volume distribution at steady state (V_{dss}), $V_{u,brain}$ does not describe a real volume. $V_{u,brain}$ is an indicator for the distribution of a compound within the brain. It indicates whether a compound is distributed merely in ISF ($V_{u,brain}$ close to 0.2-mL/g brain), throughout the brain water space, i.e., into both ISF and ICF ($V_{u,brain}$ close to 0.8-mL/g brain), or has a tendency to nonspecifically bind to brain tissue ($V_{u,brain} >$ 0.8-mL/g brain) [107].

13.4 CASE STUDY

Cole and her colleagues at Pfizer (New York, NY) have recently reported on how to use an integrated approach to screen and identify PF-3731237 as a new generation of an H3 antagonist with full efficacy that lacks CNS side effects [108]. PF-3654746, the previous leader of the H3 antagonist compounds, exhibited efficacy with visible relief in the symptoms of allergic rhinitis mediated by blocking of H3 receptors in the nasal cavity. However, a side effect of insomnia was also observed at its efficacious dose as a result of centrally mediated H3 activity. As a result, a strategy to identify a similarly potent agent that exhibited lower brain penetration was proposed: (1) design low-permeability compounds using *in silico* tools (PSA > 60 A^2); (2) target compounds that exhibit low passive membrane permeability across MDCK cell monolayer (i.e., $P_{app} < 12 \times 10^{-6}$ cm/s) and are substrates for P-gp and BCRP in first tier *in-vitro* MDCK-MDR1 and MDCK-BCRP screens (i.e., efflux ratios > 2.5); (3) identify compounds with CSF to unbound plasma concentration ratio at steady state less than 0.1 in preclinical species; (4) confirm low CNS exposures and side effects by brain receptor occupancy and electroencephalography (EGG) studies in rats; and (5) pharmacokinetic-pharmacodynamic analysis to understand the relationship between compound plasma concentration with pharmacodynamic marker level (i.e., the probability of transitioning from wake to sleep).

Optimal compounds were evaluated using the screen sequence described earlier. Eventually, PF-3731237 was selected based on the required brain penetration properties and progressed to *in vivo* studies quantifying drug in the brain and CNS effects. It exhibited low MDCK cell permeability (11×10^{-6} cm/s). Compound Z is a fully brain penetrant close analogue of PF-3654746 and was used extensively as a standard in the studies. The permeability of compound Z is 23×10^{-6} cm/s. PF-3731237 showed differentiated $K_{p,u}$ and CSF concentration to unbound plasma concentration ratio versus compound Z (0.1 vs. 1.6 and 0.05 vs. 1.10, respectively). Consequently, with a direct comparison of compound Z and PF-3731237 using the rat EEG study, a clear difference between the two compounds was observed. At a similar multiple of the free K_i in plasma (approximately 20-fold), compound Z achieved 80% receptor occupancy in brain, whereas PF-3731237 achieved only 18%. After a 7-day repeated oral dosing regimen to maintain a similar plasma exposure, this differentiation was maintained, resulting less CNS side effects [108].

13.5 CONCLUSION

Drug absorption is a complex process affected by many competing factors related to the compound, its formulation, and the biological barriers it must traverse. Accurate assessment of intestinal or brain permeability for drug candidates is crucial to making decisions according to their absorption potential and to resolving their mechanism(s) of absorption that explain limited and variable systemic or CNS exposure. The absorption of drugs across the intestine and BBB can be addressed by various computer, animal, and cell assays during discovery and development. As drugs proceed along the development pathway, the assays can be conducted in a tiered approach parallel to studies on drug efficacy and toxicity. However, given the multiple processes involved in intestinal and brain drug absorption, multiple models are used in a drug's *in vivo* permeability characteristics [4]. Absorption models can assist in drug candidate selection for *in vivo* clinical trials at early stages in drug discovery and development, along with submission of regulatory applications [5]. The preclinical models are also used to predict clinical drug–drug interactions to determine the need for *in vivo* studies [37]. Models of intestinal and brain absorption serve as tools in the decision-making process to prevent loss of drug candidates in later, more costly, clinical phases as a result of poor pharmacokinetics.

REFERENCES

1. Lennernäs, H. (2007). Intestinal permeability and its relevance for absorption and elimination. *Xenobiotica*, *37*, 1015–1051.
2. van Breemen, R.B., Li, Y. (2005). Caco-2 cell permeability assays to measure drug absorption. *Expert Opinions in Drug Metabolism and Toxicology*, *1*, 175–185.
3. Bohets, H., Annaert, P., Mannens, G., van Beijsterveldt, L., Anciaux, K., Verboven, P., Meuldermans, W., Lavrijsen, K. (2001). Strategies for absorption screening in drug discovery and development. *Current Topics in Medicinal Chemistry*, *1*, 367–383.
4. Balimane, P.V., Han, Y.H., Chong, S. (2006). Current industrial practices of assessing permeability and P-glycoprotein interaction. *The AAPS Journal*, *8*, E1–E13.
5. Matsson, P., Bergström, C.A.S., Nagahara, N., Tavelin, S., Norinder, U., Artursson, P. (2005). Exploring the role of different drug transport routes in permeability screening. *Journal of Medicinal Chemistry*, *48*, 604–613.
6. Pangalos, M.N., Schechter, L.E., Hurko, O. (2007). Drug development for CNS disorders: Strategies for balancing risk and reducing attrition. *Nature Reviews Drug Discovery*, *6*, 521–532.
7. Varma, M.V., Ambler, C.M., Ullah, M., Rotter, C.J., Sun, H., Litchfeld, J., Fenner, K.S., El-Kattan, A.F. (2010). Targeting intestinal transporters for optimizing oral drug absorption. *Current Drug Metabolism*, *11*, 730–742.
8. Englund, G., Rorsman, F., Rönnblom, A., Karlbom, U., Lazorova, L., Gråsjö, J., Kindmark, A., Artursson, P. (2006). Regional levels of drug transporters along the human intestinal tract: Co-expression of ABC and SLC transporters and comparison with Caco-2 cells. *European Journal of Pharmaceutical Sciences*, *29*, 269–277.
9. van de Waterbeemd, H., Gifford, E. (2003). ADMET *in silico* modelling: Towards prediction paradise? *Nature Reviews Drug Discovery*, *2*, 192–204.
10. Bergström, C.A.S., Strafford, M., Lazorova, L., Avdeef, A., Luthman, K., Artursson, P. (2003). Absorption classification of oral drugs based on molecular surface properties. *Journal of Medicinal Chemistry*, *46*, 558–570.
11. Norinder, U., Österberg, T. (2001). Theoretical calculation and prediction of drug transport processes using simple parameters and partial least squares projections to latent structures (PLS) statistics. The use of electrotopological state indices. *Journal of Pharmaceutical Sciences*, *90*, 1076–1085.

12. Ren, S., Lien, EJ. (2000). Caco-2 cell permeability vs human gastrointestinal absorption, QSPR analysis. *Progress in Drug Research*, *54*, 1–23.
13. Fliszar, K.L., Hill, B.T., Foster, N. (2007). Predicting human drug pharmacokinetics from *in vitro* permeability using an absorption-disposition model. *Journal of Pharmaceutical Sciences*, *96*, 2161–2170.
14. Zhao, Y.H., Abraham, M.H., Le, J., Hersey, A., Luscombe, C.N., Beck, G, Sherborne, B., Cooper, I. (2003). Evaluation of rat intestinal absorption and correlation with human intestinal absorption. *European Journal of Medicinal Chemistry*, *38*, 233–243.
15. Zhao, Y.H., Le, J., Abraham, M.H., Hersey, A., Eddershaw, P.J., Luscombe, C.N., Boutina, D., Beck, G, Sherborne, B., Cooper, I., Platts, J.A. (2001). Evaluation of human intestinal absorption data and subsequent derivization of a quantitative structure-activity relationship (QSAR) with the Abraham descriptors. *Journal of Pharmaceutical Sciences*, *90*, 749–784.
16. Kotecha, J., Shah, S., Rathod, I., Subbaiah, G. (2007). Relationship between immobilized artificial membrane chromatographic retention and human oral absorption of structurally diverse drugs. *International Journal of Pharmaceutics*, *333*, 127–135.
17. Yen T.E., Agatonovic-Kustrin, S., Evans, A.M., Nation, R.L., Ryand, J. (2005). Prediction of drug absorption based on immobilized artificial membrane (IAM) chromatography separation and calculated molecular descriptors. *Journal of Pharmaceutical and Biomedical Analysis*, *38*, 472–478.
18. Seo, P.R., Teksin, Z.S., Kao, J.P.Y., Polli, J.E. (2006). Lipid composition effect on permeability across PAMPA. *European Journal of Pharmaceutical Sciences*, *29*, 259–268.
19. Chen, X., Murawski, A., Patel, K., Crespi, C.L., Balimane, P.V. (2008). A novel design of artificial membrane for the PAMPA model. *Pharmaceutical Research*, *25*, 1511–1520.
20. Sugano, K., Takata, N., Machida, M., Ushio, H., Saitoh, K., Terado, K. (2002). Prediciton of passive intestinal absorption using bio-mimetic artificial membrane permeation assay and the paracellular pathway model. *International Journal of Pharmaceutics*, *241*, 241–251.
21. Thiel-Demby, V.E., Humphreys, J.E., St. John Williams, L.A., Ellens, H.M., Shah, N., Ayrton, A.D., Polli, J.W. (2009). Biopharmaceutics classification system: Validation and learnings of an in vitro permeability assay. *Molecular Pharmaceutics*, *6*, 11–18.
22. Volpe, D.A., Faustino, P.J., Ciavarella, A.B., Asafu-Adjaye, E.B., Ellison, C.D., Yu, L.X., Hussain, A.S. (2007). Classification of drug permeability with a Caco-2 cell monolayer assay. *Clinical Research and Regulatory Affairs*, *24*, 39–47.
23. Skolnik, S., Lin, X., Wang, J., Chen, X.H., He, T., Zhang, B. (2010). Towards prediction of *in vivo* intestinal absorption using a 96-well Caco-2 assays. *Journal of Pharmaceutical Sciences*, *99*, 3246–3265.
24. Lin, X.N., Skolnik, S., Chen, X.H., Wang, J.L. (2011). Attenuation of intestinal absorption by major efflux transporters: Quantitative tools and strategies using a Caco-2 model. *Drug Metabolism and Disposition*, *39*, 265–274.
25. Neuhoff, S., Ungell, A.L., Zamora, I., Artursson, P. (2005). pH-Dependent passive and active transport of acidic drugs across Caco-2 monolayers. *Pharmaceutical Research*, *25*, 211–220.

26. Elsby, R., Surry, D.D., Smith, V.N., Gray, A.J. (2008). Validation and application of Caco-2 assays for the in vitro evaluation of development candidate drugs as substrates or inhibitors of P-glycoprotein to support regulatory submissions. *Xenobiotica*, *38*, 1140–1164.
27. Galkin, A., Pakkanen, J., Vuorela, P. (2008). Development of an automated 7-day 96-well Caco-2 cell culture model. *Pharmazie*, *63*, 464–469.
28. Alsenz, J., Haenel, E. (2003). Development of a 7-day, 96-well Caco-2 permeability assays with high-throughput direct UV compound analysis. *Pharmaceutical Research*, *20*, 1961–1969.
29. Balimane, P.V., Patel, K., Marino, A., Chong, S. (2004). Utility of 96 well Caco-2 cell system for increased throughput of P-gp screening in drug discovery. *European Journal of Pharmaceutics and Biopharmaceutics*, *58*, 99–105.
30. Yamashita, S., Furubayashi, T., Kataoka, M., Sakane, Sezaki, H., Tokuda, H. (2000). Optimized conditions for prediction of intestinal drug permeability using Caco-2 cells. *European Journal of Pharmaceutical Sciences*, *10*, 195–204.
31. Lennernäs, H., Palm, K., Fagerholm, U., Artursson, P. (1996). Comparison between active and passive drug transport in human intestinal epithelial (Caco-2) cells in vitro and human jejunum in vivo. *International Journal of Pharmaceutics*, *127*, 103–107.
32. Tavelin, S., Milovic, V., Ocklind, G., Olsson, S., Artursson, P. (1999). A conditionally immortalized epithelial cell line for studies of intestinal drug transport. *Journal of Pharmacology and Experimental Therapeutics*, *290*, 1212–1221.
33. Volpe, D.A. (2008). Variability in Caco-2 and MDCK cell-based intestinal permeability assays. *Journal of Pharmaceutical Sciences*, *97*, 712–725.
34. Di, L., Whitney-Pickett, C., Umland, J.P., Zhang, H., Zhang, X., Gebhard, D.F., Lai, Y., Federico, J.J., Davidson, R.E., Smith, R., Reyner, E.L., Lee, C., Feng, B., Rotter, C., Varma, M.V., Kempshal, S., Fenner, K., El-Kattan, A.F., Liston, T.E., Troutman, M.D. (2011). Development of a new permeability assay using low-efflux MDCKII cells. *Journal of Pharmaceutical Sciences*, *100*, 4974–4985.
35. Greś, M.C., Julian, B., Bourrié, M., Meunier, V., Rogues, C., Berger, M., Boulenc, X., Berger, Y., Fabre, G. (1998). Correlation between oral drug absorption in humans, and apparent drug permeability in TC-7 cells, a human epithelial intestinal cell line. *Pharmaceutical Research*, *15*, 726–733.
36. Versantvoort, C.H.M., Ondrewater, R.C.A., Duizer, E., Van de Sandt, J.J.M., Gilde, A.J., Groten, J.P. (2002). Monolayers of IEC-18 cells as an in vitro model for screening the passive transcellular and paracellular transport across the intestinal barrier: Comparison of active and passive transport with the human colon carcinoma Caco-2 cell line. *Environmental Toxicology and Pharmacology*, *11*, 335–344.
37. Giacomini, K.M., Huang, S.M., Tweedie, D.J., International Transporter Consortium. (2010). Membrane transporters in drug development. *Nature Reviews Drug Discovery*, *9*, 215–236.
38. Xiao, Y., Davison, R., Smith, A., Pereira, D., Zhao, S., Soglia, J., Gebhard, D., de Morais, S., Duignan, D.B. (2006). A 96-well efflux assay to identify ABCG2 substrates using a stably transfected MDCK II cell line. *Molecular Pharmaceutics*, *3*, 45–54.
39. Förster, F., Volz, A., Fricker, G. (2008). Compound profiling for ABC (MRP2) using a fluorescent microplate system. *European Journal of Pharmacy and Biopharmaceutics*, *69*, 396–403.

40. Li, J., Volpe, D.A., Wang Y., Zhang, W., Bode, C., Owen, A., Hidalgo, I.J. (2011). Use of transporter knockdown Caco-2 cells to investigate the in vitro efflux of statin drugs. *Drug Metabolism and Disposition*, *39*, 1196–1202.
41. Kararli, T.T. (1995). Comparison of the gastrointestinal anatomy, physiology, and biochemistry of humans and commonly used laboratory animals. *Biopharmaceutics & Drug Disposition*, *16*, 351–380.
42. Chiou, W.L., Barve, A. (1998). Linear correlation of the fraction of oral dose absorbed of 64 drugs between humans and rats. *Pharmaceutical Research*, *15*, 1782–1795.
43. Akabane, T., Tabata, K., Kadono, K., Sakuda, S., Terashita, S., Teramura, T. (2010). A comparison of pharmacokinetics between humans and monkeys. *Drug Metabolism and Disposition*, *38*, 308–316.
44. Chiou, W.L., Jeong, H.J., Chung, S.M., Wu, T.C. (2000). Evaluation of using dog as an animal model to study the fraction oral dose absorbed of 43 drugs in humans. *Pharmaceutical Research*, *17*, 135–140.
45. Hoffman, D.J., Seifert, T., Borre, A., Nellans, H.N. (1995). Method to estimate the rate and extent of intestinal absorption in conscious rats using an absorption probe and portal blood sampling. *Pharmaceutical Research*, *12*, 889–894.
46. Schinkel, A.H., Wagenaar, E., van Deemter, L., Mol, C.A., Borst, P. (1995). Absence of the mdr1a P-glycoprotein in mice affects tissue distribution and pharmacokinetics of dexamethasone, digoxin, and cyclosporin A. *The Journal of Clinical Investigation*, *96*, 1698–1705.
47. Zamek-Gliszczynski, M.J., Bedwell, D.W., Bao, J.Q., Higgins, J.W. (2012). Characterization of SAGE Mdr1a (P-gp), Bcrp, and Mrp2 knockout rats using loperamide, paclitaxel, sulfasalazine, and carboxydichlorofluorescein pharmacokinetics. *Drug Metabolism and Disposition*, *40*, 1825–1833.
48. Fagerholm, U., Johansson, M., Lennernäs, H. (1996). Comparison between permeability coefficients in rat and human jejunum. *Pharmaceutical Research*, *13*, 1335–1341.
49. Kim, J.S., Mitchell, S., Kijck, P., Tsume, Y., Hilfinger, J., Amidin, G.L. (2006). The suitability of an in situ perfusion model for permeability determinations: Utility for BCS class I biowaivers requests. *Molecular Pharmaceutics*, *3*, 686–690.
50. Lacombe, O., Woodley, J., Solleux, C., Delbos, J.M., Boursier-Neyret, C., Houin, G. (2004). Localisation of drug permeability along the rat small intestine, using markers of the paracellular, transcellular and some transporter routes. *European Journal of Pharmaceutical Sciences*, *23*, 385–391.
51. Ruan, L.P., Chen, S., Yu, B.Y., Zhu, D.N., Cordell, G.A., Qui, S.X. (2006). Prediciton of human absorption of natural compounds by the non-everted rat intestinal model. *European Journal of Medicinal Chemistry*, *41*, 605–610.
52. Ungell, A.L., Nylander, S, Bergstrand, S., Sjoberg, A., Lennernäs, H. (1998). Membrane transport of drugs in different regions of the intestinal tract of the rat. *Journal of Pharmaceutical Sciences*, *87*, 360–366.
53. Perternel, L., Kristan, K., Petruševska, M., Rižner, T.L., Legen, I. (2012). Suitability of isolated rat jejunum model for demonstration of complete absorption in humans for BCS-based biowaivers request. *Journal of Pharmaceutical Sciences*, *101*, 1436–1449.
54. Kerns, E.H., Di, L., Petusky, S., Farris, M., Ley, R., Jupp, P. (2004). Combined application of parallel artificial membrane permeability assay and Caco-2 permeability assays in drug discovery. *Journal of Pharmaceutical Sciences*, *93*, 1440–1453.

55. Chu, X., Zhang, Z., Yabut, J., Horwitz, S., Levorse, J., Li, X.Q., Zhu, L., Lederman, H., Ortiga, R., Strauss, J., Li, X., Owens, K.A., Dragovic, J., Vogt, T., Evers, R., Shin, M.K. (2012). Characterization of multidrug resistance 1a/P-glycoprotein knockout rats generated by zinc finger nucleases. *Molecular Pharmacology*, *81*, 220–227.
56. Masungi, C., Mensch, J., Van Dijck, A., Borremans, C., Willems, B., Mackie, C., Noppe, M., Brewster, ME. (2008). Parallel artificial membrane permeability assay (PAMPA) combined with a 10-day multiscreen Caco-2 cell culture as a tool for assessing new drug candidates. *Pharmazie*, *63*, 194–199.
57. Pardridge, W.M. (2001). Crossing the blood-brain barrier: Are we getting it right? *Drug Discovery Today*, *6*, 1–2.
58. Smith, K. (2011). Trillion-dollar brain drain. *Nature*, *478*, 15.
59. Nicholson, C., Sykova, E. (1998). Extracellular space structure revealed by diffusion analysis. *Trends in Neuroscience*, *21*, 207–215.
60. Clark, D.E. (2003). In silico prediction of blood-brain barrier permeation. *Drug Discovery Today*, *8*, 927–933.
61. Liu, X., Tu, M., Kelly, R.S., Chen, C., Smith, B.J. (2004). Development of a computational approach to predict blood-brain barrier permeability. *Drug Metabolism and Disposition*, *32*, 132–139.
62. Guo, X., Geng, M., Du, G. (2005). Glucose transporter 1, distribution in the brain and in neural disorders: Its relationship with transport of neuroactive drugs through the blood-brain barrier. *Biochemical Genetics*, *43*, 175–187.
63. Killian, D.M., Chikhale, P.J. (2001). Predominant functional activity of the large, neutral amino acid transporter (LAT1) isoform at the cerebrovasculature. *Neuroscience Letters*, *306*, 1–4.
64. Ito, K., Uchida, Y., Ohtsuki, S., Aizawa, S., Kawakami, H., Katsukura, Y., Kamiie, J., Terasaki, T. (2011). Quantitative membrane protein expression at the blood-brain barrier of adult and younger cynomolgus monkeys. *Journal of Pharmaceutical Sciences*, *100*, 3939–3950.
65. Nishimura, M., Naito, S. (2005). Tissue-specific mRNA expression profiles of human ATP-binding cassette and solute carrier transporter superfamilies. *Drug Metabolism and Pharmacokinetics*, *20*, 452–477.
66. Cordon-Cardo, C., O'Brien, J.P., Casals, D., Rittman-Grauer, L., Biedler, J.L., Melamed, M.R., Bertino, J.R. (1989). Multidrug-resistance gene (P-glycoprotein) is expressed by endothelial cells at blood-brain barrier sites. *Proceeding of the National Academy of Science USA*, *86*, 695–698.
67. Demeule, M., Regina, A., Jodoin, J., Laplante, A., Dagenais, C., Berthelet, F., Moghrabi, A., Beliveau, R. (2002). Drug transport to the brain: Key roles for the efflux pump P-glycoprotein in the blood-brain barrier. *Vascular Pharmacology*, *38*, 339–348.
68. Breedveld, P., Pluim, D., Cipriani, G., Wielinga, P., van Tellingen, O., Schinkel, A.H., Schellens, J.H. (2005). The effect of Bcrp1 (Abcg2) on the in vivo pharmacokinetics and brain penetration of imatinib mesylate (Gleevec): Implications for the use of breast cancer resistance protein and P-glycoprotein inhibitors to enable the brain penetration of imatinib in patients. *Cancer Research*, *65*, 2577–2582.
69. Aronica, E., Gorter, J.A., Jansen, G.H., van Veelen, C.W., van Rijen, P.C., Leenstra, S., Ramkema, M., Scheffer, G.L., Scheper, R.J., Troost, D. (2003). Expression and cellular distribution of multidrug transporter proteins in two major causes of medically intractable epilepsy: Focal cortical dysplasia and glioneuronal tumors. *Neuroscience*, *118*, 417–429.

70. Liu, X., Vilenski, O., Kwan, J., Apparsundaram, S., Weikert, R. (2009). Unbound brain concentration determines receptor occupancy: A correlation of drug concentration and brain serotonin and dopamine reuptake transporter occupancy for eighteen compounds in rats. *Drug Metabolism and Disposition*, *37*, 1548–1556.

71. Goodwin, J.T., Clark, D.E. (2005). *In silico* predictions of blood-brain barrier penetration: Considerations to "keep in mind". *Journal of Pharmacology and Experimental Therapeutics*, *315*, 477–483.

72. Mensch, J., Oyarzabal, J., Mackie, C., Augustijns, P. (2009). *In vivo, in vitro* and *in silico* methods for small molecule transfer across the BBB. *Journal of Pharmaceutical Sciences*, *98*, 4429–4468.

73. Oyler, G.A., Duckrow, R.B., Hawkins, R.A. (1992). Computer simulation of the blood-brain barrier: A model including two membranes, blood flow, facilitated and non-facilitated diffusion. *Journal of Neuroscience Methods*, *44*, 179–196.

74. Kim, C.S., Ross, I.A., Sandberg, J.A., Preston, E. (1998). Quantitative low-dose assessment of seafood toxin, domoic acid, in the rat brain: Application of physiologically-based pharmacokinetic (PBPK) modeling. *Environmental Toxicology and Pharmacology*, *6*, 49–58.

75. Liu, X., Kim, C.S., Ross, I.A., Sandberg, J.A., Preston, E. (1998). Quantitative low-dose assessment of seafood toxin, domoic acid, in the rat brain: Application of physiologically-based pharmacokinetic (PBPK) modelling. *Environmental Toxicology and Pharmacology*, *6*, 49–58.

76. Liu, X., Smith, B.J., Chen, C., Callegari, E., Becker, S.L., Chen, X., Cianfrogna, J., Doran, A.C., Doran, S.D., Gibbs, J.P., Hosea, N., Liu, J., Nelson, F.R., Szewc, M.A., Van Deusen, J. (2005). Use of a physiologically based pharmacokinetic model to study the time to reach brain equilibrium: An experimental analysis of the role of blood-brain barrier permeability, plasma protein binding, and brain tissue binding. *Journal of Pharmacology and Experimental Therapeutics*, *313*, 1254–1262.

77. Fenneteau, F., Turgeon, J., Couture, L., Michaud, V., Li, J., Nekka, F. (2009). Assessing drug distribution in tissues expressing P-glycoprotein through physiologically based pharmacokinetic modeling: Model structure and parameters determination. *Theoretical Biology and Medical Modelling*, *6*, 2.

78. Mensch, J., Jaroskova, L., Sanderson, W., Melis, A., Mackie, C., Verreck, G., Brewster, M.E., Augustijns, P. (2010). Application of PAMPA-models to predict BBB permeability including efflux ratio, plasma protein binding and physicochemical parameters. *International Journal of Pharmaceutics*, *395*, 182–197.

79. Brown, R.C., Morris, A.P., O'Neil, R.G. (2007). Tight junction protein expression and barrier properties of immortalized mouse brain microvessel endothelial cells. *Brain Research*, *1130*, 17–30.

80. Mahar D.,K.M., Humphreys, J.E., Webster, L.O., Wring, S.A., Shampine, L.J., Serabjit-Singh, C.J., Adkison, K.K., Polli, J.W. (2002). Passive permeability and P-glycoprotein-mediated efflux differentiate central nervous system (CNS) and non-CNS marketed drugs. *Journal of Pharmacology and Experimental Therapeutics*, *303*, 1029–1037.

81. Kalvass, J.C., Maurer, T.S. (2002). Influence of nonspecific brain and plasma binding on CNS exposure: Implications for rational drug discovery. *Biopharmaceutics and Drug Disposition*, *23*, 327–338.

82. Liu, X., Smith, B.J., Chen, C., Callegari, E., Becker, S.L., Chen, X., Cianfrogna, J., Doran, A.C., Doran, S.D., Gibbs, J.P., Hosea, N., Liu, J., Nelson, F.R., Szewc, M.A.,

Van Deusen, J. (2006). Evaluation of cerebrospinal fluid concentration and plasma free concentration as a surrogate measurement for brain free concentration. *Drug Metabolism and Disposition*, *34*, 1443–1447.

83. Kakee, A., Terasaki, T., Sugiyama, Y. (1996). Brain efflux index as a novel method of analyzing efflux transport at the blood-brain barrier. *Journal of Pharmacology and Experimental Therapeutics*, *277*, 1550–1559.
84. Gupta, A., Chatelain, P., Massingham, R., Jonsson, E.N., Hammarlund-Udenaes, M. (2006). Brain distribution of cetirizine enantiomers: Comparison of three different tissue-to-plasma partition coefficients: K_p, $K_{p,u}$, and $K_{p,uu}$. *Drug Metabolism and Disposition*, *34*, 318–323.
85. Friden, M., Bergstrom, F., Wan, H., Rehngren, M., Ahlin, G., Hammarlund-Udenaes, M., Bredberg, U. (2011). Measurement of unbound drug exposure in brain: Modeling of pH partitioning explains diverging results between the brain slice and brain homogenate methods. *Drug Metabolism and Disposition*, *39*, 353–362.
86. Ohno, K., Pettigrew, K.D., Rapoport, S.I. (1978). Lower limits of cerebrovascular permeability to nonelectrolytes in the conscious rat. *American Journal of Physiology*, *235*, H299–307.
87. Smith, Q.R. (1996). Brain perfusion systems for studies of drug uptake and metabolism in the central nervous system. *Pharmaceutical Biotechnology*, *8*, 285–307.
88. Schinkel, A.H., Smit, J.J., van Tellingen, O., Beijnen, J.H., Wagenaar, E., van Deemter, L., Mol, C.A., van der Valk, M.A., Robanus-Maandag, E.C., te Riele, H.P., Berns, A.J.M., Borst, P. (1994). Disruption of the mouse *mdr1a* P-glycoprotein gene leads to a deficiency in the blood-brain barrier and to increased sensitivity to drugs. *Cell*, *77*, 491–502.
89. Lin, J.H., Yamazaki, M. (2003). Clinical relevance of P-glycoprotein in drug therapy. *Drug Metabolism Reviews*, *35*, 417–454.
90. Endres, C.J., Hsiao, P., Chung, F.S., Unadkat, J.D. (2006). The role of transporters in drug interactions. *European Journal of Pharmaceutical Science*, *27*, 501–517.
91. Katoh, M., Suzuyama, N., Takeuchi, T., Yoshitomi, S., Asahi, S., Yokoi, T. (2006). Kinetic analyses for species differences in P-glycoprotein-mediated drug transport. *Journal of Pharmaceutical Sciences*, *95*, 2673–2683.
92. Yamazaki, M., Neway, W.E., Ohe, T., Chen, I., Rowe, J.F., Hochman, J.H., Chiba, M., Lin, J.H. (2001). In vitro substrate identification studies for p-glycoprotein-mediated transport: Species difference and predictability of in vivo results. *Journal of Pharmacology and Experimental Therapeutics*, *296*, 723–735.
93. Zolnerciks, J.K., Booth-Genthe, C.L., Gupta, A., Harris, J., Unadkat, J.D. (2011). Substrate- and species-dependent inhibition of P-glycoprotein-mediated transport: Implications for predicting in vivo drug interactions. *Journal of Pharmaceutical Sciences*, *100*, 3055–3061.
94. Uchida, Y., Ohtsuki, S., Katsukura, Y., Ikeda, C., Suzuki, T., Kamiie, J., Terasaki, T. (2011). Quantitative targeted absolute proteomics of human blood-brain barrier transporters and receptors. *Journal of Neurochemistry*, *117*, 333–345.
95. Hsiao, P., Bui, T., Ho, R.J., Unadkat, J.D. (2008). In vitro-to-in vivo prediction of P-glycoprotein-based drug interactions at the human and rodent blood-brain barrier. *Drug Metabolism and Disposition*, *36*, 481–484.
96. Luurtsema, G., Molthoff, C.F., Windhorst, A.D., Smit, J.W., Keizer, H., Boellaard, R., Lammertsma, A.A., Franssen, E.J. (2003). (R)- and (S)-[^{11}C]verapamil as PET-tracers

for measuring P-glycoprotein function: *In vitro* and *in vivo* evaluation. *Nuclear Medicine and Biology*, *30*, 747–751.

97. Agarwal, S., Hartz, A.M., Elmquist, W.F., Bauer, B. (2011). Breast cancer resistance protein and P-glycoprotein in brain cancer: Two gatekeepers team up. *Current Pharmaceutical Design*, *17*, 2793–2802.
98. Patsalos, P.N., Alavijeh, M.S., Semba, J., Lolin, Y.I. (1992). A freely moving and behaving rat model for the chronic and simultaneous study of drug pharmacokinetics (blood) and neuropharmacokinetics (cerebrospinal fluid): Hematological and biochemical characterization and kinetic evaluation using carbamazepine. *Journal of Pharmacological and Toxicological Methods*, *28*, 21–28.
99. Lin, J.H. (2008). CSF as a surrogate for assessing CNS exposure: An industrial perspective. *Current Drug Metabolism*, *9*, 46–59.
100. Walker, M.C., Tong, X., Perry, H., Alavijeh, M.S., Patsalos, P.N. (2000). Comparison of serum, cerebrospinal fluid and brain extracellular fluid pharmacokinetics of lamotrigine. *British Journal of Pharmacology*, *130*, 242–248.
101. Liu, X., Van Natta, K., Yeo, H., Vilenski, O., Weller, P.E., Worboys, P.D., Monshouwer, M. (2009). Unbound drug concentration in brain homogenate and cerebral spinal fluid at steady state as a surrogate for unbound concentration in brain interstitial fluid. *Drug Metabolism and Disposition*, *37*, 787–793.
102. Kornhuber, J., Weigmann, H., Rohrich, J., Wiltfang, J., Bleich, S., Meineke, I., Zochling, R., Hartter, S., Riederer, P., Hiemke, C. (2006). Region specific distribution of levomepromazine in the human brain. *Journal of Neural Transmission*, *113*, 387–397.
103. Taber, M.T., Wright, R.N., Molski, T.F., Clarke, W.J., Brassil, P.J., Denhart, D.J., Mattson, R.J., Lodge, N.J. (2005). Neurochemical, pharmacokinetic, and behavioral effects of the novel selective serotonin reuptake inhibitor BMS-505130. *Pharmacology, Biochemistry and Behavior*, *80*, 521–528.
104. Schmidt, K.C., Turkheimer, F.E. (2002). Kinetic modeling in positron emission tomography. *Quarterly Journal of Nuclear Medicine*, *46*, 70–85.
105. Kelder, J., Grootenhuis, P.D., Bayada, D.M., Delbressine, L.P., Ploemen, J.P. (1999). Polar molecular surface as a dominating determinant for oral absorption and brain penetration of drugs. *Pharmaceutical Research*, *16*, 1514–1519.
106. Kalvass, J.C., Maurer, T.S., Pollack, G.M. (2007). Use of plasma and brain unbound fractions to assess the extent of brain distribution of 34 drugs: Comparison of unbound concentration ratios to in vivo p-glycoprotein efflux ratios. *Drug Metabolism and Disposition*, *35*, 660–666.
107. Hammarlund-Udenaes, M., Bredberg, U., Friden, M. (2009). Methodologies to assess brain drug delivery in lead optimization. *Current Topics in Medicinal Chemistry*, *9*, 148–162.
108. Cole, S., Bagal, S., El-Kattan, A., Fenner, K., Hay, T., Kempshall, S., Lunn, G., Varma, M., Stupple, P., Speed, W. (2012). Full efficacy with no CNS side-effects: Unachievable panacea or reality? DMPK considerations in design of drugs with limited brain penetration. *Xenobiotica*, *42*, 11–27.

14

INTEGRATED ASSESSMENT OF DRUG CLEARANCE AND CROSS-SPECIES SCALABILITY

Kevin Beaumont, James R. Gosset, and Chris E. Keefer

14.1 PHARMACOKINETIC CONSIDERATIONS OF CLEARANCE

Most drugs are not administered directly to their site of action. To show efficacy, a drug must move to the site of action, usually via some absorption process and transport in the blood. It is relatively unusual for a drug to be measured at its site of action because this may often be a tissue that is not accessible to sampling in humans. Pharmacokinetics relates the time course of drug concentration in tissues to those in blood. The clearance of a molecule is defined as the amount of blood cleared of drug molecule per unit time (units of mL/min/kg). For an intravenous administration, it is related to the dose of a drug, as shown in Equation 14.1:

$$\text{Clearance} = \frac{\text{Dose}}{\text{AUC}} \tag{14.1}$$

where AUC is the area under the plasma concentration versus the time curve.

The AUC is a measure of exposure to a drug. Combined with potency versus the target, exposure will define how well a drug will influence its target. For a given

Predictive ADMET: Integrative Approaches in Drug Discovery and Development, First Edition.
Edited by Jianling Wang and Laszlo Urban.

dose, the defining parameter for AUC is drug clearance. An extensively cleared drug exhibits lower exposure for a given dose than a slowly cleared drug.

For oral administration, clearance by the liver (hepatic clearance) plays a major role in defining oral bioavailability. All of the blood from the gut empties into the hepatic portal vein, which transports blood directly to the liver. Any drug that is absorbed from the gastrointestinal (GI) tract is subject to hepatic clearance in a process called first-pass extraction. Any drug extracted on hepatic first pass will not appear in the systemic circulation. Thus, the amount of drug removed by the liver on a single pass of blood through it will partially define the drug oral bioavailability. Consequently, the hepatic clearance ($\mathrm{CL_h}$) is a major determinant of oral bioavailability (Equation 14.2):

$$F = fa \times \left(1 - \frac{\mathrm{CL_h}}{Q_h}\right) \tag{14.2}$$

where F is the fraction bioavailable, fa is the fraction absorbed, and Q_h is hepatic blood flow.

Clearance is a parameter that can be related to the blood flow to an organ. Although organ blood flows in a particular species are fixed, drug clearance values vary widely. Clearance of a particular drug by an organ is related to organ blood flow, plasma protein (or blood) binding (ppb), and unbound intrinsic clearance ($\mathrm{CL_{int}}$) of the drug. These are related to the liver by the "well-stirred model" equation [1]:

$$\mathrm{CL_h} = Q_h \times \frac{(fu \times \mathrm{CL_{int}})}{Q_h + (fu \times \mathrm{CL_{int}})} \tag{14.3}$$

Unbound intrinsic clearance of a drug is the theoretical maximal clearance of a drug by an organ without flow and binding limitations. The units are the same as clearance (mL/min/kg). This is different from apparent intrinsic clearance ($\mathrm{CL_{intapp}}$) of a drug, which is related to the rate of compound metabolism and has the units μL/min/mg protein.

To define the clearance of a drug, a pharmacokinetic study should be completed after intravenous administration with blood concentration measurement. Such studies are rarely completed in humans. The blood is often centrifuged to produce plasma; in which case, the plasma clearance should be quoted ($\mathrm{CL_p}$). This assumes that the partition of a drug between plasma and blood cells is unity. However, this is not always the case and it is advisable to measure partitioning before comparing with blood flows. As only unbound drug is available to exert an effect [2], the clearance of unbound drug ($\mathrm{CL_{un}}$) is often quoted as the plasma or blood clearance corrected for plasma or blood binding. Finally, the oral clearance ($\mathrm{CL_o}$ or $\mathrm{CL_{po}}$) is derived from the AUC after oral administration and is the intravenous clearance divided by the fraction bioavailable.

The importance of clearance cannot be underestimated. It has implications for dose and duration of action. Its role in defining dose was recognized [3] with the steady-state dose equation (Equation 14.4):

$$\frac{\text{Dose}}{\text{tau}} = \frac{\text{Css, avg} \times \text{CL}}{F} \tag{14.4}$$

where Css, avg is the average drug concentration over the dose interval, tau is the dose interval, CL is the clearance, and F is the fraction bioavailable.

For hepatically cleared compounds given by the oral route, this can be rearranged to Equation 14.5, relating dose to steady-state unbound concentration, intrinsic clearance, dose interval, and fraction absorbed (Equation 14.5):

$$\text{Dose} = \frac{\text{Css, avg, un} \times \text{CL}_{\text{int}} \times \text{tau}}{fa} \tag{14.5}$$

The dose of a drug can be determined from potency against the pharmacological target (C_{eff}), unbound intrinsic clearance (Cl_{int}), fraction absorbed, and how often the drug will be administered.

It is a misconception that plasma protein binding (ppb) is an important parameter to modulate in a drug discovery program and that "reducing plasma protein binding" is often a goal of such projects [4]. Equation 14.5 suggests that ppb does not influence dose (no *fu* parameter) and that the important parameters to modulate are potency (unbound concentration) and unbound intrinsic clearance.

Clearance is also an important determinant of duration of action through the effect on elimination half-life (Equation 14.6):

$$\text{Half Life} = \frac{0.693 \times Vd}{\text{CL}} \tag{14.6}$$

The half-life of a drug determines the peak-to-trough ratio for a dose regimen. Avoiding high peak-to-trough ratios is an important consideration for a drug because large peak-to-trough ratios will require a significant therapeutic index. Ppb has no effect on half-life (no *fu* parameter in Equation 14.5) as lowering ppb will increase the volume of distribution and clearance to the same extent (unless *Vd* is limited to plasma volume) [4].

14.2 PHYSIOLOGICAL CONSIDERATIONS OF CLEARANCE

The liver and kidney are major organs of drug clearance. In humans, blood flow to each organ is similar [5], although the organs clear drugs very differently.

The kidney removes drugs from blood by passive and active processes. Passive removal involves filtration of blood through the glomerulus. Unbound drug passes through the glomerulus into the proximal convoluted tubule. Although drugs showing passive permeability are reabsorbed back into blood across the cells of the proximal convoluted tubule epithelium, these are not excreted unchanged in urine to any major extent. Drugs with limited passive permeability are retained inside the proximal convoluted tubule and the remaining length of the nephron to be excreted unchanged in urine at a rate that is equivalent to the glomerular filtration rate (GFR) or 1 to 2 mL/min/kg in humans [5].

Drugs can also be actively transported from blood into the proximal convoluted tubule and can exhibit renal clearance values of an unbound drug that approach renal blood flow (18 mL/min/kg) [5] exceeding GFR clearance.

Hepatic clearance can be categorized into metabolism and biliary elimination. The liver is the major organ of drug metabolism and expresses a battery of drug-metabolizing enzymes, such as cytochromes P450 (CYPs), UDP glucuronyltransferases, and sulfotransferases. The hepatocyte also expresses drug transporters on the basolateral (blood) membrane that can transport drugs into the hepatocyte. At the canalicular (biliary) membrane, further transporters can efflux drug molecules into the bile.

Drug clearance is a composite of several different processes and can be split into CL_{met}, CL_{bile}, and CL_{renal}.

14.3 CLEARANCE PATHWAYS FOR MARKETED DRUGS

Two thirds of the top 200 marketed agents in 2002 were predominantly cleared by metabolism (Figure 14.1), with 25% cleared in the urine and the remainder in bile [6]. For metabolism cleared compounds, two thirds were cleared by CYPs and half by CYP3A4. Most small-molecule drugs are administered by the oral route. The physicochemical parameters for oral drug delivery have been extensively investigated [7] and include a balance of lipophilicity, aqueous solubility, molecular weight, and hydrogen bonding potential. Pharmacological potency is improved by increasing lipophilic interactions with the target. For every methyl group added to an active

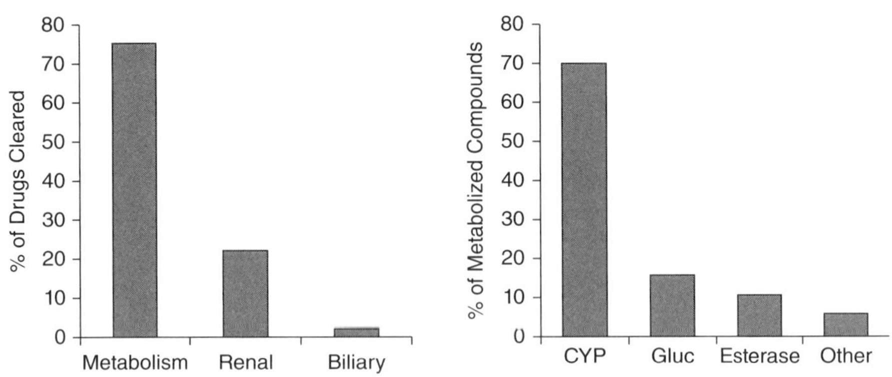

Figure 14.1 Clearance mechanisms of the top 200 marketed drugs (2002). (Adapted from Reference 6.)

scaffold, an additional 1.36 kcal/mole of binding energy (i.e., increased potency) is achieved. Thus, optimization on potency can lead to high molecular weight lipophilic molecules that are vulnerable to CYP metabolism.

Many drug discovery projects focus on addressing CYP-mediated metabolism to reduce metabolic clearance. However, CYP metabolism is one of several potential clearance mechanisms for drugs. Focusing solely on CYP metabolism has negative consequences and can lead to misprediction of human clearance. In a drug discovery project, a thorough understanding of physicochemistry and structure is required as it relates to drug clearance and is essential for success.

14.4 OVERVIEW OF THE INTEGRATION OF CLEARANCE ASSESSMENT INTO DRUG DISCOVERY

Selecting molecules with optimal human clearance is a vital component of a small-molecule discovery project. Combined with balancing potency, this will be important for dose and duration of action.

Successful discovery projects will use all the tools available for clearance optimization. The philosophy must be to optimize clearance in humans. Clearance of compound series in preclinical species is only interesting to confirm *in vitro–in vivo* correlation (IVIVC). It is important that all experiments must target understanding of human clearance.

14.5 *IN SILICO* METHODS FOR CLEARANCE IN DRUG DISCOVERY

In silico methods like metabolic site of metabolism (SOM) prediction can be used to assist in identifying the part of the molecule being metabolized and novel approaches using the pairwise analysis of large metabolic screening databases can suggest structural changes that will improve overall clearance. In addition, ideas can be prioritized via physical property trend analysis and clearance prediction using statistical modeling techniques.

14.5.1 Site of Metabolism Prediction

In vitro metabolic clearance assays measure the extent of metabolism for compounds. They do not provide any information about how to reduce the metabolic liability. One approach for doing this is to identify computationally, sites in the molecule where metabolic modification occurs. Knowledge of these sites enables strategies to address the issue like metabolic blocking or substitution with less reactive substituents. There are several computational tools and approaches to identify these sites of metabolism including Metasite [8], StarDrop [9], SMARTCyp [10], and RS-Predictor [11]. These tools use a mixture of techniques including two- (2D) and three-dimensional (3D) active site interaction estimations, chemical reactivity calculations, and quantitative structure activity relationship (QSAR) models to identify the atoms in the molecule that are most likely to undergo metabolism. The output is a rank-ordered list of vulnerable sites. Predicted sites of metabolism methods have varying degrees of success. They can identify the correct major site of metabolism in the first one to three ranked sites ~60–80% of the time [12]. They work better on well-understood mechanisms (e.g., CYP) and less well on complex metabolic mechanisms (e.g., phase I and phase II metabolism). This level of performance can be useful when it is not clear from structure alone where the major metabolism is occurring allowing one to hypothesize the SOM and most likely metabolites that can be confirmed experimentally. A limitation to predicted sites of metabolism methods is that in the case of a misprediction, they may cause project teams to focus efforts and design cycles on the wrong parts of the molecule.

14.5.2 Pairwise Analysis for Idea Generation

Although sites of metabolism prediction tools can successfully identify key parts of the molecule driving metabolic clearance, they do not offer any assistance in what changes could be made to the molecule to reduce the liability. Solutions have been based on experience, site blocking techniques, and physical property rule-of-thumb approaches (i.e., reduce overall lipophilicity). Advances in the processing of large metabolic clearance databases have led to the ability to perform full pairwise analysis of these datasets [13]. These methods identify molecular match pairs (MMPs) that are pairs of compounds that differ by only a small structural change.

Analyses of MMPs that have the same structural change can lead to the identification of transforms that can reduce clearance liabilities. Figure 14.2 shows an example of two transforms from the Pfizer (New York, NY) database that can significantly reduce CYP clearance.

Using this data, a project can start with a molecule with high clearance and identify parts of the molecule to modify, and the algorithm will then fragment that area and search the database for transforms involving the fragment that have previously led to reductions in clearance. It will then apply the transformation and run the new virtual

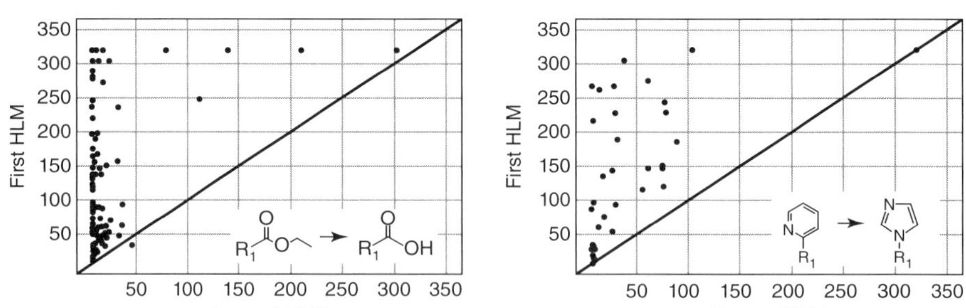

Figure 14.2 Matched molecular pair (MMP) data for two transforms that may improve HLM clearance. The *y*-axis is the HLM clearance for the compound with the first fragment in the transform, and the *x*-axis is the HLM clearance for the compound with the second fragment in the transform.

molecule through *in silico* predictive models to make a prediction of clearance. This prediction can then be used to increase confidence in the proposed change. Figure 14.3 shows an example that aims to replace the cyclopropyl group in the compound in row 2. Five transformations are highlighted with one (row 1) showing an increase in clearance and four (rows 3–6) showing a decrease in clearance. For these cases, the model predictions agree with the direction of the MMP results with some differences in the magnitude of the changes. These changes also agree with the actual clearance data, again with differences in their magnitude.

This methodology is one of the first examples of a technique that can propose structural changes based on real transforms to address clearance capitalizing on the knowledge gained from clearance measurements across different projects and chemical series.

14.5.3 Physical Property Trends

Another approach used to improve clearance for discovery projects is modification of lipophilicity. There is a relationship between lipophilicity and microsomal clearance [14]. One strategy is to reduce the lipophilicity of compounds with high clearance by adding polar groups in parts of the molecule where the change will not affect potency. Although there is a positive relationship between lipophilicity and potency, and reduction in lipophilicity can result in a decrease in clearance, there are also exceptions to the lipophilicity/clearance relationship that may be overlooked with this approach. CYP-mediated clearance is a function of general physical properties and specific molecular interaction between substrate and enzyme [15, 16]. Figure 14.4 shows the relationship between measured CL_{intapp} and calculated Log $D_{(7.4)}$ (MoKa version 1.1.0 [17]). There is a positive correlation, but there is also significant variation at any given lipophilicity.

14.5.4 Metabolic Clearance Prediction

Predictive models for clearance are used in discovery to prioritize compounds for synthesis that have desired clearance profiles. *In vitro* clearance involves a mixture of mechanisms and require statistical machine learning approaches that deal with complex nonlinear data using descriptors that capture physicochemical properties as well as specific interactions between substrates and enzymes to perform well. Some methods classify metabolic liability into a small number of categories (i.e., high, medium, or low) and others generate point estimates of an *in vitro* or *in vivo* clearance value.

An advantage of classification models is that they remove the need to learn the subtle differences in activity for compounds where those differences do not matter to the final interpretation of how the data will be used. A disadvantage is that the model is being asked to learn a boundary between classes that likely has no chemical cause. For example, if we said that compounds with CL_{intapp} values below 50 μL/min/mg are good and those ≥50 are bad, the model is being asked to make this subtle distinction.

Structure	Transform	ROW_NUMBER	Num_MMPairs	Avg_MMPChang...	Pctinc	PctDec	PctSame	Pred_HLM	Exp_HLM
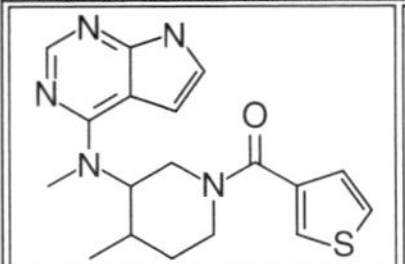		Row 1	4	67.1359	75%	0%	25%	83.4	46
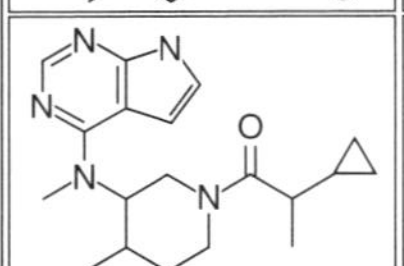		Row 2	0	0.0				93.1	70
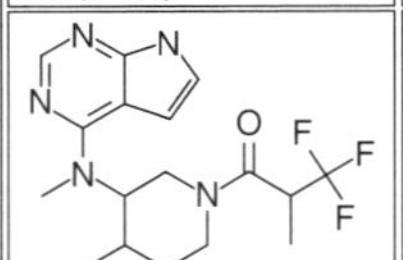	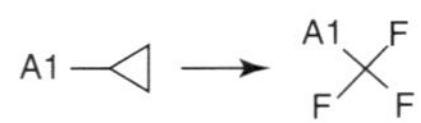	Row 3	414	−11.4401	15%	36%	47%	24.6	16
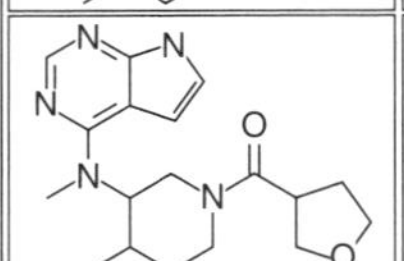		Row 4	86	−31.6044	10%	62%	26%	40.0	11
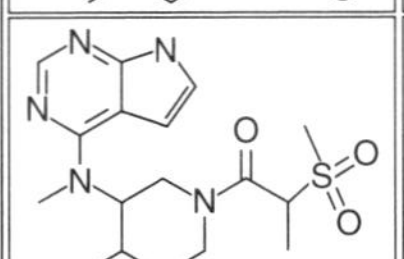	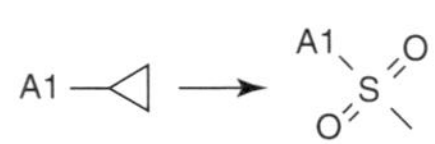	Row 5	22	−43.3355	4%	63%	27%	8.02	8
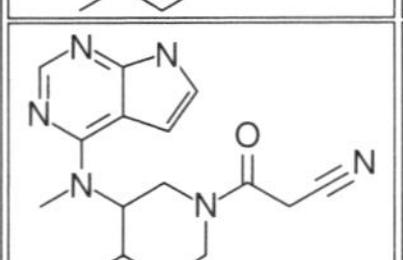		Row 6	77	−46.1378	9%	67%	23%	8.43	10

Figure 14.3 Example of pairwise transforms for the cyclopropyl moiety in the JAK inhibitor shown in row 2 ordered by pairwise change in $CLint_{app}$. *NumMMPPairs* is the number of MMPs in the database, and the *PctInc/PctDec/PctSame* columns are the percentage of pairs that increase, decrease, and make no change to the $CLint_{app}$. The *Pred_HLM* is the statistical model prediction for new compounds, and *Exp_HLM* is the experimentally determined value for the compound.

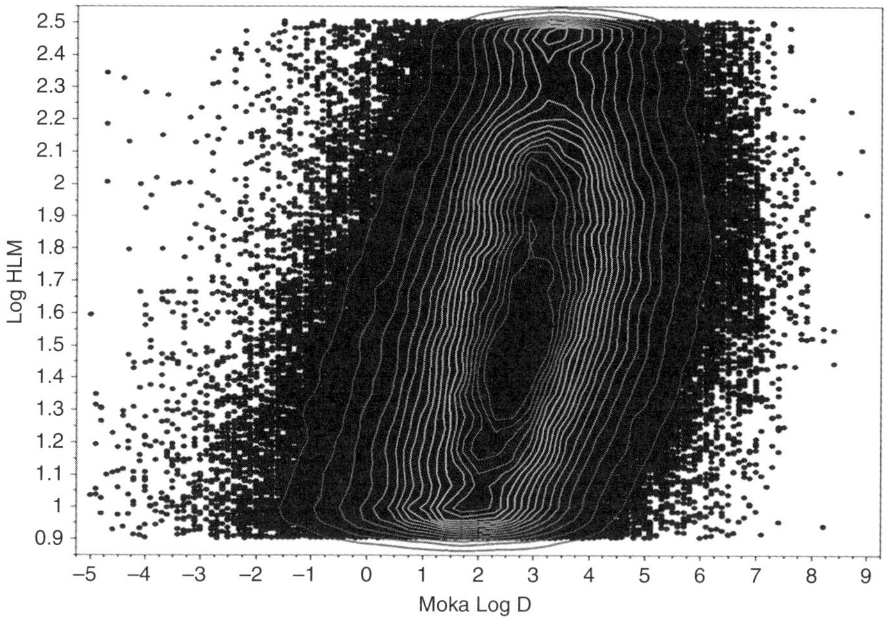

Figure 14.4 Log (HLM $CLint_{app}$) versus calculated Log $D_{(7.4)}$ values for a set of discovery compounds showing weak correlation and high variability (for any given value of HLM, the LogD spans > ±2 units). Colored lines are 5% nonparametric density lines. (*See insert for color representation of the figure.*)

And why 50, and not another value? The result are models that perform well for compounds at the extremes of CL_{intapp} values but struggle with compounds near the classification boundary. Recent examples include work by Hu et al. [18] and Lee et al. [19].

Regression models make an endpoint prediction within the range of values being modeled. They allow the relative ranking of compounds based on the predicted values, but they struggle with predictions that are near the limits of the data (overprediction of low values and underprediction of high values). Published examples include neural network and support vector machine (SVM) models [20] and partial least-squares (PLS) and random forest (RF) models for total human plasma clearance [21].

Prediction confidence for regression models typically involves applicability domain (AD) approaches or prediction error estimates. Recently, a new method has been developed that combines distance in chemistry space with distance in activity space and has been shown to work very well for HLM [22].

14.6 *IN VITRO* CLEARANCE ASSAYS

Clearance is a complicated parameter and can be made up of $CL_{metabolism}$, $CL_{biliary}$, and CL_{renal}. No single *in vitro* assay can fully evaluate the human clearance of a molecule. The choice of which *in vitro* clearance assay is a judgment based on the structure and physicochemistry of a molecule and the most likely clearance mechanism.

The use of *in vitro* assays has revolutionized drug discovery. It is now possible to evaluate significant numbers of individual molecules in predictive human and preclinical species *in vitro* clearance assays to triage and select the most appropriate compounds to take forward to *in vivo* experiments.

14.7 *IN VITRO* METABOLIC CLEARANCE ASSAYS

The major hepatic *in vitro* assay systems are isolated hepatocytes, microsomes, and S9 fractions.

Hepatocytes are liver cells with the advantage that they contain all the drug metabolizing enzymes present in the liver. S9 fraction and microsomes are formed by homogenization of liver followed by differential centrifugation. The first low-speed spin (usually 9000g) removes the large cellular organelles (such as mitochondria and the nucleus) from the homogenate. The supernatant produced is the S9 fraction. A high-speed spin (usually 100,000g) brings down the endoplasmic reticulum. These membranes tend to form into spheres called microsomes, which express all of the membrane-bound enzymes, such as CYPs and UDP-glucuronyltransferases.

14.8 HOW TO USE *IN VITRO* METABOLIC CLEARANCE ASSAYS

The use of *in vitro* metabolic clearance assays has been developed over the past few decades and is based on Michaelis–Menten kinetics. The overall method is well described by Obach [23].

In an *in vitro* assay, for a fixed enzyme concentration, if the substrate concentration is well below the Km for the enzyme (i.e., $[S] \ll Km$), then the rate of metabolism is in the linear portion of the curve and should obey first-order principles. A plot of substrate concentration versus time will be an exponential decline and log transformation will produce a straight line, with a slope of k (the rate constant of metabolism). This can be converted to the CL_{intapp} for metabolic decline using Equation 14.7:

$$CL_{intapp} = \frac{V\,\max}{\mathrm{Km}} = \frac{|\mathrm{Slope}|}{\mathrm{cell\ (protein)conc}} \qquad (14.7)$$

When using microsomes, the units of CL_{intapp} are μL/min/mg protein, whereas in hepatocytes, the units are μL/min/10^6 cells.

14.8.1 Scaling *In Vitro* Metabolic Clearance to *In Vivo* Clearance

To predict the *in vivo* clearance of a compound in a species, a method is required to convert CL_{intapp}, through to *in vivo* CL_{int} (mL/min/kg) and ultimately to blood clearance (mL/min/kg). The methodology for this and the scaling factors for each species were published approximately 20 years ago [24, 25]. The scaling factors across species are published [26].

14.8.2 Success Rates in Scaling *In Vitro* Clearance To Human Hepatic Clearance

There have been several published evaluations of the ability of human liver microsomes and hepatocyte to predict observed human *in vivo* CL_{int} or human hepatic clearance using a variety of scaling methods [27–30]. All of these analyses agree that both human liver microsomes and hepatocytes provide a reasonable approximation of the observed hepatic clearance for groups of compounds and that both methods tend to produce a bias toward underprediction.

14.8.3 IVIVC Analysis

One of the most comprehensive analyses of IVIVC has been published recently [31] showing that the rat clearance of approximately 45% of 900 Novartis (Basel, Swizerland) compounds could be well predicted from scaling of rat microsomal CL_{int}. For this group of compounds, IVIVC was established, suggesting that the major clearance mechanism for these compounds is CYP metabolism.

For a large proportion of the Novartis cohort, the observed plasma clearance was greater than that predicted from the rat liver microsomal incubations. Significant numbers of compounds exhibited plasma clearance values that were in excess of hepatic blood flow in the rat. For these compounds, IVIVC was not established. These are not uncommon findings, and in our laboratories, plasma clearance values are often not predicted by rat liver microsomal scaling. The issue is then to define the confidence of human liver microsomal scaling for human clearance prediction. There are many reasons why rat liver microsomal clearance may not predict the observed rat *in vivo* clearance.

14.8.4 *In Vitro* to *In Vivo* Scaling Requires Several Experiments and Assumptions

The fundamental assumption of using liver microsomes and hepatocytes to predict *in vivo* clearance is that clearance should be solely hepatic. CYPs, but also many other enzymes, are known to be expressed in other tissues, and there is the potential for extrahepatic metabolism.

Another assumption of the disappearance assay is that the substrate concentration is maintained below the Km for the metabolizing enzyme. For a small number of compounds, the Km value could be very low and below the initial substrate concentration. Such compounds will appear more stable than they actually are and scaling will underpredict *in vivo* clearance.

The well-stirred model shows the importance of determination of blood binding. Most groups simplify this to ppb combined with measurement of the blood-to-plasma ratio. Ppb measurement is prone to error, and it is important to control the incubation time and the pH. Compounds exhibiting extensive ppb are difficult to measure as a result of the determination of very low concentrations of drug in the buffer phase. Completing an IVIVC examination in plasma without correction for the blood-to-plasma ratio is also prone to error. Some compounds partition extensively into blood cells, and clearance values based on plasma will be much higher than from the blood concentration.

14.8.5 It Is Important that the *In Vitro* Assay Expresses the Enzyme Metabolizing the Compound

The choice of *in vitro* system is crucial to establishing IVIVC. Not all *in vitro* clearance assays express all drug metabolizing enzymes. There is a temptation within a discovery program to assume that all molecules are cleared by CYPs and that microsomal incubations are used. Many enzymes such as aldehyde oxidase and monoamine oxidase are not found in microsomes, and so heptocytes are a more appropriate system. The details for extrapolation of the *in vitro* CL_{int} of molecules metabolized by these enzymes to *in vivo* CL_{int} are not well established. Demonstration of IVIVC with such molecules is difficult.

The choice of *in vitro* system should be defined by examination of the structure of a molecule for metabolic handles for non-CYP enzymes. These have been exemplified by Obach [23].

14.8.6 Nonmetabolic Clearance

A proportion of compounds are cleared as unchanged compound in urine and bile. Some urinary and all biliary clearance is modulated by active drug transport. Most *in vitro* metabolic clearance assays do not express the transporters required to modulate the clearance of unchanged compounds into urine and bile. If a compound shows elimination of unchanged drug, the *in vitro* clearance assay will underpredict the overall clearance of that molecule and IVIVC will not be established.

14.9 *IN VITRO* ASSAYS FOR PREDICTION OF TRANSPORTER-MEDIATED CLEARANCE

Several drug transporters are expressed in the body. These include organic anion transporter polypeptides (OATPs), organic cation transporters (OCTs), multidrug resistance proteins (MRPs), and P-glycoprotein. The role of drug transport in clearance has been extensively reviewed [32]. The SAR, species specificity, and expression levels in clearing organs have not been fully established. Unlike CYP-mediated clearance, the scaling factors to *in vivo* clearance by transporters are yet to be defined and establishing IVIVC is fraught with difficulty.

14.9.1 Use of Cell Lines Expressing Transporters

In a discovery project, it is important to establish whether the compound series are substrates for particular transporters. The first assay that is used is cell lines transfected with the individual drug transport protein [33].

The use of cell lines expressing individual drug transport proteins tends to be limited to establishing whether compounds are substrates for transporters. Determination of potential clearance rates as a result of transport tends to be completed in native cells such as hepatocytes.

14.9.2 Use of Hepatocytes to Predict Biliary Clearance

Human hepatocytes are a tool to examine hepatobiliary clearance [34]. Suspended hepatocytes can be used to assess transporter-mediated uptake of compounds into cells [35]. The drawback is that suspended hepatocytes do not form bile canaliculae and cannot be used for biliary secretion.

To investigate hepatobiliary clearance more fully, plated hepatocytes are very useful [36, 37]. Sandwich cultured hepatocytes are grown on collagen-coated plates with a further layer of collagen placed on top to form the "sandwich" [38, 39].

Sandwich cultured human hepatocytes (SCHHs) have been used to predict retrospectively the human clearance of seven marketed OATP substrates [40]. SCHH underpredicted the clearance of these drugs and suggested that empirical scaling factors should be used to improve prediction accuracy.

14.10 UNDERLYING PRINCIPLES IN THE USE OF ANIMALS FOR CROSS-SPECIES SCALABILITY

14.10.1 The Use of Animals

Animal studies are essential for potential drug molecules in terms of supporting drug pharmacokinetics (PK)/ pharmacodynamics (PD)/efficacy and safety studies using preclinical pharmacological or toxicological models for the progression of such compounds toward human evaluation.

The prediction of human drug clearance from preclinical species has been used extensively, and extrapolation methods are based on science established many decades ago. From cost and ethical perspectives, the animal study should be designed with sufficient quality to answer the key question with respect to progression to humans.

Several issues make cross-species extrapolation of clearance difficult:

(a) The metabolic enzymes and drug transporters expressed in animals are related but not the same as their human isoforms. The rates of metabolism and transport can be different between animals and humans.

(b) Animals (especially rodents) tend to metabolize compounds more rapidly than humans, which is important to account for in any cross-species extrapolation of clearance.

(c) Several physiological differences exist between animals and humans. The most important variable is organ blood flows, which are higher in humans than in animals when compared on a mL/min basis. As clearance can often be related to blood flow, incorporation of these differences is key in a successful cross-species clearance prediction. Humans have a gall bladder in which bile is stored, whereas rats continually secrete bile into their gut. This can have consequences for extrapolation of biliary clearance from rats to humans.

14.10.2 Principles of *In Vivo* Cross-Species Scaling of Clearance (Allometry)

The basis of cross-species clearance scaling is that although the lives of animals are very different than humans in terms of chronological time, lives based on physiological variables, such as total number of heartbeats in a lifetime, are remarkably similar. This was translated into an examination of the half-life of ceftizoxime [41]. In minutes, the half-life of ceftizoxime shows differences between animals and humans (approximately 15 minutes in mouse to approximately 85 minutes in human). However, in terms of heartbeats, the values are very similar. The same analysis can be completed for clearance.

The basis of this observation was established many years ago [42] when it was shown that metabolic rates across species could be related to body weight, albeit with a correlation exponent that was not unity but approximately 0.75. Later it was established [43] that that blood flows and liver weights across the species could be correlated to body weight with a similar exponent. If blood flows can be related to body weight and blood flows are also a driver of clearance, the clearance of drugs may well be correlated with body weight. A cross-species correlation of drug clearance could then be used to extrapolate to human body weight to predict drug clearance in humans. This is the basis of simple allometry (Equation 14.8):

$$\mathrm{CL} = a \times \mathrm{BW}^{b} \tag{14.8}$$

where a is a constant and b is the allometric exponent.

Simple allometric profiling can be completed by measuring clearance (preferably corrected for the unbound fraction in plasma to avoid species differences in ppb) and plotting the log transformed value (in mL/min) against the log transformed body weight (kg). The slope of the correlation line is the allometric exponent. The line can be extrapolated to human body weight to predict human clearance.

14.10.3 Example of Simple Allometry for Metabolic Clearance (Sildenafil)

Sildenafil is cleared by CYP metabolism. The pharmacokinetics of sildenafil in preclinical species and humans has been published [44]. A simple allometric relationship

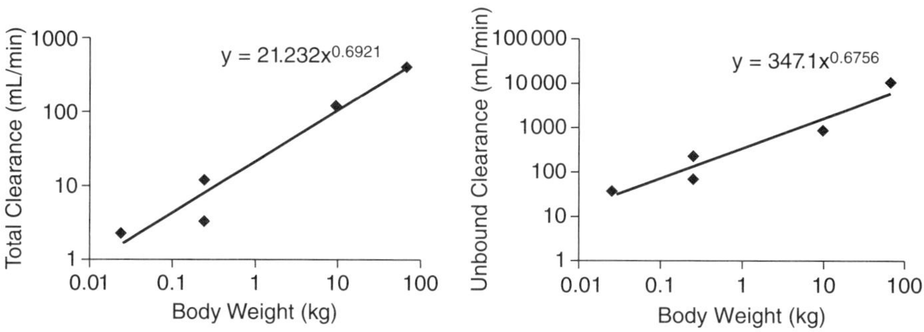

Figure 14.5 Sildenafil simple allometric scaling.

for sildenafil is shown in Figure 14.5. Simple allometry provides a reasonable approximation of the observed human clearance of sildenafil, using both total clearance and clearance corrected for the fraction unbound in plasma. The exponents are 0.69 and 0.68, respectively. This corresponds to a prediction error of approximately 23% [45].

In terms of a prediction of human clearance, the sildenafil example suggests that simple allometry can be an appropriate method to predict human clearance. There are several observations with respect to this analysis:

(a) The plasma fraction unbound values for sildenafil only shows a threefold difference across the species. Scaling total and unbound clearance values shows similar predictions of human clearance.

(b) Sildenafil shows a sex difference in rate of metabolism in rats. This translates to a sex difference in the *in vivo* clearance. If the male rat *in vitro* metabolic rate is omitted, sildenafil shows only a threefold variability in the rate of metabolism for female rats, dogs, and humans.

(c) Sildenafil shows total plasma clearance values that are moderate to high with respect to hepatic blood flow in each species.

14.10.4 Example of Simple Allometry for Renal Clearance (Fluconazole)

Fluconazole is excreted unchanged in the urine of all species. After glomerular filtration, approximately 80% of this compound is reabsorbed into the blood and fluconazole exhibits renal clearance at less than GFR in all species. Fluconazole clearance is available in six preclinical species and humans after intravenous administration [46]. It shows an allometric relationship across the species with an exponent of 0.71 that would accurately predict its human clearance (Figure 14.6). This is not surprising given that fluconazole is cleared at a constant fraction of GFR, which shows a body weight correlation.

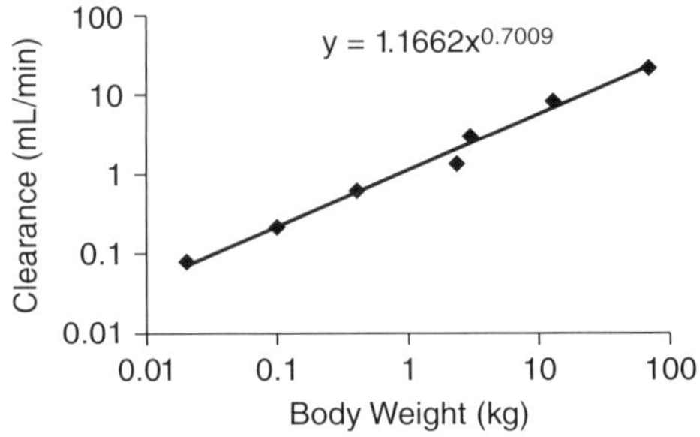

Figure 14.6 Fluconazole simple allometric scaling. (Adapted from Reference 46.) Based on mouse, rat, guinea pig, rabbit, cat, dog, and human. Total clearance used as fluconazole is not highy bound across species.

14.10.5 Example of Simple Allometry for Biliary Clearance (Susalimod)

Hepatobiliary clearance of compounds relies on active drug transport. Drug transporters show a significant species differences in expression. This complicates simple allometric relationships.

Susalimod is extensively cleared unchanged in bile. The pharmacokinetics of susalimod have been determined in six preclinical species and humans [47]. Susalimod shows a nonlinear relationship between clearance and dose. To compensate for this, only clearance after 25 mg/kg administration was taken forward to simple allometric scaling. In addition, male rats show a significant metabolic component to the clearance of susalimod, and thus, the male rat clearance was omitted. Using the clearance values from mouse, female rat, monkey, mini-pig, and dog, simple allometry predicted a human plasma clearance of 125 mL/min with an exponent of 0.82 (Figure 14.7A). The actual human plasma clearance was measured at 0.08 mL/min/kg (or approximately 5.4 mL/min, which was some 24-fold lower than the predicted human clearance (Figure 14.7B)). The prediction was improved by the incorporation

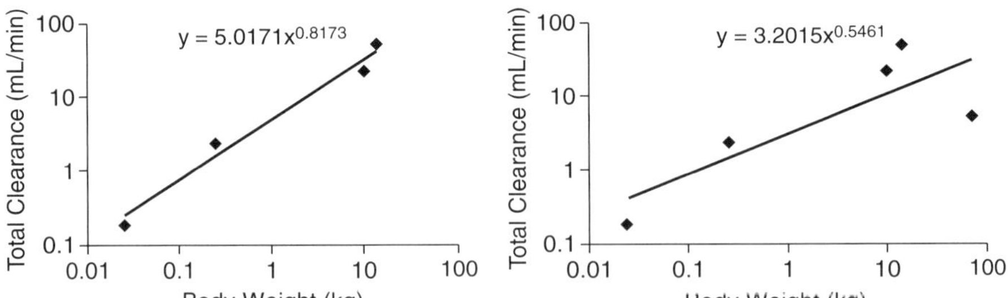

Figure 14.7 Susalimod slmple allometric scaling. Adapted from Reference 47.) Male rat omitted as a result of high metabolism. Scaling based only on 25 mg/kg dose as a result of dose-dependent pharmacokinetics.

of correction factors, but simple allometric scaling significantly mispredicted the clearance.

In terms of a prospective clearance prediction, as would be required in a discovery project, this represents an issue because such a compound may well be deselected from progression as a result of an overprediction of human clearance. Not all biliary cleared drugs are subject to such large errors in human prediction [48, 49].

14.10.6 Success Rates in Using Simple Allometry for Prediction of Human Clearance

A comprehensive analysis of the success of simple allometry for human plasma clearance has been completed for 102 small-molecule drugs where clearance after intravenous administration was published in at least three preclinical species and humans [45]:

(a) The average prediction error (irrespective of clearance route) was 254%, suggesting that the prediction accuracy from simple allometry is between two- and threefold of actual clearance. Several compounds showed accurate predictions of human clearance by this method [e.g., caffeine prediction error (PE) 9%, fluconazole PE 39%, and sildenafil PE 23%]. This was balanced by compounds with PE values of greater than 1000% (e.g., diazepam PE 3087%, reboxetine PE 1395. and susalimod PE 1676%).

(b) A more detailed analysis dividing the 102 compounds into low, moderate, and high clearance with respect to hepatic blood flow was reported. The average PEs were 598%, 120%, and 47% for the low-, moderate-, and high-clearance cohorts, respectively. These findings were rationalized on the basis of blood flows and species differences in metabolism. The high-clearance cohort shows the lowest average PE because for this cohort, clearance approaches hepatic blood flow, which is known to scale well allometrically. The low-clearance cohort will be more sensitive to species differences in CL_{int}, leading to the greater potential for misprediction.

(c) A high average PE was most likely with compounds of cLog P greater than 2 or compounds with more than a fivefold species difference in ppb. Such compounds are likely to be substrates for CYP-mediated metabolism, which is prone to significant species differences.

(d) Many exhibited allometric exponents of between 0.6 and 0.8, which is in line with expectations for body weight relationships. A proportion showed exponents outside these values. The overall range of exponents for the 102 compound set was 0.26 to 1.2.

Despite the observations that physiological processes scale with body weight across species with exponents of 0.6 to 0.8, it is not clear that drug clearance should scale with similar exponents. Some authors [50] argue that there is no rationale for drug clearance to scale with such exponents. Correction factors were proposed to improve the accuracy of prediction for allometric scaling. In their analysis, when the allometric exponent for any compound is between 0.55 and 0.7, simple allometry with no correction provides the most accuracy in prediction. When the allometric exponent is between 0.71 and 1, the most appropriate correction factor is maximum lifespan potential (MLP), whereas for exponents above 1, correction using brain weight provides the greatest accuracy. When the exponent is below 0.55 and above 1.3, these authors suggest that allometric scaling is not an appropriate method for the prediction of human clearance from animal data. These methods are termed the "rule of exponents."

Allometric scaling across preclinical species is a powerful technique for the prediction of human clearance. The choice of scaling factor is important for the use of the technique, and prediction accuracy can be improved over simple allometry using the rule of exponents. Like most prediction methods, allometric scaling is prone to large prediction errors for individual compounds. In addition, it has one major drawback for the use in discovery projects.

14.10.7 Issues with Allometric Scaling of Drug Clearance in Drug Discovery Projects

Allometric techniques are not amenable to discovery projects. The reason is that it requires pharmacokinetic studies in at least three preclinical species. As discovery projects are likely to produce several potential candidates, this multiple-species pharmacokinetic requirement would be costly and lead to significant animal use. Several groups have examined the use of single-species scaling (SSS) to predict human clearance.

14.11 SINGLE-SPECIES SCALING OF CLEARANCE

The scaling factor for SSS can be based on fixing the allometric exponent [26, 51], direct scaling from animal clearance [52], or correction for hepatic blood flows across species [53, 54].

14.11.1 Fixed Exponent Method

The fixed exponent method is based on the observation that many physiological parameters scale across species with body weight to exponents of between 0.6 and 0.8. The assumption of this method is that the human clearance of a compound can be predicted by measuring the clearance in a single preclinical species and correcting for body weight with an allometric exponent.

Several groups have examined the success rate of a single-species fixed exponent scaling in predicting human clearance. One study [51] examined using rat SSS, suggesting that for 176 drugs, the average fold error in prediction of human clearance was 2.25. This group suggested that 79% of those drugs were predicted within threefold of actual drug clearance. Another study examined the prediction of half-life (as a surrogate for clearance) for 50 proprietary compounds using SSS with a fixed exponent of 0.75 [26]. This analysis showed that using rat, 66% of compounds could be predicted within twofold of actual human half-life and 80% within threefold of actual human half-life. Similar success rates were observed using monkey data.

The use of fixed exponent SSS is controversial. Mahmood [55, 56] argues from extensive investigation that there is no reason why clearance should scale with allometric exponents of 0.6 to 0.8, and that prediction of human clearance by this method is likely to be inaccurate and misleading. However, in certain circumstances (renal clearance at GFR, clearances approaching hepatic blood flow, and compounds with low-species differences in metabolism and plasma protein binding), this method can be an effective predictor of human clearance.

14.11.2 Direct Scaling Method

Another method of SSS has been described [52]. Based on evidence from a 102 compound set, coefficients for scaling for human clearance from rat, dog, and monkey were 0.152, 0.41, and 0.407, respectively (e.g., Human CL = 0.152 × Rat Clearance). Using these fixed coefficients, the human clearance of 26 compounds not present in the 102 compound set were predicted with similar accuracy to that observed with multiple-species allometric scaling. Again this method is controversial based on the use of the fixed coefficient [56].

14.11.3 Blood Flow Method

Another method uses fixed coefficients based on the physiological variables of hepatic blood flow [53, 54]. The human clearance is predicted as the animal clearance corrected for the ratio of human to animal hepatic blood flow:

$$\text{Human CL} = \text{Animal CL} \times \frac{\text{Human Liver Blood Flow}}{\text{Animal Liver Blood Flow}} \tag{14.9}$$

Using a 103 compound set, human clearance prediction accuracy was found to be similar using this method versus multiple-species allometric scaling. In addition, the hepatic blood flow method showed slightly improved human clearance prediction accuracy using monkey over rat and dog.

14.12 PERSPECTIVE ON SINGLE-VERSUS MULTIPLE-SPECIES ALLOMETRIC SCALING

The scaling of clearance in preclinical species to predict the human clearance of a particular molecule is a powerful technique in drug discovery. Accuracy in prediction seems to be around 60% to 70% of compounds predicted within twofold of actual human clearance. However, this accuracy value does mask that all methods can produce some very large prediction errors.

The use of SSS versus multiple-species allometric scaling is controversial. The use of any particular technique is subject to judgment. At compound optimization stages, the use of multiple species is not appropriate as a result of cost and throughput. It is likely that at this stage, rodent SSS is the most appropriate method. As the project nears clinical candidate selection, multiple species allometry may well be the most appropriate method of clearance prediction.

14.13 PHYSIOLOGICALLY BASED PHARMACOKINETIC (PBPK) MODELING

The cross-species clearance scaling methods dealt with so far have been aimed solely at predicting human clearance of a compound and provide little information on the overall pharmacokinetic profile.

PBPK models are mechanistic tools that combine an understanding of the absorption, distribution, and clearance of a molecule in preclinical species to estimate the human pharmacokinetic profile.

PBPK models use the following assumptions:

1. Distribution of compound is perfusion rate limited, where the blood flow to the tissue becomes the limiting process.
2. Each tissue acts as a well-stirred compartment.
3. Membrane permeability is not a barrier to tissue distribution.
4. Distribution is governed by passive processes with no contribution of active transport.
5. Clearance is via the particular tissues selected (e.g., liver and kidney).

The PBPK framework allows the extrapolation of pharmacokinetics across different species and allows projects to validate human simulations. The following PBPK strategy has been proposed [57]. Preclinical PBPK models (routinely rat and dog), compound physicochemistry, and animal *in vitro* data are derived to generate a preclinical simulation. This simulation is then evaluated against the animal pharmacokinetic data. If these simulations agree with the observed pharmacokinetics, then it is viable to progress with the human simulations using the human PBPK model, the compound-specific physiochemical data, and the human *in vitro* data. If the simulations in preclinical species are clearly different, then this suggests that the assumptions of the model are not valid and further work must be performed. This approach can support candidate selection by integrating all available information into a framework where selection can be made based on the expected human pharmacokinetic profile. This approach has enabled reasonably accurate prediction of human PK for several compounds [58, 59].

14.14 EXAMPLES OF THE USE OF PBPK MODELING TO PREDICT PHARMACOKINETIC PROFILE OF FLUCONAZOLE AND SILDENAFIL

14.14.1 Fluconazole

Physicochemical properties, CL_{int} values scaled from HLM or preclinical data, as well as input parameters for the absorption PBPK prediction for fluconazole (and sildenafil) are shown in Table 14.1.

The prediction of *Vdss* for fluconazole in preclinical species was best described using the tissue composition equations developed by Poulin and Theil [60]. This reflects the neutral nature of fluconazole as the equations account only for lipophilicity and specific reversible binding to proteins in plasma and tissue.

When the tissue to plasma coefficients (Kp values) provided by these prediction methods were combined with the observed clearance, simulation of the intravenous concentration versus the time profiles in rats and dogs provided good agreement to the observed data (Figure 14.8A,C). These equations were then used for PBPK simulations in humans and demonstrated good agreement.

The human clearance of fluconazole was predicted to be 0.3 and 0.45 mL/min/kg when scaled from unbound clearance in rats and dogs, respectively. The observed human clearance is 0.3 mL/min/kg [46].

The absorption PBPK model was able to simulate accurately preclinical oral plasma concentration versus time profiles using the *in vitro* solubility, predicted Kp values, and observed intravenous clearance as inputs to the model as shown for fluconazole (Figure 14.8B). Therefore, the absorption model was used to simulate the oral profile in humans (Figure 14.8D). The error in the observed versus predicted

Table 14.1 Input data used in the PBPK model

Parameter	Fluconazole	Sildenafil
Molecular weight (g/mol)	306	475
LogD at pH 7.4	0.5	2.8
Charge	N	B
pKa	N/A	6.5
Effective human permeability (10^{-4}cm/s)	2.0	1.6
Buffer solubility @ pH 7.0 (mg/mL)	6.0	0.01
Particle size (μm)	25	25
C_B:C_P in rat, dog, human	0.91, 0.85, 1.0	0.9, 0.81, 0.7
Fu_p in rat, dog, human	0.94, 0.89, 0.95	0.05, 0.14, 0.04
CL in rat and dog (mL/min/kg)	1.3, 0.65	66, 11
Vdss in rat and dog (L/kg)	0.9, 0.7	1.6, 4.1
CL_{int} HLM (μL/min/mg)	<7.6	87.3
Fu, mic in HLM	N/A	0.64
Human CL_p predicted (mL/min/ kg)	0.3	3.77
Fu gut	1	1
CL mechanism	Renal	CYP

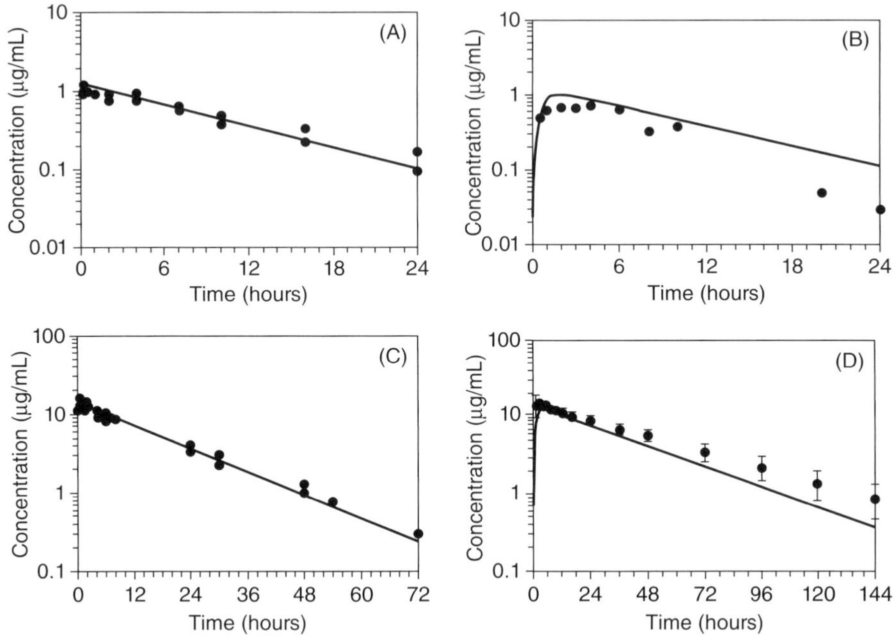

Figure 14.8 Predicted versus observed concentration time profiles in rat, dog, and human after administration of fluconazole using a physiologically based pharmacokinetic model. (a) Rat IV (1 mg/kg); CL input using observed rat CL. (b) Rat PO (1 mg/kg). (c) Dog IV (10 mg/kg); CL input using observed dog CL. (d) Human PO (800 mg); CL input using rat allometry.

profiles for fluconazole for Cmax, Tmax, and AUC were within twofold of actual (Table 14.2).

14.14.2 Sildenafil

The prediction of *Vdss* for the base sildenafil in preclinical species was best described using the tissue coefficient equations developed by Rodgers and Rowland [61] as a result of the additional consideration of ionization and charge of the molecule.

Table 14.2 Prediction of oral profile of fluconazole and sildenafil using physiologically based pharmacokinetic model versus observed data

Compound	Parameter	Observed	Predicted	Fold Error
Fluconazole (800 mg)	AUC µg·h/mL (0-last)	690	530	0.77
	Cmax µg/mL	15	12.3	0.82
	Tmax h	2.0	3.6	1.8
Sildenafil (150 mg)	AUC ng·h/mL (0-last)	3270	6723	2.1
	Cmax µg/mL	0.98	1.05	1.1
	Tmax h	1	1.7	1.7

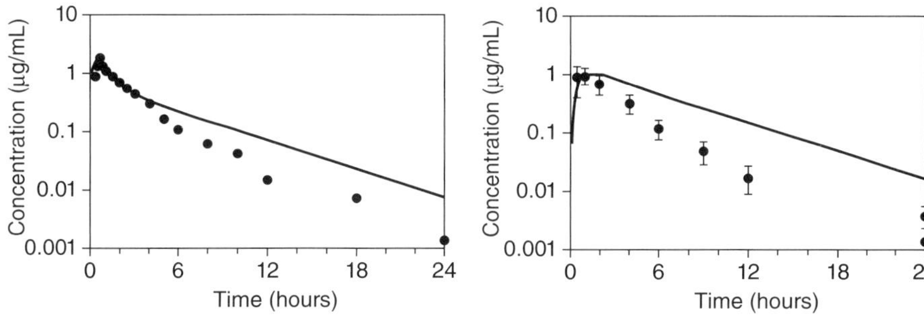

Figure 14.9 Predicted versus observed concentration time profiles in humans after intravenous and oral administration of sildenafil using physiologically based pharmacokinetic model. (a) Human IV infusion (40 min, 80 mg); CL input using human liver microsomes. (b) Human PO (150 mg); CL input using human liver microsomes.

The predicted *Vdss* for sildenafil was 0.8 L/kg versus the observed 1.2 L/kg [44] (Table 14.1).

Sildenafil is cleared by CYP metabolism, and prediction of clearance used scaled CL_{int} in HLM provides the closest prediction to the observed clearance in human (3.8 mL/min/kg vs. 6 mL/min/kg observed) [44]. Based on these data and the Kp values, the PBPK model provided a reasonable approximation of the preclinical pharmacokinetic profiles and the human intravenous and oral profiles were modeled (Figure 14.9A,B). The model predicted the intravenous and oral profiles of sildenafil within 2.1-fold of the observed human profile.

14.15 OVERALL PERSPECTIVE ON CROSS-SPECIES SCALING OF CLEARANCE FOR DRUG DISCOVERY

The purpose of a discovery project is to provide a compound with the optimal balance of pharmacology and pharmacokinetics to achieve the laboratory objectives. Clearance is a key parameter to optimize during this process, as it affects drug exposure, duration, and oral bioavailability. Throughout the discovery process, several thousand compounds can be evaluated and a strategy to select the compound with the most appropriate clearance profile is essential.

The aims of this strategy should be structured around key aspects:

1. Understanding the mechanism of clearance for a series of compounds.
2. Selecting the compound with the best balance of potency and clearance.
3. Establishing IVIVC to understand fully the clearance across species and to select the appropriate method to predict human clearance.
4. Prediction of the human clearance of the compound.
5. Revisiting the clearance prediction methodology after first time in human (FIH) evaluation to improve the properties of any backup compounds.

A clearance optimization strategy should start with the physicochemistry of the proposed molecules. This should define the predominant route of clearance and enable the appropriate *in silico* and *in vitro* tools to be applied to clearance optimization. Progression of promising compounds to *in vivo* evaluation should establish the potential for IVIVC, defining whether the appropriate *in vitro* assay has been used for optimization. In addition, at this point, it is appropriate to predict human clearance of promising molecules to define whether further optimization is required or whether the compound should be progressed to human studies. The general accuracy of prediction of most of the clearance prediction methods seems to be 60% to 80% of compounds predicted with twofold of actual [62]. This accuracy level can hide some significant prediction errors, and it is not enough to select between two closely related compounds in a compound series. However, it is true to say that clearance prediction methodology effectively "stacks the odds" in favor of a positive outcome in the clinic, and further improvements based on more science should improve that outcome.

REFERENCES

1. Pang, K.S., Rowland, M. (1977). Hepatic clearance of drugs/I. Theoretical considerations of a "well-stirred" model and a "parallel tube" model. Influence of hepatic blood flow, plasma and blood cell binding, and the hepatocellular enzymatic activity on hepatic drug clearance. *Journal of Pharmacokinetics and Biopharmaceutics*, *5*, 625–653.
2. Smith, D.A., Jones, B.C., Walker, D.K. (1996). Design of drugs involving the concepts and theories of drug metabolism and pharmacokinetics. *Medicinal Research Reviews*, *16*, 243–266.
3. Wagner, J.G., Northam, J.I., Alway, C.D., Carpenter, O.S. (1965). Blood levels of drug at the equilibrium state after multiple dosing. *Nature*, *207*, 1301–1302.
4. Smith, D.A., Di, L., Kerns, E. (2010). The effect of plasma protein binding on in vivo efficacy: Misconceptions in drug discovery. *Nature Reviews Drug Discovery*, *9*, 929–939.
5. Davies, B., Morris, T. (1993). Physiological parameters in laboratory animals and humans. *Pharmaceutical Research*, *10*, 1093–1095.
6. Williams, J.A., Hyland, R., Jones, B.C., Smith, D.A., Hurst, S., Goosen, T.C., Peterkin, V., Koup, J.R., Ball, S.E. (2004). Drug-drug interactions for UDP-glucuronosyltransferase substrates: A pharmacokinetic explanation for typically observed low exposure (AUCi/AUC) ratios. *Drug Metabolism and Disposition*, *39*, 1808–1814.
7. Lipinski, C.A., Lombardo, F., Dominy, B.W., Feeney, P.J. (1997). Experimental and computational approaches to estimate solubility and permeability in drug discovery and development settings. *Advanced Drug Delivery Reviews*, *23*, 3–25.
8. Cruciani, G., Carosati, E., De Boecck, B., Kantharaj, E., Mackie, C., Howe, T., Vianello, R. (2012). MetaSite: Understanding metabolism in human cytochromes from the perspective of the chemist. *Journal of Medicinal Chemistry*, *48*, 6970–6979.
9. StarDrop. (2012). Cambridge, UK Optibrium Ltd.
10. Rydberg, P., Gloriam, D.E., Olsen, L. (2010). The SMARTCyp cytochrome P450 metabolism prediction server. *Bioinformatics*, *26*, 2988–2989.

11. Zaretki, J., Rydberg, P., Bergeron, C., Bennett, K.P., Olsen, L., Breneman, C.M. (2012). RS-Predictor models augmented with SMARTCyp reactivities: Robust metabolic regioselectivity predictions for nine CYP isozymes. *Journal of Chemical Informatics Modeling*, *24*, 383–395.
12. Campagna-Slater, V., Pottel, J., Therrien, E., Cantin, Louis-David, Moitessier, N. (2012). Development of a computational tool to rival experts in the prediction of sites of metabolism of xenobiotics by P450s. *Journal of Chemical Informatics Modeling*, *52*, 2471–2483.
13. Keefer, C.E., Chang, G., Kaufman, G. (2011). Extraction of tacit knowledge from large ADME datasets via pairwise analysis. *Bioorganic Medicinal Chemistry*, *19*, 3739–3749.
14. Lewis, D.F.V., Modi, S., Dickins, M. (2002). Structure-activity relationship for human cytochrome P450 substrates and inhibitors. *Drug Metabolism Reviews*, *34*, 69–82.
15. Madden, J.C., Cronin, M.T.D. (2006). Structure-based models for the prediction of drug metabolism. *Expert Opinion on Drug Metabolism and Toxicology*, *2*, 545–557.
16. Sun, H., Scott, D.O. (2010). Structure-based drug metabolism predictions for drug design. *Chemical and Biological Drug Design*, *75*, 3–17.
17. MoKa. (2012). Perugia, Italy, Molecular Discovery Ltd.
18. Hu, Y., Unwalla, R., Denny, R.A., Bikker, J., Di, L., Humblet, C. (2010). Development of QSAR models for microsomal stability: Identification of good and bad structural features for rat, human and mouse microsomal stability. *Journal of Computer-Aided Molecular Design*, *24*, 23–35.
19. Lee, P.H., Cucurull-Sanchez, L., Lu, J., Du, Y.J. (2007). Development of in silico models for human liver microsomal stability. *Journal of Computer-Aided Drug Design*, *21*, 665–673.
20. Yap, C.W., Li, R., Chen, Y.Z. (2006). Quantitative structure-pharmacokinetic relationships for drug clearance by using statistical learning methods. *Journal of Molecular Graphics Modeling*, *24*, 383–395.
21. Berellini, G, Waters, N.J., Lombardo, F. (2012). In silico prediction of total human plasma clearance. *Journal of Chemical Informatics Modeling*, *52*, 2069–2078.
22. Keefer, C.E., Kauffman, G.W., Gupta, R. (2013). Probability-based confidence metric for continuous QSAR models. *Journal of Chemical Informatics and Modeling*, *53*(2), 368–383.
23. Obach, R.S. (2011). Predicting clearance in humans from in vitro data. *Current Topics in Medicinal Chemistry (Sharjah United Arab Emirates)*, *11*, 334–339.
24. Houston, J.B. (1994). Utility of in vitro drug metabolism data in predicting in vivo metabolic clearance. *Biochemical Pharmacology*, *47*, 1469–1479.
25. Obach, R.S., Baxter, J.G., Liston, T.E, Silber, B.M., Jones, B.C., MacIntyre, F., Rance, D.J, Wastall, P. (1997). The prediction of human pharmacokinetic parameters from preclinical and in vitro metabolism data. *Journal of Pharmacology and Experimental Therapeutics*, *283*, 46–58.
26. Hosea, N.A., Collard, W.T., Cole, S., Maurer, T.S., Fang, R.X., Jones, H., Kakar, S.M., Nakai, Y., Smith, B.J., Webster, R., Beaumont, K. (2009). Prediction of human pharmacokinetics from preclinical information: Comparative accuracy of quantitative prediction approaches. *Journal of Clinical Pharmacology*, *48*, 513–553.
27. Ito, K., Houston, J.B. (2005). Prediction of human drug clearance from in vitro and preclinical data using physiologically based and empirical approaches. *Pharmaceutical Research*, *22*, 103–112.

28. Riley, R.J., McGinnity, D.F., Austin, R.P. (2005). A unified model for predicting human hepatic, metabolic clearance from in vitro intrinsic clearance data in hepatocytes and microsomes. *Drug Metabolism and Disposition*, *33*, 1304–1311.
29. Stringer, R., Nicklin, P.L., Houston, J.B. (2008). Reliability of human cryopreserved hepatocytes and liver microsomes as in vitro systems to predict metabolic clearance. *Xenobiotica*, *38*, 1313–1329.
30. Hallifax, D., Foster, J.A., Houston, J.B. (2010). Prediction of human metabolic clearance from in vitro systems: Retrospective analysis and prospective view. *Pharmaceutical Research*, *27*, 2150–2161.
31. Bell, L.C., Wang, J. (2012). Probe ADME and test hypotheses: A PATH beyond clearance in vitro-in vivo correlations in early drug discovery. *Expert Opinion in Drug Metabolism and Toxicology*, *8*, 1131–1155.
32. Ayrton, A., Morgan P. (2001). Role of transport proteins in drug absorption, distribution, and excretion. *Xenobiotica*, *31*, 469–497.
33. Yamashiro, W., Maeda, K., Hirouchi, M., Adachi, Y., Hu, Z., Sugiyama, Y. (2006). Involvement of transporters in the hepatic uptake and biliary excretion of valsartan, a selective antagonist of the angiotensin II AT1-receptor, in humans. *Drug Metabolism and Disposition*, *34*, 1247–1254.
34. Fenner, K.S., Jones, H.M., Ullah, M., Kempshall, S., Dickins, M., Lai, Y., Morgan, P., Barton, H.A. (2012). The evolution of the OATP hepatic uptake transport protein family in DMPK sciences: From obscure liver transporters to key determinants of hepatobiliary clearance. *Xenobiotica*, *42*, 28–45.
35. Kitamura, S., Maeda, K., Wang, Y., Sugiyama, Y. (2008). Involvement of multiple transporters in the hepatobiliary transport of rosuvastatin. *Drug Metabolism and Disposition*, *36*, 2014–2023.
36. Poirier, A., Lave, T., Portmann, R., Brun, M.E., Senner, F., Kansy, M., Grimm, H.P., Funk, C. (2008). Design, data analysis, and stimulation of in vitro drug transport kinetic experiments using a mechanistic in vitro model. *Drug Metabolism and Disposition 36*, 2434–2444.
37. Yabe, Y., Galetin, A., Houston, J.B. (2011). Kinetic characterization of rat hepatic uptake of 16 actively transported drugs. *Drug Metabolism and Disposition*, *39*, 1808–1814.
38. Lee, J.K., Marion, T.L., Abe, K., Lim, C., Pollock, G.M., Brouwer, K.L., (2010). Hepatobiliary disposition of troglitazone and metabolites in rat and human sandwich-cultured hepatocytes: Use of Monte Carlo simulations to assess the impact of changes in biliary excretion on troglitazone sulfate accumulation. *Journal of Pharmacology and Experimental Therapeutics*, *332*, 26–34.
39. Yan, G.Z., Brouwer, K.L., Pollack, G.M., Wang, M.Z., Tidwell, R.R., Hall, J.E., Paine, M.F. (2011). Mechanisms underlying differences in systemic exposure of structurally similar metabolites: Comparison of two preclinical hepatic models. *Journal of Pharmacology and Experiemental Therapeutics*, *337*, 503–512.
40. Jones, H.M., Barton, H.A., Lai, Y., Bi, Y.A., Kimoto, E., Kempshall, S., Tate, S.C., El-Kattan, A., Houston, J.B., Galetin, A., Fenner, K.S. (2012). Mechanistic pharmacokinetic modeling for the prediction of transporter-mediated disposition in humans from sandwich-cultured human hepatocyte data. *Drug Metabolism and Disposition*, *40*, 1007–1017.
41. Mordenti, J. (1985). Forecasting cephalosporin and monobactam antibiotic half-lives in humans from data collected in laboratory animals. *Antimicrobial Agents Chemotherapy*, *27*, 887–891.

42. Kleiber, M. (1932). Body size and metabolism. *Hilgardia*, *6*, 315–353.
43. Boxenbaum, H. (1980). Interspecies variation in liver weight, hepatic blood flow, and antipyrine intrinsic clearance: Extrapolation of data to benzodiazepines and phenytoin. *Journal of Pharmacokinetics and Biopharmaceutics*, *8*, 165–176.
44. Walker, D.K., Ackland, M.J., James, G.C., Muirhead, G.J., Rance, D.J., Wastall, P., Wright, P.A. (1999). Pharmocokinetics and metabolism of slidenafil in mouse, rat, rabbit, dog and man. *Xenobiotica*, *29*, 297–310.
45. Tang, H., Mayersohn, M. (2006). A global examination of allometric scaling for predicting human drug clearance and the prediction of large vertical allometry. *Journal of Pharmaceutical Sciences*, *95*, 1783–1799.
46. Jezequel, S.G., (1994). Fluconazole: Interspecies scaling and allometric relationships of pharmacokinetic properties. *Journal of Pharmacy and Pharmacology*, *46*, 196–199.
47. Pahlman, I., Edholm, M., Kankaanranta, S., Odell, M-L., (1998). Pharmacokinetics of susalimod, a highly biliary-excreted sulphasalazine analog, in various species. Nonpredictable human clearance by allometric scaling. *Pharmacy and Pharmacology Communications*, *4*, 493–498.
48. Mahmood, I., Sahajwalla, C. (2002). Interspecies scaling of biliary-excreted drugs. *Journal of Pharmaceutical Sciences*, *91*, 1908–1914.
49. Mahmood, I. (2005). Interspecies scaling of biliary-excreted drugs: A comparison of several methods. *Journal of Pharmaceutical Sciences*, *94*, 883–892.
50. Mahmood, I., Balian, J.D. (1996). Interspecies scaling: Predicting clearance of drugs in humans. Three different approaches. *Xenobiotica*, *26*, 887–895.
51. Caldwell, G.W., Masucci, J.A., Yan, Z., Hageman, W. (2004). Allometric scaling of pharmacokinetic parameters in drug discovery: Can human CL, Vss and t1/2 be predicted from in-vivo rat data? *European Journal of Drug Metabolism and Pharmacokinetics*, *29*, 133–143.
52. Tang, H., Hussain, A., Leal, M., Mayersohn, M., Fluhler, E. (2007). Interspecies prediction of human drug clearance based on scaling data from one or two animal species. *Drug Metabolism and Disposition*, *35*, 1886–1893.
53. Ward, K.W., Smith, B.R. (2004). A comprehensive quantitative and qualitative evaluation of extrapolation of intravenous pharmacokinetic parameters from rat, dog and monkey to human. *I. Clearance. Drug Metabolism and Disposition*, *32*, 603–611.
54. Jolivette, L.J., Ward, K.W. (2005). Extrapolation of human pharmacokinetics parameters from rat, dog and monkey data: Molecular properties associated with extrapolative success or failure. *Journal of Pharmaceutical Sciences*, *94*, 1467–1483.
55. Mahmood, I. (2009). Application of fixed exponent 0.75 to the prediction of human drug clearance: An inaccurate and misleading concept. *Drug Metabolism and Drug Interactions*, *24*, 57–81.
56. Mahmood, I. (2010). Theoretical versus empirical allometry: Facts behind theories and application to pharmacokinetics. *Journal of Pharmaceutical Sciences*, *99*, 2927–2933.
57. Jones, H.M., Parrott, N., Jorga, K., Lave, T. (2006). A novel strategy for physiologically based predictions of human pharmacokinetics. *Clinical Pharmacokinetics*, *45*, 511–542.
58. Johnson, T.N., Boussery, K., Rowland-Yeo, K., Tucker, G.T., Rostami-Hodjegan, A. (2010). A semi-mechanistic model to predict the effects of liver cirrhosis on drug clearance. *Clinical Pharmacokinetics*, *49*, 189–206.

59. Zhao, P., Vieira, M.D.L.T., Grillo, J.A., Song, P., Wu, T-C., Zheng, J.H., Arya, V., Berglund, E.G., Atkinson, A.J. Sugiyama, Y., Pang, K.S., Reynolds, K.S., Abernethy, D., Zhang, L., Lesko, L.J., Huang, S.-M. (2012). Evaluation of exposure change of nonrenally eliminated drugs in patients with chronic kidney disease using physiologically based pharmacokinetic modeling and simulation. *Journal of Clinical Pharmacology*, *52*(1), 91S–108S.
60. Poulin, P., Theil, F.-P. (2002). Prediction of pharmacokinetics prior to in vivo studies. 1. Mechanism-based prediction of volume of distribution. *Journal of Pharmaceutical Sciences*, *91*, 129–156.
61. Rodgers, T., Rowland, M. (2007). Mechanistic approaches to volume of distribution predictions: Understanding the processes. *Pharmaceutical Research*, *24*, 918–933.
62. Beaumont, K., Smith, D.A. (2009). Does human pharmacokinetic prediction add significant value to compound selection in drug discovery research? *Current Opinion in Drug Discovery and Development*, *12*, 61–71.

15

PRACTICAL ANTICIPATION OF HUMAN EFFICACIOUS DOSES AND PHARMACOKINETICS USING PRECLINICAL *IN VITRO* AND *IN VIVO* DATA

Tycho Heimbach, Rakesh Gollen, and Handan He

15.1 INTRODUCTION

The anticipation of efficacious human doses is a complex process as biopharmaceutical, pharmacokinetic (PK), or pharmacodynamic (PD) properties need to be measured or projected from preclinical data [1]. The chosen starting dose in humans has to be low enough to be safe and yet high enough to avoid unnecessary dose escalation study arms in early clinical trials. Numerous prediction methodologies based on *in vitro* and preclinical *in vivo* data have been developed in recent years to identify "developable" new drug candidates with both efficacious and safe systemic exposures in humans. Predictive approaches typically use *in vivo* preclinical data along with metabolism and disposition *in vitro* data from, e.g., human microsomes and hepatocytes. The advantages and limitations of these methods in their applications to predict human pharmacokinetics have been described, and yet challenges remain

Predictive ADMET: Integrative Approaches in Drug Discovery and Development, First Edition.
Edited by Jianling Wang and Laszlo Urban.

[2–7]. One reason is that systemic exposure after oral dosing is determined by a variety of formulation and pharmacokinetic factors including drug solubility, dissolution, dosage form, intestinal permeability, efflux and first-pass effects [8–11]. Moreover, desirable molecular properties for orally administered drugs are thought to fall into a narrow range for most successfully marketed drugs [11, 12]. Another reason is that both human pharmacokinetic and pharmacodynamic (PK/PD) parameters have to be predicted from preclinical PK/PD relationships for the accurate prediction of an efficacious human dosing regimen [4, 13]. This is based on the integration of physiological and genetic information along with *in vitro* drug absorption, metabolism, and transport data to simulate and predict *in vivo* drug disposition and drug–drug interactions in patient populations. The evaluation of PK from preclinical species needs to be comprehensive enough to ensure that new compounds do not fail in clinical trials as a result of unexpected low-systemic exposure or lack of efficacy. Screening of absorption, distribution, metabolism, and excretion (ADME) properties using *in vitro* data in combination with preclinical *in vivo* data facilitates early elimination of suboptimal candidates and helps to focus drug development programs on the most promising lead candidates, thus, saving resources and time.

The prediction of efficacious human doses is performed at different developmental stages for several purposes including (1) selection of clinical candidate back-up compounds with improved absorption, distribution, metabolism, excretion, and toxicity (ADMET) properties; (2) identification of safe and efficacious doses for first-in-human (FIH) trials; (3) identification of formulation challenges; and (4) estimation of drug supply needed for toxicokinetic and FIH trials [1]. It is important to stress that the requirements for a safe maximum recommended starting dose (MRSD) for FIH trials are covered under regulatory guidance(s), which require no observable (adverse) effect levels (NO(A)ELs) in preclinical species [14–16]. The MRSD for FIH trials is chosen with a focus on safety, not on efficacy, and this approach does not include dose adjustments for interspecies differences in pharmacology, such as binding affinity and potency. Although it is useful to compare the MRSD with the human pharmacologically active dose (hPAD), the MRSD guideline does not emphasize different methods to determine the hPAD. This approach often uses a simple allometric scaling method that has an underlying assumption of a doses scale ratio of 1:1 between species when normalized to body surface area or body weight [14], using a defined safety factor. If the MRSD is far below the hPAD, this can lead to lengthy human trials with long escalation schemes, potentially exposing human cohorts to ineffective doses before reaching an active dose resulting in costly drug development processes. On the contrary, if the MRSD is higher than the pharmacologically active dose, then the complete dose–effect relationship may not be well characterized. Using a pharmacokinetically guided approach for anticipating the starting dose along with the U.S. Food and Drug Administration (FDA) guidelines, MRSD may enhance confidence in the estimation [15]. The pharmacokinetically guided approach [2, 15–17] is based on the predicted PK parameters from preclinical species and uses a correction factor with the assumption that only the parent compound is active and that the drug has equal pharmacological and toxicology activity in human as in preclinical species at a given plasma concentration. For biologics, “anticipation” of human dose (AHD) based on minimal anticipated biological effect levels (MABELs) is recommended by the European Medicines Agency (EMA) [18], which states that calculation of MABEL should use all *in vitro* and *in vivo* information available from PK/PD data including the binding affinity to molecular targets. This approach is based on the pharmacological knowledge of the compound, requiring extensive data rather than an empirical factor. This guidance implies that there should be some anticipation of human exposure before human studies can commence. Uncertainties in the anticipated human exposure as a result of, e.g., formulation variables are covered by applying safety margins and other safety factors [14].

Clinical PK/PD data from a frontrunner candidate can be helpful to validate preclinical PK/PD strategies for backup compounds using a "reverse pharmacology" approach [1]. This approach has the underlying assumptions that compounds with similar structures exhibit similar PK (e.g., CL, V_{ss}, etc.) and PD (e.g., IC_{50}, etc.), and that scaling to humans is done with similar methods for both the "backup" and the "frontrunner" drugs. This approach also requires that observed clinical frontrunner data along with preclinical data that be generated using similar assays and study designs as the backup compounds. *A priori* predictions of human oral PK or PK/PD plasma profile from preclinical data are often not reported since the identification of suitable pharmacodynamic markers and surrogates for use in preclinical and clinical PK/PD modeling can be uncertain. However, new technological advances in target validation, functional biology, and proteomics and genomics, and the correlation of these effects to the disease state or to toxic effects, may facilitate an earlier understanding of disease and toxicological processes [19].

A PK/PD model-guided approach to estimating AHD by translating animal exposure-response relationships to the human situation can minimize the uncertainties caused by interspecies differences in the exposure–response relationships [2, 20–22]. These adjustments are made using the ratio of human to nonhuman parameters as a correction factor. For example, if the affinity of receptor binding in the human is twice that of the nonhuman, e.g., *in vitro*, then the exposure–response curve observed in the nonhuman should be shifted to half the concentration to anticipate the human exposure–response. Similar adjustments can be made for the differences in (1) plasma protein binding, (2) blood cell binding, or (3) known differences in natural entities that compete for receptor binding. Application of a "reverse pharmacology" approach and step-by-step translation to human PK and PK/PD from animal exposure or response data after comparison with frontrunner clinical data is shown schematically in Figure 15.1A [1]. This approach can include "humanization" steps, such as correcting for differences in tissue and plasma protein binding, tissue distribution, blood cell binding, or differences in potency by correcting for target receptor occupancy.

Human pharmacokinetic parameters can be predicted based on ADME properties in preclinical species (e.g., CL, V_{ss}, and F), which are then integrated with PD parameters, e.g., IC_{50} (which is the plasma drug concentration that is required to maintain 50% inhibition against a target) for dose selection in phase I trials. Based on predicted human PK profiles, an appropriate dose and dosing schedule can be determined, e.g., to keep the drug concentration above the threshold of efficacy or

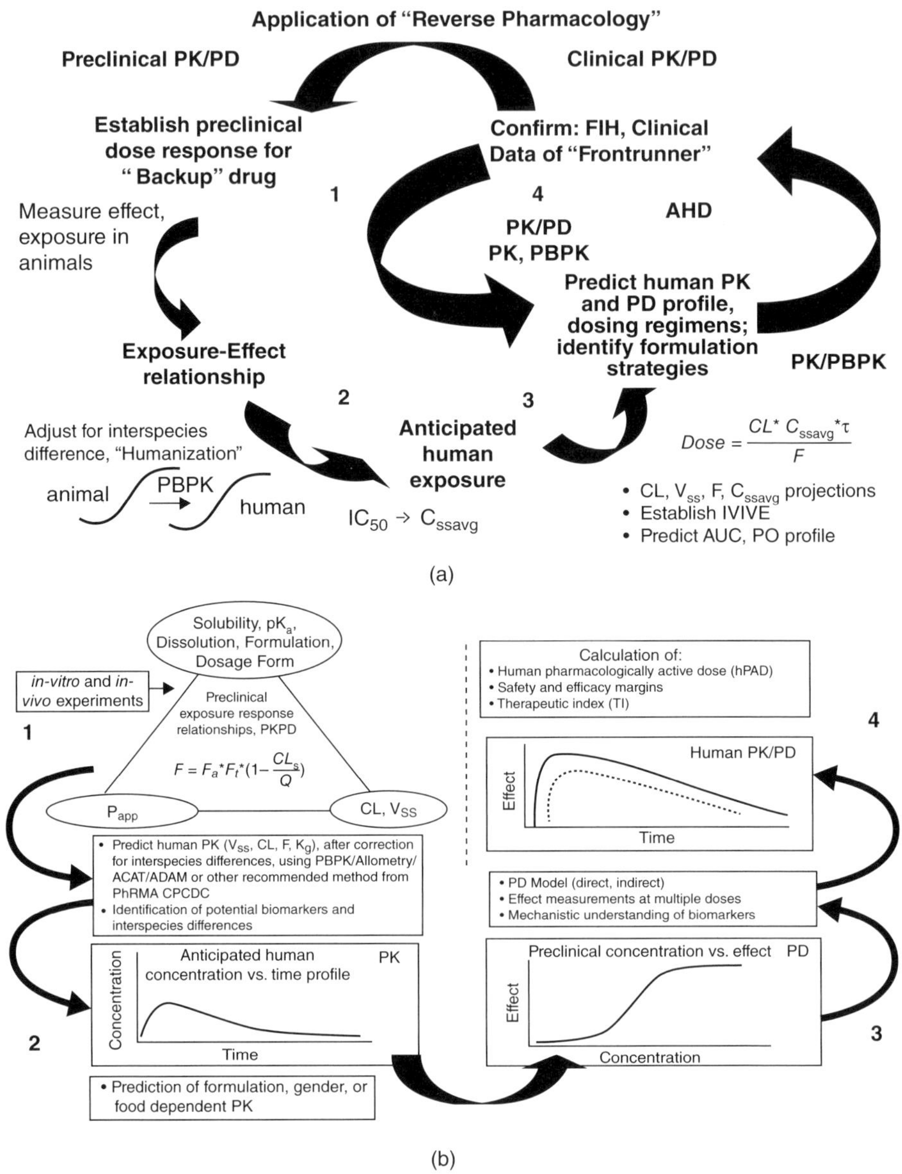

Figure 15.1 (A) Anticipation of human doses (AHD) as a "learn and refine" circular process with step-by-step translation of preclinical animal exposure and PK/PD data to human dose response and human FIH PK/PD profiles. Frontrunner clinical data allow "closing the learning loop" and allow for refinement of the preclinical-to-clinical PK/PD translation and linking steps using a reverse pharmacology approach when clinical data become available. (B) Schematic step-by-step approach for anticipating human PK and PD, by integrating physiochemical properties of drugs under investigation, after incorporating formulation dependent changes to predict several doses from the available preclinical information. Figures were modified from Heimbach et al. [1].

IC_{50} value (C_{min}) but below the threshold of adverse effect (C_{max}) on a case-by-case basis [2]. A potentially unresolved issue with this approach is the involvement of key response pathway entities, both proteins and genes. If these pathways are known to be operative in humans, but not in animals, or vice versa, an alternative approach should be used that includes an investigation in a biological system model to predict directly the human response from human *in vitro* and other prior preclinical information. This approach is applicable for well-studied classes of biotherapeutic (e.g., monoclonal antibodies) whose PK properties are uniform based on their isotypes. The receptor occupancy (RO) that is measured from *in vitro* experiments can be used as a biomarker for human PK/PD projections [23]. It has been recommended that RO values should be low for an agonist and high for an antagonist as a result of downstream biological effects for estimating the FIH doses [24].

AHD projections may also include biopharmaceutics and formulation principles, which can be critical, especially for biopharmaceutics classification system (BCS) class II to IV drugs, i.e., drugs with solubility or permeability rate-limited oral exposure. A basic projection process for the anticipation of human doses is shown schematically in Figure 15.1B [1,2], which includes the characterization of preclinical exposure–response or effect relationships from *in vitro* and *in vivo* experiments, followed by a correction for interspecies differences ("humanization") for both desired and adverse effects. This approach incorporates formulation and biopharmaceutical parameters to predict human PKPD, and it is possible to project human PK and PK/PD profiles for various dosage forms that have varied drug properties (for compounds with solubility/dissolution-dependent systemic exposure, i.e., BCS class II, IV compounds), such as different particle size, solubility, and dissolution, among others. For BCS class I compounds (high solubility and high permeability), modified release dosage formulations are sometimes pursued to optimize systemic exposure and PD response by reducing the peak-to-trough plasma concentrations [25].

15.2 METHODS FOR ANTICIPATING HUMAN PHARMACOKINETICS

Different methodologies have been previously described for predicting drug exposure in humans. The methods of choice can include (1) knowledge of PK parameters or profiles of structural analogs, (2) compound physiochemical properties and other characteristics, (3) the nature of the previously or commonly built exposure–response relationship, and (4) the clinical endpoint that needs to be predicted. To anticipate the FIH dose, preclinical PK/PD practices vary between different therapeutic areas across the world, and they are sometimes reported without establishment of the full dose–response curve. Sometimes the human pharmacologic active dose is first predicted using animal models, along with associated pharmacokinetics and biomarkers, and then it is integrated to full response-time profiles, which are then projected to anticipate human PK or PK/PD. The choice of preclinical species is determined by factors such as health authority guidelines (typically one rodent and one nonrodent) [26], and knowledge of the comparative metabolic fate of the compound in nonhuman and human *in vitro* systems [27]. Drug candidates with a desired *in vivo* pharmacokinetic profile may be further profiled *in vitro*, using assays such as metabolic stability, reaction phenotyping, CYP-450 inhibition/induction, and plasma protein binding, etc. in human microsomes, human recombinant CYP-450 enzymes, and human plasma to get an early indication of whether the compound that worked well in preclinical models would work in humans as well. This information can be particularly helpful in designing an appropriate dosing regimen for phase I studies and to guide dose escalation to achieve rapidly systemic concentrations in humans that are expected to be associated with their pharmacological effect [28]. Successful PK profile predictions have been performed by Wajima et al. and others [29–32], and they have evaluated and compared with the predictions generated using PBPK approaches [6]. However the considerations of formulation-dependent human PK/PD profile projection [4] after oral dosing have rarely been reported. Some PK/PD models have limited flexibility to evaluate the impact of formulation parameters or modified dosage forms on oral PK/PD profiles, which can be important for many reasons, especially for compounds with solubility or dissolution rate limited absorption. To minimize prediction errors, it is advised to use the same species, drug vehicle, salt form, and route of administration for both pharmacological and pharmacokinetic studies. This includes the evaluation of available toxicological, pharmaceutical, pharmacokinetics, and biopharmaceutical information, and predicting human pharmacokinetic parameters including drug systemic clearance (CL), bioavailability (F), area under the curve (AUC), and volume of distribution at steady state (V_{ss}). Most published literature describes single PK parameter projection methods that are based on different allometric interspecies scaling for predicting CL [7, 33–39], V_{ss} [2, 4, 36–41], and F [30, 39, 42]. These projected parameters serve as inputs for physiologically based pharmacokinetic (PBPK) models [29, 41, 43–45].

15.2.1 Methods for Predicting Human Clearance

CL is an important parameter in selecting the size of the dose that is needed to achieve the target concentration, as higher clearance leads to higher doses. The projections of CL by various methods based on different allometric interspecies scaling have met with varying degrees of success, as these approaches do not allow the incorporation of species difference in metabolic clearance (except those resulting from differences in body weight). This therefore increases the error in the projections when scaling these parameters across species. Furthermore, using the allometric scaling method with the rule of exponent (ROE) is proposed to decrease the error in predicting the human plasma clearance [33, 46, 47] by incorporating other species-dependent physiological parameters (e.g., life expectancy, brain weight, or *in vitro* hepatocyte intrinsic clearance (CL_{int}). Here an allometric method is selected and explained using different case examples, which include (1) simple allometry (SA), (2) product of maximum lifespan potential (MLP) and clearance (CL × MLP) versus body weight, and (3) product of brain weight (BRW) and clearance (CL × BRW) versus body weight, to predict human CL based on the exponents of simple allometric scaling with the following criteria: (1) If the exponent of the SA lies between 0.55 and 0.70, then the SA method is preferred to predict CL with the best prediction accuracy;

(2) if the exponent of the SA lies between 0.71 to 1.0, then the CL × MLP approach is superior to the other two prediction methods; and (3) if the exponent of the SA is >1.0, the CL × BRW approach is more suitable for predicting human CL when compared with other methods [33,47]. The predictive performance of all available methods of projecting human drug clearance (total = 29) and methods of calculating area under the curve (total = 106) after oral dosing from *in vivo* and *in vitro* preclinical data has been evaluated, and the fold change between predicted and observed values among different methods was discussed in the literature recently published by PhRMA CPCDC initiative [7]. Among the recommended methods to predict human clearance accurately are the fraction corrected intercept method (FCIM) [35], $TS_{rat\text{-}dog}$ [48], and quantitative structure activity relationship $(QSAR)_{rat\text{-}dog}$ method [49]. The FCIM method uses a universal equation, where the human clearance is predicted based on the intercept obtained from an allometry log–log plot, (α), and the ratio of unbound fraction in plasma between rats and humans (Rf_{up}) and is given by Equation 15.1:

$$CL_{(human)} = 33.35\ \text{mL/Min} \times \left(\alpha/Rf_{up}\right)^{0.77} \tag{15.1}$$

$TS_{rat\text{-}dog}$ and $QSAR_{rat\text{-}dog}$ approaches are based on two species, rats and dogs, and drug clearances (CLs) using allometry or QSAR, respectively. The allometry coefficient (a) for the $TS_{rat\text{-}dog}$ equation is obtained from the intercept of an SA log–log plot between rats and dogs while a fixed value of 0.628 is used as the allometry exponent, and it is given by Equation 15.2, where $BW_{(human)}$ is the average human body weight in kilograms:

$$CL_{(human)} = a_{(rat-dog)} \times BW_{(human)}^{0.628} \tag{15.2}$$

The $QSAR_{rat\text{-}dog}$ approach incorporates the systemic clearance values from rats' and dogs' PK data, molecular weight (MW) of the drug compound, and the number of hydrogen bond acceptors (NHBAs) using multiple linear regression to predict human clearance, which is given by Equation 15.3:

$$\begin{aligned} \text{LogCL}_{(human)} = {} & 0.433 \times \text{LogCL}_{(rat)} + 1.00 \times \text{LogCL}_{(dog)} - 0.00627 \times \text{MW} \\ & + 0.189 \times \text{NHBA} - 0.00111 \times \text{LogCL}_{(dog)} \times \text{MW} \\ & + 0.0000144 \times \text{MW} \times \text{MW} - 0.0004 \times \text{MW} \times \text{NHBA} - 0.707 \end{aligned} \tag{15.3}$$

Other suitable methods for predicting human clearance (CL_{human}), based on *in vivo–in vitro* extrapolation recommended by the PhRMA CPCDC initiative, include microsomal methods, where *in vitro* rates and activities in microsomes are representative of those that occur *in vivo* after being correcting for both binding in plasma and incubation medium, which is given in Equation 15.4. The predictions of CL_{human} from hepatocyte incubation studies, where *in vitro* rates and activities in hepatocytes are representative of those that occur *in vivo* without the correction for protein binding difference is commonly used, and is given by Equation 15.5:

$$CL_{(\text{int in vivo})} = CL_{(\text{int in vivo})}| \times PBSF \times fu_p \div fu_{inc} \text{ (measured)} \quad (15.4)$$

$$CL_{(\text{int in vivo})} = CL_{(\text{int in vivo})} \times PBSF \quad (15.5)$$

PBSF is a physiologically-based scaling factor; and f_{up} and f_{uinc} are correction factors for plasma protein binding and incubation medium, respectively [7].

Another methodology for predicting hepatic metabolism from *in vitro* systems (hepatocytes and microsomes) has been described by Houston and Galetin [50] and was validated in the rat model [51]. This includes the estimation of clearance from the *in vitro* metabolic rate, divided by the substrate concentration (in linear range), scaled by *in vitro* system protein levels, or hepatocyte number to the intrinsic clearance, CL_{int}, for a whole liver, which is converted to hepatic clearance (CL_h), as a function of hepatic blood flow (Q_h) [42,52], and is given by Equation 15.6:

$$CL_h = \frac{CL_{int} \times Q_h}{CL_{int} \times Q_h} \quad (15.6)$$

The total body systemic clearance is generally assumed to be mainly hepatic unless renal clearance (CL_r), i.e., renal excretion, dominates the systemic clearance ($CL_r = f_u$.GFR, where f_u is the fraction unbound of the drug and GFR is the glomerular filtration rate). For drugs that are mainly excreted by filtration through the kidneys, clearance prediction by allometry is known to be accurate. This is indeed consistent with evidence that the GFR correlates well with body weight. Another commonly used method for predicting human CL is based on allometric single-species scaling using Equation 15.7, which has the underlying assumption that unbound CL is scalable to the differences in body weight (BW), with a physiologically relevant exponent of 0.75 (for blood flow, filtration, etc.).

$$CL_{human} = CL_{animal} \times (BW_{human}/BW_{animal})b \quad (15.7)$$

where CL is the clearance, BW is body weight in kg (with typical average values of 0.25 for rats, 7.5–10 for dogs, 3.5 for monkeys, and 70 for humans), and b is the exponent. Simple allometric scaling (SA) assumes that the BW of different species is correlated with the pharmacokinetic parameter (Y) such as CL as given by Equation 15.8:

$$Y = a \times BW^b \quad (15.8)$$

where Y is CL or V_{ss} and a, b are the allometric coefficient and exponent, respectively [33,47].

If *in vitro* estimates of rodent clearance are in overall agreement with the observed value from the early rodent *in vivo* pharmacokinetics studies, then a predicted human PK parameter value can be provided from human *in vitro* metabolism, protein binding,

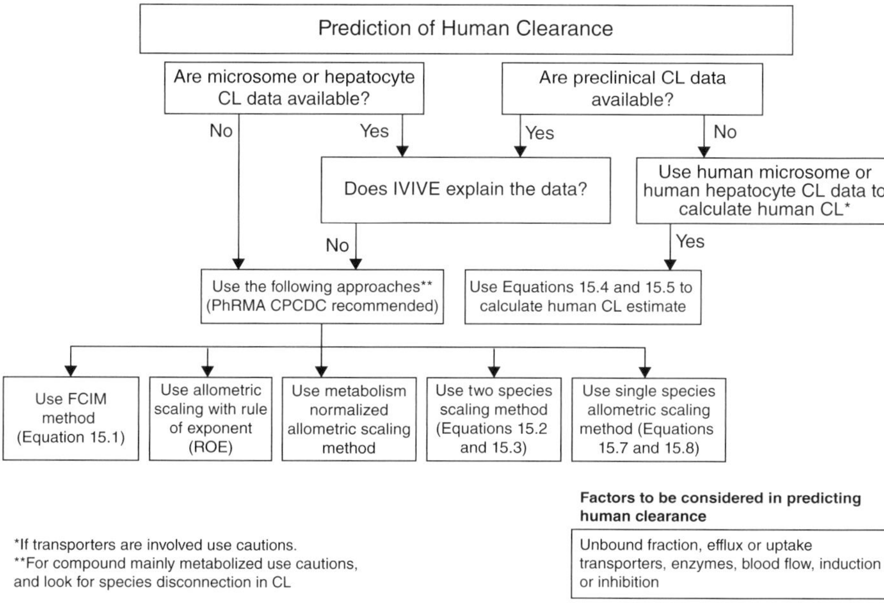

Figure 15.2 Overview flowchart with factors to predict human clearance, based on different mechanisms of elimination (renal, hepatic, or biliary), along with the other factors that need to be considered, such as fraction unbound, efflux/uptake transporters, enzyme activity, blood flow, induction or inhibition, among others.

and blood plasma distribution data. If the gap between *in vitro* and *in vivo* PK parameters is minor to moderate in rodent studies, the relevant factor should be determined, such as *in vitro* study conditions (linearity, enzyme viability, etc.), or accuracy in estimating *in vivo* clearance, which needs to be accounted for when *in vivo* clearance in humans is predicted from *in vitro* data. Alternatively, the interspecies ratio of *in vitro* metabolic rates can be used to estimate the *in vivo* clearance in humans by scaling the measured *in vivo* clearance in rats by this ratio (or other relevant scaling factors such as protein content, liver size per total body weight, etc.). Design of experiment criteria to predict the human clearance is laid out in Figure 15.2 [43] along with other factors that need to be considered in predicting human drug clearance.

15.2.2 Methods for Predicting Human Volume of Distribution

Volume of distribution determines the half-life and frequency of dosing required for maintaining the desired therapeutic effect. The different available methods (empirical, semi-mechanistic, and mechanistic, total = 24), with and without protein binding corrections for the prediction of human volume of distribution at steady state (V_{SS}) are evaluated and recommended by the PhRMA CPCDC initiative literature recently published [41]. The Oie–Tozer approach (Equations 15.9 and 15.10) was evaluated and is a recommended method for the prediction of human V_{SS} from the observed V_{ss} obtained from different preclinical species. The unbound fraction of drug in tissues (fu_t) in humans was calculated as the mean values of the fu_t values from preclinical species [37]:

$$V_{ss(human)} = V_p + fu_p \times V_e + (1 - fu_p) \times R_{E/I} \times V_p + fu_p/fu_{t(human)} \quad (15.9)$$

$$fu_{t(animal)} = (V_r \times fu_p) / [V_{ss} - V_p - (fu_p \times V_e) - (1 - fu_p) \times R_{E/I} \times V_p] \quad (15.10)$$

where V_p represents plasma volume and equals 0.0313, 0.0447, and 0.0515 L/kg for rats, monkeys, and dogs respectively; V_e represents extracellular fluid volume and equals to 0.265, 0.208, and 0.216 L/kg for rats, monkeys, and dogs, respectively; V_r represents aqueous fluid volume excluding extracellular and plasma volume and equal to 0.364, 0.485, and 0.450 L/kg for rats, monkeys, and dogs, respectively; $R_{E/I}$ represents the ratio of binding proteins in extracellular fluid to those in plasma and equal to 1.4 for all species [53, 54]. Another widely used method to calculate V_{ss} is single-species allometric scaling based on preclinical data with the assumption that the distribution process is proportional to body weight among species and is given by Equation 15.11 [37, 41, 53, 54]:

$$V_{ss\ human} = V_{ss\ animal} \times (BW_{human}/BW_{animal})^b \quad (15.11)$$

where b is the exponent of 1 ($b = 1$).

Volume of distribution, along with clearance and rate of absorption, determines the plasma concentration time profile within the dosing interval, along with C_{max} and half-life ($t_{1/2}$). This can be critical if the temporal concentrations (i.e., C_{max} or C_{trough}) are relevant to drug responses. The decision criteria used for the prediction of human volume of distribution is shown in Figure 15.3, along with other factors that need to be considered in predicting the human volume of distribution.

15.2.3 Methods for Predicting Human Oral Bioavailability and Absorption

The oral bioavailability (F) of a drug is defined as the fraction of an oral dose that reaches the systemic circulation (post-liver), as unchanged drug. F is given by the product of fraction of dose absorbed from gastrointestinal lumen, and fraction of drug that escaped first-pass elimination by liver and gut wall, and it is given by Equation 15.12:

$$F = F_a \times F_h \times F_g \quad (15.12)$$

F_a is the fraction of dose absorbed unchanged from gastrointestinal lumen, and F_h and F_g are the fractions of drugs escaping first-pass elimination by liver and gut (intestine) respectively. F_a is often determined by the solubility and stability of the drug in the gastrointestinal tract and its permeability across the intestinal membrane.

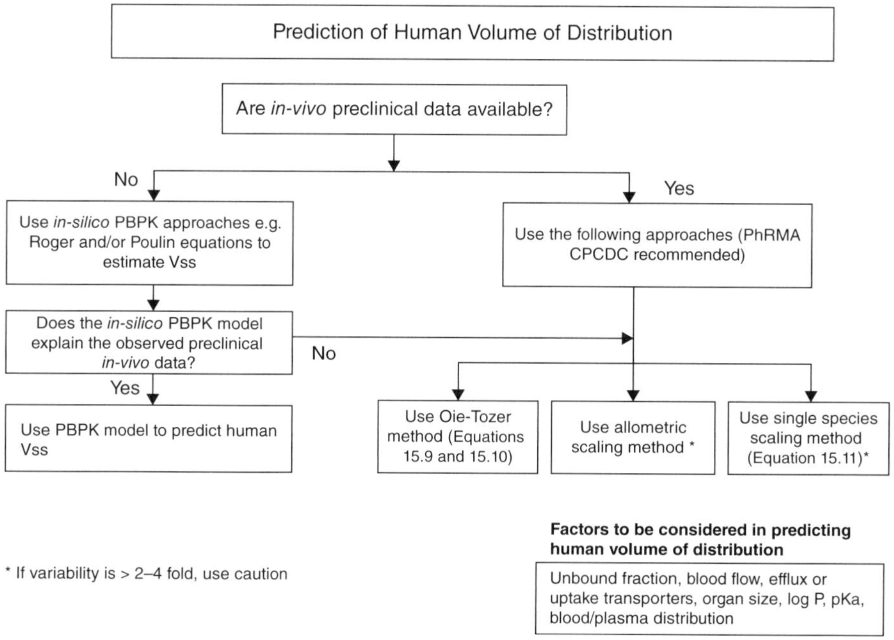

Figure 15.3 Overview flowchart of the decision criteria to predict human volume of distribution, based on different experimental criteria, along with other factors that need to be considered. For *in silico* PBPK approaches, Roger or Poulin equations or others can be obtained from the literature [41, 67, 68, 74].

F_a can also be calculated by software applications such as the ACAT physiologically based PK gut models in GastroPlus™ (Simulations Plus, Lancaster, CA)/SimCyp™. (Simcyp Ltd, Sheffield, U.K.) F_h can be calculated from the known hepatic blood flow (Q_h) [42, 52], and hepatic clearance (CL_h), as given by Equation 15.13:

$$F_h = 1 - (CL_h/Q_h) \quad (15.13)$$

F, F_a, and F_h can also be determined from intravenous and oral rat studies using radiolabeled drug and rat F_a calculated from rat ADME studies that can be used as an estimate for human F_a [55]. Often little is known about F_g at early stages, and F_g is usually assumed to be 1 unless available data indicate high intestinal metabolism. Thus, F can be calculated by Equation 15.14 [42]. The typical Q_h values for different species are 90, 70, 35, 44, and 21 mL/min/kg for mice, rats, dogs, monkeys, and humans, respectively [52].

$$F = F_a \times [1 - (CL_h/Q_h)] \quad (15.14)$$

The absorption rate constant (k_a) in preclinical species was estimated by simultaneously fitting the appropriate compartment model to the observed intravenous or oral

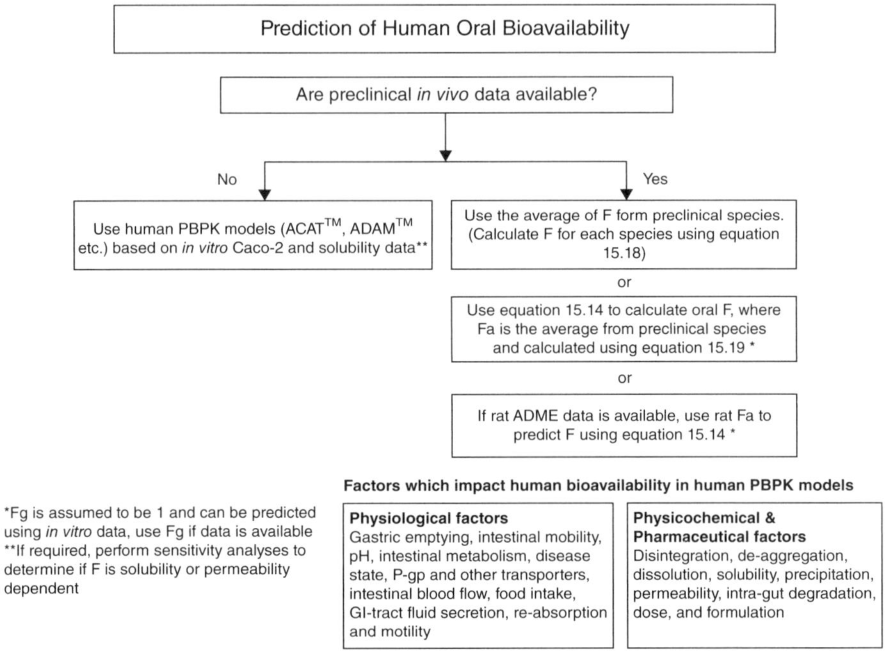

Figure 15.4 Flowchart of the decision criteria for calculating human oral bioavailability and absorption. The factors that impact human bioavailability in PBPK models including both the physiochemical and biopharmaceutical properties of drugs are also listed.

PK profiles using nonlinear regression analysis. Both human F and k_a can assumed to be the average of F and k_a from preclinical species unless there is a species disconnect, e.g., low vs. moderate or high F. Factors in the prediction of human oral bioavailability and absorption are listed in Figure 15.4, with the endpoints needed for decision making along with other physiochemical and physiological factors that also need to be considered in predicting human drug bioavailability. Collectively, F and k_a are the key parameters that need to be predicted if drug effects are assumed to be a function of the overall exposure.

15.2.4 Methods for Predicting Human Plasma Concentration-Time Profile

From the primary PK properties, i.e., V_{ss}, CL, k_a, and F of a compound, additional secondary parameters, such as the maximal concentration (C_{max}), half-life ($t_{1/2}$), and the area under the concentration-time curve (AUC), can be determined. Therefore, it is important to discuss the methods for predicting the human plasma concentration-time profile using different approaches. Successful human intravenous PK profile predictions have been performed using the Wajima method [29–32], or by using PBPK approaches [6]. Oral human PK profiles can be generated using the ACAT™ or ADAM™ models in GastroPlus or SimCyp. A flowchart for human PK profile

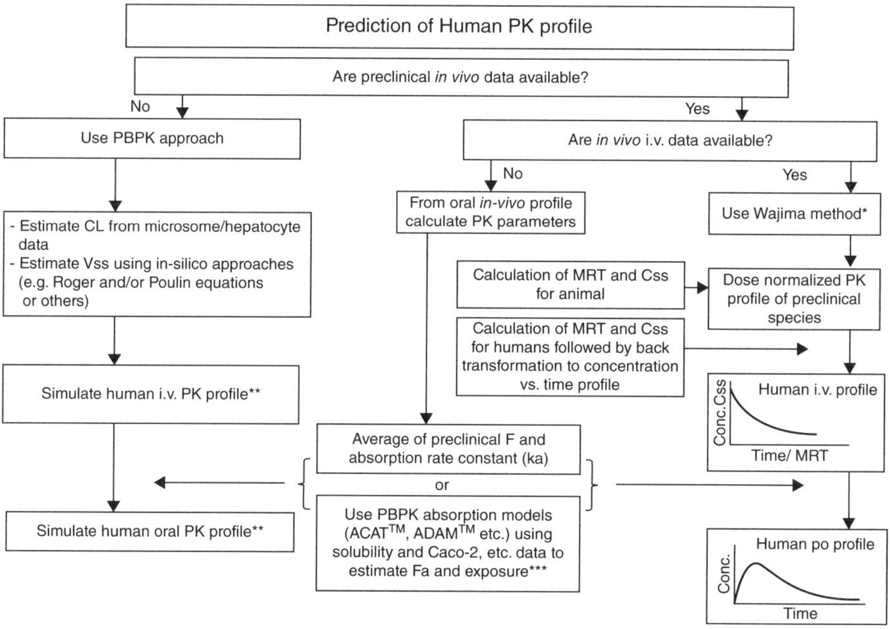

*Check for species disconnection (non-superimposable profile), and use cautions if it is observed
**Verify with the preclinical data if available
***PBPK gut models can be used along with preclinical PK to predict human PK profile

Figure 15.5 Flowchart of the decision criteria for predicting the human PK profile based on the available preclinical *in vivo* data. For *in silico* PBPK approaches, the Roger or Poulin equations are obtained from literature [67, 68, 74]. Using the Wajima method [32], caution must be used if there are species disconnects (non-superimposable profile) in predicting preclinical an intravenous PK profile.

predictions is given in Figure 15.5. Typically detailed steps for formulation-dependent human PK/PD profile projection [4] after oral dosing have not been reported extensively in the past. The discussed PK/PD models have limited flexibility to evaluate the impact of formulation parameters or modified dosage forms on oral PK/PD profiles, which can be important for many reasons in particular for compounds with solubility or dissolution rate limited absorption, i.e., BCS class II or IV compounds. It is important to note that certain factors can be a major source of uncertainty in predicting human PK. They include the following: (1) PK data generated from toxicological studies at high doses conducted with formulation that were dissimilar to FIH formulations; (2) in the case of intravenous administration of poorly soluble drugs that are aided by solubilizing excipients, caution is needed as some formulation components may affect drug disposition in animal models [56], which are not necessarily used in humans; and (3) observed nonlinear PK resulting from solubility and permeability saturation that can result in dose underproportionality, or first-pass effect saturation, drug transport saturation, and drug elimination saturation may result in dose overproportionality. These issues along with species disconnects (which may be a result of differences in interspecies metabolism or transporters), and food-dependent

PK (which may be a result of differences in biliary or pancreatic secretions or a result of differences in hepatic or intestinal metabolism), along with other physiological changes, are described. A practical projection of human PK along with PD (concentration and effect vs. time profiles) based on preclinical PK/PD and absorption model-based approaches in drug development are discussed [57, 58]. Modeling approaches that can be readily implemented by pharmaceutical drug development are illustrated using common tools such as allometric scaling [33–36, 48, 59]; the Wajima method [30, 32]; and GastroPlus [60], including its PBPK, PK/PD, ADMET Predictor™, and Simcyp [61] software modules [60, 62, 63].

GastroPlus, which evolved from Yu's compartmental, absorption, and transit (CAT) model [64], now allows physiologically based oral absorption modeling and systemic PK/PD simulations using nine gut compartments that represent different segments of the human gastrointestinal tract pharmacokinetic using Advanced Compartmental Absorption and Transit (ACAT) model [62]. PK/PD profiles based on direct and indirect PD models [65] can also be simulated using software such as GastroPlus and its integrated PDPlus™ module. Case examples demonstrate the application of the "reverse pharmacology" approach or *in vitro–in vivo* extrapolation (IVIVE) strategies for the selection of clinical backup candidates based on modeling with inclusion of data from frontrunner compounds. The examples have challenging ADME properties, such as low gastrointestinal solubility, high or unpredictable first-pass effects, unknown potency/efficacy in humans, or all of the above, for which human PK/PD profiles had to be established for decision making. For all preclinical species, the plasma concentration data obtained after the intravenous (IV) or oral (p.o.) dose administration were analyzed with noncompartmental analysis methods using WinNonlin software v5.1, (Pharsight Corporation, Mountain View, CA). The area under the plasma concentration versus time curve (AUC) was determined using the linear trapezoidal method. The terminal phase slope (λ_z) values were determined by means of log-linear regression on at least three of the latest sampling time points from the plasma concentration versus time curves. The AUC from the last measurable concentration to AUC at infinite time was estimated by C_{last} / λ_z, where C_{last} is the plasma concentration at the last sampling time point. Additionally, the area under the first moment curve (AUMC) was also determined by calculating the area under the product of time and concentration versus time curve. According to the dose administered, AUC, and AUMC values, the systemic clearance (CL), mean residence time (MRT), and steady-state volume of distribution (V_{ss}) were calculated after intravenous administration by using the following equations, respectively:

$$\mathrm{CL} = \mathrm{Dose}/\mathrm{AUC} \tag{15.15}$$

$$V_{ss} = \mathrm{MRT} \times \mathrm{CL} \tag{15.16}$$

$$\mathrm{MRT} = \mathrm{AUMC}/\mathrm{AUC} = V_{ss}/\mathrm{CL} \tag{15.17}$$

The maximum plasma concentration (C_{max}) and time to reach the maximum plasma concentrations (T_{max}) after oral administration can be determined directly from the observed pharmacokinetic profile data. The oral bioavailability (F or F_{oral}) in preclinical studies can be estimated by comparing the mean dose-normalized AUC value after oral administration with the mean dose-normalized AUC after intravenous administration by using Equation 15.18:

$$F_{oral} = \frac{AUC_{p.o.}}{AUC_{IV}} \times \frac{Dose_{IV.}}{Dose_{p.o.}} \tag{15.18}$$

From these calculated PK parameters for different preclinical species, the case examples also include human PK profile projections. Anticipation of the FIH dose (AHD) is demonstrated by incorporating the different practical aspects of drug development: (1) to show the utility of preclinical PK/PD (*Sections 15.3.1–15.3.3*); (2) by incorporating the species differences (*Sections 15.3.1–15.3.2*); (3) practical projections of human CL, $V_{ss,}$ and F parameters using an ACAT model built within GastroPlus (*Sections 15.3.1–15.3.3*), and a PBPK approach (*Section 15.3.3*).

15.3 CASE STUDIES

15.3.1 Anticipation of Human PK/PD for a Backup Compound Using a "Reverse Pharmacology" Strategy

15.3.1.1 Study goal Backup B is a follow-on candidate to frontrunner A, which has known comparable efficacy from human clinical trials and dog preclinical studies. Human PK/PD modeling was to be performed for backup B versus frontrunner A at the lead optimization stage for differentiation assessment. The following goals were set for anticipating FIH dose projections: (1) determine PK properties and profiles, (2) establish whether backup B had higher potency and PD profile over time, and (3) determine whether backup B could be administered at similar or lower daily doses when compared with frontrunner A to maintain plasma levels above the IC_{50} (drug concentration at which 50% of response is observed), while showing a similar PD profile.

15.3.1.2 Background Backup B is a base ($pK_a = 8.4$) with a MW < 500 and a gastrointestinal solubility of ~2.3 mg/mL. Given the log P of ~2 and a relatively high permeability in a Caco-2 assay (P_{app} apical to basolateral, 13×10^{-6} cm/s), backup B was classified as a potential BCS Class I drug (high solubility and high permeability). Plasma protein binding for backup B was 91.2% in rats, 81.1% in dogs, and 83.5% in humans. Human CL was predicted using "proven" PhRMA CPCDC initiative methods that had also been used to predict and describe the human oral profile and CL/F of clinical frontrunner A (<1.5-fold prediction error) and which included a single-species allometry scaling method [33, 47]. The human intravenous profile (plasma concentration vs. time profile) had been predicted by generating a profile using the Wajima method with dose-normalized rat and dog data as shown in Figure 15.6A. F_{abs} and F were obtained using the ACAT module in GastroPlus and by using

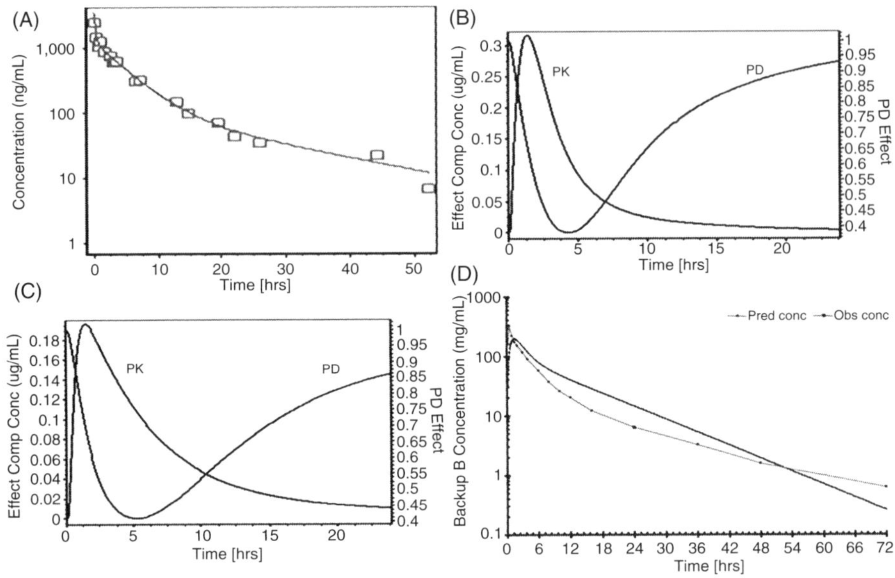

Figure 15.6 (A) Backup B predicted intravenous human plasma concentration-time profile generated by using the Wajima C_{ss}-MRT method, from observed rat and dog data (open squares) where the *x*-axis represents time (h) and the *y*-axis represents the concentration (ng/mL). (B) Frontrunner A human predicted PK and PD, as a function of time, where PK is shown on the left axis, and PD on the right axis, of the graph using an indirect response PD model. Frontrunner A observed PK and PD data are not depicted. (C) Backup B human predicted (simulated) PK and PD, as a function of time, where PK is shown on the left axis, and PD on the right axis, of the graph for Backup B using an indirect response PD model. (D) Predicted human PK profile versus the observed profile of Backup B. The human predictions for AUC and $t_{1/2}$ were <1.25-fold; PD data are not depicted.

the simulated human PK profile. Human V_{ss} was estimated using different methods including the universal Oie–Tozer method [2, 37], along with *in silico* predictions with a PBPK model for a default human [66–68]. These methods predicted the V_{ss} for backup B to be within twofold of frontrunner A (data not shown). Frontrunner A's IC_{50} was compared with that of backup B from the rat efficacy model after oral dosing.

15.3.1.3 Method As both preclinical and clinical PK/PD data were available for frontrunner A, a "reverse pharmacology" strategy was used to validate and predict the human PK/PD profiles of backup B, which included the following steps:

1. Establishment of a PK/PD link between available human and preclinical data for frontrunner A followed by application to backup B: Frontrunner A observed human oral PK profile could be described with CL and V_{ss} using allometry scaling and the ROE methods [7] combined with an ACAT human default

model. Preclinical rat PK and PD profile data were collected under identical conditions for frontrunner A and backup B (not shown), and a PK/PD rat model was established. The PD profile was modeled using an indirect I_{max} model [69] within the GastroPlus PDPlus module. For frontrunner A, the IC_{50} value was then used in combination with clinical PK data to describe the known human PK/PD profile (see step 5).

2. Determination of the potency for backup B in the rat PD model and comparison with frontrunner A: The IC_{50} values for backup B in rat model were determined and corrected for protein binding and then compared with the clinically known IC_{50} value from frontrunner A. The protein binding interspecies correction factors were near unity, as both drugs have low protein binding both in rats and humans (<90%). Protein binding corrections can be important especially for highly bound drugs [13], and they should be considered on a case-by-case basis.
3. Prediction of CL and V_{ss} parameters for backup B: CL was determined using allometric scaling and the ROE methods [7] and was found to be consistently low (~2 mL/min/kg) from preclinical species (rats and dogs) suggesting concordance between species. The scaling factor used was 0.7 among species, and all these parameters are listed in Table 15.1. For V_{ss} the Oie–Tozer method (Equations 15.9 and 15.10), the single-species scaling method (Equation 15.11, with exponent $b = 1$), and the PBPK model from Lukacova et al., were used [66–68]. However, other methods could have been employed [39,41]. Overall, the nominal human V_{ss} was calculated from these methods, and the chosen value for humans was ~ 1 L/kg.

Table 15.1 Estimation of human clearance for backup B, obtained from the single-species allometry scaling method using *in vivo* data from different species (rat, dog, and mouse), where BW is the average body weight (kg) for different species (for humans, the average BW is 70 kg); *a* and *b* are the scaling coefficients

Species	*BW* (kg)	*log BW*	*CL* (mL/min/kg)	*CL*BW* (mL/min)	*log CL*BW*	*b*	*log a*
Rat	0.25	−0.60	13	3.3	0.51	0.7	0.93
Human (estimated via rat)	70	1.85	**2.4**	168	2.23	0.7	0.93
Dog	10	1	4.4	44	1.64	0.7	0.94
Human (estimated via dog)	70	1.85	**2.5**	172	2.24	0.7	0.94
Mouse	0.02	−1.70	17	0.3	−0.45	0.7	0.72
Human (estimated via mouse)	70	1.85	**1.5**	103	2.01	0.7	0.72

$CL = a \times BW^b$.
$\log(CL) = \log(a) + b \times \log BW$.

4. Prediction of human oral PK profiles for backup B using GastroPlus: The default absorption model for a 70-kg human with a fasting gastrointestinal physiology was used to predict the human oral PK. When needed, PK profiles with an altered dosing regimen or formulations could be simulated in the ACAT model within GastroPlus.
5. Prediction of human PD profiles for backup B and comparison with human PK/PD profile of clinical frontrunner A: Here the determined preclinical IC_{50} values were used within the GastroPlus PDPlus module to simulate human plasma PK/PD profiles after oral dosing regimens for frontrunner A (Figure 15.6B) and backup B (Figure 15.6C), respectively. The human predicted PK is shown on the left axis and the PD is on the right axis of each graph for both the frontrunner A and the backup B using an indirect response model.

15.3.1.4 Results An intravenous PK profile is generated using human estimates of CL and V_{ss} from preclinical species (Figure 15.6A, for backup B), and using the same IC_{50} values for humans as obtained from preclinical rat PK/PD data after protein binding corrections, the human PK and PD profiles for frontrunner A and backup B were simulated using GastroPlus, which are shown in Figure 15.6B and Figure 15.6C, respectively. Both figures show double-Y plots with plasma levels (left axis) and the PD response (right axis). For backup B, rat PK/PD studies had predicted a human IC_{50} of ~0.07 μg/mL, and for frontrunner A, the IC_{50} prediction was ~0.14 μg/mL, respectively, suggesting possibly better potency for backup B. For frontrunner A, the preclinical results were in line with plasma concentrations achieved with efficacious clinical doses (data not shown). These PK/PD results supported frontrunner A's clinical therapeutic dose of 20 mg, yielding a ~50% (0.5 value) PD effect for ~8 h (Figure 15.6B). By comparison, backup B (Figure 15.6C) demonstrated that a slightly lower daily dose of 15 mg is sufficient to maintain a similar PD effect profile. This was mainly attributed to the higher potency of backup B, as simulated oral PK profiles are overall similar for both compounds. Figure 15.6D shows backup B's predicted human PK profile versus the observed profile. The human predictions for AUC and $t_{1/2}$ were acceptable and were within <1.25-fold.

For successful PK predictions for CL and V_{ss}, multiple methods should generally be attempted to evaluate concordance between methods (set arbitrarily to be within less than twofold) [39]. For the PD response, prediction confidence was enhanced as human data was available for frontrunner A, which allowed a "reverse pharmacology" approach, where clinical PK/PD was linked to an established preclinical rat PK/PD model.

15.3.1.5 Study Conclusion A "reverse pharmacology" approach can be successfully used to establish a link between observed clinical PK/PD data and a preclinical PK/PD model. Simulations showed that for backup B, a 15-mg dose provides a similar PD profile when compared with a 20-mg frontrunner A. The slightly lower dose for backup B is more desirable and could be attributed to its higher potency, as both compounds showed overall similar PK properties and profiles.

15.3.2 Anticipation of Human PK/PD Parameters from Preclinical PK/PD Data when Interspecies PK Disconnects are Observed

15.3.2.1 Study Goal Drug M is an inhibitor that suppresses different pathways in tumor cells. Using the available preclinical data, the specific goals were (1) to determine the anticipated average human exposure, C_{ave}, and C_{min} from preclinical efficacy models; (2) to predict human PK parameters including CL, V_{ss}, and F; and (3) to estimate the human daily dose to achieve the target exposure range (C_{ave}) based on step (1).

15.3.2.2 Background Drug M is moderately lipophilic (logP = 2.81) and exhibits low solubility at physiological pH (0.024 mg/mL at pH 6.8) and moderate-to-high permeability in Caco-2 (P_{app} apical to basolateral, 8×10^{-6} cm/s). The Caco-2 permeability was converted to a human intestinal permeability using an internally developed correlation equation. Overall, drug M was classified as a BCS II (low solubility/high permeability) or possibly as a BCS IV drug (low solubility/low permeability) as a result of its moderate permeability.

15.3.2.3 Method Pharmacokinetic studies for drug M were performed in several species. The observed systemic clearances was low in rats and mice; yet they were moderate to high in dogs and monkeys when compared with their respective liver blood flows [70] ranging from 1.1 to 21 mL/min/kg (Table 15.2). A low-to-moderate volume of distribution was observed across species (0.1 to 2.6 L/kg). The observed oral bioavailability was low to moderate (5–49%, Table 15.2). To establish metabolism IVIVE such that human CL can be estimated, *in vitro* data were obtained from rat, dog, monkey, and human hepatocytes. Although rat data showed overall

Table 15.2 Summary of observed preclinical pharmacokinetic parameters for different species for drug M, which shows a PK disconnect, low *F*, in the monkey when compared with other species

Parameters	Mouse	Rat	Dog	Monkey
Hepatocyte clearance (μL/h/10^6 cells)	Not done	123	551	686
Scaled CL_h from hepatocytes (mL/min/kg)	-	8	21.5	21
Observed *in vivo* CL (mL/min/kg)	4.3	1.4	21	17
V_{ss} (L/kg)	0.1	0.1	2.6	0.9
$AUC_{inf,po}$ (μM*h)	30	144	0.19	0.08
C_{max} (μM) (observed)	26	48	0.12	0.02
T_{max} (h) (observed)	0.5	0.5	0.3	3
$t_{1/2}$, IV (h)	0.9	2.3	5.2	1
F (%) (15.18)	43	49	42	5
F_a (%) (15.19)	45	50	100	8

Table 15.3 Predicted human PK parameters using preclinical data for drug M

Scaled CL_h from hepatocytes (mL/min/kg)	2.1
V_{ss} (Allometry method) L/kg	4.2
F_a (%) (average of preclinical F_a = human F_a, best-case scenario)	51
F_a (%) (lowest F_a from monkey = human F_a, worst-case scenario)	8
F (%) (using Equation 15.14 and average preclinical F_a, best-case scenario)	46
F (%) (using Equation 15.14 and monkey F_a, worst-case scenario)	7

low clearance both *in vitro* and *in vivo*, monkey and dog hepatocyte studies yielded higher scaled *in vitro* clearances, along with moderate-to-high observed *in vivo* total systemic clearances (Table 15.2). For humans, hepatocytes studies predicted a low clearance, as listed in Table 15.3 [39]. The V_{ss} in humans was estimated by the empirical allometric scaling of body weight of the animals (mice, rats, dogs, and monkeys) and substituting the corresponding observed animal V_{ss} in Equation 15.8 with an exponent of 1.52, without the correction of interspecies differences in plasma protein binding [41]. Human bioavailability (F) was calculated based on preclinical F_a estimates that were calculated using Equation 15.14, where F_g is assumed to be negligible (Table 15.3). The terminal human half-life was calculated based on the intravenous profile from the Wajima plot (Figure 15.8B).

15.3.2.4 Results The targeted human efficacious concentration was calculated from a preclinical model, and it was found that an ~6-mg/kg dose resulted in >50% target inhibition. Human clearance was predicted using hepatocytes data, and a low clearance in humans was predicted (Table 15.3). Hepatocytes data overall predicted the *in vivo* clearance very well (overpredicts in rats) using Equation 15.5, and there was a good *in vitro–in vivo* correlation for drug M hepatocyte data, providing a strong rationale for using human hepatocyte data, giving a predicted human clearance of 2.1 mL/min/kg, which is low as compared with human blood flow (20 mL/min/kg). Human V_{ss} was calculated to be 4.2 L/kg (Figure 15.7A). The regression coefficient of 0.98 suggests good correlation among species. Fraction absorbed (F_a) in preclinical species was calculated from the modification of Equation 15.12, where predicted F_h is calculated using Equation 15.13, F calculated from preclinical species using Equation 15.18, and assuming F_g to be 1, the F_a is given by Equation 15.19:

$$F_a = \frac{F}{\left(1 - \frac{CL_h}{Q_h}\right) \times F_g} \tag{15.19}$$

The F_a for drug M in preclinical observed data ranged from 8% to 100% among different species with an average of 51%, giving two scenarios for human fraction absorbed (worst-case scenario where F_a = 8% and best-case scenario where F_a = 51%, Table 15.3) [42]. Based on these two different scenarios of F_a, the calculated human F using Equation 15.14 has two values: 46% when F_a is 51% (best-case scenario) and 7% when F_a is 8% (worst-case scenario). The fraction absorbed can

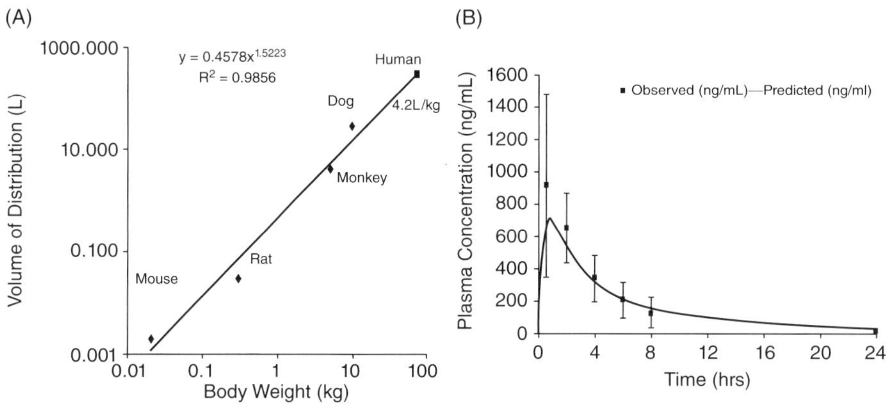

Figure 15.7 (A) Projection of human volume of distribution (V_{SS}) using allometry scaling without protein binding corrections, by plotting body weight (BW) against calculated volume of distribution (V_{SS}) for drug M, from different species. The coefficient of regression (r^2) is 0.986, and the slope is 0.457. (B) Observed and model predicted time plasma concentration profile for drug M after oral dosing, where the solid dots represent the observed concentration and the solid line represents the model predicted concentration profile.

also be derived from the mass balance studies using bile duct cannulated rats, by calculating the combined recovery of radioactivity from bile and urine after an oral administration of a radiolabeled compound [55]. The calculated terminal human half-life was ~23 hours. The rate of absorption (k_a) was calculated from the mean of rate absorption constant from preclinical species and is 3.4 per hour (1/h). Based on the predicted human PK parameters, a human efficacious dose can be predicted using Equation 15.20:

$$\text{Dose}_{\text{human}} = \frac{\text{CL}_{\text{human}} \times C_{\text{ss}} \times \tau}{F_{\text{human}}} \tag{15.20}$$

$\text{Dose}_{\text{human}}$ is the oral dose in humans, CL_{human} is the predicted clearance in humans, C_{ss} is the steady-state concentration, τ is the time to reach steady state, and F_{human} is the predicted bioavailability in humans. This yields two anticipated doses based on different predicted bioavailability in humans. For the low bioavailability scenario ($F = 7\%$), the predicted human dose is 340 mg qd, whereas for the high bioavailability scenario ($F = 46\%$), the predicted human dose is 50 mg qd. Separately, for a safe starting dose, a NO(A)EL dose from toxicology studies is used to calculate a human equivalent dose (HED). After application of another safety factor, the maximum recommended starting human dose (MRSD) was 37.8 mg and the conservatively chosen safe starting dose was 30 mg (Table 15.4). For safe starting dose calculations, the latest regulatory guidelines should always be consulted [14–16].

15.3.2.5 Study Conclusion An example of a human PK profile after a 100-mg oral dose from FIH studies is shown in Figure 15.7B. The observed AUC was within

Table 15.4 Example for the selection of the maximum recommended starting dose in humans based on NO(A)EL data [14–16]

Parameters	Rat	Dog
GLP* Toxicology Doses	30, 100, 300 mg/kg qd	10, 30, 50 mg/kg qd
NO(A)EL* (mg/kg)	100	10
HED* (mg/kg)	$100 \times 0.16 = 16$	$10 \times 0.54 = 5.4$
Conversion factor: 0.16 for rats & 0.54 for dogs		
Human total dose (mg) (× 70 kg)**	$16 \times 70 = 1120$	$5.4 \times 70 = 378$
Starting dose (mg) for humans	1120/ 10 = 112	378/ 10 = 37.8
Safety factor: 10 for rats & dogs		
Anticipated human starting dose (mg)***		30

*GLP: Good laboratory practices; NO(A)EL: no observed (adverse) effect level; HED: human equivalent dose.
**A human body weight of 70 kg is assumed.
***The human starting dose is chosen based on the preclinical species with the lower NO(A)EL (dog). 30 mg was conservatively chosen after 37.8 mg was identified via the dog, which was the more sensitive species with the lower NO(A)EL.

less than a twofold error (1.25 fold) of the predicted AUC. The observed maximum plasma concentration (C_{max}) was less than twofold (1.60 fold), indicating an overall successful prediction for drug M.

15.3.3 Anticipation of Food Effect Using Dog *In Vivo* Data and a PBPK Absorption Model

15.3.3.1 Study Goal Food can have a pronounced effect on the pharmacokinetics of various drug compounds, and it is essential to predict the effect of meals on PK in early drug development, which helps in predicting drug absorption and disposition in humans [71]. The purpose here was to determine the absorption and disposition of compound C in preclinical species and to use the preclinical data to predict human pharmacokinetic parameters and profiles using allometric scaling and other "proven" PhRMA CPCDC prediction methods [7,41].

15.3.3.2 Background Compound C is highly lipophilic (calculated LogP = ~5; ADMET Predictor, Simulation Plus Inc.) and a weak base (measured $pK_{a1} = 3.0$). The solubility is ~2 mg/mL in an acidic buffer (pH = 1) and is significantly lower in a pH 6.8 buffer (1–2 μg/mL). The apparent permeability determined in Caco-2 assays was moderate to high and was converted to a value of human effective permeability ($P_{eff} = \sim 1.5 \times 10^{-4}$ cm/s). Based on these data, compound C was classified as a potential BCS class II drug. The plasma protein binding of compound C was found to be high (>97.5), and no large differences across preclinical species were observed.

15.3.3.3 Method The PK of compound C had been evaluated in rats, monkeys, and dogs after a single intravenous or oral dose to support drug disposition and toxicology studies. These preclinical data were also used for human exposure and absorption predictions using different oral formulations. PK in rats and monkeys were used to anticipate FIH clinical doses. Human CL was estimated using allometric single-species scaling (SSS) (unit: mL/min) using Equation 15.7 [72]. Additional CL prediction approaches included allometric scaling methods with the ROE [33,46,47], FCIM [73] (Equation 15.1), $TSS_{rat\text{-}dog}$ [72] (Equation 15.2), and $QSAR_{rat\text{-}dog}$ [49] (Equation 15.3). The human V_{ss} was initially predicted by SSS with a fixed exponent of 1 ($b = 1$) from V_{ss} in rats or monkeys using Equation 15.11. As additional preclinical pharmacokinetic data became available during the late development stages, the prediction of human V_{ss} was further optimized using a simple generic allometric scaling method (Equation 15.8), and the Oie–Tozer method (Equation 15.9). The Wajima method (C_{ss}-MRT approach) [49] was used to predict the intravenous concentration-time profile in humans based on the assumption that normalized concentration-time profiles from several species are superimposable [6,49]. The normalized intravenous plasma concentrations obtained from various preclinical species (Figure 15.8A) are used to project the human intravenous profile using the Wajima method, and they are shown in Figure 15.8B. The estimated human intravenous pharmacokinetic profile was then analyzed by compartmental models using WinNonlin Phoenix v6.1 (Pharsight Incorporation, Cary, NC). The obtained micro-constants, CL, and V_{ss} were then used as input for the GastroPlus built-in compartmental PK module to simulate the human oral concentration versus time profile to estimate the fraction of drug absorption (F_a) and oral bioavailability (F). The percentage of first-pass hepatic extraction

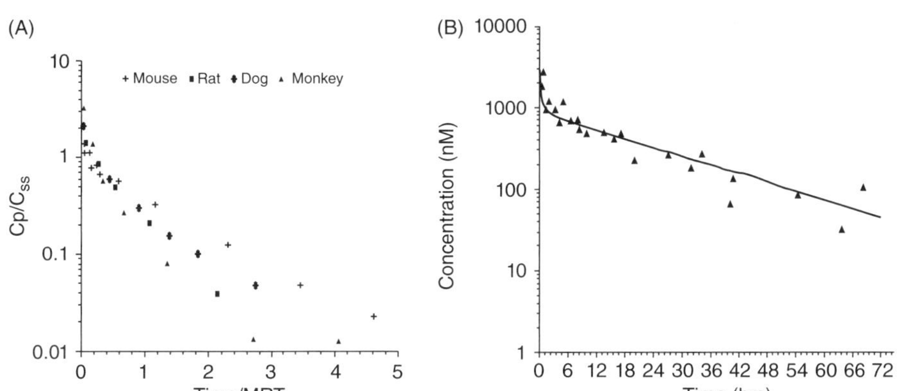

Figure 15.8 (A) The normalized intravenous plasma concentration profile from preclinical *in vivo* data from different species, using the Wajima C_{ss}-MRT approach for compound C. C_p is the observed plasma concentration, C_{ss} is steady state plasma drug concentration, Time is the time of observation, and MRT is the mean resistance time. (B) Predicted human intravenous plasma concentration-time profile after a 1-mg/kg intravenous dose from preclinical species data. This human intravenous profile serves as an input for estimating human PK performance after oral dosing in PBPK models, e.g., in GastroPlus.

(FPE %) was calculated from the ratio between predicted human CL and human hepatic blood flow (Q_h). Dog PK profiles were used to estimate the mean precipitation time (MPT) under fasting and fed conditions using GastroPlus. Higher values of MPT can be expected for compound C under the fed conditions. These values of mean precipitation time that are optimized in the dog model are used to simulate human PK with the underlying assumption that MPT has a similar influence on oral drug absorption between dogs and humans. The predicted oral PK parameters (C_{max}, T_{max}, and AUC) in humans, after applying the observed MPT value from the dog model, were then determined under fed and fasting conditions using GastroPlus and then compared with observed dogs' PK under fed and fasting conditions for immediate release (F1) and modified release (F2) dosage form. The prediction accuracy for pharmacokinetic parameters was then assessed by the prediction fold error (*FE*) using Equation 15.21 [74]:

$$FE = 10^{\left| \log \left(\frac{\text{Predicted}}{\text{Observed}} \right) \right|} \tag{15.21}$$

15.3.3.4 Results Human systemic intravenous PK parameters were predicted as follows. Human V_{ss} (2.7 L/kg) was predicted using the Oie–Tozer [7,37] method and using monkey SSS. After an intravenous dose, the total plasma clearances were low and ranged from 5 to 11 mL/min/kg in preclinical species. Using the FCIM method [7], prediction of human plasma CL was low (2.2 mL/min/kg). To predict human oral exposure under fasting and fed conditions, a GastroPlus PBPK dog absorption model [62] was used to estimate *in vivo* solubility and possible precipitation in humans. The *in vivo* predicted solubility and precipitation times from the dog model were then applied to the human situation. The observed and predicted PK profiles under fasting and fed conditions in dogs were comparable with the regression coefficient (r^2) of 0.96 and 0.92, respectively, and in humans, the r^2 between observed and model predicted was 0.90 and 0.92 under fasting and fed states, respectively, using the original F1 formulation (the comparison is shown in Figure 15.9A). The changes in exposure with formulation F1 under fasting and fed conditions in dogs (top) and humans (bottom) are depicted in Figure 15.9B. Oral exposure profiles for an alternative, follow-on formulation F2 are shown in Figure 15.10A. Observed and predicted dog PK under fasting and fed conditions (r^2 of 0.94 and 0.98, respectively) using the follow-on F2 formulation is shown in the top panel of Figure 15.10A. In humans, using the same F2 formulation, the r^2 value between observed and model predicted PK profiles are 0.90 and 0.93 under fasting and fed states, respectively (Figure 15.10A). The changes in exposure in fasting and fed conditions using the F2 formulation are shown in Figure 15.10B. Measuring *in vivo* relevant solubility remains a challenge [75], and more work is needed to refine solubility determining methods, including the generation of relevant *in vivo* precipitation kinetics [58].

15.3.3.5 Study Conclusion A PBPK model was used to describe human intravenous and oral profiles for a BCS Class II compound with low aqueous solubility and moderate-to-high permeability. Human CL was successfully estimated by allometric scaling of *in vivo* data. The precipitation time T_p obtained by convoluting dog

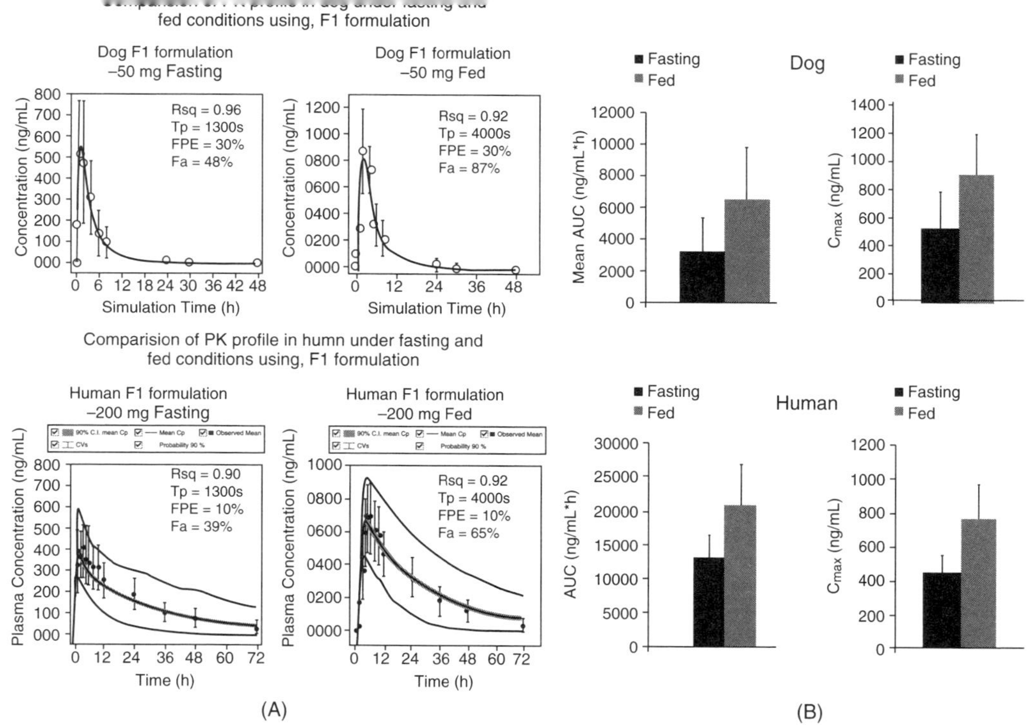

Figure 15.9 (A) Upper panel: Comparison of PK profiles in dogs after oral administration of a single dose of the original (F1) formulation under fasting (left) and fed (right) conditions. The regression coefficient (r^2) values for observed and model predicted are 0.96 and 0.92, under fasting and fed conditions, respectively. (A) Lower panel: The comparison of PK profiles in humans after a single oral dose of 200 mg of F1 formulation under fasting (left) and fed (right) conditions are given. The regression coefficient (r^2) values for observed and model predicted are 0.90 and 0.92, under fasting and fed conditions, respectively. The dots represent the observed concentrations, and the solid lines represent the model predicted concentration values. (B) Upper panel: Comparison of mean PK parameters (AUC (left) and C_{max} (right)) in dogs after oral administration of a single dose of F1 formulation under fasting and fed conditions showing an ~twofold increase (difference) in the AUC, and C_{max} under the fed condition when compared with the fasting condition. (B) Lower panel: The comparison of mean PK parameters (AUC (left) and C_{max} (right)) in humans after oral administration of 200-mg single dose of the F1 formulation under fasting and fed conditions showing an ~twofold increase (difference) in the AUC and C_{max} under the fed condition when compared with the fasting condition.

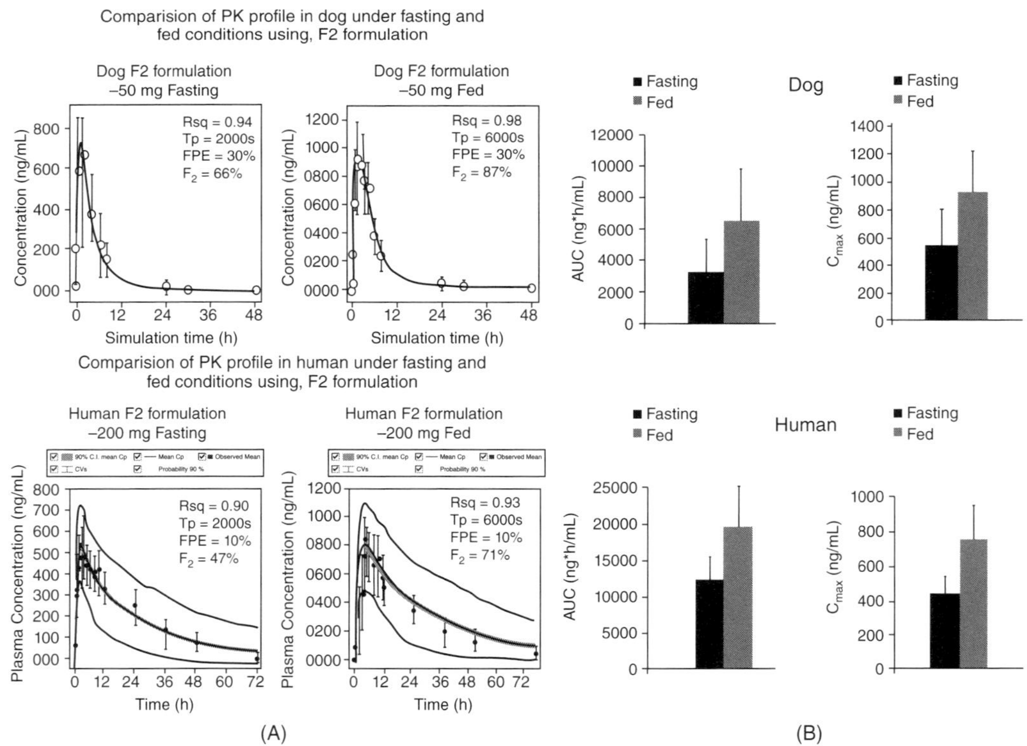

Figure 15.10 (A) Upper panel: The comparison of PK profiles in dogs after oral administration of a single dose of the F2 formulation under fasting (left) and fed (right) conditions. The regression coefficient (r^2) values for observed and model predicted are 0.94 and 0.98 under fasting and fed conditions, respectively. (A) Lower panel: The comparison of PK profiles in humans after a single oral dose of 200 mg of F2 formulation under fasting (left) and fed (right) conditions are given (lower panel). The regression coefficient (r^2) values for observed and model predicted are 0.90 and 0.93 under fasting and fed conditions, respectively. The dots represent the observed concentrations, and the solid lines represent the model predicted concentration values. (B) Upper panel: Comparison of mean PK parameters (AUC (left) and C_{max} (right)) in the dog after oral administration of a single dose of F2 formulation under fasting and fed conditions showing less than onefold increase (difference) in the AUC and C_{max} under the fed condition when compared with the fasting condition. (B) Lower panel: The comparison of mean PK parameters (AUC (left) and C_{max} (right)) in humans after oral administration of a 200-mg single dose of the F2 formulation under fasting and fed conditions showing a less than onefold increase (difference) in the AUC and C_{max} under the fasting condition when compared with the fed condition.

Table 15.5 Prediction fold error in anticipating human PK after administering two different formulations (F1 and F2) under fed and fasting conditions, respectively

Formulation	T_{max}	C_{max}	$AUC_{0\text{-inf}}$	$AUC_{0\text{-last}}$
F1 Fasting	1.38	1.05	1.09	1.06
F1 Fed	1.02	1.06	1.09	1.08
F2 Fasting	1	1.08	1.14	1.15
F2 Fed	1.02	1.04	1.01	1

in vivo data may be translated to predict human oral absorption and positive food effect for a weakly basic compound as the models were suitable and prediction accuracy expressed as *FE* was less than 1.5-fold for both fasting and fed human PK profiles and is given in Table 15.5.

15.4 DISCUSSIONS, MODELS LIMITATIONS, AND ASSUMPTIONS

Greater than 90% of all candidate compounds taken through all stages of drug development fail to reach the market [76]. Simulation and prediction of human PK/PD using *in vitro* and *in vivo* data from human tissues and animal models are of great importance in the earlier drug developmental stages. Understanding the disposition of a new compound in different species through *in vitro* and *in vivo* experiments is necessary to ensure appropriate selection of species needed to anticipate human PK/PD estimates. Characterization of ADME in animals is of fundamental importance, and extrapolation of these parameters to anticipate human PK and PD projections in early drug development is necessary to differentiate each compound and to identify efficacy along with formulation-related risk. Human PK parameter projections should be done via various approaches to compare concordance. This can be achieved by various available methods as discussed and recommended by the PhRMA CPCDC initiative [7, 41, 77, 78], along with the PD aspects of the drug candidate. The application of a "reverse pharmacology" approach can be successfully applied to predict PK for backup drug candidates by linking clinical and preclinical PK/PD data. Interspecies scaling methods are applied to determine concordance in PK parameters between species. These methods assume that the similarities in physiological, anatomical, and biochemical parameters among species are based on the fact that many physiological processes and organ sizes exhibit a power-law relationship with body weight among different species [33–36, 39]. It is noteworthy that the monkey data is overall not more predictive than the rat and dog data based on the PhRMA CPCDC initiative findings. For V_{ss}, the Oie–Tozer method along with SSS and empirical allometric scaling methods can be successfully employed [23]. For compounds where species disconnects in PK profiles have been observed, *in vivo–in vitro* metabolism correlation or extrapolation can provide a strong rationale in using human hepatocyte or microsomal data [38, 79]. For CL, preclinical IVIVE should be established whenever possible. For compounds with low CL or extraction ratios, good prediction results have been reported using metabolism normalized allometric scaling, which combines *in vivo* clearances of preclinical species with *in vitro* (hepatocyte) clearances [80]. In some cases, allometric scaling methods can fail to predict very low CL in humans (i.e., CL is overpredicted) leading to an unexpectedly long half-life, which is described as "vertical allometry." Empirical criteria have been proposed to identify vertical allometry [35], which are compounds with ClogP (calculated octanol–water partition coefficient) > 2, a ratio of unbound fraction in plasma between rats and humans Rf_u > 5, and elimination by metabolism. For F_a predictions (and F after incorporating the first-pass effect into prediction), the ACAT model or equivalent models can be used. The predicted human F should be compared and verified with the average observed F in animals.

Nonlinearities and species disconnects should not be ignored in human PK predictions. As interspecies differences in drug transporters expression levels are usually not quantitatively characterized, this can result in challenging predictions with lower confidence because transport can play a key role in tissue distribution. In such cases, a range of predictions can be made, including the uncertainty achieved either by settling an inferable boundary exposure as the worst-case scenario or by seeking more conservative approach(es). PBPK models are also recommended for certain therapeutic areas, where specific tissues are targeted. For example, drug concentrations in the lung rather than plasma concentrations are correlated with the pharmacodynamics effects of some inhaled drugs [81]. In these special scenarios, the mechanistic use of PBPK modeling is recommended to simulate the relevant local exposure of the drug. An assumption is that the molar receptor levels are low compared with the drug concentrations. If this is not the case, and target expression levels are markedly different between species, then affinities or IC_{50}/EC_{50} values determined in one species may not be the same for another. If there are differences between *in vivo* and *in vitro* IC_{50} values, then the source of the differences should ideally be ascertained, as it may be species and disease specific. Another reason for the species disconnects is a result of differences in the levels of competing entities (e.g., soluble receptors or natural antagonists), and the change in apparent affinities for receptor binding including homologies of receptors. Altered PK resulting from an altered physiology because of high doses, e.g., in toxicity studies, are not discussed here, but if the information is available, it can be included in the PBPK model analyses. In general, to maximize impact, modeling and simulation activities should be conducted prior to human studies, and methods should be kept up to date with literature, e.g., the PhRMA CPCDC initiative.

15.5 CONCLUSIONS

Human dose prediction methods aid in drug candidate selection and the identification of a clinically efficacious dosing regimen during the drug discovery, and early development stages. A multitude of prediction methods have been used to anticipate human PK parameters and profiles, including allometry, *in vitro–in vivo* extrapolation,

Wajima method, and PBPK models to predict human PK estimates (CL, V_{ss}, etc.) in early stages of drug development along with the anticipation of FIH doses. Pharmaceutical scientists can face different challenges such as formulation-dependent PK, preclinical species disconnects resulting from species-dependent clearance, or exposures that are altered by food when predicting the human PK profiles of new drug candidates. Reverse pharmacology approaches can be applied for compounds after clinical PK/PD data have become available from frontrunner compounds. Integrated simulation tools can be successfully applied in predicting the PK dependency of factors such as formulation, fasting, or fed state. Successful anticipation of the human dose requires a solid understanding of the *in vivo* and *in vitro* pharmacokinetic and pharmacodynamics behavior of the drug under investigation. Based on available approaches for predictions as discussed, e.g., the PhRMA CPCDC initiative, PK scientists are enabled to apply human PK prediction approaches, depending on the ADME, physiochemical, and other PK/PKPD properties.

REFERENCES

1. Heimbach, T., Lakshminarayana Suresh, B., Hu, W., He, H. (2009). Practical anticipation of human efficacious doses and pharmacokinetics using in vitro and preclinical in vivo data. *The AAPS Journal*, *11*(3), 602–614.
2. Lowe, P.J., Hijazi, Y., Luttringer, O., Yin, H., Sarangapani, R., Howard, D. (2007). On the anticipation of the human dose in first-in-man trials from preclinical and prior clinical information in early drug development. *Xenobiotica*, *37*(10/11), 1331–1354.
3. Thomas, V.H., Bhattachar, S., Hitchingham, L., Zocharski, P., Naath, M., Surendran, N., Stoner, C.L., El-Kattan, A. (2006). The road map to oral bioavailability: An industrial perspective. *Expert Opinion Drug Metabolism and Toxicology*, *2*(4), 591–608.
4. Miller, R., Ewy, W., Corrigan, B.W., Ouellet, D., Hermann, D., Kowalski, K.G., Lockwood, P., Koup, J.R., Donevan, S., El-Kattan, A., Li, C.S.W., Werth, J.L., Feltner, D.E., Lalonde, R.L. (2005). How modeling and simulation have enhanced decision making in new drug development. *Journal of Pharmacokinetics Pharmacodynamics*, *32*(2), 185–197.
5. Huang, C., Zheng, M., Yang, Z., Rodrigues, A.D., Marathe, P. (2008). Projection of exposure and efficacious dose prior to first-in-human studies: How successful have we been? *Pharmaceutical Research*, *25*(4), 713–726.
6. Vuppugalla, R., Marathe, P., He, H., Jones, R.D., Yates, J.W., Jones, H.M., Gibson, C.R., Chien, J.Y., Ring, B.J., Adkison, K.K., Ku, M.S., Fischer, V., Dutta, S., Sinha, V.K., Bjornsson, T., Lave, T., Poulin, P. (2011). PhRMA CPCDC initiative on predictive models of human pharmacokinetics, part 4: Prediction of plasma concentration-time profiles in human from in vivo preclinical data by using the Wajima approach. *Journal of Pharmaceutical Science*, *100*(10), 4111–4126.
7. Ring, B.J., Chien, J.Y., Adkison, K.K., Jones, H.M., Rowland, M., Jones, R.D., Yates, J.W., Ku, M.S., Gibson, C.R., He, H., Vuppugalla, R., Marathe, P., Fischer, V., Dutta, S., Sinha, V.K., Bjornsson, T., Lave, T., Poulin, P. (2011). PhRMA CPCDC initiative on predictive models of human pharmacokinetics, part 3: Comparative assessement of

prediction methods of human clearance. *Journal of Pharmaceutical Science*, *100*(10), 4090–4110.

8. Amidon, G.L., Lennernaes, H., Shah, V.P., Crison, J.R. (1995). A theoretical basis for a biopharmaceutic drug classification: The correlation of in vitro drug product dissolution and in vivo bioavailability. *Pharmaceutical Research*, *12*(3), 413–420.
9. Custodio, J.M., Wu, C.-Y., Benet, L.Z. (2008). Predicting drug disposition, absorption/elimination/transporter interplay and the role of food on drug absorption. *Advanced Drug Delivery Reviews*, *60*(6), 717–733.
10. Li, S., He, H., Parthiban, L.J., Yin, H., Serajuddin, A.T.M. (2005). IVIVC considerations in the development of immediate-release oral dosage form. *Journal of Pharmaceutical Science*, *94*(7), 1396–1417.
11. Lipinski, C.A. (2000). Drug-like properties and the causes of poor solubility and poor permeability. *Journal of Pharmacology and Toxicology Methods*, *44*(1), 235–249.
12. Kasim, N.A., Whitehouse, M., Ramachandran, C., Bermejo, M., Lennernaes, H., Hussain, A.S., Junginger, H.E., Stavchansky, S.A., Midha, K.K., Shah, V.P., Amidon, G.L. (2004). Molecular properties of WHO essential drugs and provisional biopharmaceutical classification. *Molecular Pharmaceutics*, *1*(1), 85–96.
13. Gabrielsson, J.D.H., Gillberg, P.G., Bredberg, U., Benthem, B., Duker, G. (2009). Early integration of pharmacokinetic and dynamic reasoning is essential for optimal development of lead compounds: strategic considerations. *Drug Discovery Today*, *14*(7-8), 358–372.
14. FDA Guidance for Industry - Estimating the Safe Starting Dose in Clinical Trials for Therapeutics in Adult Healthy Volunteers Center for Drug Evaluation and Research (CDER). 2005, pp. 1-30.
15. Reigner, B.G., Blesch, K.S. (2002). Estimating the starting dose for entry into humans: Principles and practice. *European Journal of Clinical Pharmacology*, *57*(12), 835–845.
16. Sharma, V., McNeill, J.H. (2009). To scale or not to scale: The principles of dose extrapolation. *British Journal of Pharmacology*, *157*(6), 907–921.
17. Mahmood, I., Green, M.D., Fisher, J.E. (2003). Selection of the first-time dose in humans: Comparison of different approaches based on interspecies scaling of clearance. *Journal of Clinical Pharmacology*, *43*(7), 692–697.
18. EMEA. *Guidline on requirements for First-in-Man clinical trials for potential high-risk medicinal products*, Committee for Medicinal Products for Human Use. London, UK, 2007, pp. 1–11.
19. Colburn, W.A. (2000). Optimizing the use of biomarkers, surrogate endpoints, and clinical endpoints for more efficient drug development. *Journal of Clinical Pharmacology*, *40*(12), 1419–1427.
20. Reigner, B.G., Williams, P.E., Patel, I.H., Steimer, J.L., Peck, C.,Van Brummelen, P. (1997). An evaluation of the integration of pharmacokinetic and pharmacodynamic principles in clinical drug development. Experience within Hoffmann La Roche. *Clinical Pharmacokinetics*, *33*(2), 142–152.
21. Iavarone, L., Hoke, J.F., Bottacini, M., Barnaby, R., Preston, G.C. (1999). First time in human for GV196771: Interspecies scaling applied on dose selection. *Journal of Clinical Pharmacology*, *39*(6), 560–566.
22. Muller, P.Y., Brennan, F.R. (2009). Safety assessment and dose selection for first-in-human clinical trials with immunomodulatory monoclonal antibodies. *Clinical Pharmacology & Therapeutics*, *85*(3), 247–258.

23. Zou, P., Yu, Y., Zheng, N., Yang, Y., Paholak, H.J., Yu, L.X., Sun, D. (2012). Applications of human pharmacokinetic prediction in first-in-human dose estimation. *The AAPS Journal, 14*(2), 262–281.

24. Agoram, B.M. (2009). Use of pharmacokinetic/ pharmacodynamic modelling for starting dose selection in first-in-human trials of high-risk biologics. *British Journal of Clinical Pharmacology, 67*(2), 153–160.

25. Tannergren, C., Bergendal, A., Lennernas, H., Abrahamsson, B. (2009). Toward an increased understanding of the barriers to colonic drug absorption in humans: Implications for early controlled release candidate assessment. *Molecular Pharmaceutics, 6*(1), 60–73.

26. FDA. S6 *Preclinical Safety Evaluation of Biotechnology-Derived Pharmaceuticals*, Center for Drug Evaluation and Research (CDER). Rockville, MD, 1997, pp. 1-14.

27. FDA. *Guidance on Nonclinical Safety Studies for the Conduct of Human Clinical Trials and Marketing Authorization for Pharmaceuticals M3(R2),* Food Drug Administration. Silver Spring, MD, 2008, pp. 1–14.

28. Chaikin, P., Rhodes, G.R., Bruno, R., Rohatagi, S., Natarajan, C. (2000). Pharmacokinetics/pharmacodynamics in drug development: An industrial perspective. *Journal of Clinical Pharmacology, 40*(12), 1428–1438.

29. De Buck, S.S., Sinha, V.K., Fenu, L.A., Nijsen, M.J., Mackie, C.E., Gilissen, R.A.H.J. (2007). Prediction of human pharmacokinetics using physiologically based modeling: A retrospective analysis of 26 clinically tested drugs. *Drug Metabolism & Disposition, 35*(10), 1766–1780.

30. Fura, A., Vyas, V., Humphreys, W., Chimalokonda, A., Rodrigues, D. (2008). Prediction of human oral pharmacokinetics using nonclinical data: Examples involving four proprietary compounds. *Biopharmaceutics & Drug Disposition, 29*(8), 455–468.

31. Sinha, V.K., De Buck, S.S., Fenu, L.A., Smit, J.W., Nijsen, M., Gilissen, A.H.J., Van Peer, A., Lavrijsen, K., Mackie, C.E. (2008). Predicting oral clearance in humans: How close can we get with allometry? *Clinical Pharmacokinetics, 47*(1), 35–45.

32. Wajima, T., Yano, Y., Fukumura, K., Oguma, T. (2004). Prediction of human pharmacokinetic profile in animal scale up based on normalizing time course profiles. *Journal of Pharmaceutical Science, 93*(7), 1890–1900.

33. Mahmood, I. (2006). Prediction of human drug clearance from animal data: Application of the rule of exponents and 'fu corrected intercept method' (FCIM). *Journal of Pharmaceutical Science, 95*(8), 1810–1821.

34. Mahmood, I., Yuan, R. (1999). A comparative study of allometric scaling with plasma concentrations predicted by species-invariant time methods. *Biopharmaceutics & Drug Disposition, 20*(3), 137–144.

35. Tang, H., Mayersohn, M. (2005). A novel model for prediction of human drug clearance by allometric scaling. *Drug Metabolism & Disposition, 33*(9), 1297–1303.

36. Mahmood, I. *Interspecies pharmacokinetic scaling: Allometric principles and applications*, Pine House Publishers. Rockville, MD, 2005.

37. Obach, R.S., Baxter, J.G., Liston, T.E., Silber, B.M., Jones, B.C., Macintyre, F., Rance, D.J., Wastall, P. (1997). The prediction of human pharmacokinetic parameters from preclinical and in vitro metabolism data. *The Journal of Pharmacology and Experimental Therapeutics, 283*(1), 46–58.

38. Obach, R.S., Lombardo, F., Waters, N.J. (2008). Trend analysis of a database of intravenous pharmacokinetic parameters in humans for 670 drug compounds. *Drug Metabolism & Disposition*, *36*(7), 1385–1405.
39. Hosea, N., Collard, W.T., Cole, S., Maurer, T.S., Fang, R.X., Jones, H., Kakar, S.M., Nakai, Y., Smith, B.J., Webster, R., Beaumont, K. (2009). Prediction of human pharmacokinetics from preclinical information: Comparative accuracy of quantitative prediction approaches. *Journal of Clinical Pharmacology*, *49*(5), 513–533.
40. Oie, S., Tozer, T.N. (1979). Effect of altered plasma protein binding on apparent volume of distribution. *Journal of Pharmaceutical Science*, *68*(9), 1203–1205.
41. Jones, R.D., Jones, H.M., Rowland, M., Gibson, C.R., Yates, J.W., Chien, J.Y., Ring, B.J., Adkison, K.K., Ku, M.S., He, H., Vuppugalla, R., Marathe, P., Fischer, V., Dutta, S., Sinha, V.K., Bjornsson, T., Lave, T., Poulin, P. (2011). PhRMA CPCDC initiative on predictive models of human pharmacokinetics, part 2: Comparative assessment of prediction methods of human volume of distribution. *Journal of Pharmaceutical Science*, *100*(10), 4074–4089.
42. Stoner, C.L., Cleton, A., Johnson, K., Oh, D.-M., Hallak, H., Brodfuehrer, J., Surendran, N., Han, H.-K. (2004). Integrated oral bioavailability projection using in vitro screening data as a selection tool in drug discovery. *International Journal of Pharmaceutics*, *269*(1), 241–249.
43. Jones, H.M., Parrott, N., Jorga, K., Lave, T. (2006). A novel strategy for physiologically based predictions of human pharmacokinetics. *Clinical Pharmacokinetics*, *45*(5), 511–542.
44. Luttringer, O., Theil, F.-P., Poulin, P., Schmitt-Hoffmann, A.H., Guentert, T.W., Lave, T. (2003). Physiologically based pharmacokinetic (PBPK) modeling of disposition of epiroprim in humans. *Journal of Pharmaceutical Sciences*, *92*(10), 1990–2007.
45. Parrott, N., Jones, H., Paquereau, N., Lave, T. (2005). Application of full physiological models for pharmaceutical drug candidate selection and extrapolation of pharmacokinetics to man. *Basics & Clinical Pharmacology & Toxicology*, *96*(3), 193–199.
46. Mahmood, I., Balian, J.D. (1996). Interspecies scaling: Predicting pharmacokinetic parameters of antiepileptic drugs in humans from animals with special emphasis on clearance. *Journal of Pharmaceutical Sciences*, *85*(4), 411–414.
47. Mahmood, I. (2002). Prediction of clearance in humans from in vitro human liver microsomes and allometric scaling. A comparative study of the two approaches. *Drug Metabolism & Drug Interaction*, *19*(1), 49–64.
48. Tang, H., Hussain, A., Leal, M., Mayersohn, M., Fluhler, E. (2007). Interspecies prediction of human drug clearance based on scaling data from one or two animal species. *Drug Metabolism & Disposition*, *35*(10), 1886–1893.
49. Wajima, T., Fukumura, K., Yano, Y., Oguma, T. (2002). Prediction of human clearance from animal data and molecular structural parameters using multivariate regression analysis. *Journal of Pharmaceutical Science*, *91*(12), 2489–2499.
50. Houston, J.B., Galetin, A. (2003). Progress towards prediction of human pharmacokinetic parameters from in vitro technologies. *Drug Metabolism Reviews*, *35*(4), 393–415.
51. Houston, J.B., Carlile, D.J. (1997). Prediction of hepatic clearance from microsomes, hepatocytes, and liver slices. *Drug Metabolism Reviews*, *29*(4), 891–922.
52. Bonate, P.L., Howard, D.R., *Pharmacokinetics in Drug Development*, AAPS. Arlington, VA, 2004, p. 451

53. Boxenbaum, H. (1984). Interspecies pharmacokinetic scaling and the evolutionary-comparative paradigm. *Drug Metabolism Reviews*, *15*(5-6), 1071–1121.

54. Sawada, Y., Hanano, M., Sugiyama, Y., Harashima, H., Iga, T. (1984). Prediction of the volumes of distribution of basic drugs in humans based on data from animals. *Journal of Pharmacokinetics and Biopharmaceutics*, *12*(6), 587–596.

55. Lin, T., Heimbach, T., Gollen, R., He, H. *Mass Balance Studies in Animals and Humans*, Wiley. New York, 2012, pp. 1–39.

56. Bittner, B., Mountfield, R.J. (2002). Intravenous administration of poorly soluble new drug entities in early drug discovery: The potential impact of formulation on pharmacokinetic parameters. *Current Opinion in Drug Discovery & Development*, *5*(1), 59–71.

57. Dannenfelser, R.-M., He, H., Joshi, Y., Bateman, S., Serajuddin, A.T.M. (2004). Development of clinical dosage forms for a poorly water soluble drug I: Application of polyethylene glycol-polysorbate 80 solid dispersion carrier system. *Journal of Pharmaceutical Sciences*, *93*(5), 1165–1175.

58. Kesisoglou, F., Wu, Y. (2008). Understanding the effect of API properties on bioavailability through absorption modeling. *The AAPS Journal*, *10*(4), 516–525.

59. Tang, H., Mayersohn, M. (2005). A mathematical description of the functionality of correction factors used in allometry for predicting human drug clearance. *Drug Metabolism & Disposition*, *33*(9), 1294–1296.

60. Rowland, M., Peck, C., Tucker, G. (2011). Physiologically-based pharmacokinetics in drug development and regulatory science. *Annual Review of Pharmacology and Toxicology*, *51*, 45–73.

61. Jamei, M., Marciniak, S., Feng, K., Barnett, A., Tucker, G., Rostami-Hodjegan, A. (2009). The Simcyp population-based ADME simulator. *Expert Opinion on Drug Metabolism & Toxicology*, *5*(2), 211–223.

62. Agoram, B., Woltosz, W.S., Bolger, M.B. (2001). Predicting the impact of physiological and biochemical processes on oral drug bioavailability. *Advanced Drug Delivery Reviews*, *50*(1), S41–S67.

63. Jamei, M., Turner, D., Yang, J., Neuhoff, S., Polak, S., Rostami-Hodjegan, A., Tucker, G. (2009). Population-based mechanistic prediction of oral drug absorption. *The AAPS Journal*, *11*(2), 225–237.

64. Yu, L.X., Lipka, E., Crison, J.R., Amidon, G.L. (1996). Transport approaches to the biopharmaceutical design of oral drug delivery systems: Prediction of intestinal absorption. *Advanced Drug Delivery Reviews*, *19*(3), 359–376.

65. Gabrielsson, J., Weiner, D., *Pharmacokinetic and Pharmacodynamic Data Analysis: Concepts and Applications*, Swedish Pharmaceutical Press. Sweden, 2007, pp. 1039–1042

66. Lukacova, V., Parrott, N.J., Lavè, T., Fraczkiewicz, G., Bolger, M.B.,Woltosz, W.S., *Role of Fraction Unbound in Plasma in Calculations of Tissue*, Plasma Partition Coefficients. Atlanta, GA, 2008.

67. Rodgers, T., Leahy, D., Rowland, M. (2007). Physiologically based pharmacokinetic modeling: Predicting the tissue distribution of moderate-to-strong bases. *Journal of Pharmaceutical Science*, *96*(11), 3151–3152.

68. Rodgers, T., Rowland, M. (2007). Physiologically-based pharmacokinetic modeling 2: Predicting the tissue distribution of acids, very weak bases, neutrals and zwitterions. *Journal of Pharmaceutical Science*, *96*(11), 3153–3154.

69. Meibohm, B., Derendorf, H. (1997). Basic concepts of pharmacokinetic/ pharmacodynamic (PK/PD) modelling. *International Journal of Clinical Pharmacology & Therapeutics*, *35*(10), 401–413.
70. Davies, B., Morris, T. (1993). Physiological parameters in laboratory animals and humans. *Pharmaceutical Research*, *10*(7), 1093–1095.
71. FDA. *Guidance for Industry - Food - Effect Bioavailability and Fed Bioequivalence Studies*, Center for Drug Evaluation and Research (CDER). Rockville, MD, 2002, pp. 1-12.
72. Tang, H., Hussain, A., Leal, M., Mayersohn, M., Fluhler, E. (2007). Interspecies prediction of human drug clearance based on scaling data from one or two animal species. *Drug Metabolism and Disposition*, *35*(10), 1886–1893.
73. Tang, H., Mayersohn, M. (2005). A novel model for prediction of human drug clearance by allometric scaling. *Drug Metabolism and Disposition*, *33*(9), 1297–1303.
74. Poulin, P., Theil, F.P. (2000). A priori prediction of tissue: Plasma partition coefficients of drugs to facilitate the use of physiologically-based pharmacokinetic models in drug discovery. *Journal of Pharmaceutical Science*, *89*(1), 16–35.
75. Dressman, J., Reppas, C. (2007). Drug solubility: How to measure it, how to improve it. *Advance Drug Delivery Reviews*, *59*(7), 531–532.
76. Selick, H.E., Beresford, A.P., Tarbit, M.H. (2002). The emerging importance of predictive ADME simulation in drug discovery. *Drug Discovery Today*, *7*(2), 109–116.
77. Poulin, P., Jones, H.M., Jones, R.D., Yates, J.W., Gibson, C.R., Chien, J.Y., Ring, B.J., Adkison, K.K., He, H., Vuppugalla, R., Marathe, P., Fischer, V., Dutta, S., Sinha, V.K., Bjornsson, T., Lave, T., Ku, M.S. (2011). PhRMA CPCDC initiative on predictive models of human pharmacokinetics, part 1: Goals, properties of the PhRMA dataset, and comparison with literature datasets. *Journal of Pharmaceutical Science*, *100*(10), 4050–4073.
78. Poulin, P., Jones, R.D., Jones, H.M., Gibson, C.R., Rowland, M., Chien, J.Y., Ring, B.J., Adkison, K.K., Ku, M.S., He, H., Vuppugalla, R., Marathe, P., Fischer, V., Dutta, S., Sinha, V.K., Bjornsson, T., Lave, T., Yates, J.W. (2011). PHRMA CPCDC initiative on predictive models of human pharmacokinetics, part 5: Prediction of plasma concentration-time profiles in human by using the physiologically-based pharmacokinetic modeling approach. *Journal of Pharmaceutical Science*, *100*(10), 4127–4157.
79. Ito, K., Houston, J.B. (2005). Prediction of human drug clearance from in vitro and preclinical data using physiologically based and empirical approaches. *Pharmaceutical Research*, *22*(1), 103–112.
80. Lave, T., Coassolo, P., Reigner, B. (1999). Prediction of hepatic metabolic clearance based on interspecies allometric scaling techniques and in vitro-in vivo correlations. *Clinical Pharmacokinetics*, *36*(3), 211–231.
81. Lopworth, B.J. (1996). Pharmacokinetics of inhaled drugs. *British Journal of Clinical Pharmacology*, *42*(6), 697–705.

16

MANAGEMENT AND MITIGATION OF HUMAN DRUG–DRUG INTERACTION RISKS IN THE DRUG DISCOVERY AND DEVELOPMENT PHASES

Heidi J. Einolf and Imad Hanna

16.1 INTRODUCTION

Pharmacokinetic drug–drug interaction (DDI) assessments are an iterative process in the pharmaceutical industry, stemming from early drug discovery and extending throughout clinical development. Clinical DDI manifests when one drug affects the distribution or clearance of another, either increasing or decreasing exposure to the co-administered drug(s). Strategies to assess DDI in early drug discovery programs may be streamlined and optimized for efficiency, choice of the enzyme system to evaluate, and robustness of the assay. The focus in discovery is to mitigate or optimize drug candidates to decrease potential DDI liability prior to the selection of a drug candidate for further preclinical and clinical development. In development, the focus is optimal *in vitro* characterization of the DDI with increased robustness of the assays and results. In this stage, more may be known about the exposure and clearance mechanisms of the drug clinically, and projections of the DDI effect in humans can

Predictive ADMET: Integrative Approaches in Drug Discovery and Development, First Edition.
Edited by Jianling Wang and Laszlo Urban.

be made with increased accuracy. After a compound is selected into development, any identified DDI liability will have to be managed including decisions to instigate clinical DDI studies, if needed, to understand the magnitude of DDI risk to aid in decisions regarding dose adjustments and labeling restrictions.

Sites of potential DDI involving pharmacokinetic changes in drug exposure include intestinal absorption, metabolism, and carrier-mediated transport across membranes (e.g., uptake into liver and kidney and efflux into bile or urine) as reviewed in Ito et al. [1]. The drug candidate could act as a perpetrator (e.g., inhibitor or inducer) of metabolic enzymes or transporters or as a victim drug (substrate). With respect to metabolism, the main focus of DDI strategies in pharmaceutical discovery and development programs are the cytochrome P450 (CYP) enzyme family, as CYP enzymes play a major role in drug clearance of marketed drugs [2]. DDIs with respect to drug transporters have become more common to evaluate in development as well as in discovery programs. This chapter will review common strategies for mitigation and managing risk with respect to both CYP-mediated and transporter-mediated DDI in drug discovery and development.

16.2 CYTOCHROME P450-MEDIATED DRUG–DRUG INTERACTIONS

DDI with respect to CYP enzymes can impact regulatory submissions, drug labeling, prescribing information, and concern for safety and efficacy. Ultimately, lack of consideration for the potential of the drug candidate to act as a perpetrator of CYP enzymes could ultimately result in drug withdrawal from the market. Such examples of drugs that have been withdrawn from the market as a result of CYP-mediated DDI include terfenadine, mibefradil, and astemizole. These drugs were pulled from the market in the late 1990s (1998–1999); however, since that time, pharmaceutical companies and regulatory authorities have put much effort into providing guidance for early risk assessment to avoid such situations [3–8]. This section will describe common *in vitro* techniques and strategies to assess the CYP DDI potential of drug candidates and to estimate clinical DDI risk.

There are several common mechanisms of CYP-mediated DDI including inhibition (reversible and irreversible) and induction. These mechanisms are typically evaluated for compounds from early drug discovery into full clinical development. The purpose of these evaluations differs in each stage of drug development. Early in discovery programs, it is the most common CYP enzymes involved in drug metabolism that are typically evaluated for CYP inhibition by the drug candidate (e.g., CYP3A4, followed by CYP2C9 and CYP2D6) [9]. In addition, higher throughput methods in discovery are important as a result of the large numbers of compounds that are screened for this potential liability. As compounds progress into preclinical development, more enzymes are evaluated as needed to inform for clinical DDI strategies and as recommended by regulatory authorities (e.g., CYP1A2, CYP2B6, CYP2C8, CYP2C9, CYP2C19, and CYP3A4) [4, 5].

The relative contribution of CYP enzymes to the test compound's total clearance is also of importance to assess whether the drug candidate could be a victim of a

DDI. Typically, however, in the early phases of drug development, the mechanism of drug clearance in humans is generally unknown. Early in discovery and development programs, conservative assumptions are likely to be made regarding other clearance pathways, whereas in later development, these pathways become more defined and projections of DDI impact will improve. This topic will be discussed in more detail later in this chapter.

16.2.1 Reversible CYP Inhibition

Typically for CYP enzymes, reversible inhibition can be described as competitive, noncompetitive, or a mixed type of inhibition. In discovery programs, reversible CYP inhibition is assessed using single-point inhibition (single-test inhibitor concentration) or a full IC_{50} curve to rank order potential DDI potency among a compound series. Generally, reversible CYP inhibition is evaluated using pooled human liver microsomes (HLMs) from many donors (to ensure ample CYP activities). IC_{50} values for competitive inhibitors (but not noncompetitive inhibitors) are dependent on substrate concentration [10]. Therefore, it is typical that the CYP-selective probe substrate concentration is at or below its K_m value for the specific reaction. For competitive inhibitors, if the substrate concentration is well below its K_m value, the IC_{50} value becomes negligibly influenced by the substrate concentration and the IC_{50} value will approximate the reversible inhibition constant (K_i) value. A list of common CYP-selective probe substrates and respective probe metabolite are listed in Table 16.1. In addition, the median K_m values reported in the literature for the CYP

Table 16.1 Common CYP-selective probe substrates used in *in vitro* CYP inhibition studies

CYP enzyme	Probe substrate	Probe metabolite	Median reported K_m value (μM)*	Number of reported *in vitro* experiments
CYP1A2	Phenacetin	Acetaminophen	27	36
CYP2B6	Buproprion	Hydroxybupropion	86	25
CYP2C8	Amodiaquine	*N*-Desethylamodiaquine	2.7	9
CYP2C9	Diclofenac	4′-Hydroxydiclofenac	5.0	45
CYP2C19	*S*-mephenytoin	4′-Hydroxy-*S*-mephenytoin	39	32
CYP2D6	Bufuralol	1′-Hydroxybufuralol	9.6	32
CYP3A4/5	Midazolam	1′-Hydroxymidazolam	3.6	73
CYP3A4/5	Testosterone	6β-Hydroxytestosterone	60	52

*Median K_m values for the CYP reaction were queried from The University of Washington Metabolism & Transport Drug Interaction Database (http://www.druginteractioninfo.org) in January 2013. The values were from incubations with HLM or recombinant CYP microsomes. If a range of values was given, an average value was taken. Reported K_m values were excluded if the data originated from incubations with variant CYP enzymes or from purified or reconstituted enzymes or if the probe metabolite was not directly measured (e.g., by substrate disappearance).

enzyme-selective reaction are also listed. It is recommended to evaluate the K_m value for the probe substrate metabolism in the test system being used (e.g., HLM) and to keep the concentration generally below this value [6]. In development, the IC_{50} curve may be followed up with determination of the K_i value for those CYP enzyme reactions that have sufficient inhibition by the test compound.

To interpret and rank order *in vitro* CYP DDI potency appropriately, the fraction of the unbound test compound in the incubation must be taken into consideration. The fraction of drug unbound in the microsomes (fu_{mic}) is dependent on the amount of microsomal protein used in the incubations. If there is considerable microsomal protein binding of the test compound in the HLM when evaluating the inhibition of individual CYP enzyme reactions, then the IC_{50} or K_i values may be overestimated and could lead to an underestimation of CYP inhibition potency. The fu_{mic} can be measured using various techniques including ultra-centrifugation, ultra-filtration, and equilibrium dialysis (as reviewed in Reference 11). These techniques are less likely to be amenable to high-throughput methods needed in discovery; however, several *in silico* models have been developed based on the physical-chemical properties of the compound to estimate the fu_{mic} value [11–13]. In general, however, if the microsomal protein concentrations in the incubations are kept low and consistent across the CYP enzymes, then the impact of fu_{mic} will be minimized.

Assessments of clinical DDI risk with respect to reversible CYP inhibition can be helpful in the discovery and development phases. Clinical risk assessment will improve as more data become available, particularly with respect to the clinical concentrations of the investigational drug. Many reported prediction approaches for reversible (and irreversible) CYP inhibition focus on the predictions of mean changes of the victim drug (or substrate) exposure at steady state. These mathematical prediction models, expressed with varying levels of complexity, incorporate the relationship of a single *in vivo* inhibitor concentration, [I], and the potency of the CYP inhibition determined from *in vitro* data. The measure of inhibition potency is expressed as the inhibition constant, K_i, for reversible inhibitors. The concentration of [I] *in vivo* is often difficult to determine accurately if drug interaction predictions are made before the first dose in humans. In early drug development, prior to knowledge of the actual clinical exposures of the test inactivator, the value of [I] to use in the risk assessment may be estimated from *in vitro* models, efficacious maximal exposures in the pharmacology animal species, or extrapolated human exposures. Refinement of the DDI predictions with actual *in vivo* [I] after the compound is in clinical development is common.

It is the intent in development to predict more accurately the magnitude of clinical DDI risk, rank order the *in vivo* DDI studies based on the CYP inhibition profile (more potent CYP inhibition likely to be evaluated first), and define co-medication inclusion/exclusion criteria in the first-in-human (FIH) studies. As the clinical [I] becomes known, DDI predictions can be refined and the projected risk resulting from the DDI will be helpful in the timing of the clinical DDI trial planning. The timing of the DDI trial, however, may also depend on the therapeutic area, i.e., known co-medications of the patient population or whether any level of DDI would be acceptable for the therapeutic indication (e.g., competitive labeling).

The simplest of drug interaction prediction models, although generally overpredictive for CYP reversible inhibition, is the $[I]/K_i$ approach [14–17]. This model describes the change in exposure of a victim drug in the presence of an inhibitor as a ratio of $AUC_{inhibited}/AUC_{control}$ (AUC_i/AUC) equal to $1 + [I]/K_i$ (Equation 16.1), where AUC is the area under the curve in the presence or absence (i.e., control) of an inhibitor drug. This model assumes competitive or noncompetitive CYP inhibition, that the victim drug is orally administered, cleared exclusively by a single metabolic pathway that is affected by the inhibitor, and the "well-stirred liver" model for hepatic clearance applies. Other assumptions in using Equation 16.1 include negligible inhibition of "first-pass" metabolism in the intestine, time-invariant inhibitor concentrations, and plasma protein binding:

$$\frac{AUC_i}{AUC} = \frac{CL_{int}}{CL_{int,i}} = 1 + \frac{[I]}{K_i} \tag{16.1}$$

The pragmatic use of the basic $[I]/K_i$ model is currently recommended as a "first tier" DDI assessment by the U.S. Food and Drug Administration (FDA) and European Medicines Agency (EMA) [4, 5]. The FDA recommends that if the value of $1 + [I]/K_i \leq 1.1$, then the drug could be labeled as a noninhibitor. If >1.1, then more mechanistic modeling can be used for the DDI assessment (see below). To be conservative, [I] is defined as the systemic total maximal concentration, $C_{max,total}$ (bound and unbound), of the inhibitor at the highest proposed clinical dose. The EMA recommendation is similar, but the [I] is defined as $C_{max,u}$ (unbound) and the cutoff value for a noninhibitor is $1 + [I]/K_i < 1.02$. For CYP3A substrates that are orally administered, both the FDA and EMA recommend calculating [I] as the maximum molar dose divided by 250 mL with a cutoff value for a noninhibitor being $1 + [I]/K_i < 11$.

There have been reported investigations using other surrogate values for [I] in the $[I]/K_i$ model other than $C_{max,total}$ [17, 18] (i.e., unbound C_{max} ($C_{max,u}$), average systemic concentration (C_{avg}) – total or unbound, or maximum hepatic inlet concentration ($C_{hep,inlet}$) – total or unbound; see Equation 16.2 [19]):

$$C_{hep,inlet} = C_{max,total} + \frac{D \times k_a \times f_a}{Q_H} \tag{16.2}$$

where $C_{max,total}$ is the maximal total systemic concentration of the inhibitor, D is the dose, k_a is the absorption rate constant, f_a is the fraction of drug escaping first-pass metabolism in the intestine, and Q_H is the human hepatic blood flow. The incorporation of protein binding may be important for better prediction accuracy and for decrease in overpredictions of actual DDI magnitude (AUC_i/AUC) whether using $C_{hep,inlet}$ or C_{max}, for instance; however there is a risk of increased false-negative predictions [17, 18, 20]. Models that incorporate the fraction of the victim drug metabolized by the inhibited enzyme ($f_{m,CYP}$) in mechanistic prediction models

(Equation 16.3) have proven to be more predictive of actual DDI magnitude than the $[I]/K_i$ approach [16, 20–22]:

$$\frac{AUC_i}{AUC} = \frac{CL_{int}}{CL_{int,i}} = \frac{1}{\sum_{j=1}^{n} \frac{f_{m,j}}{\text{fold reduction in } CL_{u,int,j}} + \left(1 - \sum_{j=1}^{n} f_{m,j}\right)} \quad (16.3)$$

where $f_{m,j}$ is the fraction of substrate clearance mediated by the inhibited metabolic pathway "j" and $CLu_{int,j}$ is the unbound intrinsic metabolic clearance of substrate down pathway "j." This mechanistic model is, like the $[I]/K_i$ approach, based on the "well-stirred liver" model, with similar assumptions, but it takes into account clearance pathways other than the inhibited pathway in the prediction of AUC change. This model, therefore, is generally less overpredictive of drug interactions than the $[I]/K_i$ approach. However, the most common use of this model is based on a simplified version involving inhibition of only one pathway, particularly for substrates with a high $f_{m,CYP}$ for one enzyme:

$$\frac{AUC_i}{AUC} = \frac{CL_{int}}{CL_{int,i}} = \frac{1}{\frac{f_{m,CYP}}{\left(1 + \frac{[I]}{K_i}\right)} + \left(1 - f_{m,CYP}\right)}. \quad (16.4)$$

For drug development programs, where an assessment of the potential clinical DDI risk of the test inhibitor compound on sensitive CYP substrates (such as midazolam for a CYP3A inhibitor) needs to be made, $f_{m,CYP}$ values have been estimated previously for some common P450 substrates [16, 22–25]. Reasonable estimations of $f_{m,CYP}$ have been typically determined by clinical exposure differences of probe substrates in extensive versus poor metabolizers of polymorphic enzymes (CYP2D6, CYP2C9, or CYP2C19). In addition, f_{mCYP} values, particularly for CYP3A and CYP1A2 substrates, have been estimated by human radiolabeled absorption, distribution, metabolism, and excretion (ADME) studies determining the urinary and fecal recovery of parent drug and metabolites along with *in vitro* information regarding the relative contributions of CYP enzymes involved in the primary clearance reactions in human liver. In addition, some estimation has been made by examining the differences of *in vivo* clearance in the presence or absence of a high dose of selective CYP enzyme inhibitor, assuming that the remaining clearance of the substrate is a result of other processes.

Figure 16.1 (adapted from Rowland and Matin [26]) emphasizes the clinical impact of the $f_{m,CYP}$ value on the predicted DDI (AUC ratio) calculated using Equation 16.4. For instance, if one pathway represents 50% of the total drug clearance (f_m value of 0.5), then 100% inhibition of that pathway will maximally result in a two-fold increase in AUC. If, on the other hand, the pathway that is being inhibited represents 90% of the drug clearance, then the maximal increase in AUC is 10-fold. The f_{mCYP}

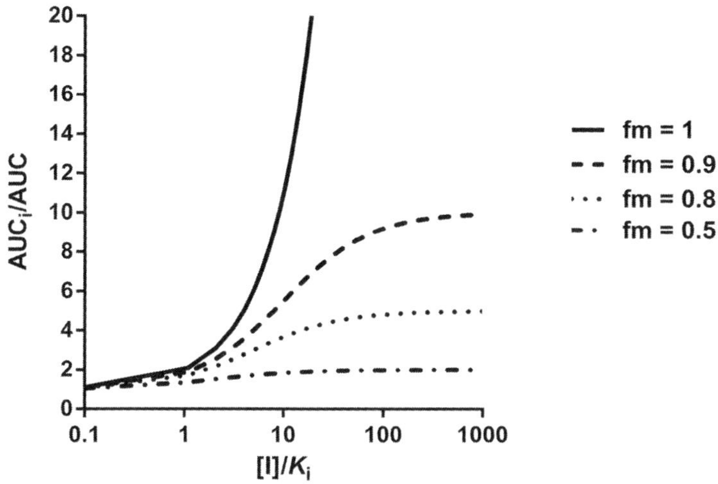

Figure 16.1 The impact of f_m on the magnitude of a drug interaction.

is a sensitive parameter in predictions of DDI; care must be given in its estimation for use in modeling DDI. More discussion regarding the estimation of individual CYP enzyme contributions to total hepatic metabolic clearance can be found later in this chapter.

For drug interactions involving CYP3A substrates, to take into account intestinal clearance of the substrate and inhibition of this pathway, Equation 16.4 can be expressed as a product of the change in hepatic clearance and intestinal availability ($F_{g,i}/F_g$), as described by Rostami-Hodjegan and Tucker [21] and incorporated in reports by Obach et al. [23] and Wang et al. [27]:

$$\frac{\mathrm{AUC_i}}{\mathrm{AUC}} = \frac{\mathrm{CL_{int}}}{\mathrm{CL_{int,i}}} = \frac{1}{\frac{\mathrm{f_{m,CYP}}}{\left(1+\frac{[\mathrm{I}]}{K_\mathrm{i}}\right)} + \left(1-\mathrm{f_{m,CYP}}\right)} \times \frac{F_{g,\mathrm{i}}}{F_g}. \tag{16.5}$$

F_g and $\mathrm{F_{g,i}}$ are the fractions of an oral dose of the victim drug that passes through the intestine unchanged in the absence and presence of an inhibitor, respectively, where:

$$\frac{F_{g,\mathrm{i}}}{F_g} = \frac{1}{\mathrm{F_g} + \left(1-\mathrm{F_g}\right) \times (\mathrm{CL_{int,gi}}/\mathrm{CL_{int,g}})}. \tag{16.6}$$

$$\frac{\mathrm{CL_{int,gi}}}{\mathrm{CL_{int,g}}} = \frac{1}{1+[\mathrm{I}]_g/K_I} \tag{16.7}$$

and the inhibitor concentration is assumed to be a function of its dose (D), absorption rate constant (k_a), and blood flow to enterocytes (Q_{ent}):

$$[I]_g = \frac{D \times k_a \times f_a}{Q_{ent}}. \tag{16.8}$$

$CL_{int,g}$ and $CL_{int,gi}$ are the intestinal intrinsic clearances of the victim drug in the absence and presence of an inhibitor, respectively. The change in intestinal intrinsic clearance in the presence of a CYP3A inhibitor is defined by the K_i and $[I]_g$, the unbound concentration of inhibitor available in the intestinal wall at the absorption phase. $[I]_g$ can be estimated as described in Reference 21, where typically a value of 248 mL/min is used for Q_{ent} [23]. This mechanistic model can be used to assess DDI potential, as recommended in recent regulatory authority guidance. If the predicted AUC ratio (AUC_i/AUC or AUC_R) is >1.25, then a clinical DDI study is recommended to characterize the magnitude of the inhibition further [4, 5].

The model described earlier is generally described as a "static" model; i.e., the predictions are made based on an invariable [I]. Physiologically based pharmacokinetic (PBPK) models, which incorporate time-varying [I], to predict DDI with respect to reversible (and irreversible) CYP inhibition, have also been described in the literature and are commercially available in platforms such as the SimCyp™ (Simcyp Ltd, Sheffield, U.K.) model [1, 28–32]. PBPK approaches (also referred to as "dynamic" models) are informative in the prediction of time-dependent DDI effects, such as CYP inactivation and induction, which are described later. These models are particularly valuable when the actual clinical PK of the compound is known and used to ensure proper modeling of the concentration-time profiles. Alternatively, when clinical data are not available, input parameters would have to be estimated, e.g., human CL, volume of distribution, fraction, and rate of absorption. Like the "static"-model–based approaches, the modeling can be refined when actual clinical data become available. Whether using "static" or "dynamic" modeling approaches, predictions of DDI magnitude can be very useful in informing for prioritization of a specific clinical study (e.g., a mechanistic study to examine clinical CYP3A inhibition) or inclusion/exclusion criteria for co-medications in the first clinical trials.

16.2.2 Irreversible CYP Inhibition

Irreversible or mechanism-based inactivation (MBI) of CYP enzymes is generally considered to be of greater concern for clinical DDI compared with that of reversible CYP inhibition. This is, in part, a result of the propensity for increased DDI effects on multiple dosing and sustained duration of DDI effect after discontinuation of the mechanism-based inhibitor. In addition, clinical consequences of MBI may also include overproportionality in the exposure of the inactivator itself after multiple doses. This clinical auto-inactivation would, however, be dependent on the fraction of the inactivator cleared by the affected enzyme (i.e., $f_{m,CYP}$). MBI results from irreversible covalent binding or "quasi-irreversible" noncovalent tight binding of a reactive intermediate generated by the CYP enzyme that is inactivated. As a result of the dependency for CYP metabolism of the inhibitor to form the reactive metabolite, MBI is characterized by NADPH-, time-, and concentration-dependent enzyme inactivation. Recovery of CYP enzyme activity *in vivo* requires resynthesis of the new enzyme; therefore, the clinical DDI effect does not cease immediately after discontinuation and clearance of the inactivator.

Mechanism-based inhibitors are also referred to, in general terms, as time-dependent inhibitors (TDIs). MBI is essentially a subset of TDI; i.e., for a chemical to be defined as an MBI, the mechanism of the time-dependent nature of the inhibition has to be elucidated. In the practical sense, identification of a TDI with characterization of inactivation kinetics is adequate for the assessment of pharmacokinetic DDI risk assessment. However, knowledge of the mechanism of inactivation, for instance, the metabolic pathway leading to the formation of the reactive intermediate, can be helpful for improvement of the drug design.

The basic intent of assessing time-dependent CYP inhibition of compounds in discovery is to identify potential CYP inactivators. The detection of a time-dependent inhibitor would be deemed a greater risk for a project compared with that of reversible CYP inhibition. Specific chemical moieties have been frequently associated with CYP MBI. The chemical motifs common in CYP MBI include 2-alkyimidazole, alkene, alkylamine, alkyne, furan, methylenedioxyphenyl, and thiophene functionalities. The chemical mechanisms of the inactivation reactions are described in detail in previous reviews [33–37]. In drug discovery, the identification of potential liabilities in a new chemical series should be evaluated as early as possible. These types of "structural alerts" may be avoided if the substituents are deemed not to interfere with the pharmacological activity of the molecule. If such avoidance cannot be made, examination of the potential to cause time-dependent inactivation of CYP enzymes *in vitro* would be of benefit to predict the clinical risk associated with the TDI as early as possible. Particularly used in discovery, abbreviated methods to assess the potential for compounds to cause time-dependent inhibition of CYP enzymes can be made to be high throughput and even integrated into CYP reversible inhibition screens (e.g., IC_{50} determinations). It is of benefit to at least screen lead compounds in a series, particularly if there is a common "structural alert" in the pharmacophore.

The abbreviated *in vitro* methods described in the subsequent discussion (percent decline in CYP activity or IC_{50} shift experiments) are limited in the ability to rank order compounds in a series, especially if the [I] value is not known. If it is assumed that [I] would be similar across a series, rank ordering based on a shifted-IC_{50} or rate if inactivation at a single test compound concentration may be a viable approach to rank order TDI [38, 39]. As a result of the large involvement of CYP3A in the metabolism of marketed drugs, it is not surprising that time-dependent inhibition of CYP3A4 is most commonly assessed in industry [8]. Other CYP enzymes that have been found to be susceptible to MBI such as CYP2D6, CYP2C9, CYP1A2, CYP2C19, and CYP2C8 are also commonly assessed for TDI, particularly in the development stage [8].

If a compound progresses into development, more definitive *in vitro* studies are run to define the kinetic parameters for the CYP inactivation, k_{inact} (maximal rate of inactivation) and K_I (inactivator concentration at one half k_{inact}). As mentioned earlier for reversible CYP inhibitors, the basic intent in development is to predict accurately the magnitude of DDI risk to provide information for the safe administration of the investigational drug in the FIH trials and form a clinical DDI strategy. Definitive assessments of time-dependent inactivation are needed for the most accurate assessments of inactivation potency and predictions of clinical DDI. This assay is much more involved than the abbreviated methods for detection of a TDI and may be performed in the late discovery or early development stages (i.e., selected leads or development candidates). The kinetic parameters k_{inact} and K_I are used in DDI prediction models or for rank ordering the inactivation potency of a series of compounds based on the efficiency of the inactivation (k_{inact}/K_I). If used for early rank ordering of compounds (i.e., prior to the availability of clinical concentration data of the inactivator), an assumption that the inactivation concentration *in vivo* is the same across the series of compounds is generally made, unless other information is available (e.g., differences in pharmacological potency, protein binding, etc.). Early predictions of clinical DDI magnitude (or risk) using an estimate of human efficacious maximal exposures are important for decisions regarding the timing of the clinical trials. In later development, when actual clinical concentrations are known, risk assessments can be refined.

The conventional method for determining the inactivation parameters includes a pre-incubation period followed by a dilution step into an "activity assay" to measure the remaining CYP activity at several time-points and to test inactivator concentrations. In the pre-incubation step, typically pooled HLM are pre-incubated for various times with several test inactivator concentrations (including a vehicle control) in the presence (or absence, as a control) of NADPH. The reactions are then diluted by least 10-fold into the "activity assay" [8]. It has been recommended to use at least five time-points, and five concentrations of test inactivator are used to establish pseudo–first-order rate constants for the inactivation (k_{obs}) [8]. The natural logarithm of the percentage of CYP activity remaining at each time-point for the individual inactivator concentrations is plotted against the pre-incubation time (Figure 16.2A). The absolute value of the slope of the linear-regression line is the observed inactivation rate constant (k_{obs}). It must be noted here that only the initial rate of inactivation should be used. If there is nonlinearity or a biphasic decline in the activity, these points must not be included in the calculation of k_{obs}. An apparent decrease in the rate of inactivation may be a result of consumption of the inactivator during the time of the pre-incubation reaction, and earlier time-points would be more dependable than later ones [40]. Depletion of the test inactivator in the pre-incubation experimental conditions can be monitored by liquid chromatography–mass spectrometry (LC-MS)/MS analysis. In some cases, a decrease in CYP activity with increasing inactivator concentration at the "zero" time-point is observed, and this is likely a reflection of reversible CYP inhibition by the test compound in the "activity assay," particularly if it is a potent CYP inhibitor and was not effectively "diluted out." There are two main assumptions in this method: (1) that there is negligible metabolism of the test inactivator during the pre-incubation stage and (2) that there is negligible inactivation (or reversible inhibition) of the CYP enzyme by the test compound in the "activity assay." Therefore, selection of the appropriate pre-incubation time and effective dilution of the test inactivator in the "activity assay" are important. In

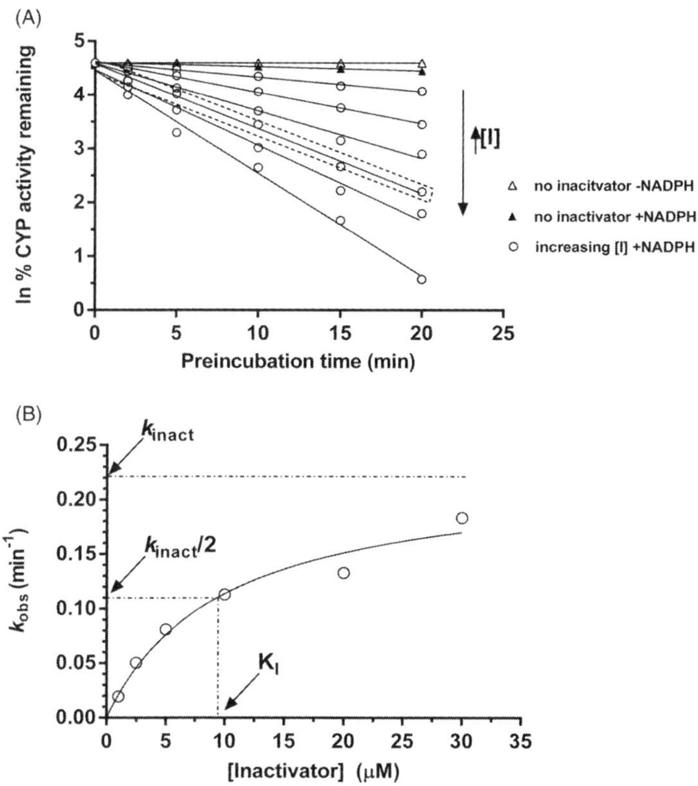

Figure 16.2 Kinetic determination of inactivation parameters. (A) Time- and concentration-dependent inactivation of a CYP enzyme by an MBI (or a TDI). The dotted line illustrates an abbreviated method where only one inactivator concentration was tested (e.g., 10 μM) at several time-points. (B) The absolute value of the slope of the lines (k_{obs} values) in (A) are plotted against the inactivator concentration, and the kinetic parameters (k_{inact} and K_I) are determined by nonlinear regression of the data.

addition to the magnitude of the dilution of the pre-incubation reaction, use of a high CYP probe substrate concentration also aids in the effective dilution of the test inactivator in the "activity assay." Up to this point, it has been recommended to use a concentration of the CYP probe substrate that is ≥four-fold the K_m for its metabolism [8]. To determine the inactivation parameters, the k_{obs} values, normalized by the control (vehicle + NADPH) inactivation rate, are plotted against their respective inactivator concentration (Figure 16.2B) and the k_{inact} and K_I values are then determined by a nonlinear regression fit to Equation 16.9:

$$k_{obs} \text{ or } \lambda = \frac{k_{inact} \times [I]}{K_I + [I]} \qquad (16.9)$$

Two commonly abbreviated approaches are used in discovery to determine whether a compound has the potential to have time-dependent CYP inhibition characteristics.

These assays are considered to be abbreviated based on a limited number of time-points or test inactivator concentrations examined. The two abbreviated approaches are generally categorized as (1) a percent decline in CYP activity or inactivation rate determined at a single test inactivator concentration or (2) an IC_{50} shift assay [8]. These assays are designed to detect time-dependent inhibition; they are not designed to determine the mechanism of the inactivation per se nor are they used to predict the magnitude of clinical DDI. Experimental details of the IC_{50} shift assay can be found in other publications [8, 41]. Essentially, two IC_{50} curves for CYP inhibition are produced after the test inactivator is pre-incubated with HLM in the presence or absence of NADPH. TDI is observed if there is a leftward shift of the IC_{50} curve (more potent inhibition) of the NADPH-pre-incubated samples compared with the samples that were pre-incubated in the absence of NADPH. An example of the other abbreviated approach is illustrated in Figure 16.2A (dotted lines), where only one inactivator concentration was tested (e.g., 10 μM) at several time-points. From the slope of the linear regression line, the k_{obs} can be determined. The shifted IC_{50} as well as the k_{obs} value has been shown to have good correlation to the efficiency of the inactivation (k_{inact}/K_I) and therefore is likely to be useful in rank ordering compounds in discovery for further development [38, 39].

From a PhRMA survey, most companies use the definitive TDI (k_{inact} and K_I) for the prediction of *in vivo* DDI and for the planning of clinical DDI studies [8]. Early DDI predictions, prior to FIH studies, are made to assess the potential risk resulting from the TDI based on anticipated clinical exposures. Mathematical models to predict the pharmacokinetic effect of TDI minimally incorporate the inactivation kinetic parameters (k_{inact} and K_I) in relation to the degradation rate constant of the affected CYP enzyme (k_{deg}) and *in vivo* inactivator concentration [I]. This type of mathematical model was first described by Stephen Hall and colleagues in 2000 [42]. In simple terms, the mathematical model described in this report assumes that the victim drug is solely cleared *via* hepatic metabolism by the enzyme being inactivated ($f_{m,CYP} = 1$) and that the fold increase in exposure is equal to the fold decrease in hepatic intrinsic clearance (CL_{int}) after oral administration [42]. It is also assumed that the change in the hepatic CL_{int} in the presence and absence of an inactivator is directly proportional to the change in the steady-state concentration of the affected enzyme $[E]_{ss}$. This model also assumes that the steady-state concentration of the enzyme is in equilibrium with the zero-order synthesis of the enzyme and the first-order rate of degradation (rate constants of k_{sys} and k_{deg}, respectively). In the presence of a CYP inactivator, there is an additional first-order loss of enzyme as a result of the inactivation, k_{obs} (or λ), as defined by Equation 16.10. Therefore, the change in exposure resulting from a CYP inactivator is described by the following relationship [42]:

$$\frac{AUC_i}{AUC} = \frac{CL_{int}}{CL_{int,i}} = \frac{[E]_{ss}}{[E]_{i,ss}} = \frac{k_{syn}/k_{deg}}{k_{syn}/k_{deg} + \lambda}$$
$$= \frac{k_{deg} + \lambda}{k_{deg}} = \frac{k_{deg} + \frac{[I] \times k_{inact}}{[I] + K_I}}{k_{deg}} \quad (16.10)$$

Similar to the [I]/K_i approach for reversible inhibition, this equation is currently recommended to use as a first tier DDI assessment by the regulatory authorities [4, 5]. For instance, the FDA recommends that if the AUC ratio of the predicted DDI is >1.1, then more mechanistic modeling can be used for the DDI assessment (see the subsequent discussion); otherwise the drug candidate could be labeled as a noninhibitor. The inactivator concentration, [I], is defined in this assessment as the systemic total maximal concentration, $C_{max,total}$ (bound and unbound), of the inhibitor at the highest proposed clinical dose. Use of this equation for risk assessment is generally thought to be capable of conservatively rank ordering the clinical DDI risk as weak (>two-fold), moderate (≤two-fold and <five-fold), and strong (≥five-fold) [8,41].

Although this equation is useful for quick risk assessments and rank ordering, the magnitude of the clinical DDI by a TDI is also dependent on the fraction of the substrate that is cleared by the affected enzyme ($f_{m,CYP}$). In addition, for CYP3A substrates, intestinal availability and inhibition of this pathway are important considerations. To take into account the fraction of the substrate metabolized by the affected CYP enzyme into the prediction model, the change in intrinsic clearance of a substrate in the presence of a mechanism-based inhibitor described in Equation 16.10 can be incorporated into Equation 16.4 to describe the change in AUC with an inactivator at steady state:

$$\frac{\mathrm{AUC_i}}{\mathrm{AUC}} = \frac{\mathrm{CL_{int}}}{\mathrm{CL_{int,i}}} = \frac{1}{\mathrm{f_{m,CYP}} \times \left(\dfrac{k_{deg}}{k_{deg} + \frac{[\mathrm{I}] \times k_{inact}}{[\mathrm{I}] + \mathrm{K_I}}} \right) + \left(1 - \mathrm{f_{m,CYP}}\right)} \tag{16.11}$$

For mechanism-based inhibitors of CYP3A, to take into account intestinal clearance and inhibition of this pathway, a similar term to express the intestinal availability of the substrate as in Equation 16.6 can be incorporated into Equation 16.11:

$$\frac{\mathrm{AUC_i}}{\mathrm{AUC}} = \frac{\mathrm{CL_{int}}}{\mathrm{CL_{int,i}}} = \frac{1}{\mathrm{f_{m,CYP}} \times \left(\dfrac{k_{deg}}{k_{deg} + \frac{[\mathrm{I}] \times k_{inact}}{[\mathrm{I}] + \mathrm{K_I}}} \right) + \left(1 - \mathrm{f_{m,CYP}}\right)} \times \frac{F_{g,i}}{F_g} \tag{16.12}$$

where $F_{g,i} / F_g$ is as described in Equation 16.6 and $\mathrm{CL_{int,gi}}/\mathrm{CL_{int,g}}$ is defined as:

$$\frac{\mathrm{CL_{int,gi}}}{\mathrm{CL_{int,g}}} = \left(\frac{k_{deg,CYP3Ag}}{k_{deg,CYP3Ag} + \frac{[\mathrm{I}]_g \times k_{inact}}{[\mathrm{I}]_g + \mathrm{K_I}}} \right) \tag{16.13}$$

The $k_{deg,CYP3A,g}$ term is the *in vivo* degradation rate constant for intestinal CYP3A. A value of 0.03 h^{-1} is commonly used, based on the time-course of CYP3A activity recovery after inactivation by grapefruit juice [38,43]. This value is generally thought to be based on the kinetics of enterocyte turnover rather than on CYP enzyme degradation [44]. The $[I]_g$ value is estimated from the dose, oral absorption rate constant (k_a), fraction of dose absorbed (f_a), and enterocytic blood flow (Q_{ent}), as described earlier.

The important parameters for DDI predictions with respect to TDI, in addition to the *in vitro* inactivation parameters, are the actual *in vivo* inactivator concentration [I], the rate constant for hepatic CYP enzyme degradation k_{deg}, and the $f_{m,CYP}$ of the affected CYP substrate. The importance of $f_{m,CYP}$ and surrogate values of [I] used for reversible CYP inhibition prediction models have been mentioned earlier. Further discussions regarding [I], with respect to TDI modeling, and the hepatic CYP k_{deg} value are included in the subsequent discussion.

The use of unbound systemic concentrations ($C_{max,u}$ or $C_{avg,u}$) in the model has been found to be appropriate for the assessment of risk with respect to TDI [38] [27, 45–47]. The value for the hepatic CYP k_{deg} cannot be directly obtained from humans *in vivo*, and it must be estimated indirectly, as reviewed in Yang et al. [48]. The range of turnover half-lives estimated for the different CYP enzymes can vary widely; therefore, this value can be a source of uncertainty in the model [48, 49]. For instance, the k_{deg} value for CYP3A has been estimated to range from 0.02 to 0.005 h^{-1} based on a $t_{1/2}$ range of 36–140 hours, as reviewed in Grimm et al. [8] and Yang et al. [48]. As a result of the current lack of consensus regarding the k_{deg} value, sensitivity analysis around this value to predict the magnitude of DDI caused by known CYP inactivators is one recommended approach to establish confidence in the performance of the model [8]. An initial estimate that has been deemed reasonable for human hepatic CYP3A half-life is within the 24–72-hour range [45].

Sensitivity analysis can be useful to understand what parameters are the most impactful in predicting a DDI and, in turn, the most important in accurately determining for the prediction. An example of the sensitivity analysis that can be performed is illustrated in Figure 16.3. In this graph, DDI predictions were made using the PBPK modeling software, SimCyp. The simulations were based on an actual DDI study of erythromycin (200-mg oral dose four times a day for four days) and midazolam (2.5-mg oral dose on day 4, 1 h after the erythromycin dose) [50]. The parameters used in the original simulation were as follows: K_I = 13.4 μM, k_{inact} = 1.49 h^{-1}, and k_{deg} = 0.0077 h^{-1}. The actual clinical DDI was 3.38-fold, and the simulated DDI was predicted to be 4.22-fold (closed and open circles on the graph, respectively). Changes were made in either of the inactivation parameters of erythromycin or the CYP3A4 k_{deg} value. Particularly for the inactivation parameters, variations of these values even by two-fold led to substantial over/underpredictions in the DDI. The AUC ratio, however, was less affected by a difference in a k_{deg} value of two-fold. This example illustrates the impact of potential experimental variability in input parameters on prediction outcomes. It is important, from this type of assessment, to know which parameters are the most sensitive and, therefore, most imperative to obtain with the best accuracy.

One example of how early risk assessment of TDI influenced the development plan of a drug candidate involved LCL161, an antagonist of IAP (inhibitor of apoptosis

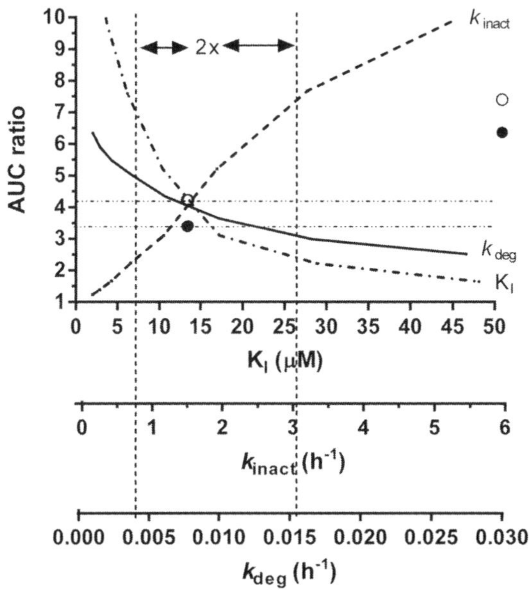

Figure 16.3 Sensitivity analysis of inactivation parameters and k_{deg} value on predictions of DDI. Change in AUC of midazolam in the presence of erythromycin was predicted with initial input parameters (as discussed in the text). The simulated AUC ratio is presented as the open circle on the graph, and the actual AUC ratio is the closed circle. The DDI was also predicted (curved dotted or solid lines on the graph) after varying the inactivation parameters (K_I, k_{inact}) or CYP3A4 k_{deg}. The vertical dotted lines represent a two-fold variance from the initial input value of the inactivation parameters or CYP3A4 k_{deg} to the DDI model.

proteins) for oncology indication [51]. The clinical TDI effect was first predicted using the fundamental equation shown in Equation 16.10. As this compound progressed into development, and actual clinical concentrations were known, the DDI was predicted to be strong (>10-fold) with oral doses > 40 mg ($C_{max,\ unbound}$ 0.049 μM). Further simulations using the PBPK model, SimCyp, found that one dose of LCL161 had a sustained DDI effect over several days (current dosing regimen of LCL161 is once a week). This initial modeling led to the design of the clinical study to assess clinical CYP3A activity at several time-points. The predicted and actual DDI on the day of co-administration was in line with each other, 12- vs. 9.3-fold, respectively, with no DDI after six days [51]. In this case, the DDI will have to be further managed. Clinical management of such DDI may include dose reduction or staggered dosing of CYP3A cleared co-medications, the need to monitor exposures of the co-administered drugs, or both. Compounds that make it to development with strong TDI potential will likely result in early assessments of clinical DDI to understand the risk for later trials. If clinical DDI is observed, this could result in more complicated dosing regimens, labeling restrictions, and increased patient risk for those co-medications that could be affected.

16.2.3 CYP Induction

CYP induction can lead to failure of therapeutic efficacy by reduction of the exposure of another drug or of the drug itself if it is metabolized by the induced enzyme (autoinduction). Sometimes it may also raise safety concerns on the significantly increased concentration of toxic or reactive metabolite(s) as a result of the extensive metabolism of the victim drugs caused by CYP induction. The most common mechanism of drug metabolizing enzyme induction is by activation of nuclear receptors and transcriptional upregulation. Three nuclear receptors are implicated in the majority of drug metabolizing enzyme gene regulation, and these are the aryl hydrocarbon receptor (AhR), pregnane X receptor (PXR), and constitutive androstane receptor (CAR). The AhR upregulates the CYP1A genes, whereas activation of PXR and CAR results in upregulation of CYP3A, CYP2B, and CYP2C genes.

As a result of the importance of CYP3A in drug metabolism and DDI, liabilities with respect to the induction of the CYP3A gene may be identified early by screening assays to measure binding or activation of the PXR receptor. Most commonly, PXR transactivation assays (reporter gene assays) are used in pharmaceutical companies compared with PXR ligand binding assays [7]. These assays are amenable to high-throughput screening methods used in discovery. One disadvantage of the ligand binding assays, however, is that they cannot distinguish between agonists and antagonists and do not measure the subsequent gene transactivation. The PXR binding assay does not seem to be a routine assay used by pharmaceutical companies, likely because of a lack of strong correlation between the receptor binding and transactivation [7]. The reporter gene assay is a cell-based method in which a cell-line is co-transfected with a PXR expression plasmid and a reporter gene construct, such as the luciferase gene downstream of the CYP3A4 promoter [52]. Luo et al. [53] found a correlation between PXR activation and the induction potential of CYP3A4 observed in human hepatocyte cultures. However, care must be taken in cases when the mechanism of CYP3A4 induction is a result of activation of the CAR receptor and not PXR activation. Although CYP3A4 induction may not be detected in a PXR reporter gene assay, it would be caught later in a human hepatocyte *in vitro* induction study. Typically in the development stage (or sometimes evaluated earlier), CYP induction is assessed using primary human hepatocytes and at least three donors [7]. Current regulatory agency guidance suggests evaluation of CYP1A2, CYP2B6, and CYP3A4 to capture the common mechanisms of induction by the nuclear receptors, AhR, PXR, and CAR [4, 5]. The evaluation of mRNA is thought to be the most sensitive measurement of CYP enzyme induction, although CYP activity is also commonly measured [54]. In regard to predictions of clinical outcomes, mRNA data are more amendable, as the dynamic range of the mRNA induction response generally is greater than activity measurements. In addition, if the test compound is a TDI or has potent reversible inhibition and has high metabolic stability in hepatocytes, measurements of the activity in cells *in situ* may lead to a false-negative result [7].

Models to predict the clinical magnitude of DDI as a result of CYP induction have been increasing over the last several years. These models have focused on CYP3A primarily, as a result of the larger number of clinical DDI studies evaluating CYP3A induction to qualify the models with. Those models that have been reported are based on a similar concept, that being the law of mass action for receptor binding [44,55–59]. The E_{max} model (Equation 16.14) is generally used to describe an induction effect based on EC_{50} (concentration at one half E_{max}) and E_{max} (maximum induction response) values obtained in *in vitro* assays in relation to an inducer concentration ([I]), i.e., a therapeutic effective *in vivo* concentration. In this case, full dose-induction response data are needed from an *in vitro* study:

$$\text{Effect} = \frac{E_{max} \times [\text{I}]}{\text{EC}_{50} + [\text{I}]} \tag{16.14}$$

Recent reviews on predictions of DDI with respect to CYP3A have been published [7,55,60]. The common prediction methods include "calibration" approaches, "static" mechanistic models, and PBPK models (such as those described earlier, e.g., SimCyp). Calibration approaches require some *in vitro* induction parameters (e.g., EC_{50} or E_{max} values) to be established for known inducers/noninducers of CYP3A in the same test system as the unknown clinical inducer. The *in vitro* parameters are plotted graphically against the clinical DDI magnitude (e.g., % decrease in AUC of a sensitive CYP3A substrate, such as midazolam) to produce a calibration curve to use to estimate the risk of the unknown clinical inducer. Various algorithms have been used in the calibration to predict a DDI (e.g., the relative induction score (RIS), which is the equation shown in Equation 16.14, versus % decrease in AUC). The "static" mechanistic models and PBPK models may require less experimentally (calibration curve not needed) but do require some type of "scalar" value (*d*). Although a full calibration set of data could be used to establish the scalar value (i.e., a value optimized to give the best predictions of DDI of the calibration dataset), just one positive control for CYP3A induction, rifampicin, has also been used to scale *in vitro* induction data.

Similar to reversible and irreversible DDI assessments, there is also a recommended first tier DDI assessment by the regulatory authorities [4, 5]. This simple basic model is shown in Equation 16.15:

$$\frac{\text{AUC}_\text{i}}{\text{AUC}} = \frac{1}{1 + \frac{d \times E_{max} \times [\text{I}]}{\text{EC}_{50} + [\text{I}]}} \tag{16.15}$$

For instance, the FDA recommends that if the AUC ratio of the predicted DDI is < 0.9, then more mechanistic modeling can be used for the DDI assessment (see the subsequent discussion); otherwise the drug candidate could be labeled as "noninducer." The inducer concentration [I] is defined conservatively by regulatory authorities as $C_{max,total}$ of the inhibitor at the highest proposed clinical dose.

The mechanistic "static" model for CYP induction is shown as follows. For CYP3A induction modeling, the equation also includes the intestinal availability term

($F_{g,i}/F_g$). The regulatory authority recommended [I] value is the unbound $C_{hep,inlet}$ concentration [4]:

$$\frac{AUC_i}{AUC} = \frac{CL_{int}}{CL_{int,i}} = \frac{1}{f_{m,CYP} \times \left(1 + \frac{d \times E_{max} \times [I]}{EC_{50}+[I]}\right) + \left(1 - f_{m,CYP}\right)} \times \frac{F_{g,i}}{F_g} \quad (16.16)$$

where $F_{g,i}/F_g$ is as described in Equation 16.6 and $CL_{int,gi}$ /$CL_{int,g}$ is defined as:

$$\frac{CL_{int,gi}}{CL_{int,g}} = \left(1 + \frac{d \times E_{max} \times [I]_g}{EC_{50} + [I]_g}\right) \quad (16.17)$$

$[I]_g$ value is estimated from the dose, oral absorption rate constant (k_a), fraction of dose absorbed (f_a), and enterocytic blood flow (Q_{ent}), as described earlier.

For all DDI assessment methods mentioned here, induction parameters (EC_{50} and E_{max}) are prerequisites for the predictions of clinical DDI as well as a defined clinical exposure ([I]). Some potential hurdles in obtaining these *in vitro* induction parameters experimentally are limitations of test compound solubility and cellular toxicity. This may limit the ability to reach E_{max} in the hepatocytes. In these cases, evaluation of the fold-change with respect to positive controls may have to be used as decision criteria. This approach is empirical but still can be found in regulatory authority guidance [5]. For example, if the mRNA fold-change is less than two-fold and <20% of the positive control inducer, the test compound can be considered a noninducer, according to the EMA [5]. The main drawback to resorting to fold-change criteria is the inability to put the data into perspective of clinically relevant concentrations ([I]) and DDI. It can be helpful, however, for rapidly identifying noninducers and for eliminating further more resource-intensive EC_{50}/E_{max} determinations.

In this example, a drug candidate was determined to be both a reversible inhibitor (K_i 1.3 μM) and an inducer (mRNA EC_{50} 60 μM, E_{max} 8.5-fold) of CYP3A4 *in vitro*. The potential magnitude of clinical CYP3A induction was predicted using the calibration approach (RIS) in Figure 16.4A. Based on a range of expected clinical efficacious exposures ($C_{max,unbound}$), the compound was predicted to decrease the exposure of the CYP3A sensitive substrate, midazolam, by 36–50% (RIS range of 0.07–0.14). The induction parameters of seven other CYP3A inducers/noninducers were used to create the calibration curve (Figure 16.4A). Using another approach incorporating induction parameters only, the SimCyp model predicted a 31% decrease in AUC, which is similar to the calibration approach. However, as a result of the concomitant CYP3A4 inhibition by this compound, the inclusion of inhibition properties of the test compound in the model resulted in a net effect of inhibition. This example illustrates the complexity of modeling perpetrators of CYP enzymes with multiple mechanisms. Most accurate predictions depend on how well the test compound has been characterized for overall DDI potential.

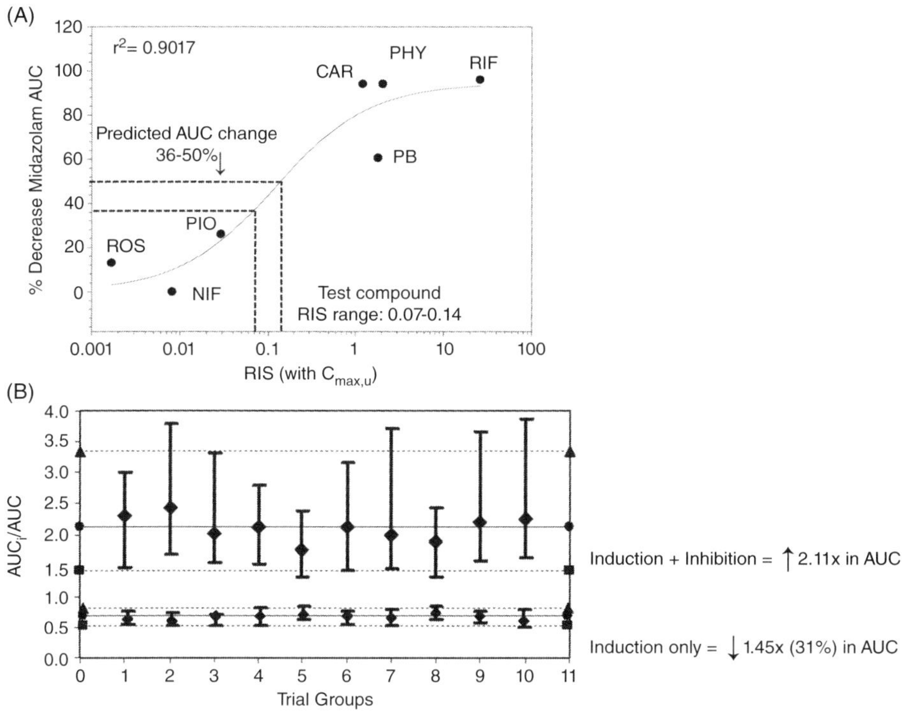

Figure 16.4 Examples of prediction approaches for clinical induction of CYP3A4. Predictions of DDI using a relative induction score (RIS) approach (A) or SimCyp modeling using models for CYP3A induction only or induction plus inhibition (B). The calibrator compounds used in the calibration curve in (A) included CAR, carbamazepine; NIF, nifedipine; PHY, phenytoin; PB, phenobarbital; PIO, pioglitazone; RIF, rifampin; and ROS, rosiglitazone.

16.2.4 CYP Reaction Phenotyping

The identification of metabolic enzymes (e.g., individual CYP enzymes), and estimation of the relative contributions of those enzymes to the total hepatic clearance of a drug, is typically described as "enzyme reaction phenotyping." The estimation of the relative contributions of individual enzymes to total drug clearance is important for drug–drug interaction assessments of the compound as a victim (substrate) drug. This information is important for decisions regarding inclusion/exclusion criteria for co-medications, exclusion of special patient populations (e.g., poor metabolizers) in the first clinical trials, and planning clinical DDI studies. If metabolism by a single CYP enzyme represents a significant proportion of a drug's clearance, then inhibition or induction of this enzymatic pathway could result in potentially substantial increases or decreases, respectively, in exposure to the drug. The clinical result of such an increase or decrease in drug exposure may be in an overdose (increased toxicity) or lead to therapeutic failure, respectively. In addition, if the CYP enzyme is polymorphic (e.g., CYP2C9, CYP2C19, and CYP2D6), this could result in potentially high variability in exposure in the clinic, including high exposure to the drug in poor metabolizers.

In the research phase, care must be taken when interpreting results with limited information about contributions of metabolism to total drug clearance. At this stage, little is known about the contributions of different clearance pathways to the total elimination of the investigational drug in humans or even animals. Besides CYP-mediated oxidative metabolism, there may be other mechanisms contributing to the total drug clearance, e.g., other enzyme systems, renal or biliary clearance of the unchanged drug, etc. Prior to the first clinical studies, knowledge about the major clearance mechanisms may be known from preclinical ADME studies using radiolabel compound. These studies can be helpful in the estimation of the relative importance of metabolism, as well as in specific metabolic pathways, to the total clearance of the compound in animals. In addition, radiolabel metabolism studies in human hepatocytes can also give a more quantitative picture of the relative importance of oxidative metabolism versus other reaction pathways (direct conjugation, hydrolysis, etc.) involved in the compound's clearance in the hepatocyte. Ultimately, human radiolabel ADME studies provide the most definitive assessment of the routes of drug clearance in humans. In addition, designated clinical DDI trials using potent and selective CYP inhibitors can define more succinctly that enzyme's involvement in the total clearance (assuming complete inhibition). However, these types of studies are typically run later in clinical development (phases 2 and 3). The knowledge about potential DDI, however, is needed earlier for phase 1 clinical development; *in vitro* studies and estimation from preclinical animal studies are important for these initial assessments.

Relative contributions of individual CYP enzymes to the total metabolic clearance can be assessed *in vitro* by several methods. Some noted techniques include correlation of the test compound metabolism activity to specific CYP enzyme activities in a panel of individual HLM preparations, kinetic analysis of individual recombinant CYP enzyme metabolism and scaling of the contributions to total HLM intrinsic clearance, and inhibition of the test compound metabolism by selective CYP inhibitors (chemicals or inhibitory antibodies) in HLM [6]. In a drug discovery setting, the latter technique may be most amenable to a more high-throughput assay, where the use of a substrate depletion approach may be sufficient to assess the change in total HLM metabolic clearance in the presence of a CYP-selective inhibitor [61–63]. Use of a low test substrate concentration that is below its K_m value is preferred when determining an IC_{50} for the inhibition. Therefore, the IC_{50} value is close to the expected K_i value for inhibition of that individual CYP enzyme. The limitations of the substrate depletion method are sensitivity for low-turnover compounds, as a substantial percentage of the initial substrate concentration during the incubation time must occur. In this case, linear initial reaction velocities cannot be measured. In more of a development setting, with the radiolabel test compound, formation and inhibition of individual metabolites can be monitored. This method is more robust and can provide the sensitivity needed for linear initial velocities to be measured.

Table 16.2 lists some inhibitors known to be CYP enzyme selective. Based on a literature search of reported K_i or IC_{50} values for the inhibition of individual CYP

Table 16.2 Some CYP-selective inhibitors and recommended concentration range to use in *in vitro* phenotyping studies

CYP enzyme	Selective inhibitor	Recommended experimental concentration range (μM)	Median K_i (or IC_{50}) value*	Number of reported *in vitro* experiments*
CYP1A2	Furafylline**	0–10	1.9	29
CYP2B6/2C19	Ticlopidine**	0–10	0.17 (CYP2B6)	8
			1.2 (CYP2C19)	23
CYP2C8	Montelukast***	0–2	0.019	15
CYP2C9	Sulfaphenazole	0–5	0.50	97
CYP2D6	Quinidine	0–2	0.058	120
CYP3A4/5	Ketoconazole	0–1	0.10	425
CYP3A4/5	Azamulin**	0–5	0.15	3

*The University of Washington Metabolism & Transport Drug Interaction Database (http://www.druginteractioninfo.org) was queried in the category of "In Vitro Parameters" for studies providing K_i or IC_{50} values with indicated precipitants. If more than one CYP was inhibited by the selected precipitant for a particular substrate metabolism, the corresponding K_i or IC_{50} values were excluded from the analysis. If individual K_i or IC_{50} values were listed for one study, the average was taken. If a range of values was given, the lowest value was taken. All queries were of data reported in the Database as of January 2013.
**Time-dependent inhibitor, pre-incubation recommended.
***IC_{50} dependent on microsomal protein concentration resulting from substantial microsomal protein binding of montelukast [64].

enzymes and compilation of the information, a recommended concentration range of inhibitor to use is listed. This range is intended to maximize the selectivity of the inhibitor for the specific CYP enzyme(s) listed. If a full IC_{50} curve is produced, the IC_{50} value obtained should be similar to those reported in Table 16.2 for inhibition of that particular enzyme. How complete the inhibition of the HLM metabolism by the CYP-selective inhibitor is can also give some information regarding the contribution of the inhibited enzyme to the total HLM oxidative metabolism.

Another common technique to assess the contribution of specific CYP enzymes to total microsomal intrinsic clearance is to determine recombinant human (rh) CYP enzyme kinetic parameters for the metabolism of the drug and to scale the rhCYP intrinsic clearance (CL_{int} or V_{max}/K_m) to HLM CL_{int}. When the contributions of specific CYP enzymes are scaled to HLM CL_{int}, then the percent contribution of each CYP enzyme with respect to the total HLM CL_{int} can be estimated. The CL_{int} values from recombinant CYP enzymes need to be scaled to HLM clearance to account, primarily, for differences in the expression/activity of the CYP enzyme in the different *in vitro* systems. As illustrated in Figure 16.5, there are varying expression levels of individual CYP enzymes in HLM.

Two techniques of scaling the rhCYP CL_{int} to HLM CL_{int} are the use of a relative activity factor (RAF) or intersystem extrapolation factor (ISEF) [66]. The RAF accounts for differences in intrinsic activity (per mg microsomal protein) between rhCYP and HLM. Equation 16.18 describes the RAF as a ratio of the maximal CYP

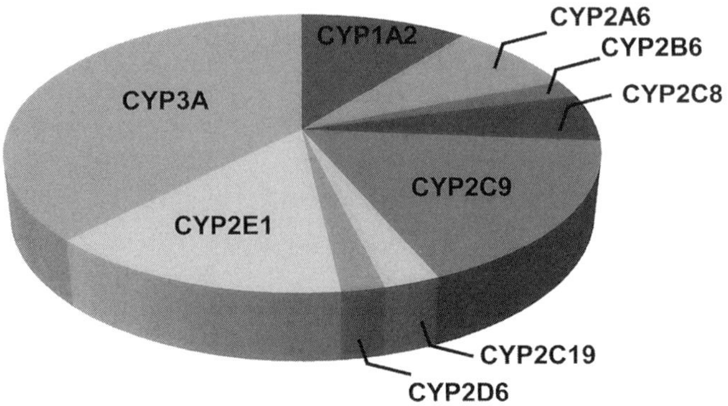

Figure 16.5 Relative abundance of individual CYP enzymes in human liver. The geometric mean of the CYP abundance in Caucasians (pmol CYP/mg microsomal protein) from Reference 65 was plotted as a percentage of the total CYP amount reported.

metabolizing activity (V_{max}, or activity at saturating substrate concentrations) in the rhCYP preparation over the maximal metabolizing activity in HLM using the same CYP-selective probe substrate (see Table 16.1):

$$\mathrm{RAF} = \frac{V_{\mathrm{max,rhCYP}}\left(\mathrm{pmol}\cdot\mathrm{min}^{-1}\cdot\mathrm{mg\ protein}^{-1}\right)}{V_{\mathrm{max,HLM}}\left(\mathrm{pmol}\cdot\mathrm{min}^{-1}\cdot\mathrm{mg\ protein}^{-1}\right)} \tag{16.18}$$

The units of the activities are both per mg protein for rhCYP and HLM; therefore, this RAF is unit-less. The ISEF accounts for differences in intrinsic activity (per unit CYP) between rhCYP and HLM and is used to scale CYP contributions to total liver after accounting for CYP abundance in liver (pmol CYP · mg protein^{-1}). The ISEF can be based on either V_{max} or CL_{int} (V_{max}/K_m) for metabolism of a CYP probe substrate in each system, like the RAF approach (see Equations 16.19 and 16.20, respectively). Some CYP abundance data from a meta-analysis can be found in the literature by Inoue et al. [65]; however, it may be more appropriate to measure this in the HLM used in the laboratory:

$$(\mathrm{V})\,\mathrm{ISEF} = \frac{V_{\mathrm{max,HLM}}\left(\mathrm{pmol}\cdot\mathrm{min}^{-1}\cdot\mathrm{mg\ protein}^{-1}\right)/\mathrm{pmol\ CYP}\cdot\mathrm{mg\ protein}^{-1}}{V_{\mathrm{max,rhCYP}}\left(\mathrm{pmol}\cdot\mathrm{min}^{-1}\cdot\mathrm{pmol\ CYP}^{-1}\right)}. \tag{16.19}$$

$$(\mathrm{CL})\,\mathrm{ISEF} = \frac{\mathrm{CL}_{\mathrm{int,HLM}}\left(\mathrm{mL}\cdot\mathrm{min}^{-1}\cdot\mathrm{mg\ protein}^{-1}\right)/\mathrm{pmol\ CYP}\cdot\mathrm{mg\ protein}^{-1}}{\mathrm{CL}_{\mathrm{int,rhCYP}}\left(\mathrm{mL}\cdot\mathrm{min}^{-1}\cdot\mathrm{pmol\ CYP}^{-1}\right)} \tag{16.20}$$

The calculation of the "HLM-scaled" $CL_{int,u}$ for the individual CYP enzyme is shown as follows using the RAF or (V)ISEF method, as an example (Equations 16.21 and 16.22, respectively).

Calculation of the "HLM-scaled" $CL_{int,u}$ using RAF:

$$CL_{int,u} = \frac{V_{max}\left(\text{pmol} \cdot \text{min}^{-1} \cdot \text{mg protein}^{-1}\right)/\text{RAF}}{K_{m,u}} = \text{mL} \cdot \text{min}^{-1} \cdot \text{mg protein}^{-1}. \tag{16.21}$$

Calculation of the "HLM-scaled" $CL_{int,u}$ using (V)ISEF:

$$\begin{aligned} &CL_{int,u} \\ &= \frac{V_{max}\left(\text{pmol} \cdot \text{min}^{-1} \cdot \text{pmol CYP}^{-1}\right) \times \text{(V) ISEF} \times \text{CYP abundance}\left(\text{pmol CYP} \cdot \text{mg protein}^{-1}\right)}{K_{m,u}} \\ &= \text{mL} \cdot \text{min}^{-1} \cdot \text{mg protein}^{-1} \end{aligned} \tag{16.22}$$

The $CL_{int,u}$ value is the unbound intrinsic clearance or ($V_{max}/K_{m,u}$), where $K_{m,u}$ is the unbound K_m value, i.e., corrected for microsomal protein binding (fu_{mic}). The "HLM-scaled" $CL_{int,u}$ is the rhCYP intrinsic clearance corrected for the RAF or ISEF value to scale the clearance to the appropriate level of CYP activity/abundance in the HLM. If the same abundance values are used to calculate the (V)ISEF value as well as the $CL_{int,u}$ parameter, then the use of the RAF and the (V)ISEF approaches are equivalent. The contribution of the individual CYP enzymes to the total CYP oxidative clearance can then be calculated by the following equation:

$$\text{Contribution of CYP enzyme to HLM } CL_{int,u}\ (\%) = 100\ \frac{CL_{int,u}\text{CYP enzyme}}{\sum \text{CYP } CL_{int,u}} \tag{16.23}$$

Use of the rhCYP enzyme kinetic scaling method in addition to other approaches, such as CYP-selective inhibition in HLM, can provide good guidance regarding the important CYP enzyme(s) involved in the hepatic oxidative clearance. Integrating this information with other data, such as the clearance pathways found in primary human hepatocytes cultures and preclinical animals, will give a better perspective of the importance of oxidative clearance to the total systemic drug clearance. This information can generally be available prior to the first dose in humans, and as mentioned earlier, it is critical to safety (inclusion/exclusion criteria), DDI trial planning, and drug labeling. As more information becomes available, such as the data from the radiolabel human ADME study as well as designated clinical DDI studies investigating the mechanism of drug clearance (i.e., DDI study with a potent CYP inhibitor), the contributions of the CYP enzymes will be refined. As shown earlier in Figure 16.1, the magnitude of a DDI is highly dependent on knowing the contribution of the affected enzyme to the total clearance of the drug. Even a compound that is cleared only 50% by one enzyme, if this pathway is completely inhibited, it could result in a two-fold increase in exposure. Depending on the therapeutic window, this change could be deemed substantial if this window is narrow.

16.3 TRANSPORTER-MEDIATED DRUG–DRUG INTERACTIONS

Drug transporters are membrane-bound proteins that catalyze the movement of drug molecules or endogenous substances from one biological environment to the other. Transporters derive the energy required for catalysis either by chemical means (e.g., ATPase activity) or by using chemical or electrochemical gradients in viable cells. As such, these proteins may regulate the identity and the amount of the molecules that enter into a tissue and may have effects on the pharmacokinetics, efficacy of therapeutics, and potential for DDI [67, 68]. The *in vivo* role that transporters can play has been demonstrated in loss-of-function mutations in humans and in knock-out animal models [69–73]. Additional knock-out animal models continue to be developed that include knock-out animals with a single or multiple disrupted transporters or those where the native transporters are replaced with human counterparts [74, 75].

Despite recent advances in understanding how transport processes can impact pharmacokinetics, the overall knowledge still lags behind that of understanding the impact of drug metabolizing processes. Therefore, under most circumstances, detailed drug transporter evaluations are not included in early drug discovery efforts. Such activities in most cases are restricted to an assessment of membrane permeability in cell-based assays that express efflux transporters (e.g., Caco-2 cell monolayers) in support of oral drug formulation development. Additionally, early research efforts on central nervous system (CNS)-targeted drugs often include evaluations to rule out an interaction with P-glycoprotein (P-gp) that may limit compound distribution at the pharmacological target [76]. Other promising targeted research efforts include the design of drug-peptide [77, 78] or drug-bile acid conjugates [79] to enhance oral bioavailability, and tissue-centric compounds to maximize distribution to target tissue examples include potential treatments for diabetes, hypercholesterolemina, and anti-hepatitis C agents [80, 81]. More detailed assessments of transporter-mediated DDIs can subsequently be done as compounds progress in the development process that are ideally guided by *in vitro* and *in vivo* observations from preclinical and early human studies. It is widely accepted that interactions as a result of transporter inhibition are reversible (i.e., competitive) in nature, which like competitive inhibition of drug metabolizing enzymes, has less of an impact on the pharmacokinetics of the victim drug when compared with irreversible (i.e., time-dependent) inhibition. Additionally, the risk of adverse event as a result of transporter inhibition is further minimized because it is a rare event to find a drug currently in development that relies entirely on a single drug transporter for its disposition. Therefore, the resultant fold increase in systemic exposure is in most cases expected to be lower in magnitude when compared with the inhibition of drug metabolizing enzymes.

However, even small changes in exposure to narrow therapeutic index drugs (e.g., digoxin) or cytotoxic anticancer agents may have severe consequences, and therefore, transporter interactions have to be considered. Moreover, because in many cases drug transporters work in concert with drug metabolizing enzymes, the magnitude of change in exposure as a result of interactions with transporters may be significantly higher than anticipated. As the major goal of the pharmaceutical industry is to develop therapies that can be safely administered to humans, such considerations need to be addressed. In particular, transporters that have been demonstrated to be clinically relevant (i.e., influence the pharmacokinetic properties of the newly developed drugs) and therefore can be potential sites of DDI in the clinic need to be studied in detail preferably prior to large-scale clinical trials. These transporters include efflux (P-gp and Breast Cancer Resistance Protein, BCRP) and uptake transporters [organic anion transporter polypeptide (OATP) 1B1, OATP1B3, organic anion transporter (OAT) 1, OAT3, and the organic cation transporter (OCT) 2] that are expressed in the tissues of drug disposition. By virtue of the nature of the activity of these transporters, different considerations and *in vitro* assays need to be put in place to investigate their ability to transport drugs or to study their inhibition. Efflux transporters such as P-gp and BCRP have historically been investigated by measuring the flux across polarized cell monolayers where the expression of either efflux transporter is localized to the apical membranes. On the other hand, the activity of uptake transporters such as OATPs, OATs, and OCTs is investigated by monitoring the accumulation of respective substrates within cells that express individual transporters or isolated cells (or tissue slices) derived from the tissue of interest. More complex systems have been described to assess simultaneously drug uptake and efflux such as doubly transfected mammalian cells and hepatocytes cultured in a manner that allows them to form structures that are analogous to canalicular bile ducts [82]. Each one of these systems offers advantages to address individual transporter-related aspect, but it is unlikely to provide a comprehensive picture of how these processes interact with each other or with other potential clearance mechanisms.

16.3.1 Efflux Transporter Inhibition

P-gp and BCRP catalyze the transport of a wide variety of drugs with diverse structural features belonging to several therapeutic areas (e.g., antivirals, immunosuppressant therapy, and cytotoxic anticancer agents) [83]. P-gp in particular shares considerable substrates with human CYP3A4, and therefore, the involvement of P-gp in causing DDI is often difficult to prove with certainty [69, 84]. On the other hand, clinical DDI among digoxin and quinidine [85], verapamil [86], or ritonavir [87] can be to a certain extent explained by the inhibition of P-gp in the gastrointestinal tract and kidney epithelia because digoxin is not metabolized by CYP3A4. These clinical observations were verified using *in vitro* models and *in vivo* studies with P-gp knock-out rodents. P-gp can also be induced by agents that also induce CYP3A4 such as rifampicin. Pretreatment with rifampicin has been shown to increase the expression level of P-gp in intestinal enterocytes and as a result has reduced the human exposure to P-gp substrates such as talinolol, fexofenadine, and digoxin [88–90].

When evaluating compounds as inhibitors of P-gp or BCRP, the site of expression of either transporter has to be taken into consideration. For example, to assess the potential for compounds to inhibit gastrointestinal P-gp and BCRP efflux, the drug concentration in the gastrointestinal lumen would be considered. Alternatively, when systemic P-gp or BCRP inhibition is assessed (e.g., at the blood-brain barrier (BBB), renal epithelia, or hepatic canalicular membrane), the concentration of the drug that is *unbound* to plasma- or intracellular-proteins should be considered. Therefore, to minimize the likelihood of potentially adverse reactions resulting from the inhibition of P-gp or BCRP, regulatory agencies have suggested evaluation of the relevant drug concentration and comparison of it with the estimated IC_{50} (or K_i) of a drug toward either transporter. To eliminate potential false-negative findings, compounds are considered as potential *in vivo* inhibitors of P-gp or BCRP if the ratio of the intestinal lumen concentration after the administration of the highest oral dose (amount of drug divided by 250 mL) or the maximum *total* plasma concentration to the P-gp or BCRP IC_{50} (or K_i) value is greater than 10 or 0.10, respectively [91, 92]. Clinical DDI studies are subsequently suggested to investigate whether such *in vitro* findings result in a clinically relevant increase in the systemic exposure of potential co-administered P-gp or BCRP substrates. Such cutoff values have often been considered to be conservative, but they have been put in place to minimize the potential for false-negative findings. Naturally, the largest discrepancy between regulatory recommendations and the scientific community practices is centered on which inhibitor concentration is to be used when assessing the risk of DDI. Sugimoto et al. [93] evaluated the magnitude of P-gp-mediated DDIs at the human BBB with the use of *in vitro* P-gp inhibition data and confirmed the appropriate cutoff limit value. No significant DDI was observed in cases where ($[I]_{unbound}/K_i$) was less than 0.1. Similarly the drug–drug interaction at the human BBB can be predicted from *in vitro* investigations and from *in situ* rat studies when the unbound inhibitor concentrations are considered [94].

16.3.2 Uptake Transporter Inhibition

A change in the systemic exposure of compounds as a result of the inhibition of uptake transport in the main tissue of clearance indicates that uptake clearance plays a role in the overall systemic clearance and tissue intrinsic clearance of this particular drug. The relationship between the intrinsic clearance and the uptake clearance was described by Maeda and Sugiyama [95] as follows:

$$CL_{int,tissue} = CL_{uptake} * \frac{CL_{metab} + CL_{efflux}}{\left(CL_{metab} + CL_{efflux}\right) + CL_{back}} \tag{16.24}$$

where $CL_{int,tissue}$, CL_{uptake}, CL_{metab}, CL_{eff}, and CL_{back} represent the whole tissue intrinsic clearance, uptake clearance, metabolic clearance, efflux clearance, and clearance back into the systemic circulation, respectively. The tissue intrinsic clearance of compounds with very high passive diffusion across individual cells (i.e., $CL_{uptake} \approx CL_{back}$) is determined by the sum of the metabolic and efflux clearances. However, the tissue intrinsic clearance of compounds that undergo appreciable uptake clearance (i.e., $CL_{uptake} \gg CL_{back}$) is determined by the combined activity of all processes.

Uptake clearance, just like metabolic clearance, can be a determinant of overall clearance, and therefore, its inhibition would result in an increase in the systemic exposure of implicated compounds. The fold-change in exposure resulting from transport inhibition could be described as indicated in Equation 16.1. Moreover, because multiple transporters may be capable of contributing to the uptake clearance, the fold-change in exposure would depend on a fraction of the overall transport that was inhibited in a manner analogous to that described in Equation 16.3 for the inhibition of individual CYP enzymes. Inhibition of uptake transport processes is also implicated in increased systemic exposure to endogenous compounds such as bilirubin and bilirubin glucuronides (OATP1B1 and OATP1B3) [96] and, to a certain extent, to creatinine (OCT2) [97]. In the case of bilirubin, the increase is often dose limiting and may be additionally exacerbated in combination drug regimens such as those given to combat hepatitis C or HIV infections [98, 99].

Clinical DDIs have been attributed to the inhibition of uptake transporters; however, most reported cases are more related to OATP inhibition compared with the inhibition of OATs or OCT2. This is not surprising because OATPs transport a wide variety of compounds that include many widely used medicines such as HMG-CoA reductase inhibitors and antihypertension agents [100–104]. The timing of studies aimed at investigating transporter inhibition liabilities should ideally be done in accordance with the clinical development strategy. This should take into consideration the compound profile that includes the disease area, patient population, likelihood of co-administration with transporter substrates, and most importantly based on prior knowledge of liabilities associated with the same class of compounds. However, as a general strategy and for patient safety considerations, OATP inhibition assessments should preferably be investigated prior to FIH studies. Additional transporter inhibition assessments (i.e., OATs and OCT2) would be done further down the clinical development path prior to the initiation of larger scale clinical trials.

When investigating transporter inhibition and downstream liability, the ratio of the maximum inhibitory concentration in plasma to the estimated IC_{50} (or K_i) value would be calculated to estimate the potential DDI liability. For intravenously administered drugs, the inhibitory concentration would be the maximum total plasma concentration and the anticipated fold-change exposure resulting from transporter inhibition would be estimated as described in Equation 16.1. This assumption provides the most conservative estimate of the inhibition liability and unsurprisingly overestimates the DDI risk. If the calculated C_{max}/IC_{50} ratio is ≥ 0.1, this implies that inhibition of the uptake transport activity may be observed *in vivo*. Incorporation of the fraction of the drug unbound to plasma proteins is a more realistic estimate of the concentration of the inhibitor at the transporter's active site. In fact, the unbound C_{max} concentration of the perpetrator drug is used when assessing potential *in vivo* inhibition of the uptake activity of the renal transporters OAT1, OAT3, and OCT2 [105]. Hepatic uptake transporters OATP1B1 and OATP1B3 are exposed to higher inhibitor concentrations at the hepatic inlet after oral absorption [1]. The maximum hepatic inlet concentration is calculated as described in Equation 16.2 and used to estimate the potential increase in systemic exposure resulting from the inhibition of OATP transport activity as described in the following equation:

$$\textit{Fold change in systemic exposure} = 1 + \left(\frac{fu * [I]_{inlet,max}}{IC_{50}}\right) \tag{16.25}$$

where the ($fu \times [\mathrm{I}]_{\mathrm{inlet,max}}$) term represents the unbound maximum hepatic inlet inhibitor concentration that comes in contact with OATPs at the sinusoidal membrane of hepatocytes. An estimated value in excess of 1.25 (represents the upper limit of equivalence) may require additional investigations to assess the inhibition liability and may include a clinical DDI study with a sensitive OATP substrate as shown in Table 16.3.

16.3.3 Efflux Transporter Phenotyping

As mentioned, compounds may be characterized as substrates for efflux drug transporters in early discovery activities to evaluate the potential effects on the extent of gastrointestinal absorption and distribution to tissues containing the pharmacological or toxicological target (e.g., CNS). The extent by which such an efflux process may impact oral drug bioavailability or distribution can be estimated from *in vitro* transport studies across polarized cell monolayers that express the efflux transporter. Such experiments can be used to estimate the passive permeability of the new drug candidate by measuring the increase in uptake permeability when the efflux transporters are inhibited. Alternatively, transport rates across cell monolayers derived from parental cell lines or those from manipulated cells where the expression of a single (or multiple) efflux transporters is obliterated either genetically or using RNAi knockdown approaches [132–134]. Experiments are typically done in both directions across the polarized monolayers (apical-to-basolateral and basolateral-to-apical) under sink conditions [135]. Transport studies across polarized monolayers can be conducted at a range of concentrations, and the permeability in either direction across the polarized cell monolayer can be estimated as a function of the substrate concentration. The contribution of efflux to the observed flux in either direction can be measured, and the kinetic parameters of the transport activity (i.e., J_{max} and K_{m} values) can be estimated. Ultimately, scaling these parameters taking into consideration the difference in expression level between the *in vitro* test system and the expression in the biological tissue can be used, in theory, to estimate the contribution to the efflux activity the disposition of efflux substrates *in vivo*.

The characterization of the nature of interaction between new clinical candidates and efflux transporters is preferably carried out prior to first-in-human studies. Moreover, the clinical co-administration with efflux inhibitors would need to be evaluated depending on the safety margin of the new clinical candidate. Definitive evidence implicating efflux drug transporters in influencing the *in vivo* disposition of the drug candidate will ultimately need to be verified from human mass balance studies.

Table 16.3 Extended list of clinically relevant transporters including probe substrates and model inhibitors that can be used clinically to assess transporter-mediated interaction in individual tissues of drug disposition

Transporter [a]	Site of interaction	Clinical drug substrate	Clinical inhibitors	References
P-gp	Gastrointestinal enterocytes	Digoxin, talinolol, loperamide	Verapamil, quinidine	[85, 106–108]
	Hepatic canalicular membrane [b]	Imatinib, digoxin, (Tc99-Sestamibi)	Elacridar, (rifampicin, PSC833)	[72, 109, 110]
	Renal brush border membrane	Digoxin, (Tc99-Sestamibi)	Cyclosporine, quinidine, (PSC833)	[85, 111]
	Blood-brain barrier	Verapamil, loperamide, gefitinib	Quinidine, cyclosporin, ritonavir, saquinavir	[112, 113]
BCRP	Gastrointestinal enterocytes	sulfasalazine, rosuvastatin	Cyclosporine, pantoprazole	[114]
	Hepatic canalicular membrane	methotrexate	pantoprazole	[115]
	Renal brush border membrane	methotrexate	pantoprazole	[115]
	Blood-brain barrier	Gefitinib	— [c]	[116]
OATP1B1	Hepatic sinusoidal membrane	Atorvastatin acid, Pravastatin, Pitavastatin, Rosuvastatin	Rifampicin, Atazanavir	[117–121]
OATP1B3	Hepatic sinusoidal membrane	Valsartan, Telmisartan	Rifampicin	[103, 122]
OATP1A2	Gastrointestinal enterocytes	Aliskiren, Fexofenadine, Celiprolol	Grape fruit juice, verapamil	[123–125]
OAT1	Renal brush border membrane	Cidofovir	Probenecid	[126]
OAT3	Renal brush border membrane	Penicillin G	Probenecid	[127]
OCT2	Renal proximal tubule cells	Metformin, Varenicline	cimetidine	[128, 129]
MATE1/ MATE2K	Renal brush border membrane	Metformin	Cimetidine	[128]

[a] The transport list excludes the MRP2 transporter as it plays a prominent role in the hepatic transport of drug metabolites [130, 131] and not of the active drug itself.

[b] Clinical studies aimed at demonstrating the unequivocal participation of P-gp in the biliary secretion of drug or its inhibition are extremely challenging and are in almost all cases based on observations in either knock-out animal models or *in vitro* specialized cell culture studies.

[c] Observation based on data from BCRP knock-out mice [116].

16.3.4 Uptake Transporter Phenotyping

To determine whether a candidate drug is a substrate of a specific uptake transporter, uptake of the candidate drug can be determined in cells expressing the individual uptake transporter and compared with the uptake of the candidate drug into the parental control cells, i.e., those that do not express appreciable levels of the transporter. If the candidate drug is a substrate of the transporter, its accumulation should be considerably higher in the cells heterologously expressing the transport protein. When a candidate drug is deemed to be a substrate, then it may be appropriate to determine the Michaelis constant (K_m) and the maximal rate of transport (J_{max}). For kinetic experiments, the uptake of increasing concentrations of a candidate drug into parental control cells is subtracted from its uptake into cells expressing the uptake transporter at each concentration of candidate drug tested. It is important to subtract uptake into the parental control cells to account for endogenous transport activity (if present) as well as nonspecific binding of the drug candidate to either the outside of the cell or the transport apparatus.

As a result of the overlapping substrate specificity, clinical drug candidates are often substrates for more than one uptake transporter in the same biological tissue. For example, both hepatic OATPs (i.e., OATP1B1 and OATP1B3) are capable of carrying out the transport of pravastatin. To estimate the fraction of the overall pravastatin hepatic transport attributed to OATP1B1 and OATP1B3, additional *in vitro* studies would need to be conducted beyond simply characterizing the kinetic parameters of the transport activity in the recombinant systems as described by Hirano et al. [136]. These scaling methods can provide an estimation of the fraction contributed by an individual transporter to the overall transport clearance based on relative transport activity (RAF) or relative expression levels (REF). In the RAF approach (Figure 16.6), the uptake clearance of pravastatin in cryopreserved human hepatocytes would be conducted in addition to that in the recombinant cells that individually express OATP1B1 and OATP1B3. The fraction contributed by an individual transporter to the overall transport clearance by either transporter would then be estimated by scaling pravastatin uptake activity into hepatocytes by that of probe substrates that are exclusively transported into the hepatocytes as shown in Figure 16.6. In the relative expression factor (REF) approach, the fraction of pravastatin uptake contributed by either transporter would be estimated directly from the uptake activity in the recombinant system after applying the relative expression factor as described in Figure 16.7. The choice of either approach will ultimately depend on the availability of the proper tools and on the scaling parameters [136]. The RAF method provides a direct estimation based on transport activity; however, it demands a considerable amount of resources and is limited by the availability of selective probe transport substrates. On the other hand, the REF method demands a considerable amount of up-front investment to characterize the expression of individual transporters in the recombinant systems and in human hepatocytes [137]. Once the expression levels have been characterized, they can be directly applied to estimate the fraction contributed to the overall transport clearance without the need to measure the actual transport activity of the new drug candidate in cryopreserved human hepatocytes. An estimation

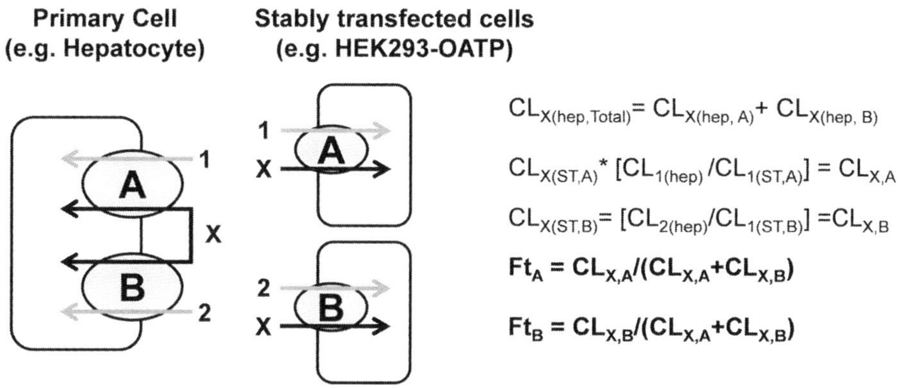

Figure 16.6 Estimation of the relative contribution of individual transporters to the overall transport activity in the tissue of interest using the relative activity factor approach. Schematic representation of the RAF *in vitro* method to estimate the fraction of overall transport attributed to individual uptake transporters. Application of this method applies to compounds (X) that have been shown to be substrates of multiple transporters. and therefore. their systemic clearance would be influenced by this activity. Successful application of this method requires specific probe substrates (represented as 1 and 2 in the figure) for individual transporters, and the estimation of the uptake clearance values for the test compound and the probe substrates in all cells as indicated in the equations.

of transporter expression level is commonly performed by LC-MS/MS approaches using a unique polypeptide derived from the transporter of interest after proteolytic digestion [137–140]. The preparation of the cellular plasma membrane fraction is critical because only those proteins localized in this fraction would contribute to the transport activity [141, 142].

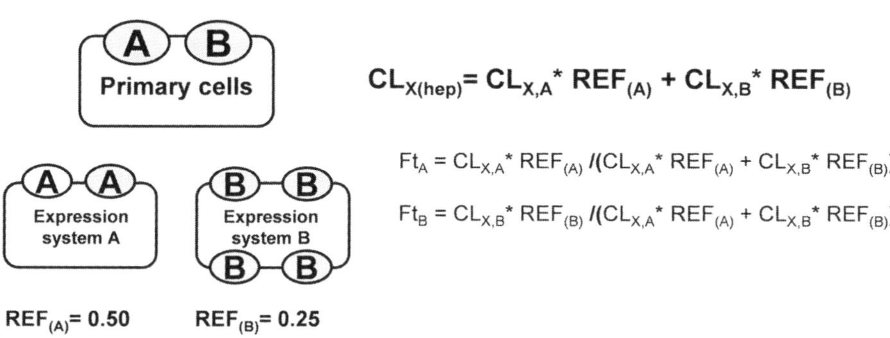

Figure 16.7 Estimation of the relative contribution of individual transporters to the overall transport activity in the tissue of interest using the relative expression factor approach. Schematic representation of the REF method to estimate the contribution of individual transporters to overall transport. This method takes into consideration the ratio of expression levels between the expression system and primary cells. This method overcomes the overlapping substrate specificity and eliminates the need to measure the uptake of test compounds into the primary cells as indicated in the equations.

The timing of the conduct of either one of the previous approaches would likely be after sufficient evidence has been gained to prove that such transport pathways contribute to the *in vivo* clearance of the drug candidate. Such evidence can be gleamed from human mass balance studies or in many cases from prior experience with compounds in the same chemical class. For example, the observation of appreciable amounts of unchanged drug in urine can be considered as evidence for the involvement of transporters in the renal secretion. In this case, the estimated renal clearance of the new drug candidate (i.e., amount of unchanged drug excreted in the urine divided by the plasma AUC) is compared with the product of the human glomerular filtration rate (GFR) by the unbound plasma fraction (f_u). An indication of active secretion is observed when the renal clearance is appreciably higher than the GFR* f_u product. Similarly the observation of considerable amounts of unchanged drug in human feces can be an indication of active hepatobiliary transport that may be mediated in part by OATP1B1 and OATP1B3 activity. Inherent challenges with the accurate interpretation of fecal extraction results include the possible incomplete absorbance of the orally administered dose, or the conversion of labile metabolites back to the parent drug by gastrointestinal microflora (e.g., glucuronide conjugates and *N*-oxide metabolites). Having sufficient information implicating the active transport process in the systemic clearance would ideally be used to design targeted clinical DDI interaction studies. Such studies would be able to assess *in vivo* risk, guide clinical dose selection, and potentially establish patient exclusion/inclusion criteria.

16.3.5 Emerging Drug Transporters of Potential Clinical Relevance

Several transporters mentioned in Table 16.3 have shown potential to play a role in drug disposition, and their activity (and inhibition) may need to be characterized. However, because insufficient clinical evidence exists to implicate them with common pharmacokinetic interactions, such characterization activities are ideally performed on a case-by-case basis and driven by the need to provide a mechanistic explanation to an observed clinical behavior. For example, the reduction in Aliskiren, Celiprolol, or Fexofenadine bioavailability when taken with grapefruit juice could be explained by possible interaction at the level of intestinal OATP1A2 activity [123–125, 143]. In many cases such as with Aliskiren, the apparent reduction in bioavailability in the presence of grapefruit juice was not shown to be of clinical significance [123].

Finally, the human organic cation transporter 1 (OCT1) was not included in the list of relevant drug transporters because of the absence of evidence implicating its involvement in causing clinical DDIs. Several studies have shown potential association between basal OCT1 expression level in lymphocytes and the clinical efficacy of imatinib [144, 145]. Interestingly, no evidence for OCT1-mediated imatinib transport could be demonstrated (in-house unpublished data). This indicated that the clinical associations may be a result of additional factors but are unlikely related to the direct participation of OCT1 in imatinib transport.

16.4 CONCLUSIONS

The importance of mitigating drug interactions starting in the early discovery stages is to decrease potential DDI liability prior to the selection of a drug candidate for further preclinical and clinical development. In development, the focus is optimal *in vitro* characterization of the DDI and to project the DDI effect in humans with increased accuracy as the drug candidate proceeds into clinical development. After a compound is selected into development, any identified DDI liability will have to be managed including decisions to instigate clinical DDI studies, if needed, to understand the magnitude of DDI risk to aid in decisions regarding dose adjustments and labeling restrictions. The DDI liability of drug candidates can impact regulatory submissions, drug labeling, prescribing information, and concern for safety and efficacy. The general strategies that were overviewed in this chapter included streamlined processes for efficiency in discovery as well as more thorough and robust assays in development. Predictive models of DDI, whether the more established methods for CYP enzyme involvement and DDI or emerging models for transporter effects, aid in the iterative process of risk assessment to optimize drug candidates in discovery and inform for clinical DDI strategies.

REFERENCES

1. Ito, K., Iwatsubo, T., Kanamitsu, S., Ueda, K., Suzuki, H., Sugiyama, Y. (1998). Prediction of pharmacokinetic alterations caused by drug-drug interactions: Metabolic interaction in the liver. *Pharmacological Reviews*, *50*, 387–412.
2. Williams, J.A., Hyland, R., Jones, B.C., Smith, D.A., Hurst, S., Goosen, T.C., Peterkin, V., Koup, J. R., Ball, S.E. (2004). Drug-drug interactions for UDP-glucuronosyltransferase substrates: A pharmacokinetic explanation for typically observed low exposure (AUCi/AUC) ratios. *Drug Metabolism and Disposition*, *32*, 1201–1208.
3. FDA. (2006). Guidance for Industry: Drug Interaction Studies- Study Design, Data Analysis, and Implications for Dosing and Labeling.
4. FDA. (2012). Guidance for Industry: Drug Interaction Studies- Study Design, Data Analysis, and Implications for Dosing and Labeling Recommendations.
5. EMA. (2012). European Medicines Agency: Guideline on the Investigation of Drug Interactions.
6. Bjornsson, T.D., Callaghan, J.T., Einolf, H.J., Fischer, V., Gan, L., Grimm, S., Kao, J., King, S. P., Miwa, G., Ni, L., Kumar, G., McLeod, J., Obach, R.S., Roberts, S., Roe, A., Shah, A., Snikeris, F., Sullivan, J.T., Tweedie, D., Vega, J.M., Walsh, J., Wrighton, S.A. (2003). The conduct of in vitro and in vivo drug-drug interaction studies: A Pharmaceutical Research and Manufacturers of America (PhRMA) perspective. *Drug Metabolism and Disposition*, *31*, 815–832.
7. Chu, V., Einolf, H.J., Evers, R., Kumar, G., Moore, D., Ripp, S., Silva, J., Sinha, V., Sinz, M., Skerjanec, A. (2009). In vitro and in vivo induction of cytochrome p450: A survey of the current practices and recommendations: A pharmaceutical research and manufacturers of america perspective. *Drug Metabolism and Disposition*, *37*, 1339–1354.

8. Grimm, S.W., Einolf, H.J., Hall, S.D., He, K., Lim, H.K., Ling, K.H., Lu, C., Nomeir, A.A., Seibert, E., Skordos, K.W., Tonn, G.R., Van Horn, R., Wang, R.W., Wong, Y.N., Yang, T.J., Obach, R.S. (2009). The conduct of in vitro studies to address time-dependent inhibition of drug-metabolizing enzymes: A perspective of the pharmaceutical research and manufacturers of America. *Drug Metabolism and Disposition*, *37*, 1355–1370.
9. Guengerich, F.P. (2008). Cytochrome p450 and chemical toxicology. *Chemical Research in Toxicology*, *21*, 70–83.
10. Segal, I.H. *Enzyme Kinetics: Behavior and Analysis of Rapid Equilibrium and Steady-State Enzyme Systems*, Wiley. New York, 1975.
11. Gao, H., Steyn, S.J., Chang, G., Lin, J. (2010). Assessment of in silico models for fraction of unbound drug in human liver microsomes. *Expert Opinion in Drug Metabolism and Toxicology*, *6*, 533–542.
12. Austin, R.P., Barton, P., Cockroft, S.L., Wenlock, M.C., Riley, R.J. (2002). The influence of nonspecific microsomal binding on apparent intrinsic clearance, and its prediction from physicochemical properties. *Drug Metabolism and Disposition*, *30*, 1497–1503.
13. Hallifax, D., Houston, J.B. (2006). Binding of drugs to hepatic microsomes: Comment and assessment of current prediction methodology with recommendation for improvement. *Drug Metabolism and Disposition*, *34*, 724–726.
14. Tucker, G.T. (1992). The rational selection of drug interaction studies: Implications of recent advances in drug metabolism. *International Journal of Clinical Pharmacology, Therapy and Toxicology*, *30*, 550–553.
15. Tucker, G.T., Houston, J.B., Huang, S.M. (2001). Optimizing drug development: Strategies to assess drug metabolism/transporter interaction potential—towards a consensus. *British Journal of Clinical Pharmacology*, *52*, 107–117.
16. Brown, H.S., Ito, K., Galetin, A., Houston, J.B. (2005). Prediction of in vivo drug-drug interactions from in vitro data: Impact of incorporating parallel pathways of drug elimination and inhibitor absorption rate constant. *British Journal of Clinical Pharmacology*, *60*, 508–518.
17. Blanchard, N., Richert, L., Coassolo, P., Lave, T. (2004). Qualitative and quantitative assessment of drug-drug interaction potential in man, based on Ki, IC50 and inhibitor concentration. *Current Drug Metabolism*, *5*, 147–156.
18. Ito, K., Brown, H.S., Houston, J.B. (2004). Database analyses for the prediction of in vivo drug-drug interactions from in vitro data. *British Journal of Clinical Pharmacology*, *57*, 473–486.
19. Kanamitsu, S., Ito, K., Sugiyama, Y. (2000). Quantitative prediction of in vivo drug-drug interactions from in vitro data based on physiological pharmacokinetics: Use of maximum unbound concentration of inhibitor at the inlet to the liver. *Pharmaceutical Research*, *17*, 336–343.
20. Einolf, H.J. (2007). Comparison of different approaches to predict metabolic drug-drug interactions. *Xenobiotica*, *37*, 1257–1294.
21. Rostami-Hodjegan, A., Tucker, G. (2004). 'In silico' simulations to assess the 'in vivo' consequences of 'in vitro' metabolic drug-drug interactions. *Drug Discovery Today*, *1*, 441–448.
22. Ito, K., Hallifax, D., Obach, R.S., Houston, J.B. (2005). Impact of parallel pathways of drug elimination and multiple cytochrome P450 involvement on drug-drug interactions: CYP2D6 paradigm. *Drug Metabolism and Disposition*, *33*, 837–844.

23. Obach, R.S., Walsky, R.L., Venkatakrishnan, K., Gaman, E.A., Houston, J.B., Tremaine, L.M. (2006). The utility of in vitro cytochrome p450 inhibition data in the prediction of drug-drug interactions. *Journal of Pharmacology and Experimental Therapeutics*, *316*, 336–348.

24. Yao, C., Levy, R.H. (2002). Inhibition-based metabolic drug-drug interactions: Predictions from in vitro data. *Journal of Pharmaceutical Sciences*, *91*, 1923–1935.

25. Rodrigues, A.D., Winchell, G.A., Dobrinska, M.R. (2001). Use of in vitro drug metabolism data to evaluate metabolic drug-drug interactions in man: The need for quantitative databases. *Journal of Clinical Pharmacology*, *41*, 368–373.

26. Rowland, M., Matin, S.B. (1973). Kinetics of drug-drug interactions. *Journal of Pharmacokinetics and Biopharmaceutics*, *1*, 553–567.

27. Wang, Y.H., Jones, D.R., Hall, S.D. (2004). Prediction of cytochrome P450 3A inhibition by verapamil enantiomers and their metabolites. *Drug Metabolism and Disposition*, *32*, 259–266.

28. Yamano, K., Yamamoto, K., Katashima, M., Kotaki, H., Takedomi, S., Matsuo, H., Ohtani, H., Sawada, Y., Iga, T. (2001). Prediction of midazolam-CYP3A inhibitors interaction in the human liver from in vivo/in vitro absorption, distribution, and metabolism data. *Drug Metabolism and Disposition*, *29*, 443–452.

29. Kanamitsu, S., Ito, K., Green, C.E., Tyson, C.A., Shimada, N., Sugiyama, Y. (2000). Prediction of in vivo interaction between triazolam and erythromycin based on in vitro studies using human liver microsomes and recombinant human CYP3A4. *Pharmaceutical Research*, *17*, 419–426.

30. Ito, K., Ogihara, K., Kanamitsu, S., Itoh, T. (2003). Prediction of the in vivo interaction between midazolam and macrolides based on in vitro studies using human liver microsomes. *Drug Metabolism and Disposition*, *31*, 945–954.

31. Yang, J., Jamei, M., Heydari, A., Yeo, K.R., de la, T.R., Farre, M., Tucker, G.T., Rostami-Hodjegan, A. (2006). Implications of mechanism-based inhibition of CYP2D6 for the pharmacokinetics and toxicity of MDMA. *Journal of Psychopharmacology*, *20*, 842–849.

32. Jamei, M., Marciniak, S., Feng, K., Barnett, A., Tucker, G., Rostami-Hodjegan, A. (2009). The Simcyp population-based ADME simulator. *Expert Opinion on Drug Metabolism & Toxicology*, *5*, 211–223.

33. Hollenberg, P.F., Kent, U.M., Bumpus, N.N. (2008). Mechanism-based inactivation of human cytochromes p450s: Experimental characterization, reactive intermediates, and clinical implications. *Chemical Research in Toxicology*, *21*, 189–205.

34. Kalgutkar, A.S., Obach, R.S., Maurer, T.S. (2007). Mechanism-based inactivation of cytochrome P450 enzymes: Chemical mechanisms, structure-activity relationships and relationship to clinical drug-drug interactions and idiosyncratic adverse drug reactions. *Current Drug Metabolism*, *8*, 407–447.

35. Riley, R.J., Grime, K., Weaver, R. (2007). Time-dependent CYP inhibition. *Expert Opinion on Drug Metabolism & Toxicology*, *3*, 51–66.

36. Fontana, E., Dansette, P.M., Poli, S.M. (2005). Cytochrome p450 enzymes mechanism based inhibitors: Common sub-structures and reactivity. *Current Drug Metabolism*, *6*, 413–454.

37. Correia, M.A., Ortiz de Montellano, P.R. Inhibition of Cytochrome P450 Enzymes, P.R. Ortiz de Montellano (ed.), *Cytochrome P450: Structure, Mechanism, and Biochemistry*, Kluwer Academic/Plenum Publishers, New York, NY, 2005, pp. 247–322.

38. Obach, R.S., Walsky, R.L., Venkatakrishnan, K. (2007). Mechanism-based inactivation of human cytochrome p450 enzymes and the prediction of drug-drug interactions. *Drug Metabolism and Disposition*, *35*, 246–255.
39. Zimmerlin, A., Trunzer, M., Faller, B. (2011). CYP3A time-dependent inhibition risk assessment validated with 400 reference drugs. *Drug Metabolism and Disposition*, *39*, 1039–1046.
40. Silverman, R.B. (1995). Mechanism-based enzyme inactivators. *Methods Enzymology*, *249*, 240–283.
41. Nettleton, D.O., Einolf, H.J. (2010). Assessment of cytochrome P450 enzyme inhibition and inactivation in drug discovery and development. *Current Topics in Medicinal Chemistry*, *11*, 382–403.
42. Mayhew, B.S., Jones, D.R., Hall, S.D. (2000). An in vitro model for predicting in vivo inhibition of cytochrome P450 3A4 by metabolic intermediate complex formation. *Drug Metabolism and Disposition*, *28*, 1031–1037.
43. Greenblatt, D.J., Von Moltke, L.L., Harmatz, J.S., Chen, G., Weemhoff, J.L., Jen, C., Kelley, C.J., LeDuc, B.W., Zinny, M.A. (2003). Time course of recovery of cytochrome p450 3A function after single doses of grapefruit juice. *Clinical Pharmacology & Therapeutics*, *74*, 121–129.
44. Kato, M., Chiba, K., Horikawa, M., Sugiyama, Y. (2005). The quantitative prediction of in vivo enzyme-induction caused by drug exposure from in vitro information on human hepatocytes. *Drug Metabolism and Pharmacokinetics*, *20*, 236–243.
45. Venkatakrishnan, K., Obach, R.S. (2007). Drug-drug interactions via mechanism-based cytochrome P450 inactivation: Points to consider for risk assessment from in vitro data and clinical pharmacologic evaluation. *Current Drug Metabolism*, *8*, 449–462.
46. Ernest, C.S., Hall, S.D., Jones, D.R. (2005). Mechanism-based inactivation of CYP3A by HIV protease inhibitors. *Journal of Pharmacology and Experimental Therapeutics*, *312*, 583–591.
47. Galetin, A., Burt, H., Gibbons, L., Houston, J.B. (2006). Prediction of time-dependent CYP3A4 drug-drug interactions: Impact of enzyme degradation, parallel elimination pathways, and intestinal inhibition. *Drug Metabolism and Disposition*, *34*, 166–175.
48. Yang, J., Liao, M., Shou, M., Jamei, M., Yeo, K. R., Tucker, G.T., Rostami-Hodjegan, A. (2008). Cytochrome p450 turnover: Regulation of synthesis and degradation, methods for determining rates, and implications for the prediction of drug interactions. *Current Drug Metabolism*, *9*, 384–394.
49. Ghanbari, F., Rowland-Yeo, K., Bloomer, J.C., Clarke, S.E., Lennard, M.S., Tucker, G.T., Rostami-Hodjegan, A. (2006). A critical evaluation of the experimental design of studies of mechanism based enzyme inhibition, with implications for in vitro-in vivo extrapolation. *Current Drug Metabolism*, *7*, 315–334.
50. Okudaira, T., Kotegawa, T., Imai, H., Tsutsumi, K., Nakano, S., Ohashi, K. (2007). Effect of the treatment period with erythromycin on cytochrome P450 3A activity in humans. *Journal of Clinical Pharmacology*, *47*, 871–876.
51. Dhuria, S., Einolf, H., Mangold, J., Sen, S., Gu, H., Wang, L., Cameron, S. (2013). Time-dependent inhibition and induction of human cytochrome P4503A4/5 by an Oral IAP antagonist, LCL161, in vitro and in vivo in healthy subjects. *Journal of Clinical Pharmacology*, *53*, 642–653.

52. Raucy, J., Warfe, L., Yueh, M.F., Allen, S.W. (2002). A cell-based reporter gene assay for determining induction of CYP3A4 in a high-volume system. *Journal of Pharmacology and Experimental Therapeutics*, *303*, 412–423.

53. Luo, G., Cunningham, M., Kim, S., Burn, T., Lin, J., Sinz, M., Hamilton, G., Rizzo, C., Jolley, S., Gilbert, D., Downey, A., Mudra, D., Graham, R., Carroll, K., Xie, J., Madan, A., Parkinson, A., Christ, D., Selling, B., LeCluyse, E., Gan, L.S. (2002). CYP3A4 induction by drugs: Correlation between a pregnane X receptor reporter gene assay and CYP3A4 expression in human hepatocytes. *Drug Metabolism and Disposition*, *30*, 795–804.

54. Fahmi, O.A., Kish, M., Boldt, S., Obach, R.S. (2010). Cytochrome P450 3A4 mRNA is a more reliable marker than CYP3A4 activity for detecting pregnane X receptor-activated induction of drug-metabolizing enzymes. *Drug Metabolism and Disposition*, *38*, 1605–1611.

55. Fahmi, O.A., Ripp, S.L. (2010). Evaluation of models for predicting drug-drug interactions due to induction. *Expert Opinion in Drug Metabolism and Toxicoogy*, *6*, 1399–1416.

56. Shou, M., Hayashi, M., Pan, Y., Xu, Y., Morrissey, K., Xu, L., Skiles, G.L. (2008). Modeling, prediction, and in vitro in vivo correlation of CYP3A4 induction. *Drug Metabolism and Disposition*, *36*, 2355–2370.

57. Kanebratt, K.P., Andersson, T.B. (2008). HepaRG cells as an in vitro model for evaluation of cytochrome P450 induction in humans. *Drug Metabolism and Disposition*, *36*, 137–145.

58. Fahmi, O.A., Boldt, S., Kish, M., Obach, R.S., Tremaine, L.M. (2008). Prediction of drug-drug interactions from in vitro induction data: Application of the relative induction score approach using cryopreserved human hepatocytes. *Drug Metabolism and Disposition*, *36*, 1971–1974.

59. Ripp, S.L., Mills, J.B., Fahmi, O.A., Trevena, K.A., Liras, J.L., Maurer, T.S., De Morais, S.M. (2006). Use of immortalized human hepatocytes to predict the magnitude of clinical drug-drug interactions caused by CYP3A4 induction. *Drug Metabolism and Disposition*, *34*, 1742–1748.

60. Almond, L.M., Yang, J., Jamei, M., Tucker, G.T., Rostami-Hodjegan, A. (2009). Towards a quantitative framework for the prediction of DDIs arising from cytochrome P450 induction. *Current Drug Metabolism*, *10*, 420–432.

61. Emoto, C., Murayama, N., Rostami-Hodjegan, A., Yamazaki, H. (2010). Methodologies for investigating drug metabolism at the early drug discovery stage: Prediction of hepatic drug clearance and P450 contribution. *Current Drug Metabolism*, *11*, 678–685.

62. Emoto, C., Murase, S., Iwasaki, K. (2006). Approach to the prediction of the contribution of major cytochrome P450 enzymes to drug metabolism in the early drug-discovery stage. *Xenobiotica*, *36*, 671–683.

63. Emoto, C., Iwasaki, K. (2007). Approach to predict the contribution of cytochrome P450 enzymes to drug metabolism in the early drug-discovery stage: The effect of the expression of cytochrome b(5) with recombinant P450 enzymes. *Xenobiotica*, *37*, 986–999.

64. Walsky, R.L., Obach, R.S., Gaman, E.A., Gleeson, J.P., Proctor, W.R. (2005). Selective inhibition of human cytochrome P4502C8 by montelukast. *Drug Metabolism and Disposition*, *33*, 413–418.

65. Inoue, S., Howgate, E.M., Rowland-Yeo, K., Shimada, T., Yamazaki, H., Tucker, G.T., Rostami-Hodjegan, A. (2006). Prediction of in vivo drug clearance from in vitro data. II: Potential inter-ethnic differences. *Xenobiotica*, *36*, 499–513.

66. Proctor, N.J., Tucker, G.T., Rostami-Hodjegan, A. (2004). Predicting drug clearance from recombinantly expressed CYPs: Intersystem extrapolation factors. *Xenobiotica 34*, 151–178.

67. de Vries, N.A., Buckle, T., Zhao, J., Beijnen, J.H., Schellens, J.H., van, T.O. (2012). Restricted brain penetration of the tyrosine kinase inhibitor erlotinib due to the drug transporters P-gp and BCRP. *Investigational New Drugs*, *30*, 443–449.

68. Breedveld, P., Pluim, D., Cipriani, G., Wielinga, P., van Tellingen, O., Schinkel, A.H., Schellens, J.H. (2005). The effect of Bcrp1 (Abcg2) on the in vivo pharmacokinetics and brain penetration of imatinib mesylate (Gleevec): Implications for the use of breast cancer resistance protein and P-glycoprotein inhibitors to enable the brain penetration of imatinib in patients. *Cancer Research*, *65*, 2577–2582.

69. van Waterschoot, R.A., Schinkel, A.H. (2011). A critical analysis of the interplay between cytochrome P450 3A and P-glycoprotein: Recent insights from knockout and transgenic mice. *Pharmacological Reviews*, *63*, 390–410.

70. van de Steeg, E., Stranecky, V., Hartmannova, H., Noskova, L., Hrebicek, M., Wagenaar, E., van, E.A., de Waart, D.R., Oude Elferink, R.P., Kenworthy, K.E., Sticova, E., al-Edreesi, M., Knisely, A.S., Kmoch, S., Jirsa, M., Schinkel, A.H. (2012). Complete OATP1B1 and OATP1B3 deficiency causes human Rotor syndrome by interrupting conjugated bilirubin reuptake into the liver. *Journal of Clinical. Investigation*, *122*, 519–528.

71. Iusuf, D., Sparidans, R.W., van, E.A., Hobbs, M., Kenworthy, K.E., van de Steeg, E., Wagenaar, E., Beijnen, J.H., Schinkel, A.H. (2012). Organic anion-transporting polypeptides 1a/1b control the hepatic uptake of pravastatin in mice. *Molecular Pharmacology*, *9*, 2497–2504.

72. Oostendorp, R.L., Buckle, T., Beijnen, J.H., van, T.O., Schellens, J.H. (2009). The effect of P-gp (Mdr1a/1b), BCRP (Bcrp1) and P-gp/BCRP inhibitors on the in vivo absorption, distribution, metabolism and excretion of imatinib. *Investigational New Drugs*, *27*, 31–40.

73. Generaux, G.T., Bonomo, F.M., Johnson, M., Doan, K.M. (2011). Impact of SLCO1B1 (OATP1B1) and ABCG2 (BCRP) genetic polymorphisms and inhibition on LDL-C lowering and myopathy of statins. *Xenobiotica*, *41*, 639–651.

74. van de Steeg, E., van, E.A., Wagenaar, E., Kenworthy, K.E., Schinkel, A.H. (2013). Influence of human OATP1B1, OATP1B3, and OATP1A2 on the pharmacokinetics of methotrexate and paclitaxel in humanized transgenic mice. *Clinical Cancer Research*, *19*, 821–832.

75. Iusuf, D., van de Steeg, E., Schinkel, A.H. (2012). Functions of OATP1A and 1B transporters in vivo: Insights from mouse models. *Trends in Pharmacological Sciences*, *33*, 100–108.

76. Taylor, E.M. (2002). The impact of efflux transporters in the brain on the development of drugs for CNS disorders. *Clinical Pharmacokineics*, *41*, 81–92.

77. Varma, M.V., Ambler, C.M., Ullah, M., Rotter, C.J., Sun, H., Litchfield, J., Fenner, K. S., and El-Kattan, A.F. (2010). Targeting intestinal transporters for optimizing oral drug absorption. *Current Drug Metabolism*, *11*, 730–742.

78. Orioli, M., Vistoli, G., Regazzoni, L., Pedretti, A., Lapolla, A., Rossoni, G., Canevotti, R., Gamberoni, L., Previtali, M., Carini, M., Aldini, G. (2011). Design, synthesis, ADME properties, and pharmacological activities of beta-alanyl-D-histidine (D-carnosine) prodrugs with improved bioavailability. *ChemMedChem*, *6*, 1269–1282.

79. Rais, R., Fletcher, S., Polli, J.E. (2011). Synthesis and in vitro evaluation of gabapentin prodrugs that target the human apical sodium-dependent bile acid transporter (hASBT). *Journal of Pharmaceutical Sciences*, *100*, 1184–1195.

80. Powell, D.A., Black, W.C., Bleasby, K., Chan, C.C., Deschenes, D., Gagnon, M., Gordon, R., Guay, J., Guiral, S., Hafey, M.J., Huang, Z., Isabel, E., Leblanc, Y., Styhler, A., Xu, L.J., Zhang, L., Oballa, R.M. (2011). Nicotinic acids: Liver-targeted SCD inhibitors with preclinical anti-diabetic efficacy. *Bioorganic & Medicinal Chemistry Letters*, *21*, 7281–7286.

81. Oballa, R.M., Belair, L., Black, W.C., Bleasby, K., Chan, C.C., Desroches, C., Du, X., Gordon, R., Guay, J., Guiral, S., Hafey, M.J., Hamelin, E., Huang, Z., Kennedy, B., Lachance, N., Landry, F., Li, C.S., Mancini, J., Normandin, D., Pocai, A., Powell, D.A., Ramtohul, Y.K., Skorey, K., Sorensen, D., Sturkenboom, W., Styhler, A., Waddleton, D.M., Wang, H., Wong, S., Xu, L., Zhang, L. (2011). Development of a liver-targeted stearoyl-CoA desaturase (SCD) inhibitor (MK-8245) to establish a therapeutic window for the treatment of diabetes and dyslipidemia. *Journal of Medicinal Chemistry*, *54*, 5082–5096.

82. Chandra, P., Brouwer, K.L. (2004). The complexities of hepatic drug transport: Current knowledge and emerging concepts. *Pharmaceutical Research*, *21*, 719–735.

83. Gottesman, M.M., Pastan, I., Ambudkar, S.V. (1996). P-glycoprotein and multidrug resistance. *Current Opinion in Genetics & Development*, *6*, 610–617.

84. Wacher, V.J., Wu, C.Y., Benet, L.Z. (1995). Overlapping substrate specificities and tissue distribution of cytochrome P450 3A and P-glycoprotein: Implications for drug delivery and activity in cancer chemotherapy. *Molecular Carcinogenesis*, *13*, 129–134.

85. Leahey, E.B., Jr., Reiffel, J.A., Heissenbuttel, R.H., Drusin, R.E., Lovejoy, W.P., Bigger, J.T., Jr. (1979). Enhanced cardiac effect of digoxin during quinidine treatment. *Archives of Internal Medicine*, *139*, 519–521.

86. Hedman, A., Angelin, B., Arvidsson, A., Beck, O., Dahlqvist, R., Nilsson, B., Olsson, M., Schenck-Gustafsson, K. (1991). Digoxin-verapamil interaction: Reduction of biliary but not renal digoxin clearance in humans. *Clinical Pharmacology & Therapeutics*, *49*, 256–262.

87. Kirby, B.J., Collier, A.C., Kharasch, E.D., Whittington, D., Thummel, K.E., Unadkat, J.D. (2012). Complex drug interactions of the HIV protease inhibitors 3: Effect of simultaneous or staggered dosing of digoxin and ritonavir, nelfinavir, rifampin, or bupropion. *Drug Metabolism and Disposition*, *40*, 610–616.

88. Westphal, K., Weinbrenner, A., Zschiesche, M., Franke, G., Knoke, M., Oertel, R., Fritz, P., von Richter, O., Warzok, R., Hachenberg, T., Kauffmann, H.M., Schrenk, D., Terhaag, B., Kroemer, H.K., Siegmund, W. (2000). Induction of P-glycoprotein by rifampin increases intestinal secretion of talinolol in human beings: A new type of drug/drug interaction. *Clinical Pharmacology & Therapeutics*, *68*, 345–355.

89. Hamman, M.A., Bruce, M.A., Haehner-Daniels, B.D., Hall, S.D. (2001). The effect of rifampin administration on the disposition of fexofenadine. *Clinical Pharmacology & Therapeutics*, *69*, 114–121.

90. Greiner, B., Eichelbaum, M., Fritz, P., Kreichgauer, H.P., von Richter, O., Zundler, J., Kroemer, H.K. (1999). The role of intestinal P-glycoprotein in the interaction of digoxin and rifampin. *Journal of Clinical Investigation*, *104*, 147–153.

91. Zhang, L., Zhang, Y.D., Strong, J.M., Reynolds, K.S., Huang, S.M. (2008). A regulatory viewpoint on transporter-based drug interactions. *Xenobiotica*, *38*, 709–724.

92. Fenner, K.S., Troutman, M.D., Kempshall, S., Cook, J.A., Ware, J.A., Smith, D.A., Lee, C.A. (2009). Drug-drug interactions mediated through P-glycoprotein: Clinical relevance and in vitro-in vivo correlation using digoxin as a probe drug. *Clinical Pharmacology & Therapeutics*, *85*, 173–181.

93. Sugimoto, H., Hirabayashi, H., Amano, N., Moriwaki, T. (2013). Retrospective analysis of P-glycoprotein-mediated drug-drug interactions at the blood-brain barrier in humans. *Drug Metabolism and Disposition*, *41*(4), 683–688.

94. Hsiao, P., Unadkat, J.D. (2012). P-glycoprotein-based loperamide-cyclosporine drug interaction at the rat blood-brain barrier: Prediction from in vitro studies and extrapolation to humans. *Molecular Pharmacology*, *9*, 629–633.

95. Maeda, K., Sugiyama, Y. (2010). The use of hepatocytes to investigate drug uptake transporters. *Methods in Molecular Biology*, *640*, 327–353.

96. van de Steeg, E., Stranecky, V., Hartmannova, H., Noskova, L., Hrebicek, M., Wagenaar, E., van, E. A., de Waart, D.R., Oude Elferink, R.P., Kenworthy, K.E., Sticova, E., al-Edreesi, M., Knisely, A.S., Kmoch, S., Jirsa, M., Schinkel, A.H. (2012). Complete OATP1B1 and OATP1B3 deficiency causes human Rotor syndrome by interrupting conjugated bilirubin reuptake into the liver. *Journal of Clinical Investigation*, *122*, 519–528.

97. Reznichenko, A., Sinkeler, S.J., Snieder, H., van den Born, J., de Borst, M.H., Damman, J., van Dijk, M.C., van, G.H., Hepkema, B.G., Hillebrands, J.L., Leuvenink, H.G., Niesing, J., Bakker, S.J., Seelen, M., Navis, G. (2013). SLC22A2 is associated with tubular creatinine secretion and bias of estimated GFR in renal transplantation. *Physiological Genomics*, *45*, 201–209.

98. von, H. N. (2008). Atazanavir/ritonavir: A review of its use in HIV therapy. *Drugs Today (Barc.)* *44*, 103–132.

99. Gallay, P.A., Lin, K. (2013). Profile of alisporivir and its potential in the treatment of hepatitis C. *Drug Design, Development and Therapy*, *7*, 105–115.

100. Lau, Y.Y., Okochi, H., Huang, Y., Benet, L.Z. (2006). Multiple transporters affect the disposition of atorvastatin and its two active hydroxy metabolites: Application of in vitro and ex situ systems. *Journal of Pharmacology and Experimental Therapeutics*, *316*, 762–771.

101. Shitara, Y., Itoh, T., Sato, H., Li, A.P., Sugiyama, Y. (2003). Inhibition of transporter-mediated hepatic uptake as a mechanism for drug-drug interaction between cerivastatin and cyclosporin A. *Journal of Pharmacology and Experimental Therapeutics*, *304*, 610–616.

102. Shitara, Y., Sugiyama, Y. (2006). Pharmacokinetic and pharmacodynamic alterations of 3-hydroxy-3-methylglutaryl coenzyme A (HMG-CoA) reductase inhibitors: Drug-drug interactions and interindividual differences in transporter and metabolic enzyme functions. *Pharmacology & Therapeutics*, *112*, 71–105.

103. Yamashiro, W., Maeda, K., Hirouchi, M., Adachi, Y., Hu, Z., Sugiyama, Y. (2006). Involvement of transporters in the hepatic uptake and biliary excretion of valsartan, a

selective antagonist of the angiotensin II AT1-receptor, in humans. *Drug Metabolism and Disposition*, *34*, 1247–1254.

104. Nakagomi-Hagihara, R., Nakai, D., Kawai, K., Yoshigae, Y., Tokui, T., Abe, T., Ikeda, T. (2006). OATP1B1, OATP1B3, and mrp2 are involved in hepatobiliary transport of olmesartan, a novel angiotensin II blocker. *Drug Metabolism and Disposition*, *34*, 862–869.
105. Giacomini, K.M., Huang, S.M., Tweedie, D.J., Benet, L.Z., Brouwer, K.L., Chu, X., Dahlin, A., Evers, R., Fischer, V., Hillgren, K.M., Hoffmaster, K.A., Ishikawa, T., Keppler, D., Kim, R.B., Lee, C.A., Niemi, M., Polli, J.W., Sugiyama, Y., Swaan, P.W., Ware, J.A., Wright, S.H., Yee, S.W., Zamek-Gliszczynski, M.J., Zhang, L. (2010). Membrane transporters in drug development. *Nature Reviews Drug Discovery*, *9*, 215 236.
106. Verschraagen, M., Koks, C.H., Schellens, J.H., Beijnen, J.H. (1999). P-glycoprotein system as a determinant of drug interactions: The case of digoxin-verapamil. *Pharmacological Research*, *40*, 301–306.
107. Spahn-Langguth, H., Baktir, G., Radschuweit, A., Okyar, A., Terhaag, B., Ader, P., Hanafy, A., Langguth, P. (1998). P-glycoprotein transporters and the gastrointestinal tract: Evaluation of the potential in vivo relevance of in vitro data employing talinolol as model compound. *International Journal of Clinical Pharmacology, Therapy*, *36*, 16–24.
108. Skarke, C., Jarrar, M., Schmidt, H., Kauert, G., Langer, M., Geisslinger, G., Lotsch, J. (2003). Effects of ABCB1 (multidrug resistance transporter) gene mutations on disposition and central nervous effects of loperamide in healthy volunteers. *Pharmacogenetics* *13*, 651–660.
109. Hendrikse, N.H., Kuipers, F., Meijer, C., Havinga, R., Bijleveld, C.M., van der Graaf, W.T., Vaalburg, W., de Vries, E.G. (2004). In vivo imaging of hepatobiliary transport function mediated by multidrug resistance associated protein and P-glycoprotein. *Cancer Chemotherapy and Pharmacology*, *54*, 131–138.
110. Drescher, S., Glaeser, H., Murdter, T., Hitzl, M., Eichelbaum, M., Fromm, M.F. (2003). P-glycoprotein-mediated intestinal and biliary digoxin transport in humans. *Clinical Pharmacology & Therapeutics*, *73*, 223–231.
111. Chen, C.C., Meadows, B., Regis, J., Kalafsky, G., Fojo, T., Carrasquillo, J.A., Bates, S.E. (1997). Detection of in vivo P-glycoprotein inhibition by PSC 833 using Tc-99m sestamibi. *Clinical Cancer Research*, *3*, 545–552.
112. Bendayan, R., Lee, G., Bendayan, M. (2002). Functional expression and localization of P-glycoprotein at the blood brain barrier. *Microscopy Research and Technique*, *57*, 365–380.
113. Sadeque, A.J., Wandel, C., He, H., Shah, S., Wood, A.J. (2000). Increased drug delivery to the brain by P-glycoprotein inhibition. *Clinical Pharmacology & Therapeutics*, *68*, 231–237.
114. Urquhart, B.L., Ware, J.A., Tirona, R.G., Ho, R.H., Leake, B.F., Schwarz, U.I., Zaher, H., Palandra, J., Gregor, J.C., Dresser, G.K., Kim, R.B. (2008). Breast cancer resistance protein (ABCG2) and drug disposition: Intestinal expression, polymorphisms and sulfasalazine as an in vivo probe. *Pharmacogenetics and Genomics 18*, 439–448.
115. Breedveld, P., Zelcer, N., Pluim, D., Sonmezer, O., Tibben, M.M., Beijnen, J.H., Schinkel, A.H., van Tellingen, O., Borst, P., Schellens, J.H. (2004). Mechanism of the

pharmacokinetic interaction between methotrexate and benzimidazoles: Potential role for breast cancer resistance protein in clinical drug-drug interactions. *Cancer Research*, *64*, 5804–5811.

116. Agarwal, S., Sane, R., Gallardo, J.L., Ohlfest, J.R., Elmquist, W.F. (2010). Distribution of gefitinib to the brain is limited by P-glycoprotein (ABCB1) and breast cancer resistance protein (ABCG2)-mediated active efflux. *Journal of Pharmacology and Experimental Therapeutics*, *334*, 147–155.
117. Lau, Y.Y., Huang, Y., Frassetto, L., Benet, L.Z. (2007). Effect of OATP1B transporter inhibition on the pharmacokinetics of atorvastatin in healthy volunteers. *Clinical Pharmacology & Therapeutics*, *81*, 194–204.
118. Deng, S., Chen, X.P., Cao, D., Yin, T., Dai, Z.Y., Luo, J., Tang, L., Li, Y.J. (2009). Effects of a concomitant single oral dose of rifampicin on the pharmacokinetics of pravastatin in a two-phase, randomized, single-blind, placebo-controlled, crossover study in healthy Chinese male subjects. *Clinical Therapeutics*, *31*, 1256–1263.
119. Kyrklund, C., Backman, J.T., Neuvonen, M., Neuvonen, P.J. (2004). Effect of rifampicin on pravastatin pharmacokinetics in healthy subjects. *British Journal of Clinical Pharmacology*, *57*, 181–187.
120. Zhang, W., Deng, S., Chen, X.P., Zhou, G., Xie, H.T., He, F.Y., Cao, D., Li, Y.J., Zhou, H.H. (2008). Pharmacokinetics of rosuvastatin when coadministered with rifampicin in healthy males: A randomized, single-blind, placebo-controlled, crossover study. *Clinical Therapeutics*, *30*, 1283–1289.
121. Hirano, M., Maeda, K., Shitara, Y., Sugiyama, Y. (2006). Drug-drug interaction between pitavastatin and various drugs via OATP1B1. *Drug Metabolism and Disposition*, *34*, 1229–1236.
122. Ishiguro, N., Maeda, K., Kishimoto, W., Saito, A., Harada, A., Ebner, T., Roth, W., Igarashi, T., Sugiyama, Y. (2006). Predominant contribution of OATP1B3 to the hepatic uptake of telmisartan, an angiotensin II receptor antagonist, in humans. *Drug Metabolism and Disposition*, *34*, 1109–1115.
123. Rebello, S., Zhao, S., Hariry, S., Dahlke, M., Alexander, N., Vapurcuyan, A., Hanna, I., Jarugula, V. (2012). Intestinal OATP1A2 inhibition as a potential mechanism for the effect of grapefruit juice on aliskiren pharmacokinetics in healthy subjects. *European Journal of Clinical Pharmacology*, *68*, 697–708.
124. Dresser, G.K., Kim, R.B., Bailey, D.G. (2005). Effect of grapefruit juice volume on the reduction of fexofenadine bioavailability: Possible role of organic anion transporting polypeptides. *Clinical Pharmacology & Therapeutics*, *77*, 170–177.
125. Lilja, J.J., Backman, J.T., Laitila, J., Luurila, H., Neuvonen, P.J. (2003). Itraconazole increases but grapefruit juice greatly decreases plasma concentrations of celiprolol. *Clinical Pharmacology & Therapeutics*, *73*, 192–198.
126. Cundy, K.C., Petty, B.G., Flaherty, J., Fisher, P.E., Polis, M.A., Wachsman, M., Lietman, P.S., Lalezari, J.P., Hitchcock, M.J., Jaffe, H.S. (1995). Clinical pharmacokinetics of cidofovir in human immunodeficiency virus-infected patients. *Antimicrobial Agents and Chemotherapy*, *39*, 1247–1252.
127. Overbosch, D., Van, G.C., Hermans, J., Mattie, H. (1988). The effect of probenecid on the renal tubular excretion of benzylpenicillin. *British Journal of Clinical Pharmacology*, *25*, 51–58.

128. Wang, Z.J., Yin, O.Q., Tomlinson, B., Chow, M.S. (2008). OCT2 polymorphisms and in-vivo renal functional consequence: Studies with metformin and cimetidine. *Pharmacogenetics and Genomics*, *18*, 637–645.

129. Feng, B., Obach, R.S., Burstein, A.H., Clark, D.J., De Morais, S.M., Faessel, H.M. (2008). Effect of human renal cationic transporter inhibition on the pharmacokinetics of varenicline, a new therapy for smoking cessation: An in vitro-in vivo study. *Clinical Pharmacology & Therapeutics*, *83*, 567–576.

130. Horikawa, M., Kato, Y., Tyson, C.A., Sugiyama, Y. (2002). The potential for an interaction between MRP2 (ABCC2) and various therapeutic agents: Probenecid as a candidate inhibitor of the biliary excretion of irinotecan metabolites. *Drug Metabolism and Pharmacokinetics*, *17*, 23–33.

131. Zamek-Gliszczynski, M.J., Day, J.S., Hillgren, K.M., Phillips, D.L. (2011). Efflux transport is an important determinant of ethinylestradiol glucuronide and ethinylestradiol sulfate pharmacokinetics. *Drug Metabolism and Disposition*, *39*, 1794–1800.

132. Graber-Maier, A., Gutmann, H., Drewe, J. (2010). A new intestinal cell culture model to discriminate the relative contribution of P-gp and BCRP on transport of substrates such as imatinib. *Molecular Pharmacology*, *7*, 1618–1628.

133. Watanabe, T., Onuki, R., Yamashita, S., Taira, K., Sugiyama, Y. (2005). Construction of a functional transporter analysis system using MDR1 knockdown Caco-2 cells. *Pharmaceutical Research*, *22*, 1287–1293.

134. Celius, T., Garberg, P., Lundgren, B. (2004). Stable suppression of MDR1 gene expression and function by RNAi in Caco-2 cells. *Biochemical and Biophysical Research Communications*, *324*, 365–371.

135. Adachi, Y., Suzuki, H., Sugiyama, Y. (2001). Comparative studies on in vitro methods for evaluating in vivo function of MDR1 P-glycoprotein. *Pharmaceutical Research*, *18*, 1660–1668.

136. Hirano, M., Maeda, K., Shitara, Y., Sugiyama, Y. (2004). Contribution of OATP2 (OATP1B1) and OATP8 (OATP1B3) to the hepatic uptake of pitavastatin in humans. *Journal of Pharmacology and Experimental Therapeutics*, *311*, 139–146.

137. Ohtsuki, S., Ikeda, C., Uchida, Y., Sakamoto, Y., Miller, F., Glacial, F., Decleves, X., Scherrmann, J.M., Couraud, P.O., Kubo, Y., Tachikawa, M., Terasaki, T. (2013). Quantitative targeted absolute proteomic analysis of transporters, receptors and junction proteins for validation of human cerebral microvascular endothelial cell line hCMEC/D3 as a human blood-brain barrier model. *Molecular Pharmacology*, *10*, 289–296.

138. Kimoto, E., Yoshida, K., Balogh, L.M., Bi, Y.A., Maeda, K., El-Kattan, A., Sugiyama, Y., Lai, Y. (2012). Characterization of Organic Anion Transporting Polypeptide (OATP) expression and its functional contribution to the uptake of substrates in human hepatocytes. *Molecular Pharmacology*,*9*, 3535–3542.

139. Karlgren, M., Vildhede, A., Norinder, U., Wisniewski, J.R., Kimoto, E., Lai, Y., Haglund, U., Artursson, P. (2012). Classification of inhibitors of hepatic organic anion transporting polypeptides (OATPs): Influence of protein expression on drug-drug interactions. *Journal of Medicinal Chemistry*, *55*, 4740–4763.

140. Liu, L., Mugundu, G.M., Kirby, B.J., Samineni, D., Desai, P.B., Unadkat, J.D. (2012). Quantification of human hepatocyte cytochrome P450 enzymes and transporters induced by HIV protease inhibitors using newly validated LC-MS/MS cocktail assays and RT-PCR. *Biopharmaceutics & Drug Disposition*, *33*, 207–217.

141. Kipp, H., Arias, I.M. (2002). Trafficking of canalicular ABC transporters in hepatocytes. *Annual Review of Physiology*, *64*, 595–608.

142. Kipp, H., Arias, I.M. (2000). Intracellular trafficking and regulation of canalicular ATP-binding cassette transporters. *Seminars in Liver Disease*, *20*, 339–351.

143. Dresser, G.K., Bailey, D.G., Leake, B.F., Schwarz, U.I., Dawson, P.A., Freeman, D.J., Kim, R.B. (2002). Fruit juices inhibit organic anion transporting polypeptide-mediated drug uptake to decrease the oral availability of fexofenadine. *Clinical Pharmacology & Therapeutics*, *71*, 11–20.

144. Sacha, T., Czekalska, S., Foryciarz, K., Zawada, M., Florek, I., Cwynar, D., Wator, G., Balwierz, W., Skotnicki, A.B. (2011). [H-oCT1 gene expression as a predictor of major and complete molecular response to imatinib of chronic myeloid leukemia. Single center experience]. *Przegl. Lek. 68*, 191–195.

145. Zach, O., Krieger, O., Foedermayr, M., Zellhofer, B., Lutz, D. (2008). OCT1 (SLC22A1) R61C polymorphism and response to imatinib treatment in chronic myeloid leukemia patients. *Leukemia and Lymphoma*, *49*, 2222–2223.

17

INTEGRATED ASSESSMENT AND CLINICAL TRANSLATION OF *IN VITRO* OFF-TARGET SAFETY PHARMACOLOGY RISKS

Patrick Y. Muller and Christian F. Trendelenburg

17.1 INTRODUCTION

The declining efficiency of pharmaceutical research and development (R&D) [1] is driving the search for improved strategies to identify unsuitable drug candidates earlier in the process and, hence, reduce the number of expensive failures in late-stage clinical trials. Historically, early discovery activities were largely focused on assessing the *in vitro* and *in vivo* efficacy of lead compounds and drug candidates. However, safety is increasingly recognized as a key differentiation criterion and competitive advantage for successful new drugs, based on changing societal [2], legal [3], and regulatory [4] expectations. Consequently, the industry faces the challenge that improved efficiency in demonstrating drug efficacy may have little impact on the time and cost of drug development unless there is a comparable or better efficiency in demonstrating drug safety [5].

Predictive ADMET: Integrative Approaches in Drug Discovery and Development, First Edition.
Edited by Jianling Wang and Laszlo Urban.

17.2 ON- VERSUS OFF-TARGET PHARMACOLOGY/TOXICITY

In general, drugs can cause adverse/toxic effects either by acting via the primary pharmacological target to elicit on-target toxicity (often referred to as mechanistic toxicity) or by acting via a known or unknown off-target (also referred to as an antitarget) receptor to elicit off-target pharmacology/toxicity. Recognizing the underlying mechanism of toxicity is essential for any kind of safety-related decision to be taken during drug development. On-target toxicity is the dominant mode of toxicity for monoclonal antibody (mAb) drugs as a result of their high selectivity for the primary target. On-target toxicity may include both exaggerated pharmacology in the therapeutic target tissue(s) as well as on-target pharmacology in tissues other than the therapeutic target tissue(s) (secondary pharmacology); the latter is often the case even for very selective small-molecule drugs. Off-target toxicity is a frequent mode of toxicity for small-molecule drugs, often resulting from their pharmacological promiscuity [6], their chemical reactivity with biomolecules [7] (either reactivity of the parent drug or its metabolite(s)), or both.

On-target toxicity of drug candidates against (novel) targets is frequently predictable based on both genetic or pharmacological knowledge of the target or its biological signaling pathway. Strategies to improve the therapeutic index (TI) [8] for on-target toxicities may include drug candidates acting as partial antagonists/partial agonists, having more selective distribution into the tissue of intended pharmacological action and/or optimizing the dosing regimen. To assess and translate on-target (off-target) pharmacology/toxicity from animals to humans, the relative *in vitro* on-target (off-target) potency of the drug candidate needs to be known for both preclinical models and humans.

17.2.1 Important Off-targets and Associated Adverse Effects

In vitro off-target receptor profiling should be considered for both parent and major metabolites as early as the lead optimization stage of small-molecule drugs, with hits in radioligand binding assays being confirmed in functional assays to distinguish agonism from antagonism [21]. The off-target pharmacology/toxicity profile of a compound can be optimized by changing the chemical structure while retaining its on-target activity [9]. An overview of important off-targets regularly screened in *in vitro* preclinical safety pharmacology profiling panels used by Novartis Pharmaceuticals (Basel, Switzerland) and validated with regard to human impact has been published previously [21, 22] and is provided in an extended version in the recent review of Bowes et al. [10]. Quantitative *in vitro* off-target pharmacology data (IC_{50}) need to be available before first-in-human studies.

17.3 EXPOSURE-CENTRIC APPROACH: CONSIDERING PLASMA PROTEIN BINDING AND FREE TISSUE EXPOSURE

In general, it is the exposure of a given tissue to a drug (that is, the drug concentration over time), rather than dose, that underlies its pharmacological and toxicological

Table 17.1 Selectivity, safety margin, and therapeutic index

Term	*In vivo*/ *In vitro*	Calculation	Influence of % PPB
Selectivity	*In vitro* ÷ *In vitro*	Off-target IC_{50} ÷ Target IC_{50}	No influence (for protein free assays)
Safety Margin	*In vitro* ÷ *In vivo*	Off-target IC_{50} ÷ C_{max} @ efficacious dose	Use free C_{max}
Therapeutic Index (TI)	*In vivo* ÷ *In vivo*	C_{max} or AUC @ NOAEL dose ÷ C_{max} or AUC @ efficacious dose	Use free Cmax & AUC (Different TI compared with total exposure in case % PPB differs across species)

effects. Toxicity often occurs in tissues other than the target tissue for efficacy. For feasibility reasons, plasma exposure to the drug is usually used as a surrogate for tissue exposure. *In vivo*, it is the free drug concentration at the site of action that exerts the biological activity [11]. Using plasma exposure as surrogate for tissue exposure assumes that free tissue exposure, in equilibrium, is similar to plasma exposure, which is usually the case for drugs with reasonable membrane permeability.

In a translational research setting, TI can be considered the quantitative ratio of the exposure level at the chosen safety end point divided by the exposure level at the chosen efficacy end point. Based on data from *in vivo* (animal or human) studies, TI can be calculated as the ratio of the highest exposure not leading to toxicity and the exposure required to elicit the desired pharmacological effect. Interpreting and making decisions based on TI numbers has been extensively reviewed by Muller and Milton [12] and depends on many factors characterizing an acceptable risk/benefit profile for the targeted indication of a drug candidate.

At the early stages of lead identification, when only *in vitro* data are available, an initial assessment of the TI of a drug candidate can be estimated based on its on/off-target selectivity (off-target IC_{50} divided by on-target IC_{50}) (Table 17.1). There are many examples where an increase in target selectivity (in particular among close-target homologues) leads to an improved *in vivo* TI [13]. However, there are also cases known where relatively high target selectivity did not prevent a drug from having a challenging or even unacceptable clinical safety profile (Figure 17.1). In the latter case, this is often a result of the relatively high systemic exposure required for certain drugs to drive the pharmacological effect, despite the drug having a high affinity (low target IC_{50}) for its pharmacological target *in vitro*.

Exposure-based TI assessments are usually focused on the parent molecule as the main compound eliciting the *in vivo* effects. However, pharmacology and toxicity as well can be driven by metabolites. If it is suspected that a metabolite is responsible for a particular toxicity, the metabolite exposure needs to be considered for calculation of

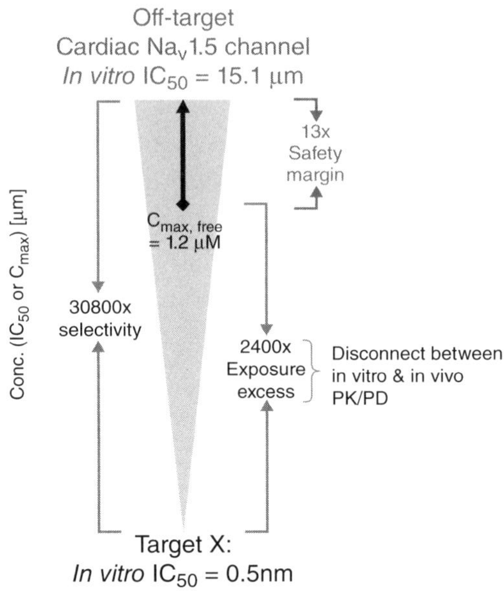

Figure 17.1 Drug candidate X failed as a result of off-target toxicity in humans. On- vs. off-target selectivity *in vitro* (calculated as Off-target IC_{50}/Target IC_{50}, outlined in Table 17.1) is reflected by the length of the vertical triangle with the upper end of the triangle at the IC_{50} value of the off-target and the lower end at the IC_{50} value of the target. The starting point of the vertical arrow corresponds to $C_{max,free}$ at efficacious dose. The length of the vertical arrow reflects the off-target safety margin (calculated as Off-target IC50/$C_{max,free}$ at efficacious dose, outlined in Table 17.1). For further details, refer to text. (*See insert for color representation of the figure.*)

TI, in particular if the parent:metabolite ratio is not constant across studies, species, or individual subjects/populations.

Dosing regimen/schedule and duration may affect pharmacology or toxicity considerably (for example, the use of an intermittent dosing schedule to decrease toxicity of some oncology drugs). For some drug candidates, the envisaged clinical dosing regimen is twice daily (as opposed to once daily), in particular, for drugs with short (predicted) half-life in humans, oral drugs with low solubility, or in cases when lower doses more frequently are used in an attempt to reduce (expected) C_{max}-driven toxicities (for example, drugs known to inhibit cardiac ion channels, which might be anticipated to have different C_{max}-based TIs compared with area under concentration (AUC)-based TIs). In situations with different preclinical compared with clinical dosing frequency, the same AUC intervals should be considered for exposure comparisons.

The determination of C_{max} concentrations relies heavily on the time at which the plasma samples are collected and as such may underestimate the true C_{max}.

The free (that is, unbound) drug hypothesis states that it is the free drug concentration at the site of action that exerts the biological activity, such as on-target pharmacology, off-target/secondary pharmacology, toxicity, or all of the above, not the total drug concentration or the concentration of drug that is bound to plasma or tissue proteins [11]. On the basis of the free drug hypothesis, designing drug candidates aimed at minimizing plasma protein binding (PPB) is not a logical approach, for reasons extensively discussed by Smith et al. [11].

In equilibrium, free tissue exposure in well-perfused tissues reflects free plasma exposure for membrane-permeable drugs, with total tissue exposure being potentially very different from total plasma exposure [11]. For this reason, membrane-permeable compounds do not need to be optimized/selected for high total exposure in tissue(s) expressing the pharmacological target or low total exposure in target tissue(s) of toxicity. For drugs having poor membrane permeability, actual free tissue concentrations/gradients can be influenced by transporters. Tissues with abundant transporter expression include brain [14], liver and kidney [15], testes [16], retina [17], as well as tumor cells [18]. Determination of drug levels in such tissues might need to be considered for drugs known or suspected to be transporter substrates to understand the possible disconnects between plasma and effective tissue concentrations.

Some *in vitro/ex vivo* assays (for example, some cell-based assays, serum supplemented assays, or whole-blood *ex vivo* assays) may have high protein concentrations, and in such cases, the free drug concentration in these assays should be considered when calculating apparent potency. As good practice, we suggest considering free drug exposure for calculation of exposure ratios such as TI or safety margins. In particular, in case there are significant interspecies differences in PPB, systemic exposure (AUC, C_{av}, and C_{max}) at steady state should be corrected for free fraction for interspecies comparison.

17.4 QUANTITATIVE *IN VITRO/IN VIVO* EXTRAPOLATION

High *in vitro* target selectivity (calculated as outlined in Table 17.1) by itself may not be a good predictor of a high therapeutic index of a drug in humans. In Figure 17.1, the example of drug candidate X is given for which high *in vitro* target selectivity did not translate into human safety. Drug candidate X (as published in Reference 10) was terminated during P-R interval (PR) prolongation observed in humans at pharmacological dose (as a result of off-target inhibition of the cardiac $Na_v1.5$ channel). An important factor to be considered is that a high *in vitro* on-target potency (low IC_{50}) may not necessarily correlate with a low *in vivo* exposure required to drive pharmacodynamics. A good example for that is drug candidate X requiring 2400x higher systemic exposure *in vivo* ($C_{max,free}$ at pharmacological dose) compared with the *in vitro* on-target potency (IC_{50}), i.e., requiring 2400x target exposure excess (Equation 17.1). As a consequence, the off-target safety margin (calculated as outlined in Table 17.1) shrinks to 13x, despite high *in vitro* selectivity of the target over the off-target (30800x). In the case of drug candidate X, the tissue expressing the target is different from the tissue expressing the off-target (i.e., the heart). Therefore, potential differences in exposure to drug in-between these tissues may contribute to the situation where the *in vitro* target selectivity (calculated as outlined in Table 17.1) may not be enough to translate into a safe drug clinically.

Table 17.2 Target exposure excess

	In vitro IC_{50} (inhibition of FXa[a]) [nM]	In vitro IC_{50} (human plasma assay) [nM]	$C_{max,free}$ (therapeutic dose) [nM]	Therapeutic dose [mg/day]
Rivaroxaban	2.1	21	14	10
Apixaban	0.62	37	17	2.5 (bid)

[a]Prothrombinase assay.

In contrast, examples for a good correlation between *in vitro* potency and *in vivo* exposure are the Factor-Xa inhibitors Rivaroxaban and Apixaban. For both drugs, the free plasma C_{max} at (relatively low) therapeutic doses is within approximately only 10- to 20-fold of the *in vitro* IC_{50} values (prothrombinase assay), i.e., requiring low target exposure excess, and even within the same range of the *in vitro* IC_{50} values determined in the human plasma assay (Table 17.2).

In general, the extent of target exposure excess required to drive pharmacodynamics depends on target, assay type, and compound class. As with any correlations involving *in vitro* data, careful judgment needs to be given to potential adsorption of compound to glass/plastic ware potentially leading to erroneously high IC_{50} values.

Off-target safety margins are calculated as off-target IC_{50}/C_{max} at efficacious exposure, typically for free plasma concentration (as outlined in Table 17.1). To err on the side of caution, plasma C_{max} should be preferred as the reference concentration *in vivo* over C_{av} as (repeated) short-term modulation of off-targets (in particular around the time C_{max} is achieved *in vivo*) may lead to (long-term) *in vivo* effects, depending on the properties of the drug and the off-target. This assessment should be done as soon as data on pharmacologically efficacious plasma concentration from preclinical *in vivo* pharmacology studies are available and should be repeated later once the efficacious plasma concentration (both predicted and actually measured) in humans is known.

Equation 17.1 ("Muller theorem") connects selectivity (as defined in Table 17.1), target exposure excess, and off-target safety margin (as defined in Table 17.1). A high *in vivo* target exposure excess (see the example in Figure 17.1) will lead to a low *in vivo* off-target safety margin, and vice versa. Pictorially, a high *in vivo* target exposure excess is eating up the *in vitro* selectivity leading to a low off-target safety margin *in vivo*. Using Equation 17.1, at an early lead stage, the *in vitro* selectivity required to avoid specific off-target pharmacology *in vivo* can be calculated based on the anticipated extent of the *in vivo* target exposure excess required to drive 1° pharmacodynamics (which depends on target and compound class) and the *in vivo* off-target safety margin required to avoid off-target pharmacology for the specific off-target (e.g., an off-target safety margin of ≥30 for human ether a go-go related gene (hERG); see the subsequent discussion).

$$\text{Selectivity}_{\text{target vs off-target}} = \text{Exposure excess}_{\text{target}} \times \text{Safety margin}_{\text{off-target}} \qquad (17.1)$$

Table 17.3 Off-target-related factors and drug-related factors modulating *in vivo* outcome of off-target activities

(Off-)target related factors	Drug-related factors
• High off-target receptor density *in vivo* (leading to a potential disconnect between *in vitro* and *in vivo* potencies) • Low turnover of the off-target receptor (slow reconstitution of inactivated off-target receptors) • Co-morbidities in off-target expressing tissues (making it more susceptible to adverse effects) • Possibility of, for example, IC_{10} rather than IC_{50} being the appropriate indicator of off-target pharmacodynamic effects (as accepted for cardiac ion channels)	• Systemic exposure (C_{max}, C_{av}) • High penetration or accumulation into off-target expressing tissues (for example, as a result of active uptake) • Cellular permeability of the drug for intracellular off targets • Mode of action at the off-target receptor (partial vs. full agonists/antagonists; fast vs. slow off-rate) • Width of the C_{max} section of concentration/time curve *in vivo* (flat curve versus sharp peak; that is, time above a certain concentration threshold)

The best known off-target is the cardiac potassium channel hERG; several drugs have been withdrawn from the market or limited in their use as a result of hERG inhibition [19], which can lead to delayed ventricular repolarization/Q-T interval (QT) prolongation and potentially fatal arrhythmia. Consequently, regulatory authorities require all drug candidates to be tested for hERG inhibition prior to entering clinical trials [20]. Redfern et al. [19] concluded that a ≥30-fold safety margin between hERG IC_{50} and free plasma C_{max} in humans, based on retrospective analysis of marketed drugs that inhibit hERG, provides a reasonable safety margin regarding arrhythmogenesis for drugs in clinical evaluation and, thus, tends to be indicative of a reasonable TI in most cases. However, there are also several examples where a ≥30-fold clinical safety margin was not enough to prevent arrhythmogenesis [18]. Therefore, these authors recommended aiming for higher preclinical safety margins, particularly for drug candidates targeting nondebilitating diseases.

Similarly, for the cardiac sodium channel $hNa_v1.5$, a ≥30-fold safety margin between IC_{50} and free plasma C_{max} in humans seems to confer an acceptable degree of safety for Q-R-S interval (QRS) prolongation and associated arrhythmia [21]. However, the magnitude of off-target safety margins required to avoid modulation of a specific off-target and, thus, indicative of sufficient TI, likely depends on many off-target and drug-specific factors. Currently, little is available on this relationship with the exception of hERG and $hNa_v1.5$.

Based on our experience, total plasma C_{max} close to or above the off-target IC_{50} (so called off-target coverage) is in many cases leading to off-target effects *in vivo*, whereas free plasma C_{max} can be (well) below the off-target IC_{50}.

As outlined, the drug exposure at the tissue(s)/compartment(s) in which the off-target is present relative to the off-target IC_{50} (off-target coverage) drives off-target effects, not just the off-target IC_{50}. Additional off-target–related factors and

Table 17.4 Strategies to increase off-target safety margins[a]

Target-related parameters	Off-target–related parameters

• Lower therapeutic exposure / C_{max}
  • By increasing potency to 1° pharmacological target
  • By changing dosing regimen
  • By changing formulation
• Increase distribution to target tissue
	• Decrease potency to off target (increase IC_{50})
	• Decrease distribution to off-target tissue
	  • e.g., decrease BBB permeability for CNS-based off-targets

[a]To be considered for both parent and metabolites. (BBB = blood-brain barrier; CNS = central nervous system).

drug-related factors determining whether a certain exposure level is actually driving specific off-target effects *in vivo* are outlined in Table 17.3.

Mitigation strategies to increase off-target safety margins are outlined in Table 17.4.

At Novartis, small-molecule drug candidates are screened preclinically in *in vitro* safety pharmacology profiling panels for activity on defined off-target receptors [22,23]. In general, the following sequence of questions should be answered to determine the potential *in vivo* impact of pharmacological/toxicological effects elicited by potential activity at an off-target receptor by a drug candidate *in vitro*.

1. Will *in vivo* exposure at the pharmacologically active dose be high enough to modulate (inhibit/activate) the off target?
2. Will the magnitude and duration of the off-target activity *in vivo* lead to a biologically relevant effect (to be assessed in *in vivo* pharmacology/toxicology studies by including tailored endpoints)?
3. Will the off-target–driven *in vivo* effect negatively affect the safety profile of the drug candidate for the targeted indication?

The answers to these questions largely depend on experience with the particular off target, closely related off targets, or both. In particular, interpretation of off-target safety margins needs to consider the available quantitative experience with a particular off target and is largely done on case-by-case basis. It is a reasonable strategy to aim for higher off-target safety margins preclinically (off-target $IC_{50}/C_{max,free}$ at efficacious exposure) than those accepted clinically, as the therapeutic plasma exposure in humans for certain drugs can be higher than initially assumed based on preclinical models [8].

REFERENCES

1. Paul, S.M., Mytelka, D.S., Dunwiddie, C.T., Persinger, C.C., Munos, B.H., Lindborg, S.R., Schacht, A.L. (2010). How to improve R&D productivity: The pharmaceutical industry's grand challenge. *Nature Reviews Drug Discovery*, *9*(3), 203–214.

2. Ray, A. (2009). Beyond debacle and debate: Developing solutions in drug safety. *Nature Reviews Drug Discovery*, *8*, 775–779.
3. Kesselheim, A.S., Avorn, J. (2007). The role of litigation in defining drug risks. *Journal of the American Medical Association*, *297*, 308–311.
4. Kesselheim, A.S. (2010). Permitting product liability litigation for FDA-approved drugs and devices promotes patient safety. *Clinical Pharmacology & Therapeutics*, *87*(6), 645–647.
5. Watkins, P.B. (2011). Drug safety sciences and the bottleneck in drug development. *Clinical Pharmacology & Therapeutics*, *89*(6), 788–790.
6. Bender, A., Scheiber, J., Glick, M., Davies, J.W., Azzaoui, K., Hamon, J., Urban, L., Whitebread, S., Jenkins, J.L. (2007). Analysis of pharmacology data and the prediction of adverse drug reactions and off-target effects from chemical structure. *ChemMedChem*, *2*(6), 861–873.
7. Park, B.K., Boobis, A., Clarke, S., Goldring, C.E., Jones, D., Kenna, J.G., Lambert, C., Laverty, H.G., Naisbitt, D.J., Nelson, S., Nicoll-Griffith, D.A., Obach, R.S., Routledge, P., Smith, D.A., Tweedie, D.J., Vermeulen, N., Williams, D.P., Wilson, I.D., Baillie, T.A. (2011). Managing the challenge of chemically reactive metabolites in drug development. *Nature Reviews Drug Discovery*, *10*(4), 292–306.
8. Muller, P.Y., Milton, M.N. (2012). The determination and interpretation of the therapeutic index in drug development. *Nature Reviews Drug Discovery*, *11*(10), 751–761.
9. Hamon, J., Whitebread, S., Techer-Etienne, V., Le Coq, H., Azzaoui, K., Urban, L. (2009). *In vitro* safety pharmacology profiling: What else beyond hERG? *Future Medicinal Chemistry*, *1*(4), 645–665.
10. Bowes, J., Brown, A.J., Hamon, J., Jarolimek, W., Sridhar, A., Waldron, G., Whitebread, S. (2012). Reducing safety-related drug attrition: The use of in vitro pharmacological profiling. *Nature Reviews Drug Discovery*, *11*, 909–922.
11. Smith, D.A., Di, L., Kerns, E.H. (2010). The effect of plasma protein binding on *in vivo* efficacy: Misconceptions in drug discovery. *Nature Reviews Drug Discovery*, *9*(12), 929–939.
12. Muller, P.Y., Milton, M.N. (2012). The determination and interpretation of the therapeutic index in drug development. *Nature Reviews Drug Discovery*, *11*(10), 751–761.
13. Lee, C.H., Zhu, C., Malysz, J., Campbell, T., Shaughnessy, T., Honore, P., Polakowski, J., Gopalakrishnan, M. (2011). α4β2 neuronal nicotinic receptor positive allosteric modulation: An approach for improving the therapeutic index of α4β2 nAChR agonists in pain. *Biochemical Pharmacology*, *82*(8), 959–966.
14. Urquhart, B.L., Kim, R.B. (2009). Blood-brain barrier transporters and response to CNS-active drugs. *European Journal of Clinical Pharmacology*, *65*(11), 1063–1070.
15. International Transporter Consortium, Giacomini, K.M., Huang, S.M., Tweedie, D.J., Benet, L.Z., Brouwer, K.L., Chu, X., Dahlin, A., Evers, R., Fischer, V., Hillgren, K.M., Hoffmaster, K.A., Ishikawa, T., Keppler, D., Kim, R.B., Lee, C.A., Niemi, M., Polli, J.W., Sugiyama, Y., Swaan, P.W., Ware, J.A., Wright, S.H., Yee, S.W., Zamek-Gliszczynski, M.J., Zhang, L. (2010). Membrane transporters in drug development. *Nature Reviews Drug Discovery*, *9*(3), 215–236.
16. Mruk, D.D., Cheng, C.Y. (2012). Drug transporters in spermatogenesis: A re-evaluation of recent data on P-glycoprotein. *Spermatogenesis*, *2*(2), 70–72.

17. Tomi, M., Hosoya, K. (2010). The role of blood-ocular barrier transporters in retinal drug disposition: An overview. *Expert Opinion on Drug Metabolism & Toxicology*, *6*(9), 1111–1124.
18. Moitra, K., Lou, H., Dean, M. (2011). Multidrug efflux pumps and cancer stem cells: Insights into multidrug resistance and therapeutic development. *Clinical Pharmacology & Therapeutics*, *89*(4), 491–502.
19. Redfern, W.S., Carlsson, L., Davis, A.S., Lynch, W.G., MacKenzie, I., Palethorpe, S., Siegl, P.K., Strang, I., Sullivan, A.T., Wallis, R., Camm, A.J., Hammond, T.G. (2003). Relationships between preclinical cardiac electrophysiology, clinical QT interval prolongation and torsade de pointes for a broad range of drugs: evidence for a provisional safety margin in drug development. *Cardiovascular Research*, *58*(1), 32–45.
20. ICH guideline on non-clinical evaluation of the potential for delayed ventricular repolarization (QT interval prolongation) by human pharmaceuticals. ICH S7b. May 2005. http://www.ich.org/fileadmin/Public_Web_Site/ICH_Products/Guidelines/Safety/S7B/Step4/S7B_Guideline.pdf.
21. Harmer, A., Valentin, J.P., Pollard, C. (2011). On the relationship between block of the cardiac Na(+) channel and drug-induced prolongation of the QRS complex. *British Journal of Pharmacology*, *164*(2), 260–273.
22. Whitebread, S., Hamon, J., Bojanic, D., Urban, L. (2005). Keynote review: in vitro safety pharmacology profiling: an essential tool for successful drug development. *Drug Discovery Today*, *10*(21), 1421–1433.
23. Lounkine, E., Keiser, M.J., Whitebread, S., Mikhailov, D., Hamon, J., Jenkins, J.L., Lavan, P., Weber, E., Doak, A.K., Côté, S., Shoichet, B.K., Urban, L. (2012). Large-scale prediction and testing of drug activity on side-effect targets. *Nature*, *486*(7403), 361–367.

18

INTEGRATED RISK ASSESSMENT OF CARDIOVASCULAR SAFETY IN DRUG DISCOVERY

Gül Erdemli and Ruth L. Martin

18.1 INTRODUCTION

Drug-induced off-target effects on cardiac physiology, such as contractility, heart rhythm, blood pressure, and end organ perfusion, can lead to life-threatening adverse reactions. In the United States, cardiovascular safety is reported to be the leading cause of drug discontinuation in all phases of drug discovery and development as well as in the postapproval stage [1–4] (Table 18.1). In one example that led to a complete overhaul of the regulatory landscape, terfenadine (Seldane) was removed from the U.S. market as a result of many documented cases of serious cardiac events (dysrythmias, Torsade de Pointes, ventricular fibrillation, sudden death) and 17 deaths associated with the drug [5]. The sequence of events that led to removal of this drug also set into motion a series of regulations and guidelines that aimed to help protect the public from potential adverse cardiac events. In 1997, the Committee for Propriety Medicinal Products of the European Medicines Agency (EMA) issued the first regulatory guideline for the assessment of drug-induced QT-prolongation

Predictive ADMET: Integrative Approaches in Drug Discovery and Development, First Edition.
Edited by Jianling Wang and Laszlo Urban.

Table 18.1 Evidence, prevalence, and occurrence of safety liabilities relating to the cardiovascular and hepatic systems (*See insert for color representation of the figure.*)

Phase	Nonclinical	Phase I	Phase I-III	Phase III/ postapproval	Postapproval	Postapproval	Post approval
Information	Causes of attrition	Serious ADRs	Causes of attrition	ADRs on label	Serious ADRs	Withdrawal from sale	Withdrawal from sale
Source	Car 2006 [111]	Sibille et al. 1998 [112]	Olson et al. 2000 [113]	BioPrint® 2006	Budnitz et al. 2006 [114]	Fung et al. 2001 [115]	Stevens & Baker 2009 [116]
Sample size	88 CDs stopped	1,015 subjects	82 CDs stopped	1,138 drugs	21,298 patients	121 drugs	47 drugs
Cardiovascular	27%	9%	21%	36%	15%	9%	45%
Hepatotoxicity	8%	7%	21%	13%	0%	26%	32%

0% 1-10% 1-20% >20%

Note: Information was collated from the commercially available database, BioPrint®, and published articles [111–116]. ADRs, Adverse Drug Reactions. CD, Candidate drug. (Reprinted from Laverty et al. [2] with permission.)

of all new drugs. This guideline was followed by the International Conference on Harmonization (ICH) guideline for preclinical safety pharmacology studies in 2001 [6] and two more guidelines in 2005: one for preclinical studies [7] specifically designed to identify the potential risk of delayed ventricular repolarization of a developing drug candidate and its metabolites and one for clinical studies [8]. Based on experience over the past several decades with compounds that block the human *ether-à-go-go* related gene (hERG) channel and the regulatory emphasis on the subject, the pro-arrhythmia risk of drug candidates has been a significant concern throughout drug development [9, 10]. More recently, it is also being recognized that drug-induced effects on other cardiac ion channels, cardiac contractility, and the vasculature could generate serious adverse effects [11–13]. As a result, it is now a routine practice in the pharmaceutical industry to profile drug candidates routinely in a range of *in vitro* and *in vivo* assays in various stages of discovery and development [13–15].

A recent study showed that drug candidates that are shown to be safe in preclinical and phase I studies in healthy volunteers, however, may still induce serious cardiovascular adverse reactions when administrated in larger groups of patients for longer periods of time [2, 16]. Analysis of the late-stage cardiovascular adverse reactions indicated that some of these adverse reactions were either not observed in preclinical studies or not considered clinically relevant [2, 16]. This implies that routine preclinical methods are efficient to identify acute cardiovascular effects and to eliminate high-risk compounds but may not be as effective for more subtle adverse reactions or for those effects that may emerge with chronic treatments or in high-risk patient groups. This is rather worrisome as late-stage adverse reactions may result in prescription restrictions, additional pre- and post-approval monitoring, and ultimately discontinuation of clinical trials or withdrawal from the market that will have significant financial consequences for the industry and delay or prevent patients' access to the medicines. These results emphasize the importance of integrated risk assessment for each drug candidate taking into consideration all available data as well as intended patient group, co-morbidities, underlying cardiovascular risk, co-medications, and treatment duration [3]. Additionally, better understanding of the translational value of existing preclinical models and development of additional preclinical and clinical biomarkers are of paramount importance for successful comprehensive cardiovascular safety assessment to decrease the overall attrition rate and to improve the safety of future medicines.

In this chapter, we reviewed preclinical models including *in silico*, *in vitro*, and *in vivo* techniques to evaluate the cardiovascular risk of new drug candidates as well as application of these results in the integrated risk assessment at different stages of drug discovery. We focused on functional adverse reactions on the heart and hemodynamic parameters as a result of off-target interactions. However, cardiovascular adverse reactions could also be structural in nature and involve blood constituents, hemostasis, and the vasculature for which preclinical models and translational biomarkers may not be readily available. In some cases, cardiovascular adverse reactions can also be a result of on-target toxicity, and the safety assessment as well as management and mitigation of these adverse reactions may substantially differ from off-target toxicities. Comprehensive coverage of such adverse reactions is not within the scope of this chapter, and readers are referred to recent reviews [2, 17, 18].

18.2 OVERVIEW OF PRECLINICAL MODELS

18.2.1 *In Vivo* Models

18.2.1.1 In Vivo Models Used to Meet Regulatory Requirements and Guide Decisions The S7B guideline outlines the expectation of having an *in vitro* rapid component of a delayed rectifier potassium channel (I_{Kr}, also known as hERG channel) assessment and an *in vivo* QT interval (a measure of the time between the start of the Q wave and the end of the T wave on the surface electrocardiogram (ECG)) assessment. Typically these *in vivo* studies are done in conscious large animals (e.g., dogs or nonhuman primates) dependent on the tolerance of the test compound. In addition, the studies use compound doses that meet and exceed the anticipated human exposure and are given by the intended route of human administration. Typical parameters assessed include mean arterial pressure (MAP), heart rate (HR), and ECG, mainly the QT interval. As the QT interval is heavily influenced by heart rate changes, correction formulas are used to minimize these influences [19–21]. Some well-known correction factors include the following: Bazett [$QTc = \sqrt{QT(RR)}$]; Fridericia [$QTc = \sqrt[3]{QT(RR)}$]; Van de Water [$QTc = QT - 0.087(RR - 1)$]; and Framingham [$QTc = QT + 0.154(1 - RR)$] (see References 20, 22, and 23 for discussions), where QTc is the corrected QT interval and RR is the RR interval, the interval from the peak of one QRS complex (the combination of Q, R and S waves on the ECG, corresponding to depolarization of the right and left ventricle) to the peak of the next. Bazett and Fredericia assume an exponential relationship between the QT and RR interval, while Framingham and Van de Water assume a linear relationship. The main problem with all of these formulas is the accuracy of their corrections over large datasets. For example, Bazett is the most accurate with a heart rate of 60 bpm or RR interval of 1 s and tends to overcorrect for rapid heart rates and undercorrect for slow heart rates. Because of these shortcomings, work is still being done to optimize methods for QT correction for both preclinical and clinical studies [23, 24].

18.2.1.2 Supplemental Non-GLP* In Vivo *Models Used for Early Decision Making Before commencing these good laboratory practice (GLP) studies, non-GLP studies are performed on several drug candidates to advance a compound with no or relatively small risk. These non-GLP studies can be done in either conscious or anesthetized animals. Under anesthetized conditions, many more parameters can be accessed, including systolic arterial pressure, diastolic arterial pressure, left ventricular end diastolic pressure, central venous pressure, cardiac output, systemic vascular resistance, pulmonary vascular resistance, contractility, and ECG parameters in addition to those assessed in GLP studies. These additional parameters allow for further refinement of the potential risks between advancing compounds.

Prior to large animal *in vivo* studies, small animal studies can also be employed. These studies can be done in guinea pig or in other preclinical species with intravenous (IV) infusion in the anesthetized state or with oral dosing in the conscious state using telemetry [25,26]. The advantage of these small animal *in vivo* models is cost effectiveness; animal costs (acquisition, housing, etc.), lesser compound requirements, and availability of early pharmacokinetic data all combine to reduce the costs of this model relative to large animal studies. Typical parameters measured include MAP, HR, contractility, and ECG parameters. If rats are used, the assessment will be limited to drug-induced effects on hemodynamic parameters and cardiac ion channels other than the hERG as the rodent heart does not rely on the rapid component of the delayed rectifier potassium current, I_{Kr} (hERG) for repolarization.

When sorting out mechanisms that contribute to changes observed in the *in vivo* model, it is useful to determine whether the effects are direct, indirect, or compensatory. For example, if a compound were to produce simultaneous, similar effects on both HR and MAP (either both increasing or both decreasing), it is possible that these changes are caused by direct modulation of the sympathetic nervous system rather than by (or in addition to) direct effects on the cardiovascular system. This hypothesis can be easily tested with a sympathetically blocked (simultaneous hexamethonium infusion) *in vivo* model [27,28]. In another example, when MAP decreases, followed by an increase in HR, the data would suggest a possible direct effect on the vasculature followed by a compensatory change in heart rate to attempt to return blood pressure to a normal range [29].

Most *in vivo* studies are done in healthy adult animals. Because drug-induced Torsades de Pointes (TdP) is a rare event, many laboratories have developed arrhythmogenic experimental animal models to improve assessment of cardiac risk. These models typically employ chronic atrioventricular (AV) blockade in anesthetized or awake dogs or anesthetized rabbits challenged with methoxamine [25]. Because the models are resource-intense, they are employed only at the final stages of drug development and then not on a routine basis.

Other *in vivo* models that have been employed to gain insight into mechanisms of cardiotoxicity use zebrafish to monitor heart rate. Milan et al. [30] have shown that bradycardia and AV block in zebrafish embryos are correlated with QT-prolongation in humans after screening a set of 100 compounds. Since then, zebrafish larvae/embryos have been used to assess the cardiac risk of antipsychotic drugs [31] and in high-throughput risk assessment screens [32, 33]. In addition, Peal et al. [32] used KCNH2 mutant *breakdance* zebrafish embryos to identify compounds that could rescue the long QT type 2 phenotype. Zebrafish embryos have also been used to determine the cardiotoxicity of engineered nanomaterials by assessing heart rate and pericardial edema [34]. A recent report described the development of a cardiovascular toxicity screen using *ex vivo* adult zebrafish hearts to monitor heart rate as a surrogate marker for risk identification [35]. However, the zebrafish assay seems to be less sensitive than hERG patch clamp assays requiring 8–10-fold higher concentrations to cause AV block than those required to block the hERG channel in a patch clamp assay and higher exposure levels than required for QT interval prolongation in human [36]. Another limitation of zebrafish assays is that uptake of compounds into zebrafish can be variable particularly for compounds with poor solubility. Compound concentrations in the whole larvae should be measured to aid interpretation of results. Additionally, advantages and limitations of zebrafish assays in relation to translation to humans have not been carefully assessed. To use zebrafish assays for decision making in preclinical cardiovascular safety assessment, extensive validation studies with diverse and comprehensive pharmacological compound sets will be required.

18.2.2 *In Vitro* Models

18.2.2.1 In Vitro Models Used to Guide Early Decisions and Study Mechanisms of Action Given the need to screen many compounds for cardiac risk, *in vitro* models are typically used. These models have distinct advantages in that they can provide a direct investigation of the underlying mechanism of potential cardiac risk. Also, these assays typically require much less compound (a few mg) and are not bound to the availability of species specific pharmacokinetic data prior to commencing the assay. That said, *in vitro* assays by their very nature have distinct disadvantages as well. First, because of the isolated system employed, no feedback or compensatory loops will be present, and thus, compensatory changes such as an increase in heart rate that would typically follow a decrease in blood pressure would be missed. Also, because of the lack of neuronal input, secondary effects on the nervous system that affect cardiovascular parameters will be lost. In addition, effects resulting from metabolites will be missed unless the metabolites are synthesized and studied in parallel assays. Finally *in vitro* assays are typically designed to look for antagonistic signals and may miss an agonist effect. In addition to the disadvantages listed earlier, one has to pay careful attention to the many inherent variables in *in vitro* assays, including those associated with the compound. First, stability of the compound in a salt solution, unprotected from light, needs to be considered [37]. Delivery of the compound whether through a perfusion system or through plate format may induce artifacts resulting from interactions with that system: leaching of components or binding of the compound to those components, leading to an incorrect or inaccurate conclusion. As a result, verification of compound concentrations is critical. Even if concentrations are verified with traditional solubility methods, the possibility of artifacts still exists as a result of potential compound aggregate formation [38]. At high concentrations, some compounds can form soluble aggregates in aqueous solutions. These compound aggregates (dimers, trimmers, and even larger complexes), even though appearing soluble, may not be available to interact with the target. Finally, the lack of plasma proteins in most *in vitro* assays needs to be considered. Typically, a simple calculation to determine the expected free fraction of a compound is done after plasma protein binding is determined in a parallel assay. For compounds that are highly plasma protein bound, caution must be exercised when employing this technique [39]. The effects observed with these highly bound compounds can be more potent and are not always well correlated with the calculated free fraction.

18.2.2.2 Nonroutine In Vitro Assays When direct effects on the vasculature are suspected as a cause of changes in blood pressure in *in vivo* assays, an isolated vascular ring assay may be used to quickly screen many compounds. These assays can employ aortic or mesenteric artery isolated rings obtained from normal rats or mice to measure compound effects on contraction or relaxation of precontracted tissue to investigate either increases or decreases in blood pressure [40,41]. Vascular rings can also be obtained from diseased animals. In one study, Mendizabal et al. [42] used vascular rings from obese spontaneously hypertensive rats to investigate the effects of glitazones on vascular function.

An *in vitro* contractility assay may be used for screening compounds after a change in contractility is noted in an *in vivo* assay. Although contractility changes can be a result of compensatory mechanisms, they can also be caused by direct interactions of a compound on the cardiac myocyte. One *in vitro* contractility assay uses isolated canine ventricular myocytes and optical measurements of changes in sarcomere length to determine cardiac iontropic risk [11].

A data-rich *in vitro* assay for assessing cardiac risk is the Langendorff heart model [43]. Hearts isolated from normal guinea pigs or rabbits are used to investigate compound effects on both hemodynamic function and cardiac electrophysiology, including ECG parameters and monophasic action potentials. Hearts can also be isolated from diseased animals and interventions employed such as ischemia/reperfusion on healthy hearts. Although this type of assay provides a more complete picture of the cardiac risk of a compound, it is very labor intensive and not suitable for routine compound assessment.

Historically, repolarization assays have been used to assess cardiac risk prior to the availability of automated electrophysiology platforms (see below) [44]. More recently, these assays have been relegated to an as-needed basis because of their animal requirements, resource, and time intensity [21]. Repolarization assays typically use isolated Purkinje fibers, papillary muscle, or wedge preparations from either dog, rabbit, or guinea pig [45]. These assays provide an integrated risk assessment with analysis of action potential upstroke velocity, duration, and changes in the action potential shape (triangulation). The effects of compounds on cardiac action potentials have also been analyzed with measurements from isolated ventricular myocytes [46].

A relatively new substrate for evaluating cardiac risk is stem cell technology. Cardiomyocytes derived from human [47,48] or mouse [49] embryonic stem cells have been used to evaluate compound effects on action potentials and isolated currents. One drawback from the use of these cells is the immature phenotype presented. Typically, cells are depolarized as a result of a decrease or lack of the inward rectifier current. Therefore, caution is required when drawing conclusions about potential cardiac risk. A report using adult human induced pluripotent stem cell (iPSC)-derived cardiomycytes and field potentials investigated the effects of cardiac glycosides and found the model to be a suitable surrogate assay for predicting QT effects [50]. In another study, Itzhaki et al. [51] used iPSC-derived cardiomycytes from a patient with long QT syndrome to evaluate reference compounds using both action potentials and current measurements. One disadvantage of the iPSC-derived cardiomyocytes is the heterogeneity of the preparations. This currently limits the utility of these preparations as a routine screening tool for cardiac risk evaluation.

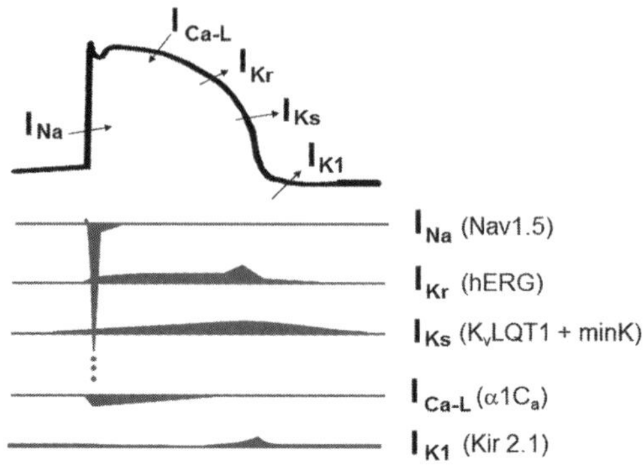

Figure 18.1 Cardiac ventricular action potential primary currents. Arrows indicate inward or outward currents. Not pictured are $I_{Ca,T}$, I_{to}, I_{Kur}, I_{KATP}, I_{KAch}, I_F, or I_{Cl}. *(See insert for color representation of the figure.)*

18.2.2.3 Cardiac Ion Channel Assays The cardiac action potential is composed of ionic currents exquisitely timed and tuned to allow for a forceful cardiac contraction. Alterations of any of these currents through drug-induced block, enhancement, or modulation will have effects on cardiac function. Although many currents impact cardiac repolarization (see Figure 18.1), most attention in drug discovery and development is given to interactions with the hERG channel [52, 53]. hERG (or KCNH2) is the gene that encodes the pore-forming subunit of the rapid component of the delayed rectifier potassium channel in humans. When expressed in the heart, the hERG channel contributes an increasing degree of repolarizing current over the duration of the cardiac action potential plateau, thus, terminating the action potential and defining the action potential duration. A block of hERG current (Kv 11.1) is associated clinically with delayed repolarization and has been implicated in (and is a surrogate marker for) pro-arrhythmia with noncardiovascular drugs.

The experimental conditions that are used to measure ionic current can have profound effects on the IC_{50}s that are generated for compounds. hERG current is typically measured with a standard step protocol with a block of tail current elicited from a step to –50 mV (after application of a positive activating voltage step) being assessed [53]. The duration of the activating step and voltage applied can influence the IC_{50} determination as much as 10-fold in the case of BeKm-1 [54]. Others have used step/ramp protocols to mimic more closely a physiologic cardiac action potential [54, 55]. In the study by Kirsch et al. [55], most compound IC_{50}s showed good agreement with acquisition via either a step pulse or a step/ramp protocol, with the exception of erythromycin, verapamil, and loratadine. With these compounds there was a two- to threefold difference in potency obtained with the different voltage protocols. In addition to voltage protocol, temperature can also have an effect on the IC_{50} generated for a compound [54, 55]. Although Yao et al. [54] found no significant differences with temperature for the compounds they investigated, Kirsch et al. [55] found significant differences for several compounds, most notably, erythromycin (eightfold more potent at 35°C vs. 22°C) and d,l-sotalol (threefold more potent at 35°C).

As companies focus on the propensity of compounds to block hERG current, other cardiac ion channels must also be considered [56]. Experimental conditions (choice of voltage clamp protocols and temperature) for investigating compound effects on these channels are equally important as for hERG channel investigations. Because hERG is only one of many currents that comprise the cardiac action potential (see Figure 18.1), relying solely on this assay would be an oversimplification of cardiac risk. An hERG current block is not always an accurate predictor of action potential duration (APD) and QT interval prolongation [57]. For example, compounds such as verapamil or fluoxetine that are potent hERG current blockers do not cause APD prolongation or QT interval prolongation because of their concomitant block of cardiac calcium current. Because compounds can block multiple ion channels, Brown [58] has suggested that a more appropriate screening method for cardiac risk would be to screen a panel of cardiac ion channels, including calcium (Cav3.2, Cav1.2), sodium (Nav1.5), pacemaker (HCN4), and five additional potassium (Kv1.5, Kv4.3, KvLQT1/minK, Kir2.1, Kir3.1/3.4) currents in addition to screening hERG current. Although investigating all 10 ion channels may not be necessary for routine compound investigation, many companies will focus on 3 channels (hERG, Nav1.5, and Cav1.2) and reserve the rest of the cardiac panel for use on an as-needed basis when *in vivo* results cannot be explained by interactions on these channels.

18.2.2.4 Enabling Technologies for Investigating Cardiac Ion Channels

Patch clamp methods measure ion channel function directly, and the whole cell voltage clamp method is considered the best approach for studying the effects of compounds. This method requires a skilled operator and is labor intensive with at best several compounds screened against a single ion channel a day. In 2002, the first automated patch clamp technology became commercially available. Since that time, several companies have launched devices, each with their own advantages and disadvantages that are briefly discussed here. This technology has now allowed pharmaceutical companies to investigate cost effectively ion channels for both efficacy and cardiac safety studies [59–62].

The platforms can be divided into two categories based on the seal quality provided. Platforms such as IonWorks™, IonWorks Quattro™, and Barracuda™ (Molecular Devices, Sunnyvale, CA, USA) all provide megaOhm perforated patch configurations. The lack of high seal quality is compensated for with recording the average response from a population of cells. These devices operate in voltage clamp mode at room temperature and have limited solution exchange capabilities. The main advantage of these platforms is throughput. For the IonWorks systems, the need for intermediate compound plate dilutions in buffer makes the assays on these systems more sensitive to compound solubility. Additionally, these systems (with the exception of Barracuda) allow only discontinuous voltage clamp, have only 48 amplifiers recording 384 wells, and lack a fluid outlet making them a single point rather than a kinetic reader [60]. To study inhibition kinetics and reversibility, follow-up manual patch clamp assays may be required. Barracuda, however, has been based on an array of 384-amplifiers allowing recordings from all-wells-at-a-time, and the improvements in the liquid handling design allow for rapid compound screening without a need for intermediate compound dilutions. Platforms that provide gigaOhm whole cell patch configurations include PatchXpress^{TM} (Molecular Devices, Sunnyvale, CA, USA), QPatch^{TM} (Sophion, Balerup, Denmark), Patchliner^{TM}, and SynchroPatch^{TM} (Nanion, Munich, Germany). In addition, IonFlux^{TM} (Fluxion, San Francisco, CA, USA) can operate in either a megaOhm or a gigaOhm configuration. This set of platforms, with regard to throughput, range from 16 to 96 simultaneous recordings. Although all platforms operate in voltage clamp mode, current clamp capabilities have recently been added to PatchXpress and QPatch. Patchliner can also operate in current clamp mode and has the distinct advantage of temperature control, making it suitable to record action potentials from cardiac stem cells. IonFlux also has temperature control capabilities [63]. The choice of automated platform depends on the goal of the experiment, whether it be rapid screening of hundreds of compounds, cardiac safety screening during lead optimization, or mechanism studies. These platforms are not limited to just screening of compounds. They can also be used to investigate ion channel biophysical properties (for example, see References 64 and 65).

18.2.3 *In Silico* Models for Cardiac Safety

In an effort to reduce animal use and refine predictability of cardiac risk assessment, *in silico* models have been developed (see Reference 66 for a review). These models are typically dependent on the robustness of the training set used in addition to the parameters used to build the model for their predictivity of cardiac risk. When developing an *in silico* hERG model, the underlying dataset, including experimental conditions (voltage protocol and temperature [see previous discussion in Section 18.2.2.3]), is crucial for the quality and performance of the model [67]. Some models have used chemical structure as one determinant of adverse drug reactions [68] or as a way to mitigate hERG current block liability [69]. In another example of the *in silico* approach, a mid-myocardial APD model was developed using IC_{50} data obtained from a panel of five ion channels (Nav1.5, Cav1.2, Kv4.3/KChIP2.2, Kv7.1/minK, and Kv11.1[hERG]) and a training set of compounds [70]. Their model had a >80% predictivity and was found useful for virtual screening of APD prolongation.

18.3 INDIRECT MODULATION OF CARDIAC ION CHANNELS AND IMPLICATIONS FOR CARDIAC SAFETY

Indirect modulation of cardiac ion channels may significantly contribute to drug-induced QT interval prolongation and pro-arrhythmia, and it may be the most common underlying reason for discrepancies observed between *in vitro* and *in vivo* studies [71, 72]. Some of these interactions may not be so obvious in preclinical studies for various reasons and may be overlooked until the drug candidate progresses to the clinic.

Numerous possible mechanisms (e.g., transcriptional, translational, and posttranslational) could influence the expression and functional properties of cardiac ion channels. One of the most studied examples of indirect channel modulation is drug-induced inhibition of hERG channel trafficking to the plasma membrane. The inhibition of hERG channel trafficking has been proposed as a contributing mechanism for the delayed QT interval prolongation observed with arsenic trioxide and pentamidine treatments in animals and humans [73–75]. Examples of drug-induced indirect modulation of cardiac ion channels other than the hERG channel are also emerging. Taniguchi et al. [76] have recently reported that probucol, a hypocholesterolemic agent used for the treatment of hyperlipidemia, decreased I_{Ks} after chronic but not acute treatment *in vitro*. The authors argued that probucol inhibits I_{Ks} by decreasing the number of functional I_{Ks} channel complexes expressed on the plasma membrane, and they proposed that this mechanism may contribute to the QT interval prolongation and ventricular arrhythmias observed with probucol treatment in clinical practice [76]. Alterations in hNav1.5 expression levels have been associated with acquired cardiac disorders, such as heart failure, and cardiac sodium channel blockers have been shown to facilitate trafficking of mutant hNav1.5 proteins in patients with long QT syndrome type 3 [77]. The inhibition of the protein kinase A-mediated trafficking of sodium channels by the antimalarial drug chloroquine is proposed to enhance the functional inhibition of the channel and to contribute to its pro-arrhythmic effects observed in preclinical species and in humans [78–80]. Apelin is an endogenous peptide shown to have strong inotropic, chronotropic, and vasodilation effects in rodents, canines, and humans acting via its angiotensin like receptor (APJ-R) [81–83]. It is also shown to potentiate cardiac sodium channels via a protein kinase C dependent manner that is thought to contribute to its positive inotropic effects [84].

It is also known that drugs could indirectly induce corrected QT interval (QTc) prolongation by influencing autonomic tone [85, 86]. Ventricular arrhythmias can be triggered by physical or emotional stress as a result of modulation of cardiac ion channel activity by increased adrenergic stimulation [87]. I_{Ks} and I_{Kr} have been shown to be regulated by α and β adrenergic stimulation, which may relate to the increased incidence of ventricular arrhythmias in long QT syndrome-1 (LQTS-1) patients and the electrical remodeling of the heart in chronic cardiac diseases, respectively [88].

Changes in blood glucose and insulin levels have been shown to delay ventricular repolarization, prolong QTc, and induce ventricular arrhythmias in healthy subjects and dogs [89, 90]. Type I and II diabetes are also associated with increased QT interval duration and dispersion that are linked to the "dead in bed" syndrome [90]. Zhang et al. [91] showed that the hERG channel is modulated by extracellular glucose levels, and it is likely that impairment of the hERG function may contribute to the ECG changes observed in hypo- or hyperglycemia.

A decrease in serum K^+ levels, hypokalemia, increases the cardiac action potential duration and prolongs QTc by decreasing I_{Kr} and is a major risk factor for arrhythmia [92–94]. Furosemide is a loop diuretic that can cause hypokalemia and produce QTc prolongation [92, 93]. The inhibition of the hERG channel by low extracellular K^+ is attributed to the accelerated inactivation or cell surface density of the hERG channel [95, 96].

Despite the ample evidence of indirect modulation of cardiac ion channels, the overall contribution of such mechanisms to QT interval prolongation or arrhythmogenesis *in vivo* is not yet well understood. Most cases of indirect ion channel modulation may present as *in vitro* and *in vivo* disconnect with an unknown mechanism. The degree of safety risk resulting from indirect cardiac ion channel modulation is usually difficult to quantify. This is at least partly because routine screening methods widely used in preclinical settings are not suitable methodologies to identify such interactions. Patch clamp assays recognized as gold standard methods to assess cardiac liabilities as a result of ion channel block are designed for acute functional assessments of channel activity. They are not suited to study indirect modulation of the ion channel protein that may take much longer to develop. Additionally, patch clamp studies are routinely performed in noncardiac host cells overexpressing the recombinant channel of interest, in most cases, only the subunits essential for functional activity (e.g., pore-forming alpha subunit of the hERG channel expressed in HEK293 or CHO cells). These host cells may not have the second messenger systems, pathways, and associated proteins subject to indirect modulation, further limiting the suitability of these assays to study indirect modulation of ion channels.

It is, however, essential to recognize that unexpected QTc prolongation could occur in preclinical telemetry studies or in human trials even with compounds showing no known direct effects on cardiac ion channels. A prudent strategy to avoid such unpleasant findings may be to invest more time and resources early in candidate profiling to identify the potential mechanisms and to design targeted experiments to help clarify impact. Figure 18.2 illustrates a schematic of how indirect ion channel modulation may be assessed as part of a safety analysis strategy. If in an *in vivo* model QTc prolongation is observed for a compound that does not directly block cardiac ion channels, human metabolites should be profiled in cardiac ion channel assays. Compounds blocking cardiac ion channels by an indirect mechanism likely to display pharmacokinetic (PK)-pharmacodynamic (PD) disconnect; e.g., PD effects are observed after the C_{max}. PK-PD modeling of *in vivo* findings should be performed to understand the mechanisms involved to help identify mechanism of action [97] (Figure 18.2). *In silico* approaches can also be helpful to propose off-target hypotheses that could be tested in additional relevant *in vitro* assays [72]. Once the mechanism of action is identified, relevant assays with sufficient specificity and throughput could be incorporated into the discovery phase to guide medicinal chemistry efforts to mitigate off-target activities.

18.4 PRECLINICAL STRATEGIES TO IDENTIFY CARDIAC ION CHANNEL–RELATED LIABILITIES

Cardiovascular safety risk assessment strategies in various stages of drug discovery have been extensively discussed in the literature [2, 4, 98–100]. The current strategies tend to focus on the hERG channel. This is mainly for two reasons: First, the hERG channel is a well-validated target for TdP, and second, it is the most frequently hit cardiac ion channel because of its pharmacological promiscuity [99, 100]. Although much of the cardiovascular safety concerns are based on the experience with compounds that block the hERG potassium channel, it is also recognized that cardiac ion channels other than the hERG channel play key roles in cardiac physiology and that

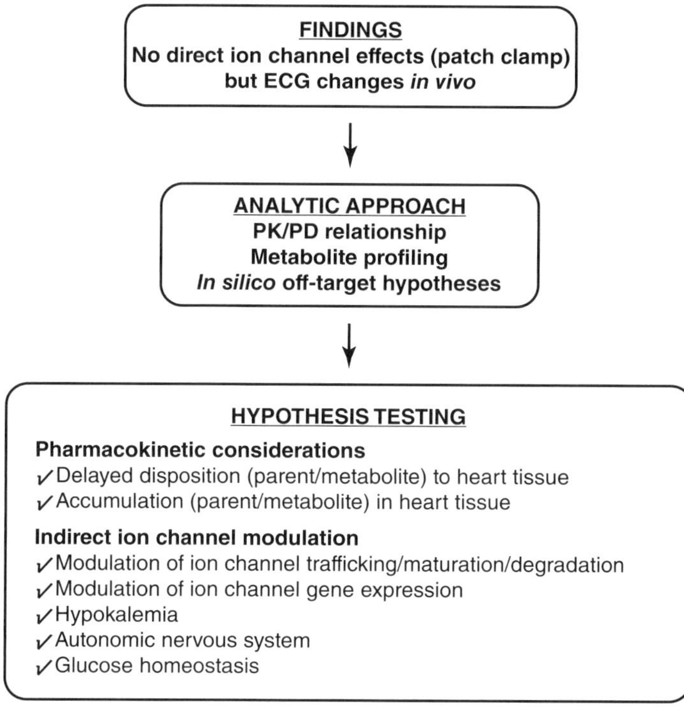

Figure 18.2 Cardiac safety schematic integrating indirect ion channel modulation analysis. PD: pharmacodynamic, PK: pharmacokinetic, ECG: electrocardiogram. (Reprinted from Erdemli et al. [72] with permission.)

drug-induced effects at these channels can result in serious adverse effects [101]. The human cardiac sodium channel (hNav1.5), for example, is critical for action potential generation and propagation in the heart. Drug-induced sodium channel inhibition decreases the conduction velocity and can have serious implications for cardiac safety [102–104]. Genetic mutations in hNav1.5 have been linked to a number of cardiac diseases [105]. Therefore, off-target hNav1.5 inhibition is considered to be a risk marker for a drug candidate. Given the potential safety implications and the costs of late-stage drug development, detection and mitigation of hNav1.5 liabilities early in drug discovery and development becomes important [102,104]. This has led to implementation of routine screening of drug candidates against the hNav1.5 channel early in drug development [60]. Figure 18.3 illustrates a preclinical strategy incorporating *in silico*, *in vitro*, and *in vivo* techniques to identify the potential of small molecules for hNav1.5 blockade and its electrophysiological consequences [104]. This is an integrated, step-wise approach starting with a routine *in vitro* assessment of ion channel activity based on a patch clamp electrophysiology assay in recombinant expression systems during early drug discovery [60]. If hNav1.5 blockade is observed, an automated patch clamp assay is used to support medicinal chemistry efforts to mitigate the hNav1.5 activity with the guidance of *in silico* tools. A manual patch clamp is

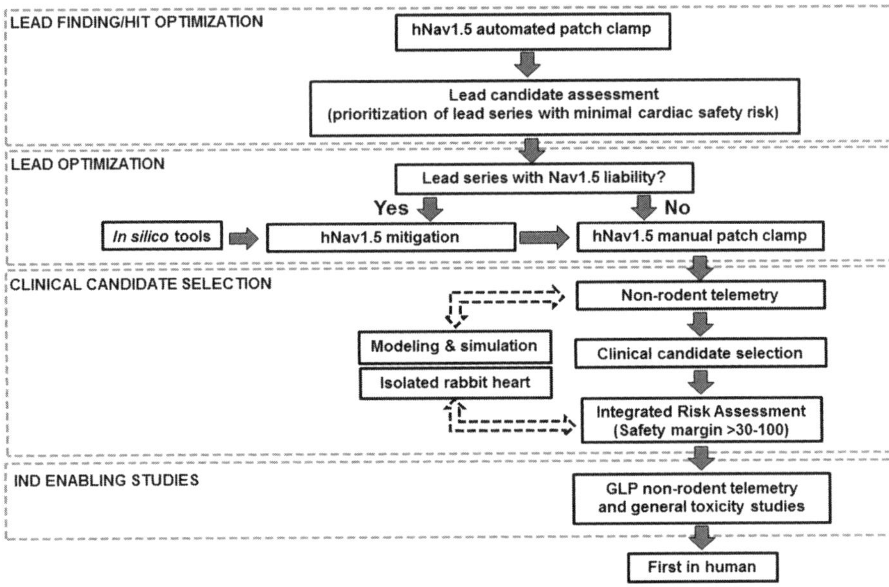

Figure 18.3 Cardiac sodium channel safety assessment scheme. Nonrodent telemetry includes dog or nonhuman primates. Safety margins are calculated by dividing hNav1.5 IC_{50}, lowest effect concentration in the rabbit heart or free C_{max} for lowest efficacious dose in nonrodent telemetry by free C_{max} at estimated efficacious plasma concentration in humans. IND: Investigational New Drug, GLP: Good Laboratory Practice. Dotted lines indicate case-by-case use. Reprinted from Erdemli et al. [104] with permission.

combined with drug concentration measurements and used during the late preclinical phase to verify automated patch clamp results and for more accurate definition of the concentration-channel inhibition relationship. Nonrodent *in vivo* telemetry studies in ascending dose general toxicity studies are performed during clinical candidate selection. Additional preclinical assays such as isolated heart models may be used as needed on a case-by-case basis at any stage (Figure 18.3). Mathematical models can be used to understand how pharmacological perturbations affect electrophysiology from the single ion channel level to the whole heart particularly when the compound affects multiple channels [106].

18.5 TRANSLATIONAL VALUE OF PRECLINICAL ASSESSMENT AND INTEGRATED RISK ASSESSMENT

The integrated risk assessment is an essential part of drug discovery and development. Together with a risk–benefit analysis, an integrated risk assessment should be performed for each molecule including all the available data from *in silico*, *in vitro*, and *in vivo* assays. The results of each assay should be judged on the basis of a therapeutic index calculated by dividing the concentration of the drug at which the off-target related/undesired effects are observed by the concentration at which primary target-mediated/desired effects are observed (see Section 18.6). As none of the preclinical assays are fully predictive of the clinical situation, an integrated risk assessment using all the existing information increases the sensitivity and specificity of the predictions [107].

To avoid late-stage failures, it is advised that lead candidates are profiled in *in vitro* cardiac ion channel assays to prioritize lead series according to cardiac ion channel liability. An integrated risk assessment prior to lead candidate selection ideally would allow only the progression of compounds with no or minimal liabilities to lead optimization. As a result of its pharmacological promiscuity, cardiac ion channel profiling in early drug discovery phases tends to focus on the hERG channel. At Novartis Institutes for Biomedical Research (Cambridge, MA, USA), we also experienced significantly high hit rates for the cardiac sodium channel and recommend profiling lead candidates *in vitro* in an hNav1.5 assay as well as hERG [60]. Profiling of other cardiac ion channels can be postponed to later stages, e.g., clinical candidate selection stage as part of an integrated risk assessment of clinical candidates.

In lead optimization, each chemical series should be carefully profiled using *in vitro* as well as *in silico* approaches to identify trends and to clarify whether off-target cardiac ion channel activity can be removed without significantly compromising the primary target activity. The selection of assay formats with sufficient specificity and throughput is the key factor to support medicinal chemistry efforts successfully. For cardiac ion channels, integration of automated patch clamp technologies in early drug discovery safety profiling has revolutionized the traditional preclinical cardiac safety assessment where the evaluations were performed at later stages when it was too late for optimization. However, automated systems may have several limitations (e.g., recordings are performed at ambient temperature, shorter compound incubation time, etc.) and complementing the cardiac ion channel profiling flowcharts with manual patch clamp recordings for carefully selected compounds could certainly add value [60, 104].

When available, pharmacokinetic data and therapeutically relevant plasma concentrations should be included in the integrated risk assessment to establish an exposure–response relationship and for therapeutic index calculations. A special consideration should be given to the anticipated target population, particularly patient populations with preexisting cardiac risk factors (including known cardiac sodium channel gene variants), co-morbidities (including electrolyte disturbances), co-medication, age, gender, duration of treatment, dosing scheme, as well as unmet medical need and the competitive landscape.

18.6 THERAPEUTIC INDEX

Redfern et al. [108], based on analyses of preclinical and clinical data from 100 drugs covering a range of therapeutics, recommended a safety margin of 30-fold between hERG IC_{50} and maximum therapeutic free (unbound) plasma concentration (C_{max}) for an acceptable degree of safety from arrhythmogenesis. More recently Harmer et al. [103] examined the relationship between hNav1.5 IC_{50} and QRS widening in humans and concluded that a QRS widening may be observed at a free plasma drug concentration 15-fold less than the corresponding Nav1.5 IC_{50}. No systematic analyses are reported to guide therapeutic window calculations for the modulation of other cardiac ion channels and corresponding drug-induced ECG changes and/or arrhythmia propensity. In the absence of such analyses, general rules of therapeutic index calculations for hERG and hNav1.5 channels may be applied to other cardiac ion channels. Considering the key role ion channels play in cardiomyocyte excitability and the analysis of Redfern et al. [108] for hERG inhibition, IC_{10} or IC_{20} values may be more appropriate for therapeutic index calculation than IC_{50} values. However, as a result of the practical reason that IC_{50} values can be obtained much more accurately from sigmoidal concentration response curves than IC_{10} or IC_{20} values, the use of IC_{50} values became a common practice.

When the free plasma concentrations in humans are unknown, then it is acceptable to use maximum therapeutic free plasma concentration in preclinical disease models. When *in vivo* nonrodent telemetry data are available, the safety margin calculations can be performed using the ratio between C_{max} at the lowest observed effect level (LOEL) and maximum therapeutic free plasma concentration.

During the early stages of drug discovery when only *in vitro* data are available, calculation of a safety window using the primary target *in vitro* potency (IC/EC_{50}) and cardiac ion channel potency (IC_{50}) has been a common practice (e.g., hERG IC_{50} divided by primary target IC/EC_{50}). Although in some cases this approach has been successful, there are also many examples that *in vitro* selectivity-based safety window calculations do not translate to *in vivo* [109]. Ideally, for this calculation, IC_{50} values for both primary target and the cardiac ion channels should be obtained from similar *in vitro* assay settings for "like with like comparisons," e.g., cell-based assays. Comparisons of binding or biochemical assay values (e.g., for primary target) with cell-based cardiac ion channel assays may be misleading. The physicochemical properties of compounds (e.g., solubility, permeability, and lipophilicity) and assay conditions (e.g., dimethylsulfoxide (DMSO) levels, protein content, and recording temperatures) should be taken into consideration when interpreting the *in vitro* data as they may heavily influence the results.

A therapeutic index value less than 30-fold would be considered a strong signal [99, 108]. The translational value of therapeutic index calculations depends on various factors including the accuracy and relevance of the information used in these calculations as well on as our knowledge at the time of the assessment. It is essential to have strong confidence in the quality of the data that are used in these calculations. Additionally, even with a therapeutic index $\geq$100, it should only be interpreted as low risk or safe to progress but never as a guarantee for a clean cardiac safety profile in the clinic or in the market. Therefore, the cardiac safety assessment of a new drug is duly completed with the inclusion of clinical and postmarketing information.

18.7 CARDIAC SAFETY EVALUATION OF BIOLOPHARMACEUTICALS

Biolopharmaceuticals encompass a broad range of biological agents from monoclonal antibodies to hormones and enzymes. Current safety pharmacology guidelines (ICH S7A and ICH S7B) address safety pharmacology studies mainly for small molecules and whether these guidelines could be uniformly applied to biopharmaceuticals, particularly monoclonal antibody-like large molecules is debatable [6, 7]. Vargas et al. [110] reviewed the preclinical safety evaluation for biopharmaceuticals and concluded that routine *in vitro* cardiac ion channel safety profiling and dedicated *in vivo* cardiovascular safety studies as described in S7B guidelines are unnecessary and recommended to integrate ECG endpoints to general toxicology studies.

18.8 CONCLUSIONS

A comprehensive cardiovascular safety assessment starts with *in vitro* assays for cardiac ion channels and expands to integrate the judicious use of *in silico*, *in vitro*, and *in vivo* tools. Although the translational experience is still limited, a systematic approach incorporating *in vitro* assays, *in vivo* models, and *in silico* modeling can provide a better understanding of the effects of dug candidates on the cardiovascular system and enable better decision making in early drug development ultimately decreasing the attrition in late stages and facilitating development of safe and efficacious medicines.

REFERENCES

1. Piccini, J.P., Whellan, D.J., Beridge, B.R., Finkle, J.K., Pettit, S.D., Stockbridge, N., Valentin, J.-P., Vargas, H.M., Krucoff, M.W. (2009). Current challenges in the evaluation of cardiac safety during drug development: Translational medicine meets the Critical Path Initiative. *American Heart Journal*, *158*, 317–326.
2. Laverty, H., Benson, C., Cartwright, E., Cross, M., Garland, C., Hammond, T., Holloway, C., McMahon, N., Milligan, J., Park, B., Pirmohamed, M., Pollard, C., Radford, J., Roome, N., Sager, P., Singh, S., Suter, T., Suter, W., Trafford, A., Volders, P., Wallis, R., Weaver, R., York, M., Valentin, J. (2011). How can we improve our understanding of cardiovascular safety liabilities to develop safer medicines? *British Journal of Pharmacology*, *163*, 675–693.
3. Pettit, S.D., Berridge, B., Sarazan, R.D. (2010). A call for more integrated cardiovascular safety assessment. *Journal of Pharmacological and Toxicological Methods*, *61*, 1–2.
4. Sarazan, R.D., Mittelstadt, S., Guth, B., Koerner, J., Zhang, J., Pettit, S. (2011). Cardiovascular function in nonclinical drug safety assessment: Current issues and opportunities. *International Journal of Toxicology*, *30*, 272–286.
5. Josefson, D. (1997). Hay fever drug to be banned by the FDA. *British Medical Journal*, *314*, 248.

6. ICH S7A. (2001). Safety pharmacology studies for human pharmaceuicals S7A. Web page: http://www.ich.org/fileadmin/Public_Web_Site/ICH_Products/Guidelines/Safety/S7A/Step4/S7A_Guideline.pdf
7. ICH S7B. (2005). The non-clinical evaluation of the potential for delayed ventricular repolarization (QT interval prolongation) by human pharmaeuticals S7B. Web page: http://www.ich.org/fileadmin/Public_Web_Site/ICH_Products/Guidelines/Safety/S7B/Step4/S7B_Guideline.pdf
8. ICH E14. (2005). The clinical evaluation of QT/QTc interval prolongation and proarrhythmic potential for non-antiarrhythmic drugs. Web page: http://www.ich.org/fileadmin/Public_Web_Site/ICH_Products/Guidelines/Efficacy/E14/E14_Guideline.pdf
9. Shah, R.R. (2005). Drug-induced QT interval prolongation–regulatory guidance and perspectives on hERG channel studies. *Novartis Foundation Symposium*, *266*, 251–280.
10. Van-Noord, C., Eijgelsheim, M., Stricker, B.H.C. (2010). Drug- and non-drug-associated QT interval prolongation. *British Journal of Clinical Pharmacology*, *70*, 16–23.
11. Harmer, A.R., Abi-Gerges, N., Morton, M.J., Pullen, G.F., Valentin, J.P., Pollard, C.E. (2012). Validation of an in vitro contractility assay using canine ventricular myocytes. *Toxicology and Applied Pharmacology*, *260*, 162–172.
12. Marks, L., Borland, S., Philp, K., Ewart, L., Lainée, P., Skinner, M., Kirk, S., Valentin, J.P. (2012). The role of the anaesthetised guinea-pig in the preclinical cardiac safety evaluation of drug candidate compounds. *Toxicology and Applied Pharmacology*, *263*, 171–183.
13. Hanton, G. (2007). Preclinical cardiac safety assessment of drugs. *Drugs in R and D*, *8*, 213–228.
14. Hoffmann, P., Warner, B. (2006). Are hERG channel inhibition and QT interval prolongation all there is in drug-induced torsadogenesis? A review of emerging trends. *Journal of Pharmacological and Toxicological Methods*, *53*, 87–105.
15. Picard, S., Goineau, S., Guillaume, P., Henry, J., Hanouz, J-L., Rouet, R. (2011). Supplemental studies for cardiovascular risk assessment in safety pharmacology: A critical overview. *Cardiovascular Toxicology*, *11*, 285–307.
16. Redfern, W.S., Ewart, L., Hammond, T.G., Bialecki, R., Kinter, L., Lindgreen, S., Pollard, C.E., Roberts, R., Rolf, M.G., Valentin, J.P. (2010). Impact and frequency of different toxicities throughout the pharmacological life cycle. *Toxicologist*, *114*, 1081.
17. Louden, C., Brott, D., Katein, A., Kelly, T., Gould, S., Jones, H., Betton, G., Valetin, J.P., Richardson, R.J. (2006). Biomarkers and mechanisms of drug-induced vascular injury in non-rodents. *Toxicologic Pathology*, *34*, 19–26.
18. Roth, B.L. (2007). Drugs and valvular heart disease. *New England Journal of Medicine*, *356*, 6–9.
19. Malik, M. (2001). Problems of heart rate correction in assessment of drug-induced QT interval prolongation. *Journal of Cardiovascular Electrophysiology*, *12*, 411–420.
20. Hosmane, B., Locke, C., Morris, D. (2006). QT interval: Correction for heart rate. *Journal of Applied Research*, *6*, 288–299.
21. Salvi, V., Karnad, D.R., Panicker, G.K., Kothari, S. (2010). Update on the evaluation of new drug for effects on cardiac repolarization in humans: Issues in early drug development. *British Journal of Pharmacology*, *159*, 34–48.

22. Spence, S., Soper, K., Hoe, C-M., Coleman, J. (1998). The heart rate-corrected QT interval of conscious beagle dogs: A formula based on analysis of covariance. *Toxicological Sciences*, *45*, 247–258.

23. Wang, D., Taubel, J., Arezina, R. (2012). Comparison of six commonly used QT correction models and their parameter estimation methods. *Journal of Biopharmaceutical Statistics*, *22*, 148–1161.

24. Markert, M., Shen, R., Trautmann, T., Guth, B. (2011). Heart rate correction moels to detect QT interval prolongation in novel pharmaceutical development. *Journal of Pharmacological and Toxicological Methods*, *64*, 25.41.

25. De Clerck, F., Van de Water, A., D'Aubioul, J., Lu, H.R., van Rossem, K., Hermans, A., Van Ammel, K. (2002). *In vivo* measurement of QT prolongation, dispersion and arrhythmogenesis: Application to the preclinical cardiovascular safety pharmacology of a new chemical entity. *Fundamental & Clinical Pharmacology*, *16*, 125–140.

26. Hess, P., Rey, M., Wanner, D., Steiner, B., Clozel, M. (2007). Measurements of blood pressure and electrocardiogram in conscious freely moving guineapigs: A model for screening QT interval prolongation effects. *Labratory Animals*, *41*, 470–480.

27. Ferguson, A.V., Smith, P. (1991). Autonomic mechanisms underlying area postrema stimulation-induced cardiovascular responses in rats. *American Journal of Physiology*, *261*, R1–R8.

28. Santajuliana, D., Hornfeldt, B.J., Osborn, J.W. (1996). Use of ganglionic blockers to assess neurogenic pressor activity in conscious rats. *Journal of Pharmacological and Toxicological Methods*, *35*, 45–54.

29. Berne, R.M., Levy, M.N. *Cardiovascular Physiology*, 5th Ed., C.V. Mosby Company. St. Louis, MO, 1986, pp. 153–196.

30. Milan, D.J., Peterson, T.A., Ruskin, J.N., Peterson, R.T., MacRae, C.A. (2003). Drugs that induce repolarization abnormalities cause bradycardia in zebrafish. *Circulation*, *107*, 1355–1358.

31. Lee, S.H., Kim, H.R., Han, R.X., Oqani, R.K., Jin, D.I. (2011). Cardiovascular risk assessment of atypical antipsychotic drugs in a zebrafish model. *Journal of Applied Toxicology*, *33*(6), 466–470.

32. Peal, D.S., Peterson, R.T., Milan, D. (2010). Small molecule screening in zebrafish. *Journal of Cardiovascular Translational Research*, *3*, 454–460.

33. Letamendia, A., Quevedo, C., Ibarbia, I., Virto, J.M., Holgado, O., Diez, M., Belmonte, J.C.I., Callol-Massot, C. (2012). Development and validation on an automated high-throughput system for zebrafish in vivo screenings. *PLoS One*, *7*, e36690.

34. George, S., Xia, T., Rallo, R., Zhao, Y., Ji, Z., Lin, S., Want, X., Zhang, H., France, B., Schoenfeld, D., Damoiseaux, R., Liu, R., Lin, S., Bradley, K.A., Cohen, Y., Nel, A.E. (2011). Use of a high-throughput screening approach coupled with *in vivo* zebrafish embryo screening to develop hazard ranking for engineered nanomaterials. *ACS Nano*, *3*, 1805–1817.

35. Kitambi, S.S., Nilsson, E.S., Sekyrova, P., Ibarra, C., Tekeoh, G.N., Andang, M., Emfors, P., Uhlen, P. (2012). Small molecule screening platform for assessment of cardiovascular toxicity on adult zebrafish heart. *BMC Physiology*, *12*, 3–10.

36. Barros, T.P., Alderton, W.K., Reynolds, H.M., Roach, A.G., Berghmans, S. (2008). Zebrafish: An emerging technology for in vivo pharmacological assessment to identify

potential safety liabilities in early drug discovery. *British Journal of Pharmacology, 154*, 1400–1413.

37. Brimecombe, J.C., Kirsch, G.E., Brown, A.M. (2009). Test article concentrations in the hERG assay: Losses through perfusion, solubility and stability. *Journal of Pharmacological and Toxicological Methods, 59*, 29–34.
38. Wang, J., Matayoshi, E. (2012). Solubility at the molecular level: Development of a critical aggregation concentration (CAC) assay for estimating compound monomer solubility. *Pharmaceutical Research, 29*, 1745–1754.
39. Limberis, J.T., McDermott, J.S., Salmen, H.J., Su, Z., Mikhail, A., Green, J.R., Cox, B.F., Gintant, G.A., Martin, R.L. (2007). The effects of plasma proteins on delayed repolarization in vitro with cisapride, risperidone and d, l-sotalol. *Journal of Pharmacological and Toxicological Methods, 56*, 11–17.
40. Haulica, I., Bild, W., Mihaila, C., Serban, D.N., Serban, L., Boisteanu, D., Ionita, T., Radasanu, O. (2004). Comparative study of the inhibitory effects of adrenomedullin on angiotensin II contraction in rat conductance and resistance arteries. Journal of the *Renin-Angiotensin-Aldosterone System: JRAAS, 5*, 79–83.
41. Bridges, L.E., Williams, C.L., Pointer, M.A., Awumey, E.M. (2011). Mesenteric artery contraction and relaxation studies using automated wire myography. *Journal of Visualized Experiments, 55*, e3119.
42. Mendizabal, Y., Liorens, S., Nava, E. (2011). Reactivity of the aorta and mesenteric resistance arteries from the obes spontaneously hypertensive rat: Effects of glitazones. *American Journal of Physiology, 301*, H1319–H1330.
43. Guo, L., Dong, Z., Gurthrie, H. (2009). Validation of a guinea pig Langendorff heart model for assessing potential cardiovascular liability of drug candidates. *Journal of Pharmacological and Toxicological Methods, 60*, 130–151.
44. Valentin, J-P., Pollard, C., Lainee, P., Hammond, T. (2010). Value of non-clinical cardiac repolarization assays in supporting the discovery and development of safer medicines. *British Journal of Pharmacology, 159*, 25–33.
45. Gintant, G.A., Limberis, J., McDermott, J., Wegner, C., Cox, B. (2001). The canine Purkinje fiber: An in vitro model system for acquired long QT syndrome and drug-induced arrhythmogenesis. *Journal of Cardiovascular Pharmacology, 37*, 607–618.
46. Johnson, D.M., Hussein, L., Spatjens, R.L., Valentin, J-P., Volders, P.G., Abi-Gerges, N. (2011). Measurement of action potential generation in isolated canine left ventricular midmyocardial myocytes. *Current Protocols in Pharmacology, 55*, 1–23.
47. Peng, S., Lacerca, A.E., Kirsch, G.E., Brown, A.M., Bruening-Wright, A. (2010). The Action potential and comparative pharmacology of stem cell-derived human cardiomyocytes. *Journal of Pharmacological and Toxicological Methods, 61*, 277–286.
48. Jonsson, M.K., Vos, M.A., Mirams, G.R., Duker, G., Sartipy, P., de Boer, T.P., van Veen, T.A. (2012). Application of human stem cell-derived cardiomyocytes in safety pharmacology requires caution beyond hERG. *Journal of Molecular and Cellular Cardiology, 52*, 998–1008.
49. Stoelzle, S., Haythornthwaite, A., Kettenhofen, R., Kolossov, E., Bohlen, H., George, M., Bruggemann, A., Fertig, N. (2011). Automated patch clamp on mESC-derived cardiomyocytes for cardiotoxicity prediction. *Journal of Biomolecular Screening, 16*, 910–916.
50. Guo, L., Quan, J.Y., Abrams, R., Tang, H.M., Weiser, T., Sanders, M.J., Kolaja, K.L. (2011). The electrophysiological effects of cardiac glycosides in human iPSC-derived

cardiomyocytes and in guinea pig isolated hearts. *Cellular Physiology and Biochemistry*, *27*, 453–462.

51. Itzhaki, I., Maizels, L., Huber, I., Zwi-Dantsis, L., Caspi, O., Winterstern, A., Feldman, O., Gepstein, A., Arbel, G., Hammerman, H., Boulos, M., Gepstein, L. (2011). Modeling the long QT syndrome with induced pluripotent stem cells. *Nature*, *471*, 225–229.
52. Priest, B.T., Bell, I.M., Garcia, M.L. (2008). Role of hERG potassium channel assays in drug development. *Channels*, *2*, 87–93.
53. Goineau, S., Legrand, C., Froget, G. (2012). Whole-cell configuration of the patch-clamp technique in the hERG channel assay to predict the ability of a compound to prolong QT interval. *Current Protocols in Pharmacology*, *57*, 1–14.
54. Yao, J-A., Du, X., Lu, D., Baker, R.L., Daharsh, E., Atterson, P. (2005). Estimation of potency of HERG channel blockers: Impact of voltage protocol and temperature. *Journal of Pharmacological and Toxicological Methods*, *52*, 146–153.
55. Kirsch, G.E., Trepakova, E.S., Brimecombe, J.C., Sidach, S.S., Erickson, H.D., Kochan, M.C., Shyjka, L.M., Lacerda, A.E., Brown, A.M. (2004). Variability in the measurement of hERG potassium channel inhibition: Effects of temperature and stimulus pattern. *Journal of Pharmacological and Toxicological Methods*, *50*, 93–101.
56. Crumb, W.J., Cavero, I. (2003). Patch-clamp studies of human cardiac ion channels in the evaluation of cardiac electrophysiological effects of compounds. *Current Protocols in Pharmacology*, *20*, 1–18.
57. Martin, R.L., McDermott, J.S., Salmen, H.J., Palmatier, J., Cox, B.F., Gintant, G.A. (2004). The utility of hERG and repolarization assays in evaluating delayed cardiac repolarization: Influence of multi-channel block. *Journal of Cardiovascular Pharmacology*, *43*, 369–379.
58. Brown, A.M. (2009). High throughput functional screening of an ion channel library for drug safety and efficacy. *European Biophysics Journal*, *38*, 273–278.
59. Mathes, C., Friis, S., Finley, M., Liu, Y. (2009). QPatch: The missing link between HTS and ion channel drug discovery. *Combinatorial Chemistry & High Throughput Screening*, *12*, 78–95.
60. Cao, X., Lee, T., Holmqvist, M., Lin, Y., Ni, Y., Mikhailov, D., Zhang, H., Hogan, C., Zhou, L., Lu, Q., Digan, M.E., Urban, L., Erdemli, G. (2010). Cardiac ion channel safety profiling on the IonWorks Quattro™ Automated Patch Clamp System. *Assay and Drug Development Technologies*, *8*, 766–780.
61. Moller, C., Slack, M. (2010). Impact of new technologies for cellular screening along the drug value chain. *Drug Discovery Today*, *15*, 384–390.
62. Bebarova, M. (2012). Advance inpatch clamp technique: Towards higher quality and quantity. *General Physiology and Biophysics*, *31*, 131–140.
63. Golden, A.P., Li, N., Chen, Q., Lee, T., Nevil,l T., Cao, X., Johnson, J., Erdemli, G., Ionescu-Zanetti, C., Urban, L., Holmqvist, M. (2011). IonFlux: A microfluidic patch clamp system evaluated with human Ether-à-go-go related gene channel physiology and pharmacology. *Assay and Drug Development Technologies*, *9*, 608–619.
64. Limberis, J.T., Su, Z., Cox, B.F., Gintant, G.A., Martin, R.L. (2006). Altering extracellular potassium concentration does modulate drug block of human ether-a-go-go-related gene (hERG) channels. *Clinical and Experimental Pharmacology and Physiology*, *33*, 1059–1065.

65. Estacion, M., Choi, J.S., Eastman, E.M., Lin, Z., Li, Y., Tyrrell, L., Yang, Y., Dib-Hajj, S.D., Waxman, S.G. (2010). Can robots patch-clamp as well as humans? Characterization of a novel sodium channel mutation. *Journal of Physiology*, *588*, 1915–1927.
66. Dumotier, B.M., Georgieva, A.V. (2007). Preclinical cardio-safety assessment of torsadogenic risk and alternative methods to animal experimentation: The inseparable twins. *Cell Biology and Toxicology*, *23*, 293–302.
67. Polak, S., Wisniowska, B., Brandys, J. (2008). Collation, assessment and analysis of literature in vitro data on hERG receptor blocking potency for subsequent modeling of drugs' cardiotoxic properties. *Journal of Applied Toxicology*, *29*, 183–206.
68. Bender, A., Scheiber, J., Glick, M., Davies, J.W., Azzaoui, K., Hamon, J., Urban, L., Whitebread, S., Jenkins, J.L. (2007). Analysis of pharmacology data and the prediction of adverse drug reactions and off-target effects from chemical structure. *ChemMedChem*, *2*, 861–873.
69. Davenport, A.J., Moller, C., Heifetz, A., Mazanetz, M.P., Law, R.J., Ebneth, A., Gemkow, M.J. (2010). Using electrophysiology and in silico three-dimensional modeling to reduce human ether-a-go-go related gene K^+ channel inhibition in a histamine H_3 receptor antagonist program. *Assay and Drug Development Technologies*, *8*, 781–789.
70. Davies, M.R., Mistry, H.B., Hussein, L., Pollard, C.E., Valentin, J-P., Swinton, J., Abi-Gerges, N. (2012). An *in silico* canine cardiac midmyocardial action potential duration model as a tool for early drug safety assessment. *American Journal of Physiology*, *302*, H1466–H1480.
71. Lu, Z., Wu, C.Y., Jiang, Y.P., Ballou, L.M., Clausen, C., Cohen, I.S., Lin, R.Z. (2010). Suppression of phosphoinositide 3-kinase signaling and alteration of multiple ion currents in drug-induced long QT syndrome. *Science Translational Medicine*, *4*, 131–150.
72. Erdemli, G., Mikhailov, D., Kim, A. (2010). Indirect modulation of cardiac ion channels and implications for preclinical safety assessment. *European Pharmaceutical Review*, *4*, 39–45.
73. Dennis, A., Wang, L., Wan, X., Ficker, E. (2007). hERG channel trafficking: Novel targets in drug-induced long QT syndrome. *Biochemical Society Transactions*, *35*, 1060–1063.
74. van der Heyden, M.A., Smits, M.E., Vos, M.A. (2008). Drugs and trafficking of ion channels: A new pro-arrhythmic threat on the horizon? *British Journal of Pharmacology*, *153*, 406–409.
75. Ficker, E., Kuryshev, Y.A., Dennis, A.T., Obejero-Paz, C., Wang, L., Hawryluk, P., Wible, B.A., Brown, A.M. (2004). Mechanisms of arsenic-induced prolongation of cardiac repolarization. *Molecular Pharmacology*, *66*, 33–44.
76. Taniguchi, T., Uesugi, M., Arai, T., Yoshinaga, T., Miyamoto, N., Sawada, K. (2012). Chronic probucol treatment decreases the slow component of the delayed-rectifier potassium current in CHO cells transfected with KCNQ1 and KCNE1: A novel mechanism of QT prolongation. *Journal of Cardiovascular Pharmacology*, *59*, 377–386.
77. Ruan, Y., Denegri, M., Liu, N., Bachetti, T., Seregni, M., Morotti, S., Severi, S., Napolitano, C., Priori, S.G. (2010). Trafficking defects and gating abnormalities of a novel SCN5A mutation question gene-specific therapy in long QT syndrome type 3. *Circulation Research*, *106*, 1374–1383.
78. Zhou, J., Shin, H.G., Yi, J., Shen, W., Williams, C.P., Murray, K.T. (2002). Phosphorylation and putative ER retention signals are required for protein kinase A-mediated potentiation of cardiac sodium current. *Circulation Research*, *91*, 540–546.

79. Sánchez-Chapula, J.A., Salinas-Stefanon, E., Torres-Jácome, J., Benavides-Haro, D.E., Navarro-Polanco, R.A. (2001). Blockade of currents by the antimalarial drug chloroquine in feline ventricular myocytes. *Journal of Pharmacology and Experimental Therapeutics*, *297*, 437–445.

80. Traebert, M., Dumotier, B. (2005). Antimalarial drugs: QT prolongation and cardiac arrhythmias. *Expert Opinions in Drug Safety*, *4*, 421–431.

81. Farkasfalvi, K., Stagg, M.A., Coppen, S.R., Siedlecka, U., Lee, J., Soppa, G.K., Marczin, N., Szokodi, I., Yacoub, M.H., Terracciano, C.M. (2007). Direct effects of apelin on cardiomyocyte contractility and electrophysiology. *Biochemical and Biophysical Research Communications*, *357*, 889–895.

82. Japp, A.G., Cruden, N.L., Barnes, G., van Gemeren, N., Mathews, J., Adamson, J., Johnston, N.R., Denvir, M.A., Megson, I.L., Flapan, A.D., Newby, D.E. (2010). Acute cardiovascular effects of apelin in humans: Potential role in patients with chronic heart failure. *Circulation*, *121*, 1818–1827.

83. Szokodi, I., Tavi, P., Földes, G., Voutilainen-Myllylä, S., Ilves, M., Tokola, H., Pikkarainen, S., Piuhola, J., Rysä, J., Tóth, M., Ruskoaho, H. (2002). Apelin, the novel endogenous ligand of the orphan receptor APJ, regulates cardiac contractility. *Circulation Researc*, *91*, 434–440.

84. Chamberland, C., Barajas-Martinez, H., Haufe, V., Fecteau, M.H., Delabre, J.F., Burashnikov, A., Antzelevitch, C., Lesur, O., Chraibi, A., Sarret, P., Dumaine, R. (2010). Modulation of canine cardiac sodium current by Apelin. *Journal of Molecular and Cellular Cardiology*, *48*, 694–701.

85. Berger, E., Patel, K., Anwar, S., Davies, W., Sheridan, D.J. (2005). Investigation of the effects of physiological and vasodilation-induced autonomic activation on the QTc interval in healthy male subjects. *British Journal of Clinical Pharmacology*, *60*, 17–23.

86. Shusterman, V., Beigel, A., Shah, S.I., Aysin, B., Weiss, R., Gottipaty, V.K., Schwartzman, D., Anderson, K.P. (1999). Changes in autonomic activity and ventricular repolarization. *Journal of Electrophysics*, *32*, 185–192.

87. Thomas, D., Kiehn, J., Katus, H.A., Karle, C.A. (2004). Adrenergic regulation of the rapid component of the cardiac delayed rectifier potassium current, I_{Kr}, and the underlying hERG ion channel. *Basic Research in Cardiology*, *99*, 279–87.

88. Kurokawa, J., Bankston, J.R., Kaihara, A., Chen, L., Furukawa, T., Kass, R.S. (2009). KCNE variants reveal a critical role of the beta subunit carboxyl terminus in PKA-dependent regulation of the I_{Ks} potassium channel. *Channels*, *3*, 16–24.

89. Eckert, B., Agardh, C.D. (1998). Hypoglycaemia leads to an increased QT interval in normal men. *Clinical Physiology*, *18*, 570–575.

90. Gill, G.V., Woodward, A., Casson, I.F., Weston, P.J. (2009). Cardiac arrhythmia and nocturnal hypoglycaemia in type 1 diabetes—the "dead in bed" syndrome revisited. *Diabetologia*, *52*, 42–45.

91. Zhang, Y., Han, H., Wang, J., Wang, H., Yang, B., Wang, Z. (2003). Impairment of human ether-à-go-go-related gene (hERG) K^+ channel function by hypoglycemia and hyperglycemia. Similar phenotypes but different mechanisms. *Journal of Biological Chemistry*, *278*, 10417–10426.

92. Chvilicek, J.P., Hurlbert, B.J., Hill, G.E. (1995), Diuretic-induced hypokalaemia inducing torsades de pointes. *Canadian Journal of Anesthesia*, *42*, 1137–1139.

93. Cohen, J.D., Neaton, J.D., Prineas, R.J., Daniels, K.A. (1987). Diuretics, serum potassium and ventricular arrhythmias in the Multiple Risk Factor Intervention Trial. *American Journal of Cardiology*, *60*, 548–554.
94. Sanguinetti, M.C., Jurkiewicz, N.K. (1992). Role of external Ca^{2+} and K^+ in gating of cardiac delayed rectifier K^+ currents. *Pflugers Arch*, *420*, 180–186.
95. Yang, T., Snyders, D.J., Roden, D.M. (1997). Rapid inactivation determines the rectification and $[K^+]_o$ dependence of the rapid component of the delayed rectifier K^+ current in cardiac cells. *Circulation Researc*, *80*, 782–789.
96. Guo, J., Massaeli, H., Xu, J., Jia, Z., Wigle, J.T., Mesaeli, N., Zhang, S. (2009). Extracellular K^+ concentration controls cell surface density of I_{Kr} in rabbit hearts and of the HERG channel in human cell lines. *Journal of Clinical Investegation*, *119*, 2745–2757.
97. Friberg, L.E., Isbister, G.K., Duffull, S.B. (2005). Pharmacokinetic-pharmacodynamic modeling of QT interval prolongation following citalopram overdoses. *British Journal of Clinical Pharmacology*, *61*, 177–190.
98. Pugsley, M.K., Authier, S., Towart, R., Gallacher, D.J., Curtis, M.J. (2009). Beyond the safety assessment of drug-mediated changes in the QT interval...what's next? *Journal of Pharmacological and Toxicological Methods*, *60*, 24–27.
99. Mikhailov, D., Traebert, M., Lu, Q., Whitebread, S., Egan, W. Should cardiosafety be ruled by hERG inhibition? Early testing scenarios and integrated risk assessment. In B. Faller and L. Urban (eds.), *Hit and Lead Profiling*, Wiley-VCH. Weinheim, 2009, pp. 387–413.
100. Traebert, M., Dumotier, B. Cardiac Safety. In J.J. Xu and L. Urban (eds.), *Predictive Toxicology in Drug Safety*. Cambridge University Press. New York, NY, 2010, pp. 34–53.
101. Chen, M.X., Helliwell, R.M., Clare, J.J. (2009). *In vitro* profiling against ion channels beyond hERG as an early indicator of cardiac risk. *Current Opinion in Molecular Therapeutics*, *11*, 269–281.
102. Gintant, G.A., Gallacher, D.J., Pugsley, M.K. (2011). The "overly-sensitive" heart: Sodium channel block and QRS interval prolongation. *British Journal of Pharmacology*, *164*, 254–259.
103. Harmer, A., Valentin, J.P., Pollard, C. (2011). On the relationship between block of the cardiac Na^+ channel and drug-induced prolongation of the QRS complex. *British Journal of Pharmacology*, *164*, 260–273.
104. Erdemli, G., Kim, A., Ju, H., Springer, C., Penland, R.C., Hoffmann, P. (2012). Preclinical evaluation of cardiac safety liabilities mediated by hNav1.5 blockade. *Front Pharmacology*, *3*, 1–9.
105. Tfelt-Hansen, J., Winkel, B., Grunnet, M., Jespersen, T. (2010). Inherited cardiac diseases caused by mutations in the Nav1.5 sodium channel. *Journal of Cardiovascular Electrophysiology*, *21*, 107–115.
106. Bottino, D., Penland, R.C., Stamps, A., Traebert, M., Dumotier, B., Georgiva, A., Helmlinger, G., Lett, G.S. (2006). Preclinical cardiac safety assessment of pharmaceutical compounds using an integrated systems-based computer model of the heart. *Progress in Biophysics and Molecular Biology*, *90*, 414–443.
107. Wallis, R.M. (2010). Integrated risk assessment and predictive value to humans of nonclinical repolarization assays. *British Journal of Pharmacology*, *159*, 115–121.
108. Redfern, W.S., Carlsson, L., Davis, A.S., Lynch, W.G., MacKenzie, I., Palethorpe, S., Siegl, P.K., Strang, I., Sullivan, A.T., Wallis, R., Camm, A.J., Hammond, T.G. (2003).

Relationships between preclinical cardiac electrophysiology, clinical QT interval prolongation and torsade de pointes for a broad range of drugs: Evidence for a provisional safety margin in drug development. *Cardiovascular Research*, *58*, 32–45.

109. Muller, P.Y., Milton, M.N. (2012). The determination and interpretation of the therapeutic index in drug development. *Nature Reviews Drug Discovery*, *11*, 751–761.

110. Vargas, H.M., Bass, A.S., Breidenbach, A., Feldman, H.S., Gintant, G.A., Harmer, A.R., Heath, B., Hoffmann, P., Lagrutta, A., Leishman, D., McMahon, N., Mittelstadt, S., Polonchuk, L., Pugsley, M.K., Salata, J.J., Valentin, J-P. (2008). Scientific review and recommendations on preclinical cardiovascular safety evaluation of biologics. *Journal of Pharmacological and Toxicological Methods*, *58*, 72–76.

111. Car, B. (2006). Enabling technologies in reducing drug attrition due to safety failures. *American Drug Discovery*, *1*, 53–56.

112. Sibille, M., Deigat, N., Janin, A., Kirkesseli, S., Durand, D.V. (1998), Adverse events in phase-I studies: A report in 1015 healthy volunteers. *European Journal of Clinical Pharmacology*, *54*, 13–20.

113. Olson, H., Betton, G., Robinson, D., Thomas, K., Monro, A., Kolaja, G., Lilly, P., Sanders, J., Sipes, G., Bracken, W., Dorato, M., Van Deun, K., Smith, P., Berger, B., Heller, A. (2000). Concordance of the toxicity of pharmaceuticals in humans and in animals. *Regulatory Toxicology and Pharmacology*, *32*, 56–67.

114. Budnitz, D.S., Pollock, D.A., Weidenbach, K.N., Mendelsohn, A.B., Schroeder, T.J., Annest, J.L. (2006). National surveillance of emergency department visits for outpatient adverse drug effects. *Journal of American Medical Association*, *296*, 1858–1866.

115. Fung, M., Thornton, A., Mybeck, K., Wu, J.H., Hornbuckle, K., Muniz, E. (2001). Evaluation of the characteristics of safety withdrawal of prescription drugs from worldwide pharmaceuticals markets 1960 to 1999. *Drug Information Journal*, *35*, 293–317.

116. Stevens, J.L., Baker, T.K. (2009). The future of drug safety testing: Expanding the view and narrowing the focus. *Drug Discovery Today*, *14*, 162–167.

19

DRUG-INDUCED HEPATOTOXICITY: ADVANCES IN PRECLINICAL PREDICTIVE STRATEGIES AND TOOLS

Donna M. Dambach

19.1 INTRODUCTION: THE ISSUE OF IDIOSYNCRATIC DRUG-INDUCED LIVER INJURY (IDILI)

Drug-induced liver injury (DILI) has remained one of the top two causes of clinical drug attrition for the last 40 years, and as such, identification of preclinical models to improve mitigation of this adverse event has continued to be a key focus area of pharmaceutical safety scientists [1, 2]. DILI accounts for ~30–50% of acute liver failure cases in the Western world and is the major cause of acute liver failure, with a mortality rate of up to 10% [3,4]. It is likely that hepatic injury will remain a key clinical adverse finding for orally administered, small-molecule pharmaceuticals because of the anatomical location predisposing the liver to high transient drug concentrations ("first-pass effect") and the physiological function of xenobiotic metabolism and elimination. Therefore, continued efforts to improve our preclinical model prediction and refine translational information regarding risk factors will remain a significant focus.

Predictive ADMET: Integrative Approaches in Drug Discovery and Development, First Edition.
Edited by Jianling Wang and Laszlo Urban.

Direct or intrinsic DILI occurs at a high incidence, will manifest in animals and humans when a drug is taken at sufficient quantities, and is apparent soon after administration (acute). Thus, drugs causing intrinsic DILI are commonly identified in preclinical models: Severely toxic drugs are discontinued, and those advanced to the clinic have safety margins that are considered acceptable. Idiosyncratic DILI (iDILI) is rare (1 in 10,000 or greater), does not follow predictable dose–response relationships, is not related to the targeted pharmacology, does not occur with a similar incidence across animal species and humans, and has an unpredictable acute or latent onset often occurring after weeks to months of dosing; it is not reliably detected in preclinical models and thus is the major cause of marketed drug attrition [5,6]. The pathogenesis of iDILI is not understood; however, based on epidemiological studies and mechanistic investigations of failed drugs, the leading hypothesis portends there is an initial, intrinsic insult caused by the drug and a subsequent adaptive response [7,8]. In the average population, the initial insult is either minimal and subclinical or transient, whereas in susceptible individuals, either the insult is amplified or the adaptive response is inappropriate leading to severe toxicity [6,9]. Clinical manifestation of iDILI is believed to be multifactorial and includes drug-specific properties (drug exposure, inherent chemical properties) and patient-specific factors (combination of physiological, environmental, and genetic risk factors) that enhance susceptibility, and precipitation of iDILI is related to a threshold concurrence of these independent factors [10,11].

In particular, drug-specific factors are related to the physicochemical and structural features that influence metabolic bioactivation and covalent binding to cellular components leading to cellular dysfunction or an immune response; noncovalent inhibition/alteration of cellular functions including unintended pharmacological activity; and cellular drug/metabolite accumulation. Affected cellular processes commonly associated with DILI that have been identified through mechanistic investigation include mitochondrial functional impairment (including initiation of apoptosis); alterations in redox status; inflammation or immune response activation; and alteration of protein function (e.g., enzymes or transporters) [7,8,12–15].

Patient-related factors are those that influence the normal adaptive responses and include age, gender, diet/nutritional status, administration of co-medications (drug–drug interactions), concurrent disease effects, and epigenetic or genetic variability that can affect drug metabolism and disposition or immunological responses. There is no definitive clinical diagnostic tool, nor set of risk factors, that uniquely defines or predicts iDILI. Various clinical causality scoring criteria have been established, but they are inconsistently used and cannot prospectively predict development of iDILI. [16] Nonetheless, epidemiological examination of clinical features of drugs causing hepatotoxicity has identified some common factors that are useful for developing preclinical assessment strategies:

1. Metabolism: Drugs with extensive hepatic metabolism ($\geq$50%) have a greater association with elevated alanine transferase (ALT) values (>3 × ULN), hepatic failure, and mortality [17].

2. Dose: Greater than 75% of drugs with a daily dose ≥50 mg were associated with DILI; the risk was increased for highly metabolized drugs [17–20].
3. Biliary Elimination: Drugs eliminated via biliary clearance have a higher incidence of jaundice, including drugs with minimal hepatic metabolism [17].
4. Gender and Age: There is no gender or age association for overall DILI. However, older age was associated with cholestatic DILI with a slight male predominance; younger age was associated with development of hepatocellular necrosis with a female predominance and a higher progression to liver failure; autoimmune-type DILI has been reported to occur exclusively in women [19, 20].
5. Hepatocellular DILI: The most common form to progress to liver failure [20].
6. Genetic Polymorphisms (metabolic, inflammatory/immunological, mitochondrial): Associations have been made with DILI, but often multiple polymorphisms in a metabolic and elimination pathway are identified versus a single determinant [20].
7. Preexisting Liver Disease: Diabetes and viral infections have been associated with enhanced susceptibility [16].

When considering the ability to predict iDILI preclinically, if the pathogenic complexity of the "multideterminant hypothesis" holds true, then no single factor would predict the development of toxicity. Instead, in addition to an initiating factor, a series of risk factors related to the physiological handling of the drug needs to be considered. Therefore, it is highly likely that preclinical hazard identification and risk assessment will require the integrated evaluation of several endpoints. However, the clinical risk factors and drivers of toxicity are largely unknown and, therefore, hamper the development of predictive models. The continued evolution of preclinical assessment of drug-induced hepatotoxicity, in particular iDILI, will require continued mechanistic investigation both preclinically and clinically. This chapter will provide an overview of the currently available preclinical surrogate models; key challenges to assess iDILI risk associated with the use of these surrogate models; recommended strategic considerations for integrated risk assessment and mitigation; and a forward-looking perspective of the remaining gaps in our tools and data integration.

19.2 OVERVIEW OF PRECLINICAL SURROGATE MODELS AVAILABLE TO PREDICT DILI: *IN VIVO*, *IN VITRO*, AND *IN SILICO* TOOLS

In vivo and *in vitro* models remain the primary preclinical surrogate models used for hazard identification and mechanistic investigation of DILI, whereas *in silico* models are the most nascent. These platforms will continue to remain the most practical tools available for preclinical risk assessment. Typically, "advances" or evolution in the strategic use of these models results from characterization of new mechanisms of toxicity, e.g., mitochondrial dysfunction, and from improvements of identified limitations of the models, e.g., metabolic activity, through technological advances. DILI manifests as a variety of histological phenotypes, and the complex and varied pathogenesis of DILI has resulted in the application of a variety of surrogate assays and endpoints. Each model platform has recognized limitations, which will be discussed in a later section. Currently, there is no universally accepted or validated strategy for preclinical assessment of iDILI.

19.2.1 Animal Models

Animal models remain the regulatory standard for decisional analysis regarding hazard identification and risk assessment, and they are the most physiologically integrated models to assess the systemic effects of drug administration. Therefore, animal models will likely remain a mainstay of safety and risk assessment analyses using single- and repeat-dosing strategies. Drugs causing severe hepatic injury in animals, i.e., intrinsic hepatotoxicants, will continued to be removed from the portfolio [1,21]. However, based on the available public literature, ~50% of clinical hepatotoxicants are not detected in animal studies [22]. Additionally, there are currently no definitive animal models of idiosyncratic liver toxicity [23]. Nonetheless, there are opportunities to leverage specialized animal models for mechanistic evaluation or to identify susceptibility factors, which can be used to generate more robust risk assessments to impact candidate decision-making or clinical trial designs.

Animal models are commonly used to elucidate the underlying mechanisms of toxicity of specific drugs. These include genetically engineered mouse models (transgenic and knock-out mice), specific disease models, cross-strain genome-wide association studies (GWAS) in rodents, and chemical manipulation to increase susceptibility to hepatotoxicants [23–36]. Transgenic and disease models have demonstrated sensitivity to idiosyncratic hepatotoxicants when normal, healthy animals did not [25, 28–32]. The use of disease models as part of an integrated, evidence-based approach to improve characterization of hepatic risk has also been advocated by several health authorities [24].

19.2.2 *In Vitro* Models

In vitro platforms contribute significantly to candidate safety lead optimization strategies; detection of underlying drivers of toxicity; and investigation of toxicities identified *in vivo* [37, 38]. Health authorities have specifically embraced the use of *in vitro* models to improve characterization of identified hepatobiliary toxicities [24]. Drug-induced hepatic injury may manifest as a myriad of histological phenotypes, but most often iDILI is classified as a primary hepatocellular injury, primary cholestatic, or mixed hepatocellular-cholestatic. Historically, most preclinical screening models have focused on predicting hepatocellular injury because this phenotype of iDILI is the most life-threatening, potentially leading to fulminant liver failure and death. More recently, models have been developed to evaluate the potential for inhibition of bile acid or xenobiotic transporters that may predict cholestatic iDILI risk or intrahepatic drug/metabolite accumulation [39–41]. There are currently no definitive *in vitro* models of iDILI, including models to evaluate complex, multicellular inflammatory or immune interactions. Immortalized cell lines and primary human monocellular culture systems remain the most commonly used.

In vitro models developed for detecting potential hepatotoxicants have relied heavily on general cytotoxicity as an endpoint using static cultures of either immortalized cell lines or primary hepatocytes. In the last 10 years, the endpoints measured have expanded from assessing nonspecific markers of cellular viability (e.g., adenosine triphosphate (ATP) levels or lactate dehydrogenase (LDH) release) to including endpoints associated with identified mechanisms of hepatotoxicity to determine the cellular processes affected by a drug and potentially improve prediction. This is apparent when reviewing the literature associated with hepatotoxicity detection, which is admixed with general cytotoxicity assessments of several drug or chemical classes and those dedicated to the qualification of cell-based assays for specific "hepatotoxicity" detection [41–55]. However, no standard assay method, defined endpoints, nor standard positive criteria have been established for cytotoxicity assessments and the reported approaches and assessments are quite varied. Therefore, attempting to determine accurately the predictivity of these models by comparing the reported data is difficult. In most studies, multiple endpoints were evaluated; however, typically a "positive" outcome was designated if any single endpoint achieved the defined threshold. Nonetheless, the overall data suggest that regardless of whether an immortalized cell line or primary hepatocyte is used, the positive predictive value (PPV) for clinical translation of these cytotoxicity assays is quite high, $\geq$94% (Table 19.1). This indicates that a positive outcome in a cytotoxicity study is highly predictive of a drug having cytotoxicity (e.g., DILI) risk. On the other hand, the negative predictive value (NPV) for clinical translation of cytotoxicity assays is highly variable and generally much lower; immortalized cells and primary hepatocytes occupy a similar predictive space: immortalized cell NPV ranges from 25% to 69% and primary hepatocytes from 31% to 58%. This indicates that a negative outcome in a cytotoxicity study misses a significant number (40–75%) of drugs with potential DILI risk, which confirms the limitations of simple, monoculture assay systems.

Recently, Lin and Will demonstrated that immortalized cells derived from specific organs (i.e., liver, heart, and kidney) have an equal ability to assess general cytotoxicity, but they cannot preferentially predict specific organ toxicity [54]. This finding was also demonstrated by other researchers when independently comparing several immortalized cell lines [44,45,55]. These data suggest that immortalized cell lines cannot specifically detect hepatotoxicity, but they do have value for screening potential cytotoxicity, especially of the non-metabolized, parent drug.

The major value added by primary hepatocytes versus immortalized cells is their enhanced ability to detect potential cytotoxic drugs that require metabolic activation because primary hepatocytes more closely retain the genotypic/phenotypic makeup and functional activity present *in vivo* [47]. Primary hepatocytes are considered the "gold standard" of hepatotoxicity detection specifically because of their metabolic capabilities that allow for detection of toxicity associated with effects on enzyme activity (inhibition; induction), metabolic bioactivation, and drug elimination. Importantly, drugs known to cause iDILI, but that were not detected or weakly

Table 19.1 Predictivity data for *in vitro* cytotoxicity detection by hepatic cell type

Cell Type	Number of Compounds Evaluated	Toxicant Classes	Endpoints Evaluated	Predictivity Indicators: Sensitivity	Specificity	Positive Predictive Value	Negative Predictive Value
HepG2							
	110 (44, 45)	mixed	NAD(P)H levels; ATP levels; DNA content	67*	ND	ND	ND
	100 (44,45)	mixed	reactive oxygen species formation; reduced glutathione levels; membrane stability	69*	ND	ND	ND
	119 (46)	mixed	DNA synthesis; protein synthesis; reduced glutathione levels; superoxide production; caspase-3 activation; membrane integrity; cell viability	56*	90*	94*	45*
		hepatotoxins only		57*	90*	95*	46*
	240 (46)	mixed	High Content Endpoints: calcium; mitochondrial membrane potential; cell number; nuclear area; membrane integrity	95*	97*	98*	92*
		hepatotoxins only		96*	93*	97*	94*
	21 (55)	hepatotoxins	electrical impedance	52*	100*	100*	25*
	212 (54)	Hepatotoxins	ATP	~80	~40	ND	ND
HepaRG							
	21 (55)	Hepatotoxins	electrical impedance	53*	100*	100*	26*
THLE							
	679 (43)	Hepatotoxins	MTS/ATP	78*	99.6*	98*	97*
	212 (53)	Mixed (animal)	ATP	56*	95*	98*	32*
	36 (41)	Hepatotoxins	MTS	80*	90*	94*	69*
Primary Hepatocytes							
	344 (49)	Hepatotoxins	mitochondrial membrane potential; reduced glutathione levels; reactive oxygen species formation	67*	100*	100*	58*
	21 (55)	Hepatotoxins	electrical impedance	43–67*	100*	100*	31–38*

*Calculated from reported positive and negative cellular endpoint data: sensitivity (TP/(TP+FN); specificity (TN/(TN+FP); PPV (TP/(TP+FP); NVP (TN/(TN+FN); TP, true positive; FN, false negative; TN, true negative; FP, false positive; PPV, positive predictive value; NVP, negative predictive value; ND, not determined. Reference numbers indicated parenthetically.

detected using immortalized cell lines or in animal models, have caused cytotoxicity in primary human hepatocyte cultures [48, 49, 55]. Additionally, species differences in metabolic pathways and resultant susceptibility to cytotoxicity have been demonstrated by using primary cell cultures [56–58]. So, there is evidence that metabolically competent primary hepatocytes can identify a subset of potential idiosyncratic hepatotoxicants.

The technological advances in cryopreservation techniques have resulted in more consistent availability of high-quality cells resulting in expanded use of primary hepatocytes for toxicity assessments [48]. The use of two-dimensional sandwich cultures in matrix gels has enabled intercellular-matrix connectivity to reestablish cellular polarity that has resulted in improved morphology, including formation of bile canalicular structures, and metabolic and transporter function, as well as improved overall functional longevity and viability [47, 48]. Sandwich systems are now used to evaluate drug effects on bile and xenobiotic transporters as a surrogate for detecting cholestatic hepatotoxicants [39]. There has been continual and exciting progress in the development of bioengineered two-dimensional and three-dimensional microfabricated, organotypic co-culture platforms that allow homotypic and heterotypic cellular interactions and flow-through perfusion to create shear stress and concentration gradients to improve recapitulation of the physiological microenvironment [59, 60]. These platforms have demonstrated genotypic and phenotypic features similar to freshly isolated hepatoctyes, improved cellular functions over conventional culture systems, and longevity over several weeks [60, 61]. The breadth of activity in this area is beyond the scope of this chapter, but it offers some of the most promising and innovative tools to study the complex pathogenesis of toxicities; for a comprehensive review, see LeCluyse et al. [59]. The anticipated advancements of these bioengineered platforms toward improved hepatotoxicity prediction are that *in vitro* refinement of cellular function will result in cell culture-based outcomes that are more physiologically relevant that will enable (1) deconvolution of the complex primary, initiating events; secondary, adaptive events; and multicellular interactions in the development of hepatotoxicity; and (2) *in vitro–in vivo* dose–effect extrapolations and modeling. These systems can take advantage of advances in human stem cell biology. Induced pluripotent stem cell (iPSC)-derived cells from individuals with clinically identified DILI should enable an understanding of the patient-related causes of idiosyncratic toxicity; permit personalized toxicity evaluation; model the effects of drugs on patients with specific diseases; and examine the underlying risk factors that may precipitate adverse effects. That said, currently there are numerous systems in development, but at this time, none have not been extensively evaluated for toxicology applications [62].

Finally, well-established *in vitro* absorption, distribution, metabolism, and excretion (ADME) model systems (e.g., intrinsic clearance, metabolite profiles, trapping studies, and enzyme and transporter studies) offer important, complementary information to help characterize the potential risk of hepatotoxicity, and the data from these analyses should be actively incorporated into these risk assessments.

19.2.3 *In Silico* Models

Computational models are the most nascent of the preclinical predictive hepatotoxicity platforms. A variety of models have been developed to predict global hepatotoxicity (nonspecific, intrinsic, idiosyncratic), as well as to link hepatotoxicity to underlying histological phenotypes, transcriptional profiles, and underlying mechanisms, and to compare species-specific effects.

19.2.3.1 Global Hepatotoxicity Models Computational models that have been developed to predict global DILI are commonly built with chemical-structural and physicochemical descriptors and are statistics based (quantitative structure activity relationship, (QSAR)) or expert-based methods. Excellent reviews of the models developed in the past decade have been recently published, and the models are summarized in Table 19.2 [63, 64]. The modeling methods described are numerous and predominantly statistics based. Similarly, the specific study objectives and data types used are diverse. In Particular, several models use *in vitro* or animal data to predict intrinsic hepatotoxicity [64–68]. Other models use a mixture of *in vitro*, animal, and human data to predict DILI and do not distinguish between intrinsic and idiosyncratic categories, or they use human-specific data and specify models for idiosyncratic toxicity [69–76]. The overall results have demonstrated a range of ability to predict hepatotoxicity with several models demonstrating sensitivity and specificity values of >70%. For example, one QSAR model developed from the data mining of published literature for hepatotoxicity descriptors derived from *in vitro* and *in vivo* (animal and human) data demonstrated concordance (predictivity) values that were comparable with the published literature estimations for specific *in vitro* or *in vivo* surrogate models, suggesting the potential value in this approach to predict outcomes for currently employed preclinical models [69].

19.2.3.2 Mechanism-Specific Models Various mechanism-specific models have also been developed to predict drug effects on cellular targets, e.g., mitochondrial toxicity or transporter inhibition. [77,78] Other models have focused on ADME-related contributions to liver toxicity including tissue distribution, biotransformation and detoxification, extrapolation of in vitro to in vivo exposure–effect relationships, and prediction of differences in species sensitivities to hepatotoxicants [79, 80]. Several groups have developed models to recapitulate the cellular-level zonal differences inherent in the hepatic lobule (“virtual livers”) to be used to refine in vitro–in vivo extrapolation of ADME and pharmacokinetic (PK) properties and whole-body exposure–effect relationships. [81] These modeling systems could potentially be applied to elucidating species-specific susceptibility to hepatotoxicity and in vitro–in vivo extrapolation of exposure–effect relationships. There is also an opportunity to leverage more traditional physiologically based pharmacokinetic (PBPK) models to evaluate potential species differences in biotransformation and detoxification [82–84].

Although computational models to predict hepatotoxicity are not yet widely used, the continued development and testing of these models is creating a foundation

Table 19.2 Summary of computation models of hepatotoxicity

Model Objective	Method	Training Set Compounds (data source)	Specificity (%)	Sensitivity (%)	Concordance (%)
General (intrinsic) (65)	Statistical: recursive partitioning (1D molecular & 2D structural descriptors)	382 (*in vivo* data) (animal and human)	75–90	76–78	76–85
General (66)	Statistical: Soft Independent Modeling of Class Analogy and Comparative Molecular Field Analysis	654 (*in vitro* data)	NR	NR	52%
Intrinsic and iDILI (94)	Statistical: SVM, discriminant analysis, decision tree, neural network, logistical regression (comparison of models)	7 (*in vitro* data) (human or non-human primate hepatotoxicants)	NR; Positive Predictive Values: 82–100	NR; Negative Predictive Values: 75–83	NR
iDILI (model comparison) (71)	3 Statistical: Linear Discriminant Analysis; Artificial Neural Networks; OneR machine learning (3D radial distribution function descriptors)	74 (human data)	84–93; 80–90; 98–100	81–88; 75–92; 76–77	86–91; 78–91; 81–84
Hepatobiliary injury (72, 73)	Statistical software: MC4PC; BioEpisteme; MDL-QSAR; LPDM (fragment and whole molecule descriptors)	1660 (human data)	avg 87; two program consensus	avg 39; two program consensus	89
	Expert: Derek for Windows	78	56		NR
General (intrinsic) Hepatotoxicity (model comparison) (67)	Statistical: weighted feature significance (WFS); naive Bayesian; sequential minimal optimization (SMO) (fragments)	1755 ((RTECS) – animal data)	NR	NR	67–83; 62–86; 61–99

(*Continued*)

Table 19.2 (*Continued*)

Model Objective	Method	Training Set Compounds (data source)	Specificity (%)	Sensitivity (%)	Concordance (%)
DILI (human versus rodent species-specific effects) (69)	Statistical: hierarchical clustering & support vector machines (2D fragments & molecular descriptors)	951 (*in vitro* & *in vivo*) (rodent and human)	NR	NR	56–73
iDILI (75)	Statistical: machine learning, Bayesian classification (2D descriptors & extended connectivity fingerprints)	295 (human data)	62–65	54	59
General Hepatotoxicity (74)	Statistical: k-nearest neighbor (topological and molecular descriptors)	490 (human serum biomarkers)	AST:96; ALT: 89; composite: 74	AST: 88; ALT: 60; composite: 74	NR
General Hepatotoxicity (70)	Expert: structural alerts	626 (*in vivo*) (animal and human)	73	46	56
DILI (76)	Statistical: k-nearest neighbor; support vector machines; naïve Bayesian (2D structure and molecular descriptors)	1087 (human data)	30–81	80–96	55–88
General Hepatotoxicity (model comparison with transcriptional profiling data) (68)	Statistical: k-nearest neighbor; support vector machines; random forest; distance weighted discrimination (chemical, molecular and toxicogenomic (TGX) descriptors)	127 (*in vivo* rat)	60–77; with TGX 77–84	45–56; with TGX 57–67	55–61; with TGX 68-77
General Hepatotoxicity (Histological Subtypes) (64)	Statistical: recursive partitioning	1380 (*in vivo*) (rat and mouse)	73	88	81

NR, not reported; RTECS, Registry of Toxic Effects of Chemical Substances; Reference numbers are indicated parenthetically.

of experience that demonstrates their theoretical feasibility and identifies areas for continued refinement.

19.3 KEY CHALLENGES TO PREDICT IDIOSYNCRATIC DRUG-INDUCED LIVER INJURY IN THE DRUG DISCOVERY AND DEVELOPMENT PHASES

Idiosyncratic drug-induced liver injury is a diagnosis by exclusion. There are no definitive biomarkers, and the underlying pathogenesis is largely unknown, so it cannot be easily predicted clinically. Without clear understanding of the mechanism of iDILI or definitive biomarkers, it is not possible to develop robust, predictive preclinical models. As such, there are no definitive preclinical models of iDILI. Thus, the key challenge is that iDILI cannot be easily predicted and, therefore, reliably mitigated either preclinically or clinically.

Regarding preclinical models, a core assumption is that surrogate models are by nature imperfect and wholly dependent on the knowledge and data at hand to create and refine them, and the available technology to implement them. Therefore, although each preclinical model platform has unique technical limitations, they all have the same fundamental need for accurate and quality data and mechanistic understanding of physiology and underlying disease/toxicity processes to improve their predictive value.

19.3.1 Data Quality and Accuracy Challenges

Accurate assessment of translatable risk to humans from results obtained with preclinical models is difficult to determine based on the limited publically available data. For example, when evaluating clinical attrition resulting from toxicity, it is often not clear whether a particular toxicity was identified in either *in vitro* or *in vivo* preclinical studies, but the drug advanced into clinical trials on the basis of a benefit–risk analysis or a wide safety margin obtained in those studies; both are common scenarios. Access to clinical trial data is highly limited, and gathering data is laborious and involves cobbling together information from various sources. As a result, conclusions drawn are based on incomplete data and therefore are, in part, inferential or empirical. Finally, as recognized by the U.S. Food and Drug Administration (FDA) Critical Path Initiative, there is the need for more standardization of data collection to allow development of effective databases that contain preclinical and clinical data, and better tools to interrogate large datasets across different preclinical and clinical studies through meta-analysis [85]. These databases are needed both within pharmaceutical companies and across regulatory agencies. The greatest impact in advancing the prediction of iDILI will come from evaluating larger datasets versus on a drug-by-drug basis; therefore, establishing sustainable and codified mechanisms to share data across industry, academia, and government will be necessary. Greater accessibility to more complete preclinical and clinical trial data for a particular toxicity can be used to identify the possible risk factors and to formulate testable hypotheses. The data generated would be used to improve existing models or to develop new models. The more that our models demonstrate the ability to accurately predict a variety of known mechanisms of toxicity or physiology, the more confidence we will gain in preclinical models to predict and characterize mechanisms that have yet to be formally identified in humans.

Coupled with greater access to data is the need for robust and quantitative measurements that will enable movement from empirical to evidence-based evaluation of the predictive and translatable utility of preclinical models. Currently, there is tremendous variability in not only the assay methods (particularly *in vitro*) used but also the criteria used to define "hepatotoxicants" versus "non-hepatotoxicants" and "positive" versus "negative" outcomes. Equally important are the weaknesses in the statistical analyses used to assess predictive value [86–89]. Together, this makes utilization of the entire body of published investigative work for such assessments difficult and hampers the ability to draw robust and accurate conclusions needed to reveal gaps and provide clear direction toward improving models. Establishing baseline best assay practices and statistical analysis standards is essential to enable robust determination of the predictive value of preclinical models [90].

Another reality is the imprecise method of diagnosing DILI clinically. There are no specific biomarkers for DILI, and there is a lack of widely accepted and used causality assessments, which further hinders the possible identification of more refined risk factors and biomarkers for DILI [16, 91, 92]. Furthermore, it is common for traditional biomarkers of liver injury (e.g., ALT and aspartate transaminase (AST)) to be elevated during clinical trials with drugs that eventually have outcomes defined as idiosyncratic, but only a small subset of patients (e.g., 1/10,000–100,000) are susceptible to developing severe, life-threatening liver injury. Currently, there is no established or routine process to prospectively investigate early clinical elevations in liver injury markers to gain insight into an underlying mechanism driving this susceptibility. This limits the development of predictive preclinical models. The incorporation of prospective, minimally invasive sampling into clinical trial designs when there is an identified hepatotoxicity risk or when one is identified clinically will help to build datasets from controlled trials that can be used to evaluate underlying drivers of toxicity.

19.3.2 Preclinical Model Platform Challenges

19.3.2.1 Animal Models Species differences in physiology, in particular, metabolism/elimination pathways, and both target and off-target sensitivity and effector pathways, are well established and considered the major driver of low concordance for predicting human hepatotoxicity. An accurate assessment of the predictivity of animal models for hepatotoxicity outcome in humans is not clear based on the limited literature and limitations of the studies to generate statistical outcomes. Olson et al. evaluated 150 drugs and reported a positive concordance of ~50% [22]. Analysis of another dataset of 69 drugs reported a positive concordance of >80%, with a false-negative rate of ~30% [93, 94]. However the true positive predictive (positive likelihood) and negative predictive (negative likelihood) values cannot be properly calculated based on missing data or inconclusive data [89]. Therefore, the true power of animal studies to predict human hepatotoxicity remains unknown.

Similarly, various animal models, e.g., disease models, may not accurately recapitulate the phenotypic changes identified in humans, and therefore, the underlying mechanisms may be different and uninformative if applied toward improving predictions or identifying risk factors of iDILI [23]. Lastly, idiosyncratic DILI is a rare event, and neither animal studies nor clinical trials are powered to detect such rare events. These studies can only detect overt toxicities, and more rare or subtle toxicities may be classified as "outliers" or "inherent animal/human variability." This limitation is difficult to overcome; however, in a situation where there is a dose-related increase in the "variability" of a liver injury signal without overt liver injury, further analyses may provide clues related to risk factors. For example, in this instance, using either specific animal models, e.g., transgenic mice or comparing outcomes in different strains, or *in vitro* models may help to identify underlying differences that can be used to formulate mechanistic hypotheses. This type of enhanced characterization can also be applied to drugs for which a clear signal for hepatic injury was identified (1) only at higher doses, (2) in one species, or (3) *in vitro* but not *in vivo*. In these instances, further characterization may lead to a more refined and translationally relevant risk assessment versus simple target organ hazard identification.

19.3.2.2 In Vitro Cytotoxicity Models The technical limitations of these artificial models are well recognized [10,59]. Traditional, static culture systems are maintained under nonphysiological conditions and are composed of single-cell populations; as such, they will never have the ability to recapitulate complex, integrated multicellular interactions and mechanisms of toxicity, such as immunological responses. Immortalized cells do not have the full complement of physiological functions present *in vivo*, in particular, metabolic activities, and primary cells lose metabolic activity and viability over time. The anticipated promise of newer bioengineered organotypic platforms is to recapitulate physiologically relevant microenvironments that may address, in part, many of these issues.

With regard to cytotoxicity assessments, no standard assay methods, defined endpoints, nor standard positive criteria have been established and the reported approaches and assessments are quite varied. Therefore, it is difficult to compare across the reported *in vitro* models to assess their respective predictive value or that may be the most useful method (Table 19.1). For example, the measurement of multiple endpoints has been an important advancement in elucidating mechanistic drivers of toxicity and has added great value, but is it unclear how to best use multi-endpoint data to improve the prediction of hepatotoxicity over single-endpoint methods. A variety of endpoints have been examined, and compelling arguments have been put forward on the most useful; however, there is no consensus on which endpoints add the most value for predicting hepatotoxicity. Several published studies report the output data as binary (toxic or nontoxic) based on a positive outcome in one of the parameters; this approach alone may not provide added predictive value. Others have attempted to integrate multiple endpoints and to consider effects over a concentration range to enhance rigor and to predict exposure–effect relationships [46,94,95]. With regard to drug lead optimization, there seems to be great variability in how the data are interpreted and used for decision making. Given the diversity of approaches, there would be value in collectively and objectively assessing best practices.

It is quite common practice to extrapolate *in vitro–in vivo* outcome relationships by comparing cytotoxicity outcome in primary hepatocytes from animal species with *in vivo* outcomes and then subsequently projecting human outcome (i.e., hazard identification). Although some investigators are beginning to evaluate alternative and more refined methods to extrapolate *in vitro–in vivo* exposure–effect relationships, for all practical purposes, this ability is still largely lacking and remains an exceptionally important gap toward informing risk translation [52,53,95].

19.3.2.3* In Silico *Models Development of computational models to predict iDILI is hampered by highly available, quality data, including reliable and specific biomarkers of iDILI both preclinically or clinically [63]. The incomplete mechanistic understanding of iDILI and paucity of predictive models for hepatotoxicity also limit the ability to create useful models. For example, all reported computational models rely on currently available endpoints or biomarkers to define a toxicity outcome; however, a multiplicity of diverse structures can result in the same toxicity outcome, thereby limiting the ability to refine predictive structure-activity relationships. In addition, numerous descriptors cannot easily be taken into consideration to refine models because their role in the development of iDILI is not fully understood, e.g., genetic polymorphisms.

With regard to structure-based models, many have narrow applicability domains that are limited by the physicochemical and structural data available for making predictions, and therefore, they may be more useful when established for a similar compound class or selective mechanisms versus as global models [66,70]. Furthermore, based on the currently reported models, it seems unlikely that global models to predict DILI based on structural properties alone, including expert-based structural alerts, will be extremely useful [63].

Another inherent limitation of computational models is their inability to detect toxicants with unknown or novel mechanisms; therefore, without a more complete understanding of the components of the underlying pathogenesis of toxicity, the predictive power of computational models to predict DILI is blunted.

19.4 RECOMMENDED STRATEGY FOR INTEGRATED IDILI RISK ASSESSMENT AND MITIGATION PROCESSES

When considering an integrated risk assessment strategy for iDILI, the following caveats need to be considered: (1) There are no standardized preclinical models for iDILI; (2) existing preclinical models are based on the current understanding of risk factors and identified cellular mechanisms of toxicity (e.g., multifactorial), and there is no consensus on endpoints; and (3) the accurate predictive value for *in vivo* and *in vitro* models is unknown. Thus, attempting to de-risk iDILI will continue to be a significant challenge and a simple solution will not be forthcoming. Instead, we safety scientists will need to shift our thinking to integrate new approaches and, equally importantly, be committed to continued iterative and prospective evaluation of new models and strategies. The suggested strategies detailed as follows are based on the current evidence that iDILI is multifactorial. These strategies should be achievable for both larger pharmaceutical companies and smaller biotechnology companies because many of the suggested models are well established and are already available within companies or through commercial laboratories. The goals of a drug discovery and development iDILI risk assessment and mitigation strategy include (1) enabling rational decision making related to drug advancement based on robust issue characterization, and (2) providing information toward improving the overall field of pharmaceutical safety assessment to minimize the safety risks for all patients.

A successful risk assessment and mitigation strategy will evaluate drug-related factors that can contribute to toxicity and identify risk factors for iDILI, and as a result, it will not only remove the drugs with the most risk but also will create an "*iDILI susceptibility/risk profile*" for each compound that is advanced clinically. The latter profile can be used to create more detailed safety monitoring plans and, more importantly, help to influence clinical designs and promote ethically-based clinical hypothesis testing regarding identification of hepatotoxicity risk factors, which is currently not a typical feature of clinical trials. Data obtained by these hypothesis-driven clinical trials can be used to refine further preclinical models.

Activities to enable this strategy can be grouped into three categories: lead optimization and candidate identification, candidate characterization and risk profile determination, and clinical considerations. The strategy draws from currently reported approaches that have demonstrated value in mitigating toxicity and define the profile of an optimal target candidate (Figure 19.1).

19.4.1 Lead Optimization and Candidate Identification

The early discovery phase is the period when there is intense medicinal chemistry design activity to identify the drug with the optimal characteristics related to pharmacology, pharmaceutics, ADME, and safety. At this early stage, target candidate criteria are established to guide the lead optimization screening strategy for the significant characteristics desired. *In silico* and *in vitro* assays are most commonly used as first and second tier assessments that result in selection of lead scaffolds or compounds-of-interest. With regard to selecting drugs with minimal DILI risk, the major screening goals are to reduce promiscuity, AMDE risks, and intrinsic cytotoxicity (Figure 19.1, Table 19.3). These safety lead optimization goals are not unique to DILI but can be applied to reducing a variety of other toxicities as well, with the optimal candidate having fewer or less severe issues.

19.4.1.1 Reducing Promiscuity The purpose of minimizing promiscuity is to reduce the potential for off-target effects, which are most commonly implicated in iDILI. Highly promiscuous drugs have a higher failure rate when compared with successfully marketed drugs [96, 97]. The physicochemical parameter lipophilicity

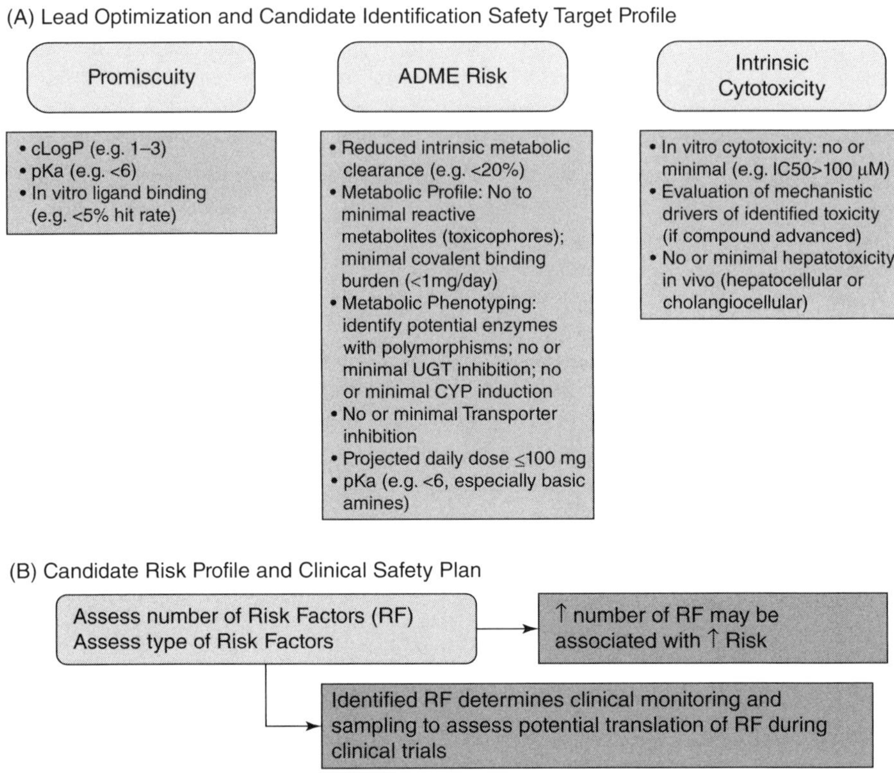

Figure 19.1 iDILI risk assessment and mitigation strategy components. (*See insert for color representation of the figure.*)

(cLogP) and, to a lesser degree, ionization state (basic, pKa > 6) and molecular size (>500 Daltons) have been correlated with increased promiscuity and *in vivo* toxicity [98, 99]. Review of the literature indicates that, with respect to safety, the optimal range for lipophilicity is a cLogP between 1 and 3 [99]. Leeson and Springthorpe suggested the application of a "ligand-lipophilicity efficiency" (LLE) calculation as part of lead optimization activities [98]. This LLE is the difference between the potency (IC_{50} or K_i) and the lipophilicity (cLogP or LogD) with a value of >5–7 as the suggested target.

Other parameters to consider are associated with ionization state. Compounds with high pKa values, e.g., >6, are highly ionized at physiological pH and tend to interact with membrane phospholipids and become trapped in acidic organelle compartments, i.e., mitochondrial intermembrane space and lysosomes, where they can cause dysfunction [100]. Compounds with low topological polar surface area (TPSA) can more readily cross membranes and distribute into tissues. Hughes et al. identified a TPSA of <75 (low charge) associated with increased risk of adverse events *in vivo*. These physicochemical parameters should be part of the chemistry design considerations for safety optimization [101].

Table 19.3 Suggested candidate profile targets to mitigate iDILI

Endpoint Evaluated	Targeted Value	Rationale	Clinical Risk
cLogP	1–3	Reduce promiscuity: lipophilicity strongly associated with promiscuous receptor binding	Off-target toxicity leading to iDILI
Ligand-Lipophilicity Efficiency (LLE)	>5–7		
pKa	<6	Reduce cell membrane and acidic cellular organelle drug accumulation	Basic amines often associated with toxicity
In vitro Ligand Binding Screens (e.g., safety pharmacology, kinases)	target hit rate: ≤5% of all targets tested	Reduce promiscuity: panel serves as a surrogate for off-target binding	Off-target toxicity leading to iDILI
Cytotoxicity Assay (e.g., ATP levels)	≥100 μM	Reduce intrinsic cytotoxicity	Direct hepatotoxicity
In Vitro Mechanistic Endpoints (e.g., mitochondrial dysfunction; reduced glutathione)	no to minimal change from controls	Identify underlying mechanistic drivers of toxicity	Identify potential underlying risk factors for iDILI
Covalent Binding	GSH Trapping: 0 to minimal Covalent Binding: 0 to <200 pmol eq/mg protein Covalent binding burden (dose adjusted) <1 mg/day Mechanism-dependent CYP450 inhibition: 0 to minimal	Surrogate for the formation of reactive metabolites and covalent binding to cellular proteins	Direct hepatotoxicity or immune-mediated iDILI
In vitro (and *in vivo*) Intrinsic Clearance	stable in liver microsomes/hepatocytes	Reduce potential for generating reactive metabolites & high doses; determine possible non-CYP450-mediated metabolism	Direct hepatotoxicity; immune-mediate iDILI

(*Continued*)

Table 19.3 (*Continued*)

Endpoint Evaluated	Targeted Value	Rationale	Clinical Risk
Metabolic Profile Evaluation	No potentially reactive metabolites (toxicophores) identified	Evaluate metabolic pathway and potential reactive metabolite formation; identify potential species differences	Direct hepatotoxicity
Metabolic Phenotyping	Identify major metabolizing enzymes	Evaluate whether single or multiple CYP450 enzymes involved; characterize potential species with polymorphisms	Alteration in major metabolizing enzyme may predispose to hepatotoxicity
Bile Salt/Xenobiotic Transporter Inhibition Assays (NTCP, OATP, BSEP, BCRP, MRP2, MRP3)	No or minimal inhibition	Reduce potential cholestatic effects or hepatocellular accumulation of reactive drug or metabolites	Cholestasis or direct hepatoxicity
UGT Inhibition	No or minimal	Reduce potential for altered bilirubin elimination and cholestasis	Cholestasis
Animal Studies	No or minimal hepatic toxicity (hepatocellular or cholestatic)	Reduce potential for direct hepatoxicity	Direct hepatotoxicity
Projected Daily Dose	<100 mg	Greater than 75% of drugs with a daily dose ≥50 mg were associated with DILI; the risk was increased for highly metabolized drugs	High drug exposure

GSH, reduced glutathione.

In vitro ligand-binding screens are composed of endogenous receptors, ion channels, and enzymes (including kinases) with known physiological roles. These screens can be used to obtain a general assessment of promiscuity based on the overall binding ("hit") rates across the assay panel, as well as insight into potential issues related to specific endogenous ligands that are bound. Although thresholds for promiscuity are somewhat arbitrarily defined, Azzaoui et al. defined a compound's target hit rate (THR) as the ratio of the number of targets bound (at >50% inhibition at 10 μM) to the total number of targets tested, with "highly promiscuous" compounds having a THR of ≥20% and "selective" compounds having a THR of ≤5% [97]. Databases containing ligand–drug interaction information and computational models (e.g., CEREP BioPrint^{TM}) have been developed based on structural similarity to predict off-target binding of drugs, and although potentially biased toward a particular chemical space (e.g., G-protein coupled receptor (GPCR) targets), they may be useful for informing chemical design [96]. An additional computational approach was recently described that expands the biological target space for potential off-target ligands that are unrelated to conventional off-target assessments [102].

19.4.1.2 Reducing ADME Risks Metabolism and the formation of reactive metabolites (RMs) are recognized as important risk factors for the development of iDILI [6, 15]. Recent studies evaluating the potential contribution of covalent binding and dose to hepatotoxicity outcome provide evidence for these risk factors [103–105]. The findings from these studies can be applied toward compound lead optimization and risk assessment for hepatotoxicity. For example, based on the work of Sakatis et al., drugs with a dose of ≥100 mg/day that also demonstrated either glutathione adduct formation (medium or high intensity), metabolism-dependent cytochrome P450 (CYP) inhibition (>5 × decrease in IC_{50}), or covalent binding (>200 pmol eq/protein) predicted >80% of clinical hepatotoxicants [104]. Thompson et al. distinguished hepatotoxicants from non-hepatotoxicants (78% sensitivity; 100% specificity) through an integrated assessment using the combination of a covalent binding burden (the product of maximum dose and fraction metabolized leading to covalent binding) of >1 mg/day coupled with two or more positive *in vitro* mechanistic (mitochondrial dysfunction; bile salt export transporter (BSEP) or multidrug resistance-associated protein 2 (MRP2) inhibition) or cytotoxicity endpoints [105]. Thus, evaluation of covalent binding potential coupled with projected dose and hepatotoxicity/cytotoxicity endpoints shows utility as part of an integrated risk assessment for DILI.

To screen for cholestasis potential, compounds can be assessed for inhibition of various hepatic transporters such as BSEP, MRP2, Na^+-taurocholate cotransporting polypeptide (NTCP), and organic acid-transporting polypeptide (OATP) [106]. Morgan et al. identified potent BSEP inhibition as IC_{50} of ≤25 μM associated with hepatotoxicants and with IC_{50} > 100 μM distinguished non-hepatotoxicants [14]. Functional effects of drugs on transporters can be evaluated in primary hepatocyte sandwich cultures [39]. Additionally, inhibition of xenobiotic efflux transporters may result in accumulation of a drug within hepatocytes, resulting in hepatotoxicity for an intrinsically cytotoxic drug that would not have been predicted based on plasma drug concentrations. Inhibitors of uridine diphosphate-glucuronyl transferase (UGT) can potentially cause hyperbilirubinemia as this is the conjugating enzyme for the endogenous processing of bilirubin [107, 108]. Evaluating whether drugs are inhibitors of UGT or transporters as potential risk factors supports the clinical findings that drugs eliminated via biliary clearance have a higher incidence of jaundice, including drugs with minimal hepatic metabolism, and that transporters and phase II genetic polymorphisms have been identified in some cases of iDILI [17, 20].

In vitro intrinsic clearance estimated by primary hepatocytes or liver microsomes can indicate the rate of metabolism of a drug and can be used to generate the metabolic profile. Epidemiological data indicate that drugs with extensive hepatic metabolism ($\geq$50%) have a higher association with clinically elevated ALT values (>3 × ULN), hepatic failure, and mortality associated with iDILI [17]. Thus, a high intrinsic clearance may indicate an added risk factor for iDILI. Furthermore, disconnects between intrinsic clearance estimated from liver microsomes versus hepatocytes may indicate other, non-CYP450 enzyme involvement in drug metabolism. Likewise, *in vitro–in vivo* clearance disconnects may also indicate non-CYP450 enzyme involvement in drug metabolism. Determining which non-CYP450 enzymes may be involved may result in the identification of metabolizing enzymes with known polymorphisms or with species differences, such as epoxide hydrolase, aldehyde oxidase, *N*-acetyltransferase, or flavin-containing monooxygenase (FMO), that could impact outcomes of toxicity studies or clinical trials [56, 109, 110].

Compound structure, metabolic profile, and metabolic pathway prediction may provide clues for potential reactive substructures (toxicophores) and for assessing the potential impact of altering a major metabolic pathway, which may result in accumulation of a metabolite or shunting to another metabolic pathway that may be predicted to be toxic. Metabolic phenotyping information to determine which CYP450 enzymes are important in metabolism and, importantly, whether a particular CYP450 may be responsible for the major metabolic route, will help determine the risk impact of inhibiting or inducing that particular enzyme, or whether there is a polymorphism that should be considered as a possible risk factor.

Minimizing metabolism and the formation of reactive metabolites that form covalent bonds, as well as minimizing the overall drug dose, seem to be important features for reducing hepatotoxicity risk. Likewise, minimizing the potent inhibition of key transporters may also reduce risk for hepatic injury especially with regard to drugs expected to be eliminated by biliary clearance or for intended co-administration with drugs that are eliminated through the bile. The full metabolic profile of compounds-of-interest should be evaluated from a safety perspective to identify additional potential risk factors related to the metabolic pathway and other effects on metabolizing enyzmes that would alter drug or metabolite concentrations.

19.4.1.3 Reducing Intrinsic Cytotoxicity Determining whether a compound is intrinsically cytotoxic is one of the key parameters for assessing DILI risk. Cell viability (e.g., ATP levels) can be used as a first tier assessment for comparing chemical scaffolds or candidate compounds. Use of primary hepatocytes provides information on the cytotoxicity of the parent and its metabolites. Use of immortalized cells, with limited metabolic capacity, usually does not provide information on the cytotoxic potential of metabolites (except for THLE cell lines), and potential antiproliferative effects observed with immortalized cells (cytostatic) need to be differentiated from direct cytotoxicity. Compounds-of-interest that have some degree of cytotoxicity can be further characterized in primary hepatocytes using various mechanistic endpoints. This tiered approach enables scaffold or compound selection and investigation of underlying drivers of cytotoxicity. Primary hepatocytes can also be used to investigate hepatotoxicity observed in animal studies and to evaluate the *in vitro*-to-*in vivo* translation and cross-species sensitivity to an identified hepatotoxicant.

Extrapolation of *in vitro* outcomes, e.g., IC_{50} values, to *in vivo* exposure–effect relationships that can inform either animal studies or clinical trials is a limitation of cytotoxicity assays, and currently there are no commonly used methods to accomplish this *in vitro–in vivo* extrapolation. Typically, the output of these *in vitro* models is binary, i.e., toxic or not toxic, and is most commonly used for ranking compounds. However, methods have been proposed to predict useful *in vitro–in vivo* exposure–effect extrapolations that may be useful to predict *in vivo* risk [52, 53, 95]. For example, an exposure relationship comparing the ratio of a projected *in vivo* concentration, to an *in vitro* LC_{50} could be used to predict a high risk potential (i.e., $C_{max}/LC_{50} > 1$) for an adverse outcome versus a low risk potential (i.e., $C_{max}/LC_{50} < 0.01$) [53]. Using an alternative method, drugs with an $LC_{50} \leq 50$ μM were shown to have fivefold increase of demonstrating toxicity *in vivo* if a C_{max} of 10 μM (total drug) was achieved when compared with a drug with an $LC_{50} > 50$ μM [53]. Using these approaches, the LC_{50} can be used to predict at what C_{avg} (average concentration for $AUC_{0\text{-}24}$) or C_{max} toxicity may be observed compared with projected efficacious C_{max} values, i.e., projected safety margin. This analysis can be used either to prioritize compounds or to set dose ranges in toxicology studies. The lead optimization profile of the selected candidate can also inform the design of the *in vivo* animal studies or the choice of *in vivo* models to improve characterization of risk. For example, if a compound is selected that has evidence of cytotoxicity associated with mitochondrial dysfunction, a transgenic mouse model that has enhanced susceptibility to mitochondrial toxicants (i.e., superoxide dismutase (SOD) +/−) can be used to evaluate translatable risk to humans [28–31].

In summary, lead optimization efforts to mitigate the potential for iDILI should minimize promiscuity, thus, reducing potential off-target hits; minimize metabolic risks (especially reactive metabolite formation); and minimize intrinsic cytotoxicity. Furthermore, the lead optimization profile for iDILI can be incorporated to inform candidate selection and design of IND-enabling animal studies. It is possible that there may be a disconnect between the *in vitro* lead optimization profile of a selected candidate and the identification of hepatotoxicity in animal studies. In cases where animals do not develop hepatotoxicity, the *in vitro* data should not be discounted as not predictive of an *in vivo* outcome because many of the *in vitro* assays use human reagents and therefore may be relevant to human risk. Similarly, if compounds with a clean *in vitro* lead optimization profile cause hepatic injury in animals, the potential discrepancy should be further investigated to characterize the injury and to determine whether there is a species-specific effect. It is important that all of these parameters be captured and used as part of the "susceptibility/risk profile" for a compound especially because the underlying pathogenesis for iDILI has not been fully elucidated.

19.4.2 Candidate Summary iDILI Susceptibility/Risk Profile and Clinical Considerations

For each candidate chosen for advancement, a "Summary iDILI Susceptibility/Risk Profile" could be compiled and discussed with the clinicians and drug safety scientists to help inform the clinical trial design. In particular, each of the endpoints listed in Table 19.3 would be reviewed and those endpoints outside an identified target value "flagged" and the liability explained as it relates to a potential risk factor for iDILI. For example, a decision may be made to advance a compound that has multiple, identified risk factors but no hepatotoxicity identified in animal studies (e.g., troglitazone) (see an example illustration in Table 19.4). In this case, the Summary iDILI Susceptibility/Risk Profile would indicate the potential risk for iDILI based on several risk factors: dose, high metabolic rate, reactive metabolite formation and high covalent binding, high lipophilicity, intrinsic cytotoxicity, and potential for cholestatic effects identified from *in vitro* findings. The potential value of this approach is that it is based on the current multideterminant hypothesis for iDILI and represents a more comprehensive, transparent, and integrated risk assessment that provides both qualitative (type of risk factors identified) and quantitative (number of risk factors identified) information. These data can be used by clinicians to design studies that may incorporate prospective, minimally invasive sampling (e.g., for pharmacogenomic or metabonomic analyses) of clinical trial participants to evaluate iDILI risk factors. Additionally, because all data used to assess iDILI risk is captured, over the longer term, this more complete dataset has the potential to be leveraged for meta-analysis to determine the predictive value of this approach and these preclinical models. Currently, these types of data are not routinely included as part of the overall risk assessments for drugs submitted for IND/CTAs, and so this information is lost. It is likely that many pharmaceutical companies use these integrated assessments with theoretical risk endpoints for internal decision making; however, there may be reluctance to include them in health authority documents without some "agreed upon safe haven" to collect these data for prospective assessment of value. This scenario is unfortunate as it results in continuation of current practices and a lack of advancement of scientific knowledge.

19.5 CONCLUSIONS AND FORWARD-LOOKING CONSIDERATIONS

Testing in the preclinical phase will always be limited as a result of the use of surrogate models and their inherent imperfections. Therefore, it is of paramount importance to identify and codify processes accepted by each company, the pharmaceutical industry as a whole, and health authorities to enable continued and sustainable joint evaluation of preclinical models through the transparent review of preclinical and clinical data, as a mechanism to understand the true concordance and enable translational

Table 19.4 Example comparison of candidate iDILI susceptibility/risk profiles for iDILI susceptibility/risk

Endpoint Evaluated	Troglitazone	Rosiglitazone	Pioglitazone
cLogP (Micromedex 2.0)	5.3	2.8	3.3
pKa (Micromedex 2.0)	6.34	6.8	6.4
promiscuity (selectivity) (*In Vitro* Ligand Binding Screens, e.g., pharmacology and kinase)	UNK	UNK	UNK
Cytotoxicity Assay (41) (e.g., ATP levels)	Yes	No	No
In Vitro Mechanistic Endpoints (13, 26) (e.g., mitochondrial dysfunction; reduced glutathione)	mitochondrial permeability & potential; oxidative stress	None	minimal, mitochondrial potential
Covalent Binding (41)	118 pmol/mg protein	11 pmol/mg protein	11 pmol/mg protein
RM Body Burden (CVB × Dose)	70,800	88	495
In Vitro (and *in vivo*) Intrinsic Clearance (Micromedex 2.0)	High	High	High
Metabolic Profile Evaluation (Micromedex 2.0)	quinone metabolite (reactive metabolite); conjugated species	N-demethylation and hydroxylation of parent; conjugated species	hydroxyl- and keto-derivatives of parent; conjugated species
Metabolic Phenotyping (Micromedex 2.0)	CYP3A4	CYP2C8	CYP3A4 and CYP2C8
Bile Salt/Xenobiotic Transporter Inhibition Assays (105) (NTCP, OATP, BSEP, BCRP, MRP2, MRP3)	Yes, BSEP	Yes, BSEP	Yes, BSEP
UGT Inhibition	UNK	UNK	UNK
Animal Studies	no hepatotoxicity	no hepatotoxicity	no hepatotoxicity
Projected Daily Dose (41)	600	8	45

risk assessments and assay improvements. It is highly likely these data are already collected but not reported as an integrated assessment of iDILI risk or in an available format for meta-analysis. Without question, data sharing needs to be conducted in a manner that protects patient identity, pharmaceutical proprietary information, and is not perceived by the public as collusion. In fact, this model exists and is exemplified by numerous consortia (e.g., ILSI/HESI, Critical Path Institute, DILI Networks, and MRC Centre for Drug Safety Science and Innovative Medicine Initiative) and could be leveraged to create a process for reviewing nonclinical and clinical data related to iDILI as a common practice. Similarly, a drug withdrawn for iDILI should automatically prompt full review of nonclinical and clinical data by the sponsor and health authority or, alternatively, a dedicated consortium. The outcomes could ultimately be made publically available to improve nonclinical surrogate models, and these consortia could be formally empowered to influence industry practices and regulatory guideline refinements. Lastly, a codified process should be initiated for a systematic and documented review of the nonclinical–clinical concordance of DILI findings for each drug in development. For example, such updates might be part of annual reviews or at significant phases of drug development, such as end of phase II meetings. These data would be included in health authority databases and regulatory documentation processes. There is a similar urgent need for pharmaceutical companies to develop a philosophy and codified process for collaboration among nonclinical and clinical safety scientists and clinicians to perform these same detailed analyses for their drug portfolios.

The current focus to overcome technical limitations, including leveraging of a diverse approach that couples toxicology with other scientific disciplines, such as bioengineering, to bring about innovative solutions to long-held issues is a recognized necessity, but alone it cannot fully impact the needed improvements in prediction. It is also important to continue to investigate and characterize mechanisms of hepatotoxicity so that we can identify risk factors and build-in appropriate endpoints or refine preclinical models or assessment strategies. This too can only come from thorough review of both preclinical and clinical data and investment in follow-up hypothesis-driven scientific investigation, including leveraging knowledge of species differences in target biology or metabolism. Likewise, there is an need to design clinical trials toward more prospective DILI safety assessments. For drugs advancing into clinical trials, prospective procurement and banking of tissue samples, such as blood or buccal swabs that involve minimally invasive procedures, from healthy volunteers and patients enrolled in those trials provides an opportunity to investigate mechanisms or risk factors associated with DILI identified during these controlled trials. This type of prospective sampling is now commonly performed in clinical trials for potential analysis of genetic polymorphisms of enzymes that may alter metabolism (pharmacogenomics) and to improve efficacy of cancer drugs by characterizing tumor tissue from nonresponding patients, which has resulted in significant therapeutic impact [111, 112]. There are examples of prospective trials to assess the incidence and risk factors of DILI led by DILI consortia that have added valuable epidemiological information [20]. The Drug Induced Liver Injury Network (DILIN), a cooperative between the National Institutes of Health (NIH) and academic clinical centers, has established prospective clinical studies enrolling patients with clinically identified DILI to create a registry of clinical information and a repository of biological samples that can be used to evaluate the mechanisms and risk factors of DILI [113]. Finally, closer interaction with DMPK scientists is needed to help develop toxicology-specific strategies to enhance our understanding of *in vitro–in vivo* exposure–effect relationships and PKTD relationships that will not only facilitate *in silico* model development but also provide more useful information for clinical monitoring and safety evaluations.

Applying these approaches to one of the most significant causes of attrition, hepatic injury, could have a significant impact on patient safety and successful drug progression.

REFERENCES

1. Stevens, J.L., Baker, T.K. (2009). The future of drug safety testing: Expanding the view and narrowing the focus. *Drug Discovery Today*, *14*, 162–167.
2. MacDonald, J.S., Robertson, R.T. (2009). Toxicity testing in the 21st century: A view from the pharmaceutical industry. *Toxicological Sciences*, *110*, 40–46
3. Andrade, R.J. Camergo, R., Lucena, I., Gonzalez-Grande, R. (2004). Causality assessment in drug-induced hepatotoxicity. *Expert Opinion in Drug Safety*, *3*, 329–344.
4. Bell, L.N., Chalasani, N. (2009). Epidemiology of idiosyncratic drug-induced liver injury. *Seminars in Liver Disease*, *29*, 337–347.
5. Zimmerman, H. (2000). Drug-induced liver disease. *Clinics Liver Disease*, *4*, 73–96.
6. Kaplowitz, N. (2005). Idiosyncratic drug hepatotoxicity. *Nature Reviews Drug Discovery*, *4*, 489–499.
7. Russmann, S., Kullak-Ublick, G.A., Grattagliano, L. (2009). Current concepts of mechanisms of drug-induced hepatotoxicity. *Current Medicinal Chemistry*, *16*, 3041–3053.
8. Greer, M.L., Barber, J., Eakins, J., Kenna, J.G. (2010). Cell based approaches for evaluation of drug-induced liver injury. *Toxicology*, *268*, 125–131.
9. Li, J., Uetrecht, P.J. Danger hypothesis applied to idiosyncratic drug reactions, in J. Uetrecht (ed.), *Adverse Drug Reactions, Handbook of Experimental Pharmacology*, Springer, New York, NY, 2010, pp. 493–509.
10. Li. A.P. (2004). An integrated, multidisciplinary approach for drug safety assessment. *Drug Discovery Today*, *9*, 687–693.
11. Ulrich, R.G. (2007). Idiosyncratic toxicity: A convergence of risk factors. *Annual Reviews in Medicine*, *58*, 17–34.
12. Boelsterli, U.A., Lim, P.L.K. (2007). Mitochondrial abnormalities–a link to idiosyncratic drug hepatotoxicity? *Contemporary Issues in Toxicology*, *220*, 92–107.
13. Dykens, J.A., Will, Y. (2007). The significance of mitochondrial toxicity testing in drug development. *Drug Discovery Today*, *12*, 777–785.
14. Morgan, R.E., Trauner, M., van Staden, C.J., Lee, P.H., Ramachandran, B., Eschenberg, M., Afshari, C.A., Qualls, C.W., Lightfoot-Dunn, R., Hamadeh, H.K. (2010). Interference with bile salt export pump function is a susceptibility factor for human liver injury in drug development. *Toxicological Sciences*, *118*, 485–500.

15. Park, B.K., Boobis, A., Clarke, S., Goldring, C.E.P., Jones, D., Kenna, J.G., Lambert, C., Laverty, H.G., Naisbitt, D.J., Nelson, S., Nicoll-Griffith, D.A., Obach, R.S., Routledge, P., Smith, D.A., Tweedie, D.J., Vermeulen, N., Williams, D.P., Wilson, I.D., Baillie, T.A. (2011). Managing the challenge of chemically reactive metabolites in drug development. *Nature Reviews Drug Discovery*, *10*, 292–306.
16. David, S., Hamilton, J.P. (2010). Drug-induced liver injury. *Gastroenterology and Hepatology Reviews*, *6*, 73–80.
17. Lammert, C., Bjornsson, E., Niklasson, A., Chalasani, N. (2010). Oral medications with significant hepatic metabolism at higher risk for hepatic adverse events. *Hepatology*, *51*, 615–620.
18. Lammert, C., Einarsson, E., Saha, C., Niklasson, A., Bjornsson, E., Chalasani, N. (2009). Relationship between daily dose of oral medications and idiosyncratic drug-induced liver injury: Search for signals. *Hepatology*, *47*, 2003–2009.
19. Lucena, M.I., Andrade, R.J., Kaplowitz, N., Garcia-Cortes, M., Fernandez, M.C., Romero-Gomez, M., Bruguera, M., Hallal, H., Robles-Diaz, M., Rodriguez-Gonzaleaz, J.F., Navarro, J.M., Salmeron, J., Martinez-Odriozola, P., Perez-Alvarez, R., Borraz, Y., Hidalgo, R. (2009). Phenotypic characterization of idiosyncratic drug-induced liver injury: The influence of age and sex. *Hepatology*, *49*, 2001–2009.
20. Chalasani, N., Bjornsson, E. (2010). Risk factors for idiosyncratic drug-induced liver injury. *Gastroenterology*, *138*, 2246–2259.
21. Peters, J. (2005). Do preclinical testing strategies help predict hepatotoxic potential? *Toxicologic Pathology*, *33*, 146–154.
22. Olson, H., Betton, G., Robinson, D., Thomas, K., Monro, A., Kolaja, G., Lilly, P., Sanders, J., Sipes, G., Bracken, W., Dorato, M., Van Deun, K., Smith, P., Berger, B., Heller, A. (2000). Concordance of the toxicity of pharmaceuticals in humans and animals. *Regulatory Toxicology and Pharmacology*, *32*, 56–67.
23. Ng, W., Lobach, A.R., Zhu, X., Chen, X., Liu, F., Metushi, I.G., Sharma, A., Li, J., Cai, P., Ip, J., Novalen, M., Popovic, M., Zhang, X., Tanino, T., Nakagawa, T., Li, Y., Uetrecht, J. (2012). Animal models of idiosyncratic drug reactions. *Advances in Pharmacology*, *63*, 81–135.
24. Morgan, S.J., Elangbam, C.S., Berens, S., Janovitz, E., Vitsky, A., Zabka, T., Conour, L. (2012). Use of animal models of human disease for nonclinical safety assessment of novel pharmaceuticals. *Toxicological Pathology*, *1*, 1–11.
25. Boelsterli, U.A. (2003) Animal models of human disease in drug safety assessment. *Journal of Toxicological Sciences*, *28*, 109–121.
26. Jaeschke, H. (2007). Troglitazone hepatotoxicity: Are we getting closer to understanding idiosyncratic liver injury. *Toxicological Sciences*, *97*, 1–3.
27. Roth, R.A., Ganey, P.E. (2012). Intrinsic versus idiosyncratic drug-induced hepatotoxicity: Two villans for one? *Journal of Pharmacology and Experimental Therapeutics*, *332*, 692–697.
28. Ong, M.M.K., Latchoumycandane, C., Boelsterli, U.A. (2007). Troglitazone-induced hepatic necrosis in an animal model of silent genetic mitochondrial abnormalities. *Toxicological Sciences*, *97*, 205–213.
29. Ong, M.M.K., Wang, A.S., Loew, K.Y., Khoo, Y.M., Boelsterli, U.A. (2006). Nimesulide-induced hepatic mitochondrial injury in heterozygous SOD+/- mice. *Free Radical Biology & Medicine*, *40*, 420–429.

30. Kashimshetty, R., Desai, V.G., Kale, V.M., Lee, T., Moland, C.L., Branham, W.S., New, L.S., Chan, E.C., Younis, H., Boelsterli, U.A. (2009). Underlying mitochondrial dysfunction triggers flutamide-induced oxidative liver injury in a mouse model of idiosyncratic drug toxicity. *Toxicology and Applied Pharmacology*, *238*, 150–159.

31. Ramachandran, A., Lebofsky, M., Weinman, S.A., Jaeschke, H. (2011). The impact of partial manganese superoxide dismutase (SOD2)-deficiency on mitochondrial oxidant stress, DNA fragmentation and liver injury during acetaminophen hepatotoxicity. *Toxcicology and Applied Pharmacology*, *251*, 226–233.

32. Caparro-Assef, S.M., Salguerio-Pagadigorria, C.L., Bersani-Amado, C.A., Bracht, A., Kelmer-Bracht, A.M., Ishii-Iwamoto, E.L. (2001). The uncoupling of the non-steroidal anti-inflammatory drug nimesulide in liver mitochondria from adjuvant-induced arthritis rats. *Cell Biochemistry and Function*, *19*, 117–124.

33. Harrill, A.H., Watkins, P.B., Su, S., Ross, P.K., Harbourt, D.E., Stylianou, I.M., Boorman, G.A., Russo, M.W., Sackler, R.S., Harris, S.C., Smith, P.C., Tennant, R., Bogue, M., Paigen, K., Harris, C., Contractor, T., Wiltshire, T., Rusyn, I., Threadgill, D.W. (2009). Mouse population-guided resequencing reveals that variants in CD44 contribute to acetaminophen-induced liver injury in humans. *Genome Research*, *19*, 1507–1515.

34. Zheng, M., Shafter, S., Liao, G., Liu, H.H., Peltz, G. (2009). Computational genetic mapping in mice: The ship has sailed. *Science Translational Medicine*, *1*, 3ps4.

35. Zheng, M., Dill, D., Peltz, G. (2012). A better prognosis for genetic association studies in mice. *Trends in Genetics*, *28*, 62–68.

36. Harrill, A.H., Ross, P.K., Gatti, D.M., Threadgill, D.W., Rusyn, I. (2009). Population-based discovery toxicogenomics biomarkers for hepatotoxicity using a laboratory strain diversity panel. *Toxicological Sciences*, *110*, 235–243.

37. Kramer, J.A., Sagartz, J.E., Morris, D.L. (2007). The application of discovery toxicology and pathology towards the design of safer pharmaceutical lead candidates. *Nature Drug Discovery*, *6*, 636–649.

38. Farkas, D., Tannenbaum, S.R. (2005). In vitro methods to study chemically-induced hepatotoxicity: A literature review. *Current Drug Metabolism*, *6*, 111–125.

39. Ansede, J.H., Smith, W.R., Perry, C.H., St. Claire III, R.L., Brouwer, K.R. (2010). An in vitro assay to assess transporter-based cholestatic hepatotoxicity using sandwich-cultured rat hepatocytes. *Drug Metabolism and Disposition*, *38*, 276–280.

40. Dawson, S., Stahl, S., Paul, N., Barber, J., Kenna, J.G. (2012). In vitro inhibition of the bile salt export pump correlates with risk of cholestatic drug-induced liver injury in humans. *Drug Metabolism and Distribution*, *40*, 130–138.

41. Thompson, R.A., Isin, E.M., Li, Y., Weidolf, L., Page, K., Wilson, I., Swallow, S., Middleton, B., Stahl, S., Foster, A.J., Dolgos, H., Weaver, R., Kenna, J.G. (2012). In vitro approach to assess the potential for risk of idiosyncratic adverse reactions caused by candidate drugs. *Chemical Research in Toxicology*, *25*, 1616–1632.

42. Guillouzo, A. (1998). Liver cell models in in vitro toxicology. *Environmental Health Perspectives*, *106*, 511–532.

43. Dambach, D.M., Andrews, B.A., Moulin, F. (2005). New technologies and screening strategies for hepatotoxicity: In vitro models. *Toxicological Pathology*, *33*, 17–26.

44. Schoonen, W.G.E.J., de Roos, J.A.D.M., Westerink, W.M.A., Debiton, E. (2005a). Cytotoxic effects of 110 reference compounds on HepG2 and of 60 compounds on HeLa,

ECC-1 and CHO cells. II mechanistic assays on NAD(P)H, ATP and DNA contents. *Toxicology*, *19*, 491–503.

45. Schoonen, W.G.E.J., Westerink, W.M.A., de Roos, J.A.D.M., Debiton, E. (2005b). Cytotoxic effects of 100 reference compounds on HepG2 and HeLa cells and of 60 compounds on ECC-1 and CHO cells. I mechanistic assays on ROS, glutathione depletion and calcein uptake. *Toxicology*, *19*, 505–516.
46. O'Brien, P.J., Irwin, W., Diaz, D., Howard-Cofield, E., Krejsa, C.M., Slaughter, M.R., Gao, B., Kaludercic, N., Angeline, A., Bernardi, P., Brain, P., Houghham, C. (2006). High concordance of drug-induced human hepatotoxicity with in vitro cytoxicity measured in a novel cell-based model using high content screening. *Archives of Toxicology*, *80*, 580–604.
47. Hewitt, N.J., Gomez-Lechon, M.J., Houston, J.B., Hallifax, D., Brown, H.S., Maurel, P., Kenna, J.G., Gustavsson, L., Lohmann, C., Skonberg, C., Guillouzo, A., Tuschl, G., Li, A.P., LeCluyse, E., Groothuis, G.M.M., Hengstler, J.G. (2007). Primary hepatocytes: Current understanding of the regulation of metabolic enzymes and transporter proteins, and pharmaceutical practice for the use of hepatocytes in metabolism, enzyme induction, transporter, clearance, and hepatotoxicity studies. *Drug Metabolism Reviews*, *39*, 159–234.
48. Li, A.P. (2007). Human hepatocytes: Isolation, cryopreservation and applications in drug development. *Chemico-Biological Interactions*, *168*, 16–29.
49. Xu, J.J., Henstock, P.V., Dunn, M.C., Smith, A.R., Chabot, J.R., de Graaf, D. (2008). Cellular imaging predictions of clinical drug-induced liver injury. *Toxicological Sciences*, *105*, 97–105.
50. Greer, M.L., Barber, J., Eakins, J., Kenna, J.G. (2010). Cell based approaches for evaluation of drug-induced liver injury. *Toxicology*, *268*, 125–131.
51. Gomez-Lechon, M.J., Lahoz, A., Gombau, L., Castell, J.V., Donato, M.T. (2010). In vitro evaluation of potential hepatotoxicity induced by drugs. *Current Pharmaceutical Design*, *16*, 1963–1977.
52. Benbow, J.W., Aubrecht, J., Banker, M.J., Nettleton, D., Aleo, M.D. (2010). Predicting safety toleration of pharmaceutical chemical leads: Cytotoxicity correlations to exploratory toxicity studies. *Toxicology Letters*, *197*, 175–182.
53. Greene, N., Aleo, M.D., Louise-May, S., Price, D.A., Will, Y. (2010). Using in vitro cytotoxicity assay to aid in compound selection for in vivo safety studies. *Bioorganic & Medicinal Chemistry Letters*, *20*, 5308–5312.
54. Lin, Z., Will, Y. (2012). Evaluation of drugs with specific organ toxicities in organ-specific cell lines. *Toxicological Sciences*, *126*, 114–127.
55. Gerets, H.H.J., Tilmant, K., Gerin, B., Chanteux, H., Depelchin, B.O., Dhalluin, S., Atienzar, F.A. (2012). Characterization of primary human hepatocytes, HepG2 cells, and HepaRG cells at mRNA level and CYP activity in response to inducers and their predictivity for the detection of human hepatotoxins. *Cell Biology and Toxicology*, *28*, 69–87.
56. O'Brien, P.J., Chan, K., Silber, P.M. (2004). Human and animal hepatocytes in vitro with extrapolation in vivo. *Chemico-Biological Interactions*, *150*, 97–114.
57. Tuschl, G., Lauer, B., Mueller, S.O. (2005). Primary hepatocytes as a model to analyze species-specific toxicity and drug metabolism. *Expert Opinion in Drug Metabolism and Toxicology*, *4*, 855–870.

58. Jover, R., Ponsoda, X., Castell, J.V., Gomez-Lechon, M.J. (1992). Evaluation of the cytotoxicity of ten chemicals on human cultured hepatocytes: Predictability of human toxicity and comparison to rodent cell culture systems. *Toxicology in Vitro*, *6*, 7–52.

59. LeCluyse, E.L., Witek, R.P., Andersen, M.E., Powers, M.J. (2012). Organotypic liver culture models: Meeting current challenges in toxicity testing. *Critical Reviews in Toxicology*, *42*, 501–548.

60. Khetani, S.R., Bhatia, S.N. (2008). Microscale culture of human liver cells for drug development. *Nature Biotechnology*, *26*, 120–126.

61. Wang, W.W., Khetani, S.R., Krzyzewski, S., Dulgnan, D.B., Obach, R.S. (2010). Assessment of a micropatterned hepatocyte coculture system to generate major human excretory and circulating drug metabolites. *Drug Metabolism and Disposition*, *38*, 1900–1905.

62. Dash, A., Blackburn, B.R., Wamhoff, B.R. (2012). Organotypic systems in drug metabolism and toxicity: Challenges and opportunities. *Expert Opinion in Drug Metabolism and Toxicology*, *8*, 999–1014.

63. Przybylak, K.R., Cronin, M.T.D. (2012). In silico models for drug-induced liver injury-current status. *Expert Opinion in Drug Metabolism and Toxicology*, *8*, 201–217.

64. Myshkin, E., Brennan, R., Khasanova, T., Sitnik, T., Serebriyskaya, T., Litvinova, E., Guryanov, A., Nikolsky, Y., Nikolskaya, T., Bureeva, S. (2012). Prediction of organ toxicity endpoints by QSAR modeling bsed on precise chemical-histopathology annotations. *Chemical and Biological Drug Discovery*, *80*, 406–416.

65. Cheng, A., Dixon, S.L. (2003). In silico models for the prediction of dose-dependent human hepatotoxicity. *Journal of Computationally Aided Molecular Design*, *17*, 811–23.

66. Clark, R.D., Wolohan, P.R., Hodgkin, E.E., Kelly, J.H., Sussman, N.L. (2004). Modelling in vitro hepatotoxicity using molecular interaction fields and SIMCA. *Journal of Molecular Graphics and Modeling*, *22*, 487–97.

67. Huang, R., Southall, N., Xia, M., Cho, M.H., Jadhav, A., Nguyen D.T., Inglese, J., Tice, R.R., Austin, C.P. (2009). Weighted feature significance: A simple interpretable model of compound toxicity based on the statistical enrichment of structural features. *Toxicological Sciences*, *112*, 385–393.

68. Low, Y., Uehara, T., Minowa, Y., Yamada, H., Ohno, Y., Urushidani, T., Sedykh, A., Muratov, E., Kuzmin, V., Fourches, D., Zhu, H., Rusyn, I., Tropsha, A. (2011). Predicting drug-induced hepatotoxicity using qsar and toxicogenomics approaches. *Chemical Research in Toxicology*, *24*, 1251–62.

69. Fourches, D., Barnes, J.C., Day, N.C., Bradley, P., Reed, J.Z., Tropsha, A. (2010). Cheminformatics analysis of assertions mined from literature that describes drug-induced liver injury in different species. *Chemical Reseach in Toxicology*, *23*, 171–83.

70. Greene, N., Fisk, L., Naven, R.T., Note, R.R., Patel, M.L., Pelletier, D.J. (2010). Developing structure-activity relationships for the prediction of hepatotoxicity. *Chemical Research in Toxicology*, *23*, 1215–22.

71. Cruz-Monteagudo, M., Cordeiro, M.N., Borges, F. (2008). Computational chemistry approach for the early detection of drug-induced idiosyncratic liver toxicity. *Journal of Computational Chemistry*, *29*, 533–49.

72. Matthews, E.J., Ursem, C.J., Kruhlak, N.L., Benz, R.D., Sabate, D.A., Marchant, C.A., Contrera, J.F. (2009a). Identification of structure-activity relationships for adverse effects of pharmaceuticals in humans: Part B. Use of (Q)SAR systems for early detection of drug induced hepatobiliary and urinary tract toxicities. *Regulatory Toxicology and Pharmacology*, *54*, 23–42.

73. Matthews, E.J., Kruhlak, N.L., Benz, R.D., Sabate, D.A., Marchant, C.A., Contrera, J.F. (2009b). Identification of structure-activity relationshps for adverse effects of pharmaceuticals in humans: Part C. Use of QSAR and an expert system for the estimation of the mechanism of action of drug-induced hepatobiliary and urinary tract toxicities. *Regulatory Toxicology and Pharmacology*, *54*, 43–65.

74. Rodgers, A.D., Zhu, H., Fourches, D., Rusyn, I., Tropsha, A. (2010). Modeling liver-related adverse effects of drugs using k-nearest neighbor quantitative structure-activity relationship method. *Chemical Research in Toxicology*, *23*, 724–32.

75. Ekins, S., Williams, A.J., Xu, J.J. (2010). A predictive ligand-based bayesian model for human drug-induced liver injury. *Drug Metabolism and Disposition*, *38*, 2302–2308.

76. Liew, C.Y., Lim, Y.C., Yap, C.W. (2011). Mixed learning algorithms and features ensemble in hepatotoxicity prediction. *Journal of Computationally Aided Molecular Design*, *25*, 855–871.

77. Zhang, H., Chen, Q.Y., Ma, C.Y., Huang, Q., Yang, S.Y. (2009). In silico prediction of mitochondrial toxicity using GA-CG-SVM approach. *Toxicology in Vitro*, *23*, 134–140.

78. Greupink, R., Nabuurs, S.B., Zarzycka, B., Vervweij, V., Monshouwer, M., Huisman, M.T., Russel, F.G.M. (2012). In silico identification of potential cholestasis-inducing agents via modeling of Na+-dependent taurocholate cotransporting polypeptide substrate specificity. *Toxicological Sciences*, *129*, 35–48.

79. Chan, K., Jensen, N.S., Silber, P.M., O'Brien, P.J. (2007). Structure–activity relationships for halobenzene induced cytotoxicity in rat and human hepatoctyes. *Chemico-Biological Interactions*, *165*, 165–174.

80. Howell, B.A., Yang, Y., Kumar, R., Woodhead, J.L., Harrill, A.H., Clewell, III, H.J., Andersen, M.E., Siler, S.Q., Watkins, P.B. (2012). In vitro to in vivo extrapolation and species response comparisons for drug-induced liver injury (DILI) using DILIsym^{TM}: A mechanistic, mathematical model of DILI. *Journal of Pharmacokinetics and Pharmacodynamics*, *39*, 527–541.

81. Ierapetritou, M.G., Georgopoulos, P.G., Roth, C.M., Androukalis, L.P. (2009). Tissue-level modeling of xenobiotic metabolism in the liver: An emerging tool for enabling clinical translational research. *Clinical and Translational Science*, *2*, 228–237.

82. Martati, E., Boersma, M.G., Spenkelink, A., Khadka, D.B., van Bladeren, P.J., Reitjens, I.M.C.M., Punt, A. (2012). Physiologically Based Biokinetic (PBBK) Modeling of Safrole Bioactivation and detoxification in humans as compared with rats. *Toxicological Sciences*, *128*, 301–316.

83. Punt, A., Paini, A., Boersma, M.G., Freidig, A.P., Delatour, T., Scholz, G., Schilter, B., van Bladeren, P.J., Reitjens, I.M.C.M. (2009). Use of physiologically based Biokinetic (PBBK) Modeling to study estragole bioactivation and detoxification in humans as compared with male rats. *Toxicological Sciences*, *110*, 255–269.

84. Poet, T.S., Wu, H., English, J.C., Corely, R.A. (2004). Metabolic rate constants for hydroquinone in F344 rat and human liver isolated hepatocytes: Application to a PBPK model. *Toxicological Sciences*, *82*, 9–25.

85. Food and Drug Administration, Challenges and opportunities report- innovation/stagnation: Challenge and opportunity on the critical path to new medical products (2004) http://www.fda.gov/ScienceResearch/SpecialTopics/CriticalPathInitiative/CriticalPathOpportunitiesReports/ucm077262.htm

86. Bruner, L.H., Carr, G.J., Harbell, J.W., Curren, R.D. (2002a). An investigation of new toxicity test method performance in validation studies: 1. toxicity testing methods that have predictive capacity no greater than chance. *Human and Experimental Toxicology*, *21*, 305–312.

87. Bruner, L.H., Carr, G.J., Harbell, J.W., Curren, R.D. (2002b). An investigation of new toxicity test method performance in validation studies: 2. comparison of three measures of toxicity test performance. *Human and Experimental Toxicology*, *21*, 313–323.

88. Bruner, L.H., Carr, G.J., Harbell, J.W., Curren, R.D. (2002c). An investigation of new toxicity test method performance in validation studies: 3. sensitivity and specificity are not independent of prevalence or distribution of toxicity. *Human and Experimental Toxicology*, *21*, 325–34.

89. Matthews, R.J.A. (2008). Medical progress depends on animal models-doesn't it? *Journal of the Royal Society of Medicine*, *101*, 95–98.

90. Hartung, T., Daston, G. (2009). Are in vitro tests suitable for regulatory use? *Toxicological Sciences*, *111*, 233–237.

91. Hayashi, P.H. (2009). Causality assessment in drug-induced liver injury. *Seminars in Liver Disease*, *29*, 348–356.

92. Fontana, R.J., Seeff, L.B., Andrade, R.J., Bjornsson, E., Day, C.P., Serrano, J., Hoofnagle, J.H. (2010). Standardization of nomenclature and causality assessment in drug-induced liver injury: Summary of a clinical research workshop. *Hepatology*, *52*, 730–742.

93. Greaves, P., Williams, A., Eve, M. (2004). First dose of potential new medicines to humans: How animals help. *Nature Reviews Drug Discovery*, *3*, 226–236.

94. Flynn, T.J., Ferguson, M.S. (2005). Multiendpoint mechanistic profiling of hepatotoxicants in HepG2/3CA human hepatoma cells and novel statistical approaches for development of a prediction model for acute hepatotoxicity. *Toxicology in Vitro*, *22*, 1618–1631.

95. Cockerell, G.L., McKim, J.M., Vonderrecht, S.L. (2002). Strategic importance of research support through pathology. *Toxicological Pathology*, *30*, 4–7.

96. Whitebread, S., Hamon, J., Bojanic, D., Urban, L. (2005). In vitro safety pharmacology profiling: An essential tool for successful drug development. *Drug Discovery Today*, *10*, 1421–1433.

97. Azzaoui, K., Hamon, J., Faller, B., Whitebread, S., Jacoby, E., Bender, A., Jenkins, J.L., Urban, L. (2007). Modeling promiscuity based on in vitro safety pharmacology profiling data. *ChemMedChem*, *2*, 874–888.

98. Leeson, P.D., Springthorpe, B. (2007). The influence of drug-like concepts on decision-making in medicinal chemistry. *Nature Reviews Drug Discovery*, *6*, 881–890.

99. Waring, M.J. (2010). Lipophilicity and drug discovery. *Expert Opinion in Drug Discovery*, *5*, 235–248.

100. Yokogawa, K., Ishizaki, J., Ohkuma, S., Miyamoto, K. (2002). Influence of lipophilicity and lysosomal accumulation on tissue distribution kinetics of basic drugs: A physiologically based pharmacokinetic model. *Methods and Finding in Experimental Clinical Pharmacology*, *24*, 81–93.

101. Hughes, J.D., Blagg, J., Price, D.A., Bailey, S., DeCrescenzo, G.A., Devraj, R.V., Ellsworth, E., Fobian, Y.M., Gibbs, M.E., Gilles, R.W., Greene, N., Huang, E., Krieger-Burke, T., Loesel, J., Wager, T., Whiteley, L., Zhang, Y. (2008). Physicochemical drug properties associated with in vivo toxicological outcomes. *Bioorganic & Medicinal Chemistry Letters*, *18*, 4872–4875.

102. Lounkine, E., Keiser, M.J., Whitebread, S., Mikhailov, D., Hamon, J., Jenkins, J.L., Lavan, P., Weber, E., Doak, A.K., Cote, S., Shoichet, B.K., Urban, L. (2012). Large-scale prediction and testing of drug activity on side-effect targets. *Nature*, *486*, 361–367.

103. Nakayama, S., Atsumi, R., Takakusa, H., Kobayashi, Y., Kurihara, A., Nagai, Y., Naiai, D., Okazaki, O. (2009). A zone classification system for risk assessment of idiosyncratic drug toxicity using daily dose and covalent binding. *Drug Metabolism and Disposition*, *37*, 1970–1977.

104. Sakatis, M.Z., Reese, M.J., Harrell, A.W., Taylor, M.A., Baines, I.A., Chen, L., Bloomer, J.C., Yang, E.Y., Ellens, H.M., Ambroso, J.L., Lovatt, C.A., Ayrton, A.D. Clarke, S.E. (2012). Preclinical strategy to reduce clinical hepatotoxicity using in vitro bioactivation data for >200 compounds. *Chemical Research in Toxicology*, September 19 epublication.

105. Thompson, R.A., Isin, E.M., Li, Y., Weaver, R., Weidolf, L., Wilson, I., Claesson, A., Page, K., Dolgos, H., Kenna, J.G. (2011). Risk assessment and mitigation strategies for reactive metabolites in drug discovery and development. *Chemico-Biological Interactions*, *192*, 65–71.

106. International Transporter Consortium. (2010). Membrane transporters in drug development. *Nature Reviews Drug Discover*. *9*, 215–236.

107. Fevery, J. (2008). Bilirubin in clinical practice: A review. *Liver International*, *28*, 592–603.

108. Zucker, S.D., Qin, X., Rouster, S.D., Yu, F., Green, R.M., Keshavan, P., Feinberg, J., Sherman, K.E. (2001). Mechanism of indinavir-induced hyperbilirubinemia. *Proceedings of the National Academy of Sciences*, *98*, 12671–12676.

109. Janmohamed, A., Hernandez, D., Phillips, I.R., Shephand, E.A. (2004). Cell, tissue, sex, and developmental stage-specific expression of mous flavin-containing monooxygenases (Fmos). *Biochemical Pharmacology*, *68*, 73–83.

110. Beedham, C., Bruce, S.E., Critchley, D.J., Al-Tayib, Y., Rance, D.J. (1987). Species variation in hepatic aldehyde oxidase activity. *European Journal of Drug Metabolism and Pharmacokinetics*, *12*, 307–310.

111. Kwak, E.L., Bang, Y.J., Camidge, D.R., Shaw, A.T., Solomon, B., Maki, R.G., Ou, S-H.I., Dezube, B.J., Janne, P.A., Costa, D.B., Varella-Garcia, M., Kim, W.H., Lynch, T.J., Fidias, P., Stubbs, H., Engelman, J.A., Sequist, L.V., Tan, W.W., Gandhi, L., Mino-Kenudson, M., Wei, G.C., Shreeve, S.M., Ratain, M.J., Settleman, J., Christensen, J.G., Haber, D.A., Wilner, K., Salgia, R., Shapiro, G.I., Clark, W.J., Iafrate, A.J., (2010). Anaplastic lymphoma kinase inhibition in non-small-cell lung cancer. *New England Journal of Medicine*, *363*, 1693–1703.

112. Chapman, P.B., Hauschild, A., Robert, C., Haanen, J.B., Ascierto, P., Larkin, J., Dummer, R., Garbe, C., Testori, A., Maio, M., Hogg, D., Lorigan, P., Lebbe, C., Jouary, T.,

Schadendorf, D., Ribas, A., O'Day, Sosman, J.A., Kirkwood, J.M., Eggermont, A.M.M., Dreno, B., Nolop, K., Li, J., Nelson, B., Hou, J., Lee, R.J., Flaherty, K.T., McAuthor, G.A. (2011). Improved survival with venurafenib in melanoma with BRAF V600E mutation. *New England Journal of Medicine*, *364*, 2507–2516.

113. Fontana, R.J., Watkins, P.B., Bonkovsky, H.L., Chalasani, N., Davern, T., Serrano, J., Rochon, J. (2009). Drug-induced liver injury network (DILIN) prospective study. *Drug Safety*, *32*, 55–68.

20

CARCINOGENICITY AND TERATOGENICITY ASSESSMENT

Hans-Jörg Martus, David Beckman, and Lutz Mueller

20.1 INTRODUCTION

In drug development, the full assessment of the carcinogenic and reproductive toxic risk represent relatively late activities, thanks to the significant resource involvement and the fact that critical information for both endpoints is generated earlier with several surrogate investigations (e.g., genotoxicity) in development. Both disciplines use special technologies and focus on defined scientific questions, which is different from the standard animal toxicology testing program. The peculiarities of these disciplines will be described in this chapter.

20.2 CARCINOGENICITY

Carcinogenicity, i.e., the induction of tumors, is one of the most serious side effects that can be associated with the chronic use of drugs, despite that the number of known human carcinogens is comparably low [1]. Cancer is a progressive disease, which is in most cases fatal or at least seriously debilitating. Furthermore, it is characterized by a long latency period. These features make it difficult and ethically inacceptable to rely on clinical monitoring to identify human risk. Therefore, special emphasis is placed on preclinical methods to detect an even minimal risk with the aim to exclude

Predictive ADMET: Integrative Approaches in Drug Discovery and Development, First Edition.
Edited by Jianling Wang and Laszlo Urban.

it for exposed subjects or to provide relevant information for a valuable risk/benefit consideration when making therapeutic decisions via a clear recommendation in the drug label. As a result of the complex nature of tumorigenesis, a carcinogenicity assessment will rely primarily on *in vivo* experiments, except for some aspects of tumor induction such as genotoxicity, where sensitive *in vitro* methods exist to identify problematic activity, as outlined in the subsequent discussion.

Formally, cancer induction is recognized as a multistage process consisting of initiation, promotion, and progression steps, which was first described in tumor models of the rat liver and mouse skin [2, 3]. Mechanistically, initiation is normally associated with activation of a proto-oncogene or inactivation of a tumor suppressor gene via mutational events that are usually not reversible and can be eliminated only via apoptosis or necrosis of the respective cells. Subsequent steps are characterized by evolutionary adaptive and acquired processes that provide the tumor cell population a survival and proliferation benefit. Eventually, this allows rapid growth and the ability to break immunological control and tissue borders including colonization of distant body parts. To this end, tumors and metastases are often associated with genetic instability and hypermutability of the tumor cells resulting in genetic, genomic, and epigenetic changes [4]. Therefore, stimulatory processes involved in tumor promotion and progression need to be present on a continued basis and are reversible, which also has been demonstrated in experimental initiation/promotion protocols.

20.2.1 Carcinogenicity Testing

Carcinogenicity testing in pharmaceutical development is intended for pharmaceuticals foreseen for chronic use, which is defined as continuous or intermittent application for more than 3 months [5]. Normally, those studies would be needed for marketing authorization unless there is cause for concern as a result of a known carcinogenic potential by the mode-of-action or compound class, by strong tissue retention leading to organ damage (of a potentially nonreversible nature), or by evidence of neoplasia-related findings in preceding repeated dose toxicity studies. In these cases, carcinogenicity testing may be required earlier. These standard approaches are not applicable routinely for biologics. However, when there is concern (e.g., by modulation of carcinogenesis-associated pathways), an evaluation of the carcinogenic risk can be advised, although often through methods not necessarily identical to those used for small molecular entities [6].

In general, carcinogenicity testing consists of one near lifetime (i.e., about 2 years) study in one rodent species plus either a second lifetime study in another rodent species, a transgenic model, or an initiation-promotion protocol [7]. In most cases, those two species will be the rat and the mouse, mostly technically as a result of their limited lifespan, which allows a lifelong treatment and normally excludes the use of longer-lived species. The duration of chronic toxicity studies in the nonrodent (i.e., 39-week toxicity studies in the dog, monkey, or minipig) is normally not sufficient to assess any tumorigenic potential. Considering the lifetime of 10 years of these species, 13-week subchronic studies in rodents are comparable.

Of the traditional lifetime rodent studies—besides the huge investment in time, laboratory animals, and resources—a major shortcoming lies in their limited predictivity of relevant human carcinogens. In fact, although their sensitivity (ability to discover a human carcinogen) is high, their specificity (their capacity to discriminate and identify noncarcinogens) is very low [8, 9]. Predominantly, this is a result of the large proportion of tumorigenicity findings in at least one species or gender. In many cases, those are interpreted as irrelevant for use under human therapeutic conditions and risk/benefit considerations. Very often, the effects are considered rodent-specific or a result of high-dose and long-duration treatment situations that need to be applied in the rodent studies to achieve technical feasibility. A prominent example is stomach tumorigenicity in rats treated with proton pump inhibitors, where a feedback endocrine loop results in prolonged enhanced blood levels of gastrin that subsequently leads to the development of stomach tumors, a mechanism that has been demonstrated as irrelevant for human therapeutic drug levels [10].

To overcome some of those limitations, several transgenic mouse lines were introduced recently and have undergone validation studies, and lifetime studies in wild-type mice are replaced increasingly by transgenic mouse studies using an abbreviated protocol with a 6-month treatment duration. Generally, those models are based on the partial activation of proto-oncogenes or on the inactivation of protective functions such as p53 or DNA repair genes, which renders them more sensitive to tumor induction so that tumors are induced after shorter treatment periods. Of those, two models, the TgrasH2 and the p53 heterozygous mouse, have attained wider use after a collaborative validation effort coordinated by ILSI/HESI, using six transgenic models (rasH2, p53 +/–, Tg.AC oral, Tg.AC dermal, XPA –/– and XPA –/–/ p53 +/–), and other studies [11, 12]. Thus, individual transgenic models correctly predicted carcinogenic activity for 74% to 83% of the chemicals, which were selected from the International Agency for the Research on Cancer (IARC) and National Toxicology Program (NTP) assessments of human carcinogens. Although transgenics as stand-alone assays failed to detect several known or probable human carcinogens, in combination with a rat 2-year bioassay, they achieved overall predictive accuracy of 85%, detected all known or probable human carcinogens, and had a low false-positive rate (15%). Thus, these results provided assurance that the new transgenic assays, particularly in the standard setting, when combined with a 2-year rat study, are able to detect reliably relevant human carcinogens, while they would not be overly sensitive, which had been a concern initially.

Nevertheless, even by replacing one lifetime study by a shorter term transgenic study, significant shortcomings in the reliability to predict human carcinogens remain. For those reasons, traditional two-species carcinogenicity assessment, which has been the standard for decades, has recently been challenged [8, 13]. In a collaborative effort that evaluated carcinogenicity data on 182 marketed and nonmarketed pharmaceuticals, evidence was provided that compounds that were (1) nongenotoxic, (2) not leading to hormonal perturbation, and (3) did not provide evidence of pre-neoplastic activity (e.g., hyperplasia, cellular hypertrophy, or atypical cellular foci) in chronic rodent toxicity studies were negative in 2-year rat carcinogenicity studies in 82% of the cases. The accuracy of predicting rodent carcinongenicity based on data from shorter-than-lifetime studies was even improved by the fact that most of the falsely predicted positive compounds were considered to be of questionable clinical relevance. Therefore, a proposal was put forward to refine the current approach of routinely conducting two 2-year studies or one rat and one transgenic mouse study by including these considerations as triggers for the need of the rat lifetime study, without compromising the accurate identification of relevant human carcinogens.

20.2.2 Genotoxic Carcinogenesis

For chemical and physical carcinogens including human pharmaceuticals, four distinctly different properties have been identified as critical for the induction of tumors, either alone or in combination. These are (1) genotoxicity, (2) immunosuppression, (3) hormonal modulation, or (4) chronic irritation [14]. Of those, genotoxic activity has been identified as the predominant predictor of human risk [15]. However, it is likely that this assessment is distorted by the fact that genotoxicity assessment is relatively straightforward, compared with the other previous endpoints, and that a genotoxic insult typically results in an irreversible damage to DNA, a target that is similar across species, in contrast to most of the other toxicities, so that genotoxicity is relatively easy to detect. Furthermore, as outlined in the subsequent discussion, cancer is a process driven by mutational events and progressive chromosomal instability in the tumor. Such markers are relatively easy to assess. The quantitative contribution of nonmutational events to tumor risk, incidence, and latency is technically more difficult to investigate.

Initiation, i.e., the process of converting a normal cell into a cell basically capable of forming the stem cell of a tumor, is generally associated with oncogene activation or tumor suppressor gene activation. Tumor promotion and tumor progression have been identified in experimental models as subsequent steps, characterized as processes whereby continuous growth stimulation is able to propagate an initiated cell, or this primary tumor overcomes tissue restrictions and metastasizes into other parts of the body, respectively. In the latter steps, a more continuous and generally reversible insult such as growth stimulatory signals is required. Historically, those phases were identified mainly in experimental animal models, most prominently in the mouse skin or rat liver, which allowed the identification and discrimination of the different steps. However, in most cases, the different phases of tumorigenesis are indistinguishable or interconnected so that this formal differentiation is difficult, and an identification of a true tumor initiator or promoter is not possible.

Thus, the assessment of a genotoxic activity represents one of the cornerstones of identifying a carcinogenic risk because irreversible damage to genetic material inevitably carries the risk of modulating critical growth parameters of cells, which may lead to tumor initiation. In fact, about 1% of all known human genes are associated with tumorigenesis and therefore considered to be oncogenes or tumor suppressor genes [16]. Furthermore, mutations in germline tissues can be responsible for heritable mutations, i.e., mutations that affect the health of subsequent generations. For that reason, unequivocally genotoxic compounds are considered to be trans-species carcinogens for which no extra carcinogenicity testing is required [7].

20.2.3 Genotoxicity Testing

Although the genotoxic induction of precancerous cells or other diseases results in the modification of specific genes, only rarely those lend themselves to the analysis of a mutagenic activity. Therefore, in practice, surrogate genes and test systems are used because for the purpose of genotoxicity testing, it is assumed that mutation induction is basically comparable across different parts of the genome. Whereas this certainly is a generalization, it is a valid assumption that whenever mutations are induced in a gene used for mutation testing, the risk for cancer-relevant genetic material will rise as well. Also, it is generally accepted that comparable mechanisms are responsible for mutation induction in somatic and germline cells; thus, the fact that a compound is mutagenic means that it carries similar risk for both effects and the biological consequences will depend on whether the compound or active metabolites are distributed into the respective tissue [17].

20.2.4 Regulatory Context

For any test system, two fundamental parameters need to be considered to define the appropriateness for the approach, which are (1) the genetic endpoint and (2) the experimental model. Mainly for practical reasons, genetic endpoints applied in routine testing are gene mutations and chromosome damage. In addition, primary DNA lesions or other endpoints can be used to investigate genetic damage; however, they are not used routinely. To measure gene mutations, biochemical cell functions such as enzyme activities or membrane receptors are used, whereas chromosome aberrations are mainly assessed by microscopic techniques or flow cytometry. Although investigating chromosomal integrity requires eukaryotic systems, gene mutations can also be measured in bacterial cells. In this context, it is important to understand that a spectrum of genetic damage can lead to tumor initiation, and this needs to be adequately covered by the combination of tests used (see Table 20.1). The second important parameter to be considered is the choice of the test system, i.e., the organism, cell

Table 20.1 Detection of mechanisms responsible for tumor induction by standard genotoxicity tests (modified from Reference 18)

Genetic lesion	Genotoxicity test system				Carcinogenic mechanism	
	Salmonella reverse mutation (Ames) test	Mammalian TK test	*In vitro* chromosome aberration test	Micronucleus test (*in vitro* or *in vivo*)	Oncogene activation	Tumor suppressor gene inactivation
Base substitution	X	X			X	X
Small deletion	X	X		X	X	X
Large deletion		X	X	X		X
Chromosomal translocation		X	X	X	X	X
Mitotic recombination		X				X
Chromosome loss				X		X

type or acellular experimental model. One important aspect that defines the reaction of biological systems to foreign compounds is summarized as absorption, distribution, metabolism, and excretion (ADME), which determines the fate of a compound in the organism or test system. Of those, for genetic toxicology, drug metabolism deserves particular attention because many mutagens or carcinogens are not active per se but only after metabolic activation by enzymes, which are particularly abundant in the liver. Therefore, using the appropriate metabolism system is essential to assure the biological relevance of genotoxicity data obtained in model systems [19,20].

Like other safety-related topics, genotoxicity testing of pharmaceutical ingredients and impurities present in the end product is a highly regulated field, coming from the significant public interest in assuring the safety of marketed drugs. For drug development, the guidelines as laid out by the International Conference of Harmonization of Technical Requirements for Registration of Pharmaceuticals for Human Use (ICH) are considered the most relevant and universally accepted (http://www.ich.org/). ICH guidelines S2A and S2B, which were the ones relevant for genotoxicity testing, have recently been revised and combined into one [21]. A test strategy that comprises a combination of test systems as well as genetic endpoints, i.e., a description of basic testing necessary to exclude a relevant genotoxic potential with sufficient level of safety for the various phases of drug development and marketing authorization, is proposed. In addition, OECD guidelines, issued by the Organization of Economic Cooperation and Development OECD, are in use, which primarily describe the technically correct conductance of individual tests [22].

According to ICH S2 guidance, basic genotoxicity assessment consists of a bacterial mutation (*Salmonella typhimurium* reverse mutation) test to investigate the induction of gene mutations *in vitro* [21]. In this test, like most *in vitro* tests, an extracellular metabolic system is used to provide for metabolic activation. Routinely this consists of a preparation from cytochrome P450-induced rat liver, which contains cytosolic as well as microsomal enzymes plus co-substrates to allow for mainly oxidative metabolism, in order to capture those mutagens that need metabolic activation to become DNA-reactive. Obviously, this experimental design is not sufficient to capture all relevant activating conditions, but experience has nonetheless shown a remarkable predictivity of genotoxic carcinogenesis [23]. Also, because this test system is both sensitive for genotoxic reactions as well as specific in the sense of not very prone to artifacts, it is considered a reliable predictor of relevant mutagenicity. Thus, in most indications, a clearly positive Ames test would be an insurmountable hurdle to develop a drug. If negative, ICH S2 provides two options. Option 1 consists of a mammalian cell test *in vitro*, which is either a thymidine kinase gene mutation test in L5178Y mouse lymphoma cells or a chromosome aberration or micronucleus test in basically any cell type that is amenable to those endpoints. If negative or positive only under conditions where human relevance can be convincingly excluded (e.g., excessive cytotoxicity), this package of two negative tests will allow the initiation of a clinical trial program in all ICH regions, and an *in vivo* test (chromosome aberration or micronucleus test in rodent bone marrow or peripheral blood) is not needed before clinical phase 3 studies. If positive, the relevance of the positive result(s) needs to be investigated. Relevance in this context can be demonstrated in various ways. Mostly, if an effect can be excluded under appropriate conditions in animals, this is a strong evidence of the absence of a risk in humans. For a positive effect in the mammalian tests mentioned, the applicable follow-up strategy would consist of a micronucleus or chromosome aberration test in rodent bone marrow or peripheral blood, plus a second test using a second endpoint and tissue. As a result of ease of conductance and applicability to a variety of tissues, the second endpoint will be in many cases the evaluation of primary DNA damage via the alkaline single-cell gel electrophoresis (comet) assay, a test for which OECD guidance is not available yet but under preparation. Tissue selection should be guided by scientific arguments specific to the situation, e.g., the liver if hepatic metabolic activation is suspected, the GI tract as tissue of high local concentration in the case of oral administration, or organs in which toxic effects or drug accumulation have been observed. For a positive effect in the Ames test, transgenic animals such as Big Blue® or Muta™Mouse are available to investigate gene mutations *in vivo* [24]. In any case, the conditions need to be appropriate to provide arguments to exclude a human risk with sufficient stringency, for example, with respect to doses and treatment schedules used. Accumulating experience has allowed ICH S2 to accept the combination of various endpoints into a single study, such as a micronucleus and comet test in the same animal. Whereas traditionally those tests were conducted under an acute treatment paradigm, it is now also accepted to use a subacute or a subchronic treatment, provided several criteria are fulfilled to assure sufficient sensitivity, such as a highest dose level of at least 50% of a potential acute maximum tolerated dose (MTD). Whereas those novel modifications are expected to lead to the reduction of animal numbers, an additional benefit will be the integration of genotoxicity endpoints into routine toxicity studies, which allows the combination of a wealth of information such as histopathology, clinical pathology, or toxicokinetics (traditionally not obtained in genotoxicity studies) into a very meaningful overall assessment. Alternatively, as Option 2, ICH S2 allows the initiation of clinical trials without conducting a mammalian *in vitro* test. In this case, an *in vivo* test is needed, similar to the situation when the mammalian *in vitro* test is positive under Option 1.

Routinely, in drug development, screening tests will precede those investigations intended to assess human risk, which allows an earlier prediction of the outcome of regulatory tests important for selecting the right molecules early. Also, computer-assisted systems are widely used to predict bacterial mutagenicity with acceptable precision. However, for other endpoints, they have not matured to the level that would render them reliable enough for routine use [25]. Generally, endpoints and test systems are used that resemble those of the regulatory tests to assure a sufficient predictivity, such as down-scaled versions of Ames or micronucleus tests. If other test systems are used, which can be the case for high-throughput purposes, thorough validation studies need to assure a reliable prediction of the definitive tests.

20.2.5 Case Study 1

Myfortic® is the enteric-coated mycophenolate sodium, for the prophylaxis of acute transplant rejection in patients receiving allogeneic renal transplants. Mycophenolic acid (MPA) is the active moiety of the ester mycophenolate mofetil (MMF, CellCept®). MMF is rapidly and completely cleaved by gastrointestinal and liver esterases to yield MPA and morpholino ethanol [26, 27], so it is not surprising that most properties are similar between MMF and MPA. MPA is a selective, noncompetitive, and reversible inhibitor of inosine monophosphate dehydrogenase (IMPDH), which inhibits the *de novo* pathway of guanosine nucleotide synthesis without incorporation into DNA. Because T- and B-lymphocytes are critically dependent for their proliferation on *de novo* synthesis of purines, whereas other cell types use mainly salvage pathways, MPA has more potent cytostatic effects on lymphocytes than on other cells [28].

When tested for genotoxic activity, mycophenolate sodium was mutagenic in a mouse lymphoma thymidine kinase assay, induced micronuclei in V79 Chinese hamster cells *in vitro,* and in the mouse bone marrow *in vivo*, but it was negative in a bacterial mutation assay using *Salmonella typhimurium* or a chromosomal aberration assay in human lymphocytes. Mycophenolate mofetil, another salt form in therapeutic use, generated similar genotoxic activity. This picture is consistent with the hypothesis that genotoxic activity of MPA is probably a result of the depletion of the nucleotide pool required for DNA synthesis as a result of the pharmacodynamic mode of action of MPA (inhibition of nucleotide synthesis), a mechanism that is known to induce genotoxic effects [29, 30]. Likely, bacterial homologues of mammalian IMPDH are not inhibited by MPA so that *Salmonella* is resistant to an MPA-induced pool imbalance with subsequent mutagenicity. In summary, MPA would be considered genotoxic in mammalian cells and as such a trans-species carcinogen. Remarkably, in a 104-week oral carcinogenicity study in rats, MPA at daily doses up to 9 mg/kg was not tumorigenic. The highest dose tested resulted in approximately 0.6–1.2 times the systemic exposure observed in renal transplant patients at the recommended dose of 1.44 g/day. Similar results were observed in a parallel study in rats performed with MMF. In a 26-week oral carcinogenicity assay in a P53 heterozygous transgenic mouse model, mycophenolate sodium at daily doses up to 200 mg/kg was not tumorigenic. The highest dose tested was 200 mg/kg, resulting in approximately five times the systemic exposure observed in renal transplant patients. Similarly, MMF was negative in a rat and mouse bioassay [31]. At the same time, in a teratology study performed with mycophenolate sodium in rats, at a dose as low as 1 mg/kg, malformations in the offspring were observed, including anophthalmia, exencephaly, and umbilical hernia.

In humans, reliable information on the carcinogenic risk is sparse as a result of the limited duration that either MPA or MMF have been in therapeutic use. However, preliminary data indicate a low carcinogenicity risk, which seems to be lower than that of other immunosuppressant therapies [32, 33].

Therefore, MPA and MMF are one of those uncommon examples where preclinical experiments, both *in vitro* and *in vivo*, indicated a genotoxic potential when tested *in vitro* and *in vivo*, but it was devoid of a relevant carcinogenic activity in both animals and humans, despite the fact that the mechanism of therapeutic action, immunosuppression, has been linked to an enhanced tumor risk.

20.3 NONGENOTOXIC CARCINOGENESIS

Although a tumor cannot form without mutation events, whether spontaneous or induced, nonmutational mechanisms of carcinogenesis play an important role for the speed at which a tumor can form, whether it survives immunological control, how it metastasizes, and with which chance initiated cells grow into a histologically discernible tumor. Nongenotoxic mechanisms of tumor "promotion" are of a wide variety. Some of them are acknowledged as of higher importance for many tumors found typically in rodents while others may be of lower importance for rodents but are acknowledged risk factors for humans. For example, the activation of nuclear receptors or the induction of metabolic enzymes is frequently observed in mice and rats. This leads often to a proliferative response of liver tissue, which, if sustained, can lead to liver adenoma and carcinoma formation [34]. Evaluation of the induction of drug metabolizing enzymes in humans can be good surrogate evidence for such effects, as exemplified in Case study 2, where enzyme induction by activation of nuclear receptors was observed in the rat liver but lack of human relevance was demonstrated. Conversely, impairment of immunological control of tumor growth with immunosuppressants after organ transplantation is an acknowledged risk factor for humans, whereas it is difficult to mimic these properties in rodents [35]. An early indicator for an increased cancer risk in humans with impaired immunological control is the induction of skin cancer in immunosuppressed patients [36]. This can be explained by the presence of numerous ultraviolet (UV)-initiated cells/clones on the human skin, which can escape immunological control much easier under such therapy [37].

In the meantime, molecular insights into mechanisms of tumorigenesis are getting more and more elucidated. The key role of tumor suppressor genes and tumor "activation" genes (commonly referred to as "oncogenes") is now widely acknowledged and reliably detected via genotoxicity testing. It is also recognized that what was previously referred to as "viral" oncogenesis can be traced down to the integration of mammalian host genes into the viral genome, which may become "oncogenic" with viral infection of a mammalian host as a result of mutation or insertion into a genome region that facilitates an altered regulatory control in the genome. Often, oncogenes are cellular kinases, which may become activated through mutational events. However, the network of kinases is complex and activation and inactivation of oncogenes via inhibition of kinases may represent two sides of a coin. In this context, it is intriguing that the use of the recently introduced BRAF inhibitor vemurafenib (Zelboraf®) in melanoma patients to control a specific type of a BRAF mutation is characterized by an unexpected increase in secondary skin cancers, so-called squamous cell carcinomas [38]. Recent biochemical studies have revealed a mechanism by which BRAF inhibitors can simultaneously suppress RAF-MAPK signaling while paradoxically activating this pathway in cells carrying an oncogenic mutation in the RAS gene. Mutated RAS promotes the dimerization of RAF isoforms, thus, forming BRAF-CRAF, BRAF-ARAF, and CRAF-CRAF dimers. This dimerization allows downstream signaling through isoforms of RAF not bound by small-molecule inhibitors.

In this manner, it is hypothesized that vemurafenib can induce transactivation of CRAF despite inhibition of its BRAF partner in an RAS-dependent manner. Of note, RAS mutations are detected in some actinic keratoses as well as in some squamous cell carcinomas. Via these means, the inhibitory control of the tumor promotional behavior of a mutated oncogene (BRAF in this case) can activate other oncogenic pathways. As these pathways are not necessarily under comparable control and in the same network as rodents, mouse or rat experiments will hardly elucidate these mechanisms with sufficient reliability for human risk assessment.

20.3.1 Case Study 2

One of the well-known issues in carcinogenicity studies is associated with the activation of (nuclear) receptors, which can lead to tumor promotion in several organ types. A frequently affected organ is the liver in rats and mice [34]. For such tumors, activation of nuclear receptors of the aryl hydrocarbon receptor (AHR), the constitutive androstane/pregnane X receptor (CAR/PXR), and the peroxisome proliferator-activated receptor-alpha (PPARα) has been widely described. Experimental systems have been set up to judge human relevance of such activation. For example, although fenofibrate and gemfibrozil (and other marketed PPARα agonists) induced hepatic tumors in rodent bioassays by such a nongenotoxic mode of action, comprehensive mechanism-based analysis suggested the absence of human relevance, which allowed market introduction of this type of drugs [39]. This was subsequently confirmed by human epidemiological evidence. Frequently, liver weight increase and hepatocellular hypertrophy are observed in early repeated dose toxicity studies in rodents. In those cases, there is a good chance of observing liver tumor formation in formal rodent lifetime studies. Dalcetrapib, a compound targeting the cholesteryl-ester transport protein (CETP), which has recently been investigated in phase III of clinical development, displayed such a profile in the livers of mice. Subsequently, in the full mouse carcinogenicity study, an increase in liver tumors was observed in male mice only [40]. Triggered by the histological profile in the liver, it was hypothesized that this hepato-carcinogenesis may be a result of activation of the nuclear constitutive androstane receptor (CAR), a major xenobiotic sensor in mammals and known as an important mediator of rodent-specific, nongenotoxic hepatocarcinogenesis [41]. This pathway was originally discovered through the study of the anticonvulsant phenobarbital, a prototypical inducer of cytochrome P450 (CYP) enzymes that acts via CAR activation. Although CAR mediates induction of CYPs of the 2b family and is indispensable for phenobarbital-induced hepatocyte proliferation and hepatocarcinogenesis in rodents [42, 43], there is much clinical evidence to support the absence of any hepatocarcinogenic effect of phenobarbital in humans [34, 44]. Molecular evidence for this difference in response between rodents and humans was recently provided by "humanized" mice in which endogenous CAR was replaced by the human CAR-gene under the control of the murine CAR promoter. In response to phenobarbital, these mice displayed induction of cytochrome P450 enzymes but no hepatocyte proliferation. On the contrary, wild-type counterparts showed a significant increase in hepatocellular proliferation. Additionally, CAR null mice showed no response, indicating that proliferation is a specific mouse CAR-driven effect not observed with human CAR [45]. Thus, although CYP induction is a ubiquitous downstream effect of CAR activation, hepato-carcinogenesis resulting from sustained proliferation seems only to develop with mouse CAR, which is a conclusion likely to be extrapolated to other rodent CARs. Accordingly, investigation of CAR activation by dalcetrapib was a key objective of the study published by Hoflack et al. [40]. To support the safety assessment by exploratory biomarkers, Hoflack et al. presented a mechanistic CYP2b10 analysis in tumor tissue and nontumor tissue as a follow-up to the significant increase in liver tumors found in male mice of the regulatory 2-year carcinogenesis bioassay with dalcetrapib [40]. CYP2b10 was used as a surrogate marker for activation of the nuclear receptor CAR, demonstrating the likely absence of human relevance of these tumors. Livers from about a third of the male animals from each dose group of the bioassay were analyzed for mRNA expression of CYP2b10. In addition, CYP2b10 enzymatic activity was measured in pooled samples of male animals from all dose groups of the bioassay. CYP2b10 mRNA was strongly induced by dalcetrapib, with an expected wide interindividual variation (5- to 1421-fold). Group average fold-induction versus vehicle-controls showed a dose-related increase from 48-fold (250 mg/kg/day) to 160-fold (750 mg/kg/day), which declined slightly at 2000 mg/kg/day (97-fold). CYP enzyme activities showed approximate doubling of total CYP P450 protein and a 9-fold increase in CYP2b10-selective pentyloxyresorufin O-dealkylase activity (750 mg/kg/day). These data from hepatic CYP2b10 monitoring are strongly suggestive of CAR activation by dalcetrapib, a mechanism known to be devoid of relevance toward hepato-carcinogenesis in humans. Their results demonstrate nicely the feasibility of CYP2b10 as a surrogate marker for this mechanism at cessation of a carcinogenesis bioassay.

20.4 EMBRYO-FETAL TOXICITY TESTING

Embryo-fetal toxicity testing is performed to support the assessment of the potential for human risk as a result of the administration of pharmaceuticals during pregnancy. The *in vivo* studies, which may also be termed “teratology tests” or “Segment II tests,” supporting the development of small molecules and biologics for use as human pharmaceuticals are guided by International Conference on Harmonisation of Technical Requirements for Registration of Pharmaceuticals for Human Use (ICH) M3(R2), S5(R2), S6(R1), and S9 [6, 46–48]. These studies are usually conducted in two species, rats and rabbits, but they may include mice or nonhuman primates.

In vivo embryo-fetal toxicity studies involve administration of a test article to pregnant animals from a period of embryo-fetal development encompassing implantation to closure of the hard palate and investigate adult female reproduction, embryonic development, and major organ formation [47]. Typically, pregnant rats in the main study groups are administered a test article or control article on gestation days 6–17, they are euthanized on either day 20 or day 21 of gestation, and the fetuses are obtained by cesarean section. The doses are commonly based on maternal toxicity with the high dose expected to produce at least minimal maternal toxicity and no mortality.

Macroscopic examinations are performed on the maternal animals, and histopathological evaluations may be included. Maternal reproductive parameters include the number of corpora lutea, implantation sites, and resorptions. Fetal examinations include viability, body weight, sex, and external, visceral, and skeletal evaluations for the presence of variations or malformations. Toxicokinetics and placental transfer evaluations are typically conducted on pregnant rats in satellite groups treated in a similar manner as those in the main study groups. Blood samples are collected from the satellite animals at appropriate time points after the last dose is administered, and blood samples are collected from fetuses obtained by cesarean section.

Although there are variations in the study designs, the basic elements of the embryo-fetal toxicity studies to support drug development have not changed in the last three decades as a result of the complexity of embryo-fetal development. Thus, a single *in vitro* system or a defined battery of *in vitro* tests has not been identified that emulates the series of genetic signals, cell proliferation/differentiation, organ formation, and functional maturation steps in the embryo-fetus. In addition, an *in vitro* system has not been devised that emulates the changing structure and function of the placenta during gestation or the interaction with the maternal organism as it undergoes a series of physiological changes associated with pregnancy. *In vitro* test systems also do not have the capacity to recover/repair from insults as does the embryo-fetus *in vivo* or to replicate fully the ADME characteristics for the drug.

Given that the goal of embryo-fetal toxicity testing is to support the assessment of the potential for human risk and that it is desirable to reduce the numbers of experimental animals used, there must be a known relevance to the human to justify replacing traditional testing with alternative tests. As the fundamental molecular pathways used during organogenesis are highly conserved across species and are used repeatedly during embryonic-fetal development [49], it would seem that there is an enormous opportunity to create either *in vitro* or *ex vivo* tests that could be used to characterize the embryo-fetal toxicity profile of a pharmaceutical. However, as great as the opportunity seems to develop alternative test systems, the challenge is to identify those tests that have sufficiently low false-negative characteristics to be useful as safety screens but also to have sufficiently low false-positive characteristics to avoid unjustified termination of efficacious pharmaceuticals.

At this time, only three alternative tests have been validated by the European Centre for the Validation of Alternative Methods for use in embryo-fetal toxicity testing [50, 51]; the protocols are available at http://ecvam.jrc.it (accessed 06 Aug 2012): embryonic stem cell test [52], limb bud micromass [53], and rat postimplantation whole embryo culture [54]. These tests were validated using 20 compounds [55]. The results of the validation of these tests are presented in Table 20.2. It should be noted that another evaluation of the embryonic stem cell test indicated poor predictivity [57]. Although resource intensive, the rat postimplantation whole embryo culture combined with assessments of cytotoxicity or gene expression has been used effectively for mechanistic studies [58]. The currently available validated alternative tests and variations cannot replace *in vivo* testing to support risk assessment in the human. However, they can be used effectively for prioritization during drug development [51, 59].

Table 20.2 Validation results for the embryonic stem cell test, limb bud micromass, and the whole embryo culture combined with cytotoxicity data[a]

	Embryonic stem cell test	Limb bud micromass	Whole embryo culture combined with cytotoxicity data
Predictivity for nonembryotoxic compounds	72%	57%	70%
Predictivity for weakly embryotoxic compounds	70%	71%	76%
Predictivity for strongly embryotoxic compounds	100%	100%	100%
Precision for nonembryotoxic compounds	70%	80%	80%
Precision for weakly embryotoxic compounds	83%	60%	65%
Precision for strongly embryotoxic compounds	81%	69%	100%
Accuracy	78%	70%	80%

[a]Based on Reference 51 and European Centre for the Validation of Alternative Methods (ECVAM) Scientific Advisory Committee statements for embryonic stem cell test, the limb bud micromass, and whole embryo culture combined with cytotoxicity data [56]. Definitions: Accuracy was defined as the concordance between *in vitro* data and *in vivo* data. The definitions of predictivity and precision used in *in vitro* toxicology tests were adapted, respectively, from the definitions of specificity (the proportion of negatives that are correctly identified) and sensitivity (the proportion of positives that are correctly identified).

20.4.1 Interpretation of *In Vivo* Embryo-Fetal Toxicity Studies to Support Human Risk Assessment

In addition to the regulatory guidance for embryo-fetal toxicity studies given in ICH S5(R2), it is also useful to refer to the Guidance for Industry Reproductive and Developmental Toxicities—Integrating Study Results to Assess Concerns (2011) when designing studies and interpreting the results [60]. Although the guideline pertains to recommendations for labeling based on a weight of evidence approach for the assessment of reproductive and developmental toxicities, this section will apply it to the assessment of adverse effects on embryo-fetal development resulting from administration of a pharmaceutical to the maternal organism from implantation through closure of the hard palate that can be evaluated at near-term cesarean section. Adverse effects on embryo-fetal development produced during this period are manifested primarily as mortality, dysmorphogenesis, and alterations to growth:

- Mortality can be manifested as an early or late resorption, fetal death, abortion, or stillbirth. As polytocous species are typically used in these studies, any combination of resorptions, fetal deaths, and stillbirths may be present.

Table 20.3 Examples of factors influencing the interpretation of embryo-fetal toxicity study results to support human risk assessment[a]

Interpretation of signal
Cross-species concordance of reproductive or developmental effects
Multiplicity of effects
Maternal toxicity
Dose–response relationship
Rare events
Pharmacodynamics
Risk assessment
Concordance between test species and humans
Relative exposure
Class alerts

[a]Based on Reference 60.

- Dysmorphogenesis, or structural abnormality, of the skeleton or soft tissues is classified either as a malformation (a permanent morphologic alteration that will have an adverse effect on viability, growth, or function) or a variation (a developmental delay, a temporary alteration, or a permanent change not considered to have an adverse effect on viability, growth, or function).
- Adverse effects on growth usually result in reductions in *in utero* growth, but increased growth may also occur. Growth is commonly assessed by body weight measurements, but other measurements, such as crown-rump length, may be made. As some structural abnormalities in a fetus may be a result of either a direct effect or a delay in growth, all relevant data must be taken into account to facilitate making the distinction.

For the results of an embryo-fetal toxicity study to support risk assessment for the human, the doses, exposures, route of administration, timing and duration of administration, and parameters evaluated must be relevant for the planned human use. Furthermore, the species used should demonstrate the intended pharmacodynamic effects of the compound, which is a toxicity profile compatible with that anticipated in the human, and similar ADME characteristics to those in the human.

The interpretation of embryo-fetal toxicity studies and their support for risk assessment in the human should take into consideration several additional factors (Table 20.3). Not all of the factors listed in Table 20.3 are relevant for biological pharmaceuticals or for oncology because of the different regulatory requirements for appropriate studies (refer to ICH S6(R1) and ICH S9 [6, 48]).

Interpretation of signal

- Cross-species concordance—The observation of the same or related effects in more than one test species suggests that the effects are a result of the test article. Concordance also increases the risk of an adverse effect in the human, and the effect is likely to be in the same category although not necessarily identical.

- Multiplicity of effects—When two or more adverse effects are observed in an embryo-fetal study, these results suggest that the effects are a result of the test article. Multiplicity of effects also increases the risk of an adverse effect in the human, especially if the effects occurred at different stages of *in utero* development.
- Maternal toxicity—Adverse embryo-fetal effects observed at doses that are toxic to the maternal animal decreases the risk of an adverse effect in the human, assuming that the maternal toxicity can convincingly be linked to the embryo-fetal effects. In contrast, adverse embryo-fetal effects in the absence of maternal toxicity increase the risk of an adverse effect in the human.
- Dose–response relationship—Assuming that the doses used were sufficient to assess developmental toxicity, even if those doses or exposures were less than those expected clinically, then a dose-related increase in the incidence of an adverse effect suggests that the effect is a result of the test article and that there may be an increased risk in the human.
- Rare events—The increased incidence of a rare event (e.g., a rare malformation) suggests that the effect is a result of the test article. This result also suggests an increased risk of an adverse effect in the human.
- Pharmacodynamics—If the observed effects on the embryo-fetus can be predicted by the pharmacologic mechanism of the compound, then the effects are a result of test article and there is an increased risk in the human. In cases when the mechanism is species specific, however, there is no increased risk in the human.

20.4.1.1 Risk Assessment

- Concordance between test species and humans—The greater the similarity in ADME characteristics between the animal species and the human, the greater the relevancy of embryo-fetal findings for risk assessment in the human. This aspect is particularly important when a metabolite is more toxic to the embryo-fetus than the parent compound.
- Class alerts—It is self-evident that there is an increased risk in the human when a class of pharmaceuticals is known to produce adverse effects in both humans and animals.
- Relative exposure, animal:human—A comparison of the no observed adverse effect level (NOAEL) for embryo-fetal effects in the animal study to the maximum recommended dose in the human is typically based on the area under concentration (AUC) or the body-surface-area–adjusted dose, taking into account both parent compound and active metabolites. Although the exposure ratio is typically calculated for each test species used, the risk assessment should be based on the most sensitive species. In general, the risk for adverse effects in the human is considered to be increased if the exposure ratio is <10 and to be decreased if the ratio is >25.

20.4.1.2 Risk Assessment in the Human vs. Extrapolation to an Individual

Animal studies provide a means of identifying exposures with the potential for producing embryo-fetal toxicity in the human population. As exposures are toxic to the embryo-fetus only under certain conditions of dose, route, gestational timing, etc., and given that there are species differences in placentation, pharmacokinetics, pharmacodynamics, and developmental timing, the risk assessment can be extrapolated to an individual pregnant woman only in a very general manner.

20.4.2 Case Study 3

Acting on different arms of the renin-angiotensin-aldosterone system (RAAS), both angiotensin-converting enzyme (ACE) inhibitors and angiotensin II receptor (AT1) antagonists are used in the treatment of hypertension and have been associated with fetopathy in the human [61–63]. In animals, ACE inhibitors and AT1 antagonists produced adverse fetal effects but not teratogenicity in test animals. Increased fetal loss was reported in rabbits treated with the ACE inhibitor captopril and confirmed in other species [64–66]. Animal studies have demonstrated that maternal ACE inhibitor exposure can lead to fetal hypotension in sheep [65], rabbit [66], and baboon [67], and to decreased utero-placental blood flow, low birth weight, fetal hypotension, preterm delivery, and death [63].

In the rat [68,69], there was no embryo-fetal toxicity when dams were treated with losartan, an AT1 antagonist, on gestational days 6–15, but fetal renal damage was produced when losartan was administered on gestational days 15–20. Additional evidence for fetal toxicity later in gestation was produced in the rat: candesartan, another AT1 antagonist, produced both hydronephrosis and enlargement of the renal tubules when administered during late pregnancy and lactation and while just hydronephrosis was produced when the maternal animals were treated only during lactation [70]. In the sheep, AT1 antagonist administration late in gestation had a direct effect on the fetal cardiovascular system and on renal function [71,72].

The animal data support the conclusion that embryo-fetal toxicity produced by ACE inhibitors and AT1 antagonists is a result of exposure later in gestation, at a time more equivalent to the second trimester in the human. It should be noted that differences in developmental timing contribute to differences in fetal outcome between the human and test species administered ACE inhibitors and AT1 antagonists. Taking the renal findings as an example, development of the kidney and the RAAS occurs in humans by the end of the first trimester, whereas the kidney and the RAAS develop closer to term and even into the early postnatal period in test species [73,74]. The animal embryo-fetal toxicity data for ACE inhibitors and AT1 antagonists support the increased risk of adverse effects in the second and third trimesters in the human.

The difference between embryo-fetal toxicity results in test species and human fetopathy produced by ACE inhibitors and AT1 antagonists illustrates the importance of taking all available information into account when assessing risk and in not assuming that adverse effects demonstrated in the test species will necessarily be identical to those in the human.

20.4.3 Case Study 4

As discussed in Case study 1, mycophenolate mofetil (MMF) exhibited genotoxicity in nonclinical evaluations but did not exhibit carcinogenic activity in animals or the human. In contrast, adverse effects on the fetus as a result of MMF administered during pregnancy have been demonstrated both in embryo-fetal toxicity studies and in the human [75, 76].

Based on the prescribing information for CellCept® (mycophenolate mofetil; Genentech, Inc., San Francisco, CA), MMF produced an adverse fetal effect in both rats and rabbits in embryo-fetal toxicity studies. MMF administration produced an increased incidence malformations (anophthalmia, agnathia, and hydrocephalus) in offspring of pregnant rats administered doses equivalent to 0.02-fold the clinical dose, based on body surface area. In rabbits administered 0.9-fold the clinical dose based on body surface area, MMF produced increased incidences of fetal resorptions and malformations (ectopia cordis, ectopic kidneys, diaphragmatic hernia, and umbilical hernia) in the absence of maternal toxicity.

Two validated *in vitro* tests, the rat whole embryo culture and the murine embryonic stem cell test, were used to investigate the embryotoxic potential of mycophenolic acid [77]. Rat embryos showed a concentration-dependent incidence of dysmorphogenic development (decreased growth, neural tube defects, malformations of the head, and heart anomalies) while the murine embryonic stem cell test demonstrated decreased proliferation and differentiation. These embryotoxic effects occurred at cytotoxic concentrations. The *in vitro* test results support the conclusion of an increased risk in the human, but the signal is not as strong as it is based on the *in vivo* embryo-fetal toxicity study results.

Based on a cohort of 57 prospectively followed pregnancies in the human, there is an increased risk of spontaneous abortion and major malformations with a distinctive phenotype including microtia, auditory canal atresia, cleft lip or palate, heart defects, as well as less frequent malformations, when exposures lasted longer than week 7 after the last menstrual period [75]. Thus, the preclinical data, including the use of *in vitro* tests, correctly predicted the human risk associated with the therapeutic use of MMF.

20.4.4 Case Study 5

Cystinosis is a rare autosomal recessive disorder caused by mutations of the CTNS gene on chromosome 17p13, which encodes a ubiquitous lysosomal cystine-specific transporter (cystinosin), leading to cystine accumulation in lysosomes as a result of a defect in lysosomal cystine transport and causing apoptotic cell death [78]. Cystinosis is associated with renal Fanconi syndrome, poor growth, and muscle wastage. Although renal transplantation enables children with cystinosis to survive the renal failure, some complications result from cysteine accumulation and accompanying degeneration of other body tissues [78]. Cysteamine (β-mercaptoethylamine) prevents the excess storage of cystine in lysosomes by participating in a thiol-disulfide interchange reaction, converting cystine into cysteine and cysteine-cysteamine mixed disulfide, which are excreted from the cells [79]. Early treatment with cysteamine can retard renal deterioration and reduce the complications associated with cystinosis [78, 80]. Infertility is common in men with cystinosis [81], but there have been successful pregnancies in women with cystinosis [82, 83].

An embryo-fetal toxicity study was performed to investigate the potential effects of cysteamine in the rat using a modified study design [84]. On day 6.5 through day 18.5 post-conception, cysteamine was administered to Wistar rats via oral gavage at doses of 0, 37.5, 75, 100, or 150 mg/kg/day, and animals were euthanized on day 21 and the fetuses were collected at cesarean section. Measures of adverse fetal outcome included intrauterine death, decreased fetal body weight, abnormalities of external and visceral structure, and abnormalities of skeletal development.

Adverse fetal outcome was increased in the 100- and 150-mg/kg/day groups: intrauterine death was increased at 150 mg/kg/day, and both decreased fetal body weight and increased incidence of malformations were noted at 100 and 150 mg/kg/day. The most common malformations associated with cysteamine were cleft palate (without cleft lip), kyphosis, and abnormalities of the vertebrae. An increase in variations of rib morphology was also noted. Maternal food consumption was decreased for the first 7 days of dosing at 150 mg/kg/day. Maternal weight gain was decreased at 150 mg/kg/day, which was at least partially a result of the fetal loss at this dose. Under the conditions of this study, cysteamine was shown to be teratogenic in the rat with an embryo-fetal NOAEL of 75 mg/kg/day.

Based on the prescribing information for CYSTAGON® (cysteamine bitartrate; Mylan Pharmaceuticals Inc., Morgantown, WV 26505 U.S.A.), the NOAEL for teratogenicity in the rat, 75 mg/kg/day, is 0.4-fold the recommended oral human dose based on body surface area. Although teratogenicity has not been demonstrated in the human, only a few pregnancies have been reported in cystinosis patients and it is recommended that women planning to conceive should stop taking oral cysteamine until after the pregnancy [80].

20.4.5 Embryo-Fetal Toxicity Conclusion

Well-designed animal studies support the assessment of pharmaceuticals for embryo-fetal risk in the human when all relevant information is taken into account. The alternative tests currently available cannot replace *in vivo* testing to support risk assessment, but they can be used effectively for prioritization during drug development or for mechanistic investigations.

REFERENCES

1. IARC (2012), http://monographs.iarc.fr/ENG/Classification/index.php http://monographs.iarc.fr/ENG/Classification/index.php, accessed 22nd January 2013.
2. Pitot, H.C., Dragan, Y.P., Teeguarden, J., Hsia, S., Campbell, H. (1996). Quantitation of multistage carcinogenesis in rat liver. *Toxicologic Pathology*, *24*(1), 119–128.

3. Slaga, T.J., Budunova, I.V., Gimenez-Conti, I.B., Aldaz, C.M. (1996). The mouse skin carcinogenesis model. *Journal of Investigative Dermatology-Symposium Proceedings*, *1*(2), 151–156.
4. Podlaha, O., Riester, M., De, S., Michor, F. (2012). Evolution of the cancer genome. *Trends in Genetics*, *28*(4), 155–163.
5. ICH (1995). International Conference on Harmonisation of Technical Requirements for Registration of Pharmaceuticals for Human Use ICH S1A: Need for Carcinogenicity Studies of Pharmaceuticals.
6. ICH (2011b). International Conference on Harmonisation of Technical Requirements for Registration of Pharmaceuticals for Human Use. ICH S6(R1): Preclinical safety evaluation of biotechnology-derived pharmaceuticals.
7. ICH (1997). International Conference on Harmonisation of Technical Requirements for Registration of Pharmaceuticals for Human Use ICH S1B: Testing for Carcinogenicity of Pharmaceuticals.
8. Alden, C.L., Lynn, A., Bourdeau, A., Morton, D., Sistare, F.D., Kadambi, V.J., Silverman, L. (2011). A critical review of the effectiveness of rodent pharmaceutical carcinogenesis testing in predicting for human risk. *Veterinary Pathology*, *48*(3), 772–784.
9. Friedrich, A., Olejniczak, K. (2011). Evaluation of carcinogenicity studies of medicinal products for human use authorised via the European centralised procedure (1995–2009). *Regulatory Toxicology and Pharmacology*, *60*, 225–248.
10. Dockray, G., Dimaline, R., Varro, A. (2005). Gastrin: Old hormone, new functions. *Pflügers Archiv—European Journal of Physiology*, *449*, 344–355.
11. Pritchard, J.B., French, J.E., Davis, B.J., Haseman, J.K. (2003). The role of transgenic mouse models in carcinogen identification. *Environmental Health Perspectives*, *111*, 444–454.
12. Storer, R.D., Sistare, F.D., Reddy, M.V., DeGeorge, J.J. (2009). An industry perspective on the utility of short-term carcinogenicity testing in transgenic mice in pharmaceutical development. *Toxicologic Pathology*, *210*(38), 1, 51–61.Tabacova, S. (2005). Mode of action: Angiotensin-converting enzyme inhibition—developmental effects associated with exposure to ACE inhibitors. *Critical Reviews in Toxicology, 35*, 747-755.
13. Sistare, F.D., Morton, D., Alden, C., Christensen, J., Keller, D., Jonghe, S.D., Storer, R.D., Reddy, M.V., Kraynak, A., Trela, B., Bienvenu, J.G., Bjurström, S., Bosmans, V., Brewster, D., Colman, K., Dominick, M., Evans, J., Hailey, J.R., Kinter, L., Liu, M., Mahrt, C., Marien, D., Myer, J., Perry, R., Potenta, D., Roth, A., Sherratt, P., Singer, T., Slim, R., Soper, K., Fransson-Steen, R., Stoltz, J., Turner, O., Turnquist, S., van Heerden, M., Woicke, J., DeGeorge, J.J. (2011). An analysis of pharmaceutical experience with decades of rat carcinogenicity testing: Support for a proposal to modify current regulatory guidelines, *Toxicologic Pathology*, *39*, 716–744.
14. Davies, T.S., Monro, A. (1995). Marketed human pharmaceuticals reported to be tumorigenic in rodents. *Journal of the American College of Toxicology*, *14*, 90–107.
15. Hernandez, L.G., Van Steeg, H., Luijten, M., Van Benthem, J. (2009). Mechanisms of non-genotoxic carcinogens and importance of a weight of evidence approach. *Mutation Research*, *682*, 94–109.
16. Futreal, P.A., Coin, L., Marshall, M., Down, T., Hubbard, T., Wooster, R., Rahman, N., Stratton, M.R. (2004). A census of human cancer genes. *Nature Reviews Cancer*, *4*, 177–183.

17. Pfuhler, S., Albertini, S., Fautz, R., Herbold, B., Madle, S., Utesch, D., Pothjk, A. (2007). Genetic toxicity assessment: Employing the best science for human safety evaluation part IV: Recommendation of a working group of the Gesellschaft fuer Umwelt-Mutationsforschung (GUM) for a simple and straightforward approach to genotoxicity testing. *Toxicological Sciences*, *97*(2), 237–240.
18. MacGregor, J.T., Casciano, D., Mueller, L. (2000). Strategies and testing methods for identifying mutagenic risk. *Mutation Research*, *455*, 3–20.
19. Rueff, J., Chiapella, C., Chipman, J.K., Darroudi, F., Silva, I.D. Duverger-van Bogaert, M., Fonti, E., Glatt, H.R., Isern, P., Laires, A., Léonard, A., Llagostera, M., Mossesso, P., Natarajan, A.T., Palitti, F., Rodrigues, A.S., Schinoppi, A., Turchi, G., Werle-Schneider, G. (1996). Development and validation of alternative metabolic systems for mutagenicity testing in short-term assays. *Mutation Research*, *353*, 1-2,151–176.
20. Ku, W.W., Bigger, A., Brambilla, G., Glatt, H., Gocke, E., Guzzie, P.J., Hakura, A., Honma, M., Martus, H.J., Obach, R.S., Roberts, S. (2007). Strategy for genotoxicity testing-Metabolic considerations. *Mutation Research*, *627*, 59–77.
21. ICH (2011a). International Conference on Harmonisation of Technical Requirements for Registration of Pharmaceuticals for Human Use. ICH S2(R1) Guidance on genotoxicity testing and data interpretation for pharmaceuticals intended for human use.
22. OECD. (1997). Guidelines for genetic toxicology testing (TGs No 471-486). OECD Guideline for the Testing of Chemicals. http://www.oecd-ilibrary.org/environment/oecd-guidelines-for-the-testing-of-chemicals-section-4-health-effects_20745788, accessed 22nd January 2013
23. Kirkland, D., Aardema, M., Henderson, L., Mueller, L. (2005). Evaluation of the ability of a battery of three *in vitro* genotoxicity tests to discriminate rodent carcinogens and non-carcinogens I. Sensitivity, specificity and relative predictivity. *Mutation Research*, *584*, 1–256.
24. OECD. (2011). Guideline for the testing of chemicals (TG 488) Transgenic Rodent Somatic and Germ Cell Gene Mutation Assays.
25. Naven, R.T., Louise-May, S., Greene, N. (2010). The computational prediction of genotoxicity. *Expert Opinion on Drug Metabolism & Toxicology*, *6*(7), 797–807.
26. Lee, W.A., Gu, L., Mikszhal, A.R., Chu, N., Leung, K, Nelson, P.H. (1990). Bioavailability improvement of mycophenolic acid through amino ester derivatization. *Pharmaceutical Research*, *7*, 161–166.
27. Bullingham, R.E.S., Nicholls, A., Hale, M. (1996). Pharmacokinetics of mycophenolate mofetil. RS 61443: A short review. *Tranplantation Proceedings*, *28*, 925–929.
28. Heinschink, A., Raab, M., Daxecker, H., Griesmacher, A., Muller, M.M. (2000). *In vitro* effects of mycophenolic acid on cell cycle and activation of human lymphocytes. *Clinica Chimica Acta*, *300*, 23–28.
29. Kunz, B.A., Kohalmi, S.E. (1991). Modulation of mutagenesis by deoxyribonucleotide levels. *Annual Reviews of Genetics*, *25*, 339–359.
30. Martomo, S.A., Mathews, C.K. (2002). Effects of biological DNA precursor pool asymmetry upon accuracy of DNA replication in vitro. *Mutation Research*, *499*, 197–211.
31. Hastings, K.L. (2000). Assessment of immunosuppressant drug carcinogenicity: Standard and alternative animal models. *Human and Experimental Toxicology*, *19*, 261–265.
32. Mydlarski, P.R. (2005). Mycophenolate Mofetil: A dermatologic perspective. *Skin Therapy Letter*, *10*(3), 1–6.

33. Marcen, R. (2009). Immunosuppressive drugs in kidney transplantation, impact on patient survival, and incidence of cardiovascular disease, malignancy and infection. *Drugs*, *69*(16), 2227–2243.

34. Holsapple, M.P., Pitot, H.C., Cohen, S.M., Boobis, A.R., Klaunig, J.E., Pastoor, T., Dellarco, V.L., Dragan, Y.P. (2006). Mode of action in relevance of rodent liver tumors to human cancer risk. *Toxicological Sciences*, *89*, 51–56.

35. Bugelski, P.J., Volk, A, Walker, M.R., Krayer, J.H., Martin, P., Descotes, J. (2010). Critical review of preclinical approaches to evaluate the potential of immunosuppressive drugs to influence human neoplasia. *International Journal of Toxicology*, *29*(5), 435-466.

36. Stoff, B., Salisbury, C., Parker, D., O'Reilly Zwald, F. (2010). Dermatopathology of skin cancer in solid organ transplant recipients. *Transplantation Reviews*, *24*, 172–189.

37. Jensen, P., Hansen, S., Møller, B., Leivestad, T., Pfeffer, P., Geiran, O., Fauchald, P., Simonsen, S., (1999). Skin cancer in kidney and heart transplant recipients and different long-term immunosuppressive therapy regimens. *Journal of the American Academy of Dermatology*, *40*(2), 177–186.

38. Huang, V., M.D., Hepper, D., Anadkat, M., Cornelius, L. (2012). Cutaneous toxic effects associated with Vemurafenib and inhibition of the BRAF pathway. *Archives of Dermatology*, *148*(5), 628–633.

39. Klaunig, J.E., Babich, M.A., Baetcke, K.P., Cook, J.C., Corton, J.C., David, R.M., DeLuca, J.G., Lai, D.Y., McKee, R.H., Peters, J.M., Roberts, R.A., Fenner-Crisp, P.A. (2003). PPARα agonist-induced rodent tumors: Modes of action and human relevance. *Critical Reviews in Toxicology*, *33*, 655–780.

40. Hoflack, J.C., Müller, L. Fowler, S., Braendli-Baiocco, A., Flint, N., Singer, T., Roth, A. (2012). Monitoring Cyp2b10 mRNA expression at cessation of 2-year carcinogenesis bioassay in mouse liver provides evidence for a carcinogenic mechanism devoid of human relevance: The dalcetrapib experience. *Toxicology and Applied Pharmacology*, *255*, 355–365.

41. Deguchi, Y., Yamada, T., Hirose, Y., Nagahori, H., Kushida, M., Sumida, K., Sukata, T., Tomigahara, Y., Nishioka, K., Uwagawa, S., Kawamura, S., Okuno, Y. (2009). Mode of action analysis for the synthetic pyrethroid metofluthrin-induced rat liver tumors: Evidence for hepatic CYP2B induction and hepatocyte proliferation. *Toxicological Sciences*, *108*, 69–80.

42. Yamamoto, Y., Moore, R., Goldsworthy, T.L., Negishi, M., Maronpot, R.R. (2004). The orphan nuclear receptor constitutive active/androstane receptor is essential for liver tumor promotion by phenobarbital in mice. *Cancer Research*, *64*, 7197–7200.

43. Qatanani, M., Moore, D.D. (2005). CAR, the continuously advancing receptor, in drug metabolism and disease. *Current Drug Metabolism*, *6*, 329–339.

44. Olsen, J.H., Wallin, H., Boice, J.D., Jr., Rask, K., Schulgen, G., Fraumeni, J.F., Jr. (1993). Phenobarbital, drug metabolism, and human cancer. *Cancer Epidemiology Biomarkers and Prevention*, *2*, 449–452.

45. Ross, J., Plummer, S.M., Rode, A., Scheer, N., Bower, C.C., Vogel, O., Henderson, C.J., Wolf, C.R., Elcombe, C.R. (2010). Human constitutive androstane receptor (CAR) and pregnane X receptor (PXR) support the hypertrophic but not the hyperplastic response to the murine nongenotoxic hepatocarcinogens phenobarbital and chlordane in vivo. *Toxicological Sciences*, *116*, 452–466.

46. ICH. (2009a). International Conference on Harmonisation of Technical Requirements for Registration of Pharmaceuticals for Human Use. ICH M3(R2): Guidance on nonclinical safety studies for the conduct of human clinical trials and marketing authorization for pharmaceuticals.
47. ICH. (2005). International Conference on Harmonisation of Technical Requirements for Registration of Pharmaceuticals for Human Use. ICH S5(R2): Detection of Toxicity to Reproduction for Medicinal Products & Toxicity to Male Fertility.
48. ICH. (2009b). International Conference on Harmonisation of Technical Requirements for Registration of Pharmaceuticals for Human Use. ICH S9: Nonclinical Evaluation for Anticancer Pharmaceuticals.
49. Committee on Developmental Toxicology, Board on Environmental Studies and Toxicology, National Research Council. *Scientific Frontiers in Developmental Toxicology and Risk Assessment.* National Academy Press, Washington, DC, 2000.
50. Genschow, E., Scholz, G., Brown, N.A., Piersma, A.H., Brady, M., Clemann, N., Huuskonen, H., Paillard, F., Bremer, S., Spielmann, H. (2000). Development of prediction models for three *in vitro* embryotoxicity tests in an ECVAM Validation Study. *In vitro Molecular Toxicology*, *13*, 51–65.
51. Genschow, E., Spielmann, H., Scholz, G., Seiler, A., Brown, N., Piersma, A., Brady, M., Clemann, N., Huuskonen, H., Paillard, F., Bremer, S., Becker, K. (2002). The ECVAM international validation study on *in vitro* embryotoxicity tests: Results of the definitive phase and evaluation of prediction models. European Centre for the Validation of Alternative Methods. *Alternatives to Laboratory Animals*, *30*(2), 151–176.
52. Spielmann, H., Pohl, I., Döring, B., Liebsch, M., Moldenhauer, F. (1997). The embryonic stem cell test (EST), an *in vitro* embryotoxicity test using two permanent mouse cell lines: 3T3 fibroblasts and embryonic stem cells. *In vitro Toxicology*, *10*, 119–127.
53. Flint, O.P., Orton, T.C. (1984). An *in vitro* assay for teratogens with cultures of rat embryonic midbrain and limb cells. *Toxicology and Applied Pharmacology*, *76*, 383–395.
54. New, D.A. (1976). Techniques for assessment of teratologic effects: Embryo culture. *Environmental Health Perspectives*, *18*, 105–110.
55. Brown, N. (2002). Selection of test chemicals for the ECVAM international validation study on *in vitro* embryotoxicity tests. *Alternatives to Laboratory Animals*, *30*(2), 177–198.
56. European Centre for the Validation of Alternative Methods Scientific Advisory Committee. (2002). Statement on the Scientific Validity of the Embryonic Stem Cell Test (EST)—an *In vitro* Test for Embryotoxicity; Statement on the Scientific Validity of the Micromass Test an *In vitro* Test for Embryotoxicity; Statement on the Scientific Validity of the Postimplantation Rat Whole-Embryo Culture Assay -an *In vitro* Test for Embryotoxicity. Available at: *http://ecvam.jrc.it,*accessed 6th August 2012.
57. Marx-Stoelting, P., Adriaens, E., Ahr, H.J., Bremer, S., Garthoff, B., Gelbke, H.P., Piersma, A., Pellizzer, C., Reuter, U., Rogiers, V., Schenk, B., Schwengberg, S., Seiler, A., Spielmann, H., Steemans, M., Stedman, D.B., Vanparys, P., Vericat, J.A., Verwei, M., van der Water, F., Weimer, M., Schwarz, M. (2009). A review of the implementation of the embryonic stem cell test (EST). The report and recommendations of an ECVAM/ReProTect Workshop. *Alternatives to Laboratory Animals*, *37*(3), 313–328.
58. Robinson, J.F., van Beelen, V.A., Verhoef, A., Renkens, M.F., Luijten, M., van Herwijnen, M.H., Westerman, A., Pennings, J.L., Piersma, A.H. (2010). Embryotoxicant-specific

transcriptomic responses in rat postimplantation whole-embryo culture. *Toxicological Sciences*, *118*(2), 675–685.

59. Piersma, A.H. (2006). Alternative methods for developmental toxicity testing. *Basic Clinical Pharmacology and Toxicology*, *98*, 427–431.
60. CDER. (2011). Guidance for Industry Reproductive and Developmental Toxicities—Integrating Study Results to Assess Concerns. U.S. Department of Health and Human Services Food and Drug Administration Center for Drug Evaluation and Research (CDER), September 2011. Available at: *http://www.fda.gov/downloads/Drugs/Guidance ComplianceRegulatoryInformation/Guidances/ucm079240.pdf*,accessed 06th August 2012.
61. Barr, M. Jr. (1994). Teratogen update: Angiotensin-converting enzyme inhibitors. *Teratology*, *50*, 399–409.
62. Pryde, P.G., Sedman, A.B., Nugent, C.E., Barr, M. Jr. (1993). Angiotensin-converting enzyme inhibitor fetopathy. *Journal of the American Society of Nephrology*, *3*(9), 1575–1582.
63. Quan, A. (2006). Fetopathy associated with exposure to angiotensin converting enzyme inhibitors and angiotensin receptor antagonists. *Early Human Development*, *82*, 23–28.
64. Broughton-Pipkin, F., Turner, S.R., Symonds, E.M. (1980). Possible risk with captopril during pregnancy. Some animal data. *Lancet*, *2*, 1256.
65. Broughton-Pipkin, F., Symonds, E.M., Turner, S.R. (1982). The effect of captopril (SQ 14,225) upon mother and fetus in the chronically cannulated ewe and in the pregnant rabbit. *Journal of Physiology*, *323*, 415–422.
66. Ferris, T.F., Weir, E.K. (1983). Effect of captopril on uterine blood flow and prostaglandin E synthesis in the pregnant rabbit. *Journal of Clinical Investigation*, *71*, 809–815.
67. Harewood, W.J., Phippard, A.F., Duggin, G.G. Horvath, J.S., Tiller, D.J. (1994). Fetotoxicity of angiotensin-converting enzyme inhibition in primate pregnancy: A prospective, placebo-controlled study in baboons (Papio hamadryas). *American Journal of Obstetrics and Gynecology*, *171*, 633–642.
68. Spence, S.G., Allen, H.L., Cukierski, M.A., Manson, J.M., Robertson, R.T., Eydelloth, R.S. (1995a). Defining the susceptible period of developmental toxicity for the AT1-selective angiotensin II receptor antagonist losartan in rats. *Teratology*, *51*, 367–382.
69. Spence, S.G., Cukierski, M.A., Manson, J.M., Robertson, R.T., Eydelloth, R.S. (1995b). Evaluation of the reproductive and developmental toxicity of the AT1-selective angiotensin II receptor antagonist losartan in rats. *Teratology*, *51*, 383–397.
70. Ishimura, Y., Chatani, F., Sato, S. (1999). Characterization of hydronephrosis in neonatal rats from dams receiving candesartan cilexetil. *International Journal of Toxicology*, *18*, 369–378.
71. Stevenson, K.M., Gibson, K.J., Lumbers, E.R. (1996). Effects of losartan on the cardiovascular system, renal haemodynamics and function and lung liquid flow in fetal sheep. *Clinical and Experimental Pharmacology and Physiology*, *23*, 125–133.
72. Forhead, A.J., Pipkin, F.B., Sutherland, M.F., Fowden, A.L. (1997). Changes in the maternal and fetal renin-angiotensin systems in response to angiotensin II type 1 receptor blockade and angiotensin-converting enzyme inhibition in pregnant sheep during late gestation. *Experimental Physiology*, *82*, 761–776.

73. Guron, G., Marcussen, N., Nilsson, A., Sundelin, B., Friberg, P. (1999). Postnatal time frame for renal vulnerability to enalapril in rats. *Journal of the American Society of Nephrology*, *10*, 1550–1560.
74. Hoeltzenbein, M., Elefant, E., Vial, T., Finkel-Pekarsky, V., Stephens, S., Clementi, M., Allignol, A., Weber-Schoendorfer, C., Schaefer, C. (2012). Teratogenicity of mycophenolate confirmed in a prospective study of the European Network of Teratology Information Services. *American Journal of Medical Genetics A*. *158A*(3), 588–596.
75. Merlob, P., Stahl, B., Klinger, G. (2009). Tetrada of the possible mycophenolate mofetil embryopathy: A review. *Reproductive Toxicology*, *28*(1), 105–108
76. Eckardt, K., Stahlmann, R. (2010). Use of two validated *in vitro* tests to assess the embryotoxic potential of mycophenolic acid. *Archives of Toxicology*, *84*(1), 37–43.
77. Gahl, W.A., Thoene, J.G., Schneider, J.A. (2002). Cystinosis. *New England Journal of Medicine*, *347*, 111–121.
78. Town, M., Jean, G., Cherqui, S., Attard, M., Forestier, L., Whitmore, S.A., Callen, D.F., Gribouval, O., Broyer, M., Bates, G.P., van't Hoff, W., Antignac, C. (1998). A novel gene encoding an integral membrane protein is mutated in nephropathic cystinosis. *Nature Genetics*, *18*(4), 319–324.
79. Butler, J.D., Zatz, M. (1984). Pantethine and cysteamine deplete cystine from cystinotic fibroblasts via efflux of cysteamine-cystine mixed disulfide. *Journal of Clinical Investigation*, *74*, 411–416.
80. Nesterova, G., Gahl, W. (2008). Nephropathic cystinosis: Late complications of a multisystemic disease. *Pediatric Nephrology*, *23*, 863–878.
81. Winkler, L., Offner, G., Krull, F., Brodehl, J. (1993). Growth and pubertal development in nephropathic cystinosis. *European Journal of Pediatrics*, *152*, 244–249.
82. Reiss, R.E., Kuwabara, T., Smith, M.L., Gahl, W.A. (1988). Successful pregnancy despite placental cystine crystals in a woman with nephropathic cystinosis. *New England Journal of Medicine*, *319*, 223–226.
83. Haase, M., Morgera, S., Bamberg, C., Halle, H., Martini, S., Dragun, D., Neumayer, H.-H., Budde, K. (2006). Successful pregnancies in dialysis patients including those suffering from cystinosis and familial Mediterranean fever. *Journal of Nephrology*, *19*, 677–681.
84. Beckman, D.A., Mullin, J.J., Assadi, F.K. (1998). Developmental toxicity of cysteamine in the rat: Effects on embryo–fetal development. *Teratology*, *58*, 96–102.

21

NEPHROTOXICITY: DEVELOPMENT OF BIOMARKERS FOR PRECLINICAL AND CLINICAL APPLICATION

Frank Dieterle and Estelle Marrer

21.1 INTRODUCTION

21.1.1 Definition and Prevalence of Nephrotoxicity

The kidneys are life-critical organs that have numerous functions such as removing waste materials from the blood, ensuring the homeostasis of the blood and production of several hormones.

In humans, the kidneys filter about 180 liters blood per day and reabsorb important proteins, ions, as well as a big proportion of the filtered fluid, while approximately 1.5 liters of remaining up-concentrated urine are excreted. The macroscopic anatomy of the kidney is depicted in Figure 21.1 (Cortex, Medulla, Papilla, and Pelvis) including the microscopic structures of the nephrons, which are the functional elements of the kidney. In the nephrons, blood is filtered by the glomerulus and the filtrate passes through the proximal tubules, loop of Henle, distal tubules, and collecting ducts. Hereby the filtrate is up-concentrated and important proteins and ions are reabsorbed, metabolized, and recycled, while waste material is excreted with the urine.

Predictive ADMET: Integrative Approaches in Drug Discovery and Development, First Edition.
Edited by Jianling Wang and Laszlo Urban.

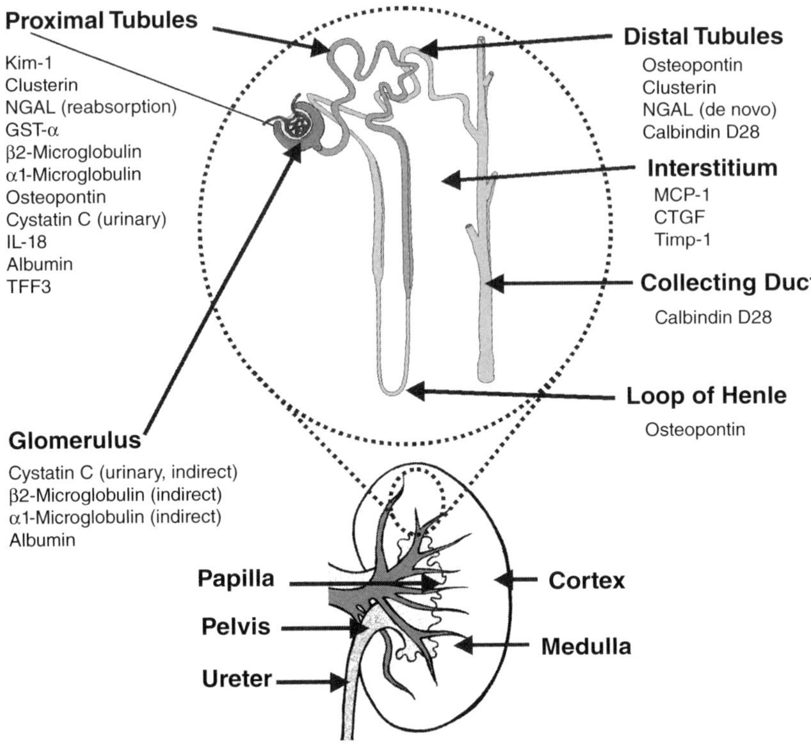

Figure 21.1 Anatomy of the kidney and of nephrons (zoomed), which are the functional units of the kidney for the filtration and production of urine. Different kidney biomarkers can be used to detect injury to different structures of the kidney and the nephrons.

Besides removing waste material, the kidney produces and regulates three important hormones: renin, which is crucial for the regulation of blood pressure; erythropoietin, which stimulates bone marrow to produce red blood cells; and calcitriol, which is important for the regulation of the body's calcium balance.

Kidney disease and renal injury are prevalent serious health conditions that significantly impact patient short- and long-term survival, as a result of alterations or even complete loss of the renal detoxification capacity, and deregulation of the salt and water balance and of kidney endocrine function. Although renal injury and impairment have a broad pathophysiological spectrum, a classification into acute kidney injury (AKI) and chronic kidney diseases (CKD) is typically performed according to the speed of progression (AKI within days, CKD within months to years). It is commonly accepted that altogether drug-induced kidney injury, also referred to as nephrotoxicity, counts for 20% of community- and hospital-acquired AKI cases. Depending on the patient-related risk factors, the incidence rates can dramatically increase. The risk factors creating negative synergies with drug-induced kidney injury are as follows: underlying chronic kidney diseases, sepsis, congestive heart failure, age, and volume depletion. Few examples of chronic drug-induced kidney injury in transplantation with calcineurin inhibitors, chronic lithium therapy for bipolar disorders, and chronic treatment with nonsteroidal anti-inflammatory drugs (NSAIDs) are known, while most of the reported drug-induced kidney injuries fall into the AKI category. As the clinical diagnosis of AKI remains a challenge, nephrotoxicity can be an insurmountable barrier at all stages of drug development.

21.1.2 Pathogenic Mechanisms and Associated Drugs

Nephrotoxicity can be the consequence of one or multiple complex pathogenic mechanisms, which is easily understandable looking at the anatomy and function of the kidney. On the one side, the kidney is basolaterally exposed to blood circulating chemicals and metabolites, and its function is the concentration and excretion of these entities, resulting in a high luminal exposure (direct exposure). For example, cisplatin accumulates in the S3 segment of tubules and the subsequent high local concentrations of the cytotoxic drug cause direct damage to the tubular epithelial cells. On the other side, the renal blood flow accounts for more than one fourth of the cardiac output. Therefore, even those drugs, which are not filtered, have a short half-life, and have low toxicity, can be potential nephrotoxicants (indirect exposure). Various drugs have been reported to be nephrotoxic as listed below by their associated modes of renal toxicity:

- Acute tubular necrosis is commonly observed with cytotoxic drugs excreted via the kidney because the role of the tubule to concentrate and reabsorb the glomerular filtrate renders it vulnerable to a direct injury of the cells. Direct tubular epithelial toxicity is one of the most frequent mechanisms leading to nephrotoxicity observed in drug development. Typical examples are aminoglycosides (neomycin, gentamicin, tobramicin, amikacin, and streptomycin), chemotherapeutics (cisplatin and carboplatin), antiviral agents (adefovir, cidofovir, and tenofovir), bisphosphonates (ibandronate and zoledronate), contrast agents, and many more.
- Acute interstitial nephritis (cell-mediated hypersensitivity reaction to a drug) is a self-limited and reversible acute inflammatory condition. Drugs associated with this condition are vancomycin, penicillins, cephalosporins, diuretics, valproic acid, and many more.
- Glomerular alterations and injury (also known as nephprotic syndrome or glomerulopathy) are marked by heavy proteinuria leading to a secondary injury of the tubules. Drugs causing the previously mentioned alterations are doxorubicin, puromycin, penicillamine, and interferon.
- Crystalline nephropathy (also known as obstructive uropathy, obstructive acute renal failure, or crystal nephropathy) manifests by the precipitation of crystals in the tubular lumens as a result of their insolubility in urine. Crystal deposition causes an obstruction of the tubular lumen and has been observed with indinavavir, tenofovir, acyclovir, and sulfadiazine.

- Hemodynamic renal failure, which is also known as prerenal nephropathy, is caused by modulations of the intrarenal blood flow. The kidney maintains intraglomerular pressure and, consequently, the glomerular filtration rate by modulating the afferent and the efferent arteriole tone. Drugs interfering with the hemodynamic system can cause a drop in the glomerular filtration rate and can result in a renal dysfunction: Angiotensin-converting enzyme (ACE) inhibitors and angiotensin receptor blockers (ARB) inhibiting angiotensin II, and calcineurin inhibitors (cyclosporin A, tacrolimus) cause vasoconstriction. NSAIDs and cyclooxygenase (COX) inhibitors cause prostaglandin-induced vasodilation. Contrast agents may impact renal perfusion in certain patients and precipitate nephropathy. Except for the calcineurin inhibitors, which induce severe nephrotoxicity in a wide population limiting their clinical use, the other agents show their nephrotoxic potential mainly in individuals with impaired renal function, with other risk factors and with co-medication of other potentially nephrotoxic drugs.

21.2 DIAGNOSIS OF NEPHROTOXICITY

21.2.1 Current Standards to Detect Kidney Injury

Preclinically, histopathology is the gold standard to detect and characterize renal injury. However, histopathology is invasive and, therefore, in clinical settings used only in very limited circumstances, such as kidney transplantation. The only remaining possibilities to detect kidney injury are peripheral parameters measured in blood or urine. For the last few decades, the key parameter has been the glomerular filtration rate (GFR) to identify and classify kidney injury. Direct measurement of GFR by renal inulin clearance is labor intensive; therefore, it has been replaced by measurement of the surrogate parameter serum creatinine. Serum creatinine is considered to be constantly produced in the body and filtered by the kidney. Therefore, its serum levels reflect directly the function of the kidney in a first approximation. To simplify and standardize the clinical definition of AKI, the Acute Dialysis Quality Initiative (ADQI) has published a consensus definition, called the RIFLE criteria (Risk for renal dysfunction, Injury to the kidney, Failure of kidney function, Loss of kidney function, and Endstage Renal disease), to diagnose and stage AKI, which has been validated in hundreds of thousands of patients [1,2]. Another parameter also constantly produced by the body and cleared by the kidney is blood urea nitrogen (BUN). It has the same diagnostic potential as serum creatinine but also the same limitations. It is a known fact in the medical community that neither marker is sensitive enough and does not give early warning. In a "healthy" volunteer experiencing AKI, up to two thirds of his or her kidney could be irreversibly injured before any of these two markers would be significantly impacted. In addition, the increases are delayed by several days at the occurrence of the first injury. Finally, the lack of specificity resulting from the dependency on protein intake, meat intake, metabolism, catabolism, liver function, age, dehydration status, muscle mass, extra-renal clearance, etc. render the picture complex. These limitations have alerted scientists and clinicians to look for new molecular and cellular parameters reflecting the loss of integrity and regeneration of the kidney, which has resulted in the use of "OMICs" technologies (genomics and proteomics) and are today referred to as sensitive kidney biomarkers (BMs).

21.2.2 New Kidney Biomarkers to Detect Kidney Injury

Progresses in both fundamental science and technologies has greatly contributed to the identification and development of kidney safety biomarkers that are primarily accessible in the urine. These new biomarkers have been demonstrated to be more sensitive in preclinical species and humans and cover exhaustively the different pathophysiological mechanisms for different regional or anatomical compartments of the kidney [3]. Thus, a combination of these markers allows a precise localization and characterization of the kidney injury (see also Figure 21.1).

Although many exploratory kidney biomarkers have been proposed, significant evidence about their preclinical and clinical utility and limitations has been compiled only for a few, which will be discussed.

- Urinary total protein is a classic "old" parameter that's utility has been reassessed recently. It is used to monitor alterations of the glomerular filtration barrier that are associated with the leakage of massive amounts of proteins into urine. Implemented in qualitative dipstick tests, it used to be a screening tool for undifferentiated glomerular or tubular injury. Nowadays, the quantitative assessment of urinary total protein is the gold standard for monitoring and predicting the progression of nephropathies. Normal urinary protein comprises low amounts of both low- and high-molecular-weight proteins (albumin, β2-microglobuin, α1-microglobulin, retinol binding protein, haptoglobin, cystatin C, and tamm horsefall glycoprotein). Today these proteins are often monitored individually and a ratio between low- and high-molecular-weight proteins allows a differential diagnosis of glomerular versus tubular injury [4–6].
- Urinary albumin is the most abundant protein in urine when proteinuria is diagnosed. It has been used for decades as a marker of glomerular and tubular injury. Albumin is slightly larger than the pores of the glomerular filter. Thus, if massive amounts are found in the ultrafiltrate, the integrity of the glomerular basement membrane is compromised. Also, a small but significant fraction of albumin is filtered and reabsorbed by proximal tubules. In the context of tubular damage, this fraction cannot be reabsorbed and consequently goes into the urine [5, 7–9].
- Serum cystatin C has gained popularity in the recent years. Similar to serum creatinine, it is a measure of the glomerular filtration rate. In contrast to serum creatinine, its production does not depend on muscle mass, age, sex, and it is not cleared extra-renally or by the tubules. Cystatin C has been demonstrated to outperform serum creatinine, and therefore, several clinical FDA-approved

assays are on the market. Nevertheless, it remains a purely functional marker with the disadvantage of a delayed increase during kidney injury [10–15].

- Urinary β2-microglobulin is a low-molecular-weight protein filtered by the glomeruli and completely reabsorbed and metabolized by the tubules. It has been primarily used to detect an impairment of the tubular reabsorption complex, for example, resulting from treatment with aminoglycosides. Increases during tubular injury are often referred to as tubular proteinuria [5, 16–20].
- Urinary and blood NGAL (neutrophil gelatinase-associated lipocalin, also known as lipocalin-2) is a small protein expressed and secreted by kidney epithelial cells during inflammation, as a tubular cell protection/regeneration mechanism. Inflammation in the kidney often results from necrosis, which is a consequence of ischemia. Therefore, urinary and serum NGAL is an early highly predictive marker for the prognosis of renal failure after cardiac surgery [21, 22]. Also, in the context of a drug-induced kidney injury, its prognostic power has been demonstrated in a clinical cisplatin study [23, 24].
- Urinary KIM-1 (kidney injury molecule 1) is a membrane protein expressed at very low levels in healthy tubules, which peaks dramatically after tubular injury and is specifically expressed in dedifferentiated tubules. Its value has been shown to detect drug-induced proximal tubular injury for various drugs preclinically and clinically (e.g., cisplatin, vancomycin, gentamycin, and contrast agents) [25–30].
- Urinary cystatin C, similar to β2-microglobulin, is filtered by the glomeruli and reabsorbed and metabolized by the tubules. Consequently, increased levels of both proteins reflect the same type of renal injury. However, the higher diagnostic value of cystatin C compared with that of β2-microglobulin is linked to its greater stability in urine [31–34].
- Urinary IL-18 is a pro-inflammatory cytokine involved in inflammation, ischemic tissue injury, T-cell mediated immunity, macrophage activation, and NK cells activation and is converted into its active form by caspase-1. In the kidney, it is induced and cleaved mainly in the proximal tubules and released into the urine after kidney injury primarily linked to ischemia. Its diagnostic value is based on early and specific increases [35–37].

21.3 PRESENT AND FUTURE OF RENAL SAFETY MANAGEMENT IN DRUG DEVELOPMENT

21.3.1 Preclinical

The identification part of the biomarker development process is performed best preclinically because of the control of conditions: ideal number of animals in mechanistic studies, their similar genetic background, the absence of co-morbidities, the easy access to the organ for histopathology readouts, and a well-controlled environment during the study.

The first step in the identification of organ specific toxicity biomarkers is the understanding of the biological and biochemical perturbations on a cellular level immediately impacting molecular pathways in adaptive or compensatory responses. These molecular events can be monitored by gene expression profiling analysis. The derived cell-related or structure-related molecular signatures can be correlated to the histopathology. Further confirmation of the observed changes involves *in situ* hybridization, immunohistochemistry, or both. As an example, we will consider puromycin, which is known to induce injury to podocytes highlighted in a mechanistic study. Analysis of time-resolved changes in gene expression profiles first identifies the induction of genes related to the podocyte structure and function (Podocin and Nephrin) and then an increased expression of tubular genes (Kim-1, Clusterin, Timp1) as a subsequent secondary injury [38, 39]. The state-of-the-art method for the investigation of kidney cell-specific toxicities is gene expression profiling of laser capture micro-dissected structures/cells allowing the concentration of these cell-type specific molecular signatures.

Kidneys exposed to toxicants experience primarily dramatic structural and functional (mostly physiological) changes that are reflected at the gene expression level by the induction of structure- and function-specific genes. As a consequence, genes involved in cellular stress (survival, death, damage/repair, mitosis, proliferation, inflammation, and remodeling) are induced and often represent prodromal signs of injury and functional loss. These molecular readouts allow a mechanistic understanding of the mode of nephrotoxicity and might help to mitigate or prevent renal damage in the clinic, as well as allow preclinical mitigation of toxicity and promote "cleaner" follow-up drugs. In addition, several toxicity-impacted genes translate into proteins, which might be cleaved or shed into the circulation and consequently can be measured in peripheral biofluids. This concept represents a crucial translational step in bringing these biomarkers into clinical renal safety management. However, several arising challenges need to be mastered. First, sensitive and reliable assays have to be developed and validated, to allow quantitative and reproducible measurements of these proteins. The fact that until now no reliable quantitative protein assays for the measurement of Podocin and Nephrin are available shows that this step can be a showstopper. Second, once assays are available, enough evidence about the performance of the biomarkers and their limitations needs to be accumulated to enable their regulatory qualification and their use for decision making in drug development. Third, the demonstration that the biomarkers can monitor the same type of kidney injury with a similar performance has to be done in human subjects, which is the last but most challenging step.

21.3.2 Translational Use

In 2009, the qualification of seven kidney biomarkers by the Predictive Safety Testing Consortium (PSTC) for preclinical and translational use set a new milestone in the biomarker history for drug development. The PSCT is a consortium of 16 pharmaceutical companies and academic institutions and regulatory authorities (U.S. Food and Drug Administration (FDA), European Medicines Agency (EMA), and

Pharmaceuticals and Medical Devices Agency (PMDA)) as advisors. The consortium collected evidence about the strengths and weaknesses of seven urinary kidney biomarkers (Albumin, Kim-1, Cystatin C, β2-Microglobulin, TFF3, Clusterin, and total protein) in numerous rat studies. The results and their interpretations from these studies presented evidence for the viability of five biomarkers for human use in kidney diseases, and a proposal was submitted to FDA, EMA, and PMDA for their regulatory biomarker qualification. This process had been introduced by the health authorities during this first safety biomarker submission. The outcome of this biomarker qualification has been published recently by the consortium in a [Nature Biotechnology Focus The Predictive Safety Testing Consortium [LINK: http://www.nature.com/nbt/focus/pstc/index.html]] and by the health authorities [Doc. Ref. EMEA/679719/2008 Rev. 1 [LINK: http://www.ema.europa.eu/docs/en_GB/document_library/Regulatory_and_procedural_guideline/2009/10/WC500004205.pdf]] with the following conclusions:

Nonclinical context:

- The urinary kidney BMs (Kim-1, Albumin, Total Protein, β2-Microglobulin, urinary Clusterin, Urinary Trefoil Factor 3, and urinary Cystatin C) are considered acceptable in the context of nonclinical drug development for the detection of acute drug-induced nephrotoxicity, either tubular or glomerular with associated tubular involvement.
- They provide additional and complementary information to BUN and serum creatinine to correlate with histopathological alterations considered to be the gold standard.
- Additional data on the correlation between the BMs and the evolution and reversibility, of acute kidney injury are needed. Also, further knowledge on species specificity is required.

Clinical context:

- It is recognized that it is worthwhile to explore further, in early clinical trials, the potential of Kim-1, Albumin, Total Protein, β2-Microglobulin, Urinary Clusterin, Urinary Trefoil Factor 3, and urinary Cystatin C as clinical BMs for acute drug-induced kidney injury. Until further data are available to correlate the BMs with the evolution of the nephrotoxic alterations, and their reversibility, their general use for monitoring nephrotoxicity in clinical setting cannot be qualified.
- The use of these renal biomarkers in clinical trials may be considered on a case-by-case basis to gather further data to qualify their usefulness in monitoring drug-induced renal toxicity in humans.

What does this mean practically for preclinical and for translational kidney safety assessment? In preclinical species, these markers can be included in non-good laboratory practice (GLP) studies to support drug candidate selection and understanding of mechanisms of toxicity. More importantly, in GLP studies, the monitoring of these

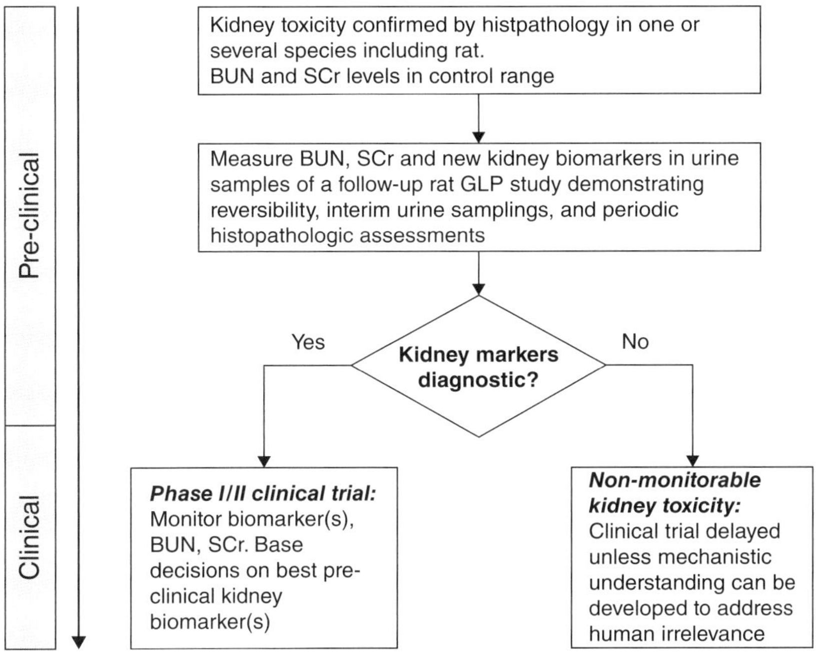

Figure 21.2 Flowchart depicting how the implementation of kidney biomarkers can enable the translation of a drug candidate that is nephrotoxic in preclinical species into first-in-human studies.

biomarkers together with the "old standards" guides safety assessment and can eventually enable the translation of drugs with preclinical glomerular or tubular toxicity findings into humans, which is illustrated in Figure 21.2.

If in preclinical toxicity studies nephrotoxicity has been identified as a safety liability, it needs to be demonstrated that kidney biomarkers can first detect renal toxicity in animals and furthermore prove whether the injury is still reversible in the presence of positive biomarker results, which means that toxicity can be managed. An ideal prospective preclinical study would encompass a real-time monitoring of the kidney biomarkers. As soon as the levels of kidney biomarkers are increased, one group of animals is sacrificed for histopathology assessment to correlate changes of biomarkers with histological findings and to quantify the level of kidney injury at the time point when kidney biomarkers are diagnostic. A second group of animals is continued while stopping dosing of the drug at the moment when the biomarkers are increased. This second group serves as a recovery group to demonstrate that the kidney injury is reversible once treatment is stopped. In summary, a positive outcome of this study means that kidney injury is manageable in animals based on the biomarkers. In case real-time measurements of the kidney biomarkers are not available, a similar study design needs to be simulated by having regular necropsy time points and peeling off different groups for recovery at different time points [40]. Assuming the safety margins established preclinically are sufficient to bring the drug into early clinical trials, the clinical study designs need to implement close monitoring of the kidney biomarkers and other kidney safety parameters. Stopping rules based on changes in those parameters, which demonstrated the best performance preclinically, have to be defined prospectively. Eventually, the toxicity observed preclinically will not translate into the clinic. In this situation, the new kidney markers helped to bring the drug candidate into humans and hereby allowed for demonstrating that nephrotoxicity is not a specific concern in humans.

21.3.3 Clinical Use

For general clinical studies, the new kidney safety biomarkers have not yet been qualified on a regulatory level. However, these markers are implemented in different clinical studies as exploratory markers together with the gold standard serum creatinine to accumulate further evidence about their utility in human subjects. Similar to the preclinical PSTC consortium, a European consortium has been founded to qualify kidney, liver, and vascular toxicity biomarkers for general regulated clinical drug development. This consortium, called IMI SAFE-T, runs numerous clinical studies encompassing several types of drug-induced injuries, diseases, as well as different clinical populations with underlying co-morbidities (e.g., diabetes, rheumatoid arthritis, and renal- and liver-impaired patients). It is expected that this consortium will submit the first clinical biomarker submission packages to FDA and EMA in 2014/2015 [41]. The new biomarkers have raised a lot of interest among academic investigators who have demonstrated in many publications that these biomarkers can surpass the current standards in many clinical contexts. Until now the most exciting advances have been shown in the utility of various kidney biomarkers in the detection of AKI after cardiopulmonary bypass surgery and in critically ill patients.

Once a critical amount of clinical evidence about the advantages and disadvantages of various kidney biomarkers to detect and manage nephrotoxicity in humans is accumulated, management of kidney safety should be revolutionized at all stages of drug development.

REFERENCES

1. Bellomo, R., Kellum, J.A., Ronco, C. (2007). Defining and classifying acute renal failure: From advocacy to consensus and validation of the RIFLE criteria. *Intensive Care Medicine*, *33*, 409–413.
2. Bagshaw, S.M. (2011). Acute kidney injury: Diagnosis and classification of AKI: AKIN or RIFLE? *Nature Reviews Nephrology*, *6*, 71–73.
3. Bonventre, J.V., Vaidya, V.S., Schmouder, R., Feig, P., Dieterle, F. (2010). Next-generation biomarkers for detecting kidney toxicity. *Nature Biotechnology*, *28*, 436–440.
4. D'Amico, G., Bazzui, C. (2003). Pathophysiology of proteinuria. *Kidney International*, *63*, 809–825.
5. Schieppati, A., Remuzzi, G. (2003). Proteinuria and its consequences in renal disease. *Acta Paediatrica Supplement*, *443*, 9–13.

6. Thielemans, N., Lauwerys, R., Bernard, A. (1994). Competition between Albumin and low-molecular-weight proteins for renal tubular uptake in experimental nephropathies. *Nephron*, *66*, 453–458.
7. Christensen, E.I., Birn, H., Rippe, B., Maunsbach, A.B. (2007). Controversies in nephrology: Renal albumin handling, facts, and artifacts! *Kidney International*, *72*, 1192–1194.
8. Greive, K.A., Nikolic-Paterson, D.J., Guimarães, M.A., Nikolovski, J., Pratt, L.M., Mu, W., Atkins, R.C., Comper, W.D. (2001). Glomerular permselectivity factors are not responsible for the increase in fractional clearance of albumin in rat glomerulonephritis. *American Journal of Pathology*, *159*, 1159–1170.
9. Russo, L.M., Sandoval, R.M., McKee, M., Osicka, T.M., Collins, A.B., Brown, D., Molitoris, B.A., Comper, W.D. (2007). The normal kidney filters nephrotic levels of albumin retrieved by proximal tubule cells: Retrieval is disrupted in nephrotic states. *Kidney International*, *71*, 504–513.
10. Mussap, M., Plebani, M. (2004). Biochemistry and clinical role of human cystatin C. *Critical Reviews in Clinical Laboratory Sciences*, *41*, 467–550.
11. Madero, M., Sarnak, M.J., Stevens, L.A. (2006). Serum cystatin C as a marker of glomerular filtration rate. *Current Opinion in Nephroly and Hypertension*, *15*, 610–616.
12. Dharnidharka, V.R., Kwon, C., Stevens, G. (2002). Serum cystatin C is superior to serum creatinine as a marker of kidney function: A meta-analysis. *American Journal of Kidney Diseases*, *40*, 221–226.
13. Herget-Rosenthal, S., Marggraf, G., Hüsing, J., Göring, F., Pietruck, F., Janssen, O., Philipp, T., Kribben, A. (2004). Early detection of acute renal failure by serum cystatin C. *Kidney International*, *66*, 1115–1122.
14. Séronie-Vivien, S., Delanaye, P., Piéroni, L., Mariat, C., Froissart, M., Cristol, J.P. (2008). Cystatin C: Current position and future prospects. *Clinical Chemistry and Laboratory Medicine*, *46*, 1664–1686.
15. Ozer, J.S., Dieterle, F., Troth, S., Perentes, E., Cordier, A., Verdes, P., Staedtler, F. et al. (2010). A panel of urinary biomarkers to monitor reversibility of renal injury and a serum marker with improved potential to assess renal function. *Nature Biotechnology*, *28*, 486–494.
16. Davey, P.G., Gosling, P. (1982). Beta 2-Microglobulin instability in pathological urine. *Clinical Chemistry*, *28*, 1330–1333.
17. Dehne, M.G., Boldt, J., Heise, D., Hempelmann, G. (1995). Tamm-Horsfall protein, alpha-1- and beta-2-microglobulin as kidney function markers in heart surgery. *Anesthesist*, *44*, 545–551.
18. Gatanaga, H., Tachikawa, N., Kikuchi, Y., Teruya, K., Genka, I., Honda, M., Tanuma, J., Yazaki, H., Ueda, A., Kimura, S., Oka, S. (2006). Urinary beta2-microglobulin as a possible sensitive marker for renal injury caused by tenofovir disoproxil fumarate. *AIDS Research and Human Retroviruses*, *22*, 744–748.
19. Schentag, J.J., Plaut, M.E. (1980). Patterns of urinary beta 2-microglobulin excretion by patients treated with aminoglycosides. *Kidney International*, *17*, 654–661.
20. Tirelli, A.S., Colombo, N., Cavanna, G., Mangioni, C., Assael, B.M. (1985). Follow-up study of enzymuria and beta 2 microglobulinuria during cis-platinum treatment. *European Journal of Clinical Pharamacology*, *29*, 313–318.
21. Mishra, J., Dent, C., Tarabishi, R., Mitsnefes, M.M., Ma, Q., Kelly, C., Ruff, S.M., Zahedi, K., Shao, M., Bean, J., Mori, K., Barasch, J., Devarajan, P. (2008). Neutrophil

gelatinase-associated lipocalin (NGAL) as a biomarker for acute renal injury after cardiac surgery. *Lancet*, *365*, 1231–1238.

22. Wagener, G., Jan, M., Kim, M., Mori, K., Barasch, J.M., Sladen, R.N., Lee, H.T. (2006). Association between increases in urinary neutrophil gelatinase-associated lipocalin and acute renal dysfunction after adult cardiac surgery. *Anesthesiology*, *105*, 485–491.
23. Mishra, J., Mori, K., Ma, Q., Kelly, C., Barasch, J., Devarajan, P. (2004). Neutrophil gelatinase-associated lipocalin: A novel early urinary biomarker for cisplatin nephrotoxicity. *American Journal of Nephrology*, *24*, 307–315.
24. Gaspari, F., Cravedi, P., Mandalà, M., Perico, N., de Leon, F.R., Stucchi, N., Ferrari, S., Labianca, R., Remuzzi, G., Ruggenenti, P. (2010). Predicting cisplatin-induced acute kidney injury by urinary neutrophil gelatinase-associated lipocalin excretion: A pilot prospective case-control study. *Nephron Clinical Practice*, *115*, 154–160.
25. Vaidya, V.S., Ozer, J.S., Dieterle, F., Collings, F.B., Ramirez, V., et al. (2010). Kidney injury molecule-1 outperforms traditional biomarkers of kidney injury in preclinical biomarker qualification studies. *Nature Biotechnology*, *28*, 478–485.
26. Han, W.K., Waikar, S.S., Johnson, A., Betensky, R.A., Dent, C.L., Devarajan, P, Bonventre, J.V. (2008). Urinary biomarkers in the early diagnosis of acute kidney injury. *Kidney International*, *73*, 863–869.
27. Liangos, O., Perianayagam, M.C., Vaidya, V.S., Han, W.K., Wald, R., Tighiouart, H., MacKinnon, R.W., Li, L., Balakrishnan, V.S., Pereira, B.J., Bonventre, J.V., Jaber, B.L. (2007). Urinary N-acetyl-beta-(D)-glucosaminidase activity and kidney injury molecule-1 level are associated with adverse outcomes in acute renal failure. *Journal of the American Society of Nephrology*, *18*, 904–912.
28. Han, W.K., Waikar, S.S., Johnson, A., Betensky, R.A., Dent, C.L, Devarajan, P., Bonventre J.V. (2008). Urinary biomarkers in the early diagnosis of acute kidney injury. *Kidney International*, *73*, 863–869.
29. Rached, E., Hoffmann, D., Blumbach, K., Weber, K., Dekant, W., Mally, A. (2008). Evaluation of putative biomarkers of nephrotoxicity after exposure to ochratoxin a in vivo and in vitro. *Toxicological Sciences*, *103*, 371–81.
30. Zhou, Y., Vaidya, V.S., Brown, R.P., Zhang, J., Rosenzweig, B.A., Thompson, K.L., Miller, T.J., Bonventre, J.V., Goering, P.L. (2008). Comparison of kidney injury molecule-1 and other nephrotoxicity biomarkers in urine and kidney following acute exposure to gentamicin, mercury, and chromium. *Toxicological Sciences*, *101*, 159–170.
31. Dieterle, F., Perentes, E., Cordier, A., Roth, D.R., Verdes, P., Grenet, O., Pantano, S., Moulin, P., Wahl, D., Mahl, A., End, P., Staedtler, F., Legay, F., Carl, K., Laurie, D., Chibout, S.D., Vonderscher, J., Maurer, G. (2010). Urinary clusterin, cystatin C, beta2-microglobulin and total protein as markers to detect drug-induced kidney injury. *Nature Biotechnology*, *28*, 463–469.
32. Herget-Rosenthal, S., van Wijk, J.A., Bröcker-Preuss, M., Bökenkamp, A. (2007). Increased urinary cystatin C reflects structural and functional renal tubular impairment independent of glomerular filtration rate. *Clinical Biochemistry*, *40*, 946–951.
33. Conti, M., Moutereau, S., Zater, M., Lallali, K., Durrbach, A., Manivet, P., Eschwège, P., Loric, S. (2006). Urinary cystatin C as a specific marker of tubular dysfunction. *Clinical Chemistry and Laboratory Medicine*, *44*, 288–291.
34. Tkaczyk, M., Nowicki, M., Lukamowicz, J. (2004). Increased cystatin C concentration in urine of nephrotic children. *Pediatric Nephrology*, *19*, 1278–1280.

35. Parikh, C.R., Jani, A., Melnikov, V.Y., Faubel, S., Edelstein, C.L. (2004). Urinary interleukin-18 is a marker of human acute tubular necrosis. *American Journal of Kidney Diseases*, *43*, 405–414.

36. Parikh, C.R., Mishra, J., Thiessen-Philbrook, H., Dursun, B., Ma, Q., Kelly, C., Dent, C., Devarajan, P., Edelstein, C.L. (2006). Urinary IL-18 is an early predictive biomarker of acute kidney injury after cardiac surgery. *Kidney International*, *70*, 199–203.

37. Parikh, C.R., Coca, S.G., Thiessen-Philbrook, H., Shlipak, M.G., Koyner, J.L., Wang, Z., Edelstein, C.L., Devarajan, P., Patel, U.D., Zappitelli, M., Krawczeski, C.D., Passik, C.S., Swaminathan, M., Garg, A.X. TRIBE-AKI Consortium. (2011). Postoperative biomarkers predict acute kidney injury and poor outcomes after adult cardiac surgery. *Journal of the American Society of Nephrology*, *22*, 1748–1757.

38. Amin, R.P., Vickers, A.E., Sistare, F., Thompson, K.L., Roman, R.J., Lawton, M., Kramer, J. et al. (2004). Identification of putative gene-based markers of renal toxicity. *Environmental Health Perspectives*, *112*, 465–479.

39. Betsholtz, C., He, L., Takemoto, M., Norlin, J., Sun, Y., Patrakka, J., Tryggvason, K. (2007). The glomerular transcriptome and proteome. *Nephron Experimental Nephrology*, *106*, 32–36.

40. Dieterle, F., Sistare, F., Goodsaid, F., Papaluca, M., Ozer, J.S., Webb, C.P., Baer, W., et al. (2010). Renal biomarker qualification submission: A dialog between the FDA-EMEA and predictive safety testing consortium. *Nature Biotechnology*, *28*, 455–62.

41. Matheis, K., Laurie, D., Andriamandroso, C., Arber, N., Badimon, L., Benain, X., Bendjama, K., Clavier, I. et al. (2011). A generic operational strategy to qualify translational safety biomarkers. *Drug Discovery Today*, *16*, 600–608.

IV

SUCCESS STORIES AND LESSONS LEARNED

22

EARLY INTERVENTION WITH FORMULATION STRATEGIES FOR MULTIDIMENSIONAL PROBLEMS TO OPTIMIZE FOR SUCCESS

Stephanie Dodd, Christina Capacci-Daniel, Christopher Towler, Riccardo Panicucci, and Keith Hoffmaster

22.1 INTRODUCTION

The poor solubility of molecules has always limited one's ability to formulate pharmaceutical molecules. Although the formation of a salt from a free form tends to improve solubility, there are limitations to this approach when seeking to improve oral exposure [1]. Formulation technologies commonly used today emerged from specific solubility-related problems. Micronization and nanosizing of pharmaceuticals were designed to improve dissolution and apparent *in vivo* solubility [2]. Solid dispersions were developed to stabilize the amorphous phase of a molecule when crystalline material did not show sufficient solubility or dissolution [3, 4]. Microemulsions of lipid and hydrophilic excipients evolved from poorly water-soluble molecules that needed complete solubilization [5]. In the past, the application of these strategies was reactive and relied on trial and error to triage solubility and exposure issues without

Predictive ADMET: Integrative Approaches in Drug Discovery and Development, First Edition.
Edited by Jianling Wang and Laszlo Urban.

looking at the root causes of these problems. Given the significant resources and extensive timelines needed to operate in this stepwise manner, proactive, parallel, and more innovative approaches are needed to develop more efficiently successful formulations for novel therapeutic agents.

Holistic understanding of a pharmaceutical candidate's physiochemical and biopharmaceutical properties can enable rapid and focused formulation development. From a formulation perspective, absorption is driven by two main concepts—solubility and permeability—whereas ultimate systemic exposure *in vivo* can also be determined by first-pass metabolism as well as by transporter-mediated efflux. *In vitro* models can be constructed to understand the behavior of a molecule with respect to expected *in vivo* solubility and permeability. Collectively, this knowledge is critical because together with an understanding of drug metabolism and the potential to interact with transporters, it impacts formulation decisions. For example, solubility should not be thought of as a static property but as a dynamic combination of kinetic solubility and supersaturation in a biorelevant environment. If one does not consider all these aspects, then potentially an insufficient or an overly complicated formulation may be proposed. Additionally, formulation strategies can improve absorption when active transport, intestinal metabolism, or a narrow gastrointestinal (GI) absorption window limits oral exposure of a molecule. As a result, a combination of formulation strategies can be employed to deliver molecules with multiple liabilities successfully that in the past would have been discarded as being too difficult to develop. The effort needed to understand a molecule's solubility characteristics is minimal compared with the iterative formulation trials required when one uses a reactive approach to formulation development based on *in vivo* preclinical or clinical outcomes. Property-based formulation development has emerged as the standard for efficient pharmaceutical molecule optimization.

Herein are presented several cases studies where understanding molecule properties contributed greatly to the successful selection of a preclinical or clinical formulation. In each instance, the pharmaceutical agent is a classic poorly water-soluble molecule, but in addition, each has unique factors limiting absorption that are not addressed by typical formulation methods. These case studies demonstrate how careful selection of excipients, novel approaches toward solubility enhancement, understanding the principles of supersaturation, determining mechanism of permeability, and *in silico* modeling can all contribute to successful formulation development prior to extensive *in vivo* evaluation and clinical trials. All of these examples highlight the importance of proactive understanding of the mechanisms limiting formulation development and subsequent implementation of a hypothesis-driven approach toward formulation optimization.

22.2 COMBINING FORMULATION STRATEGIES TO OVERCOME POOR SOLUBILITY AND INHIBIT CRYSTALLIZATION

Acidic molecules account for a smaller portion of new chemical entities developed for therapeutic purposes but offer their own challenges when developing an

Table 22.1 Equilibrium solubility of molecule A in aqueous buffers

	Free acid A mg/mL (pH*)	Salt of molecule A mg/mL (pH*)
pH 1.0	0.006 (4.12)	<0.002 (2.74)
pH 4.0	0.005 (4.15)	<0.002 (5.27)
pH 6.0	0.016 (5.92)	0.087 (6.29)
pH 6.8	0.023 (6.85)	0.450 (6.90)
pH 7.4	0.094 (7.31)	0.619 (7.28)

*Actual pH measured at the end of equilibrium solubility experiments.

in vitro–in vivo correlation to support a formulation strategy. Despite these experimental challenges, the fundamental principles of solubility- and permeability-mediated absorption still govern the oral absorption of acidic molecules. If physicochemical properties limit the *in vivo* exposure of a particular molecule, straightforward formulation strategies can be employed to attempt to overcome these limitations with deliberately designed studies.

Conducting preclinical or clinical efficacy and tolerability studies of a pharmaceutical without the most favorable form and formulation can lead to expensive failures when exposure multiples and *in vivo* biological endpoints are not achieved. Early intervention to optimize the physical form and formulation can therefore enable rapid progression from preclinical testing to the clinic. Molecule A was a carboxylic acid (pKa = 4.6) with a highly pH-dependent solubility profile. In aqueous buffers below pH 4 and simulated gastric fluids, the solubility of A was <5 ug/mL and improved to only 23 ug/mL in pH 6.8 media. Salts of A were easily produced and increased the pH 6.8 equilibrium solubility to 437 ug/mL, although very little improvement was observed at moderately acidic pH values (Table 22.1).

The free acid and salts of molecule A were highly crystalline and had poor dissolution rates. A suspension of the free acid showed 10% bioavailability in beagle dogs, whereas a salt suspension was 26% bioavailable demonstrating only modest improvement at an equivalent 10 mg/kg dose. Given the observed pH-dependent solubility for both A and its salts, the team hypothesized that the drug substances did not have sufficient time in an acceptable pH region to dissolve *in vivo*, resulting in low exposure after suspension administration. Given the expected need for a high clinically efficacious dose of molecule A, further formulation development with the salts was conducted to improve *in vivo* exposure.

In addition to the physicochemical properties of molecule A, data from transepithelial transport studies Caco-2 cell monolayers suggested that moderate-to-poor permeability and efflux could also limit the oral absorption of molecule A [6]. Unlike other preclinical species, *in vivo* data in rats showed extensive metabolism that resulted in considerably lower oral bioavailability relative to mouse and dog. As rats were the desired species to study *in vivo* toxicology, solution and optimized salt suspension formulations were dosed in hepatic portal vein cannulated rats to improve understanding of the main mechanism of low oral bioavailability. The solution formulation showed a fraction absorbed of 94%, whereas the suspension was only 14%.

These data strongly suggested that the permeability of molecule A was sufficient and that the *in vivo* solubility of molecule A limited its exposure.

To understand better the solubility of molecule A, a two-step *in vitro* dissolution method was employed before further preclinical studies were performed. Molecule A was stirred for 30 minutes in a simulated gastric environment before additional media were added to raise the pH to 6.5, thus, simulating a transition to an intestinal condition. Using this method, several formulations of molecule A were tested to prioritize further development. Interestingly, most potential formulations seemed to have an immediate 100% release with the change in pH. Solubility in pH 6.8 in this assay was thought to be indicative of *in vivo* solubility and therefore predictive of exposure. From these dissolution studies, an optimal micronized salt suspension, solid microemulsion, and solid dispersion formulation were identified and selected for *in vivo* testing. However, when these optimized formulations were dosed *in vivo*, they failed to produce a high exposure contrary to the indications from the *in vitro* dissolution data.

In this example, a rapid pH switch dissolution method as an *in vitro* model for an acidic molecule did not accurately describe the *in vivo* absorption and oral exposure. This two-step dissolution technique is recommended by the U.S. Food and Drug Administration (FDA) and is very effective for basic molecules that are highly soluble at low pH and, then once the pH switch occurs, can be observed to either stay solubilized or precipitate [7]. However, for acidic molecules, the rate of dissolution throughout a biologically relevant, moderate pH region is critical to their *in vivo* solubility. Customized dissolution procedures should be implemented for relative ranking of formulations of acidic molecules and the development of predictive *in vitro–in vivo* models.

Ultimately, dissolution enhancement was not sufficient to maximize exposure of molecule A, so a microemulsion formulation that completely solubilized molecule A was selected. *In vitro* studies indicated that this microemulsion formulation could only maintain sufficient solubilization of molecule A at dilution in simulated gastric media up to 125 mg/kg, which in turn limited the range of toxicology doses *in vivo*. Above this concentration, the formulation rapidly precipitated out as the highly crystalline free acid form of molecule A. To enable the planned toxicology design and expected high clinical doses, the formulation was therefore optimized to include an excipient to inhibit crystal growth and maintain *in vivo* supersaturation of molecule A.

A small percentage of a pH-sensitive polymer, 5% Eudragit EPO, was found to be sufficient to maintain solubility in the microemulsion vehicle at dilution in simulated gastric media. This dual-acting microemulsion formulation, which solubilized and inhibited crystallization, was found to produce dose-proportional exposure when administered above the previous precipitation threshold for molecule A. Interestingly, *in vitro* testing showed that the polymer alone would not be sufficient to improve exposure. The polymer was only capable of preventing crystallization of a supersaturated molecule A solution but did not enhance dissolution or solubility of the crystallized salt form. An *in vitro–in vivo* correlation was finally made for molecule A with the understanding that solubility should be maintained over the gastric period of the two-step dissolution assay for optimal exposure. These criteria were used to validate

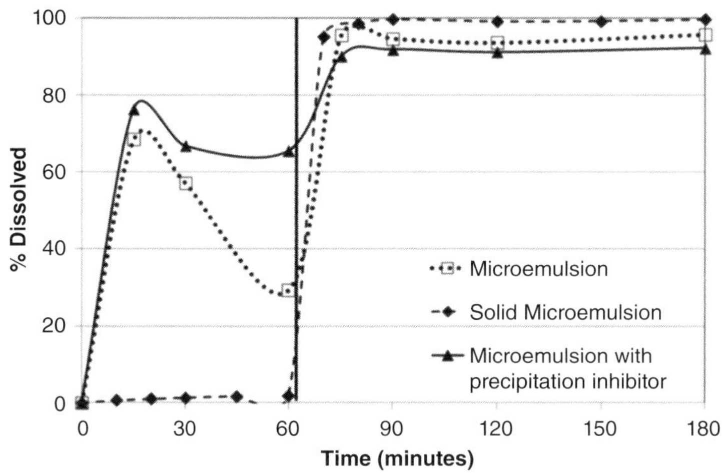

Figure 22.1 Two-step dissolution experiment showing the impact of various microemulsion formulations. Only the microemulsion with the precipitation inhibitor could best maintain the solubility of molecule A at low pH (first 60 min). The observed solubility in the *in vitro* pH 6.8 environment (60–180 min) did not correlate with exposure *in vivo*.

the microemulsion with the precipitation inhibitor against previous dissolution data that correlated to an unsuccessful formulation (Figure 22.1).

Molecule A represented an interesting case study where exposure was primarily driven by solubility and where several strategies were required to maintain solubility *in vivo* to promote adequate oral bioavailability (Table 22.2). As a result of the highly crystalline physical forms and their resulting poor solubility in low-to-moderate pH media, dissolution-enhancing strategies were not sufficient. A microemulsion preconcentrate vehicle was needed to solubilize molecule A initially and entirely. As a result of the propensity of molecule A to crystallize as its highly stable free

Table 22.2 Impact of various formulations on molecule A oral bioavailability in preclinical species at 10–30 mg/kg

	Bioavailability		
Vehicle	Rat	Mouse	Dog
Solution (30% organic content)	7%	42%	*na*
Free Acid Suspension	*na*	*na*	10%
Salt Suspension	2%	27%	26%
Solid Dispersion	11%	*na*	*na*
Solid Microemulsion	*na*	*na*	15%
Microemulsion	28%	100%	100%
Microemulsion with Precipitation Inhibitor (250 mg/kg)	32%	*na*	*na*

na = Not all formulations were tested across every species.

acid form, a precipitation inhibitor was required to maintain this microemulsion-based solubilization and supersaturation *in vivo*. It was the combination of both formulation strategies—solubility enhancement and supersaturation extension with a precipitation inhibitor—that enabled dosing of molecule A with the desired preclinical exposure. Understanding these liabilities early in preclinical development allowed for the implementation of a custom formulation such that an ideal vehicle compatible with both toxicology and clinical planned doses could be identified prior to initiation of preclinical good laboratory practice (GLP) toxicology studies.

22.3 FORMULATION STRATEGIES TO AVOID EFFLUX LIABILITY

Molecule B was under consideration for development for an infectious disease indication. The compound had a logP of 2.1, was weakly basic, and typical of a weak base, displayed a strong pH-dependent solubility profile in aqueous media. The molecule had two pKas of 8.36 and 5.8, which enabled formation of a salt of the compound to aid solubility. The isolated highly crystalline free form was anhydrous with a melting point of 155°C. Preliminary *in vivo* studies in rats showed that the molecule had low and variable plasma concentrations after oral dosing; therefore, further *in vitro* studies were aimed at identifying the potential mechanistic barriers of low exposure.

In vitro evaluation of molecule B in a Caco-2 transepithelial transport assay showed the molecule to be a potential efflux substrate, with a P_{app}(B-A)/P_{app}(A-B) ratio of 6.0. Additionally, permeability in the A→B direction, P_{app}(A-B), was low, with a value of 1.04×10^{-6} cm/s, which is equivalent to a calculated fraction absorbed % of ~35% [6]. Based on these data, the team hypothesized that efflux may be contributing to the low exposure and sought to develop a formulation strategy to use excipients to overcome the potential efflux mechanism. As a first step, transporter phenotyping studies were employed to identify the specific transporter involved in the active efflux observed in Caco-2 cell monolayers [8]. Through the use of selective inhibitors, both P-glycoprotein (P-gP) and multidrug resistance protein 2 (MRP2) were implicated as possible mechanisms of efflux; a contribution of breast cancer-resistant protein (BCRP) to the observed efflux was ruled out (Figure 22.2A). These results allowed for a focus on excipients that had been shown to modulate only P-gP and MRP2 efflux and permitted insight into possible mechanisms of molecule B disposition *in vivo*.

Several investigators have shown that in addition to their ability to aid in solubilization, surfactants can also modulate transporter activity both *in vitro* and *in vivo* [9]. Vitamin E Tocopherol Polyethylene Glycol Succinate (Vitamin ETPGS) was chosen for further investigation based on its activity against P-gP and MRP2 as well as on the potential to improve the solubility of molecule B [10]. Initial studies with Vitamin ETPGS were designed to determine the concentration of the excipient that fully inhibited efflux in the Caco-2 cell assay. As shown in Figure 22.2B, the minimum concentration required to achieve a P_{app}(B-A)/P_{app}(A-B) ratio of 1 was 0.25 mg/mL of Vitamin ETPGS. At this concentration of Vitamin ETPGS, the efflux ratio was decreased to ~1, suggesting that net efflux transporter activity was inhibited.

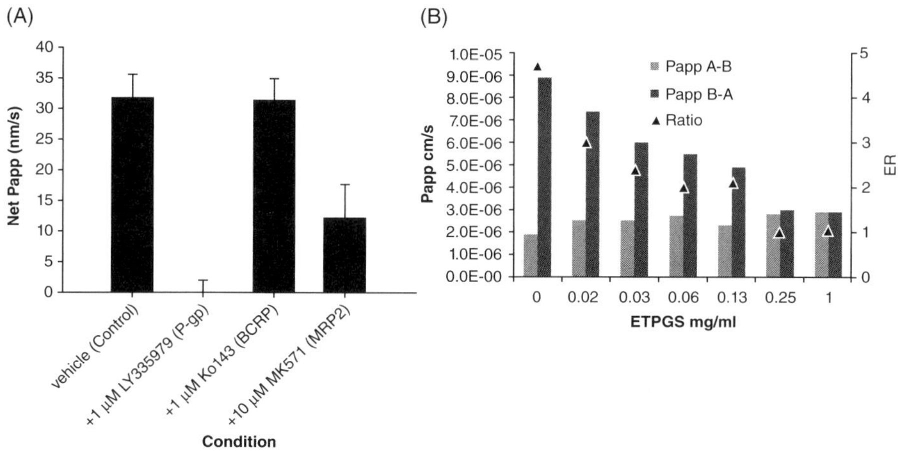

Figure 22.2 (A) Identification of the efflux transporters responsible for transport of molecule B suggests that P-gP-mediated transport may contribute to a mechanism of low oral bioavailability. The decrease in net P_{app} (apparent permeability coefficient) with MK571 also suggests that MRP2 is capable of transporting this molecule. (B) Impact of vitamin ETPGS on molecule B P_{app} A→B, P_{app} B→A, and efflux ratio. Vitamin ETPGS concentrations >0.25 mg/mL attenuated the efflux ratio to ~1, suggesting full inhibition of transporter-mediated efflux in the *in vitro* experiment.

Given these promising *in vitro* results with Vitamin ETPGS, efforts were directed to understand what concentration of excipient would be required to improve the solubility profile of the molecule. A linear relationship was established between the solubility of molecule B and increasing concentrations of Vitamin ETPGS, allowing for prediction of the amount of Vitamin ETPGS required to solubilize different amounts of the compound.

To understand how solubility and transporter-mediated efflux work together to limit oral exposure, suspension formulations were developed to evaluate the impact of dissolution via particle size reduction (down to nanosuspensions) as well as the impact of Vitamin ETPGS on overall exposure. Several stable suspensions resulted, with adequate drug loading and stability to enable further evaluation in preclinical toxicology studies. Comparative oral pharmacokinetic studies with and without ETPGS were conducted in Sprague–Dawley (SD) rats to ensure the excipient was attenuating the efflux transporters *in vivo*, as predicted from the *in vitro* Caco-2 evaluations. From these data, it was likely that efflux, rather than solubility or dissolution, was likely limiting *in vivo* exposure after oral dosing (Figure 22.3).

Preclinical confirmation that the Vitamin ETPGS formulation improved oral exposure provided a basis for a clinical formulation. A microemulsion was chosen because Vitamin ETPGS could be easily incorporated into a liquid formulation (as compared with a solid dosage form). The microemulsion supported formulation of doses from 10 to 1000 mg in humans. In a single ascending dose clinical study, the formulation was found to deliver a slightly overproportional exposure profile (based on area under concentration (AUC)), with very low interpatient variability (Figure 22.4).

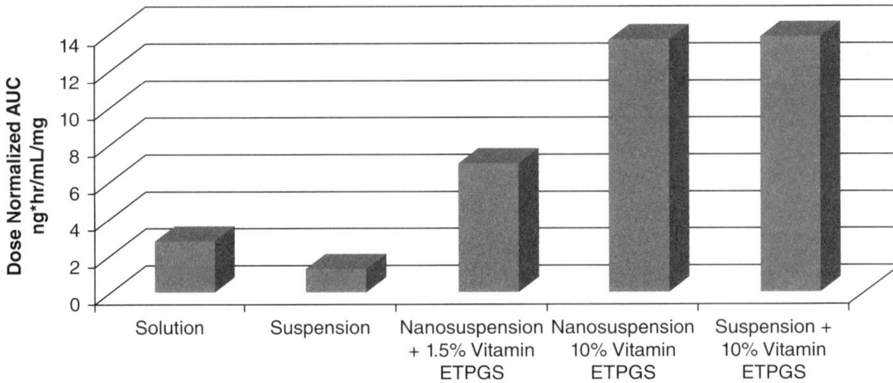

Figure 22.3 Evaluation of the effect of Vitamin ETPGS on oral exposure of molecule B in Sprague–Dawley rats. Suspension formulation contained particle sizes of >1 uM but <10 uM; nanosupsension formulation contained particle size <300 nm.

In this case study, early understanding of the potential impact of transporters together with compound solubility on oral absorption of a molecule in early preclinical development prompted design of a formulation to overcome both an efflux mechanism and possibly improve solubility. The ultimate impact was that the team saved several months of formulation development to support toxicology studies. Additionally, understanding of mechanisms that could potentially limit absorption in patients provided a clear and successful path forward in achieving desired exposures in the clinic.

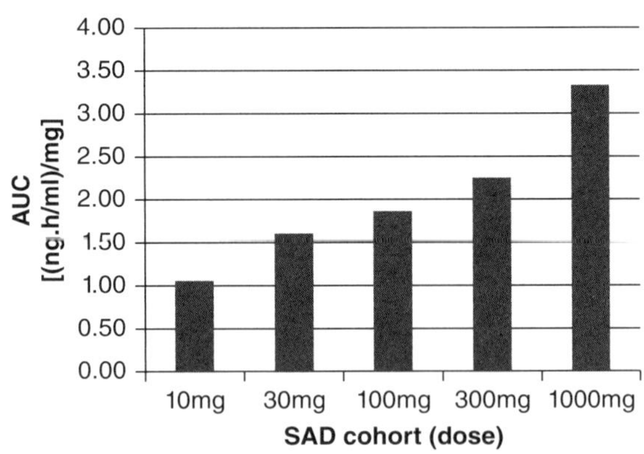

Figure 22.4 Dose-normalized AUC in the single ascending dose (SAD) clinical study shows a slightly overproportional increase in AUC from 10 to 1000 mg with microemulsion formulation containing the P-gP inhibitor Vitamin ETPGS. This observation may be explained by some minor saturation of *in vivo* clearance at higher exposures, or saturation of any noninhibited P-gP-mediated intestinal efflux at higher doses.

22.4 IDENTIFYING SOLUBILIZATION AS A KEY STRATEGY TO IMPROVE ORAL EXPOSURE

As demonstrated in the two previous examples, development of a formulation that can progress a poorly water-soluble molecule into toxicology evaluations in preclinical species and subsequently into the clinic evaluation can be a challenge. Mechanistic understanding of why a molecule shows poor exposure *in vivo* can enable scientists to apply an appropriate formulation strategy prior to the drug development phase.

Molecule C was a compound under consideration for development for a diabetic indication. The molecule was slightly acidic, with an aqueous media solubility profile that would be termed slightly soluble by USP definition in a pH range of 1–7.5. Molecule C had a measured aqueous pKa of 11.0, was acidic, and had a predicted logP of 4.033. With such a high pKa, it was not possible to make a salt form of the molecule. The isolated highly crystalline free form was anhydrous with a melting point of 221°C.

Based on the pH solubility profile and the compound's physicochemical properties, multiple formulation strategies could be applied. Understanding whether the molecule was limited by supersaturation (dictated by dissolution profile) or nucleation (time needed to form amorphous, crystalline solids, or allgomerates from drug in the dissolved state) could direct formulation techniques, whether as single strategies or in combination, to improve oral exposure.

Supersaturation and nucleation behaviors of molecule C were studied in a solvent-shift assay [7]. This method induced supersaturation of the molecule by first dissolving the molecule in dimethylsulfoxide (DMSO) and spiking the stock solution with high dilution into fasted state simulated intestinal fluid, pH = 6.5 media (FaSSIF), well above the equilibrium solubility of the molecule in FaSSIF (22 μg/mL). Initially, supersaturation was induced at a factor of 13 (294 μg/mL). At approximately 10 minutes, the concentration of molecule C remained supersaturated at a factor of 4.45 (98 μg/mL). Although the molecule showed the ability to supersaturate, nucleation tendency was very strong even at lower concentrations. These results suggested that supersaturation likely was induced by solubilization formulation. The results additionally provided an "*in vivo*" solubility number equivalent to the equilibrium solubility in FaSSIF as a result of the fast nucleating behavior of the molecule in aqueous media. These data were then incorporated into a GastroPlus™ (Simulations Plus, Lancaster, CA) model that, based on the model-based prediction of the *in vivo* pharmacokinetic profile, suggested that the observed *in vitro* equilibrium solubility was reflective of the *in vivo* solubility (Figure 22.5A).

When molecule C was administered orally as a 5 mg/kg stable solution (10% n-methyl-2-pyrrolidone (NMP), 30% polyethylene glycol (PEG) 300, 5% Cremophor EL, 55% water for injection (WFI)) *in vivo* in rats, pharmacokinetic analysis indicated complete absorption of the dose (%F ~100%). When a low-dose suspension formulation (0.5% Methyl Cellulose, 0.5% Tween 80) of amorphous molecule was dosed, oral bioavailabiliy was significantly decreased (%F < 5%). This study revealed that even in the absence of a crystalline structure, the molecule was still unable to dissolve quickly enough to achieve supersaturation. Further confirmation was obtained with

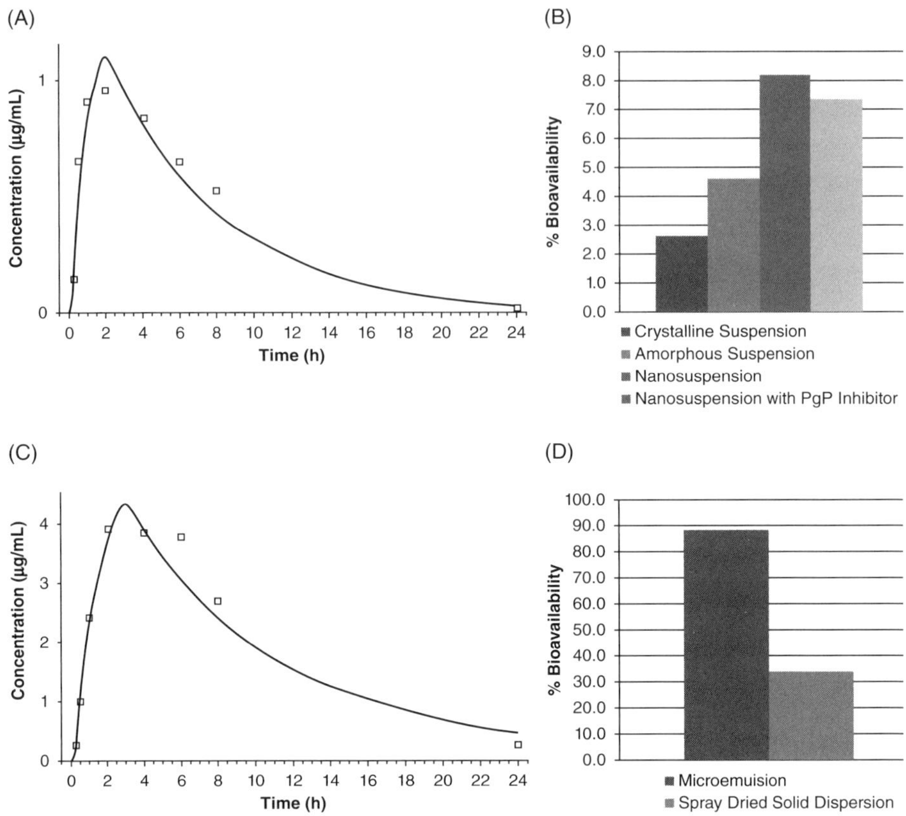

Figure 22.5 (A) Plasma concentration vs. time profile of molecule C after a 5 mg/kg dose of a stable solution in Sprague–Dawley rats. Raw data points (γ) were overlayed with predicted *in vivo* concentration vs. time profile using GastroPlus (solid line). Equilibrium solubility of the molecule in FaSSIF media (0.022 mg/mL) was used for the simulation; permeability was calculated from Caco-2 A→B permeability (2.443 × 10^4 cm/s). (B) Impact of amorphous, crystalline, and nanosuspensions on molecule C oral bioavailability at 30 mg/kg in Sprague–Dawley rats. (C) Plasma concentration vs. time profile of molecule C after a 30 mg/kg dose of a microemulsion formulation in Sprague–Dawley rats (γ) overlayed with the predicted *in vivo* concentration vs. time profile using GastroPlus (solid line). The solubility value used for the simulation was 0.055 mg/mL. The permeability was calculated from the Caco-2 assay (2.443 × 10^4 cm/s). (D) Oral bioavailability of molecule C after administration as a microemulsion and amorphous dispersion at 30 mg/kg in Sprague–Dawley rats.

a dose of a crystalline suspension, yielding a similar %F < 5%. Nanosuspension formulation showed very little improvement in oral bioavailability; a log-fold particle size reduction showed %F still less than 10% (Figure 22.5B). These results shifted the project team focus toward formulation approaches to increase the solubilization of the molecule.

The lipophilicity of molecule C supported development of a microemulsion formulation. Previous experiments with amorphous suspensions suggested that deletion of the crystal lattice would not offer much improvement in oral absorption. However, the solubilization of excipients in an amorphous dispersion could have offered some advantage in oral absorption. Therefore, both microemulsion and solid amorphous dispersions were investigated further.

Despite being lipophilic and soluble at a higher concentration in surfactants, lipids, and lipophilic excipients, molecule C showed nucleation behavior even in the microemulsion preconcentrate developed at high drug load (10%). Therefore, much lower drug load was tested *in vivo* (1%) to allow for stability in the microemulsion system itself. The exposure profile was modeled using GastroPlus based on the 30 mg/kg microemulsion dose in rats (Figure 22.5C). The same parameters were used from the previous 5 mg/kg solution GastroPlus model, with an exception that *in vivo* solubility was increased ~twofold based on results from an *in vitro* solvent shift assay solubility assessment after 1000 seconds. Approximately 88% of the dose was absorbed and orally available from the microemulsion formulation (Figure 22.5D). Results from the GastroPlus model supported that when a formulation facilitates supersaturation of the drug, the *in vitro* assessment of solubility at 1000 seconds better predicted the *in vivo* absorption of the molecule. If the formulation does not create supersaturation (e.g., suspension and nanosuspension), there is no advantage of supersaturation-induced nucleation, and therefore, the equilibrium solubility will better predict *in vivo* absorption. Through this combination of mechanistic understanding and relevant quantitative data from *in vitro* studies, PBPK modeling can well predict *in vivo* exposure.

The solid dispersion developed exhibited a glass transition (Tg) of 157°C, which is about 40°C lower than the Tg of the amorphous molecule (199°C). The solid dispersion was slightly better at improving oral bioavailability (~35%) compared with the nanosuspension and showed a 300% increase in exposure at a 30 mg/kg dose (Figure 22.5D).

Early investigation of mechanisms contributing to poor oral bioavailability of molecule C confirmed initial observations based on the supersaturation/nucleation behavior of the compound. A solubilizing formulation was required to ensure full absorption of the drug *in vivo*. In this example, modification of crystal size or creation of amorphous material was not able to create the same level of drug exposure as a formulation that completely solubilized the molecule. As a result, an optimized formulation was identified prior to toxicology studies that not only reduced the number of animal studies needed to support further development of molecule C, but also limited the demand on drug supply based on the improved exposure at lower doses.

22.5 MANAGING UNFAVORABLE PHYSCIOCHEMICAL PROPERTIES, POSSIBLE FOOD EFFECTS, AND LIKELY DRUG INTERACTIONS IN PARALLEL THROUGH FORMULATION-BASED APPROACHES

Early assessment of the physicochemical properties as it relates to the biopharmaceutical properties of new molecular entities clearly can aid in formulation development [11]. Despite an appreciation of key chemical properties that can facilitate oral bioavailability of new chemical entities, the nature of some drug targets requires

Table 22.3 Batch-to-batch variation observed in AUC, C_{max}, and T_{max} values during rat and cynomolgus monkey toxicology studies with molecule D

Species	Batch#	Dose (mg/kg)	C_{max} (ng/mL)	AUC (ng*h/mL)	T_{max} (h)
Rat	6	10	9650	154000	4
		100	63800	1440000	24
Rat	12	100	3320	45750	3
		600	8075	114500	5
Monkey	6	10	9400	182000	5.3
Monkey	12	60	480	7745	7

molecular properties for selectivity and potency that are not favorable for oral absorption. Molecule D violated several of the "rule of 5" criteria that are linked with good oral bioavailability having a molecular weight of 710, a logP of 7.1, and 10 hydrogen bond acceptors [12]. This molecule also had 12 rotatable bonds, a molecular property that has also been inversely correlated with poor oral absorption [13].

As a result of limitations in the availability of drug substance in the early development phase, a series of Sprague–Dawley rat and cynomolgus monkey toxicology studies required the administration of several different batches of molecule D to complete all of the required studies outlined in the preclinical toxicology plans. After a 100 mg/kg oral suspension dose of batch 12 to SD rats, the AUC_{plasma} of molecule D was 31-fold lower and C_{max} was 19-fold lower than the same dose of batch 6 of molecule D (Table 22.3). Dose normalization of 60 and 10 mg/kg oral doses in cynomolgus monkeys showed similar results between batch 12 and batch 6; AUC_{plasma} and C_{max} were 140-fold and 118-fold lower, respectively. Several initial hypotheses could not explain the differences in exposure between different studies (e.g., incorrect doses administered and bioanalytical errors); therefore, further investigation into the physical properties of the two different batches was initiated. Studies to characterize the solid state of the molecule showed that batch 12 displayed a higher melting, higher crystallinity polymorph that exhibited significantly lower solubility than batch 6. As a result, the straightforward development strategy identified based on the low crystallinity and the low melting nature of batch 6 could not be applied to batch 12 of molecule D. Although the team attempted to synthesize and isolate additional quantities of batch 6 for further toxicology and development studies, purification of the original polymorph was not possible so efforts were focused on formulation of the lower solubility form.

Given the high melting point, strong crystal lattice, and high lipophilicity of the new crystalline polymorph form, the solubility was observed to be very low across a wide range of physiologic pH (<0.5 μg/mL). To understand better the key mechanistic basis for this low solubility, a multiparametric solubility diagnosis tool was applied and it was suggested that lipophilicity was the primary reason for the poor solubility of this molecule [14]. The analysis considered several experimentally measured parameters (melting point, heat of fusion, pKa, intrinsic solubility, solubility at pH 6.8, and logP) as well as calculated (solubility at pH 6.8 derived from pKa and Henderson–Hasselbach equation, molar fraction, clogP, and the difference between the solubility and lipophilicity) parameters. In addition to guiding the mechanistic understanding of the observed low solubility, these parameters and data also influenced the overall formulation strategy, identified other potential limiting factors in oral absorption, and allowed the team to prioritize formulation approaches. Interestingly, a large discrepancy in the solubility in fed and fasted simulated intestinal fluids was observed with the molecule being more than 40 times more soluble in fed state simulated intestinal fluid (FeSSIF) media. Although a difference between FeSSIF and FaSSIF is typical for highly lipophilic molecules, this ratio of 40 is higher than typically observed from these authors' experience.

Based on these physicochemical data (especially the low solubility) and some understanding of similar molecules in clinical development, early *in vitro* assessment of the potential for a positive food effect *in vivo* and possible formulation approaches to address this was initiated. As bile acids can aid in compound solubilization in the GI tract *in vivo*, solubility measurements of molecule D in the presence of increasing concentrations of sodium taurocholate were conducted. Solubility in the presence of bile salts increased the aqueous solubility of molecule D 90-fold, suggesting that a food effect *in vivo* likely would be a result of interaction with food and not a pH effect. Given the observed batch dependence on molecule D exposure and the *in vitro* assessment of food effect, a formulation strategy to prevent drug precipitation was employed. *In vitro* biorelevant media precipitation tests were used to rank-order formulations by their ability to keep the molecule solubilized in both fed and fasted simulated intestinal media. As a result, a microemulsion and an amorphous solid dispersion formulation were identified for further *in vivo* evaluation. Other approaches considered include salt formation, particle size reduction, and solid microemulsions, although these were determined to be inferior to the microemulsion and solid dispersion approaches. After *in vivo* administration of the two optimized formulations, both AUC_{plasma} and C_{max} increased significantly relative to the previous results with the crystalline batch 12, and enabled continuation of preclinical toxicology studies.

Although both formulations achieved the desired exposure to enable rat and monkey toxicology studies, a single formulation needed to be selected for further clinical development. When based purely on physical properties and conventional formulation principles, both formulations were adequate for application in the clinic. However, the project team also faced the challenge that molecule D likely would be co-administered with a marketed compound known to be metabolized primarily by Cytochrome P450 3A4 (CYP3A4). To aid in final formulation selection, the team hypothesized that a potential intestinal drug interaction may occur in the clinic and assessed the ability of microemulsion and solid dispersion excipients to inhibit CYP3A4/5 *in vitro* in human liver microsomes. Excipients were examined for their effect on midazolam 1′-hydroxylation and testosterone 6β-hydroxylation. It was found that nearly all of the microemulsion components could be expected to inhibit CYP3A4/5, whereas very few of the solid dispersion components were expected to have an effect. As a result of this foresight toward a potential late-stage clinical challenge, the team selected the solid dispersion for phase 1 clinical studies. Although clinical drug interaction data between molecule D and the potential co-medication are yet to be generated, results from the phase I clinical trials showed a dose-proportional AUC_{plasma} when molecule D was administered at doses up to 600 mg/kg.

22.6 KEY LEARNINGS FROM CASE STUDIES

Given that multiple mechanisms can impact drug absorption, poor intrinsic solubility may not be the only factor limiting *in vivo* exposure. From the case studies presented earlier, three key themes emerge. First, by early incorporation of formulation strategies prior to clinical development, teams were able to assess molecules efficiently in preclinical toxicology studies, while also decreasing the risk of clinical failure as a result of inadequate exposure. Although each project team invested additional time and conducted several experiments prior to compound administration in humans, this slight increase in need for resources in the preclinical phase helped streamline formulation optimization in later development. Second, each team sought to identify the mechanism behind the observed low *in vivo* exposure. In most cases, *in silico* modeling and *in vitro* studies were able to help generate, test, and refine hypotheses, and limited *in vivo* studies were needed only to confirm these results. Finally, all of the teams faced the need to optimize multiple properties of formulations in parallel. In the case of molecule B, optimization of solubility alone may not have resulted in an orally available molecule as a result of the efflux liability. In the case of molecule D, although the microemulsion formulation likely would have shown the desired exposure in the clinic, this formulation may have caused an unfavorable drug interaction with a known co-medication, thereby limiting its clinical usage. Ultimately, a logical approach to formulation development may be rescue molecules that in the past would have been discarded as being too difficult or expensive to develop. Teams likely will also benefit from more rapid, successful progression to later stage clinical trials as well as cost savings resulting from targeted selection of optimal formulations in late discovery instead of refining formulations later in the clinical phases of drug development.

22.7 FUTURE DIRECTIONS

Understanding a molecule's physiochemical and biopharmaceutical properties and any resulting liabilities is critical for clinical success of a drug formulation. Incorporation of broader mechanistic understanding of the absorption process into strategies toward formulation development will require scientists to expand their knowledge of chemistry, biology, and drug metabolism in preclinical development. A holistic understanding of drug disposition, physical-organic chemistry, and material science should provide a strong scientific background with which scientists can diagnose a drug candidate's major liabilities and intervene with appropriate strategies earlier in the drug discovery process. Likewise, a strong understanding of fundamental principles opens opportunities to apply burgeoning technologies for novel drug delivery. In addition to the tools described herein, the next generation of computer modeling will help formulation scientists reach conclusions more quickly and can be used proactively to select molecules with the best propensity for *in vivo* exposure. Now that kinetic solubility and GI absorption can be modeled *in silico*, these same principles and approaches can be applied to other routes of administration, thus, streamlining the development of transdermal or other novel delivery routes and depot technology. Modern preclinical development is evolving such that molecule selection and drug delivery approaches will soon, if not already, be guided by earlier evaluation of rate limiting processes in absorption; the proactive, more frequent use of *in silico* and *in vitro* models to predict *in vivo* behavior; and hypothesis-driven preclinical experiments that translate to desired clinical exposures.

ACKNOWLEDGEMENTS

The authors thank Xuena Lin and Dallas Bednarczyk for conducting transporter studies to support mechanistic understanding of molecule B.

REFERENCES

1. Serajuddin, A.T. (2007). Salt formation to improve drug solubility. *Advanced Drug Delivery Reviews*, *59*(7), 603–616.
2. Merisko-Liversidge, E., Liversidge, G.G., Cooper, E.R. (2003). Nanosizing: A formulation approach for poorly-water-soluble compounds. *European Journal of Pharmaceutical Sciences*, *18*(2), 113–120.
3. Leuner, C., Dressman, J. (2000). Improving drug solubility for oral delivery using solid dispersions. *European Journal of Pharmaceutics and Biopharmaceutics*, *50*(1), 47–60.
4. Kaushal, A.M., Gupta, P., Bansal, A.K. (2004). Amorphous drug delivery systems: Molecular aspects, design, and performance. *Critical Reviews in Therapeutic Drug Carrier Systems*, *21*(3), 133–193.
5. Pouton, C.W. (2000). Lipid formulations for oral administration of drugs: Non-emulsifying, self-emulsifying and "self-microemulsifying" drug delivery systems. *European Journal of Pharmaceutical Sciences*, *9*(2), S93–S98.
6. Skolnik, S., Lin, X., Wang, J., Chen, X.H., He, T., Zhang, B. (2010). Towards prediction of in vivo intestinal absorption using a 96-well Caco-2 assay. *Journal of Pharmaceutical Sciences*, *99*(7), 3246–3265.
7. Ozaki, S., Minamisono, T., Yamashita, T., Kato, T., Kushida, I. (2012). Supersaturation-nucleation behavior of poorly soluble drugs and its impact on the oral absorption of drugs in thermodynamically high-energy forms. *Journal of Pharmaceutical Sciences*, *101*(1), 214–222.
8. Lin, X., Skolnik, S., Chen, X., Wang, J. (2011). Attenuation of intestinal absorption by major efflux transporters: Quantitative tools and strategies using a Caco-2 model. *Drug Metabolism and Disposition*, *39*(2), 265–274.

9. Rege, B.D., Kao, J.P., Polli, J.E. (2002). Effects of nonionic surfactants on membrane transporters in Caco-2 cell monolayers. *European Journal of Pharmaceutical Sciences*, *16*(4-5), 237–246.
10. Bogman, K., Zysset, Y., Degen, L., Hopfgartner, G., Gutmann, H., Alsenz, J., et al. (2005). P-glycoprotein and surfactants: Effect on intestinal talinolol absorption. *Clinical Pharmacology and Therapeutics*, *77*(1), 24–32.
11. Saxena, V., Panicucci, R., Joshi, Y., Garad, S. (2009). Developability assessment in pharmaceutical industry: An integrated group approach for selecting developable candidates. *Journal of Pharmaceutical Sciences*. *98*(6), 1962–1979.
12. Lipinski, C.A., Lombardo, F., Dominy, B.W., Feeney, P.J. (2001). Experimental and computational approaches to estimate solubility and permeability in drug discovery and development settings. *Advanced Drug Delivery Reviews*, *46*(1-3), 3–26.
13. Veber, D.F., Johnson, S.R., Cheng, H.Y., Smith, B.R., Ward, K.W., Kopple, K.D. (2002). Molecular properties that influence the oral bioavailability of drug candidates. *Journal of Medicinal Chemistry*, *45*(12), 2615–2623.
14. Faller, B., Ertl, P. (2007). Computational approaches to determine drug solubility. *Advanced Drug Delivery Reviews*, *59*(7), 533–545.

23

CYTOCHROME P450-MEDIATED DRUG INTERACTION AND CARDIOVASCULAR SAFETY: THE SELDANE TO ALLEGRA TRANSFORMATION

F. Peter Guengerich

23.1 INTRODUCTION

Metabolism can play a critical role in the safety of drugs and, consequently, whether they reach the market and remain there. Terfenadine, the active ingredient in Seldane®, has become one of the classic examples of why drug metabolism is very critical in the development process. It also serves as a case history in the development of a metabolite as a backup. Finally, this case shows how much drug metabolism has changed in the past 20 years, in terms of both the pharmaceutical industry and the regulatory agencies.

Predictive ADMET: Integrative Approaches in Drug Discovery and Development, First Edition.
Edited by Jianling Wang and Laszlo Urban.

23.2 BACKGROUND ON TERFENADINE

Terfenadine (Figure 23.1) was the first nonsedating antihistamine on the market. The drug acts as an antagonist of the H_1 receptor and is useful in the treatment of allergic rhinitis. Terfenadine does not impair psychomotor performance [1] probably because neither terfenadine nor its metabolites cross the blood-brain barrier. Because of its useful properties, it was used by ~100 million people by 1990 [2], after its introduction by Marion-Merrell-Dow (Kansas City, MO) in 1985.

23.3 BACKGROUND ON CYTOCHROME P450 (P450) ENZYMES

The metabolism of drugs in the body had been recognized since the 1800s [3], and P450 enzymes were known to catalyze many drug oxidation reactions by the 1960s. However, at the time of introduction of Seldane in 1985, the knowledge of human P450 enzymes and their involvement in the metabolism of individual drugs was very limited. Seminal observations were reported by Smith in the late 1970s, in that a single genetic locus could control the ability of the human body to metabolize a series of drugs, including the antihypertensive debrisoquine [4]. The findings suggested that at least some of the human drug-metabolizing P450s might have high selectivity for drug substrates.

Although some human liver P450s had been purified to electrophoretic homogeneity by 1983 [5, 6], the characterization of their catalytic activities was limited. An approach of purifying human liver P450s using specific drug oxidation assays yielded the proteins now known to be P450 2D6 [7], P450 1A2 [7], P450 2C8 [8], P450 2C9 [8], and P450 3A4 [9]. Antibodies prepared against these proteins, as well as a series of chemical inhibitors whose selectivity was interrogated, could be used to define the roles of individual reactions, along with correlation analyses [10, 11]. Subsequently, heterologous expression systems were also applied to the classification [12–14] and the validation of *in vivo* biomarkers for several P450s began [15, 16]. By the early 1990s, extensive data were available on the levels of expression of individual P450s in human livers [17].

Although the knowledge of the properties of human P450s was very limited before 1985, in retrospect, progress in the next five years was remarkable and the methods for evaluating roles of individual P450s (or other enzymes) were in place, at least *in vitro* [11].

Today this area of drug metabolism is rather mature, with knowledge that ~75% of drug metabolism reactions are catalyzed by P450s and that ~90% of these segments are a result of five well-characterized P450s: 1A2, 2C9, 2C19, 2D6, and 3A4 [18–21].

23.4 ISSUES ARISING WITH TERFENADINE

By 1990, ~100 million people had used terfenadine worldwide (*vide supra*) [2]. In 1989, an arrhythmia was observed in a patient with an intentional overdose,

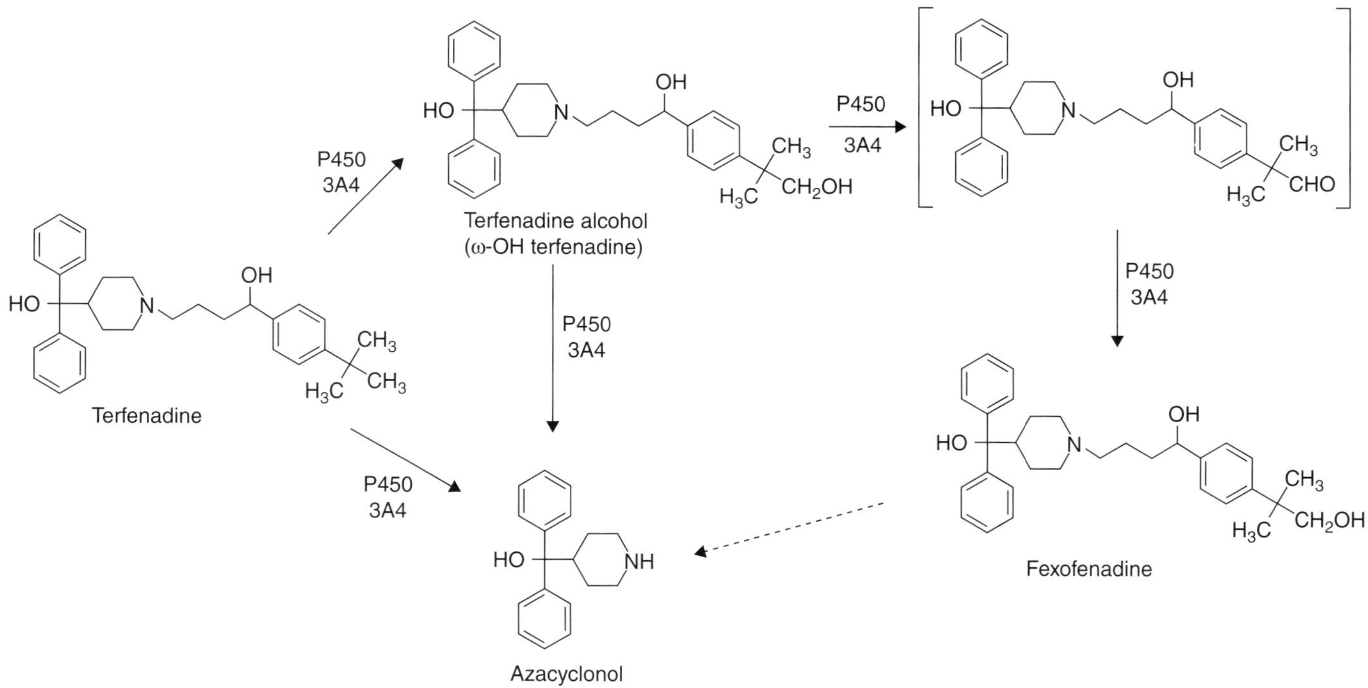

Figure 23.1 Metabolism of terfenadine [1,2]. All steps are catalyzed primarily by P450 3A4.

and by 1990, this syndrome of "torsade des pointes" was also observed in some individuals using the recommended dose [22, 23]. In June 1990, the U.S. Food and Drug Administration (FDA) released a report on the risk associated with concomitant use of erythromycin and ketoconazole and sent a letter to physicians two months later.

The problem was identified as accumulation of terfenadine in plasma. Normally the concentration of the parent drug terfenadine in plasma is very low. The plasma level of the drug rose in most of the individuals who were affected [24].

23.5 DELINEATION OF THE MECHANISM UNDERLYING THE PROBLEM

The manufacturer of Seldane®, Marion-Merrell-Dow, contacted me in late 1991 (via my late colleague Prof. Grant Wilkinson) and asked for assistance in defining the enzymes involved in terfenadine oxidation. The major two oxidation products, ω-hydroxyterfenadine (i.e., hydroxylation of a *tert*-butyl methyl group) and acyclazol, were already known (Figure 23.1). We had extensive previous experience in such "reaction phenotyping" in human liver microsomes. Although the approach was becoming somewhat monotonous and I did not consider it a role of my own group to define the P450s involved in the oxidation of every drug, we decided to accept the assignment and help the company. As I recall, we completed the work within a few weeks, thanks to the assistance of Dr. C-H. Yun, a postdoctoral fellow in my lab at the time (today this work could have been done even faster with the reagents now available).

The results were rather consistent, including chemical inhibition, antibody inhibition (Figure 23.2), and correlation with marker activities in the liver microsomes. The answer was that P450 3A4 (or 3A5) was the major oxidative enzyme. At that time we only had recombinant (human) P450 3A4 expressed in yeast [25] and these experiments were limited.

We notified Marion-Merrell-Dow of our results, finished some remaining aspects, and met with company scientists in May 1992. A manuscript was submitted in June 1992 and accepted in August; it did not actually appear in *Drug Metabolism and Disposition* until March 1993 [26].

Further studies at Marion-Merrell-Dow demonstrated that P450 3A4 was involved not only in the two reactions leading to the ω-alcohol and azacyclonal (*N*-dealkylation product) but also in the oxidation of the ω-alcohol to the corresponding carboxcylic acid, which now has the name fexofenadine [27]. All of these results were consistent with the contraindications related to the concomitant use of ketoconazole or erythroymycin with terfenadine [28, 29]. A point of note is later work showing a potential problem with terfenadine and grapefruit [30]. Although the situation was not specifically studied in this case, one might expect individuals with low intrinsic levels of P450 3A4 to be more susceptible. The level of P450 3A4 protein varies considerably among people *23* [21], and some individuals have been found to have no detectable hepatic P450 3A4 [31, 32].

More recent work has also shown that P450 2J2 can also oxidize terfenadine [33], but the level of P450 2J2 seems to be low in liver and most tissues, and the dominant

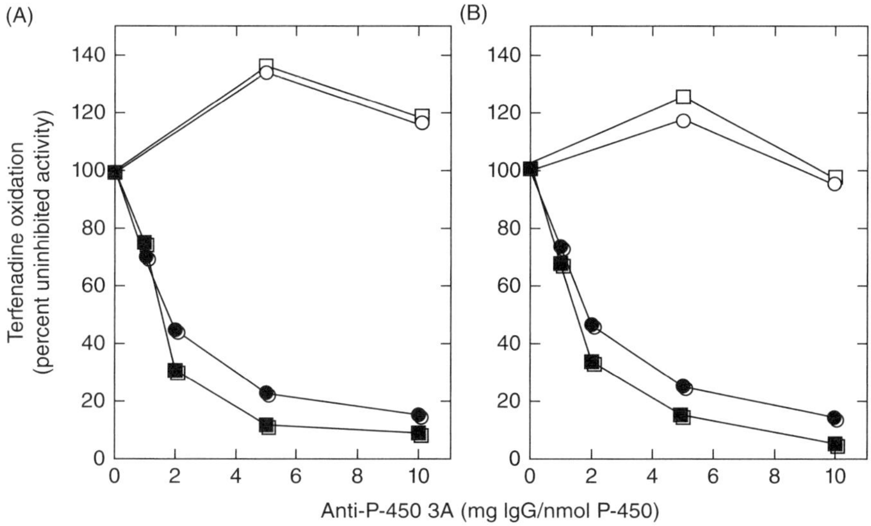

Figure 23.2 Immunoinhibition of human liver microsomal oxidation of both individual enantiomers of terfenadine by anti-P450 3A4. *A*, (*S*)-terfenadine; *B*, (*R*)-terfenadine. (*open symbols*)-results obtained with "pre-immune" immunoglobulin G; *closed symbols*, formation of azacyclonal; *squares*, formation of ω-hydroxyterfenadine. (Reprinted with permission from Reference 1.)

role of P450 3A4 is still accepted. The roles of ketoconazole and erythromycin as inhibitors of P450 3A4 are well known [21], and the mechanism of action is entirely consistent with all experimental and clinical results.

23.6 AFTERMATH OF INITIAL STUDIES

I am not exactly sure how our results on terfenadine metabolism were used by Marion-Merrell-Dow in their discussions with the FDA prior to publication of our paper [26], but in 1992, the FDA position of a "black box" warning was introduced. Subsequent pharmacological studies revealed that terfenadine bound very tightly to the hERG (human ether-a-go-go protein-related) channel, an ion channel, leading to the prolongation of the QT interval and thus torsade des pointes, i.e., the 'twisted" electrocardiogram patterns related to the observed arrhythmias [23].

As alluded to earlier, pharmacokinetic analysis had shown that very little of the parent drug was present in the plasma of individuals administered terfenadine. Most of the parent drug was accounted for by the metabolite fexofenadine. Fexofenadine was known to have desirable properties as a drug itself [34]. For instance, the H_1 receptor IC_{50} is 12 nM for fexofenadine, compared with 5 nM for terfenadine. Fexofenadine has a cLogP of 3.73, which can be considered more favorable than the value of 5.42 for terfrenadine [35] (a violation of one of Lipinski's "rules" (cLogP [calculated logarithm of the octanol-water partition coefficient]) > 5 [36], as well as earlier guidelines by Hansch) [37]. Fexofenadine did not interact with the hERG channel receptor nor cause the QT block in a feline cardioacyte model [22]. These

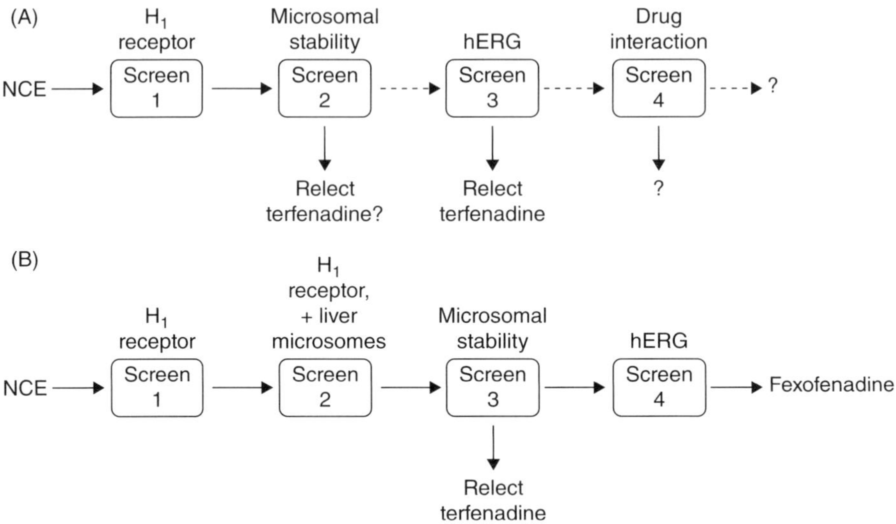

Figure 23.3 Two possible screening strategies for NCEs, including terfenadine. See text for discussion. *A*, screens commonly used today: **1**, efficacy; **2**, microsomal stability; **3**, hERG inhibition; **4**, inhibition/drug interaction. *B*, alternate screens: **1**, efficacy; **2**, efficacy in presence of microsomes; **3**, microsomal stability; **4**, hERG inhibition.

properties led to the development of fexofenadine as a drug, which has only slight oxidation (to the inactive product azacyclonal) [27].

Another early nonsedating antihistamine, astemizole, had similar features of terfenadine (despite a much different chemical structure), i.e., P450 3A4 metabolism and hERG channel inhbition [23]. At the time, another new nonsedating antihistamine, loratidine, showed none of the cardiotoxicity issues. In 1997, the FDA withdrew both terfenadine and astemizole from the market, and this market was now dominated largely by loratidine (Claritin®) and fexofenadine (Allegra®). Interestingly, a biotechnology company, Sepracor (Marlborough, MA), had obtained the patent on fexofenadine in 1993 and had sold development rights to Marion-Merrell-Dow (which became Hoechst-Marion-Roussel, Kansas City, MO, and now Sanofi-Aventis, Bridgewater, NJ). By 2004, the global sales of fexofenadine were $1.87 billion.

A brief summary of the sales of terfenadine, fexofenadine, and other nonsedating antihistamines has been published (as part of a review of strategies related to "first-in-class" drugs) [38]. Several points can be made in analyzing the graphs (Figure 23.3 of the article [38]) (recall that terfenadine was first marketed in 1985 and fexofenadine in 1996):

(i) After its introduction, terfenadine sales outpaced all other (nonsedating) antihistamines.
(ii) Sales of terfenadine peaked in 1992 and then dropped, probably related to the warnings that were issued.

(iii) Sales of fexofenadine showed a rapid increase after market entry, peaked eight years later, and then declined (probably as a result of the end of patent protection for loratidine).

(iv) Of all the (nonsedating) antihistamines in the group, the combined (drug lifetime) sales of terfenadine and fexofenadine were greater than any other, even loratidine. Cumulative sales of terfenadine and fexofenadine were approximately equal. Together, these two drugs were dominant in the field for 15 years, competing with loratidine. Fexofenadine and loratidine, now both off patent, continue to be popular.

The patent on fexofenadine has expired recently, after a long successful run since approval in 1996. An interesting aspect of this story involves the patent for fexofenadine. As mentioned, the original patent was owned by Sepracor, even though Marion-Merrell-Dow had the terfenadine patent but apparently had not applied for a patent on the metabolite. One issue was whether the fexofenadine patent covers terfenadine, in that terfenadine is converted into fexofenadine; i.e., terfenadine can be considered a pro-drug and a means of administering fexofenadine to a patient. The author was involved in legal proceedings as an expert with legal firms representing Marion-Merrell-Dow. Ultimately a judge ruled against the plaintiff, Marion-Merrell-Dow, and in favor of the (generic) defendant IVAX. However, within one month (January 1997) of initiation of the sale of generic terfenadine, the FDA withdrew terfenadine from the market, and in a sense, the legal case became a moot point. Since then some legal interpretations have led to changes in the laws regarding issues involving patenting of metabolites, what are considered “natural” processes, and the “inevitability” of certain metabolic transformations (e.g., Schering vs. Geneva case with desloratidine (Schering-Plough, Kenilworth, NJ)) [39]. (Note: the author finds this area interesting but is not offering any legal opinions.)

23.7 HOW WOULD THIS ISSUE BE DEALT WITH TODAY?

In all likelihood the outcomes with terfenadine and fexofenadine would probably not be the same today. There is even a strong probability that, depending on the pharmaceutical firm, terfenadine would have been discarded immediately because of very poor “microsomal stability.” Also, largely because of issues with terfenadine and astemizole, hERG assays have been incorporated into many developability screening assays.

For the sake of example, consider two strategies (Figure 23.3). Both strategies begin with a large set of “new chemical entities (NCEs)” or prospects for drugs (“wanna-be drugs”). The primary screen is the H_1 receptor, a major aspect of the antihistamine response. Assume that terfenadine is one of many (thousands, or at least hundreds) of NCEs in the set being considered, but fexofenadine is not. The NCEs could be coming from a combinatorial library, natural product library, semirational design set, or a mixture of these.

In the first scenario (Figure 23.3A), terfenadine would come through the first screen (**1**) with good "efficacy" (i.e., an IC_{50} of 5 nM is known, *vide infra*). It is highly unlikely that it would survive the "cut" in Screen **2** because of the poor metabolic stability in human liver microsomes. These assays are usually done by liquid chromatography–mass spectrometry (LC-MS), and the only parameter of interest is resistance of the present drug to disappearance. If it did survive the cut on this screen, it almost certainly would not survive Screen **3**, the hERG inhibition assay. If terfenadine did happen to survive Screens **1**–**3**, it would then be scrutinized for drug–drug interaction potential in Screen **4**, which would probably identify potential problems with P450 3A4 inhibitors. These issues could be addressed in preliminary human studies before proceeding with clinical trials. The likelihood that terfenadine would get to clinical trials today is quite low. Unfortunately, this process would not find fexofenadine unless it just happened to be in the initial NCE mix.

Consider an alternative, perhaps more enlightened, process in Figure 23.3B. Again assume that terfenadine is in the initial NCE mix but we know nothing about fexofenadine *a priori*. Terfenadine passes Screen **1** for efficacy, as in Figure 23.3A. In this case, Screen **2** is a repeat of Screen **1**, an H_1 receptor assay in which the assay follows a pre-incubation with human liver microsomes. Terfenadine will pass this step successfully. Screen **3**, metabolic stability, is the same as Screen **2** of Figure 23.3A, and it will yield a negative result for terfenadine. However, coupling the information of the results of Screens **2** and **3** tells the investigators that an active metabolite of terfenadine has potential use. Analysis of the LC-MS work quickly leads to the observation of fexofenadine, whose structure can rapidly be established by a combination of MS and nuclear magnetic resonance (NMR) data.

Is the scheme in Figure 23.3B realistic, and would it ever be used? There is a general pressure to (1) use microsomal stability assays (perhaps drug metabolism science has been too persuasive) and (2) to standardize screening protocols. The current industrial problem is that to filter out "problematic" drugs like terfenadine, multiple filters are implemented upfront in discovery and development. If those surrogate filters are not used wisely, one can simply do "box-checking," and we would have never found terfenadine or fexofenadine, which were billion dollar-a-year drugs that helped many people. This may account, at least in part, for concerns about the suboptimal productivity of the pharmaceutical industry, where one tends to look for "perfect" candidates but can end up "throwing babies out with the bath water."

The original approach to finding terfenadine undoubtedly involved many assays with experimental animals, both *in vitro* and *in vivo*, and probably very little with human systems prior to the IND (Investigational New Drug application to FDA)/phase I work. The only screen in Figure 23.3 that might have been used is Screen **1**, whether *in vivo* or *in vitro*.

A screening strategy of this sort shown in Figure 23.3A is common, whether or not the order is the same (and exactly which screens are included). Would a strategy such as that of Figure 23.3B be used? Obviously there would be the cost of repeating Screen **1** under these conditions. However, such a screen would avoid discarding of compounds such as terfenadine, fexofenadine, losartan, clopidogrel, desloratidine, and other valuable metabolites that became drugs [40].

23.8 WHAT WE LEARNED

Terfenadine is a very important example in the context of understanding the role of drug metabolism in drug safety and development. Some things learned include:

(i) hERG channel inhibition is a serious issue in cardiotoxicity and worthy of screening.

(ii) Knowing which P450s and other enzymes are involved in metabolism can be critical in predicting *in vivo* drug interactions.

(iii) As shown earlier, drug metabolites can be valuable backups/leads for second-generation drugs. Even with the multiple *in vitro* and *in vivo* absorption, distribution, metabolism, excretion, and toxicity (ADMET_ filters available in contemporary drug discovery and development practice, it is still very risky simply to use a "box-checking" strategy to look for perfect candidates. Integrated data interpretation is very critical to avoid throwing out promising candidates.

(iv) Even ~25 cases (1985–1992) out of >100 million people treated (i.e., <0.00003%) is enough to lead to market withdrawal.

(v) Finally, this case exemplifies how much the field has charged in the past 20 years. In 1990, a drug could reach the market and be used by >100 million people without knowledge of which enzymes (including P450s) were involved in metabolism. Today the FDA and International Conference on Harmonisation of Technical Requirements for Registration of Pharmaceuticals for Human Use (ICH) highly recommend obtaining this information prior to consideration of an IND. The experiments today are heavily focused on *in vitro* human systems, not with the view of eliminating all *in vivo* animal studies (e.g., safety assessment) but with the view of making better clinical predictions.

As a point in passing, I enjoyed presenting this example when I used to teach biochemistry to first-year medical students. Most of them did not appreciate the finer points of classic enzymology, but they were very interested in hearing about how one enzyme could have such consequences for their patients and the choice of therapies.

ACKNOWLEDGMENT

Thanks are extended to K. Trisler for assistance in preparation of the manuscript. I am also grateful, in retrospect, to Marion-Merrell-Dow, Kaye Scholer LLP, and Pennie & Edmonds for the opportunities to be involved in several aspects of this story.

REFERENCES

1. Kulshrestha, V.K., Gupta, P.P., Turner, P., Wadsworth, J. (1978). Some clinical pharmacological studies with terfenadine, a new antihistamine drug. *British Journal of Clinical Pharmacology*, *6*, 25–29.

2. Thompson, D., Oster, G. (1996). Use of terfenadine and contraindicated drugs. *Journal of the American Medical Association*, *275*, 1339–1341.
3. Guengerich, F.P. Introduction and historical perspective, in Biotransformation, F.P. Guengerich (ed.), *Comprehensive Toxciology*, 2nd ed., vol. *4*, Elsevier Science Ltd., Oxford, U.K., 2010, pp. 1–7.
4. Mahgoub, A., Idle, J.R., Dring, L.G., Lancaster, R., Smith, R.L. (1977). Polymorphic hydroxylation of debrisoquine in man. *Lancet II*, 584–586.
5. Wang, P., Mason, P.S., Guengerich, F.P. (1980). Purification of human liver cytochrome P-450 and comparison to the enzyme isolated from rat liver. *Archives of Biochemistry and Biophysics*, *199*, 206–219.
6. Wang, P.P., Beaune, P., Kaminsky, L.S., Dannan, G.A., Kadlubar, F.F., Larrey, D., Guengerich, F.P. (1983). Purification and characterization of six cytochrome P-450 isozymes from human liver microsomes. *Biochemistry*, *22*, 5375–5383.
7. Distlerath, L.M., Reilly, P.E.B., Martin, M.V., Davis, G.G., Wilkinson, G.R., Guengerich, F.P. (1985). Purification and characterization of the human liver cytochromes P-450 involved in debrisoquine 4-hydroxylation and phenacetin *O*-deethylation, two prototypes for genetic polymorphism in oxidative drug metabolism. *The Journal of Biological Chemistry*, *260*, 9057–9067.
8. Shimada, T., Misono, K.S., Guengerich, F.P. (1986). Human liver microsomal cytochrome P-450 mephenytoin 4-hydroxylase, a prototype of genetic polymorphism in oxidative drug metabolism. Purification and characterization of two similar forms involved in the reaction. *The Journal of Biological Chemistry*, *261*, 909–921.
9. Guengerich, F.P., Martin, M.V., Beaune, P.H., Kremers, P., Wolff, T., Waxman, D.J. (1986). Characterization of rat and human liver microsomal cytochrome P-450 forms involved in nifedipine oxidation, a prototype for genetic polymorphism in oxidative drug metabolism. *The Journal of Biological Chemistry*, *261*, 5051–5060.
10. Beaune, P., Kremers, P.G., Kaminsky, L.S., de Graeve, J., Guengerich, F.P. (1986). Comparison of monooxygenase activities and cytochrome P-450 isozyme concentrations in human liver microsomes. *Drug Metabolism and Disposition*, *14*, 437–442.
11. Guengerich, F.P., Shimada, T. (1991). Oxidation of toxic and carcinogenic chemicals by human cytochrome P-450 enzymes. *Chemical Research in Toxicology*, *4*, 391-407.
12. Gonzalez, F.J., Schmid, B.J., Umeno, M., McBride, O.W., Hardwick, J.P., Meyer, U.A., Gelboin, H.V., Idle, J.R. (1988). Human P450PCN1: Sequence, chromosome localization, and direct evidence through cDNA expression that P450PCN1 is nifedipine oxidase. *DNA*, *7*, 79–86.
13. Crespi, C.L., Langenbach, R., Penman, B.W. The development of a panel of human cell lines expressing specific human cytochrome P450 cDNAs, in M.L. Mendelsohn, and R.J. Albertini, (eds.), *Mutation and the Environment, Part B: Metabolism, Testing Methods, and Chromosomes*, Wiley-Liss, New York, NY, 1990, pp. 97–106.
14. Gillam, E.M.J., Baba, T., Kim, B.-R., Ohmori, S., Guengerich, F.P. (1993). Expression of modified human cytochrome P450 3A4 in *Escherichia coli* and purification and reconstitution of the enzyme. *Archives of Biochemistry and Biophysics*, *305*, 123–131.
15. Butler, M.A., Iwasaki, M., Guengerich, F.P., Kadlubar, F.F. (1989). Human cytochrome P-450_{PA} (P-450IA2), the phenacetin O-deethylase, is primarily responsible for the hepatic 3-demethylation of caffeine and N-oxidation of carcinogenic arylamines. *Proceedings of the National Academy of Science in the United States of America*, *86*, 7696–7700.

16. Butler, M.A., Lang, N.P., Young, J.F., Caporaso, N.E., Vineis, P., Hayes, R.B., Teitel, C.H., Massengill, J.P., Lawsen, M.F., Kadlubar, F.F. (1992). Determination of CYP1A2 and NAT2 phenotypes in human populations by analysis of caffeine urinary metabolites. *Pharmacogenetics*, *2*, 116–127.

17. Shimada, T., Yamazaki, H., Mimura, M., Inui, Y., Guengerich, F.P. (1994). Interindividual variations in human liver cytochrome P450 enzymes involved in the oxidation of drugs, carcinogens, and toxic chemicals: studies with liver microsomes of 30 Japanese and 30 Caucasians. *Journal of Pharmacology and Experimental Therapeutics*, *270*, 414–423.

18. Rendic, S., Di Carlo, F.J. (1997). Human cytochrome P450 enzymes: A status report summarizing their reactions, substrates, inducers, and inhibitors. *Drug Metabolism Reviews*, *29*, 413–580.

19. Williams, J.A., Hyland, R., Jones, B.C., Smith, D.A., Hurst, S., Goosen, T.C., Peterkin, V., Koup, J.R., Ball, S.E. (2004). Drug-drug interactions for UDP-glucuronosyltransferase substrates:A pharmacokinetic explanation for typically observed low exposure (AUC_i/AUC) ratios. *Drug Metabolism and Disposition*, *32*, 1201–1208.

20. Wienkers, L.C., Heath, T.G. (2005). Predicting in vivo drug interactions from in vitro drug discovery data. *Nature Reviews Drug Discovery*, *4*, 825–833.

21. Guengerich, F.P. Human cytochrome P450 enzymes, in P.R. Ortiz de Montellano (ed.), *Cytochrome P450: Structure, Mechanism, and Biochemistry*, 3rd ed., Kluwer Academic/Plenum Press, New York, NY, 2005, pp. 377–530.

22. Woosley, R.L., Chen, Y., Freiman, J.P., Gillis, R.A. (1993). Mechanism of the cardiotoxic actions of terfenadine. *Journal of the American Medical Association*, *269*, 1532–1536.

23. Woosley, R.L. (1996). Cardiac actions of antihistamines. *Annual Review of Pharmacology and Toxicology*, *36*, 233–252.

24. Honig, P.K., Woosley, R.L., Zamani, K., Conner, D.P., Cantilena, L.R., Jr. (1992). Changes in the pharmacokinetics and electrocardiographic pharmacodynamics of terfenadine with concomitant administration of erythromycin. *Clinical Pharmacology & Therapeutics*, *52*, 231–238.

25. Brian, W.R., Sari, M.-A., Iwasaki, M., Shimada, T., Kaminsky, L.S., Guengerich, F.P. (1990). Catalytic activities of human liver cytochrome P-450 IIIA4 expressed in *Saccharomyces cerevisiae*. *Biochemistry*, *29*, 11280–11292.

26. Yun, C.-H., Okerholm, R.A., Guengerich, F.P. (1993). Oxidation of the antihistaminic drug terfenadine in human liver microsomes: Role of cytochrome P450 3A(4) in N-dealkylation and C-hydroxylation. *Drug Metabolism and Disposition*, *21*, 403–409.

27. Ling, K.H., Leeson, G.A., Burmaster, S.D., Hook, R.H., Reith, M.K., Cheng, L.K. (1995). Metabolism of terfenadine associated with CYP3A(4) activity in human hepatic microsomes. *Drug Metabolism and Disposition*, *23*, 631–636.

28. Honig, P.K., Wortham, D.C., Zamani, K., Conner, D.P., Mullin, J.C., Cantilena, L.R. (1993). Terfenadine-ketoconazole interaction: Pharmacokinetic and electrocardiographic consequences. *Journal of the American Medical Association*, *269*, 1513–1518.

29. Kivistö, K.T., Neuvonen, P.J., Klotz, U. (1994). Inhibition of terfenadine metabolism: pharmacokinetic and pharmacodynamic consequences. *Clinical Pharmacokinetics*, *27*, 1–5.

30. Rau, S.E., Bend, J.R., Arnold, J.M.O., Tran, L.T., Spence, J.D., Bailey, D. G. (1997). Grapefruit juice-terfenadine single-dose interaction: magnitude, mechanism, and relevance. *Clinical Pharmacology & Therapeutics*, *61*, 401–409.

31. Guengerich, F.P., Turvy, C.G. (1991). Comparison of levels of several human microsomal cytochrome P-450 enzymes and epoxide hydrolase in normal and disease states using immunochemical analysis of surgical liver samples. *Journal of Pharmacology and Experimental Therapeutics*, *256*, 1189–1194.
32. Yang, X., Zhang, B., Molony, C., Chudin, E., Hao, K., Zhu, J., Gaedigk, A., Suver, C., Zhong, H., Leeder, J.S., Guengerich, F.P., Strom, S.C., Schuetz, E., Rushmore, T.H., Ulrich, R.G., Slatter, J.G., Schadt, E.E., Kasarskis, A., Lum, P.Y. (2010). Genetic and genomic analysis of cytochrome P450 enzyme activities in human liver. *Genome Research*, *20*, 1020–1036.
33. Lee, C.A., Neul, D., Clouser-Roche, A., Dalvie, D., Wester, M.R., Jiang, Y., Jones, J.P., 3rd, Freiwald, S., Zientek, M., Totah, R.A. (2010). Identification of novel substrates for human cytochrome P450 2J2. *Drug Metabolism and Disposition*, *38*, 347–356.
34. Garteiz, D.A., Hook, R.H., Walker, B.J., Okerholm, R.A. (1982). Pharmacokinetics and biotransformation studies of terfenadine in man. *Arz.-Forschung /Drug Research*, *32*, 1185–1190.
35. Smith, D.A., Dalvie, D. (2012). Why do metabolites circulate? *Xenobiotica*, *42*, 107–126.
36. Lipinski, C.A., Lombardo, F., Dominy, B.W., Feeney, P.J. (1997). Experimental and computational approaches to estimate solubility and permeability in drug discovery and development settings. *Advanced Drug Delivery Reviews*, *23*, 3–25.
37. Hansch, C., Bjorkroth, J.P., Leo, A. (1987). Hydrophobicity and central nervous system agents: On the principle of minimal hydrophobicity in drug design. *Journal of Pharmaceutical Sciences*, *76*, 663–687.
38. Cohen, F.J. (2006). Entryorder as a consideration for innovaiton strategies. *Nature Reviews Drug Discovery*, *5*, 285–294.
39. United States Court of Appeals for the Federal Circuit. (2003). 02-1540,-1541,-1542,-1543,-1544,-1545,-1546,-1547,-1548,-1549, 03-1021,-1022,-1023,-1025,-1027, Schering vs. Geneva (http://www.ll.georgetown.edu/federal/judicial/fed/opinions/02opinions/02-1540.html) (accessed 10 March 2012).
40. Smith, D.A., Obach, R.S. (2010). Metabolites: Have we MIST out the importance of structure and physicochemistry? *Future Science*, *2*, 1223–1233.

24

CLINICAL TOXICITY PROFILE OF VEGF INHIBITORS

Mark P. S. Sie and Ferry A. L. M. Eskens

24.1 INTRODUCTION

Angiogenesis, the formation of new blood vessels from the already established vasculature, is essential for development and subsequent growth of tumors and for the formation of metastases [1]. Tumor cells as well as various host factors release multiple pro-angiogenic factors that stimulate proliferation of endothelial cells, sprouting of microvessels, and finally the formation of new, often qualitatively poor new blood vessels. The concept of targeting tumor-related angiogenesis as a means of anticancer treatment was first described by Folkman almost half a century ago and has since then lead to an impressive amount of preclinical and clinical research activities [2]. Most research, to date, has been put into inhibiting the activity of the dominant pro-angiogenic factor, vascular endothelial growth factor (VEGF), and the currently registered and approved anti-angiogenic drugs in medical oncology all focus on inhibiting this VEGF-induced angiogenesis [3]. VEGF binds with high affinity to any of its specific transmembrane endothelial cell receptors (VEGFRs); of these, VEGFR-1 and VEGFR-2 are primarily located on blood vessels, while VEGFR-3 is primarily located on lymphatic vessels.

It has to be realized that apart from VEGF and its specific receptors, other pro-angiogenic factors, each of them also having more or less specific endothelial cell

Predictive ADMET: Integrative Approaches in Drug Discovery and Development, First Edition.
Edited by Jianling Wang and Laszlo Urban.

receptors, also exert angiogenic activity under both physiological and pathophysiological circumstances. These "other" pro-angiogenic factors, such as fibroblast growth factor 2 (FGF2/bFGF), angiopoietins 1 and 2, hepatocyte growth factor (HGF), ephrins, integrins, thrombospondin, and DLL4-mediated Notch signaling, play their role in the extremely complex process of angiogenesis. It is obvious that apart from the large number of individual ligand–receptor–pathway cascades that exert pro-angiogenic activity, extensive cross talk also comes into play. The fact that angiogenesis is such a multifactorial process probably explains the observed difficulties in achieving a complete blockade.

As currently VEGF-driven signaling remains predominant in tumor-related angiogenesis, we will focus on this pathway and its specific inhibitors in the rest of this chapter.

24.1.1 Mechanism of Action of VEGF Inhibitors

When considering the various approaches to inhibit VEGF-driven angiogenesis, ligand–receptor interaction with subsequent intracellular signal transduction pathway activation is the key process to be examined (Figure 24.1). Inhibiting this cascade can be pursued through three different mechanistic approaches; all these mechanisms have been explored, and currently compounds representing each of these approaches are either registered and approved or in advanced stages of clinical testing. Two anti-angiogenic approaches use the concept of extracellular active monoclonal antibodies, whereas small-molecule tyrosine kinase inhibitors preventing cascade activation after receptor–ligand interaction has taken place comprise the third approach.

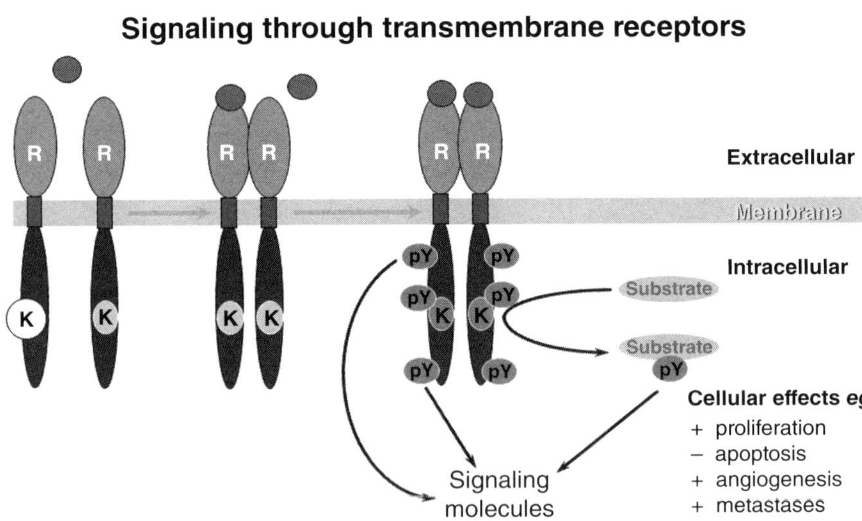

Figure 24.1 Signaling through transmembrane receptors. o = ligand; R = extracellular domain of the receptor; K = intracellular domain of the receptor; pY = phosphorylation domains of the receptor. (*See insert for color representation of the figure.*)

24.1.1.1 First Mechanism While assuming ligand–receptor interaction to be the initial step in activating the aforementioned cascade, inhibition hereof by means of monoclonal antibodies can be pursued by occupying the receptor with an inhibiting antibody, or by binding to and neutralization of the ligand prior to its attachment to the receptor.

Bevacizumab is a humanized monoclonal antibody with high binding affinity to various isoforms of VEGF and as such prevents angiogenesis by occupying the ligand before it attaches to its natural receptor(s). Bevacizumab is a humanization of the murine anti-VEGF antibody (muMab) A.4.6.1 [4,5]. It was designed by placing VEGF binding residues of a murine neutralizing antibody in the framework of normal human immunoglobulin G. Bevacizumab binds and neutralizes all human VEGF-A isoforms and bioactive proteolytic fragments, thereby resulting in reduced angiogenesis. Preclinical studies demonstrated that muMab A.4.6.1, its name at that time, was able to neutralize all isoforms of human VEGF with a dissociation constant (Kd) of 1.1 and that VEGF-induced proliferation of endothelial cells *in vitro* was inhibited with a median effective dose (ED_{50}) of 50 ± 5 ng/mL [5]. In an *in vivo* system, similar effects were observed [5]. A first clinical phase I single-agent study of bevacizumab that was performed in the early 2000s showed a highly acceptable toxicity profile (no grade 3 or 4 toxicity at doses up to 10 mg/kg) and good clinical tolerability [6]. In a simultaneously performed and presented phase Ib combination study, it was demonstrated that combining bevacizumab with various different chemotherapeutic agents (doxorubicin, carboplatin, paclitaxel, fluorouracil, and leucovorin) did not yield increased toxicity, suggesting that combining bevacizumab with cytotoxic chemotherapy was feasible [6,7]. After the first presentation of these clinical data, an extensive clinical program (*vide infra*) of studies in numerous indications ultimately led to the first approval for regular medical use in 2005 based on a randomized clinical study in metastatic colorectal cancer (CRC) showing a statistically significant and clinically relevant increase in overall survival (OS) [8]. After this registration in CRC, several other indications for malignancies followed in the years 2005–2012. In addition, local ophthalmological application of bevacizumab has become another approved indication, but it will not be discussed here.

Aflibercept (VEGF-Trap, AVE0005) is a recombinant fusion protein that contains VEGF binding portions that are fused to Fc portions of immunoglobulin (Ig) G1 [9]. This decoy receptor fusion protein targets all forms of VEGF-A, but also VEGF-B and PlGF, and has a higher affinity for VEGF-A than bevacizumab. Initially, aflibercept was developed for subcutaneous administration, but later on, the compound was developed for intravenous administration; several (phase I) studies demonstrated good tolerability and a predictable pharmacokinetic behavior [10,11]. Phase III studies also suggest clinical activity [12]. Whether the potentially superior biological activity of aflibercept translates into increased clinical activity when compared directly with bevacizumab is currently not known, as randomized studies with head-to-head comparisons have not yet been presented. In a recently presented randomized phase III study, in second-line treatment of metastatic CRC, aflibercept in combination with chemotherapy yielded statistically significant improvements in OS and progression-free survival (PFS) [13]. Currently, many other randomized phase III trials exploring the potential clinical benefit of aflibercept combined with chemotherapy in various malignant indications are ongoing or have already been closed for accrual [14]. It can therefore be anticipated that in the very near future more results defining the role of aflibercept in the treatment of cancer will become available.

24.1.1.2 Second Mechanism A second approach to inhibit VEGF-driven angiogenesis is that of monoclonal antibodies that bind to and subsequently block the VEGFR-1, -2, or -3 endothelial and lymphatic receptors. As VEGFR-2 is considered to be the most relevant receptor for VEGF-induced angiogenesis, most emphasis has been placed on this receptor. Currently, the monoclonal antibody ramucirumab, which is a fully human immunoglobulin G1 monoclonal antibody that binds with high affinity to the extracellular VEGF-binding domain of VEGFR-2, seems to be the more potent of the specific VEGF-receptors tested so far [15, 16]. Ramucurimab inhibits VEGF–VEGFR-2 interaction at inhibitory concentration 50 (IC_{50}) of 0.8–1.0 nM [17]. Phase I clinical studies revealed a good safety profile and showed anti-angiogenic and anti-tumor effects (partial response or stable disease) over a wide range of dose levels in patients with solid tumors, while currently, several phase II and III trials are assessing its activity for various indications (e.g., hepatocellular carcinoma, gastric carcinoma, breast cancer, and lung and colorectal cancer) [16, 18, 19]. Monoclonal antibodies specifically targeting VEGFR-1 and VEGFR-3 have been developed, but as these agents are currently only in early stage clinical development, they will not be discussed here.

24.1.1.3 Third Mechanism The third and by far largest category of anti-angiogenic agents is that of the VEGF tyrosine kinase inhibitors (TKIs), which act at the intracellular domain of the VEGFR (most notably, VEGFR-2) by inhibiting essential phosphorylation steps. In contrast to monoclonal antibodies, their target specificity is less pronounced, resulting in agents that sometimes have a broad (multi) target inhibitory activity. Depending on the situation and indication, this may be an advantage or a disadvantage. Another striking clinical difference between these TKIs and the monoclonal antibodies previously discussed is the observation that most TKIs exert clinical activity as monotherapy, whereas the most established and approved indications of anti-angiogenic monoclonal antibodies are in combination with chemotherapy. In contrast to the situation of the monoclonal antibodies, where currently only one agent has been registered and approved for regular clinical use, the number of TKI that has meanwhile been registered and approved for regular use approaches 10, and the number of agents to be approved will likewise increase further in the near future.

Before further discussing various compounds and classes of anti-angiogenic compounds, some general remarks should be made. VEGF is highly expressed by tumor cells and exerts its growth-stimulating effects through several specific receptors (VEGFR 1, 2, and 3) [20]. However, and as mentioned previously, angiogenesis as a process is depending on many stimulating factors and on many specific receptors. It is therefore obvious that there is no such thing as a single VEGF inhibiting "magic bullet" angiogenesis inhibitor that will be able to block this process fully. Still, as VEGF-driven angiogenesis is considered to be dominant, it can be conceived that inhibiting this process, irrespective by which mechanism, will yield biological and clinical effects. Indeed the currently registered VEGF inhibiting agents do exert clinical efficacy, but it is probably fair to state that these real-life effects are at best modest. In addition, pointing at any specific agent to be superior (or optimal) is virtually impossible, as head-to-head comparisons between compounds of the same class are just starting to appear, whereas head-to-head comparisons between agents of different classes of angiogenesis inhibitors have hardly been published. This paucity of data makes it, to date, virtually impossible to decide which approach should be considered to be optimal in a specific situation or indication or in a specific patient.

Another issue that is increasingly gaining attendance is the observation that although the target of anti-angiogenic agents has always been considered to be genetically stable, anti-angiogenic agents also seem to face the problem of primary and secondary drug resistance. For this reason, several explanations have been considered [21]. Thus, whether prolonged treatment with these agents, which is conceived to be the optimal approach when aiming to inhibit primary or secondary growth of tumor-related angiogenesis, will remain a valid paradigm, currently remains to be established.

24.2 SUCCESS STORIES

As mentioned, an ever growing number of anti-angiogenic agents has been approved for regular use, and it can be anticipated that the number of registrations will increase in the near future. In all these situations, and for all agents involved, registration followed publication of positive results obtained in large randomized clinical trials. Currently, registered and frequently prescribed anti-angiogenic agents are sunitinib, sorafenib, pazopanib, and probably best known and most frequently used, bevacizumab. These agents are registered for first-line and, albeit in only a limited number of indications, second-line treatments for several indications and have earned their place as part of global standard therapies.

24.2.1 Article I

The first registered VEGF inhibitors had the advantage of being representatives of a new class of anticancer agents, and therefore, their activity usually was demonstrated in large clinical trials comparing their (additional) activity to either best supportive care or placebo treatment. Currently, with a number of indications being covered by (or indicated for) regular use of one or sometimes even more anti-angiogenic agents, newcomers have to prove their value often in direct head-to-head comparisons with any of these already registered agents. These new compounds, in particular, the more recently developed TKI, on the one hand, often show a more broad-spectrum tyrosine kinase inhibitory activity, inhibiting more than just VEGFR tyrosine kinases, whereas on the other hand, the alleged superior activity is often based on higher target affinity (e.g., low IC_{50} values) or increased target specificity [18]. Whatever mechanism is involved in these newcomers, they have to earn their formal registration and acceptance in the current market by proving to be superior to other agents. Currently, with a rapidly increasing number of such direct comparative trials being published, shifting treatment indications from one compound to another (or just not!) do occur (e.g., AXIS trial and TIVO-1 trial) [22–24]. Additional efforts in resources, time, and patient-derived efforts will be the only way through which we will be able to raise the bar further and to prove (or, again, not!) whether the current successes of the "first-generation" VEGF inhibitors will be trumped by newer, or "second-generation" agents, and whether their apparently improved biological profile will translate into a superior clinical advantage. Preclinical data, pointing at relevant differences in biological activity, and a thorough and in-depth research into predictive biologic markers will have to be guiding the design (and the potential likelihood of success) of these trials.

24.2.2 The Monoclonal Antibodies, Bevacizumab

Although, as mentioned, several monoclonal anti-angiogenic antibodies have been developed, and (pre)clinically studied, bevacizumab is currently the only registered agent and is consistently and globally used in anticancer treatment. Bevacizumab is used in combinations with cytotoxic chemotherapy and is currently registered for metastatic renal cell carcinoma and for nonsquamous non–small-cell lung, colorectal, and breast carcinoma [8, 25].

Throughout these established treatment indications, with ovarian carcinoma probably to be added to this list in the near future, statistically significant increases in PFS have been observed, and for some indications, the effects on OS were also noted.

Table 24.1 summarizes the data of the landmark randomized clinical trials for these mentioned indications (we aim to illustrate, instead of providing an extensive or complete overview).

Bevacizumab is combined with (different) chemotherapy regimens for several different indications with generally positive (albeit often modest) effects [40]. This may be considered intrinsic proof of the efficacy of this compound, despite that we do not exactly know yet the mechanism that underlies these effects, nor the optimal dose for bevacizumab throughout these various combinations. Ongoing research and practical experience aims to elucidate this.

Of course, and based on early clinical observations, the addition of bevacizumab has increased the incidence of direct drug-attributable side effects and has initially been reason for concern. With increasing experience, however, it seems to be a safe statement to mention that even though the addition of bevacizumab to chemotherapy indeed has increased most notably the occurrence of hypertension in patients (which was generally well manageable), bevacizumab does not seem to have increased morbidity to unacceptable levels. Given its mundial application, bevacizumab may thus be, generally spoken, considered to be a "success story." In addition, the application

Table 24.1 Bevacizumab phase III studies

Indication	Author	Regimen to which bevacizumab was added	PFS	OS
Non–Small-Cell Lung Cancer (NSCLC)*	Sandler et al. [25]	Paclitaxel-carboplatin in nonsquamous NSCLC	+	+
	Reck et al. [26]	Gemcitabine + cisplatinum in nonsquamous NSCLC	+	–
Colorectal Cancer (CRC)*	Hurwitz et al. [8]	Irinotecan, fluorouracil, and leucovorin in metastastic CRC	+	+
	Saltz et al. [27]	Oxaliplatin-based chemotherapy in metastastic CRC	+	–
	Kabbinavar et al. [28]	Various fluorouracil-based regimens in metastatic CRC	+	+
	Tebbutt et al. [29]	Capecitabin in metastatic CRC	+	–
Ovarian Cancer	Aghajanian et al. [30]	Gemcitabine-carboplatin in recurrent epithelial ovarian, primary peritoneal, or fallopian tube cancer	+	–
	Perren et al. [31]	Carboplatin-paclitaxel in first-line ovarian cancer	+	–
	Burger et al. [32]	Carboplatin-paclitaxel in first-line ovarian cancer	+	–
Renal Cell Carcinoma (RCC)*	Rini et al. [33]	Interferon-alpha in metastatic renal cell carcinoma	+	–
	Escudier et al. [34]	Interferon-alpha in metastatic renal cell carcinoma	+	–
Breast Cancer	Robert et al. [35]	Several common chemotherapy regimens in Her2 negative metastatic breast cancer	+	–
	Brufsky et al. [36]	Common chemotherapy regimens in Her2 negative metastatic breast cancer	+	–
	Miller et al. [37]	Paclitaxel in metastastic breast cancer	+	–
Castration Resistant Prostate Cancer	Kelly et al. [38]	Docetaxel in metastatic prostate cancer	+	–
Pancreatic Cancer	Kindler et al. [39]	Gemcitabine in advanced pancreatic cancer	–	–

PFS = progression-free survival, OS = overall survival, * = registered for clinical application.

of bevacizumab has served as a useful and practical introduction of anti-angiogenic therapy for many clinicians.

We feel that it is important to consider the downside of this success. Clinical experience has shown benefits, but also limitations, as well as some typical safety issues. We will discuss toxicity profiles further in this chapter, but we will make the statement already here that identifying patients who may be more prone to adverse effects and who are less likely candidates for treatment with bevacizumab remains very critical in a period with wide acceptance of angiogenesis inhibitors. For example, the risk of a serious bleeding complication may vary with the histological subtype of non–small-cell lung cancer (NSCLC), and therefore, some patients may not be suitable for treatment with bevacizumab.

24.2.3 The TKIs

TKIs have been shown to be safe, active, and effective when applied as single-agent treatments, and therefore, they are generally not combined with traditional cytostatics. Broad-spectrum TKIs act on many receptors and may thus not only influence angiogenesis but also other pathways. In this respect, this last group may be considered to be a more general antitumor therapy, rather than sole inhibitors of angiogenesis.

24.2.3.1 Currently Approved TKIs Sunitinib was introduced around 2003 and is a TKI inhibiting VEGFR, PDGFRB, CSF1R, and c-Kit [41]. It has an IC_{50} value of 3.0–3.8 uM [42]. It is a multitarget TKI inhibiting signaling activity not only of VEGFR-1/2/3 but also of transmembrane receptors PDGFR-alpha and beta, KIT, FLT3, RET, and the CSF-1 receptor tyrosine kinases. It has since been registered for use in advanced renal cell carcinoma, gastrointestinal stromal tumor (GIST), and advanced pancreatic neuroendocrine tumor (NET) [43–45]. Other examples are sorafenib and pazopanib, which were developed a few years later and have since been registered for treatment of renal cell and hepatocellular carcinomas and sarcomas [46–49].

24.2.3.2 Novel TKIs Examples of newer TKIs, some of which are likely to be registered very soon, are tivozanib (a potent and selective inhibitor of VEGFR-1, -2, and -3), axitinib (a selective second-generation inhibitor of VEGFR-1, 2, and 3), vandetanib (selectively targeting VEGFR, EGFR, and RET), and the multikinase inhibitor regorafenib (which inhibits endothelial cell kinases, angiogenic kinases such as VEGFR-1 and 3, platelet-derived growth factor receptor-β and fibroblast growth factor receptor 1, and mutant kinases KIT, RET, and B-RAF) [50–52].

In addition, numerous compounds have been studied in clinical studies but either are only in early developmental phases, have failed to show meaningful success, or based on various (other) considerations, have been withdrawn from further clinical development

To illustrate current and potential future applications of TKIs, we have summarized several examples of relevant (phase III) studies (Table 24.2).

Table 24.2 TKI phase III studies

TKI	Indication	Line	Author	Compared with	PFS	OS
Sunitinib	Renal Cell Cancer (RCC)	1	Motzer et al. [43]	IFN-alpha in metastatic RCC	+	+
	Gastro Intestinal Stromal Tumor	2	Demetri et al. [44]	Placebo	+	–
	Pancreatic Neuroendocrine Tumors	1	Raymond et al. [45]	Placebo	+	+
Sorafenib	Renal Cell Cancer	2	Escudier et al. [46]	Placebo	+	+
	Hepatocellular Cancer	1	Llovet et al. [47]	Placebo	–	+
	Hepatocellular Cancer	1	Cheng et al. [53]	Placebo	+	+
	Nonsquamous Non–Small-Cell Lung Carcinoma	1	Paz-Ares et al. [54]	Gemcitabine + Cisplatinum without Sorafenib	+	–
Pazopanib	Renal Cell Cancer	1	Sternberg et al. [48]	Placebo	+	–
	Soft Tissue Sarcoma	2	Van der Graaf et al. [55]	Placebo	+	–
Vandetanib	Medullary Thyroid Cancer	1	Wells et al. [56]	Placebo	+	–
Axitinib	Renal Cell Cancer	2	Rini et al. [24]	Sorafenib	+	–
Tivozanib*	Renal Cell Cancer	≥1	Motzer et al. [57]	Sorafenib	+	N/A
Regorafenib*	Colorectal Cancer	2	Grothey et al. [58]	Placebo	+	+

PFS = progression-free survival, OS = overall survival, * = ASCO presentation.

24.3 TOXICITY PROFILE

The utopian version of systemic therapy is that effective agents that affect the tissue of interest show no adverse effects at all. Despite extensive attempts to make agents as selective, or better said, as cancer cell specific as possible, still no medication exists that lacks adverse effects completely. Treatment using angiogenesis inhibitors generally does not result in side effects associated with "classic" chemotherapy, such as alopecia, profound myelosuppression, sensory neuropathy, renal toxicity, and gastrointestinal toxicity; however, this does not mean these agents are devoid of adverse effects. It is sometimes better to refer to those effects as biological side effects, rather than as adverse effects. The character of some side effects can be considered to be an indicator of the biological activity of an agent. One may even state that such a side effect should be considered to be an "on-target" effect, as it illustrates that the agent influences an underlying pathway or pathways of interest. A befitting example of such an "on-target" effect is hypertension. Of course, side effects can also imply undesirable, unintended, and sometimes unexpected effects, better described as "off-target" effects. It is beyond the scope of this chapter to provide a definite and complete overview of all known side effects of all VEGF inhibitors. We aim to provide a more general overview of common side effects, which include hypertension, proteinuria, thromboembolic events or bleeding complications, (visceral) perforations, fatigue, hoarseness (a typical class phenomenon), mucositis, diarrhea, hypothyroidism, transaminase elevations, nausea or vomitus, hand–foot syndrome, headaches, cerebral toxicities, and rarely, myelosuppression. Probably the best known (on-target) side effect of angiogenesis inhibitors as a group is hypertension, which may occur in up to 30% [59]. VEGF exerts its effects on blood pressure through influence on the relaxation/dilation within the arterial system. Inhibiting VEGF will therefore, understandably, induce hypertension. Also rarefaction, influenced by angiogenic activity of VEGF, may play a role [60]. Hypertension after start of anti-angiogenic therapy may develop within a very short period and is often reversible after discontinuation of the agent. These characteristics argue against structural changes and support the view of a direct effect of VEGF on the blood pressure. One may even theorize that lack of a certain degree of hypertension may indicate insufficient levels or activity of the agent and consider hypertension a kind of biomarker, of which the occurrence (but not its persistence) may even be considered acceptable and, perhaps, predictive of efficacy [1, 61]. Hypertension is generally easily treatable with regular antihypertensive medication. If drug-induced hypertension is more difficult to handle or turns out to become chronic, it is conceivable that diverse sequelae on a micro- and macrovascular level will occur, and this may explain the observed "on-target" side effects such as proteinuria, cardiovascular events, and thromboembolic complications.

VEGF is also involved in tissue integrity and wound repair. Inhibition may therefore lead to decreased tissue integrity (like mucositis), disturbed wound healing, and various bleeding complications, which are examples of "off-target" effects. Treatment with some VEGF inhibitors is associated with an increased rate of gastrointestinal perforations, which may occur throughout the gastrointestinal system. For example, increased rates of gastrointestinal perforations associated with bevacizumab were reported in early studies in ovarian cancer and initially hampered the conduct of large clinical trials [62]. However, later on it was considered safe to perform these trials, and throughout these trials, enrolling large numbers of patients having undergone one or more laparotomies, the application of bevacizumab generally turned out to be safe with only a few episodes of gastrointestinal perforations. Voice changes, such as hoarseness, may also occur. Although this has a broad pathogenesis, it is not unlikely that VEGF inhibition, thereby limiting its role in mucosal integrity, may lead to local mucositis or laryngitis as an underlying cause. An important role for

VEGF in mucosal healing has been described earlier. Another frequently occurring side effect is fatigue. This may occur with chronic or long-term illness and is often found during/after regular chemotherapy. The pathophysiology of this side effect is extremely complex, but as hypothyroidism has been observed in relation to, in particular, broad-spectrum TKIs, it is conceivable that this phenomenon might, in an as yet not strictly recognized percentage, explain complaints such as fatigue and voice changes. The causal mechanism, however, still has to be further elucidated. As for transaminase elevations, VEGF has been shown to affect hepatic growth stimulation, regeneration, and cytoprotection; inhibition may in turn lead to reduced hepatic healing and, accordingly, laboratory abnormalities. Also, VEGF has been described as neuroprotective through various mechanisms. It is conceivable that anti-angiogenic agents may, in turn, induce cerebral toxicity. Although not the most common side effect, myelosuppression has been described to occur in patients on anti-angiogenic agents, particularly the multikinase inhibitors (e.g., sunitinib and sorafenib). These effects may in part be because they influence multiple angiogenic factors, such as c-Kit and Flt-3. The origin and occurrence of many side effects may be clear from the underlying pathophysiology. Preclinical data aid in understanding and explaining these effects and may also lead to ways to deal with these effects, by either preventing or limiting them, thereby enhancing the safety of the agent. For this reason, preclinical toxicity assessment may prove invaluable; however, several side effects were not immediately expected, nor could they be considered "on-target" effects.

24.4 ADMET DATA AND LIMITATIONS

Earlier we mentioned many different agents with various, sometimes overlapping, targets and pathways, eventually all related to VEGF inhibition. As with every type of medication, several agents have succeeded in obtaining a place as (part of) standard therapies for several malignancies. Other agents were developed for a similar indication, but they have not become standard therapy (or not even registered therapy). One may contemplate whether these agents will not be used definitively, or that they have not been used, yet. Of course many factors can be distinguished for why some drugs have made it, and for why some have not. ADMET data (i.e., data on absorption, distribution, metabolism, excretion, and toxicity) play a pivotal role in that process. Important issues such as drug stability, plasma half-life, and biological availability are often addressed in preclinical/ADMET studies. The selection of agents that will be developed further is largely dependent on the outcome of these studies. Although this approach is widely used in drug discovery studies, it remains important that one uses the information wisely and acts on the data. Although ADMET data are useful and necessary, literature also highlights their limitations. Suboptimal design and interpretation may limit proper discovery [63]. Also, all ADMET findings influence the outcome; e.g., a high *in vitro* potency alone is not a guarantee for success [64]. Recognizing limitations of a drug in development and lack of pharmacodynamic evidence of activity may also help in deciding to either continue or not continue further clinical development, thereby avoiding unnecessary spending in an era of limited budgets. Apart from endogenous ADMET properties, it is important also to be aware of external factors. Theoretically successful drugs may (appear to) fail or be less effective in practice if, for example, absorption is negatively influenced by dietary factors or time of ingestion. Also several agents, developed as anti-angiogenic compounds acting via the VEGF pathway, were found either to exert their effects via other pathways or to exert little or no anti-angiogenic effect at all. An example of an agent that did not come to be a regular VEGF inhibitor is ABT510. This was first described in 2004 and is a thrombospondin-1 mimetic peptide inhibitor of angiogenesis. Although first thought to be a VEGF inhibitor and act via VEGF inhibition, studies showed no relation with VEGF biology, indicating that the anti-angiogenic effect was not related to VEGF-mediated activities [65]. Also, studies on integrin inhibitors, also initially considered to act as a VEGF inhibitor, yielded conflicting results. Many multitarget TKIs may partly act through inhibition of angiogenesis, but as a result of their, sometimes, broad target inhibitory activity (or, should one say, lack of target specificity?), it is not always clear through which mechanism the observed effects are accomplished. This can make it difficult to designate those agents with certainty as, for example, a VEGF inhibitor. Of course, (pre)clinical studies, also on ADMET data, may help in distinguishing the main affected pathways.

In summary, in drug development and (pre)clinical assessment, ADMET data play an important role. However, using these data is by no means a guarantee of predicting success or failure. Some compounds have been rationally designed and developed and were applied successfully in phase I and II studies; however, they failed when they were tested in pivotal large randomized trials, such as in day-to-day clinical practice. In other cases, newly developed compounds turned out to have effects that were not clear or expected when they were developed. Therefore, and probably still somewhat disappointingly, a "trial-and-error" approach remains the gold standard to define the role of a specific new compound in day-to-day clinical care. For this reason, well-designed and rationally and scientifically based clinical trials remain essential. Well-defined pharmacological endpoints in early studies can underlie assumptions applied in these subsequent trials, even though, and merely taken as an example, for a well-recognized and frequently applied anti-angiogenic agent like bevacizumab, the truly optimal schedule/dosage in metastatic colorectal cancer in fact remains to be determined. An important issue is also the profound lack of predictive markers by which patients/diseases could be selected that, at least in theory, would benefit maximally and suffer the least adverse effects. In addition to developing new anti-angiogenic compounds, it would therefore make great sense to invest in resources to unravel these predictive markers. This would yield important information on whom to prescribe, and perhaps even more important, on whom not to prescribe certain drugs, possibly preventing unnecessary use and preventable morbidity, as well as associated health expenditure. Until then, there is no other way than to introduce an agent in a patient using the already mentioned "trial-and-error" method.

Finally, an issue that cannot be neglected is that of the currently extremely high price of almost all angiogenesis inhibitors. Especially because one cannot reliably predict which patients are or are not to benefit from therapy, treatment with these agents may form a burden on society, as these agents will sometimes be administered in vain. Hopefully in the future we may be able to select "responsive" patients better.

Nevertheless, on the whole, ADMET data are invaluable in drug development, but these data are also only part of all the information that eventually may lead to a new therapy. Knowing the advantages, but certainly also the limitations, may prove the most prudent and useful approach.

24.5 SUMMARY

When the first VEGF inhibiting agents were discovered, it probably could not have been anticipated that their impact on oncology practice would become so huge only a decade or two later.

Nowadays these agents, each with its own mechanism of action, undoubtedly have become part of the standard armamentarium for several frequently occurring indications. In addition, for some other indications, such as renal cell carcinoma and hepatocellular carcinoma, the development of these agents has meant nothing less than a major breakthrough.

With an ever increasing amount of clinical data and experience, it seems fair and safe to state that most VEGF inhibitors have shown an acceptable toxicity profile and thus can be considered to be by and large suitable for the general cancer patient population, including older persons. Specific risk factors have been identified, making patient selection doable.

When considering which patient or patients are likely to benefit most from these agents, the great challenge remains to identify predictive markers to indeed usefully and sensibly select them. Treating only selected patient who are best suited for anti-angiogenic treatment may hopefully result in a considerable decrease in the numbers of patients that do not derive any benefit and yet suffer from sometimes cumbersome and poorly predictable toxicity. In addition, in an era of increasing awareness of ever increasing health care costs, it makes sense to only get paid for success and get rid of wasting of medication in this era of decreasing and already limited funding.

In the past 20 years, anti-angiogenic therapy including VEGFR inhibitors, has come a long way: from basic experimental agents to several approved, registered, and widely applied drugs for a range of indications. Some of these agents have become part of standard therapy, while others have somewhat limited treatment applications.

It is to be foreseen that in the next two decades of drug development, further developments and refinements are to be expected, and hopefully in the years to come, this will lead us toward optimal personalized application of this class of "breakthrough" anticancer agents.

REFERENCES

1. Eskens, F.A., Verweij, J. (2006). The clinical toxicity profile of vascular endothelial growth factor (VEGF) and vascular endothelial growth factor receptor (VEGFR) targeting angiogenesis inhibitors: A review. *European Journal of Cancer*, *42*(18), 3127–3139.
2. Folkman, J. (1971). Tumor angiogenesis: Therapeutic implications. *New England Journal of Medicine*, *285*(21), 1182–1186.

3. Ferrara, N., Gerber, H.P., LeCouter, J. (2003). The biology of VEGF and its receptors. *Nature Medicine*, *9*(6), 669–676.
4. Ferrara, N., Hillan, K.J., Novotny, W. (2005). Bevacizumab (Avastin), a humanized anti-VEGF monoclonal antibody for cancer therapy. *Biochemical and Biophysical Research Communications*, *333*(2), 328–335.
5. Presta, L.G., Chen, H., O'Connor, S.J., Chisholm, V., Meng, Y.G., Krummen, L., Winkler, M., Ferrara, N. (1997). Humanization of an anti-vascular endothelial growth factor monoclonal antibody for the therapy of solid tumors and other disorders. *Cancer Research*, *57*(20), 4593–4599.
6. Gordon, M.S., Margolin, K., Talpaz, M., Sledge, G.W., Jr., Holmgren, E., Benjamin, R., Stalter, S., Shak, S., Adelman, D. (2001). Phase I safety and pharmacokinetic study of recombinant human anti-vascular endothelial growth factor in patients with advanced cancer. *Journal of Clinical Oncology*, *19*(3), 843–850.
7. Margolin, K., Gordon, M.S., Holmgren, E., Gaudreault, J., Novotny, W., Fyfe, G., Adelman, D., Stalter, S., Breed, J. (2001). Phase Ib trial of intravenous recombinant humanized monoclonal antibody to vascular endothelial growth factor in combination with chemotherapy in patients with advanced cancer: Pharmacologic and long-term safety data. *Journal of Clinical Oncology*, *19*(3), 851–856.
8. Hurwitz, H., Fehrenbacher, L., Novotny, W., Cartwright, T., Hainsworth, J., Heim, W., Berlin, J., Baron, A., Griffing, S., Holmgren, E., Ferrara, N., Fyfe, G., Rogers, B., Ross, R., Kabbinavar, F. (2004). Bevacizumab plus irinotecan, fluorouracil, and leucovorin for metastatic colorectal cancer. *New England Journal of Medicine*, *350*(23), 2335–2342.
9. Holash, J., Davis, S., Papadopoulos, N., Croll, S.D., Ho, L., Russell, M., Boland, P., Leidich, R., Hylton, D., Burova, E., Ioffe, E., Huang, T., Radziejewski, C., Bailey, K., Fandl, J.P., Daly, T., Wiegand, S.J., Yancopoulos, G.D., Rudge, J.S. (2002). VEGF-Trap: A VEGF blocker with potent antitumor effects. *Proceedings of the National Academy of Sciences of the United States of America*, *99*(17), 11393–11398.
10. Tew, W.P., Gordon, M., Murren, J., Dupont, J., Pezzulli, S., Aghajanian, C., Sabbatini, P., Mendelson, D., Schwartz, L., Gettinger, S., Psyrri, A., Cedarbaum, J.M., Spriggs, D.R. (2010). Phase 1 study of aflibercept administered subcutaneously to patients with advanced solid tumors. *Clinical Cancer Research*, *16*(1), 358–366.
11. Lockhart, A.C., Rothenberg, M.L., Dupont, J., Cooper, W., Chevalier, P., Sternas, L., Buzenet, G., Koehler, E., Sosman, J.A., Schwartz, L.H., Gultekin, D.H., Koutcher, J.A., Donnelly, E.F., Andal, R., Dancy, I., Spriggs, D.R., Tew, W.P. (2010). Phase I study of intravenous vascular endothelial growth factor trap, aflibercept, in patients with advanced solid tumors. *Journal of Clinical Oncology*, *28*(2), 207–214.
12. Ramlau, R., Gorbunova, V., Ciuleanu, T.E., Novello, S., Ozguroglu, M., Goksel, T., Baldotto, C., Bennouna, J., Shepherd, F.A., Le-Guennec, S., Rey, A., Miller, V., Thatcher, N., Scagliotti, G. (2012). Aflibercept and Docetaxel versus Docetaxel alone after platinum failure in patients with advanced or metastatic non-small-cell lung cancer: A randomized, controlled phase III trial. *Journal of Clinical Oncology*, *30*(29), 3640–3647.
13. Van Cutsem, E., Tabernero, J., Lakomy, R., Prenen, H., Prausova, J., Macarulla, T., Ruff, P., van Hazel, G.A., Moiseyenko, V., Ferry, D., McKendrick, J., Polikoff, J., Tellier, A., Castan, R., Allegra, C. (2012). Addition of Aflibercept to Fluorouracil, Leucovorin, and Irinotecan improves survival in a phase III randomized trial in patients with metastatic colorectal cancer previously treated with an oxaliplatin-based regimen. *Journal of Clinical Oncology*, *30*(28), 3499–3506.

14. Gaya, A., Tse, V. (2012). A preclinical and clinical review of aflibercept for the management of cancer. *Cancer Treatment Reviews*, *38*(6), 484–493.
15. Spratlin, J. (2011). Ramucirumab (IMC-1121B): Monoclonal antibody inhibition of vascular endothelial growth factor receptor-2. *Current Oncology Reports*, *13*(2), 97–102.
16. Spratlin, J.L., Cohen, R.B., Eadens, M., Gore, L., Camidge, D.R., Diab, S., Leong, S., O'Bryant, C., Chow, L.Q., Serkova, N.J., Meropol, N.J., Lewis, N.L., Chiorean, E.G., Fox, F., Youssoufian, H., Rowinsky, E.K., Eckhardt, S.G. (2010). Phase I pharmacologic and biologic study of ramucirumab (IMC-1121B), a fully human immunoglobulin G1 monoclonal antibody targeting the vascular endothelial growth factor receptor-2. *Journal of Clinical Oncology*, *28*(5), 780–787.
17. Lu, D., Jimenez, X., Zhang, H., Bohlen, P., Witte, L., Zhu, Z. (2002). Selection of high affinity human neutralizing antibodies to VEGFR2 from a large antibody phage display library for antiangiogenesis therapy. *International Journal of Cancer*, *97*(3), 393–399.
18. Chu, E. (2012). An update on the current and emerging targeted agents in metastatic colorectal cancer. *Clinical Colorectal Cancer*, *11*(1), 1–13.
19. Garon, E.B., Cao, D., Alexandris, E., John, W.J., Yurasov, S., Perol, M. (2012). A randomized, double-blind, phase III study of Docetaxel and Ramucirumab versus Docetaxel and placebo in the treatment of stage IV non-small-cell lung cancer after disease progression after 1 previous platinum-based therapy (REVEL): Treatment rationale and study design. *Clinical Lung Cancer*, *13*(6), 505–509.
20. Schneider, B.P., Radovich, M., Miller, K.D. (2009). The role of vascular endothelial growth factor genetic variability in cancer. *Clinical Cancer Research*, *15*(17), 5297–5302.
21. Bergers, G., Hanahan, D. (2008). Modes of resistance to anti-angiogenic therapy. *Nature Reviews Cancer*, *8*(8), 592–603.
22. Cheng, A., Kang, Y., Lin, D., Park, J., Kudo, M., Qin, S., Omata, M., Pitman Lowenthal, S.W., Lanzalone, S., Yang, L., Lechuga, M., Raymond, E. *Phase III Trial of Sunitinib (Su) Versus Sorafenib (So) in Advanced Hepatocellular Carcinoma (HCC)*, In *ASCO*. Chicago, IL: Journal of Clinical Oncology, 2011.
23. Bhargava, P., Robinson, M.O. (2011). Development of second-generation VEGFR tyrosine kinase inhibitors: current status. *Current Oncology Reports*, *13*(2), 103–111.
24. Rini, B.I., Escudier, B., Tomczak, P., Kaprin, A., Szczylik, C., Hutson, T.E., Michaelson, M.D., Gorbunova, V.A., Gore, M.E., Rusakov, I.G., Negrier, S., Ou, Y.C., Castellano, D., Lim, H.Y., Uemura, H., Tarazi, J., Cella, D., Chen, C., Rosbrook, B., Kim, S., Motzer, R.J. (2011). Comparative effectiveness of axitinib versus sorafenib in advanced renal cell carcinoma (AXIS): A randomised phase 3 trial. *Lancet*, *378*(9807), 1931–1939.
25. Sandler, A., Gray, R., Perry, M.C., Brahmer, J., Schiller, J.H., Dowlati, A., Lilenbaum, R., Johnson, D.H. (2006). Paclitaxel-carboplatin alone or with bevacizumab for non-small-cell lung cancer. *New England Journal of Medicine*, *355*(24), 2542–2550.
26. Reck, M., von Pawel, J., Zatloukal, P., Ramlau, R., Gorbounova, V., Hirsh, V., Leighl, N., Mezger, J., Archer, V., Moore, N., Manegold, C., Group, B.O.S. (2010). Overall survival with cisplatin-gemcitabine and bevacizumab or placebo as first-line therapy for nonsquamous non-small-cell lung cancer: Results from a randomised phase III trial (AVAiL). *Annals of Oncology*, *21*(9), 1804–1809.
27. Saltz, L.B., Clarke, S., Diaz-Rubio, E., Scheithauer, W., Figer, A., Wong, R., Koski, S., Lichinitser, M., Yang, T.S., Rivera, F., Couture, F., Sirzen, F., Cassidy, J. (2008). Bevacizumab in combination with oxaliplatin-based chemotherapy as first-line therapy in

metastatic colorectal cancer: A randomized phase III study. *Journal of Clinical Oncology*, *26*(12), 2013–2019.

28. Kabbinavar, F., Irl, C., Zurlo, A., Hurwitz, H. (2008). Bevacizumab improves the overall and progression-free survival of patients with metastatic colorectal cancer treated with 5-fluorouracil-based regimens irrespective of baseline risk. *Oncology*, *75*(3-4), 215–223.
29. Tebbutt, N.C., Wilson, K., Gebski, V.J., Cummins, M.M., Zannino, D., van Hazel, G.A., Robinson, B., Broad, A., Ganju, V., Ackland, S.P., Forgeson, G., Cunningham, D., Saunders, M.P., Stockler, M.R., Chua, Y., Zalcberg, J.R., Simes, R.J., Price, T.J. (2010). Capecitabine, bevacizumab, and mitomycin in first-line treatment of metastatic colorectal cancer: Results of the Australasian gastrointestinal trials group randomized phase III MAX study. *Journal of Clinical Oncology*, *28*(19), 3191–3198.
30. Aghajanian, C., Blank, S.V., Goff, B.A., Judson, P.L., Teneriello, M.G., Husain, A., Sovak, M.A., Yi, J., Nycum, L.R. (2012). OCEANS: A randomized, double-blind, placebo-controlled phase III trial of chemotherapy with or without bevacizumab in patients with platinum-sensitive recurrent epithelial ovarian, primary peritoneal, or fallopian tube cancer. *Journal of Clinical Oncology*, *30*(17), 2039–2045.
31. Perren, T.J., Swart, A.M., Pfisterer, J., Ledermann, J.A., Pujade-Lauraine, E., Kristensen, G., Carey, M.S., Beale, P., Cervantes, A., Kurzeder, C., du Bois, A., Sehouli, J., Kimmig, R., Stahle, A., Collinson, F., Essapen, S., Gourley, C., Lortholary, A., Selle, F., Mirza, M.R., Leminen, A., Plante, M., Stark, D., Qian, W., Parmar, M.K., Oza, A.M. (2011). A phase 3 trial of bevacizumab in ovarian cancer. *New England Journal of Medicine*, *365*(26), 2484–2496.
32. Burger, R.A., Brady, M.F., Bookman, M.A., Fleming, G.F., Monk, B.J., Huang, H., Mannel, R.S., Homesley, H.D., Fowler, J., Greer, B.E., Boente, M., Birrer, M.J., Liang, S.X., Gynecologic Oncology, G. (2011). Incorporation of bevacizumab in the primary treatment of ovarian cancer. *New England Journal of Medicine*, *365*(26), 2473–2483.
33. Rini, B.I., Halabi, S., Rosenberg, J.E., Stadler, W.M., Vaena, D.A., Archer, L., Atkins, J.N., Picus, J., Czaykowski, P., Dutcher, J., Small, E.J. (2010). Phase III trial of bevacizumab plus interferon alfa versus interferon alfa monotherapy in patients with metastatic renal cell carcinoma: Final results of CALGB 90206. *Journal of Clinical Oncology*, *28*(13), 2137–2143.
34. Escudier, B., Bellmunt, J., Negrier, S., Bajetta, E., Melichar, B., Bracarda, S., Ravaud, A., Golding, S., Jethwa, S., Sneller, V. (2010). Phase III trial of bevacizumab plus interferon alfa-2a in patients with metastatic renal cell carcinoma (AVOREN): Final analysis of overall survival. *Journal of Clinical Oncology*, *28*(13), 2144–2150.
35. Robert, N.J., Dieras, V., Glaspy, J., Brufsky, A.M., Bondarenko, I., Lipatov, O.N., Perez, E.A., Yardley, D.A., Chan, S.Y., Zhou, X., Phan, S.C., O'Shaughnessy, J. (2011). RIBBON-1: Randomized, double-blind, placebo-controlled, phase III trial of chemotherapy with or without bevacizumab for first-line treatment of human epidermal growth factor receptor 2-negative, locally recurrent or metastatic breast cancer. *Journal of Clinical Oncology*, *29*(10), 1252–1260.
36. Brufsky, A.M., Hurvitz, S., Perez, E., Swamy, R., Valero, V., O'Neill, V., Rugo, H.S. (2011). RIBBON-2: A randomized, double-blind, placebo-controlled, phase III trial evaluating the efficacy and safety of bevacizumab in combination with chemotherapy for second-line treatment of human epidermal growth factor receptor 2-negative metastatic breast cancer. *Journal of Clinical Oncology*, *29*(32), 4286–4293.

37. Miller, K., Wang, M., Gralow, J., Dickler, M., Cobleigh, M., Perez, E.A., Shenkier, T., Cella, D., Davidson, N.E. (2007). Paclitaxel plus bevacizumab versus paclitaxel alone for metastatic breast cancer. *New England Journal of Medicine*, *357*(26), 2666–2676.

38. Kelly, W.K., Halabi, S., Carducci, M., George, D., Mahoney, J.F., Stadler, W.M., Morris, M., Kantoff, P., Monk, J.P., Kaplan, E., Vogelzang, N.J., Small, E.J. (2012). Randomized, double-blind, placebo-controlled phase III trial comparing docetaxel and prednisone with or without bevacizumab in men with metastatic castration-resistant prostate cancer: CALGB 90401. *Journal of Clinical Oncology*, *30*(13), 1534–1540.

39. Kindler, H.L., Niedzwiecki, D., Hollis, D., Sutherland, S., Schrag, D., Hurwitz, H., Innocenti, F., Mulcahy, M.F., O'Reilly, E., Wozniak, T.F., Picus, J., Bhargava, P., Mayer, R.J., Schilsky, R.L., Goldberg, R.M. (2010). Gemcitabine plus bevacizumab compared with gemcitabine plus placebo in patients with advanced pancreatic cancer: Phase III trial of the Cancer and Leukemia Group B (CALGB 80303). *Journal of Clinical Oncology*, *28*(22), 3617–3622.

40. Strickler, J.H., Hurwitz, H.I. (2012). Bevacizumab-based therapies in the first-line treatment of metastatic colorectal cancer. *Oncologist*, *17*(4), 513–524.

41. O'Farrell, A.M., Abrams, T.J., Yuen, H.A., Ngai, T.J., Louie, S.G., Yee, K.W., Wong, L.M., Hong, W., Lee, L.B., Town, A., Smolich, B.D., Manning, W.C., Murray, L.J., Heinrich, M.C., Cherrington, J.M. (2003). SU11248 is a novel FLT3 tyrosine kinase inhibitor with potent activity in vitro and in vivo. *Blood*, *101*(9), 3597–3605.

42. Oechsle, K., Honecker, F., Cheng, T., Mayer, F., Czaykowski, P., Winquist, E., Wood, L., Fenner, M., Glaesener, S., Hartmann, J.T., Chi, K., Bokemeyer, C., Kollmannsberger, C. (2011). Preclinical and clinical activity of sunitinib in patients with cisplatin-refractory or multiply relapsed germ cell tumors: A Canadian urologic oncology group/german testicular cancer study group cooperative study. *Annals of Oncology*, *22*(12), 2654–2660.

43. Motzer, R.J., Hutson, T.E., Tomczak, P., Michaelson, M.D., Bukowski, R.M., Oudard, S., Negrier, S., Szczylik, C., Pili, R., Bjarnason, G.A., Garcia-del-Muro, X., Sosman, J.A., Solska, E., Wilding, G., Thompson, J.A., Kim, S.T., Chen, I., Huang, X., Figlin, R.A. (2009). Overall survival and updated results for sunitinib compared with interferon alfa in patients with metastatic renal cell carcinoma. *Journal of Clinical Oncology*, *27*(22), 3584–3590.

44. Demetri, G.D., van Oosterom, A.T., Garrett, C.R., Blackstein, M.E., Shah, M.H., Verweij, J., McArthur, G., Judson, I.R., Heinrich, M.C., Morgan, J.A., Desai, J., Fletcher, C.D., George, S., Bello, C.L., Huang, X., Baum, C.M., Casali, P.G. (2006). Efficacy and safety of sunitinib in patients with advanced gastrointestinal stromal tumour after failure of imatinib: A randomised controlled trial. *Lancet*, *368*(9544), 1329–1338.

45. Raymond, E., Dahan, L., Raoul, J.L., Bang, Y.J., Borbath, I., Lombard-Bohas, C., Valle, J., Metrakos, P., Smith, D., Vinik, A., Chen, J.S., Horsch, D., Hammel, P., Wiedenmann, B., Van Cutsem, E., Patyna, S., Lu, D.R., Blanckmeister, C., Chao, R., Ruszniewski, P. (2011). Sunitinib malate for the treatment of pancreatic neuroendocrine tumors. *New England Journal of Medicine*, *364*(6), 501–513.

46. Escudier, B., Eisen, T., Stadler, W.M., Szczylik, C., Oudard, S., Siebels, M., Negrier, S., Chevreau, C., Solska, E., Desai, A.A., Rolland, F., Demkow, T., Hutson, T.E., Gore, M., Freeman, S., Schwartz, B., Shan, M., Simantov, R., Bukowski, R.M., Group, T.S. (2007). Sorafenib in advanced clear-cell renal-cell carcinoma. *New England Journal of Medicine*, *356*(2), 125–134.

47. Llovet, J.M., Ricci, S., Mazzaferro, V., Hilgard, P., Gane, E., Blanc, J.F., de Oliveira, A.C., Santoro, A., Raoul, J.L., Forner, A., Schwartz, M., Porta, C., Zeuzem, S., Bolondi, L., Greten, T.F., Galle, P.R., Seitz, J.F., Borbath, I., Haussinger, D., Giannaris, T., Shan, M., Moscovici, M., Voliotis, D., Bruix, J., Group, S.I.S. (2008). Sorafenib in advanced hepatocellular carcinoma. *New England Journal of Medicine*, *359*(4), 378–390.

48. Sternberg, C.N., Davis, I.D., Mardiak, J., Szczylik, C., Lee, E., Wagstaff, J., Barrios, C.H., Salman, P., Gladkov, O.A., Kavina, A., Zarba, J.J., Chen, M., McCann, L., Pandite, L., Roychowdhury, D.F., Hawkins, R.E. (2010). Pazopanib in locally advanced or metastatic renal cell carcinoma: Results of a randomized phase III trial. *Journal of Clinical Oncology*, *28*(6), 1061–1068.

49. van der Graaf, W.T., Blay, J.Y., Chawla, S.P., Kim, D.W., Bui-Nguyen, B., Casali, P.G., Schoffski, P., Aglietta, M., Staddon, A.P., Beppu, Y., Le Cesne, A., Gelderblom, H., Judson, I.R., Araki, N., Ouali, M., Marreaud, S., Hodge, R., Dewji, M.R., Coens, C., Demetri, G.D., Fletcher, C.D., Dei Tos, A.P., Hohenberger, P., Tissue, E.S., Bone Sarcoma, G., Group, P.S. (2012). Pazopanib for metastatic soft-tissue sarcoma (PALETTE): A randomised, double-blind, placebo-controlled phase 3 trial. *Lancet*, *379*(9829), 1879–1886.

50. Eskens, F.A., de Jonge, M.J., Bhargava, P., Isoe, T., Cotreau, M.M., Esteves, B., Hayashi, K., Burger, H., Thomeer, M., van Doorn, L., Verweij, J. (2011). Biologic and clinical activity of tivozanib (AV-951, KRN-951), a selective inhibitor of VEGF receptor-1, -2, and -3 tyrosine kinases, in a 4-week-on, 2-week-off schedule in patients with advanced solid tumors. *Clinical Cancer Research*, *17*(22), 7156–7163.

51. Lee, J.S., Hirsh, V., Park, K., Qin, S., Blajman, C.R., Perng, R.P., Chen, Y.M., Emerson, L., Langmuir, P., Manegold, C. (2012). Vandetanib versus placebo in patients with advanced non-small-cell lung cancer after prior therapy with an epidermal growth factor receptor tyrosine kinase inhibitor: A randomized, double-blind phase III trial (ZEPHYR). *Journal of Clinical Oncology*, *30*(10), 1114–1121.

52. Wilhelm, S.M., Dumas, J., Adnane, L., Lynch, M., Carter, C.A., Schutz, G., Thierauch, K.H., Zopf, D. (2011). Regorafenib (BAY 73-4506): A new oral multikinase inhibitor of angiogenic, stromal and oncogenic receptor tyrosine kinases with potent preclinical antitumor activity. *International Journal of Cancer*, *129*(1), 245–255.

53. Cheng, A.L., Kang, Y.K., Chen, Z., Tsao, C.J., Qin, S., Kim, J.S., Luo, R., Feng, J., Ye, S., Yang, T.S., Xu, J., Sun, Y., Liang, H., Liu, J., Wang, J., Tak, W.Y., Pan, H., Burock, K., Zou, J., Voliotis, D., Guan, Z. (2009). Efficacy and safety of sorafenib in patients in the Asia-Pacific region with advanced hepatocellular carcinoma: A phase III randomised, double-blind, placebo-controlled trial. *Lancet Oncology*, *10*(1), 25–34.

54. Paz-Ares, L.G., Biesma, B., Heigener, D., von Pawel, J., Eisen, T., Bennouna, J., Zhang, L., Liao, M., Sun, Y., Gans, S., Syrigos, K., Le Marie, E., Gottfried, M., Vansteenkiste, J., Alberola, V., Strauss, U.P., Montegriffo, E., Ong, T.J., Santoro, A. (2012). Phase III, randomized, double-blind, placebo-controlled trial of Gemcitabine/Cisplatin alone or with Sorafenib for the first-line treatment of advanced, nonsquamous non-small-cell lung cancer. *Journal of Clinical Oncology*, *30*(25), 3084–3092.

55. van der Graaf, W.T., Blay, J.Y., Chawla, S.P., Kim, D.W., Bui-Nguyen, B., Casali, P.G., Schoffski, P., Aglietta, M., Staddon, A.P., Beppu, Y., Le Cesne, A., Gelderblom, H., Judson, I.R., Araki, N., Ouali, M., Marreaud, S., Hodge, R., Dewji, M.R., Coens, C., Demetri, G.D., Fletcher, C.D., Dei Tos, A.P., Hohenberger, P. (2012). Pazopanib for metastatic soft-tissue sarcoma (PALETTE): A randomised, double-blind, placebo-controlled phase 3 trial. *Lancet*, *379*(9829), 1879–1886.

56. Wells, S.A., Jr., Robinson, B.G., Gagel, R.F., Dralle, H., Fagin, J.A., Santoro, M., Baudin, E., Elisei, R., Jarzab, B., Vasselli, J.R., Read, J., Langmuir, P., Ryan, A.J., Schlumberger, M.J. (2012). Vandetanib in patients with locally advanced or metastatic medullary thyroid cancer: A randomized, double-blind phase III trial. *Journal of Clinical Oncology*, *30*(2), 134–141.
57. Motzer, R.J. *Tivozanib Versus Sorafenib as Initial Targeted Therapy for Patients with Advanced Renal Cell Carcinoma: Results from a Phase III Randomized, Openlabel, Multicenter Trial, in ASCO*, Chicago, IL: Journal of Clinical Oncology, 2012.
58. Grothey, A. *Results of a Phase III Randomized, Double-Blind, Placebo-Controlled, Multicenter Trial (CORRECT) of Regorafenib Plus Best Supportive Care (BSC) Versus Placebo Plus BSC in Patients (Pts) with Metastatic Colorectal Cancer (Mcrc) Who Have Progressed After Standard Therapies, in ASCO*, Chicago, IL: Journal of Clinical Oncology, 2012.
59. Nazer, B., Humphreys, B.D., Moslehi, J. (2011). Effects of novel angiogenesis inhibitors for the treatment of cancer on the cardiovascular system: Focus on hypertension. *Circulation*, *124*(15), 1687–1691.
60. Steeghs, N., Gelderblom, H., Roodt, J.O., Christensen, O., Rajagopalan, P., Hovens, M., Putter, H., Rabelink, T.J., de Koning, E. (2008). Hypertension and rarefaction during treatment with telatinib, a small molecule angiogenesis inhibitor. *Clinical Cancer Research*, *14*(11), 3470–3476.
61. Dienstmann, R., Brana, I., Rodon, J., Tabernero, J. (2011). Toxicity as a biomarker of efficacy of molecular targeted therapies: Focus on EGFR and VEGF inhibiting anticancer drugs. *Oncologist*, *16*(12), 1729–1740.
62. Cannistra, S.A., Matulonis, U.A., Penson, R.T., Hambleton, J., Dupont, J., Mackey, H., Douglas, J., Burger, R.A., Armstrong, D., Wenham, R., McGuire, W. (2007). Phase II study of bevacizumab in patients with platinum-resistant ovarian cancer or peritoneal serous cancer. *Journal of Clinical Oncology*, *25*(33), 5180–5186.
63. Leeson, P.D., St-Gallay, S.A. (2011). The influence of the ‘organizational factor’ on compound quality in drug discovery. *Nature Reviews Drug Discovery*, *10*(10), 749–765.
64. Gleeson, M.P., Hersey, A., Montanari, D., Overington, J. (2011). Probing the links between in vitro potency, ADMET and physicochemical parameters. *Nature Reviews Drug Discovery*, *10*(3), 197–208.
65. Hawighorst, T., Oura, H., Streit, M., Janes, L., Nguyen, L., Brown, L.F., Oliver, G., Jackson, D.G., Detmar, M. (2002). Thrombospondin-1 selectively inhibits early-stage carcinogenesis and angiogenesis but not tumor lymphangiogenesis and lymphatic metastasis in transgenic mice. *Oncogene*, *21*(52), 7945–7956.

25

CARDIOMYOPATHY: DRUG INDUCED AND PREDISPOSED

Shirley A. Aguirre and Eileen R. Blasi

25.1 INTRODUCTION

Several morphologic changes within the rat heart have been observed as spontaneous events in control populations [1,2]. One of the most common spontaneous rat myocardial changes reported in the literature is cardiomyopathy [1, 2]. Cardiomyopathy is described as a condition of aging rats and is characterized by the increased prevalence and severity of focal-to-multifocal areas of necrosis, inflammation, and fibrosis and may be secondary to ischemic change resulting from myocardial vascular dysfunction [1, 2]. Myocardial changes reminiscent of the condition of cardiomyopathy in aging rats can also be observed in young control rats [3]. In contrast to the aging rats, the incidence of myocardial changes in young rats occurred with high variability in control populations [3]. The relationship between the well-characterized cardiomyopathy of the aging rat and the myocardial changes observed in the young rat is currently not understood [3]. In an attempt to avoid confusion with the highly variable myocardial changes observed in young rats with the well-documented cardiomyopathy condition in aging rats, the use of descriptive terms such as myocardial degeneration/necrosis or inflammation in the young rat was encouraged [3]. Importantly in preclinical studies using young rats between 3 and 6 months of age, these highly variable spontaneous myocardial changes must be distinguished from those that are drug induced.

Predictive ADMET: Integrative Approaches in Drug Discovery and Development, First Edition.
Edited by Jianling Wang and Laszlo Urban.

This chapter will describe the preclinical studies undertaken for the development of mesenchymal-epithelial transition factor (c-MET) inhibitors, also called hepatocyte growth factor receptor (HGFR) inhibitors, during the late phase of lead optimization, candidate selection, and the subsequent investigational studies that were conducted to understand the translation of *in-vitro*-to-*in-vivo* off-target phosphodiesterase (PDE) inhibition [4]. Because perturbation of PDE3 is an established alert for possible serious cardiac side effects [5], safety pharmacology and toxicology profilings were used to select the best c-MET/HGFR prototype inhibitor for progression into good laboratory practice (GLP) regulatory toxicology studies required to support first-in-human (FIH) studies.

A relatively high rate of adverse cardiac events in the clinic has been associated with anticancer therapies targeted to inhibit the enzymatic activity of certain tyrosine kinase signaling pathways [6–11]. Tyrosine kinases catalyze the transfer of phosphate from adenosine triphosphate (ATP) to the tyrosine residues in polypeptides [12, 13]. Phosphorylated polypeptides are under tight control and regulate cellular proliferation, differentiation, and motility and are abundantly expressed in cancer cells [12, 13]. Tyrosine kinase signaling pathway activity associated with cancer cell survival may also perform a similar but beneficial function in the myocardium. Inhibition of some tyrosine kinase pathways in the heart may be detrimental and lead to untoward cardiotoxicity, which currently remains a leading safety concern in drug development [6]. Off-target myocardial toxicity associated with inhibitors of tyrosine kinases and receptors of tyrosine kinases (RTK) may be a result of their nonselectivity [4, 6]. The c-MET/HGFR is an attractive oncology target because of its essential role in oncogenesis and tumor metastasis [14]. The c-MET/HGFR inhibitor PF-04254644 with exquisite potency for the intended RTK target was also associated with off-target inhibitory activity for phosphodiesterases 3 (PDE3) and several other PDE subfamilies [4]. Mechanistic *in vitro* and *in vivo* investigations with PF-04254644 suggested that cardiotoxicity was likely related to activation of adenosine monophosphate (c-AMP) signaling and, possibly, to disruption of intracellular calcium and oxidative stress signaling pathways [15].

25.2 *IN VITRO* SAFETY PROFILING FOR THE PROTOTYPE C-MET/HGFR INHIBITORS

Two classes of small-molecule c-MET/HGFR inhibitors, PF-04254644 a triazolopyridazine prototype [16] and PF-04217903 a triazolopyrazine prototype [17], with similar potency but different binding modes and selectivity profiles were screened at Pfizer (New York, NY) and submitted to CEREP (Poitiers, France) BioPrint for *in vitro* binding and functional assays (68 in total) to assess off-target pharmacology as well as screened for potential activity against 150 kinases. Of the two lead prototype c-MET/HGFR inhibitors, only PF-04254644 inhibited the phosphodiesterase-3 (PDE3) ligand by 98% at a concentration of 10 μM with an IC_{50} of 170 nM (Table 25.1).

Follow-up screening of PF-04254644 for PDE enzyme activity was evaluated in a PDE enzyme panel developed at Pfizer using scintillation proximity assay (SPA)

Table 25.1 Select Tier 0 and 1 Ligand Profiling for PF-04254644[a]

	%inhibition	IC_{50} (nM)	K_i (nM)
5-HT7 (h)(agonist site	43		
5-HT7 (h)	31		
GABA A BZD	97	180	150
Kappa (KOP)	37		
Muscarinic 1	47		
Muscarinic 2	59	7000	4900
PDE3 (h)	98	170	
PDE4 (h)	63	3700	

[a]CEREP BioPrint = CEREP – Poitiers, France Laboratories Le bois l'Evêque 86600 Celle l'Evescault, France. BioPrint® positions a new drug candidate in the context of marketed drugs, anticipating potential *in vivo* liabilities, predicting off-target activities, and ADME characteristics (Drug profile interpretation). In another application, BioPrint® serves to identify secondary targets that are not genetically parented to a test target (Target profile design).

technology. PF-04254644 was evaluated in a 10-point IC_{50} curve ranging from 1 nM to 10 mM. The assay uses recombinant human phosphodiesterase proteins partially purified from insect cells. The results for PF-04254644 in the PDE assay panel demonstrated strong enzymatic inhibitory activity at multiple PDE families including 1A, 1C, 2, 3A, 3B, 4, 5, 10, and 11 (Table 25.2).

25.3 TOXICOKINETIC AND PRECLINICAL SAFETY STUDY PROFILES FOR PROTOTYPE C-MET/HGFR INHIBITORS

The relationship between oral dosing and systemic exposure for the two lead c-MET/HGFR inhibitors (PF-04254644 and PF-04217903) was evaluated in a rat toxicokinetic (TK) study. Rats were given a single oral dose of PF-04254644 or PF-04217903 at 0, 50, 150, or 500 mg/kg and euthanized at 24 h after dosing. Toxicity was determined by assessing changes in pre- and postdrug body weight, clinical signs, and behavior. Blood samples for the assessment of systemic exposure of the two lead c-MET/HGFR inhibitor compounds were collected temporally from all treatment groups on day 1 of the study. Laboratory examinations (hematology and clinical chemistry) were performed at 24 h postdosing. Necropsy was performed at the end of treatment; organ weights were recorded, and histopathologic examinations were performed.

Systemic exposure, as defined by area under the curve (AUC), to PF-04254644 increased with increasing dose, whereas the increase in maximum serum concentration (C_{max}) was less than dose proportional. Systemic exposure of PF-04217903 did not increase from 150 to 500 mg/kg, and C_{max} and AUC values were variable at all doses and were not proportional. Time of maximum plasma concentration (T_{max}) for PF- 04254644 and PF-04217903 was in the range of 2 to 4 h postdose. The ranges for the estimated free plasma concentrations on the first and last day of dosing were

Table 25.2 PDE Profile of PF-04254644[a]

PDE subfamily	IC_{50} (μM)
1A	0.6
1B	>10
1C	1.4
2	4.6
3A[a]	3.5
3B[b]	0.2
4A	>10
4B	>10
4C[c]	6.7
4D	9.6
5	4.9
6	>10
7A	>10
7B	>10
8A	>10
8B	>10
9	>10
10	0.1
11	3.0

[a]Milrinone IC50 2.1 μM.
[b]Milrinone IC50 0.32 μM.
[c]Milrinone IC50 1.45 μM.

6.2 to 12 μM. Importantly, the estimated free plasma concentrations of PF-04254644 exceeded the IC_{50} values for PDEs 1A, 1B, 1C, 2, 3A, 3B, 4A, 4B, 4C, 4D, and 5 expressed in cardiac tissue and for PDEs 6, 7A, 7B, 8A, 9, 10, and 11 expressed in other tissues (Table 25.2).

Histopathologic evaluation revealed that only rats given ≥150-mg/kg PF-04254644 developed a dose-related, mild-to-moderate myocardial degeneration at the apex, interventricular septum, and in the left and right ventricular free walls (Figure 25.1B). The moderate change was characterized by the presence of thin and fragmented hypereosinophilic myofibers with loss of cytoplasmic striations and infiltration of neutrophils in areas of degeneration (Figure 25.1B). No myocardial changes were observed by light microscopy after a single dose of PF-04254644 at the 50-mg/kg dose level. Because myocardial changes in rats given PF-04254644 were more severe than the typical spontaneous myocardial changes observed in young rats and surprisingly appeared within 24 h after a single dose, a 7-day repeat-dose oral exploratory toxicity study was conducted in rats to confirm PF-04254644-induced myocardial damage and to select the final lead candidate for the c-MET/HGFR inhibitor prototype.

Results from the 7-day repeat-dose experiment in rats with PF-04254644 confirmed the presence of myocardial toxicity. However, in the 7-day repeat-dose exploratory toleration study with PF-04217903 in rats, no treatment-related

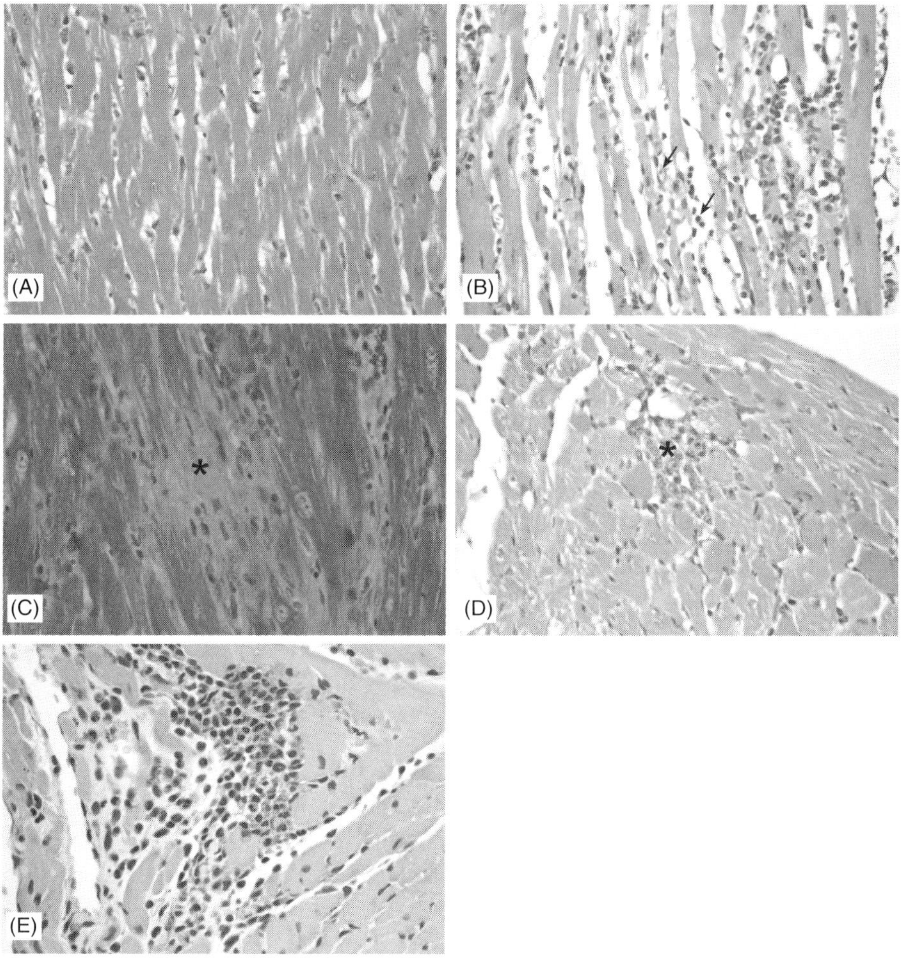

Figure 25.1 Microscopic findings of hematoxylin and eosin stained or Masson's Trichrome stained sections of rat myocardial interventricular tissue. (A) Hematoxylin and eosin staining of myocardial sections from a control rat in the TK study (400X). (B) Hematoxylin and eosin staining of myocardium sections from a rat in the TK study with 500-mg/kg PF-04254644 24 h after dosing, 400X. Moderate myocardial degeneration in the interventricular septum (arrows). (C) Masson's Trichrome staining of myocardial sections of a rat with 320/160-mg/kg PF-04254644 in the 7-day repeat dose toleration study after 7 days of dosing (400X). Fibrosis of the myocardial interventricular septum (asterisk). (D) Hematoxylin and eosin staining of myocardial sections of a rat given 80-mg/kg PF-04254644 in the 7-day repeat dose toleration study after 7 days of dosing (400X). Minimal myocardial degeneration in the interventricular septum (asterisk). (E) Hematoxylin and eosin staining of myocardial tissue taken from a rat 24 h after dosing with 10.5-mg/kg milrinone (400X). Note the minimal myocardial degeneration with macrophage and few neutrophil infiltrates. (*See insert for color representation of the figure.*)

findings were observed and this compound was advanced as the lead prototype candidate for drug development.

As CEREP screening and internal PDE panel data demonstrated an off-target alert for PDEs with PF-04254644, the 7-day repeat-dose exploratory study included additional endpoints for myocardial injury and myocardial function. In this study, five male rats per group were given PF-04254644 at 0, 40, 80, and 320 mg/kg/day for 7 days. The lowest dose selected in this study was based on the predicted therapeutic estimated free plasma concentration in humans, and two- to eightfold doses were selected to calculate safety margins over any toxic effects. Laboratory evaluations (cardiac troponin I [cTnI]), hematology, and clinical chemistry were performed 24 h postdose and on day 8. Blood samples for the assessment of systemic exposure to PF-04254644 were collected temporally from all treatment groups on days 1 and 7 of the study. Necropsy was performed on day 8; organ weights were recorded, and histopathologic examinations were performed. Cardiac function was evaluated *in vivo* by collecting echocardiography parameters on lightly anesthetized rats during the predose period (baseline values) and on day 6 of the 7-day repeat-dose study. Rats were anesthetized with 1.5% isofluorane (AErrane; Baxter, Inc., Deerfield, IL); underwent hair removal from the thorax by shaving, followed by treatment with a depilatory agent; and then were placed on their left side to collect echocardiographic images in the parasternal long axis, parasternal short axis, and apical four-chamber view. Echocardiograms were performed using the Philips SONOS 5500 system equipped with a 15-MHz linear-array transducer (Philips, Andover, MA) [18], and the parameters analyzed were end systolic volume (ESV), end diastolic volume (EDV), stroke volume (SV), and cardiac output (CO). Ejection fraction (EF), an indirect measure of cardiac contractility and overall cardiac performance, was reported and calculated by the formula EF = EDV − ESV/EDV × 100%. Cardiac output (SV × heart rate, HR) was used as an indicator of overall cardiovascular health. All measurements were performed according to the recommendations of the American Society for Echocardiography leading-edge method from three consecutive cardiac cycles [19, 20].

TK analysis revealed that the estimated free peak plasma concentrations for PF-04254644 ranged from 3.1 to 15.0 μM and exceeded the IC_{50} of multiple PDEs listed in Table 25.2.

Treatment-related morbidity occurred in two of five male rats receiving 320-mg/kg/day PF-04254644. Because of the poor condition, treatment with 320 mg/kg was stopped on day 2, and these animals were euthanized. Other animals in that dose group subsequently received a reduced dose of 160 mg/kg for the remainder of the study.

Serum cTnI level was increased (0.288 ng/mL) in one rat given 320 mg/kg at 24 h postdose, and cTnI levels were at baseline levels ($\leq$0.02 ng/mL) in the other two rats given 320 mg/kg and in rats given 40 or 80 mg/kg at 24 and 168 h postdose.

Echocardiography evaluation revealed a dose-dependent decrease in cardiac contractility (measured by changes in EF and ESV) that occurred in rats treated with 80 or 320/160 mg/kg. EF was decreased 6% and 11% from baseline, and ESV was increased 67% and 113%, in rats given 80- and 320/160-mg/kg PF-04254644, respectively. Rats treated with 80- or 320/160-mg/kg PF-04254644 had preserved

CO because of small increases in diastolic chamber volume. Readers interested in reviewing parameter means for all echocardiography studies described in this chapter are referred to p. 423 of Aguirre et al. [4].

Myocardial histopathology findings in moribund rats given 320 mg/kg were similar to those observed in the TK study and were characterized by mild-to-moderate multifocal areas of degeneration in the apex, interventricular septum, and left and right ventricular free walls. After 7 days of repeat dosing with 320/160-mg/kg, fibrosis replaced the degenerative myofibers (Figure 25.1C). Minimal focal myocardial degeneration in the same areas of the heart occurred in two of five rats given 80-mg/kg PF-04254644 (Figure 25.1D). Minimal myocardial degeneration occurred in one of five rats given 40 mg/kg.

In summary, increases in serum TnI, decreases in cardiac contractility, and myocardial degeneration in rats given PF-04254644 suggested that off-target inhibition of multiple PDE families contributed to the myocardial toxicity. As a result of these findings in this study, PF-04254644 drug development was stopped but further evaluated in two additional investigative studies to improve characterization of the onset and reversibility of myocardial damage by evaluating myocardiocyte ultrastructure, cTnI responses, and changes in myocardial function using safety pharmacology endpoints (contractility, HR, and blood pressure, BP).

25.4 *IN-VITRO-TO-IN-VIVO* TRANSLATION OF PHOSPHODIESTERASE INHIBITION

In the first investigative oral dose study using PF-04254644, we accurately assessed the onset and reversibility of myocardial changes after a single high dose of 500-mg/kg PF-04254644 in 12 rats by collecting echocardiography parameters, serum cTnI levels, hearts for ultrastructural examination, and blood for TK analysis at various time points. Echocardiography endpoints were collected predose and at 6 and 24 h postdose. Serum cTnI and hearts for ultrastructural evaluation were examined at 2, 8, 24, and 168 h postdose in three rats per time point.

TK analysis showed that the estimated free peak plasma concentration ranges for PF-04254644 were 6.2 to 10.4 μM and exceeded the IC_{50} of multiple PDEs (Table 25.2).

The results of the echocardiographic evaluation revealed that EF was increased by 12% over baseline values at 6 and 24 h, and significant increases in HR (19%) and CO (33%) were also observed at 6 and 24 h postdose in rats given 500-mg/kg PF-04254644.

At the ultrastructural level, myocardial changes in rats given a single 500 mg/kg dose of PF-04254644 were observed as early as 2 h postdose and were characterized by myofiber lysis with visibly intact mitochondria, disruption of the intercalated disk, and deposition of lipid material (Figure 25.2B). At 2 h postdose, ultrastructural changes in the heart were more consistently observed than histopathologic changes. At 8 h after dosing, myocardial changes progressed to coagulative necrosis of myofibers with the presence of intramitochondrial densities (Figure 25.2C) and

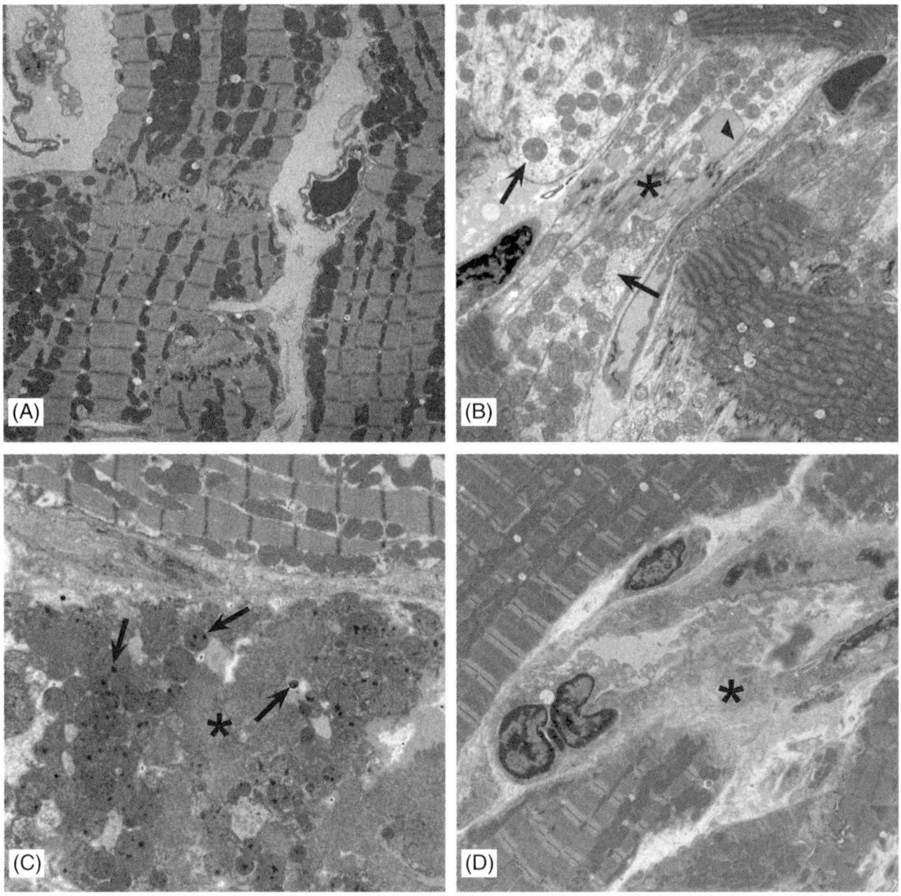

Figure 25.2 Ultrastructural findings in rat myocardial tissue. (A) Myocardial tissue from a control rat (2000X). (B) Myocardial tissue from a rat in the 7-day single-dose investigative study 2 h after dosing with 500-mg/kg PF-04254644 (2000X). Note the loss of myofiber actin and myosin filaments and intact mitochondria (arrows), deposition of flocculent material near disrupted intercalated disk (asterisk), and lipid deposition (arrowhead). (C) Myocardial tissue from rat in the 7-day single-dose investigative study 8 h after dosing with 500-mg/kg PF-04254644 (2000X). Note the coagulative necrotic myofibers (asterisk) and intra-mitochondrial densities (arrows). (D) Myocardial tissue from a rat in the 6-day repeat-dose investigative study 24 h after dosing with 10.5-mg/kg milrinone (2000X). Note the coagulative necrosis of myofibers (asterisk) and macrophage infiltration.

were consistently correlated with histopathologic findings. Coagulatively necrotic myofibers and macrophage infiltrates were consistently observed by ultrastructural and histopathologic examination at 24 h after dosing.

Serum cTnI levels in control rats ranged from 0.126 to <0.02 ng/mL. The serum cTnI levels ranged from 0.564 to 2.158 ng/mL at 2 h after a single dose of 500-mg/kg PF-04254644. At the 2-h time point, the increased cTnI levels were more consistently correlated with findings observed in the myocardium of rats at the ultrastructural level than findings observed in the heart by light microscopic assessment of H&E stained sections. At 8 h postdose, serum cTnI levels ranged from 0.421 to 1.472 ng/mL in 4 rats and correlated with mild-to-moderate myocardial degeneration observed by ultrastructural or histopathologic examination. By 24 h, serum cTnI levels ranged from 0.236 to 0.253 ng/mL or returned to background levels and correlated with a mild-to-moderate myocardial degeneration observed by ultrastructural or histopathologic examination. Comparable cTnI levels at 24 h and myocardial changes at 24 to 48 h were observed in rats given 500 mg/kg in the 7-day single-dose investigative study or 320 mg/kg in the 7-day repeat-dose toleration study. Collectively, the data suggested that PF-04254644 systemic exposure levels were sufficient to exceed the IC_{50} of multiple PDE subfamilies and induced off-target myocardial injury as early as 2 h postdose. Serum cTnI levels peaked at 2 h postdose concomitant with evidence of myocardial damage at 2 h at the ultrastructural level. These ultrastructural and biomarker changes led to the next series of investigative evaluations that focused on examining the effect of inhibition of multiple PDE subfamilies on HR and BP.

In the second investigational study with PF-04254644, we conducted a rat telemetry study to determine whether other myocardial *in vivo* functional alterations (HR and BP) could be contributing to the myocardial degeneration and included TK, histopathology, and ultrastructural evaluation endpoints. A 6-day repeat-dose study was conducted in 74 male Sprague–Dawley (SD) rats given 0-, 40-, or 80-mg/kg PF-04254644 or 10.5-mg/kg milrinone, a known PDE3 inhibitor [21, 22]. Doses of PF-04254644 were selected to affirm the induction of myocardial degeneration at the predicted human therapeutic dose (40 mg/kg) previously observed in the 7-day exploratory toxicity study. The dose for milrinone (10.5 mg/kg) was selected because it was predicted selectively to inhibit the PDE3 subfamily and attained an estimated free serum concentration predicted to exceed PDE3 IC_{50} values. Rats were segregated into three assessment groups: *in vivo* cardiovascular analysis including echocardiography and telemetry ($n = 28$), histo- and ultrastructural pathology with cTnI ($n = 40$), and satellite toxicokinetics ($n = 6$). Rats intended for *in vivo* cardiovascular assessment were instrumented with PA-C40 radio telemetry units (Data Sciences Inc., St. Paul, MN) for conscious, unrestricted blood pressure and heart rate measurements [23]. Arterial BP and HR were measured and analyzed with the DATAQUEST A.R.T. Version 3.0- Gold software (Data Sciences International, St. Paul, MN). The values represent the average of all data points collected from each animal, every 15 min for a 10-s interval over a 24-h period. Data were further consolidated and analyzed using 1-h averages. Echocardiography was performed on telemetry instrumented rats as follows: on day 1 at baseline predose, at times of maximal drug concentration (C_{max}; 0.5–1 h postdose for milrinone and 1–2 h postdose for PF-04254644), and at 24 h; on day 6 prior to dosing (0 h) and at C_{max}. Hearts for histologic and ultrastructural pathology were collected from three rats per group at various time points (2, 6, 24, and 144 h) postdose and were bisected longitudinally through both ventricles immediately adjacent to both atria. Half of the heart was retained in Karnovsky's fixative, and the other half of the myocardium was saved in RNA later. Samples of left ventricle, right ventricle, and interventricular septum were retained in Karnovsky's fixative for ultrastructural examination. The remaining heart tissue was processed for histopathologic examination. Analysis of serum cTn1 was conducted at 2, 6, 24, and 144 h postdose.

TK analysis revealed that maximal estimated free plasma concentrations of PF-04254644 exceeded the IC_{50} values for PDE 1A, 3A, 3B, 10, and 11 (Table 25.2). The T_{max} for milrinone occurred at 2 h postdose (data not shown). The maximal estimated free plasma concentration of milrinone was 2.5 μM and exceeded the IC_{50} values for PDE3A, 3B, and 4D (Table 25.2).

Histopathology results in this second investigational study were negative in rats given 40- or 80-mg/kg PF-04254644 or 10.5-mg/kg milrinone at 2 h postdose. Minimal multifocal myocardial degeneration occurred in the apex, interventricular septum, left and right ventricular free walls of rats at ≥6 h after dosing with 40- or 80-mg/kg PF-04254644; this change progressed to fibrosis on day 6. Similar to PF-04254644 at the 40-mg/kg dose level, no myocardial changes were observed in rats given 10.5-mg/kg milrinone at 2 h postdose; and minimal multifocal myocardial degeneration occurred in the apex, interventricular septum, and left and right ventricular free walls of rats at ≥6 h after dosing (Figure 25.1E).

Serum cTnI levels did not increase above background levels in rats dosed with PF-04254644 at 40- or 80-mg/kg or milrinone at 10.5 mg/kg. In this case, the assay used to detect release of cTnI may have been of insufficient sensitivity to detect release of cTnI in the presence of minimal myocardial injury.

Ultrastructurally, myocardial changes in rats given 10.5-mg/kg/day milrinone were undetectable at 2 h and minimal at ≥6 h postdose and characterized by coagulative necrosis of myofibers with macrophage infiltration (Figure 25.2D).

The results of the echocardiography study revealed an increase in EF (2% and 10%), CO (17% and 25%), and HR (5% and 17%) observed at the C_{max} in rats given 40- and 80-mg/kg PF-04254644, respectively, and all parameters returned to normal or baseline levels 24 h postdose. However, a significant decrease in EF (5%) and SV (6%) was evident before dosing with 80-mg/kg PF-04254644 on day 6. Despite this decreased contractility during the time of minimal drug concentrations, treatment with 80-mg/kg PF-04254644 increased EF, CO, and HR (6%, 25%, and 17%, respectively) on day 6 at C_{max}. No significant changes in echocardiography parameters occurred on the 6th day of dosing in rats given 40-mg/kg PF-04254644. Milrinone treatment in rats on day 1 induced an increase over baseline in EF (7%), CO (37%), and HR (21%) at C_{max} and no significant changes in these parameters at 24 h postdose. Similarly, on day 6 at 0 h before dosing with milrinone, EF and SV were not significantly different from vehicle-treated animals, and EF, CO, and HR increased (5%, 37%, and 21%, respectively) at C_{max}. Across all studies, echocardiographic changes in rats given vehicle were consistent with historical data from nontreated and vehicle-treated animals (data not shown).

After a single dose of 80-mg/kg PF-04254644, cardiac contractility and HR increased as early as 3 h postdose and returned to normal baseline levels at 24 h. However, after repeat dosing and before the last treatment on day 6 (time 0 h) with 80 mg/kg, there was a significant decrease in cardiac contractility. Despite this decreased contractility, rats that were given the last dose of 80 mg/kg on day

6 were able to respond with an increase in cardiac contractility (EF), CO, and HR at C_{max}. Rats given 40-mg/kg PF-04254644 or 10.5-mg/kg milrinone responded with an increased contractility, HR, and CO at C_{max} on day 1 and returned to baseline levels 24 h later. On day 6, prior to the last dose (time 0 h), no changes in cardiac function were detectable in rats given 40-mg/kg PF-04254644 or 10.5-mg/kg milrinone. However, on day 6 at C_{max}, rats given milrinone responded with an increased cardiac contractility, HR, and CO, while rats given 40-mg/kg PF-04254644 did not respond with any detectable changes in cardiac function.

Overall, soon after dosing rats with 500-mg/kg PF-04254644, cardiac contractility measured by EF, CO, and HR as measured by echocardiography R-R interval (late diastole) remained increased over a 24-h time period. After 7 days of repeat dosing in rats with ≥80-mg/kg PF-04254644, there was an overall reduction in cardiac function with dose-dependent decreases in cardiac contractility but CO was preserved.

In the rat telemetry study, HR changes were continuously monitored in rats given 40- or 80-mg/kg PF-04254644 or 10.5-mg/kg milrinone on day 1 through day 6, as shown in Figure 25.3. On day 1, an increase in HR or tachycardia (HR change from vehicle ≥25 bpm) was first observed at 2 h postdose in rats given 40- or 80-mg/kg PF-04254644 and was elevated for a duration of 5 h. In comparison, tachycardia was first observed 1 h postdose in rats given milrinone and reached a maximum effect 4 h postdose. On day 2, the magnitude and duration of the milrinone-induced tachycardia was greater than that observed with PF-04254644. However, over days 3 to 5, although the initial magnitude of the tachycardia postdose was greater for milrinone than PF-04254644, the overall duration of the PF-04254644-induced tachycardia was longer (at least 8 h) than that observed for milrinone (<4 h). On day 6, a maximum tachycardia (120 bpm) relative to vehicle controls occurred 1 h postdose in rats given 80-mg/kg PF-04254644 or 2 h postdose in rats given 40 mg/kg and was sustained for 8 h. In contrast, milrinone on day 6 induced a maximum tachycardia (100 bpm) relative to vehicle controls that lasted for less than 1 h and returned to baseline levels in <3 h. Therefore, over the course of 6 days of repeat dosing in rats, PF-04254644 or milrinone induced a tachycardia (>100 bpm) 1 to 2 h after dosing, but the tachycardia (>25 bpm) in rats given PF-04254644 lasted 1 to 6 h longer in duration compared with rats given milrinone.

Systolic blood pressure (SBP), diastolic blood pressure (DBP), and mean blood pressure (MBP) were also monitored for 6 days in rats given 40- or 80-mg/kg PF-04254644 or 10.5-mg/kg milrinone (data not shown). Prior to dosing, BP measurements ranged between –5 and +5 mmHg of group averages of 118-mmHg SBP, 100-mmHg MBP, and 80-mmHg DBP. The most remarkable drop in blood pressure was observed in rats given milrinone. Over the course of 6 days, a 15–20-mmHg decrease in SBP, DBP, and MBP at time of maximal drug concentration (C_{max}) was observed approximately 1 h postdose and was sustained for 4 to 6 h in rats given milrinone. The BP decrease in rats given milrinone was consistent with internal Pfizer data (unpublished data). After 6 h of dosing, BP values returned to baseline levels. Upon repeat dosing with milrinone, the decrease in BP diminished to a level similar to PF-04254644-treated rats. The maximal decrease in SBP, DBP, and MBP in rats given 40- or 80-mg/kg PF-04254644 was 8 to 10 mm Hg at 1 h postdose (or C_{max})

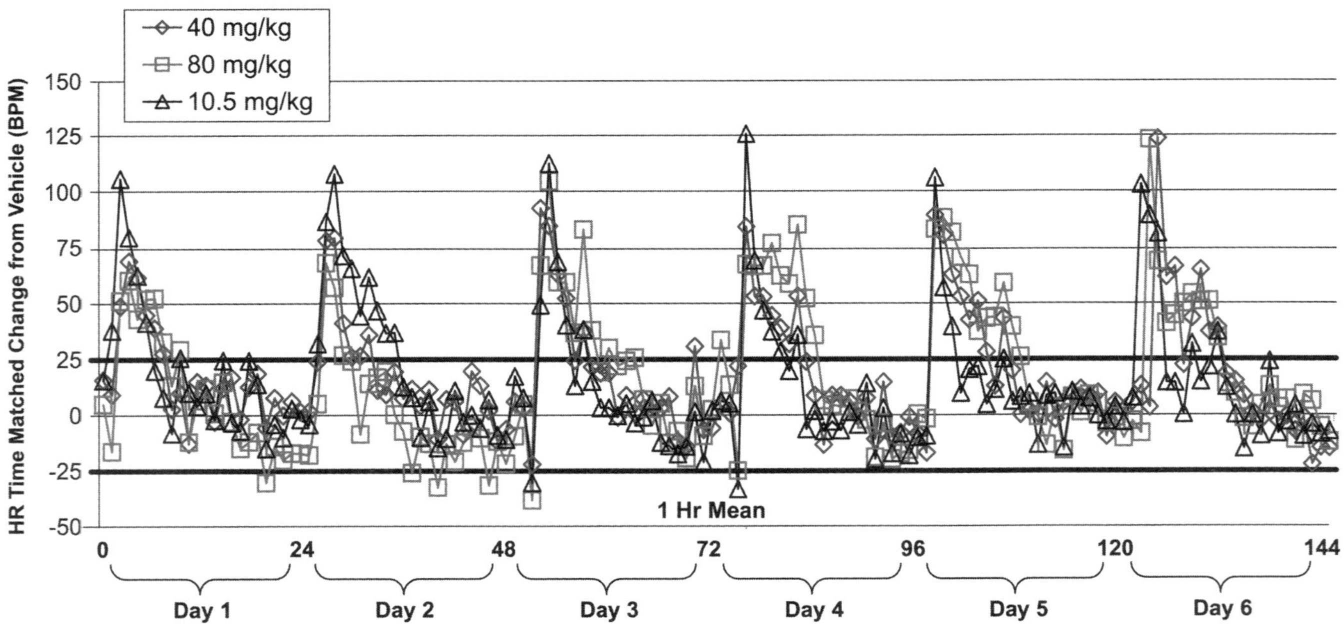

Figure 25.3 Heart rate profile in telemetry instrumented rats from the 6-day repeat-dose investigative study. Rats were given 40- or 80-mg/kg PF-04254644 or 10.5-mg/kg milrinone. Heart rate data represented as 1 h means of time-matched change from rats given vehicle. Dosing occurred at or during hours 2, 25, 49, 73, 96, and 121. Heavy black bars at 25 and -25 depict baseline HR range prior to dosing for male Sprague–Dawley rats ages 7 to 9 weeks. (*See insert for color representation of the figure.*)

and was sustained for 4 to 6 h and then returned to baseline values after 6 h over the 6-day treatment period. With PF-04254644 repeat dosing, the BP values remained within the range of baseline values after 6 h of dosing and were consistently higher than baseline BP values observed with milrinone treatment. Therefore, the overall pattern of BP decrease induced by milrinone or PF-04254644 treatment in rats was somewhat similar, but the overall decrease in BP was greater in milrinone-treated rats compared with PF-04254644-treated rats.

In summary, the c-MET/HGFR inhibitor, PF-04254644, provided an excellent example where the generation of early absorption, distribution, metabolism, excretion, and toxicology (ADMET) data in the preclinical environment helped to mitigate ADMET risks associated with off-target PDE3 inhibition and related cardiac toxicities. This work further affirmed the cardiovascular alert for PDE3 and suggested that other PDEs expressed in the heart may have contributed to untoward cardiac toxicity.

REFERENCES

1. Greaves, P. *Cardiovascular System. Histopathology of Preclinical Toxicity Studies*, 3rd ed. Elsevier, Amsterdam, 2007, pp. 277–278.
2. King, W.M., Russell, S.P. Metabolic, traumatic and miscellaneous diseases: cardiovascular system, in: M.A. Suckow, S.H. Weisbroth, C.L. Franklin (eds.), *The Laboratory Rat*, 2nd ed. Boston, MA: Elsevier, 2006, pp. 515–517.
3. Keenan, C.M., Hughes-Earle, A.R., Maleeff, B.E., et al. (2010). Industry survey of approaches to examination and terminology of spontaneous changes in the heart of young rats. *Toxicologic Pathology*, *38*, 995–998.
4. Aguirre, S.A., Heyen, J.R., Collette, W., III, Bobrowski, W., Blasi, E.R. (2010). Cardiovascular effects in rats following exposure to a receptor tyrosine kinase inhibitor. *Toxicologic Pathology*, *38*, 416–428.
5. Whitebread, S., Hamon, J., Bojanic, D., Urban, L. (2005). Keynote review: In vitro safety pharmacology profiling: An essential tool for successful drug development. *Drug Discovery Today*, *10*, 1421–1433.
6. Force, T., Krause, D.S., Van Etten, R.A. (2007). Molecular mechanisms of cardiotoxicity of tyrosine kinase inhibition. *Nature Reviews Cancer*, *7*, 332–344.
7. Kerkela, R. (2006). Cardiotoxicity of the cancer therapeutic agent imatinib mesylate. *Nature Medicine*, *12*, 908–916.
8. Chen, M.H., Kerkela, R., Force, T. (2008). Mechanisms of cardiac dysfunction associated with tyrosine kinase inhibitor cancer therapeutics. *Circulation*, *118*, 84–95.
9. Schmidinger, M., Zielinski, C.C., Vogl, U.M., et al. (2008). Cardiac toxicity of sunitinib and sorafenib in patients with metastatic renal cell carcinoma. *Journal of Clinical Oncology*, *26*, 5204–5212.
10. Orphanos, G.S., Ioannidis, G.N., Ardavanis, A.G. (2009). Cardiotoxicity induced by tyrosine kinase inhibitors. *Acta Oncology*, *48*, 964–970.
11. Mellor, H.R., Bell, A.R., Valentin, J.P., Roberts, R.R.A. (2011). Cardiotoxicity associated with targeting kinase pathways in cancer. *Toxicological Sciences*, *120*, 14–32.

12. Van Etten, R.A., Shannon, K.M. (2004). Focus on myeloproliferative diseases and myelodysplastic syndromes. *Cancer Cell*, *6*, 547–552.

13. Krause, D.S., Van Etten, R.A. (2005). Tyrosine kinases as targets for cancer therapy. *New England Journal of Medicine*, *353*, 172–187.

14. Stellrecht, C.M., Gandhi, V. (2009). MET receptor tyrosine kinase as a therapeutic anti-cancer target. *Cancer Letters*, *280*, 1–14.

15. Hu, W., Hirakawa, B., Jessen, B., Lee, M., Aguirre, S.A. (2012). A tyrosine kinase inhibitor induced myocardial degeneration in rats through off-target phosphodiesterase inhibition. *Journal of Applied Toxicology*, *32*, 1008–1020.

16. Zou, H., Li, Q., Lee, J., et al. *Receptor tyrosine kinase cross talk affects tumor cell sensitivity to RTK inhibitors*. AACR, 99th Annual Meeting American Association for Cancer Research. Late Breaking Abstracts & Posters, San Diego, 2008, LB-198.

17. Zou, H.Y., Li, Q., Lee, J.H., et al. (2012). Sensitivity of selected human tumor models to PF-04217903, a novel selective c-Met kinase inhibitor. *Molecular Cancer Therapy*, *4*, 1036–1047.

18. Heyen, J.R., Blasi, E.R., Nikula, K., et al. (2002). Structural, functional, and molecular characterization of the SHHF model of heart failure. *American Journal of Physiology, Heart and Circulatory Physiology*, *283*, 1775–1784.

19. Schiller, N.B., Shah, P.M. Crawford, M., et al. (1989). Recommendations for quantitation of the left ventricle by two-dimensional echocardiography. *Journal of American Society of Echocardiography*, *2*, 358–367.

20. Weyman, A.E. *Principles and Practice of Echocardiography*, Lea & Febinge, Philadelphia, PA, 1994.

21. Earl, C.Q., Linden, J., Weglicki, W.B. (1986). Biochemical mechanisms for the inotropic effect of the cardiotonic drug milrinone. *Journal of Cardiovascular Pharmacology*, *8*, 864–872.

22. Earl, C.Q., Linden, J., Weglicki, W.B. (1986). Inhibition of cyclic AMP-dependent protein kinase activity by the cardiotonic drugs amrinone and milrinone. *Life Sciences*, *39*, 1901–1908.

23. Blasi, E.R., Rocha, R., Rudolph, A.E., Blomme, E.A., Polly, M.L., McMahon, E. (2003). Aldosterone/salt induces renal inflammation and fibrosis in hypertensive rats. *Kidney International*, *63*, 1791–1800.

26

SAFETY MANAGEMENT BY PHARMACOKINETIC CONSIDERATIONS: RANIBIZUMAB (LUCENTIS) AND BEVACIZUMAB (AVASTIN)

Nicole H. Siegel and Manju L. Subramanian

26.1 INTRODUCTION

The vascular endothelial growth factor (VEGF) plays a pivotal role in a wide array of physiologic processes in the body. Its upregulation is involved in the development of numerous ocular diseases, including the neovascularization observed in exudative age-related macular degeneration and proliferative diabetic retinopathy, as well as macular edema in diabetes and retinal vein occlusions. VEGF inhibition, namely in the form of intravitreal injections of bevacizumab and ranibizumab, has revolutionized the treatment and prognosis of exudative age-related macular degeneration. Given the importance of VEGF and its involvement in both physiologic and pathologic angiogenesis, the ocular and systemic safety of administering VEGF antagonists are a primary concern [1]. There is ample evidence to suggest adequate safety of both bevacizumab and ranibizumab, when administered via intravitreal injection, although there are key differences between the two drugs, which will be reviewed in this chapter.

Predictive ADMET: Integrative Approaches in Drug Discovery and Development, First Edition.
Edited by Jianling Wang and Laszlo Urban.

Age-related macular degeneration (AMD) is a progressive disease of the retina and is the leading cause of blindness in persons older than 50 years of age in developed nations [2]. AMD has significant economic and social implications [3]. The non-neovascular or nonexudative (dry) form of AMD, the more common form occurring in 80–90% of cases, is characterized by macular pigmentary changes, drusen, and atrophy of the retinal pigment epithelium (RPE) and choriocapillaris. Its treatment centers on nutritional supplementation and antioxidants to slow disease progression, as concluded in The Age-Related Eye Disease Study (AREDS) [4]. The neovascular or exudative (wet) form of AMD, which often causes more significant visual decline and is implicated in a greater proportion of visual loss related to AMD, is defined by the growth of abnormal blood vessels under the retina, or choroidal neovascular membranes (CNVMs). These are observed clinically by the presence of subretinal fluid, subretinal hemorrhages, exudates, and inflammation [4–6]. Without treatment, subretinal fibrosis develops with damage to the outer retinal layers, RPE, and choriocapillaris resulting in permanently decreased vision and visual distortion secondary to irreversible scarring.

Previous treatments for age-related macular degeneration included submacular surgery, photodynamic therapy (PDT) with verteporfin (Visudyne) (established in both the Verteporfin in Photodynamic therapy (VIP) and Treatment of Age Related Macular Degeneration with Photodynamic Therapy (TAP) trials) [7, 8], and thermal laser photocoagulation (established in the Macular Photocoagulation Study). These treatments delayed visual decline but seldom improved final visual acuity [9, 10]. In 2004, the Vascular Endothelial Growth Factor (VEGF) Inhibition Study in Ocular Neovascularization (VISION) trials supported the benefit of VEGF inhibition by Pegaptanib sodium (Macugen, Pfizer/Eyetech, New York, NY) compared with sham-injected controls. Macugen is an RNA aptamer that binds the primary isoform of VEGF, $VEGF_{165}$ [9].

The advent of vascular endothelial growth factor A (VEGF-A) inhibition therapy has revolutionized the treatment and prognosis of neovascular age-related macular degeneration. It has also had a major impact on visual prognoses for vascular occlusions, proliferative diabetic retinopathy, and macular edema stemming from diabetic retinopathy and retinal vein occlusions. AMD, diabetes, and retinal vascular occlusions cause vision loss secondary to VEGF-associated macular edema and neovascularization [1]. Diabetic macular edema and clinically significant macular edema are characterized by the accumulation of extracellular fluid within the retina in the setting of diabetic retinopathy [11]. The arsenal of treatment for clinically significant macular edema includes focal and grid laser as well as the intravitreal use of VEGF inhibitors.

The purpose of this chapter is to review comprehensively the advent, utility, pharmacokinetics, and safety of inhibiting vascular endothelial growth factor in treating ocular diseases.

26.2 VASCULAR ENDOTHELIAL GROWTH FACTOR (VEGF)

The importance of angiogenesis, defined as the growth of new capillaries in blood vessel networks [9], in the pathogenesis of disease is well known. Rudolf Virchow and others reported the association between tumor growth and angiogenesis in the late 1800s [12]. The discovery of the role of vascular endothelial growth factor by Ferrara and Henzel in 1989 [13] and its subsequent utility in treating exudative age-related macular degeneration began a whirlwind in drug development to optimize the treatment for this previously visually devastating disease [14]. By the 1990s, the theoretical role of VEGF and its relevance to retinal and choroidal neovascularization was elucidated [15].

Although other members of the VEGF family exist that interact with the VEGF tyrosine kinase receptors, VEGF receptors (VEGFR) 1, 2, and 3, including VEGF-A, VEGF-B, VEGF-C, VEGF-D, VEGF-E, and placental growth factor (PlGF) [14], VEGF-A has the greatest relevance in ocular disease by contributing to maintaining baseline ocular functions as well as to increasing vascular permeability and producing angiogenesis [9]. Systemically, VEGF-A promotes endothelial cell proliferation and vascular permeability and is responsible for systemic homeostasis and precise embryologic development [16].

The VEGF-A gene comprises eight exons, and different splicing patterns of VEGF pre-messenger RNA produce its six discrete isoforms, which include 121-, 145-, 165-, 183-, 189- and 206-amino acid molecular species, which manifest different properties [14, 17]. $VEGF_{121}$ is more diffusible without heparin binding. Alternatively, $VEGF_{189}$ is bound to the extracellular matrix and cell surface. $VEGF_{165}$, which is the predominant form of VEGF, has some properties of both and is more diffusible than $VEGF_{189}$ resulting in its greater efficacy and bioavailability, which is further demonstrated in Figure 26.1 [17, 18].

In vivo, VEGF-A contributes to angiogenesis, and *in vitro,* VEGF-A results in vascular endothelial cell growth and is upregulated in the setting of hypoxia and retinal ischemia [19]. It is derived from numerous cell lines in the retina, including the RPE, pericytes, retinal ganglion cells and neurons, Müller cells, and vascular endothelial cells [14]. VEGF protects against apoptosis aiding in the survival of endothelial and tumor cells, and it is theorized be involved in retinal leukostasis as well as having neurotrophic and neuroprotective effects [20, 21]. The retinal pigment epithelium, which expresses VEGF, is responsible for maintaining both the photoreceptor outer segments and the choriocapillaris. VEGF inhibition, and AMD can both cause damage to the RPE, photoreceptors, and choriocapillaris [22]. Therefore, the long-term and dose effect on retinal tissues from VEGF inhibition has been evaluated and must be weighed against the potential beneficial role of VEGF both systemically and in the eye.

The pathogenesis of increased VEGF expression varies in different disease states. In exudative age-related macular degeneration, retinal pigment epithelium damage results in increased VEGF expression, which contributes to the formation of choroidal neovascularization that enters the subretinal space and causes subretinal fluid and hemorrhages [23]. In proliferative diabetic retinopathy, the pathologic retinal neovascularization occurs secondary to hypoxia-induced VEGF upregulation [23]. In retinopathy of prematurity (ROP), hyperoxia suppresses VEGF during retinal development. VEGF suppression is also implicated in treatment of ROP, as examined in the Bevacizumab Eliminates the Angiogenic Threat of Retinopathy of Prematurity (BEAT-ROP) Study [24].

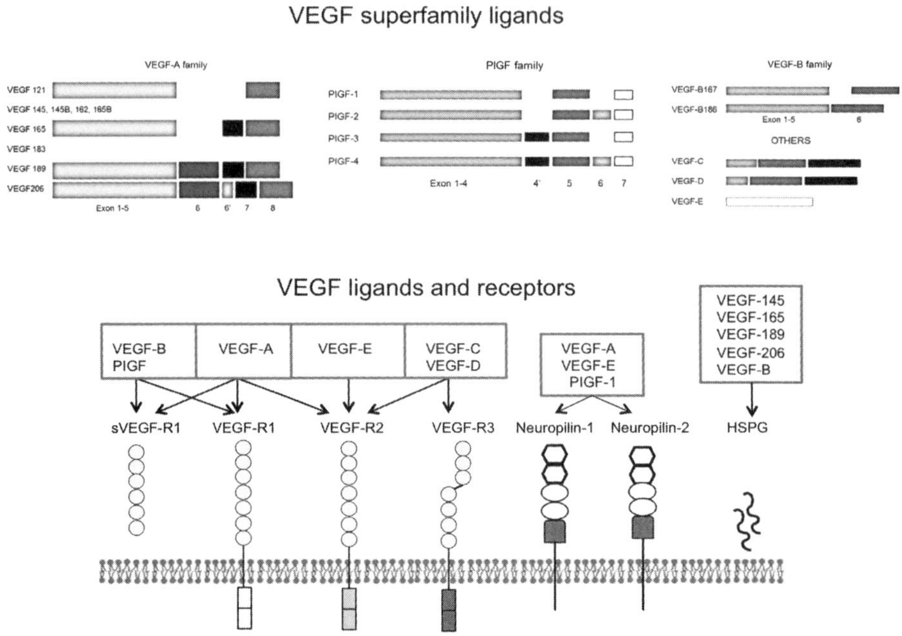

Figure 26.1 VEGF superfamily ligands and receptors. A schematic of VEGF ligands and their interactions with VEGF receptors is shown. (*See insert for color representation of the figure.*)

26.3 INTRAVITREAL ADMINISTRATION OF VEGF ANTAGONISTS

The intraocular use of VEGF antagonists began with intravitreal administration of Pegaptanib sodium (Macugen) after its approval by the U.S. Food and Drug Administration (FDA) in 2004. Pegaptanib is an oligonucleotide RNA aptamer that solely binds to the heparin binding domain of the $VEGF_{165}$ isoform [4]. The success of pegaptanib sodium, which was administered via intravitreal injection every 6 weeks, was noted in its ability to decrease the chance of moderate-to-severe visual loss, although it did not result in significant visual improvement, as demonstrated in the VISION trials [25]. The success of Pegaptanib confirmed the value of VEGF inhibition, as had been previously theorized, in the treatment of angiogenesis observed in AMD.

Bevacizumab (Avastin, Genentech, San Francisco, CA), a 149-kDa full-length antibody composed of two light chains and two heavy chains with a murine-derived Fc segment, is a humanized version of an anti-VEGF monoclonal antibody that targets all human VEGF-A isoforms [1]. It received approval from the FDA in 2004 as a systemic adjuvant for the treatment of metastatic colorectal cancer [15, 26]. In patients receiving intravenous bevacizumab at 5 mg/kg of body weight every 2 weeks, there was an observed increased rate of thromboembolic disease and systemic hypertension [27]. After studies proved the beneficial effect of intravenous bevacizumab to treat exudative AMD, a dose was determined for intravitreal use in an attempt to reduce the adverse events that occur with systemic administration [28–30]. The currently administered doses of 1.25 or 2.5 mg/0.05 mL are found to have a relative lack of systemic complications compared with controls after intravitreal injection [27, 28]. However, its intravitreal administration remains off-label in the treatment of several ocular neovascular diseases including AMD and proliferative diabetic retinopathy [31]. Most of the data supporting its intravitreal efficacy to treat ocular disease comes primarily from a wealth of noncontrolled trials [32–34].

In 2006, shortly after its FDA approval, Rosenfeld et al. reported on the use of ranibizumab (Lucentis; Genentech), a 48-kDa recombinant, higher affinity human monoclonal antibody fragment (Fab) that neutralizes all active forms of VEGF-A as described earlier [2, 15]. Unlike bevacizumab, ranibizumab was formulated specifically for intravitreal administration and is most commonly administered at a dose of 0.50 mg/0.05 mL.

Both bevacizumab and ranibizumab led to an improvement in visual acuity and to a decrease in central retinal thickness as evaluated by spectral domain optical coherence tomography (OCT) in a large percentage of patients who underwent treatment. Figures 26.2 and 26.3 represent examples of normal and abnormal OCT images. This was the first time in the history of AMD management that patients could hope for visual improvement with treatment, rather than slowing visual loss [35]. Early studies of intravitreal ranibizumab, including the Minimally Classic/Occult Trial of the Anti-VEGF Antibody Ranibizumab in the Treatment of Neovascular Age-Related Macular Degeneration (MARINA) and Anti-VEGF Antibody for the Treatment of Predominantly Classic Choroidal Neovascularization in Age-Related Macular Degeneration (ANCHOR) studies, revealed that after one year a higher percentage of patients who received intravitreal ranibizumab treatment gained greater than 15 letters of visual acuity compared with a sham treatment and PDT, respectively [5].

The newest VEGF antagonist aflibercept (Eylea, VEGF Trap-Eye, Regeneron, Tarrytown, NY), which received FDA approval for the treatment of wet AMD in

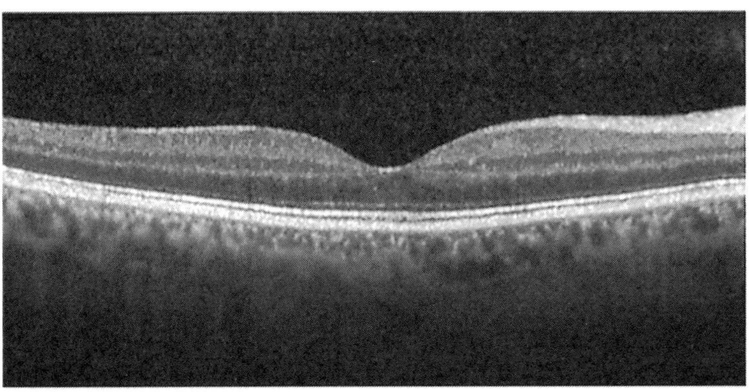

Figure 26.2 Normal macular spectral domain optical coherence tomography (OCT) scan.

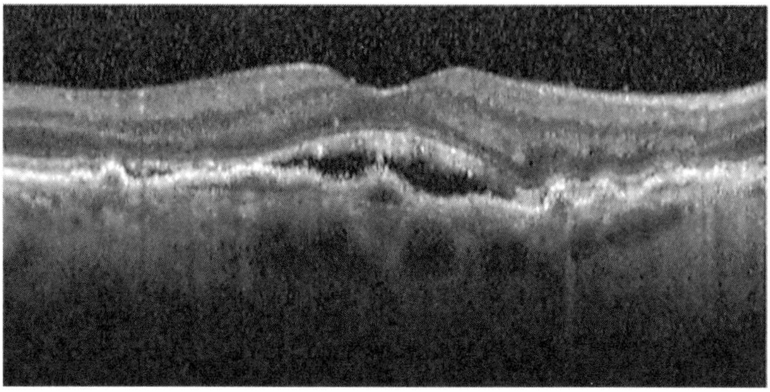

Figure 26.3 Exudative age-related macular degeneration as seen on spectral domain optical coherence tomography (OCT) scan with subretinal fluid and drusen.

November 2011, is a fully human, soluble recombinant decoy VEGFR fusion protein that contains the binding domains of VEGFR-1 and -2 and that binds both VEGF-A and another factor involved in angiogenesis, PlGF [36].

26.4 PHARMACOKINETIC AND PHARMACODYNAMIC CONSIDERATIONS

Bevacizumab and ranibizumab both neutralize all active forms of VEGF-A. Bevacizumab is a full-length antibody. Ranibizumab, a Fab, is a smaller molecule with only one antigen-binding murine domain and a nonbinding human component. It was developed to bind theoretically VEGF more strongly and be less inflammatory without the Fc segment [15, 37, 38]. The proposed benefit of the smaller molecule of ranibizumab, compared with bevacizumab, was its ability to penetrate all layers of the retina [37]. Ranibizumab has been shown in the rabbit to penetrate the full-thickness retina in 3 days, compared with 7 days for bevacizumab [31].

Bakri et al. investigated the pharmacokinetics of intravitreal bevacizumab and ranibizumab using the standard dose of 0.5 mg of ranibizumab versus 1.25 mg of bevacizumab in rabbit eyes [31, 39]. Although the inherent structure of the rabbit eye differs from the human eye, specifically that rabbit eyes have a larger lens, a smaller vitreous cavity, and are less vascular, rabbit eyes have been effectively used to study the pharmacokinetics of drugs administered by intravitreal injection [39]. In the rabbit eye, drug clearance from the vitreous is related to molecular size [37]. In the Bakri et al. study, the vitreous and aqueous humor in the treated eyes and the untreated fellow eyes were measured, along with the serum concentration at different time points, which is further demonstrated in Table 26.1 [31, 39]. The vitreous half-life of 4.32 days was noted to be longer after intravitreal injection of the full-length antibody bevacizumab, compared with 2.88 days with the antibody fragment ranibizumab. Gaudreault et al. studied the vitreous half-life in cynomolgus

Table 26.1 Comparison of the pharmacokinetics of Bevacizumab and Ranibizumab*

Intravitreal Drug	Compartment	Half-life (Days)	MRT (Days)	T_{max} (Days)	C_{max} (μg/ml)	% of Vitreous C_{max}	Exposure to Drug (% of Vitreous Exposure)
Bevacizumab, 1.25 mg	Vitreous	4.32	5.92	1	400		
	Aqueous	4.88	6.77	3	37.7	9.4	8.9
	Serum	6.86	12.8	8	3.33	0.8	1.6
Ranibizumab, 0.5 mg	Vitreous	2.88	4.03	1	162		
	Aqueous	2.84	6.77	3	17.9	11.0	14.2
	Serum				None detected		

C_{max} = maximum concentration; MRT = mean residence time (time to clear >69% of the drug);
T_{max} = time to attain maximum concentration.
Samples were taken from the aqueous and vitreous of the injected eye.

*Reprinted with permission from Reference 31.

monkeys and found comparable results to the study of rabbit eyes. They found that in the monkey the vitreous half-life varied between 2.6 and 4.0 days and that the peak vitreous concentration occurred between 6 and 24 hours depending on the administered intravitreal dose of ranibizumab, with a fast delivery from the vitreous into the retina and anterior chamber [37].

In the studies by Bakri et al., the concentration in the serum and untreated fellow eyes were also evaluated. Ranibizumab was noted to be absent systemically, while there was a small concentration of bevacizumab [31]. The intravitreal concentration of bevacizumab peaked at 400 μg/mL 1 day postinjection of 1.25 mg and sustained a concentration >10 μg/mL for 30 days posttreatment. The serum concentration reached a maximum of 3.3 μg/mL 8 days posttreatment, and bevacizumab was detected in the fellow eye at a peak concentration of 11.17 ng/mL at 4 weeks posttreatment [39]. By comparison, the intravitreal concentration of ranibizumab peaked 1 day postinjection at 162 μg/mL and sustained a concentration of >0.1 μg/mL for 29 days posttreatment (Table 26.1). The concentration of both bevacizumab and ranibizumab in the aqueous humor corresponded to that of the vitreous humor; however, the ranibizumab was notably absent in the serum concentration at all measured time points after injection [31].

Bevacizumab was initially used for the treatment of the CNVM, the hallmark of the exudative form of age-related macular degeneration. To be effective in treating the upregulation of VEGF observed in the setting of CNVM, namely VEGF-A, the dose must adequately inhibit the choroidal vascular endothelial cells. The toxicity to various retinal tissues at a variety of doses of bevacizumab was studied [40]. Brar et al. evaluated the concomitant damage to the retinal pigment epithelial and ganglion cells at a dose effective to treat the CNVM [40]. The clinically used doses of intravitreal bevacizumab are 1.25 mg and 2.5 mg, which correspond to a concentration of 0.3125 and 0.625 mg/mL assuming a 4-mL vitreous cavity size and 0.25 and 0.5 mg/mL with a vitreous cavity size of 5 mL [40,41]. At commercially available doses of 0.1, 1.0, and 2.0 mg/mL of bevacizumab, there was no noted toxicity to the human retinal pigment epithelium and rat ganglion cells (RGC-5) after a 24-hour exposure [40].

In a similar experiment, Sharma and Chalam failed to reveal any cytotoxicity to RGC-5 from the *in vitro* administration of 0.1-, 1.0-, and 2.0-mg/mL doses of bevacizumab. Furthermore, bevacizumab had an excitatory, dose-dependent effect on cell proliferation, including RGC-5 proliferation [41]. It was theorized that this effect might be secondary to increasing the soluble protein concentration in the vitreous. Additional *in vitro* studies are needed to classify further this observed association and determine its relevance and impact in ocular diseases.

Avci et al. studied the effect of intravitreal VEGF inhibition on photoreceptors by comparing photoreceptor apoptosis of bevacizumab at intravitreal doses of 0.25, 0.625, 1.25, and 2.5 mg and pegaptanib sodium at intravitreal doses of 0.15, 0.3, and 0.6 mg [42]. Using 48 rabbit eyes, the early effect was measured after 14 days of one intravitreal injection at all the listed doses and the late effect was evaluated after 90 days with three intravitreal injections given monthly of 0.625 mg of bevacizumab or 0.15 mg of pegaptanib sodium. These doses in the rabbit vitreous volume of 2.5 mL correspond to the clinically accepted doses of 1.25 mg and 0.3 mg in the human vitreous. There was no evidence of retinal toxicity resulting from either medication observed in H&E stained sections.

Evaluating DNA fragments in the nuclei of apoptotic cells in the outer retinal layers using the *in situ* terminal-deoxynucleotidyl transferase-mediated biotin-deoxyuridine triphosphate nick-end labeling (TUNEL) technique, it revealed that the TUNEL-positive apoptotic cells in the outer nuclear layer (ONL) were only observed in the study groups, while they were observed in both the study and control groups in the ganglion cell layer, inner nuclear layer, and retinal pigment epithelium [42]. Comparing the mean percentage of apoptotic TUNEL-positive cells, there was a statistically significant higher percentage in the 0.625-mg, 1.25-mg, and 2.5-mg day 14 bevacizumab groups and the 0.625-mg bevacizumab three-injection group compared with controls with similar levels of increased apoptosis in all pegaptanib sodium groups. Although apoptosis was observed in the control eyes, there was a significant increase in apoptotic activity in rabbit photoreceptor cells at higher doses with bevacizumab [42].

A similar study using mouse retina failed to reveal any toxic effects of bevacizumab [43]. Studies of bevacizumab in different animal populations cannot be directly compared. Although it is known that bevacizumab can bind both primate and rabbit VEGF, controversy exists as to the ability of bevacizumab to bind to murine VEGF [44]. The differences stem from the ability and affinity for species-related VEGF to bind humanized anti-VEGF [42].

26.5 CLINICAL DATA

The Comparison of Age-Related Macular Degeneration Treatments Trials (CATT), a large, single-blind, noninferiority trial, compared the change in visual acuity at one year after intravitreal treatments of bevacizumab versus ranibizumab administered on either a monthly schedule or as needed with monthly evaluation [34]. It was noted that the mean gain of visual acuity was greater for intravitreal injections given monthly than for treatment administered as needed, for both the first and the second year of the CATT trial [34, 45]. Given the significantly lower cost of intravitreal bevacizumab versus ranibizumab, bevacizumab continues to be used off-label for treatment of VEGF-mediated ocular diseases. After one year in the CATT trial, the visual acuity measurements in the bevacizumab and ranibizumab groups were similar when the same dosing schedule was compared [34].

The results were similar to a previous prospective, double-masked randomized clinical trial, which failed to show a statistically significant difference between the visual and anatomic outcomes in the bevacizumab and ranibizumab groups after one year of treatment for exudative AMD although the number of injections required earned statistical significance with the ranibizumab group requiring fewer injections [46]. In the CATT trial, there was no statistically significant difference in the rate of death, arteriothrombotic events, or venous thrombotic events between the groups in the first or second year of the trial [34, 45]. However, the rate of serious systemic adverse events was greater ($P = 0.04$) in the bevacizumab-treated groups, the as-needed group more than the monthly group, resulting mainly from hospitalizations for systemic infections and gastrointestinal disorders. However, these systemic adverse events were not observed with intravenous administration of the drug in cancer trials at 500 times the dose given in the eye [34]. The adverse events associated with the bevacizumab-treated group continued to be observed over the second year of the study [45].

A Medicare claims–based study evaluating more than 40,000 patients concluded that ranibizumab had a lower risk for myocardial infarction and cerebrovascular accidents, as compared with bevacizumab [47,48]. Despite the similar VEGF inhibition bevacizumab and ranibizumab and the small intravitreal doses administered, the noted increased incidence of systemic adverse events with bevacizumab is thought to occur secondary to its larger molecule and longer half-life, leading to more longstanding VEGF suppression [49]. Additionally, animal studies have noted an increased systemic absorption of intravitreal administration of bevacizumab as compared with ranibizumab, which can contribute to the adverse events given the fundamental role that VEGF plays in maintaining key physiologic processes throughout the body [50].

26.6 CONCLUSION

The advent of VEGF inhibition via intravitreal administration has been one of the greatest achievements in the delivery of targeted intraocular treatment. The clinical indications for VEGF inhibitors and number of drugs on the market continue to expand for the treatment of a variety of ocular diseases, including exudative age-related macular degeneration and macular edema from multiple causes. These medications have been instrumental in being able to maintain and potentially improve visual function in several ocular diseases, which previously had poor visual prognosis.

Additional large-scale studies are needed to characterize further the systemic adverse events observed with VEGF inhibition.

REFERENCES

1. Stewart, M.W. (2012). The expanding role of vascular endothelial growth factor inhibitors in ophthalmology. *Mayo Clinic Proceedings*, *87*(1), 77–88.
2. Rosenfeld, P.J., Brown, D.M., Heier, J.S., Boyer, D.S., Kaiser, P.K., Chung, C.Y., et al. (2006). Ranibizumab for neovascular age-related macular degeneration. *New England Journal of Medicine*, *355*(14), 1419–1431.
3. Adamis, A.P., Aiello, L.P., D'Amato, R.A. (1999). Angiogenesis and ophthalmic disease. *Angiogenesis*, *3*(1), 9–14.
4. Prasad, P.S., Schwartz, S.D., Hubschman, J.P. (2010). Age-related macular degeneration: Current and novel therapies. *Maturitas*, *66*(1), 46–50.
5. Vedula, S.S., Krzystolik, M.G. (2008). Antiangiogenic therapy with anti-vascular endothelial growth factor modalities for neovascular age-related macular degeneration. *Cochrane Database System Review*, *2*, CD005139.
6. Subramanian, M.L., Ness, S., Abedi, G., Ahmed, E., Daly, M., Feinberg, E., et al. (2009). Bevacizumab vs ranibizumab for age-related macular degeneration: Early results of a prospective double-masked, randomized clinical trial. *American Journal of Ophthalmology*, *148*(6), 875–882 e871.
7. Photodynamic therapy with verteporfin for age-related macular degeneration. American Academy of Ophthalmology. (2000). *Ophthalmology*, *107*(12), 2314–2317.
8. Brown, D.M., Michels, M., Kaiser, P.K., Heier, J.S., Sy, J.P., Ianchulev, T., et al. (2009). Ranibizumab versus verteporfin photodynamic therapy for neovascular age-related macular degeneration: Two-year results of the ANCHOR study. *Ophthalmology*, *116*(1), 57–65 e55.
9. Bressler, S.B. (2009). Introduction: Understanding the role of angiogenesis and antiangiogenic agents in age-related macular degeneration. *Ophthalmology*, *116*(10), S1–7.
10. Macular Photocoagulation Study Group. (1991). Laser photocoagulation of subfoveal neovascular lesions in age-related macular degeneration. Results of a randomized clinical trial. *Archives of Ophthalmology*, *109*(9), 1220–1231.
11. Parravano, M., Menchini, F., Virgili, G. (2009). Antiangiogenic therapy with anti-vascular endothelial growth factor modalities for diabetic macular oedema. *Cochrane Database System Review*, *4*, CD007419.
12. Ferrara, N. (2002). VEGF and the quest for tumour angiogenesis factors. *Nature Reviews Cancer*, *2*(10), 795–803.
13. Ferrara, N., Henzel, W.J. (1989). Pituitary follicular cells secrete a novel heparin-binding growth factor specific for vascular endothelial cells. *Biochemical and Biophysical Research Communications*, *161*(2), 851–858.
14. Chong, V. (2012). Biological, preclinical and clinical characteristics of inhibitors of vascular endothelial growth factors. *Ophthalmologica*, *227*(1), 2–10.
15. Keane, P.A., Sadda, S.R. (2012). Development of anti-VEGF therapies for intraocular use: A guide for clinicians. *Journal of Ophthalmology*, *2012*, 483034.

16. Tolentino, M. (2011). Systemic and ocular safety of intravitreal anti-VEGF therapies for ocular neovascular disease. *Survey of Ophthalmology*, *56*(2), 95–113.
17. Ferrara, N., Gerber, H.P., LeCouter, J. (2003). The biology of VEGF and its receptors. *Nature Medicine*, *9*(6), 669–676.
18. Ferrara, N. (2011). From the discovery of vascular endothelial growth factor to the introduction of avastin in clinical trials - an interview with Napoleone Ferrara by Domenico Ribatti. *International Journal of Developmental Biology*, *55*(4-5), 383–388.
19. Adamis, A.P., Shima, D.T., Tolentino, M.J., Gragoudas, E.S., Ferrara, N., Folkman, J., et al. (1996). Inhibition of vascular endothelial growth factor prevents retinal ischemia-associated iris neovascularization in a nonhuman primate. *Archives of Ophthalmology*, *114*(1), 66–71.
20. Ding, X.M., Mao, B.Y., Jiang, S., Li, S.F., Deng, Y.L. (2005). Neuroprotective effect of exogenous vascular endothelial growth factor on rat spinal cord neurons in vitro hypoxia. *Chinese Medical Journal (Engl)*, *118*(19), 1644–1650.
21. Jin, K.L., Mao, X.O., Greenberg, D.A. (2000). Vascular endothelial growth factor: Direct neuroprotective effect in in vitro ischemia. *Proceedings of the National Academy of Science in the United States of America*, *97*(18), 10242–10247.
22. Ford, K.M., Saint-Geniez, M., Walshe, T., Zahr, A., D'Amore, P.A. (2011). Expression and role of VEGF in the adult retinal pigment epithelium. *Investigative Ophthalmology & Visual Science*, *52*(13), 9478–9487.
23. Saint-Geniez, M., Maharaj, A.S., Walshe, T.E., Tucker, B.A., Sekiyama, E., Kurihara, T., et al. (2008). Endogenous VEGF is required for visual function: Evidence for a survival role on muller cells and photoreceptors. *PLoS One*, *3*(11), e3554.
24. Mintz-Hittner, H.A., Kennedy, K.A., Chuang, A.Z., Group, B.-R.C. (2011). Efficacy of intravitreal bevacizumab for stage 3+ retinopathy of prematurity. *New England Journal of Medicine*, *364*(7), 603–615.
25. Mitchell, P. (2011). A systematic review of the efficacy and safety outcomes of anti-VEGF agents used for treating neovascular age-related macular degeneration: Comparison of ranibizumab and bevacizumab. *Current Medical Research and Opinion*, *27*(7), 1465–1475.
26. Fernando, N.H., Hurwitz, H.I. (2004). Targeted therapy of colorectal cancer: Clinical experience with bevacizumab. *Oncologist*, *9*(1), 11–18.
27. Wu, L., Martinez-Castellanos, M.A., Quiroz-Mercado, H., Arevalo, J.F., Berrocal, M.H., Farah, M.E., et al. (2008). Twelve-month safety of intravitreal injections of bevacizumab (Avastin): Results of the Pan-American Collaborative Retina Study Group (PACORES). *Graefes Archive for Clinical and Experimental Ophthalmology*, *246*(1), 81–87.
28. Michels, S., Rosenfeld, P.J., Puliafito, C.A., Marcus, E.N., Venkatraman, A.S. (2005). Systemic bevacizumab (Avastin) therapy for neovascular age-related macular degeneration twelve-week results of an uncontrolled open-label clinical study. *Ophthalmology*, *112*(6), 1035–1047.
29. Moshfeghi, A.A., Rosenfeld, P.J., Puliafito, C.A., Michels, S., Marcus, E.N., Lenchus, J.D., et al. (2006). Systemic bevacizumab (Avastin) therapy for neovascular age-related macular degeneration: Twenty-four-week results of an uncontrolled open-label clinical study. *Ophthalmology*, *113*(11), 2002 e2001–2012.
30. Geitzenauer, W., Michels, S., Prager, F., Rosenfeld, P.J., Kornek, G., Vormittag, L., et al. (2008). Comparison of 2.5 mg/kg and 5 mg/kg systemic bevacizumab in neovascular

age-related macular degeneration: Twenty-four week results of an uncontrolled, prospective cohort study. *Retina*, *28*(10), 1375–1386.

31. Bakri, S.J., Snyder, M.R., Reid, J.M., Pulido, J.S., Ezzat, M.K., Singh, R.J. (2007). Pharmacokinetics of intravitreal ranibizumab (Lucentis). *Ophthalmology*, *114*(12), 2179–2182.
32. Rich, R.M., Rosenfeld, P.J., Puliafito, C.A., Dubovy, S.R., Davis, J.L., Flynn, H.W., Jr., et al. (2006). Short-term safety and efficacy of intravitreal bevacizumab (Avastin) for neovascular age-related macular degeneration. *Retina*, *26*(5), 495–511.
33. Gregori, N.Z., Gaitan, J., Rosenfeld, P.J., Puliafito, C.A., Feuer, W., Flynn, H.W., Jr., et al. (2008). Long-term safety and efficacy of intravitreal bevacizumab (Avastin) for the management of central retinal vein occlusion. *Retina*, *28*(9), 1325–1337.
34. Group, C.R., Martin, D.F., Maguire, M.G., Ying, G.S., Grunwald, J.E., Fine, S.L., et al. (2011). Ranibizumab and bevacizumab for neovascular age-related macular degeneration. *New England Journal of Medicine*, *364*(20), 1897–1908.
35. Dafer, R.M., Schneck, M., Friberg, T.R., Jay, W.M. (2007). Intravitreal ranibizumab and bevacizumab: A review of risk. *Seminars in Ophthalmology*, *22*(3), 201–204.
36. Brown, D.M., Heier, J.S., Ciulla, T., Benz, M., Abraham, P., Yancopoulos, G., et al. (2011). Primary endpoint results of a phase II study of vascular endothelial growth factor trap-eye in wet age-related macular degeneration. *Ophthalmology*, *118*(6), 1089–1097.
37. Gaudreault, J., Fei, D., Rusit, J., Suboc, P., Shiu, V. (2005). Preclinical pharmacokinetics of Ranibizumab (rhuFabV2) after a single intravitreal administration. *Investigative Ophthalmology & Visual Science*, *46*(2), 726–733.
38. Zou, L., Lai, H., Zhou, Q., Xiao, F. (2011). Lasting controversy on ranibizumab and bevacizumab. *Theranostics*, *1*, 395–402.
39. Bakri, S.J., Snyder, M.R., Reid, J.M., Pulido, J.S., Singh, R. J. (2007). Pharmacokinetics of intravitreal bevacizumab (Avastin). *Ophthalmology*, *114*(5), 855–859.
40. Brar, V.S., Sharma, R.K., Murthy, R.K., Chalam, K.V. (2009). Evaluation of differential toxicity of varying doses of bevacizumab on retinal ganglion cells, retinal pigment epithelial cells, and vascular endothelial growth factor-enriched choroidal endothelial cells. *Journal of Ocular Pharmacology and Therapeutics*, *25*(6), 507–511.
41. Sharma, R.K., Chalam, K.V. (2009). In vitro evaluation of bevacizumab toxicity on a retinal ganglion cell line. *Acta Ophthalmology*, *87*(6), 618–622.
42. Avci, B., Avci, R., Inan, U.U., Kaderli, B. (2009). Comparative evaluation of apoptotic activity in photoreceptor cells after intravitreal injection of bevacizumab and pegaptanib sodium in rabbits. *Investigative Ophthalmology & Visual Science*, *50*(7), 3438–3446.
43. Kim, J.H., Kim, C., Lee, B.J., Yu, Y.S., Park, K.H., Kim, K.W. (2008). Absence of intravitreal bevacizumab-induced neuronal toxicity in the retina. *Neurotoxicology*, *29*(6), 1131–1135.
44. Lin, Y.S., Nguyen, C., Mendoza, J.L., Escandon, E., Fei, D., Meng, Y.G., et al. (1999). Preclinical pharmacokinetics, interspecies scaling, and tissue distribution of a humanized monoclonal antibody against vascular endothelial growth factor. *Journal of Pharmacology and Experimental Therapeutics*, *288*(1), 371–378.
45. Comparison of Age-related Macular Degeneration Treatments Trials Research, G., Martin, D.F., Maguire, M.G., Fine, S.L., Ying, G.S., Jaffe, G.J., et al. (2012). Ranibizumab and Bevacizumab for Treatment of Neovascular Age-related Macular Degeneration: Two-Year Results. *Ophthalmology*, *119*(7), 1388–1398.

46. Subramanian, M.L., Abedi, G., Ness, S., Ahmed, E., Fenberg, M., Daly, M.K., et al. (2010). Bevacizumab vs ranibizumab for age-related macular degeneration: 1-year outcomes of a prospective, double-masked randomised clinical trial. *Eye (London)*, *24*(11), 1708–1715.
47. Wong, D., Joussen, A.M. (2011). The safety of using anti-VEGF: Is there strength in numbers? Curtis LH, Hammill BG, Schulman KA, Cousins SW (2010). Risks of mortality, myocardial infarction, bleeding, and stroke associated with therapies for age-related macular degeneration. *Archives of Ophthalmology*, *128*(10), 1273–1279. *Graefes Archive for Clinical and Experimental Ophthalmology, 249(2)*, 161-162.
48. Curtis, L.H., Hammill, B.G., Schulman, K.A., Cousins, S.W. (2010). Risks of mortality, myocardial infarction, bleeding, and stroke associated with therapies for age-related macular degeneration. *Arch Ophthalmol*, *128*(10), 1273–1279.
49. Lim, L.S., Cheung, C.M., Mitchell, P., Wong, T.Y. (2011). Emerging evidence concerning systemic safety of anti-VEGF agents—should ophthalmologists be concerned? *American Journal of Ophthalmology*, *152*(3), 329–331.
50. Sharma, S., Johnson, D., Abouammoh, M., Hollands, S., Brissette, A. (2012). Rate of serious adverse effects in a series of bevacizumab and ranibizumab injections. *Canadian Journal of Ophthalmology*, *47*(3), 275–279.

INDEX

Predictive ADMET: Integrative Approaches in Drug Discovery and Development, First Edition.
Edited by Jianling Wang and Laszlo Urban.